D1797498

Model Validation

Model Validation: Perspectives in Hydrological Science

Edited by

MALCOLM G. ANDERSON
University of Bristol, UK

and

PAUL D. BATES
University of Bristol, UK

JOHN WILEY & SONS, LTD
Chichester · New York · Weinheim · Brisbane · Singapore · Toronto

National   01243 779777
International   (+44) 1243 779777
e-mail (for orders and customer service enquiries): cs-books@wiley.co.uk
Visit our Home Page on http://www.wiley.co.uk
                http://www.wiley.com

Other Wiley Editorial Offices

John Wiley & Sons, Inc., 605 Third Avenue,
New York, NY 10158-0012, USA

WILEY-VCH Verlag GmbH, Pappelallee 3,
D-69469 Weinheim, Germany

John Wiley & Sons Australia, Ltd, 33 Park Road, Milton,
Queensland 4064, Australia

John Wiley & Sons (Asia) Pte Ltd, 2 Clementi Loop #02-01,
Jin Xing Distripark, Singapore 129809

John Wiley & Sons (Canada) Ltd, 22 Worcester Road,
Rexdale, Ontario M9W 1L1, Canada

Library of Congress Cataloging-in-Publication Data
Model validation: perspectives in hydrological science / edited by M.G. Anderson and P.D. Bates.
    p. cm.
  Includes bibliographical references.
  ISBN 0-471-98572-4 (alk. paper)
  1. Hydrologic models.   I. Anderson, M.G.   II. Bates, Paul D.
GB656.2.H9 M59 2001
551.48′01′1 —dc21   00-047753

British Library Cataloguing in Publication Data

A catalogue record for this book is available from the British Library

ISBN 0-471-98572-4

Typeset in 10/12pt Times from the author's disks by Vision Typesetting, Manchester
Printed and bound in Great Britain by Antony Rowe Ltd, Chippenham, Wiltshire
This book is printed on acid-free paper responsibly manufactured from sustainable forestry, in which at least two trees are planted for each one used for paper production.

To Rebecca, Joseph and Samuel

Contents

List of contributors   ix

Chapter 1   Hydrological Science: Model Credibility and Scientific Integrity   1
            Malcolm G. Anderson and Paul D. Bates

Chapter 2   Kinds of Models   11
            Adam Morton and Mauricio Suárez

Chapter 3   Philosophical Issues in Model Assessment   23
            Naomi Oreskes and Kenneth Belitz

Chapter 4   Calibration, Validation and Equifinality in Hydrological Modelling:
            a Continuing Discussion . . .   43
            Keith Beven

Chapter 5   Models in the Courtroom   57
            E. Scott Bair

Chapter 6   On Simulation, Calibration and Ill-conditioning with Application to
            Environmental System Modelling   77
            Claude R. Dietrich

Chapter 7   Data-Based Mechanistic Modelling and Validation of Rainfall–Flow
            Processes   117
            Peter Young

Chapter 8   The Use of Remote Sensing to Validate Hydrological Models   163
            D. Pearson, M.S. Horritt, R.J. Gurney and D.C. Mason

Chapter 9   Soil Water Relations   195
            H.W.G. Booltink

Chapter 10 A Hydromechanical Approach to Preferential Flow   233
            P. Germann

Chapter 11   Validation of Snow Models   261
                 Robert E. Davis, Rachel Jordan, Steven Daly and George Koenig

Chapter 12   Groundwater   293
                 Hubert J. Morel-Seytoux

Chapter 13   Validation of Hydraulic Models   325
                 Paul D. Bates and Malcolm G. Anderson

Chapter 14   Modelling Water Quality Processes in Riverine Systems   357
                 R.A. Falconer, B. Lin and S.M. Kashefipour

Chapter 15   Modelling Sediment Entrainment into Suspension, Transport and
                 Deposition in Rivers   389
                 Marcelo H. Garcia

Chapter 16   The 'Validation' of Hydrodynamic Models: Some Critical
                 Perspectives   413
                 Stuart N. Lane and Keith S. Richards

Chapter 17   The Validation of Ice-sheet Models   439
                 Martin J. Siegert and Antony J. Payne

Chapter 18   Discussion of Model Validation in Relation to the Regional and
                 Global Scale   461
                 Jens Christian Refsgaard

Author Index   485
Subject Index   494

List of Contributors

M.G. Anderson — School of Geographical Sciences, University of Bristol, University Road, Bristol, BS8 1SS, UK
m.g.anderson@bristol.ac.uk

E.S. Bair — Department of Geological Sciences, Ohio State University, 231 Mendenhall Lab, 125 South Oval Mail, Columbus, Ohio, 43210-1398, USA
bair.1@osu.edu

P.D. Bates — School of Geographical Sciences, University of Bristol, University Road, Bristol, BS8 1SS, UK
paul.bates@bristol.ac.uk

D.K. Belitz — NAWQA Project Chief, U.S. Geological Survey, Water Resources Division, 5735 Kearny Villa Road, Suite O, San Diego, California 92123, USA
kbelitz@usgs.gov

K.J. Beven — Institute of Environmental and Natural Sciences, University of Lancaster, Bailrigg, Lancaster, LA1 4YQ, UK
k.bevan@lancaster.ac.uk

H.W.G. Booltink — Laboratory of Soil Science and Geology, Wageningen Agricultural University, Duivendaal 10, PS Box 37, 6700 AA Wageningen, The Netherlands
harry.booltink@bodlan.beng.wau.nl

S. Daly — Cold Regions Research and Engineering Laboratory, 72 Lyme Road, Hanover, New Hampshire 03755-1290, USA
Steven.F.Daly@crrel.usace.army.mil

R.E. Davis — Cold Regions Research and Engineering Laboratory, 72 Lyme Road, Hanover, New Hampshire 03755-1290, USA
bert@crrel.usace.army.mil

C.R. Dietrich — Centre for Resources and Environmental Studies, Integrated Catchment Assessment & Management, Australian National University, Canberra, ACT 0200, Australia
claude@cres.anu.edu.au

R.A. Falconer — School of Engineering, University of Wales, Cardiff, PO Box 686, Cardiff, CF2 3TB, UK
falconerra@cardiff.ac.uk

M.H. Garcia — Hydrosystems Laboratory, Department of Civil and Environmental Engineering, University of Illinois at Urbana Champaign, 205 North Matthews Avenue, Urbana, IL 61801, USA
mhgarcia@uiuc.edu

P. Germann   Universitat Bern, Geographisches Institut, Hallerstrasse 12, CH-3012
                      Bern, Switzerland
                      germann@giub.unibe.ch

R.J. Gurney   Earth Systems Science Centre, Harry Pitt Building, University of
                      Reading, 3 Earley Gate, Whiteknights, Reading RG6 6AL, UK
                      rjg@mail.nerc-essc.ac.uk

M.S. Horritt   School of Geographical Sciences, University of Bristol, University
                      Road, Bristol, BS8 1SS, UK
                      matt.horritt@bristol.ac.uk

R. Jordan   Cold Regions Research and Engineering Laboratory, 72 Lyme Road,
                      Hanover, New Hampshire 03755-1290, USA
                      Rachel.E.Jordan@crrel.usace.army.mil

S. M. Kashefipour   School of Engineering, University of Wales, Cardiff, PO Box 686,
                      Cardiff, CF2 3TB, UK
                      kashefipoursm@cardiff.ac.uk

G. Koenig   Cold Regions Research and Engineering Laboratory, 72 Lyme Road,
                      Hanover, New Hampshire 03755-1290, USA
                      George.G.Koenig@crrel.usace.army.mil

S.N. Lane   School of Geography, University of Leeds, Leeds LS2 9JT, UK
                      s.lane@geog.leeds.ac.uk

B. Lin   School of Engineering, University of Wales, Cardiff, PO Box 686,
                      Cardiff, CF2 3TB, UK
                      linbl@cardiff.ac.uk

D.C. Mason   Earth Systems Science Centre, Harry Pitt Building, University of
                      Reading, 3 Earley Gate, Whiteknights, Reading RG6 6AL, UK
                      Dave.Mason@mail.nerc-essc.ac.uk

H.J. Morel-Seytoux   Hydrology Days Publications, 57 Selby Lane, Atherton, CA 94027-
                      3926, USA
                      Hydroprose@batnet.com

A. Morton   Department of Philosophy, University of Bristol, Bristol, BS8 1TB,
                      UK
                      Adam.Morton@bristol.ac.uk

N. Oreskes   Department of History and Program in Science Studies, University of
                      California, 9500 Gilam Drive, La Jolla San Diego, California,
                      92093-0104, USA
                      noreskes@ucsd.edu

A.J. Payne   Department of Geography, University of Southampton, Highfield,
                      Southampton, SO17 1BJ, UK
                      A.J.Payne@soton.ac.uk

D. Pearson   Earth Systems Science Centre, Harry Pitt Building, University of
                      Reading, 3 Earley Gate, Whiteknights, Reading RG6 6AL, UK
                      dwcp@mail.nerc-essc.ac.uk

J.C. Refsgaard   Department of Hydrology, Geological Survey of Denmark and
                      Greenland (GEUS), Thoravej 8, DK-2400 Copenhagen NV, Denmark
                      jcr@geus.dk

K.S. Richards   Department of Geography, University of Cambridge, Downing Place,
                   Cambridge, CB2 3EN, UK
                   ksr10@cam.ac.uk

M.J. Siegert   School of Geographical Sciences, University of Bristol, University
                   Road, Bristol BS8 1SS, UK
                   m.j.siegert@bristol.ac.uk

M. Suárez   Department of Philosophy, University of Bristol, 9 Woodland Road,
                   Bristol, BS8 1TB, UK
                   m.suarez@bristol.ac.uk

P. Young   Institute of Environmental and Natural Sciences, University of
                   Lancaster, Bailrigg, Lancaster, LA1 4YQ, UK
                   p.young@lancaster.ac.uk

1 Hydrological Science: Model Credibility and Scientific Integrity

MALCOLM G. ANDERSON AND PAUL D. BATES
Geographical Sciences, University of Bristol, UK

1.1 WHY WE (DON'T) MEAN MODEL 'VALIDATION'

1990   Absolute validity of a model is never determined

                                                              (National Research Council 1990)

1992   What is usually done in testing the predictive capability of a model is best characterised as calibration or history matching; it is only a limited demonstration of the reliability of the model. We believe the terms validation and verification have little or no place in ground-water science; these terms lead to a false impression of model capability. More meaningful descriptors of the process include model testing, model evaluation, model calibration, sensitivity testing, benchmarking, history matching, and parameter estimation. Use of these terms will help to shift emphasis towards understanding complex hydrogeological systems and away from building false confidence into model predictions.

                                                          (Konikow and Bredehoeft 1992)

1996   According to the terminology defined in Chapter 2 of this book a modelling system can, in principle, never be validated. Instead of a full validation we can think of the degree of validity as the credibility of a given modelling system. The degree of validity of a modelling system is expressed in the first and most immediate place by the sum of all successful validation of all models that have been constructed and operated to date using the modelling system. As the number of such successful models increases, so the credibility of the system itself grows in strength.

        Behind this, most superficial of views, lies the assumption that the modelling system is in fact being improved on the basis of the operating experience; that it is functioning within its market, tracking the needs of that market and thus learning from this market. From this point of view, the development of a modelling system is not one that leads directly to a finished, rounded and complete product, but it is rather a process of adaptation through evolution. Thus, although the modelling system is indirectly a product, it is one that is constantly evolving, so that its evolution corresponds to a process.

                                                                 (Refsgaard and Storm 1996)

2001   The inherent uncertainties of models have been widely recognised, and it is now commonly acknowledged that the term 'validation' is an unfortunate one, because its root – valid – implies a legitimacy that we may not be justified in asserting (Tsang 1991, 1992; Anderson and Woessner 1992; Konikow 1992; Konikow and Bredehoeft 1992; Oreskes et al. 1994; Oreskes 1998; Beck et al. 1997; Steefel and van Cappellen 1998). But old habits die hard, and the term persists. In formal documents of major national and international agencies that sponsor modelling efforts and in the work of many modellers, 'validation' is still widely used in ways

Model Validation: Perspectives in Hydrological Science. Edited by M.G. Anderson and P.D. Bates.
© 2001 John Wiley & Sons, Ltd.

that assert or imply assurance that the model accurately reflects the underlying natural processes, and therefore provides a reliable basis for decision-making. This usage is misleading and should be changed. Models cannot be validated.

(Oreskes and Belitz, this volume, chapter 3)

There can be few books that begin with an initial commentary upon the title suitability in respect of its appropriateness within the science that the text contains. Over the last ten years or so there has been increased awareness within hydrological science that there is a need to have assessments made of model performance, as the above quotations testify. Simultaneously, there is awareness that the root of the term 'validation' has implications that may be inappropriate. The term 'model validation', although widely used, should best be interpreted as a *stimulus* for a wide community of both scientists and decision-makers to discuss underlying fundamental issues. It is this context that drives the title for this text.

On the one hand over the last decade it has become widely recognised that model validation in an absolute sense is not possible. On the other is the fact that:

Modellers practically never declare their models to be 'invalidated', primarily because ground water models nearly always have enough adjustable parameters to fit a limited set of field observations. This leads us to ask how we can distinguish a good fit that is based on artificial manipulation of an over-parameterised model from a good fit that is based on an accurate description of the processes.

(NRC 1990)

The notion of a constantly evolving model embodying physical theories that are provisional is the true context in which we have to develop procedures for assessing model credibility. Refsgaard and Storm (1996) outline the need for multi-criteria, multi-scale validation criteria (Table 1.1). The table illustrates the point that for different model structures the model credibility criteria will be different. However, as we have stated, it must be assumed that the model is evolving, and that evolution must imply a *process*. Whether or not the process and associated credibility criteria are adequately researched is, in large measure, the function of this text to explore. In attempting to do so there are two relevant contexts to bear in mind. Firstly, the notion that hydrological science has an element of immaturity and, secondly, that it is not a data-rich science. We highlight both of these aspects below, because in our view they contribute to the evolution of the validation argument over the last decade.

1.2   THE REQUIREMENTS FOR SUSTAINING HYDROLOGICAL SCIENCE

The context for this text stems from the belief that there exists a pressing need to appraise, integrate and direct views on systems we should use to evaluate performances of models within hydrological science. This, of course, should be a process in which we are constantly engaged, but it is an area of research that has not received either the level nor integration of attention that we believe is necessary. It is worthwhile briefly examining certain background issues that are relevant and perhaps contributory to the absence of substantive texts relating to model validation.

1.2.1   The Relative Immaturity of Hydrological Science

A number of researchers have observed that hydrological science has not yet reached maturity as a geoscience. Most notably perhaps this is a comment emanating from an overview of a set of

Table 1.1 An illustration of the need for the incorporation of multi-criteria and multi-scale aspects in methodologies for the validation of distributed models (from Refsgaard and Storm 1996, p. 52)

	Lumped conceptual	Distributed physically based
Output	At one point: • Runoff	At many points: • Runoff • Surface water level • Groundwater head • Soil moisture
	$= > single\ variable$	$= > multi\ variable$
Success criteria (excluding problem of selecting which statistical criteria to use)	Measured $< = >$ simulated Runoff, one site	Measured $< = >$ simulated • Runoff, multi sites • Water levels, multi sites • Groundwater heads, multi sites • Soil moisture, multi sites
	$= > single\ criteria$	$= > multi\ criteria$
Typical model application	Rainfall–runoff • stationary conditions • calibration data exist	Rainfall–runoff, unsaturated zone, groundwater, basis for subsequent water quality modelling Impacts of man's activity • non-stationary conditions sometimes • calibration data do not always exist
Validation test	Usually 'Split-sample test' is sufficient	More advanced tests required: • Differential split sample test • Proxy basin test
	$= > well-defined\ practice\ exists$	$= > need\ for\ rigorous\ methodology$
Modelling scale	Model: catchment scale Field data: catchment scale	Model: depends on discretisation Field data: many different scales
	$= > single\ scale$	$= > multi\ scale$ problems

invited lecturers organised by the US National Research Council in 1997 'Hydrologic Science – Taking Stock and Looking Ahead'. Here Dunne (1998) implies that while much excellent hydrological research is undertaken, the science is fragile and relatively immature. The overview of that report summarises the evidence for this immaturity as including a diversity of approaches as opposed to a consistent set of theoretical constructs, lack of communication (and even respect) between groups such as modellers and field observers, engineers and scientists, and the relative lack of hydrologists fully participating in the design of data collection campaigns. Finally the argument is advanced for the need for institutional foci to support and mark hydrological science as a mature geoscience.

Most hydrologists have entered hydrology from another discipline and there is thus an in-built tendency to ossify pre-existing labels rather than fully integrate the science across the relevant disciplines. Additionally, and perhaps consequentially, hydrologists are not as influential as other groups in specifying data needs to major international programmes, despite exciting new data acquisition possibilities. As a consequence, Dunne argues that there is a vital need to improve communication across multiple disciplines, background and approaches, and that this maturity in the discipline 'needs to be developed by a convergence of approaches and a consistent set of theoretical constructs'.

A specific example of these concerns is the 1993 flooding of the Mississippi and Missouri basins. Dunne (1998) recounts that in what was the largest ever hydrological event in the United States (Figure 1.1), the debate on future responses to flooding was not guided by hydrological science. Moreover, in top-ranked journals in hydrology he comments that one can find virtually no analysis of the flood.

In 1986 Klemes commented that 'hydrology has not yet consolidated itself as a science in its own right', and discussed the possibility of mediocrity in hydrology being driven by 'water management organisations'. Since that time, a key element in how hydrological science is communicated to many groups is through the huge numbers of available hydrological models.

1.2.2   A Proliferation of Models

The relative vulnerability of hydrological science among the geosciences is compounded by the proliferation of hydrological and hydraulic models available from commercial sources. Further, certain of these models are accorded an enhanced status by government agency endorsements. Taken with the exponential growth of environmental consultancy companies in recent years, commercially driven demand is being met by commercially available software. The likelihood of the results of such models being 'accepted' in a complete and relatively unquestioning sense is thus increased, while those closely involved with modelling are well aware of uncertainty in both parameterisation and model structure. How fundamental research in hydrological science should be configured to best develop and have a strengthening influence against such a background is an interesting question to pose.

Abbott and Refsgaard (1996) recognised the strong link (and influence) that exists between model developers, institutional–political factors, administration and broader social contexts. These interconnected influences need increasingly to be managed and to be the focus of those who wish to drive hydrological science. As Abbott and Refsgaard observe, 'in a world of increasingly commercial pressures on most organisations, few of these like to admit that it is they that do not have enough data or it is their organisation that cannot mobilise enough knowledge'. Whatever the impact of such pressures in absolute terms, their existence is another important driver for us to present as coherent a picture as possible of issues relating to model validation; or rather, and perhaps more precisely, for us to present a complete specification of model performance. As Oreskes and Belitz (this volume, chapter 3) point out, however, we are in a position where in many cases a scientist can derive greater professional benefit from building a flawed model than from undertaking further empirical investigation.

Somewhat parallel discussions are taking place in groundwater research. Narasimhan (1998) points to the fact that decision-makers find it easier to make decisions when 'numbers are fed back to them'. He is simultaneously concerned with the fact that:

> We have created a group of people labelled modellers . . . this is troublesome. Without a grasp of the fundamental cause, effect and the process limitations the mere exercise of mathematical models does not necessarily imply sophistication.

There are pressures, therefore, on the rigour and creditability of model development, utilisation and interpretation that are widespread across the geosciences. These pressures are enhanced by the sparsely available data.

1.3   ISSUES OF DATA HEALTH IN HYDROLOGICAL SCIENCE

Although in many respects measurement and observations are central to hydrology, Wood (1998) argues that specific quality and quantity criteria still do not exist against which to evaluate collection efforts. A notable issue here is the role of major field facilities, such as Coweeta Hydrologic Laboratory, the H.J. Andrews Experimental Forest in Oregon, and Oak Ridge National Laboratory Hydrology field sites. While, ideally, there should be a highly co-ordinated programme of consolidated theory associated with a range of major field installations, reality is probably a little different. Franklin (1993) observes that 'continued existence of facilities such as Coweeta must be credited to one or two individuals. Maintaining such sites and databases have cost individuals and programmes, sometimes dearly'. In short, Franklin remarks 'someone's curriculum vitae is going to be the thinner for the time and cost involved'. Although there are frequent calls for a sustained interaction between field research and modelling, such interaction, as we have seen, is not always easy.

It is important to recognise that major field facilities and associated experiments do have a sustained and potentially long-standing impact on this linkage. For example, Steenhuis et al. (1999) utilise the Coweeta field data from Hewlett and Hibbert (1963) – some 36 years earlier – to undertake modelling work, just as Sloan and Moore (1984) did to examine macropore flow issues. Evidence can thus be accumulated of the impact of major facilities and associated experiments and their role in 'linking' field and modelling efforts.

There are echoes here of the 1990 NRC report on hydrological science which commented on the fact that designing and executing data collection programmes are frequently reviewed as mundane. 'It is therefore difficult for agencies to be doggedly persistent about the continuity of high-quality hydrologic data sets . . . The scientific community tends to allow data collection programmes to erode'. Of course this issue is sharply in focus with model simulation periods of greatest interest often being orders of magnitude in excess of the time frame for which relevant data (either for parameterisation purposes or split sample testing and the like) are available.

Wood (1998) concludes that because the hydrological community has failed to establish data needs (for operational, management or scientific hydrology purposes), hydrology has been subordinate to climate field programmes. It can, and should be strongly argued that major field laboratories are essential for comparison, collaboration and long-term databases. Collaboration between hydrologists of different initial backgrounds is vital for the interface science that we know to be the source of greatest advances. Such 'diverse' teams appropriately configured and with a much enlarged range of dedicated sites is what Franklin has argued is necessary to drive the Suprasite synthesis generating new paradigms and fresh approaches. Such a commentary underpins views of those engaged in distributed modelling specifically. Beven (1996), for example, highlights the fact that we know relatively little about the value of different data sets for distributed modelling and thus advocates the creation of test catchment databases. Fawcett et al. (1995) outline 'blind' runs on a highly instrumented catchment in Switzerland, using fully distributed hydrological models. In respect of internal and external (bulk flow) validation the results are particularly concerning – calibration was required to achieve satisfactory predictions, and this with model parameterisation levels that would exceed all but the very best of test catchment databases. The requirement for such databases is thus urgent and absolute in respect of the evaluation of model credibility.

A decade ago the US National Research Council (1990) in a thorough review of the scientific and regulatory application of groundwater models stated the following:

> Modellers tend to go their own way, building impressive computer codes, while experimentalists tend to gather data for purposes other than evaluating models. The resolution of this problem must eventually come from a close interaction among modellers, experimentalist, and field scientists. We have probably reached the point at which it is now imperative to gather laboratory and field data to evaluate the validity and utility of the geochemical codes.
>
> (NRC 1990)

1.4 PRESSURES TO ESTABLISH MODEL CREDIBILITY

Over the last decade we have begun to have an appreciation of the need to be much more rigorous in establishing procedures for defining model credibility. However, the context in which this argument has evolved and is progressing is, as we have seen, one in which the discipline, hydrological science, has an element of immaturity and is limited by data. The demand side is, however, increasing from policy makers and agencies for standards, which can be used to certify model results, or the utilisation of a specific code. Such standards, it is argued, reduce the risk that a model will lead to inappropriate decisions (NRC 1990). In short, modellers are under increasing pressure to show their longer term predictions are 'worthwhile', when the real context for such predictions may typically include unknown boundary conditions and lack historical data (Bair, chapter 5).

The regulatory push to standardisation in environmental fields is also associated with simplification (Elsenheimer 1999) – this can therefore educe a 'disconnect' between research scientists and a user community. The complication is that the user community, in becoming more environmentally aware, is becoming also more litigious. This change is aided by the US EPA, for example, which has set standards that *mandate* use of numerical *models* in the assessment of certain environmental system *acceptability* (see Morel-Seytoux, chapter 12). Oreskes and Belitz (chapter 3) go further and argue that such a trend in fact leads to the increased use of models that do not work. The frequency of model failure creates a positive incentive to use bad models if those models have already been used by others. A further issue of concern that has been expressed in terms of the litigation aspect of models is one that 'experts' may act more for the client than as a disinterested specialist seeking the truth (Griffiths 1999). However, Bair (chapter 5) observes that the primary function of the (US) legal system, 'unlike the scientific method, is not to find the truth. It is designed to resolve conflicts'. Such a fact can be thought of as another dimension of the 'disconnection' between scientist and user.

Against this general background, the current academic structures, for the most part, are perhaps less helpful than they might be. Credibility needs to be based upon a specific mode of scientific inquiry. Critical rationalism ('hypothesis – experiment – test – falsify) focuses on observation while the 'realist' approach outlined by Richards (1990, 1994) advocates a more holistic approach. This latter approach, in which acceptance of a theory is based not upon prediction but explanatory power, is worthy of some discussion. The diverse origins of those researching hydrological science provide a powerful pool from which 'interdisciplinary' focused research can be undertaken.

In the realist approach a key element is that any phenomena can be structured into three levels: the generating mechanism, the events and the empirical observation (Richards et al. 1997). Richards et al. discuss the practical issues involved in 'validating' the theoretical findings of research programmes typical of realist science. They argue that generally there is a need to invoke relationships with a number of phenomena, at a physically appropriate scale, and that seeking explanatory power involves a strong interdisciplinary approach. The argument is developed by proposing that realist research is best undertaken as an in-depth study at a single site – focusing on multiple interacting processes without reliance on a single mode of experimentation. However, most academic institutions are organised in a manner that, at best, is not readily conducive to parallel research into related structures (Richards et al. 1997). According to this analysis we see critical rationalism to depend heavily upon observations, but we see hydrological science to be relatively poor in terms of the relevant data acquisition techniques. We see realism to be holistic but hydrological science as requiring of more integration across the discipline (see Refsgaard, chapter 18).

However, because of site-uniqueness, the above analysis does not actually offer explanatory power additional to that which exists in a perceptual sense. The basic question is how we can utilise the perceptual models we have (which themselves contain a significant amount of qualitative explanatory power) to develop predictions given site uniqueness. In this regard scaling is one of the major outstanding challenges (Blöschl and Sivapalan 1995). There is demand to use models when conditions for predictions are different from those used to calibrate/validate them. The circumstance in which simulations of interest are at scales significantly different from those of the relevant processes is an element of the ill-conditioning of environmental systems (see Dietrich, chapter 6) – a situation complicated by the existence of catchment heterogeneity. In summary, an illustration of heterogeneity in catchments and processes at a range of space and time scales is shown in Figure 1.1 – observational limitations and the requirement for upscaling and downscaling are thus clearly evident.

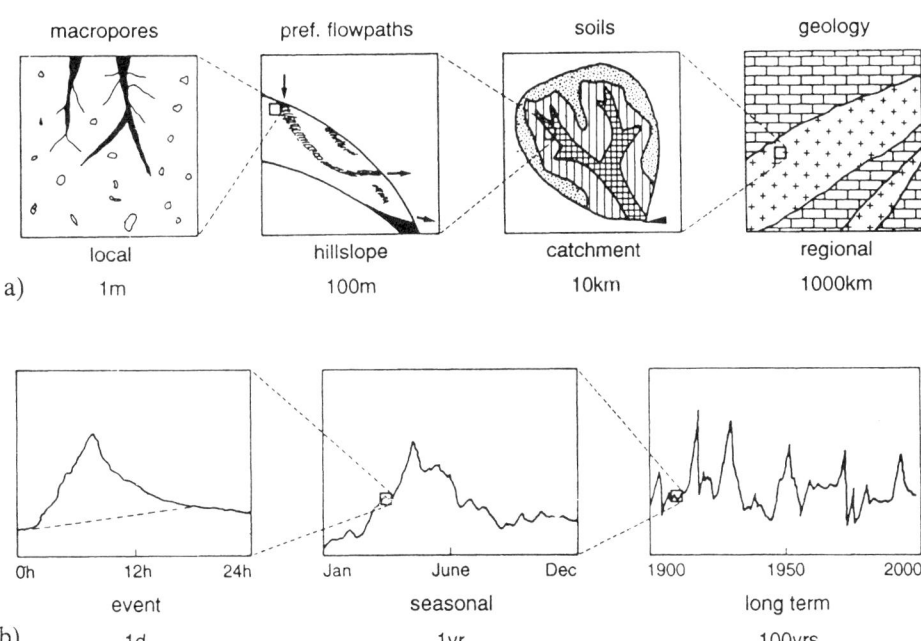

Figure 1.1   Heterogeneity of catchments and hydrological processes at a range of (a) spatial and (b) time scales (from Blöschl and Sivapalan 1995)

The observational limitations that are inevitable in hydrological science result in models with differing parameter sets yielding similar model efficiencies (Pinol et al. 1997) and simple models with only a small number of parameters providing a satisfactory empirical fit to observed data – models that lack sufficient parameter numbers to fully characterise the space–time variability (Sivapalan et al. 1997). Equifinality has not tempered the fundamental research need to examine scaling problems (e.g. using large-scale models and data for small-scale predictions) – see Blöschl and Sivapalan (1995).

A thorough review of a framework and analysis for scaling approaches – one model orientated (scaling the state variables, model parameters, inputs and so forth), the other based on dimensional analysis and similarity concepts – has been given by Blöschl and Sivapalan (1995). We are reminded here, among other things, of the upscaling and downscaling choices, the critical position that scaling theory has in terms of model validation issues, and the potential opportunities afforded by the possibilities of classifying heterogeneity in terms of future predictive frameworks. Hydrology is hardly unique in this context – scaling issues are of obvious importance in meteorology, hydraulics, climatology etc. In all such instances, although perceptual understanding of processes at different scales may be good, the observational limitations demand scaling theory developments and similarly a focused recognition by field research within hydrology of this issue. By way of illustration of recent developments in this regard we can cite the work of Freer et al. (1997). In observing recent use of Digital Terrain Model (DTM) based maps of soil saturation, based on flowpaths assumed to be related to hydraulic gradients derived from surface topography, they comment on the need to validate the resultant spatial pattern. The DTM upscaling thus stimulated Freer et al. to thoroughly review the spatial representation of topographically derived indices for saturation prediction. For two catchments where suitable data were available they concluded that bedrock topography may be critical to an understanding of hillslope flow processes and thus be relevant in terms of appropriate indices for saturation prediction. This then is but one targeted avenue by which a carefully conducted experimental design can aid upscaling, thereby helping the implementation of our perceptual knowledge and providing strong underpinning of modelling efforts that *implicitly* attempt to incorporate three-dimensional heterogeneity of catchment processes and properties without having to model them *explicitly*.

Formally embracing uncertainty in a post Konikow and Bredehoeft era is increasingly being advocated. Beven (2000), for example, argues 'why not . . . start to recognise such uncertainty at the very start and take account of the very different models that might be acceptable to the experts (and consistent with any observational data)'. The strengthening links between model developers and socio-political factors to which we have already made reference is evident in structures such as quantitative risk assessments. Risk assessments are but one example of the widening exposure hydrological science is being afforded and *must* participate in.

The chapters in this book provide a commentary on, and amplification of, these issues. Readers will note a number of areas of correspondence and departure in the various chapters and in introducing the volume one should be careful of summarising too far a debate that is both complex and fluid. With this in mind we have deliberately avoided setting out here our own interpretation of the views expressed in the various chapters of this book. Nevertheless, we can perhaps note agreement on a number of basic issues. For example, there is reasonable consensus between authors that all knowledge is provisional and thus models can never be conclusively validated, only falsified. Further, the Realist call for explanatory rather than predictive power in models is also widely invoked. Debate exists too as to, for example, whether models are truly propositional and can embody hypotheses and theory in a formal way or whether in the limit even falsification may prove impossible. Readers will find their own debates and truisms in considering the chapters contained herein and it is this process of opening up validation issues to wider consideration that this volume seeks to encourage. In hydrological science there are a large number of issues relating to scaling, equifinality, linkages to socio-political frameworks, legal frameworks, philosophical interpretations and so forth that make the need for discussing and establishing model credibility criteria both urgent and compelling.

REFERENCES

Abbott, M.B. and Refsgaard, J.C. (eds). 1996. *Distributed Hydrological Modelling*. Kluwer, Dordrecht, 321 pp.

Beven, K. 1996. Response to comments on 'A discussion of distributed hydrological modelling' by J.C. Refsgaard et al. In Abbott, M.B. and Refsgaard, J.C. (eds), *Distributed Hydrological Modelling*. Kluwer, Dordrecht, 289–295.

Beven, K. 2000. On model uncertainty, risk and decision making. *Hydrological Processes*, **14**, 2605–2606.

Blöschl, G. and Sivapalan, M. 1995. Scale issues in hydrological modelling: a review. In Kalma, J.D. and Swapalan, M. (eds), Scale issues in Hydrological Modelling, John Wiley, Chichester, 9–48.

Dunne, T. 1998. Wolman Lecture: hydrologic science . . . in landscapes . . . on a planet . . . in the future. *Hydrologic Sciences – Taking Stock and Looking Ahead*. National Research Council, Washington, 10–43.

Elsenheimer, D. 1999. The commoditization of hydrogeology. *Groundwater*, **37**, 641–642.

Fawcett, K.R., Anderson, M.G., Bates, P.D., Jordan, J-P. and Bathurst, J.C. 1995. The importance of internal validation in the assessment of physically based distributed models. *Transactions of the Institute of British Geographers*, **20**, 248–265.

Franklin, J.F. 1993. Past and future of ecosystem research – construction of dedicated experimental sites. In: Frank, W.T. and Crossley, D.A. (eds), *Forest Hydrology and Ecology at Coweeta*. Spinger-Verlag, Berlin, 415–427.

Freer, J., McDonnell, J., Beven, K.J., Brammer, D., Burns, D., Hooper, R.P. and Kendal, C. 1997. Topographic controls on subsurface storm flow at the hillslope scale for two hydrologically distinct small catchments. *Hydrological Processes*, **11**, 1347–1352.

Griffiths, J.S. 1999. Proving the occurrence and cause of a landslide in a legal context. *Bulletin of Engineering Geology and the Environment*, **58**, 75–85.

Hewlett, J.D. and Hibbert, A.R. 1963. Moisture and energy conditions within a slope mass during drainage. *Journal of Geophysical Research*, **68**, 1081–1087.

Klemes, V. 1986. Dilettantism in hydrology: transition or destiny? *Water Resources Research*, **22**, 1775–1885.

Konikow, L.F. and Bredehoeft, J.D. 1992. Groundwater models cannot be validated. *Advances in Water Resources*, **15**, 75–83.

Narasimhan, T.N. 1998. Quantification and groundwater hydrology. *Groundwater*, **36**, 1.

National Research Council, 1990. *Groundwater Models; Scientific and Regulatory Applications*. National Academy Press, Washington, 303 pp.

Pinol, J., Beven, K. and Freer, J. 1997. Modelling the hydrological response of Mediterranean catchments, Prades, Catalonia. The use of distributed models as aids to hypothesis formulation. *Hydrological Processes*, **11**, 1287–1306.

Refsgaard, J.C. and Storm, B. 1996. Construction, calibration and validation of hydrological models. In: Abbott, M.B. and Refsgaard, J.C. (eds), *Distributed Hydrological Modelling*. Kluwer, Dordrecht, 41–45.

Richards, K.S. 1990. Real geomorphology. *Earth Surface Processes and Landforms*, **15**, 195–196.

Richards, K.S. 1994. Real geomorphology revisited. *Earth Surface Processes and Landforms*, **19**, 277–281.

Richards, K.S., Brooks, S., Clifford, N., Harris, T. and Lowe, S. 1997. Theory, measurement and testing in 'real' geomorphology and physical geography. In: Stoddart, D.R. (ed), *Process and Form in Geomorphology*. Routledge, London. 265–292.

Sivapalan, M., Woods, R.A. and Kalma, J.D. 1997. Variable bucket representation of TOPMODEL and investigation of the effects of rainfall heterogeneity. In: Beven, K.J. (ed), *Distributed Hydrological Modelling – Applications of the TOPMODEL Concept*. John Wiley, Chichester. 265–288.

Sloan, P.G. and Moore, I.D. 1984. Modelling subsurface storm flow on steeply sloping forested watersheds. *Water Resources Research*, **20**, 1815–1822.

Steenhuis, T.S., Parlange, J-Y., Sanford, W.E., Heilig, A., Stagnitti, F. and Walter, M.F. 1999. Can we distinguish Richards' and Boussinesq's equations for hillslopes?: The Coweeta experiment revisited. *Water Resources Research*, **35**, 589–593.

Wood, E.F. 1998. Hydrologic measurements and observations: an assessment of needs. In: *Hydrologic Sciences – Taking Stock and Looking Ahead*. National Research Council, Washington, 67–86.

2 Kinds of Models

ADAM MORTON AND MAURICIO SUÁREZ
University of Bristol, UK

2.1 THE M WORD

'Model' is a term of the working scientist's self-explanatory and self-justifying vocabulary. 'Here is my model of the phenomenon', 'it follows from our model that . . .', 'our model does not capture the following aspects of the data, but this is no problem since it is just a model . . .', 'to model the data we have made the following assumptions . . .'. ('We are modelling the river as a drunken snake', 'the 3000% discrepancy between predicted sedimentation and observation is very satisfactory for a purely theoretical model.') In some such assertions the word 'model' could be replaced with 'theory' or 'hypothesis' with no big loss of meaning. But in many it could not. There are examples of both in the chapters in this volume, as we argue in Section 2.3. When scientists describe their creations as models they often intend to take advantage of some of the following features of models, as opposed to theories:

- Two models can be inconsistent with each other, and both can be good models.
- A model can contradict some aspects of the observed phenomena and not be refuted.
- A model can contain assumptions which there are theoretical reasons to believe to be false.
- A model can contain assumptions which observation shows to be false.

That makes it sound as if making models is just doing sloppy science. But models are also supposed to have these further features:

- Models are evaluated in accordance with the available data, and rejected if they are inadequate.
- A good model informs us about important properties of its subject matter.

How can we have the features on the second list and also those on the first? How do we get our cake and eat it? It may not be as difficult as it seems, depending on how we interpret the ideas of evaluation (and its variants confirmation and validation or vindication – see Section 2.2) and explanation (and its cousins prediction and derivation). One aspect of modelling builds on the familiar idea of a harmless idealisation, with objects as point masses in Newtonian mechanics, gases as homogeneous fluids or as collections of randomly distributed point masses. It is clear that we can use such idealisations in formulating hypotheses which have explanatory value and can be tested. But some of the false consequences of such hypotheses are not to count as refuting them: those that are direct results of the idealisation rather than of those aspects of the hypothesis that is intended as a description of the subject matter.

Model Validation: Perspectives in Hydrological Science. Edited by M.G. Anderson and P.D. Bates.
© 2001 John Wiley & Sons, Ltd.

The danger now is that all theories will be models, that the theory–model distinction will collapse. Instead of an interesting category of intellectual constructs called 'models' we will have an account of the conditions under which a false consequence does not refute a hypothesis. (This would be worth having, but it could be had without any use of the M word.)

We strongly agree with Bates and Anderson's arguments in this volume (chapter 13) that models are typically propositional and truth-valued – and thus that they can and must be empirically tested and verified. It is very important to remind ourselves of that. An old philosophical tradition used to take it that models were merely interpretations of theories, and therefore could neither be verified nor falsified empirically. Models could not be tested, only theories could. This is a tradition that we oppose, and that we have independently criticised elsewhere. Typically, in putting forward a model, we claim, a scientist (among other things) puts forward a hypothesis – i.e. a set of claims that are subject to empirical scrutiny. But we do not believe that 'model' *just* means 'theory' (or 'hypothesis', or 'assumption'). Modelling is a distinct activity from mere theorising: in addition to putting forward hypotheses, in talking of models, scientists usually are signalling that they are making some very specific uses of idealising assumptions. The problem is that there are many different such signals they can be sending. Here are some worthwhile non-redundant claims that can be intended when a scientists describes her creation as a model:

(a) *Models as tamed theories.* Often there are theoretical reasons for thinking that the laws governing a certain domain should take a certain form, but the result is a theory that is very hard to work with. The difficulty can take the form of problems in deriving practical consequences or problems deriving the kinds of observable predictions which could be evidence for the theory. Then simplifications or idealisations are needed. Fluid mechanics is the obvious example. Newtonian physics conceives of physical systems as complexes of discrete particles, so that when we apply this way of thinking to continua it is not surprising that we get such intractable monsters as the Navier–Stokes equations. (So modelling in hydrology inevitably runs into the hard questions.) A model in this case will be obtained by taking the background theory and changing some of the assumptions that make it badly behaved or hard to understand, hoping that the particular changes made will not produce inaccurate results in the particular application that one has in mind.

In fact there are two distinct directions in which one can tame a theory. One direction is towards intelligibility or intuitive understanding. For example, the equations of general relativity permit mind-bogglingly many solutions. So expositions of the theory usually subtly restrict the range of solutions so that only those that are cosmologically plausible or geometrically manageable are considered. The other direction is towards the deduction of observable consequences. This can be motivated by the need to test the background theory or by practical applications. Most of the models in this book that are tamed models are of this second kind, which we shall refer to as *theory-based models.*

(b) *Models as analogies with other (real) systems.* We model the atom as a planetary system (warily); we model gases as fluids; we model an economy as a collection of independent self-interested perfectly rational agents. In each of these cases we know perfectly well that our assumptions are false. Electrons are much more unlike planets than they are like them, and electrostatic force and gravitation are fundamentally different. Gases are composed of molecules with empty space between them. Economies consist of people of limited rationality who group into coalitions and care very much about the welfare of particular others. We make these false assumptions in part to reduce the possibilities enough that we can begin to describe and explain, in part to make it possible to apply known theories or theory-making techniques, and in part to stimulate our imaginations. The result is then sometimes a fairly simple account, too simple to account for all the data. So as with the models discussed under (a) there is a crucial and problematic distinction between discon-firming evidence which merely shows the obvious fact that the modelling assumptions are not literally true, and more worrying evidence which shows that even taken on its own terms there is something wrong with the model.

Sometimes, though, the attempt to model one kind of phenomena in terms of ideas taken from a different domain produces a result that is far from simple. Thinking in Newtonian terms about fluids results in the Navier–Stokes equations. So then a model of the (a) kind may be needed to tame the results of (b)-type modelling.

(c) *Models as images of data.* Just as a theory can be intractable taken on its own, so can the data that a theory is supposed to explain. A macroeconomic theory, for example, may have as its empirical base the entire statistical corpus of a nation's financial data. There is too much there to begin to explain. So one has to pre-digest the data into a manageable form. Two forms of data models which are so familiar that they might escape notice are generalisations and statistics. A theory might be asked to explain why all businesses over a certain size in a certain sector of the economy were subject to take-over bids, or why the variance of the life-expectancy of new firms in a sector had declined. At first these look like raw data, but in fact they are several steps removed from it. In fact, such assertions are rarely literally true. Thus we say that elementary physics explains why all unsupported bodies near the surface of the earth accelerate downwards at 980 m/sec^2, quite ignoring birds and balloons. We are not lying; we do not really mean that absolutely literally all bodies behave in this way. We mean that for the purposes of understanding the world through physical theory it is a not too misleading summary of the data to represent it in this way. Similarly, *mutatis mutandis*, for sophisticated statistical summaries of data.

Some data models work in a much less obvious way. Quantities that cannot be directly observed, organised in ways that beg questions about their relations, are essential to much of science (see Suppes 1969).

(d) *Models as instances of theories.* Only the very simplest theories are ever completely stated. Put the equations or the axioms, even assuming they exist in a standard form, in front of an intelligent person who knows nothing about the area, and they will usually fall far short of grasping the theory, without all the necessary background facts and unstated assumptions. And there is no uniform way of stating these. They may not all even be stateable in any familiar language. So what is a theory? One fashionable suggestion – see Giere (1988) – is to take a theory to be a collection of abstract objects, such that the theory would be true if the real world resembles any of them. So, for example, we can take a theory in dynamics to consist of a set of trajectories in an abstract space, which is true of real objects in real space if they actually follow trajectories corresponding to those in one model of (or, constituting) the theory. (The model in question may not be one that anyone has described or can imagine, perhaps resulting from initial conditions and forces that are beyond our ken.)

These models are like (a)-models in that they represent reality (and like all representa-tions are selective in what they represent), and like (c)-models in that they are abstract objects, typically mathematical entities, rather than material objects. But in fact (d)-models are very unlike the other kinds of models. They are not shaped by theories – they are (constitutive of) theories. They do not abstract from data – they represent the causal facts not the available observations.

Who cares about a word? These four kinds of models are very different, and used for very different scientific purposes. So there is room for a verbal debunking strategy. That line would run: there are at least four different ways to use the word 'model', with very little in common; it is a linguistic accident that we use the one word in these different ways, and it would be less confusing if we used four different words.

This is a line to take seriously. Science consists of a lot more than theories bearing explanatory relations to bare data. And each of the purposes that one of the model types (a) to (d) serves is a real and necessary one. But that does not mean that the *same* objects do all these things, or even that it is possible for the same objects to do them. The constraints are too different.

It is possible that this is the end of the story. We should stop studying modelling techniques and validation of models in general, perhaps, and start differentiating between the techniques and validations appropriate to different kinds of models. We think there is a lot of truth in this moral. But it is also an exaggeration. To bring out the ways in which it is exaggerated consider the connections between the kinds of models.

One connection has already been mentioned. (b)-type modelling (model-as-analogy) may result in a complexity which needs to be simplified by an (a)-type modelling (model-as-tamed-theory). Moreover, a need for (c)-type modelling (data models) can be created by (a)- or (b)-type modelling. The structures that result from such modelling will by their very nature simplify, abstract away from, or simply ignore some of the causal processes at work in the given domain. As a result there will almost always be evidence with which they are not consistent. So it would be inappropriate to apply to them a 'one strike and you're out' evidential methodology (such as one might get from the writings of Karl Popper or other classic philosophers of science). Instead, the evidence will have to be massaged before it can fairly be compared with the model. It has to be structured in such a way that some of the phenomena, that the model aims to capture, are highlighted, and others hidden. So starting with (a) or (b) we are pushed towards (c).

There are links between other kinds of modelling and the rather dissimilar (d)-type modelling (models-as-instances), too. The range of models that constitute a theory is typically enormous and hard to grasp. So as an approximation to understanding the theory (rather than using it as in (a) or testing it as in (a) or (c)) one takes a small number of relatively easily described models and uses them as paradigms. So with mechanics one uses planetary systems and harmonic oscillators. But even these may be hard to handle or grasp, or more importantly hard to apply to a specific subject matter, so one can make explicit simplifications or idealisations, with a particular application or form of evidence in mind. Therefore, thinking of theories in terms of (d)-type models invites us to make (a)-type models.

Modelling is like sin. Once you begin with one form of it you are pushed to others. In fact, as with sin, once you begin with one form you *ought* to consider other forms. (You have lied to your best friend, to avoid hurting his feelings. Now you have to break a promise to tell him what a third person said about him, in order to keep up the benevolent deception.) But unlike sin – or at any rate unlike sin as a moral purist conceives of it – modelling is the best reaction to the situation in which we find ourselves. Given the meagreness of our intelligence in comparison with the complexity and subtlety of nature, if we want to say things which are true, as well as things that are useful and things which are testable, then we had better relate our bids for truth, application and testability in some fairly sophisticated ways. This is what modelling does.

2.2 MODEL VALIDATION

We now turn to the concept of validation. Oreskes and Belitz suggest in this volume (chapter 3) that the notion is unhelpful, and that the term 'validation' should be abandoned in favour of more neutral terms such as 'evaluation' and 'assessment'. All scientific knowledge is provisional, and at most we can aspire to link a particular model to current modelling practice elsewhere, or to justify the model by appeal to our best theories of natural processes, or to ground the model on a rich and highly constraining set of abundant data. (See Oreskes and Belitz, this volume, chapter 3).

Our claims in this paper are in a sense complementary to theirs: we agree that scientific knowledge is provisional, and that this makes the notion of validation initially suspect. But instead of rejecting the notion, we prefer to fine-grain, by analysing the notion of validation down into five different categories, and showing all of them to be in one way or another legitimate.

Broadly speaking, validation has two different meanings. Validating a model is sometimes taken to involve calibrating it to fit a particular set of data, or physical system, and sometimes it is taken to involve providing a vindication of the model by appeal to high-level theory. Then model calibration can take one of two forms: fixing the parameters, and refining the description. And there are three types of vindication of a model by theory: logical deduction from theory, physical derivation from theory and consistency with theory.

2.2.1 Calibration

A model may be calibrated in order to apply it to either (i) a concrete physical system, or (ii) a specific phenomenon. Examples of calibration in this volume are provided by, among others, Dietrich, Germann and Refsgaard (chapters 6, 10 and 18). Calibration may take one or both of two mathematical forms. The first consists of fixing the parameters in the model in order to fit the data that result from experiments or observations on the system, or to fit the main characteristics of interest of the phenomenon in question. This is typically just a matter of specifying the boundary conditions for the equations in the model, and it is relatively straight-forward. It is arguable that the data can never suffice to fix all parameters, especially so when the data is statistical, but it nonetheless constrains the values that the parameters can take. So this type of calibration, let us call it *calibration-1*, consists of explicitly introducing formal constraints on the values of the parameters in the model.

The second form that calibration may take is the introduction of correction factors in the model to account for causal variables known to be part of the physical system, or to be efficacious in the phenomenon. This involves actually tinkering with the equations in the model, or at least making additional assumptions beyond those strictly speaking contained in the model about the real system to be modelled. The goal of this process is a *refinement* of the model, in order to bring it closer to the physical situation at hand.

This second form of calibration, or *calibration-2*, is philosophically and methodologically the most interesting. It involves a process of concretisation: an abstract model is made more concrete by introducing causal factors that permit its application to a particular system or specific phenomenon. The important fact to note, however, is that for a correct calibration these correction factors need not be suggested by any well-established theory; they may instead be suggested by tacit knowledge of past model-building practice, or specific knowledge of the causal properties of the system or phenomenon at hand. (For example, Germann – this volume, Section 10.3 – in discussing the 'validation of momentum dissipation approach' introduces a number of causal hypothesis, such as 'recognising that some of the input water may get stuck between the surface and the depth of observation').

It is important to distinguish here between idealisation and abstraction (see Cartwright 1989, chapter 5). A model is idealised if it says something literally false about some of the features of the system that it is intended for, and it is abstract if it remains silent about a number of its features, and says nothing literally false. Everything turns on the presuppositions: from the formal point of view these two models may look identical. Consider some classical mechanical models of a pendulum. The simple harmonic oscillator, for instance, has no term to represent the slowing down of the pendulum due to the friction in the pendulum's motion through air. If, in entertaining this model, we presuppose that a pendulum suffers no air-friction at all, we idealise. If, on the other hand, we simply ignore the question of friction, and refuse to make any assumptions regarding it, we abstract.

A useful way to characterise this difference is due to Ronald Giere (see Giere 1988, chapter 3, and Giere 1999, chapter 6). We say that an idealised model *speaks about the real world*, and that it makes some false claims about it. The classical mechanical model of a simple harmonic oscillator, the SHO, if an idealisation, makes claims about real-life pendulums, but not all the claims it makes are true – for instance it makes the false claim that real-life pendulums are subject to no air-friction. An abstract model, on the other hand, does not lie. But this is *because it does not speak about any object in the real world*, only about an abstract object, the one implicitly defined by the equations in the model. Considered as an abstraction, thus, the SHO model would speak truly about an abstract object, namely the simple harmonic oscillator, defined implicitly by its equation of motion.

We can employ Giere's framework to make further headway on the relation between models and theories, and to understand the kind of procedure involved in calibration. Idealisation requires that there be a (true) theory, available to us, describing the corrections needed to apply the model to every concrete system and specific phenomenon. For only then can we be sure that the model makes *false* claims about the system, and can we try to correct these claims. Classical mechanics, for instance, recommends modelling air-friction as a linear function of velocity. So it tells us exactly the way in which the SHO idealises. It is because of this that the SHO is, properly speaking, an idealisation, not an abstraction.

But in most cases of calibration – such as those discussed by Dietrich, Germann and Refsgaard in this volume – no such theory is available. The corrections come from a varied mixture of theoretical and non-theoretical modelling techniques. We have no established body of knowledge to compare the model with, and thus we have no means to tell if the model is speaking falsely about the world. It is the availability of such background knowledge that turns an equation into an idealisation; as we have seen there is nothing in the formal description a model M that will tell us that M is an idealisation of some phenomenon P. So, properly speaking, most of the models that require validation in the calibration-2 sense are not idealised, but abstract. And calibration-2 is a process that makes an abstract model concrete.

2.2.2   Theory Vindication

A model can also be said to be validated if it can be shown to be a consequence (logical or a physical consequence) of some established theory, or if it can be shown to be consistent with some established theory.

Let us carefully define these terms. A logical consequence of a theory is a statement that is consistent with the theory and whose contrary is inconsistent with the theory. In other words, as a theory logically dictates all of its logical consequences, a contradiction can only be a logical consequence of an inconsistent theory. Logical deduction provides the strongest type of theoretical vindication; let us call this *vindication-1*. But genuine examples of vindication-1 in science are surprisingly hard to find. Astronomy provides some: Kepler's laws, which describe the ratios of the planets' periods and distances from the sun, for instance, follow deductively from Newton's theory, and they are thus validated by Newtonian mechanics.

On the other hand, the contrary of a theory's physical consequence, as we wish to define the notion here, is not necessarily inconsistent with the theory. The theory cannot logically dictate, but merely suggest, its physical consequences, and a contradiction may be a 'physical consequence' of a perfectly logically consistent theory.

This may seem at first counterintuitive, but it is in fact only counterintuitive in light of the assumptions of determinism and completeness that many of us have been schooled into. The assumption of determinism states that at its most fundamental level nature is always deterministic; statistics only enters the picture in order to account for our lack of knowledge. But consider genuinely irreducible statistical theories. Quantum theory, for instance, even if complete, is only able to tell us that the electron just created in a cascade experiment in the laboratory will be found at end E of the lab with probability p, or at end E* with probability p*; where, say, p = 1 − p*. Then the statements 'the electron will be at E' and 'the electron will be at E*' are both physical consequences of the theory, even if, of course, when the experiment is performed, the electron appears either at E or at E*, but never simultaneously at E and E*. Thus a model of the electron in E may be vindicated by showing it to be a physical consequence of the fundamental quantum theory, even if another model of the inconsistent claim that the electron is at E* can similarly be vindicated.

We claim that this type of vindication of a model as 'physical consequence' of a theory (let us call it *vindication-2*) is not unique to statistical theories. It is by now common lore that the relation of theory and evidence is rather more subtle and complex than simple logical deduction, even for deterministic theories. There is a gap between theory and evidence that, we claim, the notion of physical consequence can fill. This is because the theory does not describe all the facts that are conditions for the occurrence of its physical consequences. The theory, in other words, is not complete. As a result of this, the theory cannot logically dictate its physical consequences but, at most, suggest them. The assumption of completeness would rule out theories that are not able to describe all the relevant facts in their domain. Combined with the assumption of determinism this gives us the result that all consequences of a theory must be consistent with each other. But the result does not follow if any of these two assumptions is abandoned. And there are good reasons to think that our best theories reject not just determinism but also completeness. (For arguments against determinism, see, for instance, Dupre (1993) and Suppes (1984). For arguments against completeness see Cartwright (1999).)

Why is this relevant? Vindication of a model by a theory of which it is shown to be a logical consequence provides confirmation for the theory. In those instances, the model, if true, confirms the theory, and, if false, falsifies it. This is a strong relation, and difficult to ever find in practice. When it is found, the model that the theory vindicates provides a resounding verification (or falsification, if the model turns out to be incorrect) of the theory, and vice versa. The observational correctness of Kepler's laws verifies Newtonian mechanics, and Newtonian mechanics lends confirmation to Kepler's laws. However, very often a theory does not logically dictate a model of a phenomenon; at most, and with some luck, it suggests it. The model may then increase the probability of the theory being true. If it does (i.e. if Prob (T/M) > Prob (T/not M)), we can still say that the model confirms the theory, although only by constituting *evidence* for T. Bayesianism explains well how this works. The theory is more likely to be true if the model is true (and vice versa). But if the model turns out to be false the theory could still be true: the theory and the contrary of the model are not logically inconsistent. (This is so unless (i) Prob (T/not M) = 0 and (ii) Prob (T) > 0.) The quantum mechanical example of the position of an electron is arguably such a case. Both the observation that 'the electron is at E' and the observation that 'it is at E*' can constitute positive evidence for the theory, even if they are contrary.

The logical consequences of a theory may increase its confirmation. But so can the physical consequences of a theory. Both can increase the degree of confirmation of the theory. These possibilities are well known to philosophers, and they are well understood. But sometimes, perhaps often, the truth of the model does not increase the probability of the theory being true even if the theory vindicates, and in this sense validates, the model. Even if a theory can be applied to a model of a phenomenon, it may not be able to suggest the corrections required for the application. (See Suárez (1999), where an example from the physics of superconductivity is discussed.) If the theory does not even suggest a model, then the success of the model will not provide any confirmation for the theory.

To many philosophers this is deeply puzzling – how is it possible? The contrast that we drew earlier between idealisation and concretisation may help here. A model M that can only be applied to a phenomenon P by introducing corrections into M that are not suggested by any established theory T is not an idealisation of P (or not an idealisation relative to P). M is, at best, an abstraction relative to P. And the process of correction of the model is therefore not a de-idealisation, but a concretisation of the model. It follows that the model does not speak about P, and therefore T cannot be confirmed by M, either by strict verification or by the Bayesian methods of probabilistic confirmation. What this entails is that, in this case, M is neither a *logical* consequence nor a *physical* consequence of T. The most we can do is to show that M and T are consistent. This proof of consistency is, we claim, a legitimate form of theoretical vindication – and in fact a most interesting and common one. Let us refer to it as *vindication-3*. Examples of vindication-3 in the present volume are the non-cohesive sediment transport models by Garcia (chapter 15), the more general among the snow models described by Davis et al. (chapter 11), and some of the simulations in Booltink (chapter 9). The conceptual and physics-based models of Young (chapter 7) may be employing vindication-3 or vindication-2.

This completes our analysis of the notion of model validation. We have seen that validation may take one of five forms: (i) calibration by fixing boundary conditions; (ii) calibration by refinement, (iii) vindication by logical deduction (from a theory), (iv) vindication by physical derivation (from a theory), and (v) vindication by proof of consistency (with a theory).

2.2.3   When Vindication is Calibration

Even if these five forms of validation are clearly distinct, presenting different practical difficulties and requiring different solutions, they have often been conflated. We think that we can diagnose why. It turns out that the most common and complex form of calibration, calibration-2, is methodologically very similar to the most common and complex form of vindication, vindication-3. Vindication-3 and calibration-2 are alike in that they do not involve idealisation, but concretisation. Even if there is a body of theoretical knowledge available to us that we can compare our model to, we are nonetheless unable to use it in order to correct our model to make it applicable to the phenomenon. The corrections required to apply the model come from other sources. This is true of calibration-2 by definition. It turns out to be true also of vindication-3. Remember that in vindicating-3 a model by appeal to a theory, we show the model and theory to be consistent because we cannot show the theory to suggest the model, or the corrections required to apply the model to a particular physical system. It follows that the corrections required to apply the model to the physical system are suggested by other sources.

In both cases the same broad methodology is used: consider the model, consider the object to be modelled, and with patience and ingenuity bring to bear on the modelling situation whatever knowledge you may have – knowledge that is typically not already contained in the high-level theory assumed to be true in the domain. So, even if fluid dynamics is the underlying true theory of hydrological phenomena, validation in hydrological models must follow a course that essentially ignores fluid dynamics, and takes it at most as one more tool in model-building. And what we have been arguing is that this is neither exceptional nor uncommon in the practice of model-building in science at large.

2.3   SIMULATION VERSUS ESTIMATION

This chapter has been an exercise in distinction-making. There are distinct things one can mean by 'model' and distinct enterprises of model validation. In fact there are five of each. So any particular model validation should find a place in one of the cells of a five by five grid. And – if our account of modelling is along the right lines – the characteristic difficulties of modelling should respect the distinctions between the cells. Difficulties characteristic of one category should – if there is anything to what we are saying – be largely absent in another. (Just about any data can be squeezed and shoved into just about any purely descriptive framework; a stricter test is whether the framework captures significant causal generalisations.) Let us consider some data: the chapters on hydrological modelling in this book. But before we confront the data, a simplification of the account is in order.

To begin, as remarked in Section 2.1, models in hydrological science are very rarely tamed theories where the taming is motivated by the need for understanding or exposition. For hydrology does not aim to challenge fluid dynamics. It assumes fluid dynamics is fundamentally right, though not much use in its pure form for predicting and explaining actual flows of water in actual geological conditions. Moreover, hydrological scientists rarely worry about models as instances of theories. That is not to say that philosophers or psychologists of science may not interpret their patterns of thought in terms of them. But the scientists themselves rarely take such things as conscious objects of concern. (Nor should they.) And while models as analogies may play an important role in the exposition and motivation of hydrological theories, hydrological scientists do not spend a lot of time trying to fix their exact form or decide between rivals. If metaphorical models do their job of suggesting or communicating, they will be used, and questions of their accuracy or uniqueness will not arouse much concern. And this, too, is as it should be, as uniqueness can be expected to reside only in what actual water flows do and in what the underlying laws of fluid motion are.

That leaves us with just two large categories of models: what in Section 2.1 we called theory-based models, models as theories tamed for evidence and prediction, and data models. If what we are saying is right, there should be different kinds of problems which arise in the four main possibilities: theory-based/calibration, theory-based/vindication, data model/calibration and data model/vindication. And indeed there do seem to be two characteristic methodological problems, each of which arises in just one cell of the grid.

The first problem is that of simulation. As argued in the chapters by Beven and by Young (chapters 4 and 7), one frequent hydrological situation is that one has a predictive model that does not generate good simulations. That is, one has a model in which the values of crucial parameters have been fixed in such a way that it predicts the output of a particular physical system quite accurately. But when one tries to generate data for different related systems problems arise. Either the predictions that emerge are clearly wrong, or there is no clear way of determining the appropriate values of the parameters. It is as if the predictively adequate model works because of a specific combination of parameter values, but once one parameter is varied suitable combinations of the others are not forthcoming.

The simulation problem is a crucial hazard of data models, as is clear from Beven and Young. With theory-based models, though simulation may be problematic, there will not typically be a prediction/simulation asymmetry. Neither need be easy, but the problems should be the same. As Germann's chapter (10) argues, the problem that arises with theory-based models is not comprehensive prediction for one system that does not transfer to others, but adequate prediction and simulation for one kind of data that does not transfer to other kinds of data.

The simulation problem is moreover a problem primarily of vindication rather than of calibration. As applied to a given physical system the given parameter values are fine, but if they do not generate reasonable *counterfactual* consequences – specifications of what the system would have done under other circumstances – then they cannot be taken as a way of describing the real causes of the phenomena. (In this connection see Section 1.1 of Dietrich's chapter (6).) Without some simulatory power a power is 'only' a description of the data. The other characteristic problem is best illustrated by Garcia's chapter (15), on sediment transport (though particular models described in several other chapters would also have made the point). Here the model consists of a highly theoretical treatment of a widespread phenomenon and the problem is to make a connection between the some undetermined theoretical quantity and some quantity whose value can be estimated from the behaviour of a particular system, using either raw data or a suitable data model. The problem is essentially a statistical one, of inferring the true value of some quantity which, if the theoretical assumptions are right, will correspond to an unknown in an equation derived by artful simplification from the underlying physics. Call this the problem of truthful estimation.

Truthful estimation problems will occur primarily in connection with theory-based models. For given that the underlying theory characterises correctly the underlying reality, the value of the unknowns to be filled in is what they actually are (or what an infinite set of them are, as when one is specifying a boundary condition). And such problems will occur primarily when the aim is calibration rather than vindication. For vindication is conceptually simple with theory-based models: it consists of showing that the model is deducible from, physically derivable from, or consistent with the underlying physics. These may not be simple tasks, but their nature is clear, and they do not involve any pure estimation.

The four cells are thus characterised as follows:

- theory-based/calibration = (+ truthful estimation, − simulation)
- theory-based/vindication = (− truthful estimation, − simulation)
- data model/calibration = (− truthful estimation, − simulation)
- data model/vindication = (− truthful estimation, + simulation)

Of the four combinations of kinds of model and kinds of validation, two receive unique characterisations in this way, and two are not distinguished. Appeal to these two kinds of problem does not distinguish 'theory-based/vindication' from 'data model/calibration'. But those two are distinguished in other ways. For, as remarked above, the characteristic problems of 'theory-based/vindication' are those of theoretical ingenuity and mathematical sophistication, while those of 'data model/calibration' are those of physical intuition, and experimental skill and ingenuity. In other words, in the first case one is looking for assurance that a model is an accurate portrayal of reality as characterised by a background theory, so the basic question is just 'does it follow?' (or, since these are models 'how closely does it follow?'), and in the second one the basic question is just 'how can we find out?'. (Note that the situation described is one where one already has a data model and is trying to calibrate its parameters. In the construction of data models, on the other hand, no end of theoretical ingenuity and mathematical sophistication can enter.) So, representing this dimension, admittedly rather crudely, as the relevance or not of empirical data, we can complete the grid as follows.

- theory-based/calibration = (+ truthful estimation, − simulation, + empirical)
- theory-based/vindication = (− truthful estimation, − simulation, − empirical)
- data model/calibration = (− truthful estimation, − simulation, + empirical)
- data model/vindication = (− truthful estimation, + simulation, + empirical)

As model of modelling, this can only be a first attempt. The most general conclusion to draw is that modelling in any scientific domain is a very varied as well as a very challenging activity. The variability is as important as the challenge, since the kinds of difficulties that modellers face are not randomly distributed. Different kinds of models lead to different kinds of challenge.

REFERENCES

Cartwright, N. 1989. *Nature's Capacities and their Measurement*. Oxford University Press, Oxford.

Cartwright, N. 1999. *The Dappled World: A Study of the Boundaries of Science*. Cambridge University Press, Cambridge.

Dupre, J. 1993. *The Disorder of Things: Metaphysical Foundations of the Disunity of Science*. Harvard University Press, Cambridge, MA.

Giere, R. 1988. *Explaining Science: A Cognitive Approach*. University of Chicago Press, Chicago.

Giere, R. 1999. *Science Without Laws*. University of Chicago Press, Chicago.

Suárez, M. 1999. The role of models in the application of scientific theories: epistemological implications. In: Morrison, M. and Morgan, M. (eds), *Models as Mediators*. Cambridge University Press, Cambridge, 168–196.

Suppes, P. 1969. Models of Data. In: P. Suppes, *Studies in the Methodology and Foundations of Science*. Reidel Publishing Company, Dordrecht, 24–35.

Suppes, P. 1984. *Probabilistic Metaphysics*. Blackwell, Oxford.

3 Philosophical Issues in Model Assessment

NAOMI ORESKES[1] AND KENNETH BELITZ[2]
[1]*Department of History and Program in Science Studies, University of California San Diego, USA*
[2]*Water Resources Division, US Geological Survey, USA*

3.1 Introduction

In recent years, there has been an explosive increase in the use of models in a wide variety of fields – pharmacology, toxicology, economics, forest ecology, climatology and psychology are some examples. With increased use has come increased concern about model assessment. How can we tell if a model of a complex system is a good model? How can we judge the relative strengths of different models? How can we test a model that we wish to use in a predictive mode? In this chapter, we consider these issues in light of examples from both hydrology and other disciplines in which similar issues comes to the fore.

In many disciplines model results can be compared with independent lines of evidence. In pharmacology, a simulation model can be compared with a clinical trial. In aeronautics, simulations can be compared with physical results in a wind tunnel or from a prototype. In chemistry, numerical output can be compared with laboratory experiments. In hydrology, however, the geographical and chronological scales of processes and the difficulty of access to the subsurface make it difficult to find a suitably comprehensive independent line of evidence against which to test model claims. So how do we judge the reliability of the knowledge the model provides? Both the makers and the users of models want to know whether a model accurately reflects the natural processes it claims to represent.

The inherent uncertainties of models have been widely recognised, and it is now commonly acknowledged that the term 'validation' is an unfortunate one, because its root – valid – implies a legitimacy that we are not justified in asserting (Tsang 1991, 1992; Anderson and Woessner 1992; Konikow 1992; Konikow and Bredehoeft 1992; Oreskes et al. 1994; Beck et al. 1997; Oreskes 1998; Steefel and van Cappellen 1998). But old habits die hard and the term persists. In formal documents of major national and international agencies that sponsor modelling efforts, and in the work of many modellers, 'validation' is still widely used in ways that assert or imply assurance that the model accurately reflects the underlying natural processes, and therefore provides a reliable basis for decision-making. This usage is misleading and should be changed. Models cannot be validated. The reasons why have been outlined in detail elsewhere (Konikow and Bredehoeft 1992; Oreskes et al. 1994). Here, we summarise these reasons before proceeding to a more detailed discussion of conceptual issues that pose problems even for models that seem to work.

Model Validation: Perspectives in Hydrological Science. Edited by M.G. Anderson and P.D. Bates.
© 2001 John Wiley & Sons, Ltd.

It is widely acknowledged that all scientific knowledge is provisional. It has to be. If scientific enquiry is a process of discovery of new knowledge and refutation of old mistakes, then the knowledge we have at any given moment must be provisional. For science to advance, we must be critical of existing knowledge. Therefore all scientific knowledge is, in some sense, uncertain, and the issues of uncertainty inherent in model validation are not unique to modelling. However, there are particular aspects of numerical simulation models that exacerbate the problem of uncertainty to a degree that may be substantially greater than in some forms of scientific endeavour. These features are non-uniqueness, the problem of temporal and spatial divergence, and the subjectivity of model assessment.

1. *Non-uniqueness.* A fundamental issue in modelling is the problem of non-uniqueness: more than one model configuration may produce the same output. The more complex the model, the more opportunity for alternative model realisations.

   There are three major forms of non-uniqueness: numerical, parametrical and conceptual. Bethke (1992) has emphasised numerical non-uniqueness: the possibility of more than one solution to the governing equations. Even a well-constrained model may give a uncertain result if the governing equations can be solved in more than one way. Konikow and Bredehoeft (1992) have emphasised parametrical non-uniqueness: there can be a wide range of possible model inputs to achieve an expected output. The more complex the problem being addressed, the greater the likelihood of significant non-uniqueness in the model solution. Below, we discuss the problem of conceptual non-uniqueness: that more than one conceptual model may prove adequate to account for the empirical evidence. Because most problems in the earth and environmental sciences are inverse problems – we know the configuration of the world, but we lack knowledge of the processes and parameters that produced it – we always face the problem of non-uniqueness.

2. *Temporal and spatial divergence.* Models are frequently calibrated or tested against historical data for a given region or time frame before using them to predict the behaviour of the system. Typically, however, our calibration or history match is smaller or shorter than the space or time frame that we want to predict. Although the model may accurately reproduce available observational data, there is no guarantee that it will perform at an equal level when applied over a larger geographic scale or a longer time frame. There may be small errors that do not impact the model fit, but which, when extrapolated over much longer time frames or much larger geographical areas, will generate significant deviations.

   A good example of this is given by Konikow and Bredehoeft (1992) in an analysis of a 40-hour pump-test in the Dakota sandstone, one of the most important aquifers in the continental United States. The results of the pump-test were consistent with two models: the Theis solution, which assumes no flow through confining layers, or the Hantush solution, which allows for transient flow through confining layers. Konikow and Bredehoeft estimate that the pump-test would have to be run for more than 1000 years for a divergence to be detected. Yet some models calibrated over short time frames attempt to predict regional flow over geological time frames. A model may also diverge from historical data because the system it describes changes – this point is discussed at length below.

3. *Subjectivity of model assessment.* The possibility of magnification of small errors over time and space leads to another source of uncertainty in modelling: the subjectivity of our judgements of what constitutes a 'good' fit. The literature of model validation is filled with terms like 'adequate', 'acceptable', 'satisfactory', and 'reasonable', even in sophisticated mathematical treatments. These are obviously subjective terms and they highlight what should be an obvious point: all models are approximations. Most modellers freely acknowledge this. But how do we determine what constitutes a good enough approximation?

Because the definition of good is inherently subjective, there is unavoidable uncertainty associated with the definition of acceptable error. As put by Konikow and Bredehoeft (1992), 'under the common operational definitions of validation, one competent and reasonable scientist may declare a model validated while another may use the same data to demonstrate that the model is invalid'. There is no way to eliminate subjectivity and value judgement when we ask ourselves what constitutes a 'reasonable' degree of uncertainty (Bredehoeft and Konikow 1992; Konikow 1992).

For these reasons, we argue here, as we have elsewhere, that the language of validation is unhelpful and should be avoided (Oreskes et al. 1994). We should prefer neutral terms like evaluation or assessment. Good science requires a critical stance, but the language of validation shifts our focus in a different direction. It implies that we can provide firm assurance of the reliability of our models. In fact, the most we can do is to say that a model is close to the state of the art (if it is), that it has been grounded in our best understanding of known natural processes (if it has), and that we built it on the basis of abundant, well-constrained empirical input (if we did).

Even if we use a neutral term like model evaluation, problems remain. How *do* we evaluate a complex model? How do we differentiate between better and worse models? What should we do if we need to use a model in a predictive capacity for social, political or economic purposes? These questions motivate the remainder of this chapter.

3.2   THE PROBLEM OF PREDICTION

Much of the demand for model assessment derives from the desire to use models in a predictive mode. A large number of models have been built in response to environmental problems – nuclear waste disposal, acid rain, water supply, groundwater contamination – which involve forecasting the behaviour of complex natural systems (Sarewitz et al. 2000). Models that involve predictive capacity raise a particular set of issues related to a basic fact: predictions are generally wrong. We might put this another way: predictions are always wrong, in the sense that we can't and don't expect any model to be precisely correct in all respects. What is more significant is that models are often wrong in systematic ways. This is true in most areas of human endeavour and the earth sciences are no exception. While the number of studies that examine the predictive accuracy of model results is modest compared to the number of models that have been built, the available studies are clear in their results. We do not know enough about complex natural (or human) systems to be able to predict them reliably. This leads to two questions: Why are we making predictions? What can we learn from past mistakes?

3.2.1   Are Predictions Necessary?

If predictions are unreliable, then why do scientists make them? When models are built in aid of public policy, scientists may feel that they have to make predictions to serve the agencies and constituencies that support them. While there may be good reasons to run a model in a predictive mode, it is important to realise that public policy does not necessarily require it. In many cases it is possible to develop sound public policy based on a general scientific understanding without specific scientific predictions (Brunner and Ascher 1992; Sarewitz and Pielke 1999; Rayner 2000). It is important for scientists to understand this so as not to feel compelled to make (bad) predictions simply because there seems to be no other choice.

A good example is global climate change. A great deal of climate research is dedicated to the construction of complex General Circulation Models (GCMs) with the goal of predicting the effects of increased atmospheric carbon dioxide. This is in part the result of a policy decision. In the early 1990s, the first Bush administration in the United States defined the primary problem of climate change to be scientific uncertainty, and made research its central policy response (Brunner 1991; Brunner and Ascher 1992; Rayner 2000). Eliminating or at least greatly reducing uncertainty was considered a prerequisite for action, and this led to greatly increased funding for climate research, particularly climate modelling. Scientists did not object; the idea of knowledge as a basis for action is eminently reasonable. And what scientist would protest better funding? Yet, despite (or perhaps because of) recent advances in understanding, it is clear that significant reductions in predictive uncertainty will be some time in coming. If global warming is real, and if it is happening now, then we may not be able to wait for predictive accuracy. By the time we have achieved it, damage will have been done and may be impossible to undo.

In response to external pressure to generate predictions, one can point to the existence of alternatives strategies and complementary courses of action. In the case of climate change, for example, society could focus on mitigation and adaptation without precise predictions. As Rayner (2000) points out, steps can be taken to protect and increase human welfare that would be beneficial to the populations involved even if the predicted climate change does not occur. Policy analysts refer to this as a 'no regrets' strategy – to act so people will benefit even if the feared event does not occur. There are positive local benefits to be accrued from reductions in atmospheric pollution (e.g. in human health, protection of ecological habitat, preservation of viewsheds, the pleasure of breathing fresh air) irrespective of whether global warming turns out to be a profound global threat. It can be rational to act despite uncertainty, and actions can be evaluated as one goes along. Finally, even if we had an accurate predictive model of the effect of atmospheric carbon dioxide on global climate, it would not in itself dictate appropriate policy response. Policy decisions hinge as much on the values of communities as they do on the facts of science.

Scientists can also create alternative avenues of study. An example is earthquake prediction. In the 1970s, hopes were high for accurate forecasting of earthquakes in California within a few years, and some seismologists went so far as to claim that short-term warning of impending earthquakes was imminent (Nigg 2000). Today few if any scientists consider this a realistic goal. The failure of earthquake prediction has led to a shift in scientific focus toward the study of seismic wave propagation and material response, which in the long run may prove efficacious in reducing seismic hazard. Meanwhile, policy-makers have focused on societal preparedness: emergency response plans, back-up medical facilities, individual family preparedness etc. (Nigg 2000). In hindsight one can see that accurate prediction of individual earthquakes may be less important to society at large than a general appreciation of seismic risk and an appropriate overall pattern of planning and preparedness.

Given the difficulty of making accurate predictions and the insight that public policy may not necessarily require them, it seems clear that model assessment need not focus solely on predictive capacity. One can gain insight and test intuitions through modelling without making predictions. One can use models to help identify questions that have scientific answers.

One approach that has been developed recently is to acknowledge that available data is often insufficient to select one model uniquely among many that might be built. Keith Beven and colleagues refer to this problem as equifinality – a form of non-uniqueness – and developed the GLUE procedure (Generalised Likelihood Uncertainty Estimation) to address it (Beven 1993, 1996; Aronica et al. 1998; Hankin and Beven 1998a, 1998b). The purpose is to identify the range of model output that is generated from a range of feasible model input. Gupta et al. (1998) also address the issue of model non-uniqueness. They develop a multi-objective optimisation procedure for identifying families of solutions that satisfy various combinations of different measures of model performance.

These efforts help to focus attention away from expectations of certainty, and underscore the fact that models of natural systems are simplifications of complex systems. By formally recognising a range of possible model input and corresponding output, these methods alert policymakers and the public that model output is not the same as an accurate prediction.

But what happens when scientists are asked to make predictions to support public policy (Beck et al. 1997)? Whether one makes a single prediction, or generates a range of predictions, there are still conceptual issues at stake. Reality may turn out to lie beyond the range of model prediction. Therefore it is worthwhile to understand how and why model predictions have gone wrong in past. The errors in models are frequently non-random. In fact, they appear to be systematic in particular and recognisable ways.

3.3 SYSTEMATIC ERROR AND BIAS

The incompleteness of our knowledge of natural systems opens the door to systematic error and bias. We can think of model input as falling into three categories: factors that are known and measured, factors that can be estimated based on informed judgement (e.g. based on prior experience in systems that are believed to be similar), and guesswork. All models involve informed judgement and typically a bit of guesswork as well. Wherever subjective judgements are required, the potential exists for systematic error and bias.

Given the diversity of human attitudes and opinion, one might hope that individual bias and idiosyncracy would tend to cancel out over the course of modelling efforts. For example, construction of more than one model of a system can reveal biases and errors in a single model. This can be an argument for a 'competing teams approach'. Different modellers or groups of modellers ought to have different subjective tendencies, and the totality of independently constructed models might converge on a correct result. But redundancy in modelling efforts is very costly. Even if we pursue it, it may not solve our problems, because systematic bias in professional communities can cause independent groups and individuals to bias their results in similar ways (Fischhoff 1982; Tversky and Kahneman 1982; Ascher and Overholt 1983; Ascher 1981, 1993).

Ascher (1993) presents a *post hoc* analysis of 21 development projects funded by the World Bank, all of which had been subject to rigorous *ex ante*, quantitative rate-of-return analyses. In each case, the analyses were performed by educated, experienced professionals, working under conditions in which the criteria for assessment were codified. Ideally such conditions should leave relatively little room for subjective bias, or at least render any such bias individual and random. But not so. The results of these analyses were systematically and seriously biased toward overly optimistic results. The true rates of return on these projects were substantially lower than predicted, and many projects were approved which in retrospect should not have been, given the evaluation criteria.

In a related study, Isham and Kaufmann (1995) analysed *ex ante* appraisals for over 1200 World Bank projects, in which the average predicted rate of return was 22%. The actual average rate of return turned out to be 12%, just skimming the 10–12% threshold value set by the World Bank for approval. Many individual projects fell well below this threshold value and several had a negative rate of return. Perhaps more important, many of these projects caused substantial environmental damage. These detrimental effects were not wholly unanticipated, but they had been justified *ex ante* on the grounds of the expected economic benefits, benefits that may have made alternative patterns of development seem ineffectual. In retrospect, however, the exaggerated economic benefit appears as a significant factor in discouraging alternatives.

Why were these analyses systematically biased, and what relevance does this example have for hydrology? One answer is human nature. While the human factors may seem obvious, they bear repeating because they apply to natural scientists as well as to social scientists: positive appraisals create work for the appraisers. The World Bank is in the business of lending money for development projects, and an appraiser who consistently rejected proposed projects would soon find him or herself under pressure. There is a parallel here with models in the natural sciences: much of the funding for modelling comes from agencies that want to act on a specific public policy or environmental problem. And who is more likely to be funded: The modeller who says he or she can produce the desired predictions, or the one who says he or she can't?

The appraisers of economic models generally share the goals of those proposing the projects being modelled: they believe in economic development and are therefore willing to provide analyses despite incomplete information. Again there is a parallel: most scientists believe that scientific information can and should form the basis of rational public policy. Scientists share the goals of agencies that want to use the available scientific knowledge as the basis for action, and this makes them susceptible to overestimating the capability of the model to provide such a basis.

Shared goals and commitment to scientific solutions may help to account for why modellers construct models with incomplete information, but do not explain why the models are biased in a particular direction. There are obvious political reasons why promoters of projects would wish to exaggerate their potential benefits, but the analysts at the World Bank are professionals who, like natural scientists, consider it their job to make informed and objective judgements based on quantitative analysis. A professional who consistently made inaccurate forecasts would seem to be placing his or her credibility at risk. Why would seasoned professionals consistently overestimate the benefits of proposed development projects? Ascher (1993) argues that two technical factors contributed to systematic error in this case: (i) inadequate incorporation of negative impacts of unknown or unlikely effects and (ii) an optimistic bias with respect to implicit conditionals.

3.3.1   Inadequate Incorporation of Unknown or Unlikely Events

All models involve parameters or processes that are unknown or poorly known, and many models involve the problem of estimating unlikely events. What do we do about it? In the case of parameters that are poorly known, modellers may take a 'best-guess' approach. Lacking quantitative data on a parameter, they make a best assessment of what it might be. Typically this is done on the basis of past experience in systems believed to be similar. In the case of parameters that are unknown, modellers may leave them out.

Ideally, we would improve our knowledge of the input parameters, and a large number of researchers are actively doing this. The methods typically draw upon relationships between model input parameters and auxiliary data, or utilise mathematical inverse methods to estimate the input parameters based on limited measurements of those values and more extensive state variables (e.g. Hill 1992). These methods improve capacity to specify model input. Nevertheless, in many modelling problems, input parameters remain incompletely known, and there may still be unknown effects.

By definition, an unknown effect cannot be incorporated into a model. But if an unrecognised factor rears its head, then the model will be in error. In highly structured, settled human societies, unanticipated events are almost always costly: they require responses (which cost money), and they may cost lives. Leaving out unknowns in models that involve human systems almost always biases a model in an optimistic direction (Ascher 1993). Reality is likely to be worse than we wish.

There is a similar pattern with low probability events. The rarer an event is, the more difficult it is to study and analyse. While considerable progress has been made in the statistical analysis of rare events, it remains a vexing domain. Many rare events leave little or no trace, and therefore are almost impossible to analyse. If you cannot analyse something, then you cannot generate a meaningful quantitative measure of it for model input. If an event has never occurred, or never occurred in this particular context, it is impossible to quantify its likelihood.

Scientists are understandably loath to make up values where none exists. Lacking adequate basis for analysis, modellers may leave out extremely low probability events. The small probability of the event discourages both its analysis and its incorporation. Yet rare events do occur, and largely because they are unanticipated their impact is generally negative. The omission of very low probability events exaggerates our capacity to make accurate predictions and biases our models in an optimistic way.

3.3.2   Optimistic Bias in Implicit Conditionals

Ascher points out that a model is a conditional proposition: if certain factors are in place, other things will follow (Ascher and Overholt 1983; Ascher 1993). Yet the proposed factors are almost never in place in exactly the way that the model proposes, and the effect of deviation is generally negative. This is analogous to the phenomenon of construction delays, well known to anyone who has renovated a kitchen or awaited an office in a new university building: there are many factors that make projects take longer but few that speed them up (Ascher 1993; see also Tversky and Kahneman 1982, p. 16; Gutierrez and Kouvelis 1991).

The idea of a model as an implicit conditional applies to natural systems as well as human behaviour, because running a model in a predictive mode presupposes that the driving forces of the system will remain within the bounds of the model conceptualisation. If the driving forces change, then even a well-calibrated model will fail. Consider the following examples.

Example One: The Salt River and Lower Santa Cruz River Basins

Konikow (1986) presents the example of the Salt River and lower Santa Cruz River basins in Central Arizona, an arid area of intense groundwater use. In the mid-1960s, a groundwater model was built and calibrated on the basis of 41 years of historical data on pumping and water levels (1923–1964), and used to predict future water level changes for the next ten years. On the basis of the model's success in matching the historical data, the modeller had concluded that the model was a valid representation of the system and therefore could 'be used to predict future ground-water conditions' (Konikow 1986, quoting Anderson 1968; see also Konikow and Patten 1985). In the early 1980s, the predictions were compared with what really happened. The model predictions were systematically wrong: real water levels were consistently higher than predicted. Groundwater levels had been falling throughout the calibration period due to groundwater pumping, and the model predicted that they would continue to fall. Major declines were predicted everywhere, but the real changes were either smaller than anticipated or in the opposite direction.

Why did the model fail? In retrospect, it can be seen that the system changed almost as soon as the model was produced. The 20-year period prior to model construction was one of relatively uniform downward trend in water levels changes driven by a consistent upward trend in groundwater pumping. But, as Konikow explains, 'a marked break in this trend occurred very soon after the end of the calibration period'. In the mid 1960s, people began to pump less. They may have done so because costs were rising as the water table was falling; because farmers took steps to increase irrigation efficiency in response to rising costs; because more surface water was made available; or because land was taken out of production in response to rising costs or encroaching urbanisation. It is unclear which of these factors was most significant, but it is clear than human factors played a large role in the inaccuracy of the predictions. Although this was a model of a physical system, human activity was decisive in undermining its predictive capacity.

Although it is not news that humans are unpredictable, many models in the natural sciences implicitly assume consistency in human behaviour. Few terrestrial systems are completely closed to human effects, and this is particularly so for the systems we are likely to be modelling. If the purpose of a model is to aid in decision-making, by definition this means that the system is of interest to humans and very likely impinged upon by human behaviour. But it goes against the grain of physical scientists to acknowledge this. We have been trained to study 'natural' systems, and only recently have we begun to assimilate the fact that there is no sharp line between human and 'natural' systems.

Beyond demonstrating the role of human factors in models of natural systems, there are two other striking features of the Salt River study. The first is that modellers may have insights that are not incorporated into their models. In this case, the modeller noted that the assumption of consistency in driving forces was unlikely: actual pumping rates would probably fall as costs continued to rise. If this happened, then the predicted declines might be exaggerated – and they were. (In hindsight, the model stands as a marker of the changes that were imminent: the situation was under pressure, people were concerned – which is why the model was built – and changes in human behaviour were likely.) But although the modeller acknowledged this in discussions of the model, he did not incorporate this insight into the model. To predict a change in the driving forces of a system may require explicit insertion of the modeller's judgement; the goal of objectivity makes modellers reluctant to do this. The result is models that fail to encompass the full extent of their builders' insight and intuition.

The second striking feature of this study is that the effects of human behaviour are not the whole story. Konikow's analysis revealed that the spatial pattern of predictive error was not explained by changes in groundwater pumping. The correlation between the spatial distribution of error in assumed pumping and the spatial distribution of error in water levels was very low (-0.086). Again, these errors were not random. Certain areas were much worse than others, and errors in nearby wells tended to be close in value. Yet Konikow was unable to correlate the pattern of error with any known factor in the model. Something else in the model was wrong, and even retrospective analysis failed to reveal what it was. This leads to an important conclusion: most complex models probably contain more than one source of error (see also Alley and Emery 1986). If so, then the various errors may cancel each other out to create a model that fits the historical data, but is still flawed. The more complex the model, the greater is the prospect for compensating errors, and the more likely it is that such errors will be undetected.

Example Two: The Cochella Valley

A second example comes from the Cochella Valley region in Riverside County, California, east of Los Angeles (Konikow and Swain 1990; Konikow and Bredehoeft 1992). Here a model was built to predict the effect of an artificial recharge programme on groundwater levels and dissolved solute concentrations. The model was calibrated on the basis of nearly 40 years of historical data (1936–1973), and used to predict water levels over the next seven years. As in the example described above, the modeller presumed that the success of the model in reproducing historical data was evidence that the model accurately represented the system, and could be used 'to predict water level changes from projected pumpage and (or) projected artificial recharge' (Konikow and Swain 1990, quoting Swain 1978).

The model predicted that groundwater levels would continue to drop despite the effect of the recharge programme. Again, the results were systematically in error. *Post hoc* comparison of 92 wells showed that the magnitude of the decline was consistently over-predicted. For individual wells, there was a wide range of error. Errors were greatest in and near tributary canyons entering the main valley. In hindsight, the period 1974–1980 was unusually wet, leading to high levels of recharge from creeks discharging from the tributary canyons.

The Cochella Valley case is similar to the Salt River case: the model was in error because of changes in the forcing function of the system, in this case climate. The model calibration implicitly assumed that the previous 40-year record covered the full range of conditions that operate in the valley, but this turned out not to be the case. The ability of the model to reproduce historical data did not translate into a capacity to predict future conditions.

3.4   LESSONS FROM EXPERIENCE?

These examples highlight a number of points. When model parameters are adjusted to obtain a best fit with historical data, it introduces a bias towards extrapolating existing trends. If those trends do not persist, the model is likely to be in error. Calibrated models are biased in favour of stasis.

The bias of stasis may operate in two ways, one explicit and one implicit. When a model assumes a particular driving force, like climate or pumping rates, then the extrapolation of that driving force into the future explicitly embeds an assumption of stasis into the model. If the driving forces change, then the model will diverge from actual conditions (Konikow and Patten 1985; see also Alley and Emery 1986). Stasis may also be implicitly embedded. A model will diverge from reality if the data gathered in the calibration period were in some way unrepresentative, even if the overall driving forces of the system remain the same. The process of calibration against historical data assumes the representivity of that data, and therefore embeds an assumption that the conditions they represent are on-going (Konikow and Person 1985). Finally, as noted above, a calibrated model may contain more than one set of errors, which cancel each other out. If conditions remain unchanged, the errors may continue to cancel each other out, and the model continues to work. But if operating conditions change, the errors may no longer cancel each other out, and the model begins to break down.

These examples show that our models may be less nuanced than our subjective understanding of the natural systems we are modelling. Scientists may have insights about systems that they are reluctant or unable to incorporate into their models. The goal of objectivity and adherence to Ockham's razor (the principle that we should prefer simpler explanations) create resistance to incorporation of nuance and subjective judgement. Yet it may be precisely within the realm of nuance and subjective judgement that our best understanding of a natural system lies.

To use William Ascher's terminology, modellers may recognise that their models are implicit conditionals – the predictions will come true if and only if there are no major changes in the forcing functions of the systems – but this insight is not often articulated, and the users of the model may be unaware of it. If they are aware, they may still choose to ignore it in order to proceed with the task at hand. The modeller may later be blamed for faulty predictions caused by limitations of which he or she was well aware, and perhaps even warned people. An obvious point here is the value of scenario development. The Salt River and lower Santa Cruz River model could have been run using a range of assumptions about future pumping rates or possible human responses to greater pumping lifts; the Cochella Valley model could have been run with 'extreme' case values for rainfall based on geological and meteorological evidence and insight rather than temporally limited historical records.

3.5 REPRODUCING THE BEHAVIOUR OF A SYSTEM VERSUS CAPTURING CAUSAL PROCESSES

The examples presented above involve retrospective analysis, but how should we evaluate a model in the present? Many modellers believe that the capacity of a model to reproduce data from the natural world – past or present – is evidence that it will be able to reproduce future behaviour. De Marsily and colleagues (1992) write, for example: 'We all know that the parameters of a model are uncertain, probably wrong in many cases, and can easily be invalidated. . . . So what? As long as they reproduce the observed behavior of the system, we can use them to make predictions.' Model post-audits refute this assertion. The examples described above successfully reproduced the observed behaviour of the modelled systems, yet failed to make accurate predictions.

The capacity to reproduce a response may or may not be sufficient grounds for prediction. Consider a proposed mine in which one needs to anticipate the day-to-day variability of the ore feed into the mill. A geostatistical model of the ore deposit can be developed that accurately represents the variability of the ore to help guide the design of the mill. The cause of the variability is irrelevant, and the model provides an adequate basis to design the mill. Now consider an exploration team, working at the same site, trying to find more ore. For this purpose, understanding the cause of the underlying variability – the geological controls on ore grade – is essential, and the model is useless for predicting where to look. In this case, no one would claim that the model addresses the cause of ore grade, and no one would make the mistake of imagining it could be used in aid of exploration. Yet there are many cases where people do make precisely this mistake: they presume that because a model accurately mimics the behaviour of a system, it must encompass an accurate representation of the underlying causal processes.

Houser et al. (1998) present a method for generating state variables, such as moisture content, which explicitly merges simulation-based estimates and observational data. As in the case of geostatistical methods, if the goal is the generation of a distributed parameter on a grid, then the method provides an efficacious means of doing that. However, the method does not seek to explain the reasons for the divergence between the simulation-based estimate and the observational data.

To generalise: if the underlying causal processes are irrelevant to the task at hand, then a calibrated model may make perfectly accurate predictions and prove highly reliable. One can design a mill without understanding the causes of ore grade variability; one can design a dam without worrying about the causes of rainfall fluctuation. But if the underlying processes are relevant – as in the case of the Salt River or Cochella Valley models – and the model fails to capture them, then the model is likely at some point to fail (see also Klemes 1982; Konikow and Patten 1985; Bredehoeft and Konikow 1992).

Validation is sometimes described as a process of confidence-building (de Marsily et al. 1992; Neumann 1992), but the most common approach to confidence building – reproduction of historical data – may be no more than mimicry. The capacity to mimic data is not evidence that you have captured underlying processes, and therefore not evidence of predictive capacity.

3.6 CONCEPTUAL ERROR

Konikow and Bredehoeft (1992) point out that the Cochella Valley example can be viewed in two ways: (i) that the historical database was not long enough, because it did not cover the full range of natural conditions, or (ii) that the model involved a flawed conceptualisation, because it inadequately appraised the highly variable nature of recharge in the desert environment. From the second perspective, the problem is not that the historical data were insufficient, but that the model conceptualisation was wrong. It embedded a faulty assumption: that the near future would closely resemble the near past.

Conceptualisation is probably the most thorny issue in modelling. It is the foundation of any model, and everyone knows that a faulty foundation will produce a faulty structure (Tsang 1991; Anderson and Woessner 1992; Konikow and Bredehoeft 1992). Yet what to do about it remains a problem. Much attention in model assessment has focused on quantification of error, but how does one quantify the error in a mistaken idea? Some refer to this as the epistemological uncertainty, to emphasise its foundational aspects (Funtowicz and Ravetz 1985, as cited in Rayner 2000). It is uncertainty rooted in the foundations of our knowledge, a function of our limited access to and understanding of the natural world. Almost by definition, conceptual error cannot be quantified. We don't know what we don't know, and we can't measure errors that we don't know we've made. In most cases, it is difficult to identify our errors at all. Not being able to identify them, we hope that they do not matter, but they may, because faulty conceptual models can lead to faulty conclusions and wrong courses of action. The following examples illustrate the point.

3.6.1 Conceptual Error, Example 1: Isostasy and Continental Drift

An example from the history of the earth sciences shows how scientists can come to radically wrong conclusions on the basis of a faulty conceptual model (Oreskes 1999).

In the early 20th century, there was a spirited debate over Alfred Wegener's theory of continental drift, which proposed that the earth's continents were mobile, and had been rearranged in various configurations during the long course of earth history. Wegener's theory was a reasoned response to a fundamental problem in the earth sciences: how to explain the palaeontological evidence that plants and animals had once freely migrated between what are now widely separated continents. In the late 19th century, this question was answered by the theory of sunken continents: that early in earth history, the entire surface of the earth was covered by a giant supercontinent, Gondwanaland, whose pieces collapsed to form the ocean basins. Organisms that had once freely migrated were subsequently isolated.

Most geologists accepted this explanation until it was refuted by the proof of isostasy: the idea that the continents sit in hydrostatic equilibrium within a denser substrate. In the early 20th century, studies of regional variations in gravitational acceleration proved that the continents are composed of less dense material than the ocean floors, and float within them. If so, then the theory of sunken continents was flawed. Wegener proposed drifting continents as an alternative explanation of the palaeontological data that did not conflict with isostasy.

Because Wegener's theory was premised on the principle of isostasy, one might expect that geodesists would have been among his strongest supporters, but in fact they were among his most vocal opponents, convincing many geologists to reject the idea despite its obvious explanatory power. Why? There were in fact two competing conceptualisations of isostasy. One, called the Airy model, supposed that mountains float as icebergs do, with large, hidden roots beneath them, and the depth at which isostatic equilibrium is achieved is proportional to the height of the topography above. The other, called the Pratt model, supposed that mass variations at the surface are compensated by density variations in the rock column below, and the depth at which isostatic equilibrium is achieved is uniform.

The two conceptual models were empirically equivalent. Both were compatible with available geodetic evidence, and they led to the same quantitative predictions with respect to surface gravitational effects. However, the Pratt model, with its assumption of a uniform depth to the base of the crust, was far easier to use. One could more readily calculate predicted surface effects using the Pratt than the Airy model. In the age before digital computers, this was an important consideration. Its mathematical simplicity was also appealing from the perspective of Ockham's razor. So a consensus formed around Pratt.

The choice of the Pratt model had serious conceptual consequences when continental drift was discussed a few years later, because it implied that there were no large-scale horizontal forces operating in the earth's crust. Most geodesists therefore rejected continental drift. Some became adamant opponents, not merely of the specifics of Wegener's model (which most contemporary scientists would say was flawed), but of the very idea that continents could move in a horizontal fashion – an idea that is now fundamental to earth science.

In later years, on the basis of seismic evidence, the Pratt model was shown to be wrong. Today geophysicists believe that isostatic compensation is primarily achieved by differences in the thickness of the crust, as argued by Airy. But because the Pratt model worked, the scientists who used it came to believe that it was true. From their perspective, their model was 'validated'. It fit the relevant data. It made accurate predictions. It *worked*. But it was not a realistic representation of the earth. Its success at mimicking surface behaviour was not evidence that it had captured the underlying causal structure.

The example of isostasy and continental drift shows how conceptual models can be under-determined by available evidence, causing an empirically adequate model to break down when applied to a problem other than the one for which it was built. A second example shows how a model based on a faulty premise may fail even at the job for which it was intended.

3.6.2 Conceptual Error, Example 2: The Limits to Growth

In the early 1970s, a group of systems analysts addressed the question we now call 'sustainability'. Their project was sponsored by the Club of Rome, a group of European industrialists, statesmen and scientists concerned about overuse of natural resources. The '*World*' model (very modestly labelled!), described in the widely-read book, *The Limits to Growth*, predicted wide-spread natural resource shortages, exponential price increases for raw materials, and possibly global economic collapse before the end of the century (Meadows et al. 1972; see also Peccei 1977; Meadows et al. 1992). The modellers concluded that the industrialised nations had to decrease their consumption of natural resources, and fast.

The end of the century has come and gone and resource use continues to grow, but real prices are level or down for virtually all commodities, and reserves of most natural resources are greater today than when the model was built (Hodges 1995; Moore 1995). While there has been economic change, much of this has had to do with falling rather than rising commodity prices (Simon and Kahn 1984; Tierney 1992).

One reason why the predictions of the *World* model have not come true is obvious in hindsight: the static way in which the model treated natural resource commodities. Resources such as copper, chromium, silver and gold were treated in the model as fixed and finite masses whose volume could only decrease over time. The greater the use rate, the faster the depletion. This seems like common sense; many people cannot imagine how it could be otherwise. On one level it is indisputable: the mass of any element in the earth is a finite (albeit unknown) number. However, from the perspective of resources management and sustainability, this view is in-adequate because the resource of something is not the same as the mass of it in the earth. By definition, a resource is something that may be used by humans (Hodges 1995; Moore 1995). This involves many variables, including the capital and technology available to find and extract it, the cost of labour, and the price people are willing to pay for it. Proven reserves are an even more constricted thing: they are only that portion of a resource that has been discovered, delineated and measured, and can be exploited under given conditions (Hodges 1995; Moore 1995; see also Oreskes 1998).

The *World* modellers made an elision between the proven reserves of a metal and its total mass in the earth as if they were the same thing. But they are not the same. Over time, the total mass of a metal in the earth must decrease or stay the same, but proven reserves can increase as a result of increased exploration, improved technology and/or decreased costs. Proven reserves of most metals *have* increased since the early 1970s, primarily because of more effective geological exploration during the subsequent decades; prices have fallen as a result (Hodges 1995).

This is not to say that the world will never run out of resources, that commodity prices will never rise, or that there are not important social, political, environmental and economic dimensions to sustainability. But it does illustrate the way in which faulty conceptual models can lead to faulty predictions. What seemed like common sense – increasing use must lead to diminishing stocks – was in error, because the definition of what constituted a 'stock' was misconceived.

The example of the *World* model also shows that a faulty conceptual model can inhibit consideration of alternative courses of action. If natural resources were fixed in the way the modellers conceived, then the industrialised nations would have had no choice but to decrease consumption in order to avoid economic crisis. But if we recognise that resource stocks are a more fluid thing, then other courses of action, such as improved technological capacity for resource recovery, or substitution, come to mind. Even if reserves *were* fixed – even if our technological capacities were utterly static – the use of proven reserves as measure of the availability of a commodity presupposes that the world is fully explored. But it isn't. The world may be known in a geographical sense – 'terra incognita' no longer appears on our maps – but from a geological point of view the world is quite unexplored. There are substantial areas that have scarcely been examined from a mineral exploration perspective, still more that have not been explored with modern technology. One need only to consider the recent discoveries in Russia, China, South America, Australia, and even Europe and North America to see this.

The assumption that the world was more explored than it really was is an example of the general problem of overconfidence in our knowledge. Studies in the social sciences have shown that most people, including experts, tend to overestimate their knowledge and are willing to make predictions even when the information supplied is inadequate to the task. This has been called the 'illusion of validity' (Tversky and Kahneman 1982, p. 9). Remarkably, this illusion persists even when the person is aware of the problem (Tversky and Kahneman 1982; see also Ascher 1989). Our cognitive biases lead us to think we know more than we do, and encourage us to act on knowledge that is inadequate or incomplete.

The two examples discussed above involve overconfidence. A third example illustrates the problem of oversimplification.

3.6.3. Conceptual Error, Example 3: Beach Processes and Erosion

Beach erosion is ubiquitous in the United States, primarily because of human efforts to prevent natural beach migration in order to maintain homes, roads and other human constructions in the near-shore area. Left to their natural state, all beaches will migrate, and under present geological and climatic conditions this migration is generally inland. To prevent shoreline retreat, communities along both the Atlantic and Pacific Coasts of the United States have constructed sea walls, jetties, groins and other forms of 'armouring' designed to stabilise the beach location and prevent sand loss. These efforts generally fail. Evidence suggests that beach-parallel constructions such as sea walls may exacerbate erosion, while beach-perpendicular constructions such as jetties help trap sand on the up-current side but accelerate erosion on the down-drift side (Hall and Pilkey 1991; see also Dean 1999).

An alternative approach that has become popular in recent years is beach nourishment, also referred to as 'soft stabilization'. In this strategy, lost sand is replaced by mined sand, usually from off-shore locations. In principle, beach nourishment is a direct response to the need for more beach sand, and it could be an environmentally and aesthetically satisfying response (Houston 1991). In practice, it is extremely expensive, and must be periodically repeated. This leads communities considering soft stabilisation to want information on how much it will cost and how long it will last. To do this, groups such as the US Army Corps of Engineers have built models. Commonly, these models produce overly optimistic assessments of the efficacy of this technique. In many documented cases, nourished beaches have lasted far shorter than predicted; in some cases erosion began even before the nourishment process was over (Pilkey 1994, 2000; see also Dean 1999).

Why are the models wrong? Orrin Pilkey and colleagues at Duke University have documented many reasons why models of coastal processes fail to make accurate predictions (Leonard et al. 1990; Pilkey 1990, 1994, 1995, 1997, 2000; Pilkey et al. 1993; Young et al. 1995; cf. Houston 1990, 1991 for counter-arguments). At root the problem is conceptual: the gap between the complexity of the shoreline systems as compared with the simplicity of the models of them.

Most models of shoreline erosion are based on the 'profile of equilibrium' concept, where the shape of the shoreline, described in terms of water depth, is expressed as a simple power law function with two empirical constants: $Y = AX^n$, where Y is the distance offshore, X is the water depth, and A and n are empirical constants. The value used for n is a worldwide average for all beaches, leaving only one parameter as a variable, 'A', which is believed to depend upon sediment characteristics. The shoreline equilibrium profile concept thus effectively asserts that the only factor controlling the profile of a beach is the sand. The complexity of beaches – the wind, weather, sand supply, coastal topography, bathymetry, subsurface geology, organismal action – is reduced to a single equation with a single degree of freedom. One need not be a coastal scientist to suspect that this simplification is an oversimplification, unlikely to account for the observed variability in coastal processes (Pilkey et al. 1993; Pilkey 1994).

Furthermore, as Pilkey and co-workers point out, the very concept of the equilibrium profile is flawed. It assumes that the shoreface is covered with a sand layer that is redistributed by wave action, and that at some depth – the closure depth – the interaction of surface waves with the sea-floor stops. Beyond the closure depth, there is no significant seaward movement of sand. Geologists and oceanographers refute this claim. Bottom currents that transport sand in a seaward direction have been known for a century, and scour channels on the sea floor attest to their widespread efficacy in moving sand beyond the zone of surface effects (Pilkey et al. 1993; Pilkey 1994).

Beach nourishment models also assume an inappropriate driving force. They predict the rate of loss of the artificial beach based on 'normal' or average waves conditions, but it is widely recognised that the bulk of coastal erosion occurs during rare, major storms. This choice is a form of simplification: averages are easier to calculate than extremes. It is notoriously difficult to predict where, when, and how often major storms will occur, much less what the details of their effects will be. The forcing function of coastal erosion – the major storm – is stochastic. Given that this simplification introduces a conceptual error, beach modellers might decide to make probabilistic predictions, as meteorologists do, but this is not generally done. As Pilkey notes, a common goal of modellers is to predict the 10-year lifespan of a nourished beach, but 'we are no closer to predicting the 10-year behavior of a beach than we are to prediction of 10 years of weather' (Pilkey 2000, p. 20).

Given the gross conceptual oversimplification and the use of an inappropriate driving force, the failure of these models to predict the efficacy of nourishment projects is hardly surprising.

3.7   WHY DO PEOPLE USE 'BAD' MODELS?

In the examples of Pratt isostasy and the *World* model, the modellers had no way to know that their models were flawed, but the models of beach nourishment raise a thornier question: why do people make and use models that they know are conceptually flawed and have perhaps already failed in a predictive mode?

An obvious reason is the political and social context of the decision-making process. In a sense, one may argue that the conceptualisation of beaches as suffering erosion is flawed: beaches left alone do not generally erode, they migrate (Pilkey and Thieler 1992). Erosion is primarily an anthropogenic effect of efforts to stabilise shorelines. An obvious solution to shoreline migration is to remove endangered structures, or let them collapse and build no new ones. A century ago, it was common practice in some areas of the United States to move cottages back from the shoreline (Dean 1999). However, the political and economic power of wealthy beach-front property owners and the presence of large resorts and high-rise hotels that cannot be moved now make yielding manoeuvres difficult.

Since hard stabilisation has been proved to be counter-productive, soft stabilisation is left as the only option when humans refuse to yield. This creates an implicit pressure on modellers to show that this option can work. As in the case of the World Bank, models of beach nourishment projects are commissioned by people who want the job done. So models are built despite past failures, and biased in an optimistic way.

And what constitutes past failure? In one study of 56 large replenishment projects, the US Army Corps of Engineers argued that their models had been successful because the costs of the projects and the amount of sand placed on beaches were close to projections. From an engineering and budgetary perspective, these projects did not fail. However, the study failed to discuss was whether the beaches were actually maintained for the predicted durations. In many cases, the replenished beach had disappeared long before the next replenishment cycle (Pilkey 1995; cf. Hillyer and Stakhiv 1997). The analysts knew what they meant by 'success' – their post-audit was designed to evaluate costs incurred and volumes of materials used – but it was not what would-be beach-goers considered success. It was not what the project was intended to *do*.

The word 'success' – like the word 'validation' – has an ordinary meaning to lay people. If modellers claim that their models were successful, this is generally translated by ordinary people into a claim that the model accurately predicted what would happen in the natural world.

Modellers sometimes argue that their predictions should really be placed in probabilistic terms, but that the decision-makers who fund these projects – members of the US Congress for example – do not understand error bars. Perhaps so, but who will be blamed if the model fails? Not the member of Congress! It is hard to see how it can be in the long-term interest of the scientific community to make predictions that are bound to fail.

Concerns over legal liability also contribute to the use of models that do not work. The frequency of model failure leads to the paradox that there is a positive incentive to use bad models if those models have already been used by others. Because failure is so common, there is pressure to follow precedent to avoid later legal liability (Pilkey 2000). If the government has an established and widely used model, there is an incentive to use it. Standardisation can be a defensive posture.

And often people do not realise or do not remember that previous modelling efforts failed to make accurate predictions. At least in the United States, there has historically been little funding for long-term monitoring (Houston 1991; Hillyer and Stakhiv 1997). Konikow and Patten (1985) point out that data collection in hydrological systems frequently ends with the study period, making it impossible to compare the model predictions with actual system conditions. (However, this may not necessarily be the case in other countries; see Smith 1990). Many factors combine to provide short-term incentive to build and use models, even if they are likely to fail in the long run.

Finally there is another point, relevant to all modelling, policy – driven or not. That is the view that modelling is cutting-edge science, and therefore inherently good. There are many cases where a scientist will reap more professional benefit from building a flawed model than from doing further empirical investigation. Who is not attracted to the promise of greater rigour in studying messy systems? But if our mathematical and computational prowess exceed our empirical understanding, we may achieve sophistication at the expense of knowledge. We may also achieve it at the expense of the open-mindedness necessary to learn from our mistakes. As Pilkey puts it, the state of the art is not necessarily close to the state of nature (Pilkey 1997; see also Ascher 1978; Ascher and Overholt 1983).

3.8   CONCLUSIONS

The problem of uncertainty is not unique to modelling, but it takes on an added dimension when scientific knowledge is used in support of public policy. Not all models in hydrology are relevant to public policy, but many are, and this makes it all the more important for modellers to be articulate about the sources of uncertainty in their models, and to think about ways to test for hidden errors.

We have raised here a number of issues that contribute to uncertainty in modelling: systematic error and bias; the difference between reproducing the behaviour of a system and capturing the underlying causal processes; and the problems of model conceptualisation. We have tried to highlight some of the ways in which models can go wrong, even when they are calibrated against large databases.

The most difficult issue to address in modelling is conceptual error. The examples provided in this chapter show that models may match available observations, yet still be conceptually flawed. Such models may work in the short run, but later fail. Recent advances in computational power may help us isolate conceptual error: if a set of simulation output, based on exhaustive sampling of parameter space, fails to encompass the observational data, this signals inadequacy in the underlying model structure. On the other hand, if the range of model outputs encompasses observational data, it is not a guarantee that the model is conceptually correct.

All models are approximations, and it is only in use over time that we discover where the model diverges most from reality. Therefore, rather than think of models as something to accept or reject in an either/or fashion – to validate or invalidate – it may be more useful to think of models as tools to be modified in response to knowledge gained through continued observation of the natural systems being represented.

The inherent uncertainties in models of complex natural systems provide a strong argument for monitoring when models are used in support of public policy. Model predictions *will* be wrong. This is inescapable. We hope that they will be wrong in inconsequential ways, but we cannot guarantee this. Therefore, any action guided by model results – be it an artificial recharge system or a nuclear waste storage facility – should be subject to continued monitoring. The more serious the consequence of error, the more important such monitoring is. Modellers can play an important role in public policy as advocates for monitoring, by emphasising modelling as a heuristic process, and by resisting the demand for predictions that are likely to be misleading, or simply wrong. Model output is not the same as a prediction of the future state of the system.

ACKNOWLEDGEMENTS

We are grateful to many colleagues for conversations that have helped to clarify the points made here, particularly William Ascher, John Bredehoeft, Dale Jamieson, Leonard Konikow, Daniel Sarewitz and Orrin Pilkey; and to an anonymous reviewer for thoughtful and helpful advice.

REFERENCES

Alley, W.M. and Emery, P.A. 1986. Groundwater model of the Blue River Basin, Nebraska – twenty years later. *Journal of Hydrology*, **85**, 225–249.

Anderson, M.P. and Woessner, W.W. 1992. The role of the postaudit in model validation. *Advances in Water Resources*, **15**, 167–173.

Anderson, T.W. 1968. Electric analog analysis of ground-water depletion in central Arizona. *US Geological Survey Water Supply Paper*, 1860, 21pp.

Aronica, G., Hankin, B. and Beven, K. 1998. Uncertainty and equifinality in calibrating distributed roughness coefficients in a flood propagation model with limited data. *Advances in Water Resources*, **22**, 349–365.

Ascher, W. 1978. *Forecasting: An Appraisal for Policy-Makers and Planners*. Johns Hopkins University Press, Baltimore, MD.

Ascher, W. 1981. The forecasting potential of complex models. *Policy Sciences*, **13**, 247–267.

Ascher, W. 1989. Beyond accuracy. *International Journal of Forecasting*, **5**, 469–484.

Ascher, W. 1993. The ambiguous nature of forecasts in project evaluation: diagnosing the over-optimism of rate-of-return analysis. *International Journal of Forecasting*, **9**, 109–115.

Ascher, W. and Overholt, W.H. 1983. *Strategic Planning and Forecasting: Political Risk and Economic Oppportunity*. Wiley, New York.

Beck, M.B., Ravetz, J.R., Mulkay, L.A. and Barnwell, T.O. 1997. On the problem of model validation for predictive exposure assessments. *Stochastic Hydrology and Hydraulics*, **11**, 229–254.

Bethke, C. 1992. The question of uniqueness in geochemical modelling. *Geochimica et Cosmochimica Acta*, **56**, 4315–4320.

Beven, K. 1993. Prophecy, reality, and uncertainty in distributed hydrological modelling, *Advances in Water Resources*, **16**, 41–51.

Beven, K. 1996. Equifinality and uncertainty in geomorphological modelling. In: Rhoads, B.L. and Thorn, C.E. (eds), *The Scientific Nature of Geomorphology: Proceedings of the 27th Binghampton Symposium in Geomorphology, 27–29 September 1996*. John Wiley, Chichester, 289–313.

Bredehoeft, J.D. and Konikow, L.F. 1992. Reply to comment (by de Marsily et al. 1992). *Advances in Water Resources*, **15**, 371–372.

Brunner, R.D. 1991. Global climate change: defining the policy problem, *Policy Sciences*, **24**, 291–311.

Brunner, R.D. and Ascher, W. 1992. Science and social responsibility. *Policy Sciences*, **25**, 295–331.

Dean, C. 1999. *Against the Tide: The Battle for America's Beaches*. Columbia University Press, New York.

de Marsily, G., Combes, P. and Goblet, P. 1992. Comment on 'Ground-water models cannot be validated', by L.F. Konikow and J.D. Bredehoeft. *Advances in Water Resources*, **15**, 367–369.

Fischhoff, B. 1982. For those condemned to study the past: heuristics and biases in hindsight. In: Kahneman, D., Slovic, P. and Tversky, A. (eds), *Judgment under Uncertainty: Heuristics and Biases*. Cambridge University Press, Cambridge, 335–351.

Gupta, H.V., Sorooshian, S. and Yapo, P.O. 1998. Toward improved calibration of hydrologic models: multiple and noncommensurable measures of information. *Water Resources Research*, **34**, 751–763.

Gutierrez, G.J. and Kouvelis, P. 1991. Parkinson's Law and its implications for project management, *Management Science*, **37**, 990–1001.

Hall, M.J. and Pilkey, O.H. 1991. Effects of hard stabilization on dry beach width for New Jersey. *Journal of Coastal Research*, **7**, 771–785.

Hankin, B.G. and Beven, K.J. 1998a. Modelling dispersion in complex open channel flows: equifinality of model structure (1). *Stochastic Hydrology and Hydraulics*, **12**, 377–396.

Hankin, B.G. and Beven, K.J. 1998b. Modelling dispersion in complex open channel flows: fuzzy calibration (2). *Stochastic Hydrology and Hydraulics*, **12**, 397–412.

Hill, M.C. 1992. A computer program (MODFLOWP) for estimating parameters of a transient, three-dimensional, ground-water flow model using non-linear regression. *US Geological Survey Report*, 91–484.

Hillyer, T.M. and Stakhiv, E.Z. 1997. Discussion of Pilkey, O.H., 1996 (sic). The fox guarding the hen house (editorial). *Journal of Coastal Research*, **13**, 259–264.

Hodges, C.A. 1995. Mineral resources, environmental issues, and land use. *Science*, **268**, 1305–1312.

Houser, P.R., Shuttleworth, W.J., Famiglietti, J.S., Gupta, H.V., Syed, K.H. and Goodrich, D.C., 1998. Integration of soil moisture remote sensing and hydrologic modelling using data assimilation. *Water Resources Research*, **34**, 3405–3420.

Houston, J.R. 1990. Discussion of: Pilkey O.H., 1990. A time to look back at beach replenishment (editorial) and Leonard L., Clayton T. and Pilkey, O.H. 1990. An analysis of beach design parameters on US east coast barrier islands. *Journal of Coastal Research*, **6**, 1023–1036.

Houston, J.R. 1991. Beachfill performance. *Shore and Beach*, **59**, 15–24.

Isham, J. and Kaufmann, D. 1995. The forgotten rationale for policy reform: the productivity of investment projects. *The World Bank Policy Research Working Paper* 1549, 35pp.

Klemes, V. 1982. Empirical and causal models in hydrology. In: *National Research Council Geophysics Study Committee* (eds), *Scientific Basis of Water-Resource Management*. National Academy Press, Washington, DC, 95–104.

Konikow, L.F. 1986. Predictive accuracy of a ground-water model: lessons from a post-audit. *Ground Water*, **24**, 173–184.

Konikow, L.F. 1992. Discussion of 'The modelling process and model validation' by Chin-Fu Tsang. *Ground Water*, **30**, 622–623.

Konikow, L.F. and Bredehoeft, J.D. 1992. Ground-water models cannot be validated. *Advances in Water Resources*, **15**, 75–83.

Konikow, L.F. and Patten, E.P. Jr 1985. Groundwater forecasting. In: Anderson, M.G. and Burt, T.P. (eds), *Hydrological Forecasting*. John Wiley, Chichester, 221–270.

Konikow, L.F. and Person, M. 1985. Assessment of long-term salinity changes in an irrigated stream-aquifer system. *Water Resources Research*, **21**, 1611–1624.

Konikow, L.F. and Swain, L.A. 1990. Assessment of predictive accuracy of a model of artificial recharge effects in the upper Cochella Valley, California. In: Simpson, E.S. and Sharp, J.M. (eds), *Selected Papers on Hydrogeology from the 28th International Geological Congress (1989), volume 1*. Heinz Heise, Hannover, 433–449.

Leonard, L., Clayton, T. and Pilkey, O.H. 1990. An analysis of replenished beach design parameters on US east coast barrier islands. *Journal of Coastal Research*, **6**, 15–36.

Meadows, D.H., Meadows, D.L. and Randers, J. 1972. *The Limits to Growth: A Report for the Club of Rome's Project on the Predicament of Mankind*. Universe Books, New York.

Meadows, D.H., Meadows, D.L. and Randers, J. 1992. *Beyond the Limits: Confronting Global Collapse, Envisioning a Sustainable Future*. Chelsea Green Publishing Co, White River Junction, VT.

Moore, S. 1995. The coming age of abundance. In: Bailey, R. (ed.), *The True State of the Planet*. The Free Press, New York, 110–139.

Neumann, S.P. 1992. Validation of safety assessment models as a process of scientific and public confidence building. In: *High Level Radioactive Waste Management, Proceedings of the Third International Conference*. Las Vegas, NV, April 12–16, 1992, published by the American Nuclear Society, La Grange Park, Illinois, and the American Society of Nuclear Engineers, New York, 1404–1413.

Nigg, J. 2000. The issuance of earthquake 'predictions': scientific approaches and strategies. In: Sarewitz, D., Pielke, R. Jr and Byerly, R. Jr (eds), *Prediction: Science, Decision-Making and the Future of Nature*. Island Press, Washington, DC, 135–156.

Oreskes, N. 1998. Evaluation (not validation) of quantitative models. *Environmental Health Perspectives*,

106 (suppl. 6), 1453–1460.

Oreskes, N. 1999. *The Rejection of Continental Drift: Theory and Method in American Earth Science*. Oxford University Press, New York.

Oreskes, N., Shrader-Frechette, K. and Belitz, K. 1994. Verification, validation, and confirmation of numerical models in the earth sciences. *Science*, **263**, 641–646.

Peccei, A. 1977. *The Human Quality*. Pergamon Press, Oxford.

Pilkey, O.H. 1990. A time to look back at beach nourishment (editorial). *Journal of Coastal Research*, **6**, iii–vii.

Pilkey, O.H. Jr 1994. Mathematical modeling of beach behaviour doesn't work. *Journal of Geological Education*, **42**, 358–361.

Pilkey, O.H. 1995. The fox guarding the hen house. *Journal of Coastal Research*, **11**, iii–v.

Pilkey, O.H. 1997. Reply to: Hillyer, T.M. and Stakhiv, E.Z., 1997. Discussion of: Pilkey, O.H. 1996 (sic). The fox guarding the hen house (editorial). *Journal of Coastal Research*, **13**, 265–267.

Pilkey, O.H. 2000. Predicting the behavior of nourished beaches. In: Sarewitz, D., Pielke, R. Jr and Byerly, R. Jr (eds), *Prediction: Science, Decision-Making and the Future of Nature*. Island Press, Washington, DC, 159–184.

Pilkey, O.H. Jr and Thieler, E.R. 1992. Erosion of the United States Shoreline. *Quaternary Coasts of the United States: Marine and Lacustrine Systems, SEPM Special Publication*, **48**, 3–7.

Pilkey, O.H., Young, R.S., Riggs, S.R., Smith, A.W.S., Wu, H. and Pilkey, W.D. 1993. The concept of shoreface profile of equilibrium: a critical review. *Journal of Coastal Research*, **9**, 255–278.

Rayner, S. 2000. Prediction and its alternatives in climate change policy. In: Sarewitz, D., Pielke, R. Jr and Byerly, R. Jr (eds), *Prediction: Science, Decision-Making and the Future of Nature*. Island Press, Washington, DC, 269–296.

Sarewitz, D, Pielke, R.A., Jr. and Byerly, R., Jr. (eds) 2000. *Prediction: Science, Decision-making, and the Future of Nature*. Island Press, Washington, DC.

Sarewitz, D. and Pielke, R. Jr 1999. Prediction in science and society. *Technology in Society*, **21**, 121–133.

Simon, J.L. and Kahn, H. (eds.) 1984. *The Resourceful Earth: A Response to Global 2000*. Blackwell Scientific, Oxford.

Smith, A.W.S. 1990. Discussion of: Pilkey O.H., 1990. A time to look back at beach replenishment (editorial) and Leonard L., Clayton T. and Pilkey, O.H. 1990. An analysis of beach design parameters on US east coast barrier islands. *Journal of Coastal Research*, **6**, 1041–1045.

Steefel, C.I. and van Cappellen, P. 1998. Reactive transport modeling of natural systems. *Journal of Hydrology*, **209**, 1–7.

Swain, L.A. 1978. Predicted water-level and water-quality effects of artificial recharge in the upper Cochella Valley, California, using a finite-element digital model. *US Geological Survey Water Resources Investigation*, **77–29**, 54pp.

Tierney, J. 1992. Betting the planet. *The New York Times Magazine*, 2 December 1992, 52–81.

Tsang, C-F. 1991. The modeling process and model validation. *Ground Water*, **29**, 825–831.

Tsang, C-F. 1992. Reply to the preceding discussion of 'The modeling process and model validation'. *Ground Water*, **30**, 622–624.

Tversky, A. and Kahneman, D. 1982. Judgment under uncertainty: heuristics and biases. In: Kahneman, D., Slovic, P. and Tversky, A. (eds), *Judgment under Uncertainty: Heuristics and Biases*. Cambridge University Press, Cambridge, 3–20.

Young, R.S., Pilkey, O.H., Bush, D.M. and Thieler, E.R. 1995. A discussion of the Generalized Model for Simulating Shoreline Change (GENESIS). *Journal of Coastal Research*, **11**, 875–886.

4 Calibration, Validation and Equifinality in Hydrological Modelling: A Continuing Discussion . . .

KEITH BEVEN

Institute of Environmental and Natural Sciences, Lancaster University, UK

This lunchtime discussion takes place in a university bar following the morning session at a workshop on Validation of Environmental Models.[1]

Taking part in the discussion are a Professor of Hydrology (P), a graduate student (G) who has recently started a research project with P, a visiting scientist from Brazil (B), a consulting engineer with particular interests in the remediation of contaminated land (C), a scientist from the Environment Agency (E) with responsibilities for the licensing of effluent discharges, and a scientist from the UK Government Department of the Environment (D) with interests in the impact of climate change on water resources.

With the exception of the graduate student and the scientist from Brazil, all the others are old friends, having studied together as undergraduates.

D   So what did you think of the presentations this morning?

E   Well, I got completely lost during those philosophical presentations at the beginning – they really didn't seem to bear much relation to what happens in practice.

D   How do you mean?

E   With the amount of data that we have for most model applications the problem is not so much whether we can verify or validate or confirm or evaluate a model but rather whether we can gather enough decent data from different sources to make any predictions at all. And we have so many sites to check that we cannot spend too much time on each one.

C   We have the same problem in knowing how to characterise a site, especially since I am dealing with what might be happening under the ground. The important thing is to make a prediction quickly, not to worry about validation.

D   Does the data available generally affect what type of model you choose to use?

C   No, not often – that is usually a question more of what variables are required to be predicted. That might vary for different projects.

P   So how do you choose a model to use?

C   Well obviously we have a degree of experience in using different types of model, some of which are well established as tools and some of which we have developed or adapted

---

[1] The audience had heard papers similar to Morton (1993), Oreskes et al. (1994), Schrader-Frechette (1989), Beck (1994), and lastly Beven (2001a).

Model Validation: Perspectives in Hydrological Science. Edited by M.G. Anderson and P.D. Bates.
© 2001 John Wiley & Sons, Ltd.

ourselves – in fact it is important that we should be able to list this experience to the client in tendering for a project.

P   But that sort of implies a process of prior evaluation and modification of models over different projects. Your reputation depends on using a model that should give results that are in some way reasonable so that if you were aware that a model does not perform for a given purpose you would have dropped it or modified it. Does that not count as validation or verification?

C   It would appear not according to the philosophers. It may be at least a *confirmation*, since at least we certainly do some evaluation of performance in each project but we are not really in the business of showing that a model is a true representation of reality, only that it gives predictions that are practically useful.

G   But surely we have to believe that it is possible to validate a model as a representation of reality, otherwise we don't have a science of hydrology.

D   I am not sure I agree with you there. There are plenty of examples from the history of science that have shown how scientists have modified models to fit new observations so as to have improved predictive capabilities only to see those theories superseded as a 'true' representation of reality. Indeed there are examples where the new theory is not, initially at least, as good a predictor of the data but appears to have other attractions or potential, if only to act as a competitor to the existing theory.[1] It seems quite possible to go on using and modifying a model or theory, within the current paradigm so to speak, as scientists with the best of intentions but without having to make a strong statement that this model is a true representation of reality. Look at how widely the Horton infiltration excess concept has continued to be used in hydrological prediction, even where overland flow is not the dominant runoff mechanism, by adjusting the infiltration losses so that the effective rainfall has a volume equal to the 'storm runoff'.

P   Is the problem in hydrology not more that it is all too easy to recognise the faults or limitations of models, especially of subsurface flow processes, so that if we were really scientists we should be rejecting all the models?

E   That wouldn't get us very far – I have to make predictions about the impact of a new effluent on river water quality every day. The models may not be totally realistic in terms of representing all the details of all the processes but I would argue that they can have useful predictive capabilities, so they must be reasonably realistic in that sense.

P   But how do you know that – how do you check or validate those predictions?

E   Well, very often of course we don't have time to do any checking – we would normally only go back for more detailed studies at a site if a problem occurs, and problems will often be the result of extreme low flow conditions where the model might be less reliable because it is outside its calibrated range.

B   That suggests, however, that if a model is only as good as the range of data over which it is calibrated it is not necessarily realistic in terms of reproducing the processes involved.

E   But remember there is a historical context here. Today's models are much better than 10 years ago and we can hope that there will be more improvements in the future.

B   Where do you think that the improvements will come from? New theory? New measurement techniques? Or just more computer power?

E   I think new measurement techniques will be very important – we are only just starting to make use of remote sensing data and continuous water quality measurements. These should eventually lead to better understanding and new theoretical developments.

P   But I cannot see how this will lead to improved confirmation or validation of models in

---

[1] See the discussions in, for example, Chalmers (1990) and Feyerabend (1978), or, for pure entertainment, Feyerabend (1991).

general practical applications. Think of a real catchment, with its patchwork of different hydrological conditions on the hillslopes. I was out in the field with some students last week, trying to help them relate what they could observe with the textbook diagrams they have seen about hydrological processes. To do so you have to filter out all the complexity. That is perhaps useful as a teaching tool but, if you dig a soil pit in wet conditions or do a surface or subsurface tracing experiment, you realise how the complexity dominates the nature of the flow processes. The water flow is usually dominated by very discrete preferential flow pathways.[1] Emmett showed the same thing for surface runoff back in the 1970s,[2] while tracer experiments in rivers show how complex the flow structures, mixing processes and exchanges within the hyporheic zone are.[3] In the field, it is only too obvious that the hydrological theory of the textbooks has its limitations. Even Darcy's law has been shown to fail a (stringent) validation test when applied to Darcy's original data set.[4]

G   But what is wrong with Darcy's law? Surely that is fundamental to hydrological theory? It is used in all the physically based models.

D   I thought the arguments were well rehearsed in that talk this morning about the sociological context of theory development.

C   Oh that! I went for coffee as soon as she mentioned Derrida and the signification of signs.

D   Well it wasn't easy to follow, for sure, but I think a crude interpretation of what she was trying to get over was that the post-modernists regard a theory as a social construct or sign that, over time, many people agree to recognise. As such, they are then reluctant to give up that theory, despite the fact that it is difficult to demonstrate its adequacy.

G   But how do we throw away Darcy's law? Surely it has been used successfully and besides what is there to put in its place?

D   Quite – that was the point (and a whole lot more about adequacy and misapplication of theory at different scales). There isn't really anything to put in its place at its correct scale of application (which would appear to be very small), and at larger scales we can get away with calibrating the parameters. Remember she posed the question as to whether a hydraulic conductivity calibrated by some inverse method could be a 'real' hydraulic conductivity.

G   How could you show it was?

C   Go and take measurements.

D   But wasn't that the point she was trying to make? The calibrated value is a large-scale effective value that also depends, through model calibration, on the model structure, the surrounding values and responses over the calibration period. The measured value is some local value in both space and time that depends on the measurement technique used. The two may not be the same quantity.[5]

C   But surely the very process of taking the measurement adds new information that can be used in calibration – using kriging of the measured values for example, or as prior information in a Bayesian inverse method.[6]

D   But there is a danger of treating that measured value as true when it may not be at the right scale for the effective values required by the model. There might not be a simple functional link between them.

P   In fact the problem may be worse than that. I remember a paper by one of David Lerner's

---

[1] See, for example, Flury et al. (1994), Binley et al. (1996), Henderson et al. (1996).
[2] See Emmett (1978).
[3] See Harvey et al. (1996).
[4] See Davis et al. (1992).
[5] See also discussion in Beven (1996b).
[6] See, for example, McLaughlin and Townley (1996).

students in WRR a while back[1] where they tried all sorts of different optimisations and came up with completely different best fit parameter sets.

B   Perhaps they just hadn't found the global optimum. We have seen that when trying to do groundwater calibrations in the past, but recently we have had a lot of success with a simulated annealing method.[2] It seems to be much more effective in converging to a consistent optimum parameter set. It does take a lot of runs but there is not such a problem with computer power these days.

G   That also depends on what you are optimising. There is a recent paper in WRR using multi-criterion optimisation that tries to take account of the fact that different parameter sets perform best on different criteria. I think they came up with what they called the Pareto optimal set of models.[3]

E   Is this not getting away from the point? Ok, so we have used optimisation as a way of fitting models to data in the past, but isn't that just like using a regression analysis? The parameters are only going to compensate for errors in the model, boundary conditions and observations. It does not have much to do with whether the model is realistic or not, just whether it has the right sort of functionality to reproduce the data. In regression analysis, it is generally easy to see if you have the right sort of functionality. If not, you try a different sort of transform. In hydrological modelling it is not nearly so easy to reject a model if you have calibrated the parameters. There are usually enough degrees of freedom to fit the data reasonably well.

C   No, surely it is not so different. If a model produces accurate predictions in some sense is that not sufficiently realistic or valid or, at the very least, fit for purpose. If it does not produce sufficiently accurate predictions then you should change the model. It is allright for you academics to argue about these things but I sometimes have to defend fitness for purpose of the model I have chosen at a public inquiry or in court.

P   And what argument do you use?

C   Past experience of successful predictions together with any validation tests if there are data available to do so.[4]

E   But if success in prediction is the only criterion, surely we have a problem – we could have as many calibrated models as there are hydrologists (it was almost like that in the days of conceptual rainfall–runoff models). If we are scientists should we not take a viewpoint that we can at least falsify some of the possible models?

G   Falsification is not the problem if you are thinking in terms of process representation. Unfortunately it is all too easy to falsify all our models in applications to real sites, so I do not think that hydrologists can afford to take a critical rationalist position here.[5] I learned a bit about Popper's ideas on falsification in my MSc course last year but I did not see how it related to the type of modelling I do. As Morton said this morning,[6] I *know* that the available process theories are not true descriptions of a catchment but I do not reject them (or I would be left with nothing to predict with) but try to improve them. That is why I am doing a PhD project.

---

[1] See Brooks et al. (1994).
[2] See Sen and Stoffa (1995).
[3] See Gupta et al. (1998).
[4] See Bair (1994).
[5] For a simple introduction to Popper, critical rationalist and modern realist philosophies, see Chalmers (1990). For a Bayesian perspective on falsification see Howson and Urbach (1993). For a discussion with specific reference to geomorphological explanation, see Richards (1990, 1994), Bassett (1994), Rhoads (1994), Beven (1996a).
[6] See Morton (1993).

D How far are improvements possible, though? Do we not need a radically different, more realistic theory in hydrology that will explain rather than simply predict. That would give me much more faith in the predictions that are being made about the hydrological impact of a changed climate in Britain for example. At the moment, since calibration with future conditions is not possible, I am very dubious about whether the models that are being used will produce adequate predictions.

P That is an interesting example. In fact, we generally have quite a good *qualitative* theory of what is going on in a catchment in our heads. Even if we cannot be sure about all the details of the spatial and temporal variability in the flow pathways, especially through the deeper subsurface pathways, we can at least be aware of the possibilities. In that sense we do, already, have a good theory of hydrology. The difficulty is putting that qualitative theory into a quantitative mathematical description.

E But will such a quantitative theory ever be possible? When it comes down to real hydrological applications, the problem is not one of generalisations, it is about the hydrology of specific places where we will inevitably have imperfect knowledge of the inputs, outputs and processes over a range of hydrologically relevant scales. Every place and every application is unique in that sense.

P That is true and, in part, what I meant about being unsure of the details of the flow pathways. Listening to the discussion this morning, I was thinking how does this relate to *my* catchment. I am not only interested in the theory, but in its description of my unique catchment. I do not see how we can get away from calibration in the end.

B And think of our example of the Parana catchment in Brazil, 1.5 million km^2. But we still need to make predictions for water resource management and to evaluate the impacts of land use and climate change. And we are not even sure of the pattern of rainfall inputs under current conditions. My own feeling is that it may not be necessary to have a completely accurate description of the processes to have a model that is useful in prediction, as long there are some data available to calibrate the particular characteristics of a catchment.

P But if we are forced always to calibrate in some way there is an implication that there will always be the possibility of multiple descriptions that will do equally well in calibration. After all, the recent trend is for models to include more processes which means more parameters which means more degrees of freedom in calibration.

E That was what I understood this morning from Beck's use of *non-identifiability* and Beven's *equifinality*.[1] Did they mean the same thing?

P I am not sure that they did. I think that non-identifiability was intended to be in respect of some 'true' description of the system, whereas equifinality accepted that we may *only* be able to find a variety of acceptable descriptions given the data available so that we are unlikely ever to be able to say that we have the true description.

B What do you mean by description? Getting the qualitative description of the dominant *processes* right, or getting accurate predictions of the volume and timing of runoff and evapotranspiration *fluxes*. It is not absolutely necessary to do the first in order to achieve the second.

G But that is to accept the traditional split between engineering hydrology and hydrological science. Surely we not only want the get the flux predictions right, but also the process descriptions.

E But I cannot see that this will be possible without a dramatic change in measurement techniques that would enable us to determine fluxes or parameters more accurately. Until

---

[1] See Beck (1987), Beck et al. (1990), Beven (1993, 1996a).

then we may have to accept that the equifinality problem is endemic to environmental models.

C   I was not happy about this equifinality idea. Surely such a relativistic viewpoint is untenable for a scientist and there are examples from the environmental sciences that suggest that real advances can be made. Look at how numerical weather prediction using Global Circulation Models has improved in recent years. They don't worry about equifinality.

D   Not in terms of parameter values no, but in weather forecasting they do now make 'ensemble' runs with different initial conditions to test the sensitivity of the predictions. And there are some very conceptual sub-grid parameterisations in those models – look at how they represent the land surface hydrology, for example – it would not be considered acceptably scientific to a hydrologist doing catchment modelling, since they are mostly one-dimensional vertical representations averaged over the GCM grid square that cannot reproduce discharges correctly.[1]

P   And when Global Circulation Models are used for climate prediction into the next century? These are nonlinear dynamic models and the predictions should be expected to diverge using different initial conditions and parameterisations if they were not con-strained by the boundary conditions of the seasonal energy budget. The idea of different nonlinear trajectories leading to much the same answer only reinforces the possibility of having different acceptable or feasible model structures or parameter sets.

C   But they do allow that there is some degree of uncertainty in the results produced by different models.

D   Yes, and surely this is a situation crying out for a critical rationalist approach. We can surely falsify some of these descriptions by some well chosen experiments in the classical scientific way – that is set up some hypotheses that will distinguish between different representations and test them. That way we are sure to move towards some more realistic models and theories as Popper suggested – and thereby reduce the resulting uncertainty.

E   That may be true up to a point but how far is it possible with current measurement techniques to make such tests and what if, when different models or parameter sets are evaluated in this way, there are several or many that pass the test of acceptability. Rainfall–runoff models are a classic example of this, if they are only evaluated on the basis of their success in predicting discharges.

P   You can set the test up within a Bayesian framework, that is you start with some prior beliefs and after testing have some posterior distribution of models or parameter sets. The test can then even be subjective.

C   But isn't that only a justification for the same relativistic idea posing as science – either a model passes the test or not, either it is acceptable or not, either it is realistic or not.

P   It is not always so easy, surely. Take all those odd dotty plots in Beven's talk this morning of goodness of fit against parameter values.[2] How does anyone decide where the level of acceptability is? Is it possible that the one that appears best is not necessarily the most 'realistic' because of the effects of errors in the data against which the model is being tested – or that none of them are 'realistic' despite fitting the data quite well.

C   That was where I started to get confused – especially when he started to talk about transforming a landscape space into model space. I didn't see what that had to do with practical prediction.

---

[1]   See Lohmann et al. (1998).

[2]   Figure 4.1, see also the applications of the GLUE methodology in Beven (1993), Freer et al. (1996), Romanowicz et al. (1996), Franks and Beven (1997a), Zak et al. (1997), Franks et al. (1998), Romanowicz and Beven (1998), Beven (2001b).

B   I think I saw what he was getting at and its application to the type of large scale problem I have to deal with. What I understood of the idea was this: let us accept for the moment that there may be many models that have similar function to what is happening in the landscape in the sense that they can be shown to predict similar fluxes to those measured. The important thing in prediction is that the functioning should be represented correctly. So the modeller would be trying to create a map of a landscape or catchment in the model space in a way that properly reflects the uncertainties associated with predicting those fluxes at a point. You can probably envisage that most easily if you think of the model space as represented by a number of parameter axes, with each patch of the landscape being represented by a rather fuzzy region in the parameter space represented by parameter sets that, to the best of our knowledge, have the same functioning as the real patch. The uncertainty could be represented by weights or fuzzy measures associated with the mapping of individual patches to different parameter sets, but the predictions, of course, are done with the different parameter sets. The transformation is a sort of conditioning of the parameter space to represent the catchment.[1]

E   But this takes us back to how important it is to have a correct model. How far does this transformation depend on having a correct representation of the processes? A simple model could provide adequate predictions of the fluxes required after such a conditioning under one set of circumstances but be completely wrong under a different set of circumstances because it doesn't correctly represent the processes.

P   That can be true of an optimised model as well, of course, or a model with measured parameter values as Keith Loague has demonstrated at Chickasha.[2] But if we have evidence that a model is invalid, then rejection of that model should be part of the process. It would represent an inappropriate mapping (if we can ever bear to reject the model that we have developed, of course). Remember we are discussing conditioning using the available data and certainly not ruling out refinement of that conditioning by carrying out experiments and collecting more data. In fact, that is one of the things I like about the idea – it focuses attention on the value of collecting data and not on theory alone. There does seem to be a tendency to think that the more physically based a model is the more accurate will be its predictions, and to ignore the fact that all models should depend on some field data to define the characteristics of a particular unique catchment or site.[3] I will accept that a process based model may be richer scientifically than, say, a transfer function model, but that does not mean it will be more accurate in reproducing the data with which it can be compared. Process based models may well have additional value in exploring the implications of different assumptions and representations of the physics and chemistry, but they also must ultimately face the same tests of being consistent with the available data, or else they should not be considered useful (except in demonstrating what is *not* a realistic description). The difference, perhaps, is that many of their predictions may well not be testable with the data currently available.

C   I am sorry to belabour the point but where does all this conditioning and relativism leave the idea of validation? What use is more data if the model is based on the wrong theory?

P   But will you not agree that any useful theory must be consistent with the data available, and that if it is not a critical rationalist like yourself would reject it.

C   Yes, if I felt the data were reliable.

---

[1] Figure 4.2, see also Beven (1995), Franks and Beven (1997b), Beven (2000).
[2] See Loague and Kyriakidis (1997) and earlier studies of the R-5 catchment referenced therein.
[3] See discussion in Beven (2000).

A

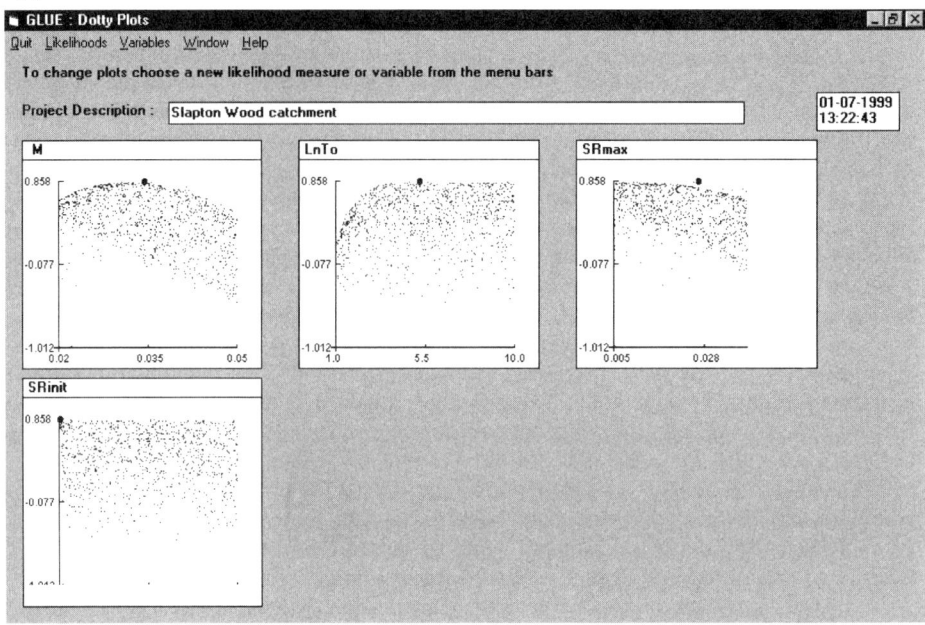

B

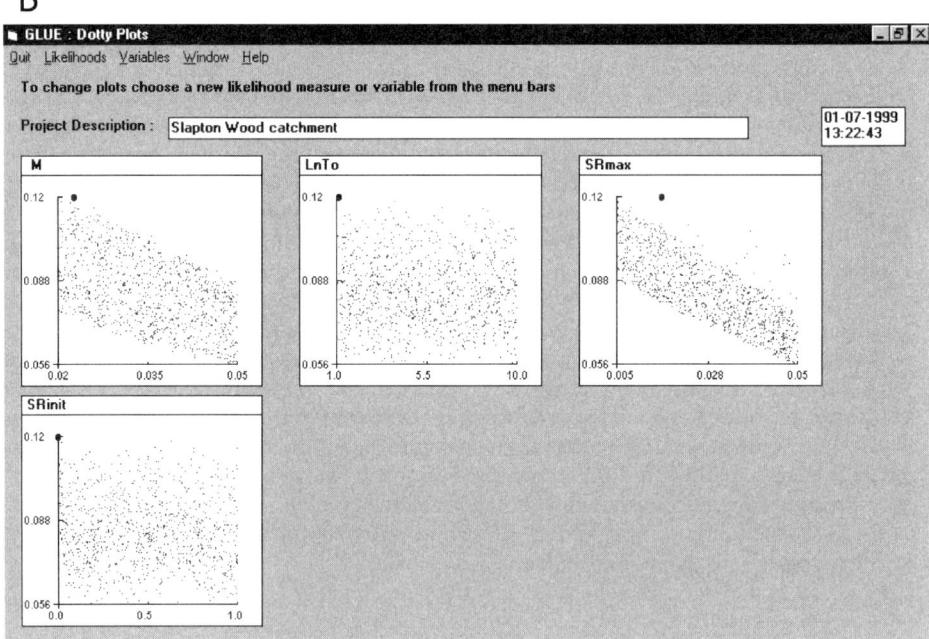

P   And that if you have two models or theories that are both consistent with the data then you have no reason to reject either?

C   Well, . . .

P   And if there were many such models?

C   Then surely you have either not yet collected the right data or have not defined consistency properly.

D   But, as far as I can see, refining the assessment of the different models is not inconsistent with these ideas – whether that be by collecting more data or by changing your definition of acceptability but it would seem that you would need to have a good reason for redefining acceptability.

E   I am still trying to think this through. Even if I accept these ideas, what am I supposed to do if I have to make a decision about whether to issue an abstraction or effluent discharge licence. If I understand correctly I should not just be running one model but a whole suite of models and different parameter sets within each model and then evaluating them all in terms of how well they predict what is happening at the site I may be investigating. I don't have the time to do that – I have to make decisions quickly and very often without any information available to evaluate the model predictions. I am also not sure that I want to have to worry about the uncertainty implied by all these different predictions in making a decision.

P   But that doesn't change the principle involved, only what might be possible in practice. And the whole idea of discussing validation of models is surely one of establishing principles to work by, whether validation of a model is actually possible or not.

E   I am sure, however, that even if we accept this multiple model framework, and make a whole range of predictions there is still plenty of scope for being wrong. It might only take an extreme flood, for example, to show that a prediction of sediment production might go way outside of any prediction limits, perhaps, due to a landslide changing the nature of the sediment supply. Allowing for multiple acceptable models and uncertainty in predictions does not mean that the predictions will be realistic.

B   But there is no implication that they are necessarily realistic – only that they are consistent with the data available (including our perception of what is acceptable). In that respect, events that produce observations that go outside the prediction limits of the models may

---

Figure 4.1   A. Dotty plots of model goodness-of-fit (here, the coefficient of determination based on a simple sum of squared errors) versus parameter value for 1000 random sets of parameter values chosen assuming prior uniform and independent distributions for each parameter. The model is the rainfall–runoff model, TOPMODEL (see Beven et al. 1995; Beven 1997, 2001b). The application is the simulation of discharges for the 1 km^2 Slapton Wood catchment over a 950 hour winter period. Each dot represents one run of the model. The larger dot represents the maximum value over all the sample of realisations. The plots represent the projection of points on the goodness-of-fit response surface onto a single parameter axis, noting that the results of each model run are controlled by the whole set of parameter values. The input rainfall and potential evapotranspiration time series used to run the model are the same in each case. B. As Figure 4.1A but with the results of each run presented in terms of a predicted variable (here, the predicted discharge summed over all time steps) rather than a goodness-of-fit measure. The dots represent exactly the same set of Monte Carlo realisations as for Figure 4.1A. An option to produce a file of such Monte Carlo realisation results is included in the TOPMODEL demonstration Windows software package. The analyses have been made using the demonstration Windows GLUE software. Both software packages can be downloaded from *http://www.es.lancs.ac.uk/hfdg/hfdg.html*

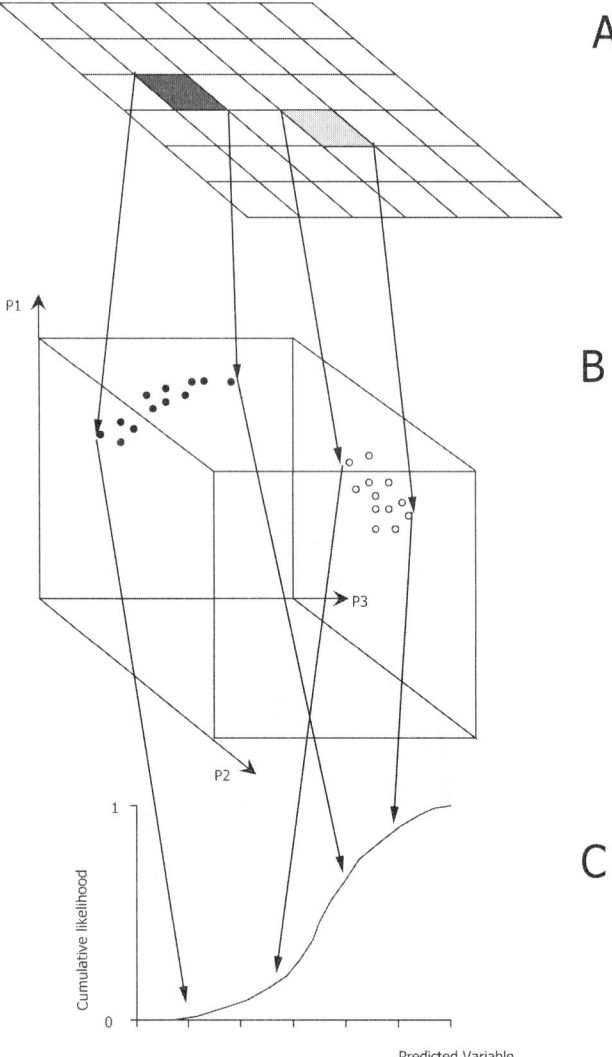

Figure 4.2 A schematic representation of mapping a 'landscape space' into a 'model space', using the same type of Monte Carlo realisations as in Figure 4.1. For any particular point or parcel of the landscape (Figure 4.2A) there may be many acceptable models in the model space (Figure 4.2B). The model space is here represented in terms of different parameter axes but an extension to multiple model structures is also feasible within this framework. The mapping from landscape space to model space, indicated by the dotted lines, is achieved using a weighting function for each model that reflects both the number of points in the landscape that model is representing and also, where appropriate, the goodness-of-fit of that model to any observational data available. These weighting functions can be used in the prediction of either local or landscape scale responses (Figure 4.2C). See Beven (2000) for further discussion of this approach

be the most valuable in refining the model space description in future, in that such failures are a driving force to extend the models to cover a wider range of conditions.

G   So do we conclude that we might never be able to validate model predictions of hydrological systems in the sense of showing that a model is a true representation of reality, but can only continue to monitor the performance of models that have been acceptable up to now?

D   That process would be much easier if we had measurement techniques that gave us better data about the process variables and fluxes – especially if they gave spatially distributed data like remote sensing of water table depths or latent heat fluxes.

E   What do you mean by easier? It might turn out that the user would get so overwhelmed by data that it would make practical prediction and decision-making more difficult. Our experience of using spatially distributed models is that it becomes more and more difficult to set the model up with all the inputs and parameters it requires. Then when you do have some observations to compare the predictions with, which parameters do you start to change to improve the fit or even to evaluate which multiple models are consistent with the spatial data?

P   And you have to remember that remote sensing data itself requires a model with parameters to interpret the digital signal into a variable that we are interested in. Such interpretative model parameters are often treated as constants even though they are known to be time variable. This must introduce additional uncertainty even for a variable like surface temperature from thermal imaging, let alone for derived variables like soil moisture values estimated from microwave sensing. So we might be trying to evaluate uncertain models with uncertain data – or even data associated with an uncertain level of uncertainty.

C   Hang on. I am not sure my brain can cope with all these layers and layers of uncertainty – how are we ever going to take them all into account? Is it necessary to take them all into account? Is it not true that even for the simplest prediction, all we need to do is to estimate the risk of being wrong, or rather the potential cost of being wrong, however we do it? Routine decisions will surely require simple and easy to understand ways of doing that, not layer upon layer of uncertainties. We don't have the money or manpower to continue monitoring to check and revise our decisions, but there is a cost if the initial decision does not prove to be adequate. And failures almost invariably involve some 'unforeseen circumstances' that are not easily assessed since they are, by definition, unforeseen.

E   But how are the levels of risk to be set without someone making some assessment in terms of the value of different types of data in constraining all these layers of uncertainty for different models – at least for some example cases?

C   But is that not equivalent to saying that if we could validate a model, say by a split record test, we might be able to have more faith in the predictions?

G   How do you mean validate? In such a framework the 'validation' period of a split record test (and any other available data) can effectively be used to condition the model space in terms of the level of acceptability of different parameter sets. Is that not the best confirmation we can have unless we go to collect more, and perhaps different types of, data?

P   If I understood correctly, that was also the conclusion of the last paper this morning . . .

ACKNOWLEDGEMENTS

Over the years, many parts of this dialogue have been rehearsed with a variety of friends over a variety of beers and in other seminar and discussion sessions and I am indebted to all those who have, perchance, participated. The origins of the chapter go back before the invitation to contribute to this volume. The first draft was scribbled on a plane to Brazil resulting from an invitation from Robin Clarke to visit Porto Allegre. I am particularly grateful to Tom Dunne, not only for his support in making a sabbatical and seminar series in Santa Barbara possible a few years back, but also for his insightful and valuable review of this manuscript. I am also grateful to Jan Feyen and the Francqui Foundation for support for a stay and extended seminar series in Leuven and other Belgian universities that also helped clarify the presentation. I know that some people still feel these ideas are counterproductive and will undermine hydrology as a science. I hope this chapter will help to show that this need not be the case, and that a continuing dialogue between modellers and measurers is essential to the future of the science. What is still needed is an appropriate framework within which such a dialogue can take place. That is what this contribution is really about!

REFERENCES

Bair, E.S. 1994. Model (in)validation – a view from the courtroom. *Ground Water*, **32**, 530–531.

Bassett, K. 1994. Comments on Richards: the problems of real geomorphology. *Earth Surface Processes and Landforms*, **19**, 273–276.

Beck, M.B. 1987. Water quality modelling: a review of the analysis of uncertainty. *Water Resources Research*, **23**, 1393–1442.

Beck, M.B. 1994. Understanding uncertain environmental systems. In: Grasman, J. and van Straten, G. (eds), *Predictability and Nonlinear Modelling in Natural Sciences and Economics*. Kluwer, Dordrecht, 294–311.

Beck, M.B., Kleissen, F.M. and Wheater, H.S. 1990. Identifying flow paths in models of surface water acidification. *Reviews of Geophysics*, **28**, 207–230.

Beven, K.J. 1993. Prophecy, reality and uncertainty in distributed hydrological modelling. *Advances in Water Resources*, **16**, 41–51.

Beven, K.J. 1995. Linking parameters across scales: subgrid parameterisations and scale dependent hydrological models. *Hydrological Processes*, **9**, 507–525.

Beven, K.J. 1996a. Equifinality and uncertainty in geomorphological modelling. In: Rhoads, B.L. and Thorn, C.E. (eds), *The Scientific Nature of Geomorphology*. John Wiley, Chichester, 289–313.

Beven, K.J. 1996b. A discussion of distributed modelling. In: Refsgaard, J.C. and Abbott, M.B. (eds), *Distributed Hydrological Modelling*. Kluwer, Dordrecht, 255–278.

Beven, K.J. 1997. TOPMODEL; a critique. *Hydrological Processes*, **11**, 1069–1086.

Beven, K.J. 2000. Uniqueness of place and process representations in hydrological modelling. *Hydrology and Earth System Science*, **4**(2) 203–213.

Beven, K.J. 2001a. Calibration, validation and equifinality in hydrological modelling: a continuing discussion, this volume.

Beven, K.J. 2001b. *Rainfall–Runoff Modelling: The Primer*. John Wiley, Chichester.

Beven, K.J. Lamb, R., Quinn, P., Romanowicz, R. and Freer, J. 1995, TOPMODEL, in Singh, V.P. (ed.), *Computer Models of Watershed Hydrology*, Water Resource Pubns. Colorado, 627–668.

Binley, A.M., Henry-Poulter, S. and Shaw, B. 1996. Examination of solute transport in an undisturbed soil column using electrical resistance tomography. *Water Resources Research*, **32**, 763–769.

Brooks, R.J., Lerner, D.N. and Tobias, A.M. 1994. Determining the range of predictions of a groundwater model which arises from alternative calibrations. *Water Resources Research*, **30**, 2993–3000.

Chalmers, A. 1990. *Science and its Fabrication*. Open University Press, Milton Keynes.

Davis, P.A., Olague, N.E. and Goodrich, M.T. 1992. Application of a validation strategy to Darcy's experiment. *Advances in Water Resources*, **15**, 175–180.

Emmett, W. 1978. Overland flow. In: Kirkby, M.J. (ed.), *Hillslope Hydrology*. John Wiley, Chichester, 145–176.

Feyerabend, P. 1978. *Against Method*. Verso, London.

Feyerabend, P. 1991. *Three Dialogues on Knowledge*. Blackwell, Oxford.

Flury, M., Flühler, H., Jury, W.A. and Leuenberger, J. 1994. Susceptibility of soils to preferential flow of water: a field study. *Water Resources Research*, **30**, 1945–1954.

Franks, S.W. and Beven, K.J. 1997a. Bayesian estimation of uncertainty in land surface–atmosphere flux predictions. *Journal of Geophysical Research*, **102**(D20), 23991–23999.

Franks, S.W. and Beven, K.J. 1997b. Estimation of evapotranspiration at the landscape scale: a fuzzy disaggregation approach. *Water Resources Research*, **33**, 2929–2938.

Franks, S.W., Gineste, Ph., Beven, K.J. and Merot, Ph. 1998. On constraining the predictions of a distributed model: the incorporation of fuzzy estimates of saturated areas into the calibration process. *Water Resources Research*, **34**, 787–797.

Freer, J., Beven, K.J. and Ambroise, B. 1996. Bayesian estimation of uncertainty in runoff production and the value of data: an application of the GLUE approach. *Water Resources Research*, **32**(7), 2161–2173.

Gupta, H.V., Sorooshian, S. and Yapo, P.O. 1998. Toward improved calibration of hydrologic models: multiple and noncommensurable measures of information. *Water Resources Research*, **34**(4), 751–763.

Harvey, J.W., Wagner, B.J. and Bencala, K.E. 1996. Evaluating the reliability of the stream tracer approach to characterise stream–subsurface water exchange. *Water Resources Research*, **32**, 2441–2451.

Henderson, D.E., Reeves, A.D. and Beven, K.J. 1996. Flow separation in undisturbed soil using multiple anionic tracers (2) steady state core scale rainfall and return flows. *Hydrological Processes*, **10**(11), 1451–1466.

Howson, C. and Urbach, P. 1993. *Scientific Reasoning, The Bayesian Approach*, 2nd edition. Open Court, Peru, IL.

Loague, K. and Kyriakidis, P.C. 1997. Spatial and temporal variability in the R-5 infiltration data set: déjà vu and rainfall–runoff simulations. *Water Resources Research*, **33**, 2883–2895.

Lohmann, D. and 27 co-authors. 1998. The project for intercomparison of land-surface parameterization schemes (PILPS) Phase 2(c) Red-Arkansas basin experiment: 3. spatial and temporal analysis of water fluxes. *Global and Planetary Change*, **19**, 161–179.

McLaughlin, D. and Townley, L.R. 1996. A reassessment of the groundwater inverse problem. *Water Resources Research*, **32**, 1131–1161.

Morton, A. 1993. Mathematical models: questions of trustworthiness. *British Journal of Philosophy of Science*, **44**, 659–674.

Oreskes, N., Schrader-Frechette, K. and Belitz, K. 1994. Verification, validation and confirmation of numerical models in the earth sciences. *Science*, **263**, 641–646.

Rhoads, B.L. 1994. On being a real geomorphologist. *Earth Surface Processes and Landforms*, **19**, 269–272.

Richards, K.S. 1990. Real geomorphology. *Earth Surface Processes and Landforms*, **15**, 195–197.

Richards, K.S. 1994. Real geomorphology revisited. *Earth Surface Processes and Landforms*, **19**, 277–281.

Romanowicz, R. and Beven, K.J. 1998. Dynamic real-time prediction of flood inundation probabilities. *Hydrological Science Journal*, **43**, 181–196.

Romanowicz, R., Beven, K.J. and Tawn, J. 1996. Bayesian calibration of flood inundation models. In: Anderson, M.G., Walling, D.E. and Bates, P.D. (eds), *Floodplain Processes*, John Wiley, Chichester, 333–360.

Schrader-Frechette, K.S. 1989. Idealised laws, antirealism and applied science: a case in hydrogeology. *Synthese*, **81**, 329–352.

Sen, M.K. and Stoffa, P.L. 1995. *Global Optimisation Methods in Geophysical Inversion*. Elsevier, Amsterdam.

Zak, S., Beven, K.J. and Reynolds, B. 1997. Uncertainty in the estimation of critical loads: a practical methodology. *Soil, Water and Air Pollution*, **98**, 297–316.

5   Models in the Courtroom

E. SCOTT BAIR
Department of Geological Sciences, Ohio State University, USA

5.1   INTRODUCTION

Most scientists and engineers will never enter a courtroom as a party in a lawsuit. Most of us may never provide expert testimony in a trial. Yet trial by jury is one of the most valued and highly defended rights guaranteed to citizens of the United States by the US Constitution. It is the hallmark of the US system of justice. Although the Anglo-American system of jurisprudence is rooted in the 'inquisitions' ordered by William the Conqueror in 11th century England to determine the countryside's wealth and population (Van Dyke 1987), the appearance of hydrologic models in courtrooms is obviously a late 20th century phenomenon.

Modern hydrologic models founded on solving partial differential equations using numerical methods also can be traced to England. Shaw and Southwell (1941) may have been the first to use a finite-difference model to address a groundwater seepage problem. Advances in numerical methods and in our theoretical and applied knowledge over the past 30 years combined with continuous improvement in the ease, cost and power of computers have enabled hydrologists to address problems previously considered intractable. The ability to model complex hydrology combined with federal and state environmental laws promulgated in the latter part of the 20th century to protect water supplies, manage waste handling, and improve waste disposal have caused US hydrologists to participate in various types of litigation. As expert witnesses, we may testify in lawsuits filed under violations of environmental statutes or tort laws (laws relating to wilful or negligent damage or injury). In 1995, US District Courts began over 45 000 new cases dealing with tort actions and environmental matters (US Department of Commerce 1997, p. 216). Although most of these cases do not involve hydrologic issues or require testimony from hydrologic experts, and only 5% actually go to trial (Grossman and Vaughn 1999), the statistics reveal the zeal we Americans have for resolving conflict by litigation.

In court, as in other applications, hydrologic models are used to evaluate historic hydrologic events or conditions, and are used to predict future hydrologic events or conditions. In both cases, the credibility of the model rests in the subjective judgment of a jury or a judge. Ideally, we, the scientists and engineers, would like the jury or judge to first decide whether the model is a reasonable representation of the real hydrologic system and then determine whether the predictive simulations are reasonable based on the ability of the model to reproduce known (measured) physical states of the flow system. It is unlikely, however, that this type of sequential, critical evaluation will occur. It is more likely that the jury or judge will simply assess the overall believability of the scientific testimony. Terms such as *validate* and *verify* that suggest or imply realism or truth are used to paint righteous images of models for the jury or judge. As a consequence, these common modelling terms and the procedures hydrologists use to substantiate their conclusions are often highly contested by the opposing party.

Model Validation: Perspectives in Hydrological Science. Edited by M.G. Anderson and P.D. Bates.
© 2001 John Wiley & Sons, Ltd.

It has been said that cross-examination is the most powerful tool in the US legal system for eliciting truth from witnesses. It has also been said that cross-examination by harassment is one of the greatest impediments to truth (Gerber 1987). For a hydrologic modeller, cross-examination is guaranteed to be a challenging experience regardless of the degree of validity, amount of veracity, or the number of confirming observations supporting the model and its predictions.

My objectives in writing this chapter are to describe issues, some substantive and some philosophic, that arise in the courtroom and bear on (1) the use and presentation of hydrologic models by expert witnesses, and (2) the evaluation of these models and their results by a jury or judge. In so doing, I will use the events surrounding a famous federal trial (*Anne Anderson et al. v. W.R. Grace & Company and Beatrice Foods, Inc.*) and the testimony of three expert hydrologists who participated in that trial to examine the interface between science and the US legal system.

5.2   SCIENTIFIC METHODS IN COURT

Hydrologists, like most other witnesses, are visitors in the legal arena. The vocabulary and rules used by lawyers may seem foreign to us, but the legal procedures that can eventually place a plaintiff's complaint before a jury are similar to the general scientific methods we use to develop and test hypotheses. As a result, the path a lawsuit takes through the American legal system is not unlike the path hydrologic research takes through the scientific community. Although scientists do not necessarily agree on all the components that comprise the scientific method, most agree that the scientific method is a set of informal rules for formulating concepts, conducting experiments, making observations, developing inferences, and then testing these inferences by further experimentation and observation (Peters 1996). Hydrologic modellers commonly follow a similar set of procedures in constructing and testing their mathematical models.

Figure 5.1 shows the general steps in the scientific method, procedures followed by hydrologic modellers, and their similarities to the path a civil lawsuit follows through the American legal system. In the latter scheme, a perceived injustice leads to identification of possible causation and the filing of a legal complaint by a plaintiff. In the scientific method, this is analogous to initial observations leading to the formulation of a question and the development of a hypothesis. The scientific hypothesis and the legal complaint both represent queries that have been framed but need to be tested and resolved. In a parallel sense, they are similar to the statement of purpose of a hydrologic model and the development of a conceptual framework of the flow system.

Pre-trial discovery, the phase of a lawsuit when information is acquired and solicited, legal strategies are formulated, and professional opinions are developed, stated and queried, corresponds to the steps in the scientific method where experiments are planned and executed, observations recorded, and the results analysed to see whether the experimental data support or refute the hypothesis. In a hydrologic model, these steps are analogous to the calibration of the model, wherein measured water levels and/or discharges are compared to simulated values to determine how well the model reproduces the measured physical state of the system.

In a trial, discovered evidence is presented by lay witnesses and expert witnesses during direct examination and then is scrutinised by the opposition during cross-examination. A jury or judge makes the final deliberations regarding acceptance or rejection of the legal complaint. At the end of the trial process, evaluation of the testimony by the jury or judge is immediate and the final decision to accept or reject the plaintiff's complaint is made by non-experts and is reached quickly.

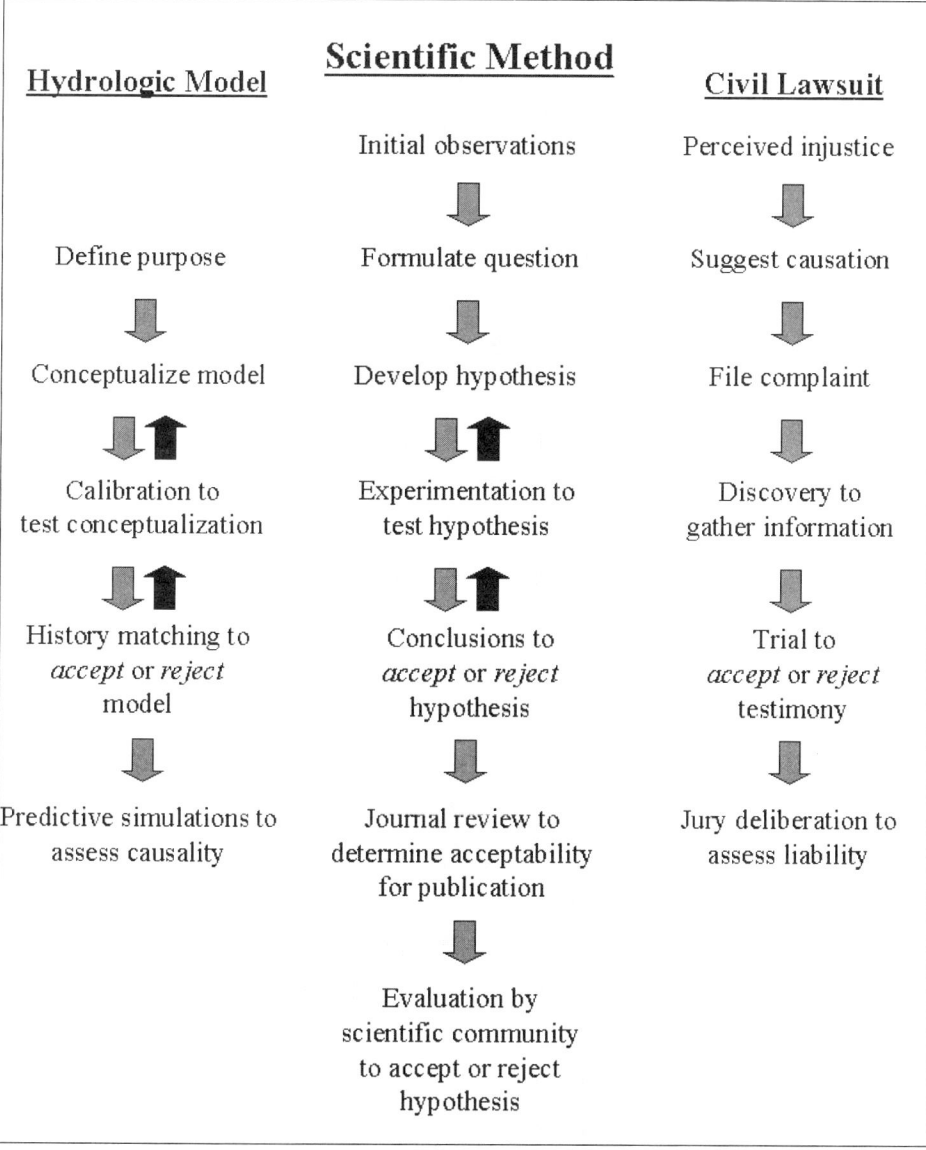

Figure 5.1 Similarities among the scientific method, trial procedures and hydrologic model construction

In the scientific arena, the experimental data are analysed and conclusions (inferences) are drawn whether to accept or reject the hypothesis. If the hypothesis is accepted, the work is commonly submitted for publication in a professional journal where it is scrutinised by peer reviewers to assess its acceptability for publication. If the work is accepted for publication, the greater scientific community makes the final acceptance or rejection of the hypothesis. In contrast to the legal process, evaluation of professional opinions through the scientific process is slow, is made by fellow experts, and the final decision may not be reached for years as other scientists try to replicate the work (Bair and Wood 1999).

Although there are similarities between science and law, there are also differences. A major difference is in the time spent evaluating whether to accept or reject the hypothesis, the testimony and the model (Figure 5.1). Deliberations by a jury over testimony presented in a trial may take hours, sometimes days, on rare occasions weeks. Evaluation by the greater scientific community of new concepts published in a journal may take months to years, sometimes decades. Another difference is that the jury is made up of non-experts whose common knowledge is relied upon to render an impartial judgment, whereas the scientific community is comprised of competing scientists whose expertise is relied upon to repeat the experiments and validate or reject the work. A third difference is that the scientific method facilitates feedback between hypothesis development, design and execution of experiments, and formulation of conclusions. The path of a civil lawsuit is one directional (see Figure 5.1), although it can be stopped by settlement or the verdict overturned on appeal.

Federal Judge Lynn N. Hughes expressed the differences as follows:

> Scientific investigation is a long-term pursuit focused at understanding our experiences, whereas jurisprudential investigation is a short-term pursuit focused at resolving a conflict between two or more parties. Science has the freedom of the eventuality of finding answers and the generality of application, whereas the law has the constraints of the immediacy of resolution and the particularity of each case... Courts are a societal mechanism for resolving disputes in which the concepts of truth and justice are significant but secondary. Although both science and the law seek the truth, both also must rely on incomplete data and both also must operate under uncertainty. Science and law do not, then, have competing visions of truth. Rather, knowledge competes with resolution... We need to fit the product of science into the task of courts.
>
> (Hughes 1999)

5.3   SCIENTISTS IN AN ADVERSARIAL LEGAL SYSTEM

Hydrologists, like most applied scientists, are accustomed to constructing models with inherent uncertainty (Voss 1998). It is highly unusual for us not to deal with sparse and/or noisy data. The general public looks to scientists to unravel and explain the physical, chemical and biological processes that operate day-to-day within our realm. The public, however, is accustomed to scientists describing these processes in terms of laws – Newton's Laws of Motion, the Laws of Thermodynamics, Ohm's Law, Snell's Law, Darcy's Law, etc. In allowing this perception of science to prevail, scientists have unwittingly encouraged the public to believe in the certainty of science. After all, science, deals with definitive laws that are tested, proven, and repeatable. As a consequence, the general public, which is sampled to form the jury pool, is unaccustomed to the methods routinely used by applied scientists to deal with uncertainty.

For hydrologists, the most uncertain aspects of model construction – the ones that often make an expert witness squirm on cross-examination – are sparse and noisy data. Using a shallow, transient contaminant transport model as an example, the following types of sparse and noisy data commonly confront the expert witness (Prickett and Pettyjohn 1995).

- surface-water/groundwater interaction
- heterogeneity, anisotropy and assigned hydraulic conductivity distributions
- spatial and temporal variations in recharge rates
- historic pumping rates and schedules of wells

- non-synoptic water-level and contaminant concentration measurements
- contaminant source history and contaminant release rates

In most cases involving site-specific application of hydrologic models, many factors are simply unknown and hence uncertain. Because of uncertainties, most hydrologists, if pressed, might admit to accepting model results that are, in a loosely quantitative sense, 75 to 90% accurate (see Figure 5.2). This assumes that a model cannot be constructed that is 100% accurate (Bredehoeft and Konikow 1993). Given a probability range from 0 to 1.0 that a hydrologic model accurately represents the flow system and accepting that no hypothesis or model can be 100% accurate, then a range of 0.75 to 0.90, or 75 to 90%, for a hydrologic field problem would demonstrate a relatively high degree of correspondence between model results and measured values. In the example of a contaminant transport model, this would be a relatively high degree of correspondence between simulated and measured values of hydraulic head, streamflow change and solute concentration. Over the past several decades, hydrologists have come to expect a higher degree of accuracy because modern finite-element and finite-difference techniques require fewer simplifying assumptions about the hydrogeologic setting and can address more difficult boundary conditions than the closed-form calculus-based analytical techniques developed in the middle of the past century.

In a civil trial, the jury bases its verdict on a preponderance of the evidence. For the jury to find for the plaintiff, a majority of the evidence must indicate that the defendant is liable. To

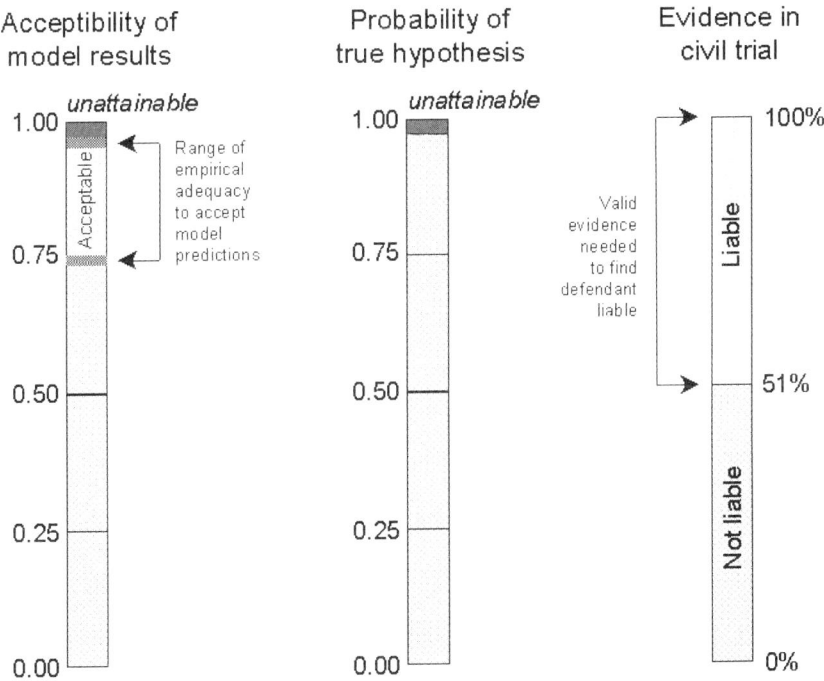

Figure 5.2 Comparison of standards between a civil trial, the probability of a correct hypothesis, and the acceptability of results from a hydrologic model

scientists and engineers accustomed to using modern numerical and computational techniques, a 51% criterion may seem low. It may seem so low that it affords some creativity for experts to use analysis techniques that produce results more favourable to their client's case. This approach is not unethical. It may be employed for a variety of reasons – scientists may approach problems differently, parties often cannot afford the most rigorous scientific approach, legal strategies may not call for a complex analysis. An overly simplistic approach can lead to as much criticism on cross-examination concerning inappropriate use of isotropic and homogeneous parameter values as can use of inappropriate ranges of heterogeneous parameter values in a complex analysis. Simplifying assumptions are always good fodder for cross-examination. It is up to the expert, in consultation with his/her lawyer, to decide whether a simple, one-dimensional analytical model is sufficiently representative of the flow system, or whether a complex, three-dimensional numerical model is required, or whether a laboratory demonstration using a sand tank is appropriate. It is difficult to predict whether, in the minds of the jurors, a simple elegant but fundamentally flawed construction will be more or less believable than a complex arcane truth.

5.4 EXPERT WITNESSING

Expert witnesses, unlike lay witnesses, can present opinions. An expert witness possesses knowledge through experience, training or education of a particular subject in greater depth than the general public (Matson 1994). As society's reliance on specialised knowledge has expanded, so too has the use of expert witnesses (Weinstein 1986). In recent years, several articles and books have been written about the proliferation of junk science in the courtroom. In these essays, expert witnesses are commonly depicted as prostitutes, who, for the right fee, will testify to any farfetched theory in an attempt to win sympathy and a huge award from a scientifically challenged jury selected from a techno-terrified society (Huber 1990; Jost 1991). Peter Huber, author of Galileo's Revenge: Junk Science in the Courtroom (1991), argues that the admission of junk science is in response to the easing of standards for admissibility of expert testimony. It is a problem that may be further heightened by judges and lawyers who are not adequately trained to assess the veracity of the scientific evidence placed before them (Blauvelt 1999).

For much of the 20th century, courts maintained that if the scientific theory or reasoning was sufficiently reliable to attain the *general acceptance* of the scientific community, then the evidence was sufficiently reliable to be admitted in court. Known as the Frye Rule, this general acceptance test was established in 1923 (*Frye v. United States*) and governed the admissibility of scientific evidence in US courts for decades. The main criticism of the Frye Rule was that it excluded new scientific advances that were too new to have established general acceptance by the scientific community (Foster et al. 1993). In the latter part of the 20th century, many states adopted the Federal Rules of Evidence, which included a broadly discretionary set of guidelines for the acceptance and restriction of scientific testimony. Rule 702, the most important of these rules, provides that expert testimony is admissible 'if scientific, technical, or other specialised knowledge will assist the trier of fact to understand the evidence or to determine a fact in issue' (Weinstein 1986). Although the 'assistance to the trier of fact' standard liberalised the admissibility of scientific evidence, it led to the admission of various types of novel expert testimony and a blurring of the distinction between expert testimony and lay testimony (Weinstein 1986).

The US Supreme Court in *Daubert v. Merrel Dow Pharmaceuticals, Inc.*, a 1993 case alleging birth defects as a result of taking the anti-nausea drug Bendectin, held that under Rule 702 scientific knowledge presented as testimony must be derived by the scientific method. The Supreme Court also held that evidentiary reliability is to be based on scientific validity. Under the Daubert ruling, the trial judge is assigned the responsibility of gatekeeper to make preliminary evaluation of whether the reasoning and methodology underlying the testimony is scientifically valid or reliable and whether it is applied properly to the facts at issue (Foster et al. 1993; Blauvelt 1999). The gatekeeper's role is not to determine which expert opinion is correct. Rather, it is to determine whether each expert used valid scientific reasoning and principles to reach his/her conclusions and to screen out professional opinions based on conjecture and speculation (Blauvelt 1999). The Supreme Court established that the initial determination regarding the admissibility of an expert's scientific testimony is to be based on (1) whether the expert opinion is founded on reliable data, (2) whether the expert opinion is supported by valid scientific reasoning and methodology, and (3) whether the expert opinion is based on information relevant to the case (Blauvelt 1999).

5.5. EXAMPLE OF *ANNE ANDERSON ET AL. v. W.R. GRACE & COMPANY AND BEATRICE FOODS, INC.*

The 1986 Woburn Toxic Trial, the informal name referring to *Anne Anderson et al. v. W.R. Grace & Company and Beatrice Foods, Inc.*, is an excellent example to demonstrate the analogy between scientific hypothesis testing and the civil trial process using a case that centres on the interpretation of hydrologic data and the predictions of hydrologic models. In this lawsuit, which is the focus of the award-winning book *A Civil Action* (Harr 1995) and the 1999 Touchstone Pictures movie of the same name, eight plaintiffs, who lived in eastern Woburn, Massachusetts, filed suit against two large corporations that operated a local manufacturing plant, the Cryovac Division of W. R. Grace & Company, and a local tannery owned by Beatrice Foods, Inc. The plaintiffs' 1982 legal complaint alleged that improper handling and disposal of five industrial chemicals, including trichloroethene (TCE), dichloroethene (DCE) and tetrachloroethene (PCE) at the two properties, entered the groundwater system, flowed to two municipal wells, and prolonged ingestion and exposure to the toxic chemicals caused leukaemias, central nervous system disorders, and other health problems.

The two municipal wells, known as wells G & H, began operating in 1964 and 1967, respectively. The wells are 2-feet in diameter, 88-feet deep, and 600 feet apart. Both wells have 10-foot screens and, when operating, pumped at average rates of 700 and 400 US gallons per minute, respectively. Wells G & H were used together only 23% of the time and primarily were used only during periods of drought to augment water produced by six other municipal wells in another part of town. The wells were built on earthen mounds within the floodplain of the Aberjona River. At normal stage, the river flows within 300 feet of well G and 100 feet of well H (Figure 5.3). At elevated stages, the river floods across a wetland and flows within a few feet of the municipal wells. The wells are constructed in highly permeable glacial outwash deposits filling a buried bedrock valley composed of fractured granodiorite (Figure 5.4). Wells G & H were closed in May 1979, by order of the Commonwealth of Massachusetts, after water samples revealed concentrations of TCE and PCE exceeding public health standards. US EPA (1985) considered TCE and PCE to be probable human carcinogens. Samples taken from wells G & H prior to this date were not analysed for volatile organic compounds.

Because of the complex nature of the case, the judge, in accordance with federal procedures, divided the case into three phases. Each phase contained a trial that focused on a specific issue. The resolution of each issue, in sequence, would progress to the determination of the ultimate issue – whether prolonged consumption of dilute concentrations of TCE, DCE and/or PCE caused the specific types of leukaemia contracted by the plaintiffs. The first of these trials was to determine which, if any, of the contaminants entered the groundwater system on the defendants' properties and, if so, whether the contaminants flowed to the municipal wells within the time the wells operated. If either defendant was found liable in this phase, then the subsequent two phases would proceed to determine whether the contaminant(s) reaching the wells could cause leukaemias and the other health effects. In trifurcating the trial, the judge produced a series of sequential subordinate hypotheses. If the jury rejected the subordinate hypothesis for a particular defendant, the lawsuit ended at that phase. If the jury accepted the subordinate hypothesis for a particular defendant, the lawsuit continued to the next phase.

From a hydrologist's perspective, the unique aspect of this trifurcated case is that the first phase focused entirely on surface-water and groundwater hydrology. From the jury's perspective, the unique aspect of the first phase is that the plaintiffs did not testify and almost all the testimony presented by lay and expert witnesses dealt with scientific issues related to contaminant source areas and concentrations, geologic characterisation, physical properties of sediment and rock, induced streambed infiltration, flood frequency, surface-water and groundwater interaction, flowpaths, retardation factors, organic chemistry, and the prediction of contaminant arrival times at the wells using hydrologic models. As a result, the trial provides an interesting view of the interface between science and the law.

Pre-trial discovery began in October 1982. No field data, including groundwater levels or monitoring well samples, were collected prior to closure of the two municipal wells in May 1979. As a result, during discovery it was necessary to design and perform field tests to reproduce the configuration of the water table when the wells were periodically operational between 1964 and 1979, and to determine the degree to which the wells were hydraulically connected to the Aberjona River and wetland. Independent of the trial, the US Geological Survey and US EPA conducted a 30-day pumping test using wells G & H. The test began in December 1985 and was done in support of federal Superfund cleanup activities at known sources of contamination near wells G & H. During the test, wells G & H were pumped at their average rates.

The plaintiffs' and defendants' experts were provided the groundwater level and streamflow measurements made during the 30-day test. These data helped them develop their professional opinions for the trial concerning the role of induced stream infiltration, flowpaths from the defendants' properties to the wells, and travel times of contaminants to the wells. Based on these and other data, the hydrologic experts developed conceptual models of the hydrogeologic setting and constructed mathematical models of the flow system that became the foundation of their opinions. The experts' opinions represent inferences drawn from the experimental data. Figure 5.3 shows the configuration of the water table after pumping wells G & H for 30 days. Water levels in the 45 monitoring wells used to construct Figure 5.3 had not quite attained steady state at the end of the pumping period (Myette et al. 1987). After the first day of pumping, periodic stream discharge measurements made at Olympia Avenue, north of the wells, and at Salem Street, south (downstream) of the wells (see Figure 5.3), showed that water in the river and wetland was being induced to flow downward into the aquifer to supply the two municipal wells. By the end of the 30 day pumping period, there was 565 US gallons per minute less stream discharge measured at Salem Street than Olympia Avenue (Myette et al. 1987).

5.5.1 Experts' Opinions on the Hydrologic System at Woburn

During the 78-day trial, the jury heard opening remarks by a lawyer representing each of the three parties, direct testimony and cross-examination from lay and expert witnesses, and closing arguments from the lawyers. The jury also visited the wells G & H area. During direct examination, the geologic experts hired by each party described the aquifer in which wells G & H are completed as a heterogeneous sequence of transmissive glacial sediments (see Figure 5.4). The main points of controversy in the opinions of the hydrologic experts concerned (1) the

Figure 5.3   Water-table map of the Aberjona River valley near wells G & H after 30 days of continuous pumping, 3 January 1986 (contours from Myette et al. 1987). WRG = W.R. Grace property. BF = Beatrice Foods property. G & H = municipal wells. T = tannery well

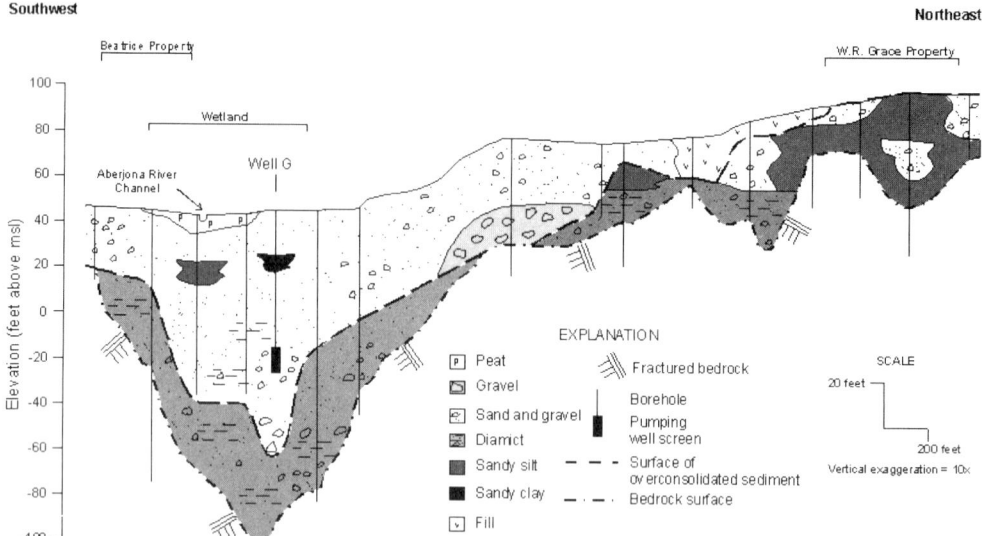

Figure 5.4 Generalized geologic cross-section from the Beatrice property, through well G, to the W.R. Grace property (after Metheny 1998)

degree to which the Aberjona River and wetland were hydraulically connected to the aquifer, (2) the hydraulic properties of interbedded peat and fine sand underlying the river and wetland, outwash sand and gravel in the buried valley, and silty clay (lodgement) till underlying the uplands (see Figure 5.4), and (3) the configuration of the water table beneath the river and wetland when wells G & H were pumping (see Figure 5.3). The opinions of the three hydrologic experts differed substantially on pathlines and travel times of contaminants to the wells, the capture zones of the wells, and whether induced infiltration of surface water entered the aquifer and flowed to the wells during pumping. Following are descriptions of the testimony of the three hydrologic experts in the case. A summary of their opinions and the bases of their opinions are given in Figure 5.5.

5.5.2   Plaintiffs' Expert

The defendants' lawyers deposed the plaintiffs' hydrologic expert on three separate occasions for a total of 16.5 hours (Grossman and Vaughn 1999). His direct testimony occurred on days 39, 40 and 41 of the trial, which was followed by seven days of cross-examination and two days of re-direct examination and re-cross examination.

The plaintiffs' expert constructed a three-dimensional flow and transport model synthesising data from well logs, soil borings and slug tests into a heterogeneous geologic framework. The numerical model was used to develop a 'cosmopolitan view' of the aquifer and flow system. However, because the model was not calibrated in a rigorous manner against measured values of head, streamflow change or solute concentrations, it was never introduced into evidence at trial and was used solely to illustrate his opinions drawn from other analyses. According to the plaintiffs' expert, the model did simulate the overall patterns of groundwater flow measured during the 30-day pumping test (Deposition #4, 13 February 1986, p. 130). The plaintiffs' expert

	Plaintiffs'	Beatrice's	W.R. Grace's
Field data	Water levels, well logs, f_{oc}, slug tests	Water levels, stream flows, well logs	Water levels, stream flows, well logs, f_{oc}, slug tests, recharge
Basis	Flow vectors, plus 1-D analytical transport model	Potentiometric maps and flowlines	3-D numerical flow and transport model
Geologic Heterogeneity	Uniform in 1-D K_{GRACE} = 75 ft/day $K_{BEATRICE}$ = 2300 ft/day	Not addressed directly	Heterogeneous 3 layers K_{RANGE} = 0.75 - 113 ft/day
River infiltration (wells pumping)	Limited infiltration due to peat, reaches wells in 10 to 20 years	Infiltration creates groundwater divide, reaches wells in 3 to 4 months	Substantial infiltration 50% of well discharge, reaches wells in 2 to 4 months

Figure 5.5 Summary of hydrologists' expert opinions in *Anne Anderson et al. v. W.R. Grace & Company and Beatrice Foods, Inc*

held the following opinions with respect to the interaction of the wetland and river with the groundwater flow system.

> . . . if I were to try and bracket the time of travel of contaminants from the river to the wells, it could be as long as 10 to 20 years.
>
> (Trial Day 43, p. 67)

> I think very little [water], if any at all was pumped out of the river during the pumping test of wells G and H].
>
> (Trial Day 43, p. 71)

The negligible amount of induced infiltration, he stated, was due to the very low hydraulic conductivity of the peat underlying the river and wetland.

As the basis of his opinion, the plaintiffs' expert used a one-dimensional analytical flow and transport equation (Ogata-Banks equation, see Domenico and Schwartz 1998, p. 373) to estimate travel times of contaminants from the W.R. Grace and Beatrice properties to the municipal wells. This one-dimensional analytical model necessitated use of uniform values of hydraulic conductivity, hydraulic gradient and porosity, and limited hydrodynamic dispersion to the longitudinal direction of groundwater flow. Based on this one-dimensional analysis, the plaintiffs' expert made the following testimony regarding contaminant travel times.

> . . . a travel time for TCE of 3 years from Grace. . . the travel time in terms of the Grace site to the wellfield would be for [DCE] – 1.03 years, [PCE] – 9.67 years . . .
>
> (Trial Day 41, pp. 108–109)

. . . in terms of Beatrice, I would like my testimony to read that TCE was 3 months, [DCE] – 1.03 months, [PCE] – 9.67 months . . .

(Trial Day 41, p. 109)

In this analysis, the plaintiffs' expert used a porosity of 0.20, a dispersivity of 67.5 feet, and different retardation factors for each contaminant that ranged between 1.5 and 8. For the calculations pertaining to contaminants at W.R. Grace, he used a hydraulic conductivity of 75 feet/day, a hydraulic gradient of 0.001 and a distance of 2840 feet, whereas he used a hydraulic conductivity of 2300 feet/day, a hydraulic gradient of 0.018 and a distance of 525 feet for the Beatrice calculations (Trial Day 41, p. 109). In his opinion the chemicals originating at Beatrice reached wells G & H within a few months, and TCE and DCE originating at W.R. Grace were present at the future locations of wells G & H before the wells were actually built, assuming the chemicals entered the flow system in 1960, the year the plant opened. The Aberjona River, in his view, was neither a source of additional contamination to the wells nor a source of dilution because of the poor hydraulic connection between the river and the aquifer.

5.5.3 Beatrice's Expert

The other parties never deposed the expert hydrologist hired by Beatrice. His entire testimony occurred during days 58, 59 and 60 of the trial. His opinions were based on his conceptualisation of the hydrogeologic setting using well logs, and groundwater level and streamflow gain/loss measurements made during the 30-day pumping test. He presented a conceptual model of the flow system, but did not construct an analytical or numerical flow model, or compute travel times. With respect to the interaction between the Aberjona River and the pumping wells, he made these statements during the trial.

I would say that river water could flow to Wells G and H in a period of 3 to 4 months.

(Trial Day 58, p. 127)

. . . the river creates this ridgepole effect, and it is the high point on the water table surface with the tarpaulin that slopes away on both sides. This is caused by the pumpage of G actually inducing water to leave the river. As the water goes out of the river channel, it forms a mound on the water table, and the mound is this ridge-shaped affair . . .

(Trial Day 59, pp. 22–23)

. . . the cone of depression from G and H, which is what we've depicted on this map, can only go up to the river. It does not cross the river. So the effect of [wells] G and H will not cross the river but will be contained and be stopped from leakage by the river itself.

(Trial Day 59, pp. 31–32)

The Beatrice expert made no calculations of contaminant travel times to wells G & H because his analysis showed that groundwater under the Beatrice property would not flow under the river to the east and downward to the well screens. He made the following statement.

What we found out was that groundwater [at Beatrice], on the contrary, moves away from Wells G and H when they were pumping. Therefore, the fact of having chemicals in that water during the 1964 to 1979 period may really be secondary because even if they were there – and I'm not saying they were – but even if they were, they would not move to Wells G and H.

(Trial Day 59, p. 57)

In his opinion, contaminants, if present on the Beatrice property, would not flow eastward toward municipal wells G & H, but would flow westward, away from the wells due to the groundwater mound formed beneath the Aberjona River by induced infiltration of river water. Hence, contamination measured in wells G & H in May 1979 could not be derived from 55-gallon drums dumped on the Beatrice property because induced infiltration from the Aberjona River formed a barrier to groundwater flow from that direction.

5.5.4   W.R. Grace Expert's Testimony

The plaintiffs' attorney deposed the expert hired by W.R. Grace for 1.5 hours. His direct examination occurred on days 66, 67, and 68 of the trial, and was followed by cross-examination on days 68, 69 and 70, and re-direct examination on day 71. The W.R. Grace expert used well logs, seismic surveys, and soil borings from the area to construct a conceptual framework of the geologic setting. He used historic pumping records, historic streamflow data, and water-level and streamflow measurements from the 30 day pumping test to develop a conceptual model of the hydrologic system and the interaction between the river and wetland with the groundwater flow system. The data were incorporated into a three-dimensional numerical flow and transport model that was used to simulate transient water level and streamflow conditions, and chemical transport in the flow system from 1960 to 1979. The model incorporated spatial variations in hydraulic conductivity, spatial and temporal variations in recharge, bedrock leakage, partial penetration of the wells and river, and historic pumping rates of the municipal wells. Hydraulic conductivity values in the three-layer model varied from 0.75 feet/day for the lodgement till underlying the bedrock uplands to 113 feet/day for the glacial outwash filling the buried valley. The flow model was calibrated to data collected for the 30 day pumping test using water levels measured immediately before wells G & H commenced pumping, transient water levels measured during the 30 day pumping period and streamflow gain and loss measurements made before and during the test. The transport model was not calibrated because of the lack of historic TCE, DCE and PCE data. A sensitivity analysis was performed using ranges of transport parameters and a conservative assumption was made placing the contaminants at the water table the day the plant opened in 1960.

In his testimony, the W.R. Grace expert made the following statements regarding the interaction between the river/wetland and wells G & H:

> But under conditions where a man-caused hydrologic stress, that is pumping of a production well for water supply, and that well is located close to the Aberjona River, it will induce water to flow out of the river into the aquifer and be a source of the supply of water to those wells.
>
> (Trial Day 67, p. 73)

> ... on a long-term period ..., the water that comes out of the river gets to the wells within two months, and that is equal to about half the well pumpage.
>
> (Trial Day 67, p. 73)

The W.R. Grace expert viewed the river as a probable source of contamination to wells G & H because of abundant induced infiltration and periodic flooding of the wetland bringing volatile organic compounds and heavy metals from known historic sources of contamination upstream in the basin into the capture zones of the municipal wells.

With respect to simulated travel times of contaminants from the W.R. Grace property to the wells, he made the following statements

... for TCE the three periods of time that I calculated were 11 years, 19 years, and 25 years ... At the end of 11 years, TCE would have moved a distance of 750 feet. For 19 years, the distance would be less than 1000 feet. And for 25 years, the distance is less than 1100 feet.

(Trial Day 68, pp. 2–3)

The distance (1,2) trans (DCE) would have moved in 11 years is less than 800 feet; in 19 years, less than 1300 feet; and in 25 years, less than 1600 feet.

(Trial Day 68, pp. 2–3)

Based on these model simulations, his opinion regarding the movement of organic solvents at the W.R. Grace property, which were dumped on the land surface and placed in 55-gallon drums in a shallow pit, was that even if the TCE, DCE and PCE were released into the groundwater flow system on the day the plant opened in 1960, the chemicals would not have reached wells G & H, over 2500 feet away, by May 1979 when the wells were shut down (Trial Day 68, pp. 2–3).

5.5.5   Jury Verdict

At the end of the trial, the judge described to the jury the legal statutes that applied to the case and read two sets of special interrogatory questions, one set of four questions pertaining to each defendant, which would elucidate their verdict. The judge in concert with the lawyers composed the special interrogatory questions. Obviously, the judge believed the jury could answer the questions based upon the testimony presented at the trial. The first special interrogatory question, shown below for W.R. Grace, asked the jurors to determine which, if any, of the contaminants substantially contributed to the contamination of wells G & H during the period of time the wells operated.

#1. Have the plaintiffs established by a preponderance of the evidence that any of the following chemicals were disposed at the Grace site after October 1, 1964 and substantially contributed to the contamination of Wells G & H by these chemicals prior to May 22, 1979?

(a)	trichloroethylene	yes_____	no_____
(b)	tetrachloroethylene	yes_____	no_____
(c)	1,2 transdichloroethylene	yes_____	no_____

The second interrogatory question, shown below, asked the jurors to determine the year and month when each contaminant selected in #1 first reached the wells.

#2. If you answered 'Yes' in question 1 as to any chemical(s), what, according to the preponderance of the evidence, was the earliest time that such chemical(s) disposed of on the Grace site after October 1, 1964 made a substantial contribution to the contamination of Wells G & H?

(a)	trichloroethylene	mo._____	yr._____
(b)	tetrachloroethylene	mo._____	yr._____
(c)	1,2 transdichloroethylene	mo._____	yr._____

According to Pacelle (1986), the jury was dumbfounded by the specificity of the special interrogatory questions, especially having to identify a date (year and month) that each contaminant made a substantial contribution to the contamination of the wells. Prior to receiving instructions from the judge, the jurors believed they would answer a singular question concerning the simple liability of each defendant, not a series of complexly worded questions about each chemical named in the lawsuit (Pacelle 1986). The third and fourth interrogatory questions addressed the legal issue of negligence and were framed and worded in a manner similar to the first two interrogatory questions.

After nearly ten days of sometimes heated deliberations (Pacelle 1986), the jury reached answers to the special interrogatory questions. W.R. Grace was found liable of contaminating wells G & H with TCE beginning in September 1973, but not with DCE or PCE. Beatrice Foods was found not liable of contaminating the wells with any industrial chemicals. The answers to the special interrogatory questions serve as the verdict and represent the final acceptance or rejection of the first subordinate hypothesis. The trial, however, did not proceed to a second phase with the remaining defendant, W.R. Grace & Company. Several months later, a settlement was reached between the plaintiffs and W.R Grace, shortly after the judge indicated he would grant the defendant's motion for a mistrial based, in part, on inconsistencies in the responses to the special interrogatory questions (Pacelle 1986).

5.5.6 Discussion of Woburn Verdict

Based on the accounts of Pacelle (1986), Kennedy (1989), and Harr (1995), it is apparent that many of the participants and observers of the Woburn Toxic Trial felt that the US legal system operated at a level below its best. A common criticism of the US legal system is its reliance on lay juries in exceptionally technical cases. Arguments are made that highly technical cases should be tried before a judge because judges are more highly educated and experienced in resolving technical issues than a lay jury. Arguments are made that technical cases should be tried before an expert jury – one comprised solely of experts whose technical disciplines are germane to the case.

Although these arguments have merit, so do the counter-arguments. Prentice H. Marshall, a US District judge, argues that

> . . . in most civil and criminal cases a carefully selected jury which has received coherently submitted evidence and accurate and succinct instructions is superior to a judge in making decisions. The jury's collective comprehension of the facts is more complete than the judge's. The jury renders its decision more quickly than the judge can. The jury's independence and integrity are superior to the judge's. The jury's cross-sectional makeup brings community values to the decision-making process.
>
> (Marshall 1990)

Lynn N. Hughes, also a US District judge, argues that it would be nearly impossible to empanel a jury of experts, who individually did not develop biases as a result of their professional experiences, whereas a lay jury, while naïve, is unbiased and can usually understand the pertinent technical issues (personal communication, October 1999).

As a society, we trust that the collective wisdom of the jury will produce the correct verdict (Marshall 1990). However, the process by which the jury derives the verdict is inherently passive (Friedland 1990). Jurors may or may not be allowed to make notes during the trial. They cannot read supplemental materials or reference books. They probably are not allowed to review the testimony transcribed by the court recorder. They cannot discuss the case with others, or among themselves except during formal deliberations following the trial. Students are active learners; jurors are not. Jurors, unlike students, do not get to benefit from homework assignments, term papers, laboratory reports, and exams that reinforce concepts covered in class.

A list of the hydrologic topics the jurors in the Woburn Toxic Trial were asked to comprehend includes geologic mapping, glacial history and glacial stratigraphy, physical properties of rocks and sediments, drawdowns and cones of depression produced by pumping wells, stream gauging, flood recurrence intervals, induced stream infiltration, aqueous and organic chemistry, regional groundwater flow, contaminant transport with retardation, and analytical and numerical models of groundwater flow and transport. The topics are nearly equivalent to the content of courses a graduate student is required to take for a master's degree in hydrology. Only one of the six jurors in the Woburn Toxic Trial attended college and none had any specific scientific training or background (Pacelle 1986).

The crux of the jury's decisions in the Woburn Toxic Trial centred on their understanding the hydrogeologic setting and its influence on groundwater pathlines and contaminant travel times, the interaction between the groundwater flow system and the wetland when the municipal wells were pumping and when they were not, and the processes affecting the fate and transport of chlorinated solvents. As a society, we hope the jury had a sufficient level of understanding of the field data and an appreciation of the necessary theory and concepts to differentiate among the divergent opinions held by the hydrologic experts. As scientists, we hope the jury was able to differentiate among the technical approaches used by the hydrologic experts. As hydrologic modellers, we hope the jury was able to decipher the limitations of each expert's predictive model. We have high hopes.

In January 1999, Harvard Law School held a conference called *Lessons from Woburn*. Many of the lawyers, several of the plaintiffs' mothers, some of the expert witnesses, and one juror from the trial participated in the conference. Many of these people led focus sessions dealing with specific aspects of the trial. (Videotapes of the sessions can be viewed on the conference website at *www.cyber.law.harvard.edu/acivilaction*.) The juror who spoke mentioned growing up on a cotton and tobacco farm where chemicals were routinely sprayed on the crops and that her family used a spray bottle of DDT in the house to kill flies. To this juror, exposure to toxic chemicals was a common, non-deleterious occurrence. The juror also mentioned that water from the Beatrice property could not flow *through* the Aberjona River to the municipal wells. To me, sitting in the audience, it sounded as if the juror considered the hydrologic system to consist of streams of water that could not pass *through* each other, as do beams of light. The juror's comment suggested that the jury may have seen the flow system in only two dimensions and did not understand the importance of the partially penetrating wells, induced infiltration out of the river and wetland downward to the wells, and the three-dimensional character of the capture zones of the wells. If this is true, as a teacher, I find it terribly disappointing – all three hydrologic experts, especially the plaintiffs' and W.R. Grace's, took several hours in their direct testimony to educate the jury on the local geology and the general concepts of surface-water and groundwater hydrology.

Hydrologists invariably are dismayed at the degree of scientific certainty the judge and the lawyers expected of the jury in order to answer the contaminant arrival time questions in special interrogatories #2 and #4 with a precision of one month. However, one of the hydrologic experts stated contaminant travel times in terms of hundredths of a month, seemingly expressing his certainty in the arrival time of the contaminants within plus or minus eight hours. This may have been the precedent for the judge and the lawyers to request the unrealistic degree of scientific certainty stated in special interrogatories #2 and #4. As potential expert witnesses, we should ask ourselves whether the unrealistic degree of scientific certainty requested in special interrogatories #2 and #4 demonstrates a general lack of scientific understanding by the courts, or does it demonstrate our disregard for acknowledging to the general public the uncertainty of our work? Outside the courtroom, we acknowledge that even the most complete model analysis of an actual field problem has limitations and contains uncertainty. Inside the courtroom, uncertainty is usually revealed on cross-examination in attempts by the opposing party to counteract and damage testimony presented during direct examination.

It is interesting to contrast the degree of scientific sophistication employed by the three hydrologic experts to develop their opinions compared with the verdict rendered by the jury.

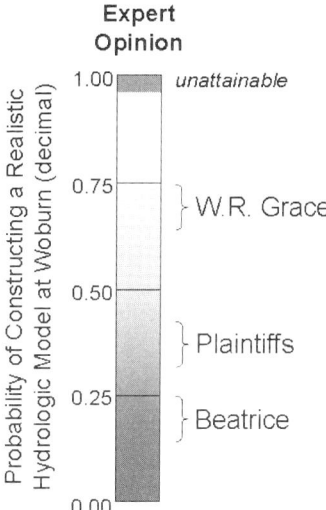

Figure 5.6 Probability of constructing a realistic model of the hydrologic system at Woburn.

Having read over 1800 pages of testimony presented by the three hydrologic experts and having helped construct three-dimensional numerical flow and transport models of the wells G & H area (see Metheny 1998), I rank the degree of sophistication of their approaches according to that shown on Figure 5.6.

The plaintiffs' expert analysed well logs from the valley and uplands, grain-size analyses and slug tests and portrayed the geology as a heterogeneous assemblage of glacial materials. He used this geologic framework to construct an uncalibrated three-dimensional numerical flow and transport model used for illustrative purposes. The one-dimensional analytical equation he used to compute contaminant travel times inadequately incorporated the heterogeneous nature of the glacial materials, converging three-dimensional flow to the pumping wells, induced infiltration from the river and wetland, leakage from bedrock, hydrodynamic dispersion, spatial and temporal variations in recharge, and temporal variations in pumping rates. Based on these limitations, I believe the probability that this analysis realistically represents the flow system when wells G & H were in use to be no more than 40% (Figure 5.6).

The Beatrice expert plotted and analysed the water level and streamflow data from the 30-day pumping test. He constructed no models and made no travel time calculations. This analysis inadequately incorporates the complexity of the interaction between the surface-water and the groundwater flow systems. It ignores the three-dimensional nature of the groundwater flow system created by the partial penetration of the wells and the river. By relying solely on limited field data and not incorporating historic data into a model of any kind, his analysis inadequately accounts for bedrock leakage, aquifer heterogeneity, temporal and spatial variations in recharge, and temporal variations in pumping rates. I believe the probability that this conceptual analysis realistically represents the flow system between 1964 and 1979 to be less than 25% (Figure 5.6).

The W.R. Grace expert used the available geologic, hydrologic and pumping data to construct a transient three-dimensional numerical flow and transport model of the area that accounted for the heterogeneous character of the glacial materials, partial penetration of the pumping wells and the river, bedrock leakage, spatial and temporal variations in recharge, historic variations in pumping rates, hydrodynamic dispersion, and chemical retardation. It was the most comprehensive analysis and the most sophisticated model. The flow model was calibrated using two sets of measured water level and streamflow gain/loss data. The transport model was not calibrated, although a sensitivity analysis was performed to determine possible ranges of contaminant arrival times at wells G & H. This model incorporated many more of the hydrologic factors affecting surface-water and groundwater flow over the 15-year period wells G & H operated. The model also included more parameters and processes and fewer unrealistic assumptions than the other approaches. I believe the probability is about 75% that this analysis represents the actual behaviour of the flow system during the period when wells G & H were operating (Figure 5.6).

The major difficulties each of the hydrologic experts faced in constructing a reliable hydrologic model that could predict water levels, streamflows and contaminant concentrations were the lack of historic source concentration values of the contaminants and the lack of historic water level, streamflow change, and chemical concentration measurements that could be used as calibration targets. The same difficulties exist today for researchers trying to further understand the hydrology of the site (Myette et al. 1987; de Lima and Olimpio 1989; Metheny 1998). Even though a sophisticated model may calibrate reasonably well to the available data, there are simply inadequate historic data to enable the probability of a reliable analysis to be much better than 70 to 80%. It is the lack of historic data and the uncertainty produced thereby that makes working in the legal arena so challenging.

From the viewpoint of scientific sophistication, the jury's verdict is the inverse of the ranking shown on Figure 5.6 – Beatrice was found not liable and W.R. Grace was found liable. The nagging questions for lawyers and hydrologists studying the Woburn Toxic Trial is whether the science itself was too complicated for the judge and jury, and/or whether the presentation of the science was too confusing.

5.6   CONCLUSIONS

Testimony from the Woburn Toxic Trial shows that predictions from hydrologic models were central to the expert opinions presented to the jury, whereas model calibration was not a major issue brought before the jury (although it was discussed heatedly in depositions and during the trial at several side bars between the lawyers and the judge). Now, fifteen years after the trial, I believe model calibration would be an important part of the direct examination and the cross-examination of each expert hydrologist. Today, the methods of model calibration are much better established and known by the hydrologic community than they were then. Textbooks (Anderson and Woessner 1992) and standards (Brown and Laase 1995) have been written and their influence on modelling studies is pervasive. Philosophic, pragmatic and guidance articles also have been written about model calibration (Bredehoeft and Konikow 1993; McCombie and McKinley 1993; Bair 1994; Oreskes et al. 1994; Woessner and Anderson 1996) and the essential procedures used in hydrologic modelling have become part of the fabric of our discipline. In the interim, expert witnesses have passed this knowledge on to their lawyers, who appear to understand the terms and the methods. This was not true in 1986 when the Woburn Toxic Trial took place. Today, model calibration would be a critical part of the scientific methods applied to the construction of a predictive model used by a hydrologic expert witness. As such, the methods used to calibrate the model would probably be evaluated by the trial judge under the Daubert ruling and model calibration would undoubtedly be highly scrutinised under cross-examination.

Models will always be attractive tools for lawyers and experts to convey information to a jury. Continued advances in image processing and visualization will bring digital animation of model simulations into the courtroom with increasing frequency (Stafford 1998). Models synthesize field data and incorporate physical, chemical and biological processes in a manner that addresses the causal relations that most legal disputes seek to resolve. Models can simulate the unobserved conditions that occurred at the crux of the conflict. The ability of hydrologic models to make reasonable predictions of these conditions, however, will always be dependent on the accuracy of the conceptual hydrogeologic framework and on having sufficient historic data to assure the jury that the model is a realistic representation of reality. Without historic data, rigorous calibration of the model is not possible and the opinions of experts are vulnerable to attack by the opposition – as well they should be. The following statement by Hugh Greenwood (1989) is worth remembering when formulating expert opinions based on model results, 'A model that has no predicted testable consequences is unverifiable and, if not useless, is sterile.' Presentation of a sterile model in the court, before a judge and jury, with the spectre of vigorous cross-examination, is not a prudent legal strategy.

Scientists and engineers need to recognise that the legal system in the United States, unlike the scientific method, is not designed to find the truth. It is designed to resolve conflicts. The primary purpose of a civil trial is to end a dispute. Many hydrologists who testify in court are disappointed by the experience, perhaps this is because their experience demonstrates that the legal system cannot attain the same goal as the scientific method. As stated by Federal Judge Lynn N. Hughes (1999), 'When our legal system works at its best, the court solves the problem by a process that respects the legal conventions, honors the ideal of truth, incorporates the force of reason, recognizes the dignity of the participants, and accepts the necessity of a decision.'

REFERENCES

Anne Anderson et al. v. Cryovac, Division of W.R. Grace & Company; W.R. Grace & Company; John J. Riley Division of Beatrice Foods, Inc.; Beatrice Foods, Inc.; United States District Court for the District of Massachusetts, Civil Action No. 82–1672-S, 1982, Boston, MA.

Anderson, M.P. and Woessner, W.W. 1992. *Applied Groundwater Modeling: Simulation of Flow and Advective Transport*. Academic Press, San Diego, CA.

Bair, E.S. 1994. Model (in)validation – a view from the courtroom. *Ground Water*, **32**(4), 530–531.

Bair, E.S. and Wood, W.W. 1999. *A Civil Action* – what will be the legacy of wells G & H? *Ground Water*, **37**(2), 161–162.

Blauvelt, S. 1999. Expert scientific testimony examined: part 1. PA GEOTECH, *Newsletter of the Pennsylvania Council of Geologists*, **9**(4), 2–6.

Bredehoeft, J.D. and Konikow, L.F. 1993. Ground water models: validate or invalidate. *Ground Water*, **31**(2), 178–179.

Brown, D.E. and Laase, A. 1995. *Standard Guide for Calibrating a Groundwater Flow Model Application C–7*, ASTM Standards on Ground Water and Vadose Zone Investigations, Philadelphia, PA.

Daubert v. Merrel Dow Pharmaceuticals, Inc., 61, U.S.L.W. 4805 (1993)

de Lima, V. and Olimpio, J.C. 1989. Hydrogeology and simulation of groundwater flow at Superfund-site Wells G and H, Woburn, Massachusetts. *US Geological Survey Water-Resources Investigations Report* 89–4059.

Domenico, P.A. and Schwartz, F.W. 1998. *Physical and Chemical Hydrogeology*. John Wiley, New York.

Foster, K.R., Bernstein, D.E. and Huber, P.W. 1993. Science and the toxic tort. *Science*, **261**, 1509, 1614.

Friedland, S.I. 1990. The competency and responsibility of jurors in deciding cases. *Northwestern University*

Law Review, **85**(8), 190–220.

Frye v. United States, 293 F. 1013 (D.C. Cir., 1923)

Gerber, R.J. 1987. Victory vs. truth: the adversary system and its ethics. *Arizona State Law Journal*, **19**(1), 3–26.

Greenwood, H.J. 1989. On models and modeling. *The Canadian Mineralogist*, **27**(1), 1–13.

Grossman, L.A. and Vaughn, R.G. 1999. *A Documentary Companion to* A Civil Action. Foundation Press, New York.

Harr, J. 1995. *A Civil Action*. Random House, New York.

Huber, P. 1990. Junk science and the jury. *University of Chicago Legal Forum, The Role of the Jury in Civil Dispute Resolution*, **1990**, 273–302.

Huber, P. 1991. *Galileo's Revenge: Junk Science in the Courtroom*. Basic Books, New York.

Hughes, L.N. 1999. Clients, cogency & candor: the complications of consulting in court. *Abstracts with Programs, Geological Society of America Meeting*, **31**(7), A181.

Jost, K. 1991. Weird science. *American Bar Association Journal*, October.

Kennedy, D. 1989. Death and Justice: Environmental tragedy and the Limits of Science. *The Boston Phoenix*, copyrighted, www.dankennedy.net/woburn_trial.html

Marshall, P.H. 1990. A view from the bench: practical perspectives on juries. *University of Chicago Legal Forum, The Role of the Jury in Civil Dispute Resolution*, **1990**, 147–160.

Matson, J.V. 1994. *Effective Expert Witnessing*, 2nd edition. CRC Press, Boca Raton, FL.

McCombie, C. and McKinley, I. 1993. Validation – another perspective. *Ground Water*, **31**(4), 530–531.

Metheny, M.A. 1998. Numerical simulation of groundwater flow and advective transport at Woburn, Massachusetts, based on a sedimentologic model of glaciofluvial deposition, unpublished Master's thesis, Department of Geological Sciences, Ohio State University.

Myette, C.F. Olimpio, J.C. and Johnson, D.G. 1987. Area of influence and zone of contribution to Superfund-Site wells G and H, Woburn, Massachusetts. *US Geological Survey Water-Resources Investigations Report 87*–4100.

Oreskes, N., Shrader-Frechette, K. and Belitz, K. 1994. Verification, validation, and confirmation of numerical models in earth sciences. *Science*, **263**(February), 641–646.

Pacelle, M. 1986. Contaminated verdict. *American Lawyer*, December, pp. 75–80.

Peters, E.K. 1996. *No Stone Unturned – Reasoning About Rocks and Fossils*. W.H. Freeman, New York.

Prickett, T.A. and Pettyjohn, W.A. 1995. Groundwater modeling and litigation. In: El-Kadi, A.I. (ed.), *Groundwater Models for Resources Analysis and Management*. Lewis Publishers, Boca Raton, FL, 137–145.

Shaw, F.S. and Southwell, R.V. 1941. Relaxation methods applied to engineering problems; VII Problems relating to the percolation of fluids through porous media. *Proceedings Royal Society*, **A178**, 1–17.

Stafford, J.M. 1998. Geoscience and the law. *Geotimes*, May, pp. 18–22.

US Department of Commerce, 1997. *Statistical Abstract of the United States – The National Data Book*, 117th Edition. Economic & Statistics Administration, US Bureau of Census, Washington, D.C.

US Environmental Protection Agency, 1985. National Primary Drinking Water Regulations: Volatile Synthetic Organic Chemicals. 40 CFR 141–142, Federal Register 50 (219).

Van Dyke, J.M. 1987. Trial Courts and Practice. In: Janosik, R.J. (ed.), *Encyclopedia of the American Justice System*, vol. 2, 734–752.

Voss, C.I. 1998. Groundwater modeling: simply powerful. *Hydrogeology*, **6**(4), A4–A6.

Weinstein, J.B. 1986. Improving expert testimony. *University of Richmond Law Review*, **20**(3), 473–498.

Woessner, W.W. and Anderson, M.P. 1996. Good model – bad model, understanding the flow modeling process. In: Ritchey, J.D. and Rumbaugh, J.O. (eds), *Subsurface Fluid-Flow (Ground Water and Vadose Zone) Modeling*, ASTM STP 1288, American Society for Testing Materials, Philadelphia, PA, 14–23.

6 On Simulation, Calibration and Ill-conditioning with Application to Environmental System Modelling

CLAUDE R. DIETRICH
Centre for Resources and Environmental Studies, Integrated Catchment Management and Assessment, Australian National University, Canberra, Australia

6.1 INTRODUCTION

In this chapter, focus is placed on quantitative issues associated with environmental system simulation, calibration and ill-conditioning. A reason for this focus is that while simulation is the bread and butter of modellers, the attendant issues of calibration and ill-conditioning are not always paid the attention they deserve. Indeed, inspection of the vast literature on system modelling shows that modellers generally focus on system simulation, often leaving unreported difficulties that may have emerged during model calibration. Paying attention to such difficulties is important in environmental modelling since environmental systems are generally character-ised by high levels of uncertainty. In this respect, given that:

- meaningful simulations can rarely be executed without prior system calibration; and
- calibration can give rise to *ill-conditioned* behaviours;

the connections between simulation, calibration and ill-conditioning are investigated in this chapter. Also, given the focus of this book, these issues are considered in the context of environmental system modelling.

While this chapter by necessity places emphasis on mathematical structures, it is self-contained and of a tutorial nature, and so should allow readers with a basic knowledge of calculus and applied linear algebra to follow ideas and concepts without having to resort to extensive references. It also extends a chapter on ill-conditioning written by Dietrich et al. (1993).

6.1.1 Calibration and Ill-conditioning

In practice, the modelling of a natural system consists in simulating the system's response to changes in driving forces. To implement such simulations, a modelling exercise relies on a mathematical abstraction of the system under study, with internal system properties encap-sulated in parameters and variables, and the system's driving forces viewed as system inputs. An illustration of this structure is in Figure 6.1. As we shall see, simulation is generally well-conditioned and follows the direction of the horizontal arrows in Figure 6.1. On the other hand, calibration somewhat follows the horizontal arrows in reverse and so can be ill-conditioned.

Model Validation: Perspectives in Hydrological Science. Edited by M.G. Anderson and P.D. Bates.
© 2001 John Wiley & Sons, Ltd.

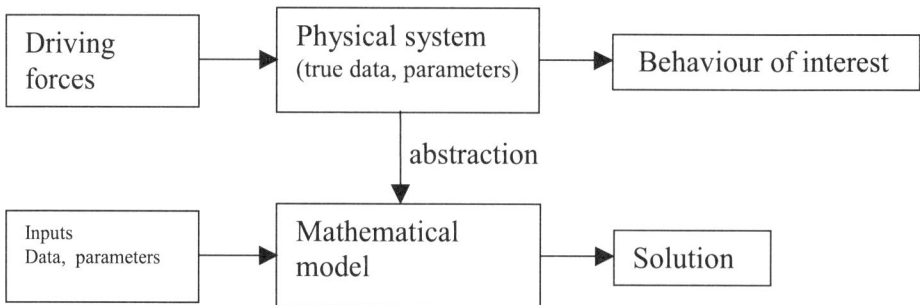

Figure 6.1 Process of abstraction from a real physical system to a mathematical representation

Ill-conditioned procedures have the property that small errors in data and/or model can significantly magnify. Thus error magnification during calibration can severely affect confidence levels in system parameters and variables estimates. It is therefore important that the presence of potentially ill-conditioned behaviours be recognised early in model development and construction.

In this regard, two questions naturally arise. These are:

1. What is ill-conditioning and what intrinsically characterises it?
2. Is there an aspect of ill-conditioning that affects environmental systems in particular?

In seeking an answer to these questions, a definition of ill-conditioning in terms of its structure and modelling implications is first required. The next step is to delineate what is meant by environmental systems and provide typical features associated with the calibration of environmental systems that can lead to ill-conditioning.

Ill-conditioning was first defined by the French mathematician Hadamard early in the 20th century. The definition is as follows (Hadamard 1932):

Definition 1 Ill-conditioning (or ill-posedness).

A problem is ill-conditioned (or ill-posed) if:

1. A solution does not exist.
2. The solution is not unique.
3. The solution does not depend continuously on the problem data.

Issues of existence, uniqueness and continuity are mathematical. Indeed, the above definition indicates that the concept of ill-conditioning, often termed *ill-posedness*, has its roots in mathematics and so requires mathematical tools for problem analysis and remedy. Nevertheless, prior to embarking on a mathematical investigation, it is possible to make a few points associated with this definition when applied to modelling, in particular items 1 and 2.

On existence, if a mathematical model and associated problem domain is a reasonable approximation to a physical system, then the existence of such a physical entity, as observed in a laboratory or in field experiments, ensures on logical grounds the existence of a solution to the mathematical problem. Going back to Figure 6.1, model and physical system closeness is measured by the distance that separates the mathematical model and physical system entities. If both are close to each other, the system's behaviour of interest will be close to the model's solution, implying that a mathematical solution must exist. Thus, in practice, existence of a solution is guaranteed if the mathematical model chosen is a sufficiently good approximation to the physical system under investigation.

As for non-uniqueness, it usually arises if the number of independent pieces of data about the system is smaller than the number of degrees of freedom in model calibration. For example, non-uniqueness may arise if l independent data measurements are available and m parameters have to be calibrated, with $l < m$. Thus non-uniqueness can most often be remedied by adding information to the system, either in the form of field data or as a priori knowledge based on judgement and expertise. The critical point here is independence of information, since adding large amounts of strongly correlated data to a system's database will only add small amounts of independent information. How to measure information content and independence of data is not pursued here; interested readers are referred to information theory such as that of Shannon and Weaver (1962).

Regarding continuity, i.e. the definition's item 3, it lies at the centre of the difficulties experienced by modellers dealing with ill-conditioned problems. It is the non-continuous dependence of solutions on data that flags a problem as truly ill-conditioned. To investigate the implication of this third item requires some mathematical analysis such as that presented in Sections 6.3 to 6.6. At this stage, suffice to say that non-continuous dependence of a solution on the data implies that small changes in the data can lead to very large oscillatory changes in the solution and so may yield estimates that are meaningless.

Having provided a definition of ill-conditioned problems, there remains to provide a clear and concise definition of environmental systems. Such a definition is not obvious as the very notion of environment encompasses a broad, rich and often fuzzy mixture of earthly systems and processes. For this reason, and given the focus of this chapter and others in the book, it is preferable to list environmental system features that can cause ill-conditioned behaviours. With such features in hand, it will be possible to establish links between modelling environmental systems and ill-conditioning and so highlight potential difficulties when embarking on the former given the potential presence of the latter.

6.2   MODELLING ENVIRONMENTAL SYSTEMS

Environmental processes are generally characterised by high levels of uncertainty and complex interactions at different scales. In this section this feature is explored together with its relation to simulation, calibration and ill-conditioning. In this respect, the focus is primarily on catchment hydrological processes from plot to basin scales. This focus does not limit the relevance of the discussion to other environmental systems, such as atmospheric and oceanographic, as all display similarities in their simulation, calibration and ill-conditioning aspects.

6.2.1   Features of Environmental Systems

Consider a water budget over a catchment. The major processes considered here are rainfall, evapotranspiration, runoff and percolation to unsaturated zones and deeper groundwater formations. These are shown in Figure 6.2. Features of such a system relevant to simulation, calibration and ill-conditioning are:

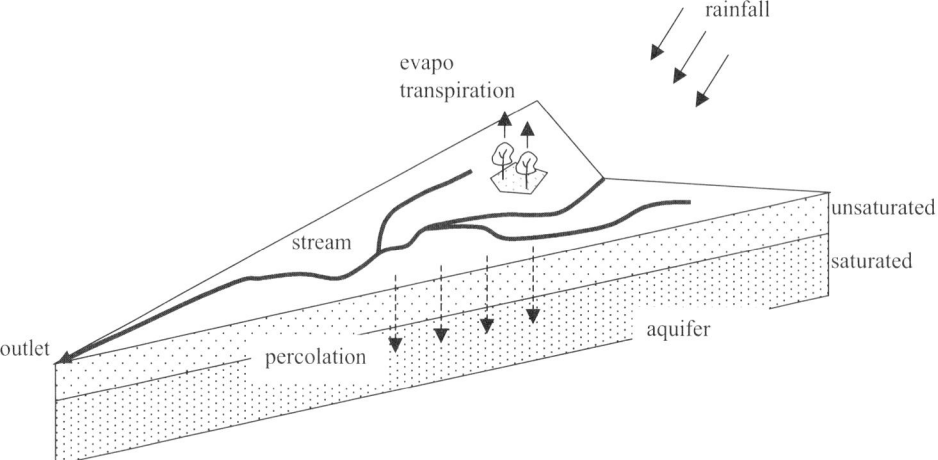

Figure 6.2 Water budget over a catchment with major processes at play being rainfall, evapotranspiration, runoff and percolation to aquifers

1. Water flow processes are local.
2. The processes occur within highly heterogeneous media.
3. Most processes are not controllable or reproducible.
4. Data cover only a small portion of the processes.
5. In most practical cases, simulations of interest are at scales significantly larger than those of the water flow processes at play.

With processes being local, detailed modelling requires that mathematical equations expressing local properties (mass, momentum, energy conservation etc.) be invoked. Furthermore, heterogeneity of the support through which processes occur implies that extensive local knowledge be available over the whole catchment. Lack of controllability and reproducibility implies that it is generally not possible to carry out experiments to verify the soundness of a simulation framework. An assessment of simulation performance is therefore often limited to checking the model's ability to reproduce and/or predict historical data. These properties and the usual lack of local data over the whole catchment (features typical of environmental systems) highlight the well-known fact that environmental systems modelling is characterised by high levels of uncertainties. In turn, these uncertainties bear on calibration procedures, since uncertainty in data can multiply by orders of magnitudes in estimates and so render calibration results meaningless. In order to explore these points further, a discussion on lumped and distributed systems is presented in the next subsection.

Before closing this subsection, it is worth indicating that there are other characteristics of environmental processes that compound the difficulties mentioned so far. For example, processes can be highly coupled with the presence of strong feedback loops leading to chaotic behaviours.

6.2.2 Lumped Systems

The interplay between data availability, model complexity and simulation goals determines in many practical problems the choice of a lumped versus a distributed approach. In a nutshell, lumped approaches average out information in both spatial and temporal dimensions to match simulation complexity and scale, with data intensity and modelling aims. On the other hand, distributed models retain local information both in time and space. For example, if the goal of a modelling exercise is to predict solely the time evolution of some bulk property of a complex system, then spatial lumping (or spatial integration) can be invoked to filter out spatial variability. Such lumping simplifies model construction and calibration and, as we shall see, counteracts the effect of ill-conditioning. In any case, such lumping may be required because of the inadequacy between data availability, system knowledge and modelling complexity and requirements. A good discussion on this is in Jakeman and Hornberger (1993).

To illustrate these issues, let us consider again the catchment example of Figure 6.2 and consider the prediction of water runoff. If the purpose of the simulation exercise is to model over time only water runoff at the catchment outlet, then the catchment can be represented simply as the input/output system illustrated in Figure 6.3. Expressing all water fluxes as volume per unit time t, we define the system input as $p(t) = r(t) - u(t) - v(t)$ where $r(t)$ is rainfall, $u(t)$ percolation to groundwater reservoirs, and $v(t)$ evapotranspiration. The output is runoff $q(t)$ with $V(t)$ representing the water volumetric storage of the catchment. With this, a water mass balance results in the first order ordinary differential equation (ODE)

$$\frac{dV}{dt} = p - q \tag{1}$$

where the dependence of V, p and q on t is omitted for the sake of notational simplicity.

With knowledge of the system's driving forces encapsulated in the input p, equation (1) has two unknowns V and q and so requires an additional equation for closure. A simple power law of the form $q = \alpha V^{\beta}$ is often invoked with α and β unknown parameters that may depend on time

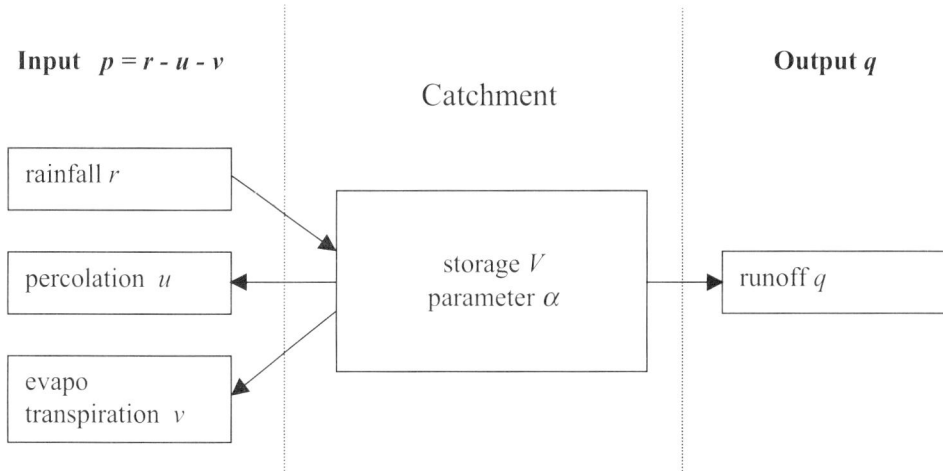

Figure 6.3 Lumped conceptualisation of the catchment illustrated in Figure 6.2. In this figure, r is rainfall, u percolation and v evapotranspiration. The output is runoff q at the catchment outlet with V representing the catchment water volumetric storage characterised by an unknown parameter α

because of seasonally, antecedent conditions etc. For the sake of simplicity, assume $\beta = 1$ so that the water mass balance (1) is

$$\frac{d(q/\alpha)}{dt} + q = p \tag{2}$$

i.e. a first order, linear ODE in q. This ODE provides a spatially lumped model where for any input p, catchment runoff q can be computed once the runoff initial conditions and the catchment parameter α are specified. For example, given the initial condition $q(0) = q_0$ and an estimate $\hat{\alpha}$, the solution $q(t)$ to (2) for $t \geq 0$ is (Zill and Cullen 1997, p. 45)

$$q(t) = q_0 e^{-\int_0^t \hat{\alpha}(\tau)d\tau} + \hat{\alpha}(t) \int_0^t e^{-\int_\tau^t \hat{\alpha}(\xi)d\xi} p(\tau)d\tau \tag{3}$$

Anticipating discussions on ill-posedness, (2) is used to make a few points regarding calibration. Given the differential equation (2) or its equivalent integral form (3), the catchment calibration procedure is that of estimating α. If at all possible, direct measurements of α are far from trivial since α summarises a bulk property of the whole catchment. On the other hand, inspection of (2) indicates that with an input/output pair (p, q) available, α is given by

$$\alpha = \frac{q}{\int(p - q)dt} \tag{4}$$

In other words, the water mass balance differential equation (2) rewritten as (4) can be used to estimate α as long as one measurement of α is available to estimate the integration constant in the denominator of (4). As discussed later, such 'inverse' procedure, whereby a mathematical model attached to a system is used to estimate unknown parameters, is often used in calibration and can often lead to ill-conditioned behaviours. In the particular case of calibrating α from input and output data via (4), as indicated at the end of Section 6.4, ill-conditioning will not arise. However, increasing model complexity by moving for example from a lumped to a distributed model, can change this. This is the object of the next section.

6.2.3 Distributed Systems

Aims of a modelling exercise may require to be spatially explicit. For example, referring again to the catchment of Figure 6.2, simulations may be required to predict flooding in the valley. In this case, a model for water flow along the stream is required. If the stream length is large compared to its width and height, the stream can be viewed as one-dimensional with modelling involving time t and distance x only. Following the development of Section 6.2.2, if $A(x, t)$, $Q(x, t)$ and $p(x, t)$ denote respectively the stream wetted cross-sectional area, flow, and water sinks and sources expressed as mass per unit time and unit length, the water mass balance (2) becomes the partial differential equation (PDE) (Dietrich et al. 1998)

$$\frac{\partial A}{\partial t} + \frac{\partial Q}{\partial x} = p(x, t) \tag{5}$$

where the sink/source term p includes various effects such as tributaries, rainfall/runoff inputs, and seepage through the stream channel. As was the case for the lumped mass balance (2), the

PDE (5) has two unknown A and Q and so requires an additional equation for closure. For this, one can use the St Venant equation, which express momentum conservation (Stepien 1984). A simpler approach often followed in practice is to assume that water flow Q is a function of A with the functional form equal to the power law $Q = aA^b$ with a and b unknown parameters depending possibly on time and location (Leopold and Maddock 1953). With this, (5) becomes

$$\frac{\partial A}{\partial t} + abA^{b-1}\frac{\partial A}{\partial x} = p(x, t) \tag{6}$$

where the chain rule $\partial Q/\partial x = (dQ/dA)(\partial A/\partial x)$ was used. For the sake of simplicity, assume that both a and b are independent of time. Then, stream calibration is the estimation of $a(x)$ and $b(x)$. Given a spatial discretisation of the stream spatial extent into a set of n nodes $\{x_i\}_{i=1}^n$, calibration can take place prior to modelling by linear least squares fitting at each node x_i of the relationship $Q(x_i, t) = a(x_i)A(x_i, t)^{b(x^i)}$ using historical time series of stage height (via $A(x_i, t)$) and discharge data. Thus, unlike calibration of the lumped catchment via its bulk storage parameter $\alpha(t)$ discussed in Section 6.2.2, stream calibration here can take place prior to simulation. However, for various reasons including the fact that measurement scales may no be those implied in the PDE (6), prior calibration may not be good enough to ensure reasonable fit to stage height historical data. Modellers may therefore use (6) and measurements A^* of A to infer information about a and b. This can be achieved in various ways, one of them being to choose for a and b the values that, upon solving (6) for A, best fit the historical cross-sectional data A^*. In other words, some measure of the distance J between A^* and computed cross-section $\bar{A}$ solution to (6) is invoked, and the a and b estimates are those values that minimises J. However, doing so leads to ill-conditioning since, as we shall see, item 3 of definition 1 is not satisfied, i.e. the dependence between a and b and cross-sectional data A via (6) involves derivatives of the data A and so is not continuous. Thus small errors in data A can magnify significantly in the best fit estimate of a and b.

The above shows the possible emergence of ill-conditioning when moving from a lumped to a spatially distributed, one-dimensional (1-D) framework. In this respect, it is worth considering an extension of the 1-D case to the full three spatial dimensions. For this, as illustrated in Figure 6.4, consider an elemental volume δV in the catchment defined by some catchment discretisation. Instead of focusing on water exclusively, let us consider the general mass balance associated with any conservative fluid flow.

If q denotes fluid density, the mass of fluid in the elemental volume δV is $\int q dV$. Now the mass of fluid flowing per unit time through an element $\mathbf{s}$ of δV's surface boundary is $q\mathbf{u} \cdot d\mathbf{s}$ where $\mathbf{u}$ is the fluid velocity vector, $\cdot$ is the vector dot product, the magnitude of the vector $d\mathbf{s}$ equals to the area of the surface element, and its direction is along the outward normal. With this, the total mass of fluid flowing out of δV per unit time is

$$\oint q\mathbf{u} \cdot d\mathbf{s}$$

where the surface integral $\oint$ is taken over the whole of the closed surface surrounding δV. Inclusion of a source of fluid flowing into the volume δV subtracts to the above the integral $\int_{\delta V} p dV$ where p is a source of fluid per unit time and unit volume. Now, the decrease per unit time of mass in the volume δV due to fluid leaving through its surface boundary is

$$-\frac{\partial}{\partial t}\int_{\delta V} q dV$$

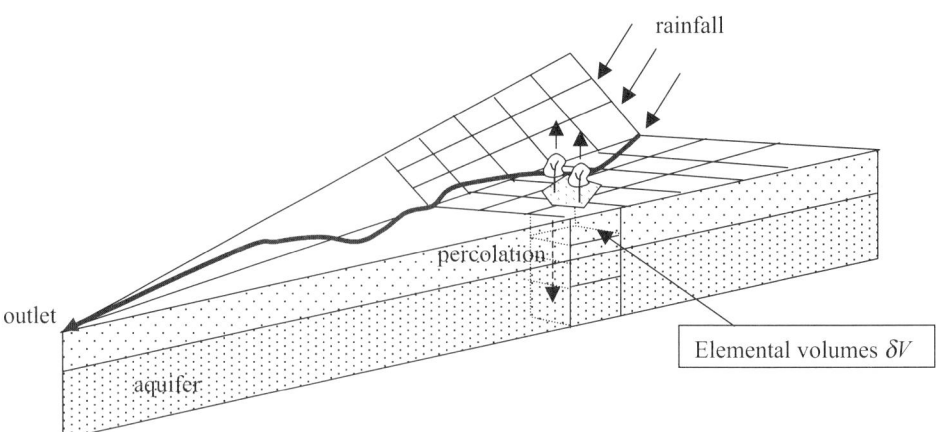

Figure 6.4 Discretisation of catchment into small volumes δV. Only the upstream portion of the discretisation grid is shown

Equating mass out per unit time through δV's boundary with mass decrease per unit time within δV yields the mass balance

$$-\frac{\partial}{\partial t}\int_{\delta V}qdV = \oint q\mathbf{u}\cdot d\mathbf{s} - \int_{\delta V}pdV$$

Using Green's formula (Stakgold 1967, 1968) to transform the surface integral into a volume integral and denoting ∇ the gradient operator $(\partial/\partial x_1, \partial/\partial x_2, \partial/\partial x_3)^T$ where x_1, x_2 and x_3 are the three independent spatial dimensions, and ^{T} denotes vector transpose, yields the relationship

$$\int_{\delta V}\left(\frac{\partial q}{\partial t} + \nabla . q\mathbf{u} - p\right)dV = 0$$

true for any volume δV. The above integrand must therefore vanish resulting in the mass balance PDE

$$\frac{\partial q}{\partial t} + \nabla\cdot q\mathbf{u} = p \tag{7}$$

As was the case for the mass balances (2) and (5), given the input $p(x_1, x_2, x_3, t)$, the PDE (7) has two unknown q and $\mathbf{u}$ and so requires additional equations for closure. For this, one can use the Euler momentum conservation equations (Landau and Lifshitz 1989). Often, for fluid flows taking place through a porous medium, as is the case in groundwater systems, it is possible to avoid the non-linear and complicated Euler equations by assuming that the mass flux density $q\mathbf{u}$ is proportional to the density gradient ∇q, i.e.

$$q\mathbf{u} = -K\nabla q \tag{8}$$

where K is a 3×3 positive definite function matrix that encapsulates the fluid carrying capacity of the medium through which the fluid flows. The negative sign accounts for the fact that fluid flows from regions of higher to lower densities. Darcy's law for groundwater systems is a typical example of (8) with K termed the groundwater transmissivity (Bear 1972). With this, the mass balance (7) become the diffusion PDE

$$\frac{\partial q}{\partial t} - \nabla \cdot K\nabla q = p \tag{9}$$

System calibration requires that the parameter K composed of up to nine 3-D functions be estimated. This is no simple task for the following two reasons:

1. K is a property of a medium that is most often out of access to modellers (e.g. transmissivity of underground formations).
2. K is a function of x_1, x_2, x_3 so that estimates are required locally over the whole catchment.

In contrast, measurements of q, or directly related quantities, are often easier to obtain. In other words, modellers may use available information on q to infer information about K via (9). Using again the example of a groundwater system, q represents piezometric heads and these can be measured far more easily and at lesser costs than transmissivity. For this reason there have been a lot of studies reported in the literature on approaches to estimate transmissivity from head measurements via so-called inverse procedures (Carrera and Neuman 1986; Yeh 1986). A difficulty with this is that the solution K depends on derivatives of the data q and so, as detailed in Section 6.3 below, does not depend continuously on q. This calibration problem is thus ill-conditioned and small errors in q measurements can lead to very large errors in K estimates. Given the uncertainties associated with environmental systems such as the catchment illustrated in Figure 6.4, the above indicates that unless ill-conditioning is counteracted, groundwater system calibration can result in parameter estimates that are meaningless.

In summary, in this section we have seen that large aggregation in system modelling leads to a mathematical representation in which system internal parameters and driving forces are lumped over spatial dimensions. A mathematical representation of such systems results in sets of ordinary differential equations (ODE) with time t as single independent variable. System calibration requires estimation of parameters that may vary only in time. On the other hand, partial aggregation over spatial dimensions results in a mathematical representation that involves (a set of) partial differential equations (PDE) with time t and at least one of the three spatial dimensions in x_1, x_2, x_3 being the independent variables. This increase in dimensionality increases significantly the difficulties associated with system calibration due to the fact that calibration may require solution of an in ill-conditioned problem in the presence of highly uncertain and scarce data.

Before moving to the next section where these issues are further investigated in the context of *forward* and *inverse* problems, it is worth mentioning here that the calibration of spatially lumped system can be placed within a time series analysis framework with system identification and/or input estimation viewed as statistical identification and estimation problems, the latter being typically much more difficult to solve than the former (Young 1984; Barker and Young 1985). The reason for this difference in difficulty levels is that system identification requires the estimation a small number of parameters only, i.e. those few parameters that determine the system's transfer function. On the other hand, time series input estimation is akin to a non-parametric ill-posed problem, or upon discretisation, to an ill-posed problem with a finite but large degree of freedom.

While such a time series analysis viewpoint is important to keep in mind, we have found useful in this chapter to retain spatial dependencies and the attendant system identification issues and difficulties.

6.3 Simulation and Calibration Procedures

The development and execution of a mathematical model requires (i) to choose an appropriate mathematical representation of the system under study, (ii) to calibrate unknown quantities, and (iii) to solve the system's mathematical equations. This procedure generally requires that two complementary problems be solved. These are the *forward* and *inverse* problems discussed in the next subsection.

6.3.1 Forward versus Inverse Problems

Each of the models (2), (6) and (9) discussed in Section 6.2 can be schematically represented by the abstract input/output box illustrated in Figure 6.5. In this representation, the system denoted $\mathscr{S}$ has internal parameter set u. The structure of $\mathscr{S}$ is determined by the physical laws governing the system under study and by the various simplifying assumptions and levels of aggregation chosen by the modellers. It is driven by an input p upon which $\mathscr{S}$ operates, resulting in an output q which is formally written in the form

$$q = \mathscr{S}p$$

The purpose of simulation is to construct the operator $\mathscr{S}$ and execute the operation $\mathscr{S}p$ to generate outputs q given input scenarios p. On the other hand, the purpose of calibration is to estimate the system parameter set u from any available information such as input/output pairs (p, q) and knowledge of the system $\mathscr{S}$.

The three water mass balance examples of the previous section clearly fall into this representation. Indeed, the lumped model (2) has output runoff q, input p and system parameter $u = \alpha$. The stream flow (6) has wetted cross-sectional area A output, input p and system parameter $u = \{a, b\}$. The 3-D fluid flow (9) has output density q, input p and parameter $u = K$.

Now observe that the three representations (2), (6) and (9) can be formally written as

$$D(u)\,\text{output} = \text{input}$$

where D(u) is a differential operator. For example, for (2) we have

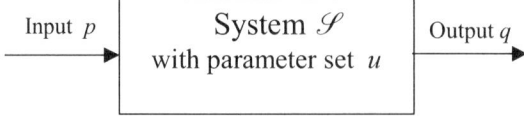

Figure 6.5   Schematic representation of an abstract input/output system with $q = \mathscr{S}p$

$$D(\alpha) = \frac{1}{\alpha}\frac{d}{dt} + g(\alpha)\mathrm{Id}$$

where $g(\alpha) = 1 + d(1/\alpha)/dt$, and Id denotes the identity operation. For (6) we have

$$D(a, b) = \frac{\partial}{\partial t} + abA^{b-1}\frac{\partial}{\partial x}$$

and for (9)

$$D(K) = \frac{\partial}{\partial t} - \nabla \cdot K\nabla$$

In other words, the model representations (2), (6) and (9) invoke a differential operation D that maps output to input. Such differential representations arise naturally when the system governing equations are derived by expressing physical laws locally.

In order to define the operator that maps input to output, the inverse operation D^{-1} has to be formally invoked. Now with D being a differential operator, D^{-1} is an integral operator denoted S. Example of an integral operator $S = D^{-1}$ is in the lumped water flow example (2). Indeed, while (2) is in the form of a differential operation mapping output to input, its solution (3) maps the input p into the output q via the integral operation $q = Sp = \hat{a}(t)\int_0^t k(t, \tau)p(\tau)d\tau$ where without loss of generality the initial condition q_0 is set to zero, and the kernel $k(t, \tau)$ in the integrand is equal to $e^{-\int_\tau^t \hat{a}(\xi)d\xi}$.

We are now in a position to define what forward and inverse problems are and link these to simulation, calibration and ill-posedness.

Definition 2 *Forward versus inverse problems.*

Let a system $\mathscr{S}$ with internal parameter u be defined by governing equations expressed in the form of the differential representation

$$D(u)q = p \tag{10}$$

where p and q are respectively the system input and output and the differential operator $D(u)$ may depend on some parameter set u. Let $S(u) = D(u)^{-1}$ be the formal inverse integral operator of $D(u)$. Then the *forward problem* associated with the simulation of the system $\mathscr{S}$ is that of constructing the operator $S(u)$ and apply it to any input p to obtain the output $q = S(u)p$. Two associated inverse problems are

1. *System identification*: given input/output pairs (p, q), estimate the system parameter set u using (10).
2. *Input identification*: given outputs q and the system parameter set u, estimate the input p using (10).

The above definition assumes that the mathematical equations in the form of the operator S describing the system under investigation $\mathscr{S}$ are accurately identified. Clearly, this is not always the case and system identification should also include those situations where the system under study is sufficiently uncertain or ill-defined to require identification of model structure (Young 1978, 1984). This problem has been extensively studied in time series analysis (Box and Jenkins

1970). Nevertheless, here we assume perfect knowledge of the system's mathematical structure S and so equate system identification with system parameter and/or input estimation only.

In term of stability, a fundamental difference between forward and inverse problems is that the former are based on an integral operation while the latter are based on a differential operation. Since integral operators are bounded while differential operators are unbounded (Griffel 1981), as we shall see, system simulation is a well-conditioned problem, while system and input identification is an ill-conditioned problem.

Boundedness of an operator is defined as follows:

Definition 3   *Boundedness of an operator.*

An operator A mapping a functional space P into a functional space Q is bounded if there exist a constant c such that for any p in P the inequality $\|Ap\|_Q \leq c\|p\|_P$ holds true where $\| \ldots \|_P$ and $\| \ldots \|_Q$ are norms in P and Q, respectively.

Boundedness of an operator A implies that small perturbations (or errors) in the input p can only lead to small perturbations (or errors) in the output $q = Ap$. Operator boundedness results in stable numerical operations, their unboundedness results in unstable numerical operations.

A simple illustration of the unboundedness of differentiation is to take $P = Q$ equal to the functional set of infinitely differentiable (smooth) real functions $p(x)$ over $[0, \pi]$ with $A = d/dx$ and norm $\|p\|^2 = \int_0^\pi p(x)^2 dx$. Choosing $p = \sin(\omega x)$ we have $q = Ap = \omega \cos(\omega x)$ and $\|p\|^2 = \int_0^\pi \sin(\omega x)^2 dx \leq \pi$ since the absolute value of sin is always less or equal to 1. However there cannot be a constant c such that $\|q\| = \|Ap\| \leq c\|p\| \leq c\pi$ since $\|q\|^2 = \omega^2 \int_0^\pi \cos(\omega x)^2 dx$ so that the norm $\|q\|$ can be made arbitrarily large by choosing arbitrarily large frequencies ω for $\|p\|$. This unboundedness is made worse by increasing the degree of differentiation.

The differences between forward and inverse problems reflect the properties of most environmental systems. In such systems, small changes in inputs or parameters result only in small changes in outputs. Furthermore, not only are most environmental systems stable, but the influence of rapid fluctuations in input or parameters is strongly attenuated in the system's output. The properties of system stability, error insensitivity and attenuation will be reflected in any reasonable mathematical model so that system simulation will share these stability properties. Note again that although stable physical systems are most often described via systems of differential operations such as in (2), (6) and (9) that map outputs to inputs, their stability is due to the fact that they map inputs to outputs via $D^{-1} = S$, i.e. via an integral operator. Thus, the usual sign of stability in a mathematical model is that the input is mapped to the output via an integral operation. As integration is a smoothing process, high frequency fluctuations in inputs will be damped in resulting outputs.

Identification problems do not share these stability properties as they are not analogues of actual stable systems. Rather they correspond to running a real system backwards. Thus if the original system contained damping, by necessity the identification process must be unstable: small errors in specification or calculation can induce arbitrarily large errors in the identified component. This is clearly true for input identification, and in most problems it is also true for system identification (although in this case the particular causes of instability may not be so obvious).

The instability of system identification has already been recognised in a number of environmental problems. One example is the identification of the impulse response function of a rainfall/runoff system from input/output data (Dietrich and Chapman 1993). Another is the system identification problem for flow in a confined groundwater system where numerical difficulties in reconstructing hydraulic conductivity, storativity and boundary conditions from piezometric head measurements have been observed and analysed in recent years (Carrera and Neuman 1986; Yeh 1986; McLaughlin and Townley 1996). Likewise the input identification problem of estimating recharge in a confined groundwater system is known to be unstable.

Further discussion about inverse problems from the point of view of indirect measurement problems and information loss is given in the next section as this standpoint sheds further light on their structure.

6.3.2 Inverse Problems as Indirect Measurement Problems

An alternative view of inverse problems is as indirect measurement problems. For example, in input identification, an output q can be viewed as an indirect measurement of the associated input p, the system acting as an imperfect instrument. In system identification, the output for a given input depends on the unknown parameter set u and therefore measures it indirectly. This view is relevant for environmental modelling, where output is often reasonably easy to measure but either system input or parameters may have to be estimated by inversion.

The important issue in the indirect measurement viewpoint is that the smoothness of integration has for effect that input or parameter information is reduced when passed to the output via integration. Furthermore, as we shall see, information losses are larger for high resolution information than they are for lower resolution. This implies that if an inverse procedure is invoked to identify a system, only low resolution (or equivalently low frequency) information can be estimated. To illustrate further the issue of information content in inverse problems, imagine a proposal to estimate local rainfall from runoff in a catchment such as that illustrated in Figure 6.4. In this catchment short local rainfall event will produce only a smeared and small runoff perturbation at the catchment outlet. In other words, the runoff will only show heavily attenuated variations of sharp changes in rainfall and any attempt to reconstruct variations in rainfall will be sensitive to small runoff measurement errors. Furthermore, it will not be possible to distinguish the runoff effects of rainfall in different regions of the catchment without monitoring runoff at numerous places.

6.3.3 Direct and Indirect Method for Inverse Problems

There are essentially two classes of approaches to solve an inverse problem. The first is the class of *direct approaches*. Direct approaches use the mathematical equations describing the system requiring calibration and solve *directly* for the unknown input or parameter. The second is the class of *indirect approaches*. Indirect approaches solve *successive forward problems* for a sequence of inputs or parameters with the sequence constructed so as to minimise some norm J between historical and computed data. To illustrate these two classes, consider again the fluid flow PDE (9). This choice is not limiting as its simulation and identification features are found in many environmental systems.

A direct method for estimating the parameter K will be based on (9) with K being the independent variable, i.e.

$$\nabla q \cdot \nabla K + (\nabla^2 q)K = \frac{\partial q}{\partial t} - p \qquad (11)$$

where $\nabla^2 = \nabla \cdot \nabla$ denotes the Laplacian operator $\partial^2/\partial x_1^2 + \partial^2/\partial x_2^2 + \partial^2/\partial x_3^2$. In terms of K, (11) is a first order PDE and any method for solving it in terms of K is called a direct method. Similarly, direct use of (9) to estimate the input p is called a direct method.

On the other hand, indirect methods estimate K from a best fit between historical and computed data, i.e. solve the non-linear optimisation problem

$$\min_{K} \|\mathbf{q}^* - \bar{\mathbf{q}}(K)\| \tag{12}$$

where $\mathbf{q}^*$ is a vector of historical output data available at discrete locations, $\bar{\mathbf{q}}(K)$ is the computed vector obtained by solving the forward problem (9) and taking the resulting q values at all historical data locations, and $\| \dots \|$ denotes a norm. Clearly, the minimisation (12) could be carried out over the input p instead of K or over a mixture of both.

As we shall see, major drawbacks of direct methods are that:

1. They require an output surface over the whole simulation domain. This surface must be reconstructed solely from noisy and scattered output data $\mathbf{q}^*$ using pertinent smoothing algorithms such as splines (Craven and Wahba 1979; Hutchinson 1984).
2. They provide estimates that generally explain poorly the available historical data $\mathbf{q}^*$.
3. They are not optimum in a statistical sense in that estimation errors generally retain a large amount of structure.

On the other hand, indirect methods are designed to best explain the available data with estimate having generally better statistical properties. Also, noise smoothing implicit in the minimisation (12) attenuates the effects of ill-posedness.

6.4   DEGREE OF ILL-CONDITIONING

In Section 6.3, we saw that a major source of ill-conditioning in inverse problems is the dependence of inputs or parameters on derivatives of the data. It was also indicated that the severity of error magnification due to differentiation increases with the degree of differentiation. In this section, these points are investigated further using again the fluid flow diffusion PDE (9).

For the sake of simplicity, assume the fluid flow to be in steady state so that instead of (9) we have

$$-\nabla \cdot K \nabla q = p \quad \text{over the system domain } \Omega \tag{13}$$

supplemented with appropriate boundary conditions. With this, assuming the system mathematical structure (13) to be exactly known, including boundary conditions

- system simulation is to compute the output q from (13) given the pair (p, K);
- system identification is to compute the system parameter K from (13) given (q, p); and
- input identification is to compute the input p from (13) given (q, K).

To comment on the stability of such fluid flow simulation and identification, let $\mathbf{x} = (x_1, x_2, x_3)$ be the 3-D spatial coordinate, and let G be the Green's function associated with (13) i.e. solution of

$$-\nabla \cdot K \nabla G = \delta(\mathbf{x} - \mathbf{v}) \quad \text{over the system domain } \Omega \tag{14}$$

where $\delta(\mathbf{x} - \mathbf{v})$ is the Dirac distribution centred at $\mathbf{v}$ (Stakgold 1967, 1968). Readers unfamiliar with Green's function can think of them as a system's response to a unit impulse. Thus Green's functions are similar to the well-known unit hydrograph concept often invoked in rainfall/runoff modelling (Chapman 1985; Dietrich and Chapman 1993).

As G depends on K, $\mathbf{x}$ and $\mathbf{v}$ it can be written $G_K(\mathbf{x}, \mathbf{v})$. With this, it can be shown that the integral representation equivalent to (13) is (Stakgold 1967, 1968)

$$q(\mathbf{x}) = \int_\Omega G_K(\mathbf{x}, \mathbf{v})p(\mathbf{v})d\mathbf{v} \tag{15}$$

which maps the input p to the output q.

In system simulation mode via (15), the input $p(\mathbf{v})$ is integrated linearly against the Green's function G_K over the whole spatial domain Ω. Thus an error Δp in the input causes an error in the output Δq equal to

$$\Delta q(\mathbf{x}) = \int_\Omega G_K(\mathbf{x}, \mathbf{v})\Delta p(\mathbf{v})d\mathbf{v} \tag{16}$$

and because the integral operator in (16) is bounded, there exist a constant c such that (see Definition 3)

$$\|\Delta q\| \leq c\|\Delta p\|$$

for any input perturbation $\|\Delta p\|$. As a consequence, input errors Δp will not magnify in Δq.

Effect on q of errors in K cannot be investigated as easily since their interdependence is not linear. Nevertheless, observe from (14) that errors in K will show in errors in the Green's function G_K only. Furthermore, inspection of (14) shows that highest derivatives in the Green's function G are second order, while highest derivatives in K are first order. Thus formally, computation of G_K requires two integrations and so depends on one integration of K. Errors in K will therefore not magnify in G_K, and so not magnify in q either.

Now to comment on stability issues associated with system identification, i.e. estimation of K, we consider a direct approach and rewrite (13) in terms of the unknown K as follows

$$\nabla q \cdot \nabla K + (\nabla^2 q)K = -p \tag{17}$$

In terms of K, (17) is a first order PDE and its solution over characteristic pathways is (see Appendix)

$$K(s) = K(0)e^{-\int_0^s \nabla^2 q(\mathbf{x}(\tau))d\tau} - \int_0^s e^{-\int_\xi^s \nabla^2 q(\mathbf{x}(\tau))d\tau} p(\mathbf{x}(\xi))d\xi \tag{18}$$

System identification is formally encapsulated in (18). First observe that (18) links K to the input p via integration. Thus errors in p will not magnify in K. On the other hand, the first term on the right hand side of (18) links $K(s)$ to the output q via two differentiations (Laplacian $\nabla^2 q(\mathbf{x}(\tau))$) followed by one integration over the interval $[0, s]$. This link is therefore formally that of one differentiation. Errors in q will thus magnify in K and the degree of magnification will correspond to one differentiation. As for input identification, (13) shows that p depends on one differentiation of K and two differentiation of q and so depends non-continuously on both.

The degree of ill-conditioning (or ill-posedness) of an inverse problem can be defined as follows.

Definition 4   Degree of ill-conditioning (or ill-posedness).

Consider an inverse problem which solution requires n differentiations of the data. Then the degree of ill-conditioning of the inverse problem is equal to n.

In view of the above definition, the ill-conditioning of input identification from q data is structurally more severe than the identification of K using the same q data, a fact perhaps not recognised often enough in groundwater investigations.

Note in passing that the degree of ill-posedness of an inverse problem can be less than one. For example, problems associated with the solution of Abel type integral equation display fractional degrees of ill-posedness (Anderssen 1980; Elliott 1980). The degree of ill-posedness can even be zero such as in the estimation of α via (4) where no data differentiation occurs. In this case, as long as the cumulative input $\int p dt$ and output $\int q dt$ are sufficiently different to ensure the integral in the denominator is not close to zero, the effects of arbitrary small changes in p and q are bounded and the estimation of α is well-posed.

6.5   INSTABILITY AND NUMERICAL ISSUES

Since in most practical problems, simulation and system identification are carried out via a discretisation of the differential governing equation over its domain of definition, details on discretisation and associated numerical issues for direct and indirect approaches to inverse problems need to be incorporated in this discussion. For this purpose, the fluid flow PDE (13) is again invoked.

Any finite difference or Galerkin (e.g. finite element) discretisation of (13) over n nodes $\{\mathbf{x}_i\}_{i=1}^n$ of the simulation domain Ω will result in a matrix equation of the form

$$D(\mathbf{k})\mathbf{q} = \mathbf{p} \tag{19}$$

where $D(\mathbf{k})$ is a non-singular $n \times n$ matrix with the vectors $\mathbf{k}$, $\mathbf{q}$ and $\mathbf{p}$ denoting the associated discretisation of K, q and p at the discretisation nodes and therefore each of dimension n. Formally, solution to (19) can be written

$$\mathbf{q} = S(\mathbf{k})\mathbf{p} \tag{20}$$

where the $n \times n$ matrix $S(\mathbf{k}) = D(\mathbf{k})^{-1}$ is a discretisation of the integral operator (15). Let us assume that we have available an n dimensional output data vector

$$\mathbf{q}^* = \mathbf{q} + \mathbf{e}$$

where $\mathbf{e}$ contains errors such as noise in the data, discretisation approximation errors etc. For simplicity, we assume that input identification is carried out with a perfect system parameter $\mathbf{k}$ and system identification is carried out with a perfect input $\mathbf{p}$.

Following the terminology specified in Section 6.3.3, input identification via a direct approach is based on the discrete differential representation (19) i.e.

$$Dq^* = Dq + De = \mathbf{p} \tag{21}$$

while input identification via an indirect approach is based on discrete integral representation (20) i.e.

$$q^* = S\mathbf{p} + \mathbf{e} \tag{22}$$

The effect of the noise $\mathbf{e}$ in the direct approach representation (21) can be assessed by comparing the norm of $\|De\|$ with that of $\|Dq\|$ where for any vector $\mathbf{v}$ the norm $\|\mathbf{v}\|^2$ is defined as $\mathbf{v}^T\mathbf{v}$. As shown below, this will be done by estimating the noise to signal ratio $r = E[\|De\|/\|Dq\|]$ where $E[...]$ denotes an expectation, with potentially large r values due to the differentiation matrix D operating on the noise $\mathbf{e}$. On the other hand, (22) indicates that an indirect approach will be sensitive to spurious large high frequency components in $\mathbf{p}$ smoothed out in the integrated term $S\mathbf{p}$ below noise level $\mathbf{e}$. These two equivalent and reciprocal features are typical of direct and indirect approaches to inverse problems and for their analysis, we invoke the singular value decomposition of D written as (Barnett 1990)

$$D = V\Lambda W^T \tag{23}$$

where the matrices V and W are orthonormal and Λ has all zero entries except for its diagonal having positive singular values $\{\lambda_i\}_{i=1}^n$ listed by increasing value $0 < \lambda_1, \ldots, \leq \lambda_n$. Note that if D is symmetric positive-definite, the singular value decomposition (23) becomes an eigenvalue decomposition. The link between eigenvalues and singular values of any symmetric positive-definite matrix A originates from the fact that the singular value decomposition $A = V\Lambda W^T$ is constructed from an eigenvalue decomposition of $A^T A$ (or AA^T) (Barnett 1990). In particular, the non-zero singular values of A are equal to the square root of the non-zero eigenvalues of $A^T A$ (or AA^T). Thus for all symmetric and positive definite matrices, eigenvalues and singular values coincide.

Recalling that D is a non-singular discretisation of a non-singular differential operator, its singular value decomposition has the following properties:

1. All singular values are strictly positive.
2. Larger singular values are associated with eigenvectors of higher frequencies.
3. If n increases so that the approximation error between D and D decreases, the larger singular value λ_n and associated singular frequency increase without bound.
4. In contrast to item 3, the smallest singular value λ_1 admits a lower bound c with $0 < c \leq \lambda_1$ and c independent of n.

To illustrate items 1 to 4 above, consider the differential operator $D = -\nabla \cdot K\nabla$ in one spatial dimension $[0, \pi]$ with $K = 1$, i.e. $D = -d^2/dx^2$, and zero boundary conditions. Then D is self-adjoint and the singular value decomposition is the eigenvalue decomposition with eigenvalues/eigenfunctions pairs $\{i^2, \sin(ix)\}_{i=1}^\infty$. Observe that (i) the eigenvalues are strictly positive, (ii) larger eigenvalues are associated with eigenfunctions of higher frequencies, and (iii) eigenvalues grow to infinity. Clearly, if the matrix D is a reasonable approximation to the operator D, its singular values will mimic those of D and λ_n will approach infinity as D approaches D. This carries through to the 3-D case with K non-constant.

In addition to the singular value decomposition of D, it is useful to define the condition number of a matrix (Barnett 1990). For this let $\| \ldots \|$ denote any norm that is consistent, i.e. with the property $\|D\mathbf{p}\| \leq \|D\| \|\mathbf{p}\|$.

Definition 5   *Condition number $\kappa(D)$.*

The condition number $\kappa(D)$ of a non-singular matrix D is $\kappa(D) = \|D\| \, \|D^{-1}\|$. If $\| \dots \|$ is the spectral norm, then $\kappa(D)$ is the ratio of the largest to smallest singular value of D.

From the above definition, it is easy to derive the following inequalities true for any non-singular system $\mathbf{q} = D\mathbf{p}$

$$\frac{\|\delta\mathbf{p}\|}{\|\mathbf{p}\|} \le \kappa(D)\frac{\|\delta\mathbf{q}\|}{\|\mathbf{q}\|} \quad \text{and} \quad \frac{\|\delta\mathbf{q}\|}{\|\mathbf{q}\|} \| \le \kappa(D)\frac{\|\delta\mathbf{p}\|}{\|\mathbf{p}\|} \tag{24}$$

where $\delta\mathbf{p}$ and $\delta\mathbf{q} = D\delta\mathbf{p}$ are perturbations in $\mathbf{p}$ and $\mathbf{q}$, respectively.

The two inequalities (24) indicate that the size of the condition number $\kappa(D)$ is a measure of error sensitivity to matrix/vector operation and matrix inversion and so is of fundamental importance in the field of matrix computation (Golub and Van Loan 1989).

Now from the Definition 5, $\kappa(S) = \kappa(D)$ so that with D being the discretisation of a differential operator, both S and D are potentially ill-conditioned. Furthermore, the properties of the singular values of D listed above indicate that the system $\mathbf{q} = S\mathbf{p}$ is sensitive to matrix inversion. On the other hand, although D is ill-conditioned, the system $\mathbf{p} = D\mathbf{q}$ *is not* sensitive to matrix inversion. This difference arises from the above items 3 and 4, i.e. from the fact that the matrix D is associated with an unbounded operator and therefore has unbounded singular values (item 3) while all singular values of S are bounded above by a constant independent of n (item 4).

With the singular value decomposition (23), it is easy to estimate how large the ratio $r = E[\|D\mathbf{e}\|/\|D\mathbf{q}\|]$ can be. For the sake of argument, assume the signal of unit sise, i.e. $\|\mathbf{q}\| = 1$, and $\mathbf{e}$ being normal $\mathcal{N}(0, \sigma^2 I)$ where I denotes an identity matrix. Since V and W in (23) are orthonormal and $\mathbf{e}$ is $\mathcal{N}(0, \sigma^2 I)$, we have $E[\|D\mathbf{e}\|^2] = E[(\mathbf{e}^T \Lambda^2 \mathbf{e})] = \sigma^2 \sum_{i=1}^{n} \lambda_i^2$. Thus

$$r = \frac{E[\|D\mathbf{e}\|]}{\|D\mathbf{q}\|} = \frac{\sigma(\sum_{i=1}^{n}\lambda_i^2)^{\frac{1}{2}}}{\|D\mathbf{q}\|} \ge \frac{\sigma\lambda_n}{\|D\mathbf{q}\|} \tag{25}$$

As for the transformed signal $D\mathbf{q}$, the Rayleigh quotient $\|D\mathbf{q}\|^2/\|\mathbf{q}\|^2 = \mathbf{q}^T D^T D\mathbf{q}/\mathbf{q}^T\mathbf{q}$ can be as small as the square of smallest singular value λ_1^2 (Golub and Van Loan 1989). Since $\|\mathbf{q}\| = 1$, this implies that $\|D\mathbf{q}\|$ can be as small as λ_1. In other words, for an output $\|\mathbf{q}\|$ of unit size the ratio r can be as large as $\sigma\lambda_n/\lambda_1 = \sigma\kappa(D)$. Now if the matrix D is a reasonable approximation to the differential operator D, the larger singular value λ_n will be very large compared to the smallest singular value λ_1 and D will have a very large condition number.

To summarise, upon a discretisation of the fluid flow equation (13) and adoption of a direct approach to estimate the input $\mathbf{p}$ from noisy output data $\mathbf{q}^*$, the noise to signal ratio affecting any estimate $\hat{\mathbf{p}}$ computed from (21) can be as large as product $\sigma\kappa(D)$. In turn, the condition number $\kappa(D)$ will be very large if D is a reasonably close approximation to D causing severe estimate corruption.

The condition number $\kappa(D)$ is clearly less informative than the full singular value decomposition of D. However, given that the computation of all singular values of a $n \times n$ matrix requires in general a number of floating point operations proportional to n^3, the feasibility of computing the full singular value decomposition of D is dictated by the size of n, i.e. by the resolution of the discretisation D of D. The closer D approximates D, the larger n will be and the more demanding is the computation of the singular decomposition. For this reason, in practice where n is usually large, potential numerical instability can be flagged through estimates of $\kappa(D) = \lambda_n/\lambda_1$ with λ_n and λ_1 generally computed via some iterative algorithms.

For the indirect approach to input identification using (22), the singular value decomposition (23) can also be used to provide a measure of the sensitivity of the approach to spurious large high frequency components in $\mathbf{p}$. Indeed, from (23) and the orthonormality of V and W we have the singular value decomposition

$$S = W\Lambda^{-1}V^T \tag{26}$$

Being the inverse of those of D, the singular values of S decay to zero as n increases, with the largest one $1/\lambda_1$ remaining bounded above independently of n. Now the effect of any input perturbation $\delta\mathbf{p}$ added to $\mathbf{p}$ in (22) can be assessed by comparing the norm of $[\|S\delta\mathbf{p}\|]$ with that of $E[\|\mathbf{e}\|]$. With $\mathbf{e}$ being $\mathcal{N}(0,\sigma^2 I)$, we have immediately $E[\|\mathbf{e}\|] = \sqrt{n}\sigma$. As for $[\|S\delta\mathbf{p}\|]$, set $\delta\mathbf{p} = \gamma V\mathbf{n}$ where $\mathbf{n}$ has all entries equal to zero except for the last equal to 1. $V\mathbf{n}$ is thus the last column of the orthonormal matrix V and so is the eigenvector of highest frequency associated with the smallest singular value $1/\lambda_n$. With this, $\|S\delta\mathbf{p}\| = (\delta\mathbf{p}^T V\Lambda^{-2}V^T\delta\mathbf{p})^{1/2} = |\gamma|(\mathbf{n}^T\Lambda^{-2}\mathbf{n})^{1/2} = |\gamma|/\lambda_n$. Therefore, as long as n is large enough to ensure $1/\lambda_n \leq \sqrt{n}\sigma/|\gamma|$, a large high frequency component $\delta\mathbf{p} = \gamma V\mathbf{n}$ can always be added to $\mathbf{p}$ in (22) with $\|S\delta\mathbf{p}\| = |\gamma|/\lambda_n$ remaining below the expected noise level $E[\|\mathbf{e}\|] = \sqrt{n}\sigma$.

So far, we have assumed that the unknown input vector $\mathbf{p}$ is of dimension n, i.e. has the same resolution as that of the modelling problem discretisation. However, as mentioned before, inverse problems are characterised by information losses. It is therefore generally necessary to reduce the resolution of unknowns to match the information content of the available data. This is often done by expressing unknowns in terms of linear combinations of a relatively small number of basis functions. Details on this are in the next subsection.

6.5.1 Reducing Resolution Using Basis Functions

Let us consider again the input p in (13) and the estimation of its discretised form $\mathbf{p}$ from $\mathbf{q}^*$ using (19) to (22). A way to reduce the dimension n of the unknown $\mathbf{p}$ is to expand linearly $p(\mathbf{x})$ in terms of m basis functions with $m \leq n$. This amounts to considering the sum

$$p(\mathbf{x}) = \sum_{j=1}^{m} u_j b_j(\mathbf{x}) \qquad m \leq n \tag{27}$$

with $\{b_j(\mathbf{x})\}_{j=1}^m$ being any set of appropriate basis functions such as orthogonal polynomials, piecewise constant functions, splines, radial basis functions etc., and $\mathbf{u} = (u_1,\ldots,u_m)^T$ being now the dimensionally reduced unknown input.

If the discretisation of the differential operator at hand is of a finite difference type, the entries p_i of the vector $\mathbf{p}$ are values of p taken at the n discretisation nodes $\{\mathbf{x}_i\}_{i=1}^n$ i.e. $p_i = p(\mathbf{x}_i) = \sum_{j=1}^m u_j b_j(\mathbf{x}_i)$ or in matrix notation

$$\mathbf{p} = B\mathbf{u} \tag{28}$$

with B being an $n \times m$ matrix having entries $b_{ij} = b_j(\mathbf{x}_i)$. Note that the expansion (27) is appropriate only if B is full rank, i.e. if $B^T B$ is invertible.

If the discretisation of the differential operator at hand is of a Galerkin type (e.g. finite element), then the available differential equation can be used to generate a set of linear equations by integrating it over its domain of definition Ω against a set of basis functions $\{\psi_i(\mathbf{x})\}_{i=1}^n$. In the case of the flow equation (13), this would take the form

$$-\int_\Omega (\nabla \cdot K\nabla q)(\mathbf{x})\psi_i(\mathbf{x})d\mathbf{x} = \sum_{j=1}^m u_j \int_\Omega \psi_i(\mathbf{x})b_j(\mathbf{x})d\mathbf{x} \qquad i = 1,\ldots,n$$

with the $n \times m$ matrix B having now entries $b_{ij} = \int_\Omega \psi_i b_j d\mathbf{x}$.

With an expansion such as (27), there arises the issue of choosing the basis functions $b_j(\mathbf{x}_i)$ and the dimension m of the unknown so that the output data information base and the resolution of the unknown match. Choosing basis functions that are too smooth and/or a value for m that is too small will not do justice to the information content present in the output data. On the other hand, choosing high resolution basis functions and/or a value for m that is too large attempts to reconstruct information that has been lost, and this will cause estimates to be corrupted by error magnifications. For linear problems such as input estimation, Newsam (1982) has shown that optimum basis functions and dimension m can be derived from a singular value decomposition of the linear operator at hand. Further details on this are in Section 6.6.3.

Having considered linear combinations of basis functions as a means to reduce the dimensionality of an inverse estimation problem, input estimation and system identification are investigated further in the next subsection, with the former first discussed due to its linear, hence more tractable, mathematical structure.

6.5.2   Input Estimation

Recall that input estimation is the inverse problem of estimating the input vector $\mathbf{p}$ from l noisy data $\mathbf{q}^*$ with l generally much smaller than the dimension n of $\mathbf{p}$. Given potential error magnifications, for this to be meaningful the dimension of the parameter vector $\mathbf{p}$ needs to be reduced. This can be done via an expansion of the form (27) expressed as (28), i.e. $\mathbf{p} = B\mathbf{u}$ with $\mathbf{u}$ of dimension $m \le l$ being now the dimensionally reduced unknown input.

As evident from (21), a direct approach to input estimation requires that an output vector be available at all n discretisation nodes. Therefore, an output surface q^o must first be reconstructed from the data $\mathbf{q}^*$. Having reconstructed such q^o, an n-dimensional vector $\mathbf{q}^o$ is obtained by taking q^o at the n discretisation nodes. With this, input identification via a direct approach will be based on the overdetermined system got from (19) and (28) i.e.

$$D\mathbf{q}^o = B\mathbf{u} + \mathbf{e} \qquad\qquad (29)$$

Note that in setting (29) we have not taken on board the fact that the vector $\mathbf{p}$ not only contains information about the input p, but will also contain flow boundary information if they arise. This was ignored as, otherwise, the notation would have been burdened without adding much value and/or insight in the discussion.

In contrast to direct approaches, indirect approaches construct the unknown input by best matching computed and historical output data. To construct such best match, a computed data vector $\bar{\mathbf{q}}$ that specifies computed output values at historical data locations needs to be defined. To do this let L be an $l \times n$ interpolation matrix that interpolates the output values $\mathbf{q}$ at the location of the historical output data $\mathbf{q}^*$, i.e.

$$\bar{\mathbf{q}} = L\mathbf{q} \qquad\qquad (30)$$

Note that if the data locations coincide with the discretisation nodes, L is a projection matrix formed of ones and zeroes with the property that $LL^T = I$ where I is an identity matrix. This implies that the first l singular values of L are equal to 1. This property will be useful in an application of the lemma below.

Using (20), (22) and (28) allows (30) to be rewritten as

$$\bar{\mathbf{q}} = LSB\mathbf{u} + \mathbf{e} \tag{31}$$

Be mindful that although the same notation is used for both direct and indirect approach error vectors $\mathbf{e}$ in (29) and (31), both are very differently structured. Nevertheless, for the sake of simplicity no notational distinction is made between both error structures. This is because to account for all major error types such as errors to surface reconstruction, data errors, model errors, discretisation errors, dimensionality reduction errors etc. requires a cumbersome notation. Suffice to say that, as already indicated, the error term in direct methods will most likely be much more structured than that in indirect methods and this results in estimates with far poorer statistical properties.

If $\mathbf{e}$ is $\mathcal{N}(0, C)$ where C is a symmetric, positive-definite covariance matrix, the statistically optimum least squares estimate associated with (29) satisfies the system

$$(B^T C^{-1} B)\hat{\mathbf{u}} = B^T C^{-1} D\mathbf{q}^o \tag{32}$$

As for an indirect approach based on minimising a norm $\|\mathbf{q}^* - \bar{\mathbf{q}}\|^2$, the statistically optimum least squares estimate associated with (31) satisfies the equation

$$F\hat{\mathbf{u}} = B^T S^T L^T C^{-1} \mathbf{q}^* \tag{33}$$

where $F = B^T S^T L^T C^{-1} LSB$.

Recalling that $S = D^{-1}$, observe the difference between (32) and (33) which reflects in the most illuminating way the difference between estimates obtained from direct and indirect inverse approaches. Focusing first on (32), observe that numerical instabilities originate from D operating directly on errors in $\mathbf{q}^o$. Thus even if the dimension m of the unknown $\mathbf{u}$ is small, error magnification will persist. In other words, for direct methods, noise stabilisation is only achievable via a smooth reconstruction of the surface q^o. However, as illustrated in Section 6.8, the surface has to be smooth and also very accurate. These two requirements are usually mutually exclusive, causing major difficulties in practice.

The picture is very different for the indirect approach estimate solution to (33). In this case, ill-conditioning can arise because of the need to invert the matrix F involving the smoothing matrix S. To quantify this, the following lemma which proof is in Dietrich and Newsam (1989) can be used:

Lemma Let A be an $n \times n$ symmetric matrix. Then for any $n \times m$ matrix X the following inequality holds true

$$\lambda_i(X^T AX) \leq \min_{r+s \leq i+1} \lambda_r(A)\lambda_s(X^T X) \qquad i = 1, \dots, m \tag{34}$$

with $\lambda(A)_i$ denoting the ith singular value of A indexed in decreasing value.

Using the fact that with $m \leq l$ the first m singular values of $L^T L$ are equal to 1, it is easy to show from (34) that the smallest singular value of F in (33) satisfies the inequality

$$\lambda_m(F) \leq \lambda_1(C^{-1})\lambda_1(B^T B)\lambda_m(S^T S)$$

Thus if m is sufficiently large, $\lambda_m(S^T S)$ will be small causing $\lambda_m(F)$ to be small. In turn F is ill-conditioned with, as a consequence, the estimate $\hat{\mathbf{u}}$ solution to (33) sensitive to errors in $\mathbf{q}^*$. This said, a reduction in the dimension m will improve the conditioning of F in (33). For example with $m = 1$, (33) is a perfectly conditioned scalar equation.

The above stability analysis was relatively straightforward because of the linear relationship between the reduced input $\mathbf{u}$ and the output data $\mathbf{q}$. In the next subsection, the non-linear and hence more complicated problem of estimating the system parameter vector $\mathbf{k}$ from output data is briefly discussed.

6.5.3   System Identification

Recall that system identification is the inverse problem of estimating the system parameter vector $\mathbf{k}$ from l noisy data $\mathbf{q}^*$ where $\mathbf{k}$ is a discretisation of K in the fluid flow PDE (13). Therefore, as done for input estimation, the dimension of $\mathbf{k}$ is reducible via an expansion of the form (27) expressed as (28), i.e.

$$\mathbf{k} = B\mathbf{u} \tag{35}$$

with $\mathbf{u}$ being now the dimensionally reduced unknown parameter vector, and B being an $n \times m$ matrix with $m \leq l$.

With (30) we have from (20)

$$\bar{\mathbf{q}}(\mathbf{u}) = LD(\mathbf{u})^{-1}\mathbf{p} \tag{36}$$

Since the relationship between $\mathbf{u}$ and the computed output $\bar{\mathbf{q}}$ is not linear, a stability analysis of any indirect approach is far from trivial. Nevertheless, some observations can be made from a linearisation of (36). To do this, consider the $m \times l$ Jacobian matrix $G_{\bar{\mathbf{q}}}$ with entries $\partial u_i / \partial \bar{q}_j$ where the link between $\mathbf{u}$ and $\bar{\mathbf{q}}$ is given implicitly by (36) with $\mathbf{p}$ assumed perfectly known. The Jacobian $G_{\bar{\mathbf{q}}}$ is useful because of the relationship

$$\hat{\mathbf{u}} - \mathbf{u} = G_{\bar{\mathbf{q}}}(\mathbf{q}^* - \bar{\mathbf{q}}) + O((\mathbf{q}^* - \bar{\mathbf{q}})^2)$$

indicating that to first order, data noise $\mathbf{q}^* - \bar{\mathbf{q}}$ can magnify in the estimate error $\hat{\mathbf{u}} - \mathbf{u}$ by factors that can be determined from a singular value decomposition of $G_{\bar{\mathbf{q}}}$.

Derivation of an expression for $G_{\bar{\mathbf{q}}}$ in terms of D is cumbersome. Considering instead $\bar{\mathbf{q}}$ as a function of $\mathbf{u}$ via the explicit relationship (36) and denoting $G_{\mathbf{u}}$ the associated $l \times m$ Jacobian with entries $\partial \bar{q}_i / \partial u_j$ yields

$$\mathbf{q}^* - \bar{\mathbf{q}} = G_{\mathbf{u}}(\hat{\mathbf{u}} - \mathbf{u}) + O((\hat{\mathbf{u}} - \mathbf{u})^2) \tag{37}$$

Differentiating (36) with regards to the m entries $(u_j)_{j=1}^m$ in $\mathbf{u}$ provides an expression for $G_{\mathbf{u}}$ equal to

$$G_{\mathbf{u}} = -LS(\mathbf{u})\left(\frac{\partial D(\mathbf{u})}{\partial k_1}, \ldots, \frac{\partial D(\mathbf{u})}{\partial k_m}\right)S(\mathbf{u})\mathbf{p} \tag{38}$$

with $S = D(\mathbf{u})^{-1}$. From (37) and the smoothing effect of $G_{\mathbf{u}}$ via that of $S(\mathbf{u})$ in (38), observe that large and high frequency deviations $\hat{\mathbf{u}} - \mathbf{u}$ may cause only small deviations $\mathbf{q}^* - \bar{\mathbf{q}}$. In other words, large fluctuations of frequency ω in $\hat{\mathbf{u}} - \mathbf{u}$ can cause fluctuations in $\mathbf{q}^* - \bar{\mathbf{q}}$ that are below noise levels as long as the frequency ω is high enough. As a consequence, small noise in the historical output data $\mathbf{q}^*$ will not allow *any estimation procedure* to ensure $\hat{\mathbf{u}} - \mathbf{u}$ small. In the next two subsections, these features are further described in the specific context of direct and indicated approaches to system identification.

Direct Approaches

A direct approach to estimating K can be based on a finite difference or Galerkin discretisation of (13) with K being the dependent variable and the reconstructed surface q^o yielding the vector $\mathbf{q}^o$ obtained by taking q^o at the n discretisation nodes. This result in the $n \times m$ overdetermined linear system $\mathbf{p} = H(\mathbf{q}^o)\mathbf{k}$ where the matrix H is a discretisation of the hyperbolic operator $\nabla q^o \cdot \nabla + (\nabla^2 q^o)\mathrm{Id}$ appearing in (17). Taking into account errors and (35) yields the $n \times m$ overdetermined linear system

$$\mathbf{p} = H(\mathbf{q}^o)B\mathbf{u} + \mathbf{e}$$

with least squares solution

$$\hat{\mathbf{u}} = ((H(\mathbf{q}^o)B)^T C^{-1}(H(\mathbf{q}^o)B))^{-1}(H(\mathbf{q}^o)B)^T C^{-1}\mathbf{p} \tag{39}$$

where C denotes the covariance matrix of the error vector $\mathbf{e}$.

Observe that errors in the historical output data $\mathbf{q}^*$ appear structurally as errors in $\mathbf{q}^o$ affecting the hyperbolic matrix $H(\mathbf{q}^o)$. A meaningful error analysis will therefore be next to impossible to carry out and the estimate (39) is likely to be far from optimal. Worse, if the above estimate $\hat{\mathbf{u}}$ is used to compute $\mathbf{q}$ from (19), there is no guarantee at all that the computed output $\bar{\mathbf{q}} = L\mathbf{q}$ will reasonably match the historical data $\mathbf{q}^*$. In other words, disadvantages of the direct approach estimate (39) for the parameter K are such that for most practical cases, a non non-linear and hence more demanding indirect approach is required.

Indirect Approaches

Indirect approaches to estimate the system reduced parameter $\mathbf{u}$ from output historical data are based on minimising a cost function

$$J(\mathbf{u}) = \|\mathbf{q}^* - \bar{\mathbf{q}}(\mathbf{u})\|^2 \tag{40}$$

between the historical $\mathbf{q}^*$ and computed $\bar{\mathbf{q}}(\mathbf{u})$ output data. In other words, system parameter identification is reduced to the non-linear optimisation problem

$$\min_{\mathbf{u}} J(\mathbf{u}) = \min_{\mathbf{u}} \|\mathbf{q}^* - \bar{\mathbf{q}}(\mathbf{u})\|^2 \tag{41}$$

The numerical sensitivity of the minimisation (41) to errors present in $\mathbf{q}^*$ will now be reflected in the gradient vector

$$\nabla_{\mathbf{u}} J(\mathbf{u}) = -2G_{\mathbf{u}}^T(\mathbf{q}^* - \bar{\mathbf{q}}(\mathbf{u})) \tag{42}$$

Note the presence of the Jacobian G_u in $\nabla_u J(\mathbf{u})$ with closed form given in (38). Because of the smoothing effect of G_u, (42) indicates that in the presence of noise in the output data, the gradient of the objective function $J(\mathbf{u})$ is likely to be very small over a wide range of $\mathbf{u}$ values causing overall flatness of $J(\mathbf{u})$. Such flatness will be a sign of ill-conditioning and a singular value decomposition of G_u can provide a measure of possible instabilities and associated error magnification.

Another way of looking at numerical instabilities associated with the estimation of the reduced parameter $\mathbf{u}$ is to use (37) to rewrite the gradient (42) as

$$\nabla_u J(\mathbf{u}) = -2G_u^T G_u(\hat{\mathbf{u}} - \mathbf{u}) + O((\hat{\mathbf{u}} - \mathbf{u})^2) \tag{43}$$

Equation (43) indicates that to first order, small eigenvalues in $G_u^T G_u$ will reduce significantly the sensitivity of the gradient $\nabla_u J(\mathbf{u})$ to the combinations of entries in $\mathbf{u}$ associated with high frequency eigenvectors. As was the case for input estimation, this can cause spurious high frequency components to affect the estimate $\hat{\mathbf{u}}$ solution to (41). These difficulties need to be remedied through stabilisation procedures such as those presented in the next section.

To conclude this section, it is worth reminding the reader that the above discussion was based specifically on the inverse problem of aquifer identification associated with the fluid flow PDE (13). While the issue of error magnification will arise in any inverse problem for which the unknown parameter dimension m is large compared to the number of independent data, it may not occur if m is small compared to the number of independent data available for calibration. Indeed, keeping m small provides a simple way to counteract the effect of ill-posedness. For example, as already indicated in Section 6.2, this occurs in time series analysis where system identification amounts to estimate only the few m parameters that determine a system's transfer function (Young 1984; Barker and Young 1985).

6.6 STABILISATION

As mentioned in Section 6.2, modelling environmental systems is often carried out with only limited information on system data and processes. For example, fluid flows modelled via the mass conservation equation (9) require detailed information on the system parameter K and the system driving forces represented by the source p, the initial state of the system, and the boundary conditions on the fluid system boundary. Clearly, in many practical situations such information is not available. If inverse procedures are used to estimate system inputs and/or parameters, ill-conditioning can arise and this is compounded by the usually high levels of uncertainties found in environmental data and processes. This will show in the form of numerical instabilities when calibrating inputs and/or system parameters. To control such instabilities, stabilisation approaches have to be implemented to limit the tendency of inverse problems to magnify high frequency noise. This said, prior to any simulation and calibration exercise, modellers need to ensure that:

1. Models have a level of complexity and resolution commensurate with the available data and the required model predictive resolution.
2. Any available prior information on the data and system structure is injected into the calibration procedure.

Item 1 states the obvious fact that one should only use high resolution models if data are sufficient and detailed simulations of the system under study are required. As seen in Section 6.2, increased model resolution can lead to increased susceptibility to ill-conditioning. Reducing resolution by decreasing the dimension of the unknown via linear expansions such as that presented in Section 6.5.1 is a simple and effective stabilising approach often used in practice.

Item 2 states that any prior (even subjective) information should be used as constraints imposed on the system's structure to control the emergence of spurious features in estimates due to the non-continuous dependence between unknown data, and narrow down the array of possible solutions. Indeed, the high level of uncertainty that often characterises environmental systems implies that there may be several solutions to a problem, all displaying high sensitivity to data errors. For example, calibrating the hydraulic conductivity K from noisy indirect head measurements will most often lead to a number of oscillatory conductivity estimates, each explaining the head data to the same degree. To narrow down the choice of possible estimate $\hat{K}$, one can use prior knowledge, e.g. over m portions of the groundwater system domain the conductivity is essentially constant. Thus instead of having to know K at any point of the groundwater domain, the modeller has to decide a priori on groundwater zones over which hydraulic conductivity has constant values and use an expansion of the form (27) where the m basis functions b_i are equal to 1 over each region of constant conductivity, and zero elsewhere. The modellers could also decide on the basis of hydrogeological evidence that over sections of the boundary the water flow is very small and so impose a priori zero flow boundary conditions there. Furthermore, as shown below, in many practical problems data about the unknown can be included in the estimation problem by requesting that its estimate be not too far away from the data. If no data are available, one can constrain the size of the unknown or some linear combination of it.

6.6.1 Prior Information Invoking Quadratic Forms

Consider input estimation based on (29) and (31). In both cases, the estimate $\hat{\mathbf{u}}$ is a solution to a least squares problem

$$\min_{\mathbf{u}} \|\mathbf{y} - X\mathbf{u}\|^2 \tag{44}$$

where for any vector $\mathbf{v}$ the norm $\|\mathbf{v}\|$ is defined by $\|\mathbf{v}\|^2 = \mathbf{v}^T C^{-1}\mathbf{v}$, and C denotes the covariance (symmetric positive definite) matrix of the zero mean error $\mathbf{e} = \mathbf{y} - X\mathbf{u}$. Rewriting (44) as

$$\min_{\mathbf{u}} \mathbf{u}^T X^T C^{-1} X\mathbf{u} - 2\mathbf{y}^T C^{-1} X\mathbf{u}$$

yields the standard least square estimate $\hat{\mathbf{u}}$ solution to

$$(X^T C^{-1} X)\hat{\mathbf{u}} = X^T C^{-1}\mathbf{y} \tag{45}$$

Now in many practical problems, modellers have prior information which can be expressed as the constraint

$$\zeta(\mathbf{u}) = (\mathbf{u} - \mathbf{a})^T M(\mathbf{u} - \mathbf{a}) \le c \tag{46}$$

where $\zeta(\mathbf{u})$ is a quadratic form with the vector $\mathbf{a}$ and the symmetric, positive-definite matrix M chosen a priori. For example:

1. If data $\mathbf{u}^*$ is available about the unknown $\mathbf{u}$, modellers often set $\mathbf{a} = \mathbf{u}^*$ and $M = I$ where I is an identity matrix. Then the quadratic form $\zeta(\mathbf{u})$ is equal to the distance $(\mathbf{u} - \mathbf{u}^*)^T(\mathbf{u} - \mathbf{u}^*) = \|\mathbf{u} - \mathbf{u}^*\|^2$ between $\mathbf{u}$ and the available data $\mathbf{u}^*$ (Carrera and Neuman 1986).

2. If it is known a priori that the size of $\mathbf{u}$ is bounded, modellers can set $\mathbf{a} = 0$ and $M = I$. With this $\zeta(\mathbf{u}) = \mathbf{u}^T\mathbf{u} = \|\mathbf{u}\|^2$. More generally, if it is a priori known that some linear combination $A\mathbf{u}$ is bounded, modellers can set $\mathbf{a} = 0$ and $M = A^TA$. With this $\zeta(\mathbf{u}) = \mathbf{u}^TA^TA\mathbf{u}$. Application of this idea with A being a finite difference matrix to limit the variability of $\mathbf{u}$ can be found in Dietrich and Chapman (1993).

Now it is well known that the minimisation (44) subject to the constraint (46) can be rewritten as the unconstrained minimisation

$$\min_{\mathbf{u}} \|\mathbf{y} - X\mathbf{u}\|^2 + \mu\zeta(\mathbf{u}) \qquad (47)$$

where the strictly positive weight μ can be determined from the constant c and properties of $\zeta(\mathbf{u})$. The interesting point in (47) is that the estimate $\hat{\mathbf{u}}$ is now solution to

$$(X^TC^{-1}X + \mu M)\hat{\mathbf{u}} = X^TC^{-1}\mathbf{y} + \mu M\mathbf{a} \qquad (48)$$

Thus inclusion of prior information in the form of the constraint (46) has changed $X^TC^{-1}X$ in (45) to $X^TC^{-1}X + \mu M$ in (48). Now for any symmetric matrices A and M, a standard result of linear algebra is that $\lambda_i(A + M) \geq \lambda_i(A) + \lambda_n(M)$ where $\lambda_i(A)_{i=1}^n$ and $\lambda_i(M)_{i=1}^n$ denote respectively the singular values of A and M both arranged in decreasing order (Barnett 1990). We thus have

$$\lambda_i(X^TC^{-1}X + \mu M) \geq \lambda_i(X^TC^{-1}X) + \mu\lambda_n(M)$$

As a consequence, inclusion of the prior information (46) into the inverse estimation problem has increased all singular values of $X^TC^{-1}X$ by an amount at least as large as $\mu\lambda_n(M)$. In view of the discussion of Section 6.5, this has the effect of reducing the ill-conditioning of input estimation from output data based on (19).

Note in passing that estimation of the optimum weight μ that trades off data explanation $\|\mathbf{y} - X\mathbf{u}\|$ with the sise of $\zeta(\mathbf{u})$ in the linear optimisation (47) can easily be computed by general cross-validation (Golub et al. 1979).

6.6.2 Application to Non-linear Inverse Problems

Stabilisation approaches such as those above mentioned can also be invoked to reduce the ill-conditioning associated with the non-linear estimation of a system parameter $\mathbf{k}$ from output information $\mathbf{q}$. For example, the cost function $J(\mathbf{u})$ given in (40) can be replaced by the stabilised cost function

$$J(\mathbf{u}) = \|\mathbf{q}^* - \bar{\mathbf{q}}(\mathbf{u})\|^2 + \mu\zeta(\mathbf{u}) \qquad (49)$$

so that the optimisation problem (41) now becomes

$$\min_{\mathbf{u}} \|\mathbf{q}^* - \bar{\mathbf{q}}(\mathbf{u})\|^2 + \mu\zeta(\mathbf{u})$$

However, because of the non-linear relationship between $\mathbf{k}$ and $\mathbf{q}$ it is generally difficult to optimally control the stabilisation process. Also it is generally not possible to derive a clean, closed form expression such as the left hand side of (48) to easily control the stabilising effect of bounding the size of the quadratic form $\zeta(\mathbf{u})$. This said, it is worth extending the discussion of Section 6.3.2 (Indirect approaches) by noticing that the gradient of the cost function $J(\mathbf{u})$ defined in (49) is

$$\nabla_{\mathbf{u}} J(\mathbf{u}) = - 2G_{\mathbf{u}}^T(\mathbf{q}^* - \bar{\mathbf{q}}(\mathbf{u})) + 2\mu M(\mathbf{u} - \mathbf{a})$$

Using (37), the above becomes

$$\nabla_{\mathbf{u}} J(\mathbf{u}) = - 2G_{\mathbf{u}}^T G_{\mathbf{u}}(\hat{\mathbf{u}} - \mathbf{u}) + 2\mu M(\mathbf{u} - \mathbf{a}) + O((\hat{\mathbf{u}} - \mathbf{u})^2)$$

or

$$\nabla_{\mathbf{u}} J(\mathbf{u}) = 2(G_{\mathbf{u}}^T G_{\mathbf{u}} + \mu M)\mathbf{u} - 2(G_{\mathbf{u}}^T G_{\mathbf{u}} \hat{\mathbf{u}} + \mu M \mathbf{a}) + O((\hat{\mathbf{u}} - \mathbf{u})^2) \tag{50}$$

The first term on the right hand side of (50) indicates that the deleterious effect of small singular values $G_{\mathbf{u}}^T G_{\mathbf{u}}$ on the independent $\mathbf{u}$ mentioned in the subsection on indirect approaches in Section 6.5.3 is now counteracted by the inclusion of the matrix μM with the effect of the smallest singular value of μM on $G_{\mathbf{u}}^T G_{\mathbf{u}}$ similar to that of μM on $X^T C^{-1} X$ on the left hand side of (48).

6.6.3   Smoothing Effect of Prior Information

It is worth concluding this section by indicating a link between injection of prior information, smoothing and expansion of the unknown in terms of basis functions. For this consider the inverse problem of input estimation stabilised by a constraint on the size of the input, i.e. with $M = I$ and $\mathbf{a} = 0$ in (46). However, here we assume no reduction in the dimension of the unknown. We also assume availability of output data at all n nodes. In other words $m = l = n$ with $B = L = I$ in (28) and (30). Furthermore, and for the sake of simplicity, the covariance matrix C is set to the identity I. In other words, we have $\mathbf{q}^* = S\mathbf{p} + \mathbf{e}$ as in (22) with the estimate $\hat{\mathbf{p}}$ solution to

$$\min_{\mathbf{p}} \|\mathbf{q}^* - S\mathbf{p}\|^2 + \mu \mathbf{p}^T \mathbf{p}$$

i.e.

$$(S^T S + \mu I)\hat{\mathbf{p}} = S^T \mathbf{q}^* \tag{51}$$

With the eigenvalue decomposition $S^T S = V\Lambda^{-2}V^T$ arising from (26), the linear system (51) becomes

$$V(\Lambda^{-2} + \mu I)V^T \hat{\mathbf{p}} = S^T \mathbf{q}^* \tag{52}$$

Consider now an expansion $\mathbf{p} = \Sigma_{i=1}^n u_i \mathbf{v}_i$ for the unknown $\mathbf{p}$ where $\mathbf{v}_i$ are the eigenvectors forming the columns of V. This expansion can be equivalently written $\mathbf{p} = V\mathbf{u}$ where $\mathbf{u}$ is now the unknown. Given the orthonormality of the matrices W and V in (26), the linear system (52) yields the estimate

$$\hat{\mathbf{u}} = (\Lambda^{-2} + \mu I)^{-1} \Lambda^{-1} W^T \mathbf{q}^* \tag{53}$$

The components r_i of the vector $\mathbf{r} = W^T \mathbf{q}^*$ are the coordinates of the data $\mathbf{q}^*$ in the eigenvector basis forming the columns of W. In other words, high frequency noise in the data $\mathbf{q}^*$ will result in non-zero components r_i even for the largest indices i. Now the diagonal matrix $(\Lambda^{-2} + \mu I)^{-1}\Lambda^{-1}$ in (53) has entries $\lambda_i/(1 + \mu\lambda_i^2)$ so that the entries $\hat{u}_i$ are given by

$$\hat{u}_i = \frac{\lambda_i}{(1 + \mu\lambda_i^2)} r_i \tag{54}$$

The relationship (54) encapsulates in a quintessential way the stabilising and smoothing effect of introducing prior information in an inverse problem. Without stabilisation, i.e. $\mu = 0$, we have $\hat{u}_i = \lambda_i r_i$. Since for n large λ_n is large and high frequency noise ensures r_n is generally non-zero, $\hat{u}_n = \lambda_n r_n$ will corrupt significantly the estimate $\hat{\mathbf{p}} = V\hat{\mathbf{u}}$. With inclusion of prior information, the deleterious effect of the larger singular values λ_i is filtered out by the denominator $(1 + \mu\lambda_i^2)$ in (54). This filtering effect also reduces the sise of the higher indices entries $\hat{u}_i$ associated with high frequency, high resolution components in the estimate $\hat{\mathbf{u}}$ implying that the estimate will contain only components associated with smooth, slowly varying eigenvectors. This carries through to the estimate $\hat{\mathbf{p}} = V\hat{\mathbf{u}}$.

In summary, prior information has smoothed out the input estimate and this without much concern about the choice of basis functions and reduced dimension m. The price to pay for this stabilisation method is a bias $1/(1 + \mu\lambda_i^2)$ introduced in all estimate components, including the smooth, low frequency ones.

Note that noise filtering such as that presented above leads to the concept of the essential dimension in an inverse problem. This concept first reported for linear inverse problems in Newsam (1982) and Newsam and Barakat (1985) provides guides on the optimum choice and number m of basis functions in which to expand an unknown. Note also that noise filtering along the lines presented above is extensively used in statistics through ridge regression estimators (Goldstein and Smith 1974; Hemmerle 1975; Marquardt and Snee 1975). Furthermore, from a time series analysis viewpoint, it can be shown that regularisation approaches such as that implemented via (51) are equivalent to recursive fixed interval smoothing (and other smoothing techniques) and that the smoothing (perhaps multi-dimensional) parameter such as μ can be optimised in a maximum likelihood sense (Young and Pedregal 1999).

The reader interested in detailed and rigorous discussions about stabilisation procedures for inverse problems is directed to the seminal work of Tikhonov and Arsenin (1977). On a more practical side, further discussion on and description of stabilisation methods can be found in Dietrich et al. (1993).

6.7 DETERMINISTIC VERSUS STOCHASTIC CALIBRATION

In this chapter, focus has so far been placed on calibration placed within a deterministic framework. In other words, system parameters, inputs and outputs were assumed deterministic with error vectors $\mathbf{e}$ being the only components having probability distributions. However, for environmental systems with components displaying a high degree of variability, there has been in the past two decades a strong interest in viewing spatially dependent parameters, inputs and/or outputs as realisation of random variates, or more generally random fields, and placing the calibration problem in a stochastic framework. For example, computation of fluid flows through porous media found in hydrogeology and petroleum engineering require estimates of the fluid conductivity parameter. Given that porous matrices such as those found in underground formations display random structures, one may view fluid conductivity as a spatial random field K with conductivity data being a realisation of K at discrete points of the flow domain. Now any system variable Q linked to K via some governing equation will inherit randomness from K. One may thus consider the joint field $Y = (Q, K)$ with probability distribution $f(q, k, \mathbf{a})$ where $\mathbf{a}$ is a vector parameter that needs to be estimated from all available system knowledge and data. The idea then is to view data on K and Q as random field samples and estimate $\mathbf{a}$ via some appropriate statistical inference technique such as maximum likelihood. With an estimate $\hat{\mathbf{a}}$ available, the unknown K can be taken as the expectation $\hat{K} = E[K] = \iint Kf(q, k, \hat{\mathbf{a}})dQdK$ with the variance $E[(K - \hat{K})^2]$ measuring confidence levels in estimates.

For inverse problems where Q is linked to K via integration, such stochastic approaches to calibration have many attractive features and this for the following reasons:

1. One can generally invoke distributions with the dimension of the parameter $\mathbf{a}$ small compared to that of the unknown K. In doing so, the dimensionality of the inverse problem is significantly reduced with ensuing stabilising benefits.
2. By setting the estimate $\hat{K}$ to be an expectation, one implicitly assumes that only smooth, slowly varying components in K can be reconstructed. This is in accordance with the fact that inverse problems are characterised by loss of information so that in the presence of noise, only low resolution components in the unknown can be reconstructed.

This said, the practical implementation of a stochastic approach to calibration is far from trivial. This is due to the fact that the choice of a probability distribution function (pdf) for K can be difficult to justify. Nevertheless, assuming some pdf for K to be available, it is generally not possible to derive the full joint distribution $f(q, k, \mathbf{a})$. An exception to this is when K is assumed to be normally distributed and the link between K and Q is linear. For this reason, stochastic approaches to calibration reported in the literature are generally based on normality assumption and rely on some linearisation of the governing equations. In this case, it can be shown that stochastic calibration results in procedures that are essentially those obtained from a deterministic framework (Dietrich and Newsam 1989). Examples of such stochastic inverse procedures can be found in aquifer identification with hydraulic conductivity estimated from piezometric head measurements and Gaussian conditional pdfs derived from a linearisation of the groundwater flow equation (Dagan 1985).

To conclude this section, it is worth mentioning some recent developments made possible by the increasing availability of powerful desk-top computers. Desk-top computers can now be used to generate quickly large numbers of random field realisations out of which distributional properties of system inputs in relation to system outputs can be investigated. Consider, for example, the estimation of the parameter $\mathbf{k}$ and input $\mathbf{p}$ from output data $\mathbf{q}^*$ based on (20), i.e. $\mathbf{q} = S(\mathbf{k})\mathbf{p}$. Set $\mathbf{y} = (\mathbf{k}, \mathbf{p})$ so that $\mathbf{q} = T(\mathbf{y})$ where T is a non-linear mapping. Consider now a large number of realisations $\{\mathbf{y}_m\}_{m=1}^M$ of $\mathbf{y}$ displaying relevant statistical properties. These could be to be uniformly distributed between acceptable bounds. Then the associated realisations $\mathbf{q}_m = T(\mathbf{y}_m)$ of $\mathbf{q}$ together with the available output data $\mathbf{q}^*$ can be used to select sub-sets of $\{\mathbf{q}_m\}_{m=1}^M$ having desired properties, out of which average and dispersion information about $\mathbf{y}$ can be derived.

Another computer-intensive stochastic approach is based on Bayesian analysis with both parameter K and data Q viewed as random variate with joint pdf $f(k, q)$. Briefly, this approach is structured as follows. Consider Bayes' theorem

$$f(k, q) = f(q|k)f(k) = f(k|q)f(q) \tag{55}$$

where $f(q|k)$ denotes the conditional distribution of q given k with $f(k) = \int f(k, q)dq$ and $f(q) = \int f(k, q)dk$ being the marginal pdfs. Bayes theorem provides a means to express the a posteriori distribution $f(k|q)$ of the parameter K given data on Q from the a priori pdfs $f(q|k), f(k)$ and $f(q)$. With $f(k|q)$ and data $\mathbf{q}^*$ available, the estimate $\hat{K}$ can be set to the conditional expectation $\hat{K} = E[K|\mathbf{q}^*] = \int kf(k|\mathbf{q}^*)dk$ with variance $E[(K - \hat{K})^2]$ measuring confidence levels in estimates.

A problem with direct use of Bayes' relationship (55) to evaluate $f(k|q)$ is that it requires the numerical evaluation of complicated integrals. As shown in the book by Sen and Stoffa (1995), this difficulty is avoided if the a posteriori distribution $f(k|q)$ is obtained numerically from Monte Carlo simulations invoking the Metropolis algorithm. For this reason, Metropolis-based approaches to estimate a posteriori pdfs for parameter inference have recently received wide attention.

An advantage of brute force approaches such as those mentioned above is that the difficult problem of deriving joint and/or conditional distributions is bypassed. The drawback is that little insight can be gained into the problem structure, as is the case when some mathematical analysis is carried out.

Finally, we note that for spatially lumped system the stochastic machinery of time series analysis can be invoked for both parameter estimation and system identification.

6.8   ILLUSTRATION

To illustrate our discussion, we use the steady-state one dimensional diffusion problem

$$-\frac{d}{dx}K\left(\frac{dq}{dx}\right) = p \qquad x \in \Omega = (0, 1) \tag{56}$$

with mixed prescribed and flow boundary conditions

$$q(0) = 0 \quad \text{and} \quad -K\frac{dq}{dx}\bigg|_{x=1} = h$$

The major issues discussed in this chapter can be illustrated with the above simple one-dimensional diffusion problem and readers interested in more realistic set-ups are directed to the water resources literature where a large number of case studies and other investigations associated with calibration of real environmental systems can be found (e.g. Cooley 1985; Carrera and Neuman 1986 for aquifer identification).

Let us start with the forward problem of calculating the output q. Although a Green's function of the form (15) that maps the input p to the output q could be invoked, the exact solution $q(x)$ to (56) is written as

$$q(x) = -\int_0^x \frac{P(\xi) + h - P(1)}{K(\xi)}d\xi$$

where

$$P(x) = \int_0^x p(\xi)d\xi$$

The ill-posedness of estimating the input p from output data q originates from the dependence of p on two differentiations of q. For example a high frequency perturbation $\delta q = \varepsilon \sin \omega x$ added to the data q magnifies into a large change $\delta p = \varepsilon \omega (K\omega \sin \omega x - K' \cos \omega x)$ where $'$ denotes derivatives with regards to x.

The ill-posedness of estimating the parameter K from output data q originates from the dependence of K on one differentiation of q as $K(x) = - P(x)/q'(x)$. For example let $1/K_1 = - q_1'/P$ and $1/K_2 = - q_2'/P$ with $q_2 = q_1 + \varepsilon \sin(\omega x)$ being a small perturbation of q_1 with $\varepsilon > 0$. Then $|1/K_1 - 1/K_2| = \varepsilon |\omega \cos(\omega x)/P| = \varepsilon |\omega/P|$ for all coordinates x_i such that $\omega x_i = i\pi$. Thus even if the output perturbation size ε is very small, the resulting perturbation $|1/K_1 - 1/K_2|$ can be very large at the points x_i if the frequency ω is large enough. Assuming there exists a constant $c > 0$ such that $c \le K(x)$ for all $x \in [0,1]$, we have $|1/K_1 - 1/K_2| = |K_1 - K_2|/|K_1 K_2| \le |K_1 - K_2|/c^2$. Thus the perturbation $|K_1 - K_2|$ in the parameter due to the perturbation in the output can be as large as $c^2 \varepsilon |\omega/P|$ at the points x_i. Note also, that $K(x)$ goes to infinity as the gradient q' goes to zero. Therefore, the sensitivity of K to errors in data q is significantly enhanced in regions of low output gradient, a fact well known in practice.

The numerical results below were obtained from Mathematica sessions (Wolfram 1991) and are available as active Mathematica notebooks to any interested reader.

6.8.1   Discretisation

To illustrate conditioning issues as they arise in practice, the problem (56) has to be discretised. This is done by discretising the ODE (56) over n nodes $\{x_i = i\delta x\}_{i=1}^n$ with $\delta x = 1/n$ with the following finite differences (FD) approximation to $D = - d/dx(Kd/dx)$ where for any variable $y(x)$, y_i is defined as $y_i = y(i\delta x)$

1. forward FD:

$$\frac{- k_i q_i + (k_i + k_{i+1})q_{i+1} - k_{i+1}q_{i+2}}{\delta x^2}$$

2. backward FD:

$$\frac{- k_{i-1}q_{i-2} + (k_{i-1} + k_i)q_{i-1} - k_i q_i}{\delta x^2}$$

3. central FD:

$$\frac{- k_{i-1}q_{i-2} + (k_{i-1} + k_{i+1})q_i - k_{i+1}q_{i+2}}{4\delta x^2}$$

The differential operator D is approximated by a forward FD on node 1, backward FD on node $n - 1$, and central FD for all nodes in between. At node n the flow boundary condition is approximated with a backward FD. With this we have the system $D(\mathbf{k})\mathbf{q} = \mathbf{p}$ with the vectors $\mathbf{k}$, $\mathbf{q}$, $\mathbf{p}$ having respectively entries k_i, q_i, p_i for $i = 1,\ldots,n$.

The input is set to $p(x) = 5 + 4x - 4x^2$ while the flow boundary constant is set to $h = 0.7$. To reduce the dimension of the parameter $\mathbf{k}$, we divide the interval $\Omega = [0,1]$ into five equal intervals and set $K(x) = u_1$ over the first interval $[0,0.2]$, $K(x) = u_2$ over the second interval $[0.2,0.4]$ etc. In other words, $K(x)$ is expanded in terms of a linear combination of the form (27) involving five piecewise constant basis functions. Thus $m = 5$ with reduced parameter here denoted $\mathbf{u}_K = (u_1, \ldots, u_5)$ with subscript K to distinguish it from the input parameter $\mathbf{u}_p$ used further below.

6.8.2   Matrix Conditioning

To illustrate the conditioning of the matrix D, Figure 6.6 plots the condition number $\kappa(D)$ as a function of n for $\mathbf{u}_K = (0.7, 1.2, 2, 3.5, 5)$ for case 1, $\mathbf{u}_K = (5, 3.5, 2, 1.2, 0.7)$ for case 2, and $\mathbf{u}_K = (5, 3, 2, 1, 2, 4)$ for case 3. Note the power law increase of $\kappa(D)$ as n increases, with $\kappa(D)$ in the million mark for $n = 150$.

The sensitivity of the matrix $G_{\mathbf{u}_K}^T G_{\mathbf{u}_K}$ appearing in the gradient $\nabla_{\mathbf{u}_K} J(\mathbf{u}_K)$ in (43) is illustrated in Table 6.1 where the values of $\kappa(\mathbf{u}_K^T G_{\mathbf{u}_K})$ for $n = 100$ and 25 output data, and changes to the first and last entries of $\mathbf{u}_K = (5, 3, 2, 1, 2, 4)$ are listed.

6.8.3   Estimation of the input p

We now illustrate the numerical instability that can originate from computing the input p from the output q via the direct and indirect approaches (32) and (33), respectively. For the sake of simplicity, the expansion for the true input $p(x) = 5 + 4x - 4x^2$ is chosen to be $p(x) = u_1 + u_2 x + u_3 x^2$. Thus the input parameter here denoted $\mathbf{u}_p$ is estimated *knowing exactly* its basis function functional form and number m. This unreasonably favourable situation is clearly never encountered in practice.

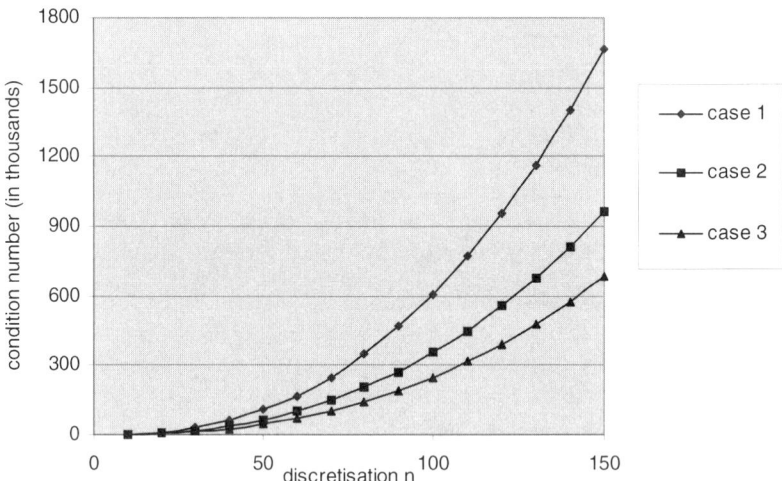

Figure 6.6   Condition number for three K parameter values and for discretisation sizes n varying from 10 to 150. The condition number values are expressed in thousands over five equispaced zones, $\mathbf{u}_K = (0.7, 1.2, 2, 3.5, 5)$ for case 1, $\mathbf{u}_K = (5, 3.5, 2, 1.2, 0.7)$, for case 2, and $\mathbf{u}_K = (5, 3, 2, 3, 4)$ for case 3

Table 6.1   For all $n = 100$ output data points, and for 25 equispaced output data points, condition number $\kappa(G_{uK}^T G_{uK})$ for a set of parameter values based on $\mathbf{u}_K = (4, 3, 2, 3, 5)$. Changes in $\mathbf{u}_K$ entries are made *only* to the first and last entry u_1 and u_5, respectively

First entry u_1	Last entry u_5	Condition number $\kappa(G_{uK}^T G_{uK})$ for 100 output data	Condition number $\kappa(G_{uK}^T G_{uK})$ for 25 output data
4	1	381	561
4	5	165 417	340 136
4	10	2 634 640	5 406 620
1	5	4 453 400	7 506 400
10	5	163 434	368 121

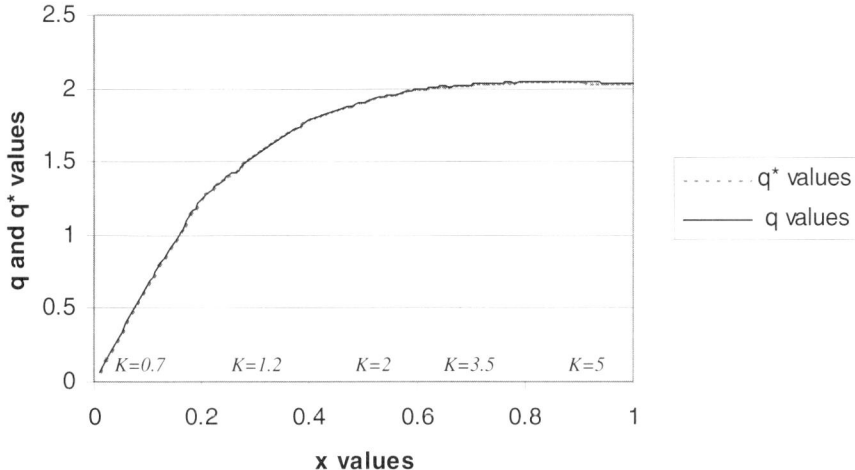

Figure 6.7   Plots of $\mathbf{q}$ and $\mathbf{q}^*$ values for the parameter K value piecewise constant and equal to 0.7 over the interval $[0, 0.2)$, 1.2 over the interval $[0.2, 0.4]$, . . ., 5 over the interval $[0.8, 0.1]$. The input is exactly known and set to $p(x) = 5 + 4x - 4x^2$. With the noise level noise σ in $\mathbf{q}^*$ set to the very low value 0.001, the computed data $\mathbf{q}$ and noisy data $\mathbf{q}^*$ on this plot are undistinguishable. Nevertheless, despite this low noise value, a direct approach to input estimate using *unsmoothed* $\mathbf{q}^*$ is $\hat{\mathbf{u}}_p = (8.3, -30, 41)$ with true value $\mathbf{u}_p = (5, 4, -4)$ indicating very strong sensitivity to errors (despite the low parameter dimension)

With the system parameter $\mathbf{u}_K$ set to $\mathbf{u}_K = (0.7, 1.2, 2, 3.5, 5)$, noisy data $\mathbf{q}^* = \mathbf{q} + \mathbf{e}$ was generated over the $n = 100$ discretisation nodes with the true output $\mathbf{q}$ solution to $D(\mathbf{u}_K)\mathbf{q} = B\mathbf{u}_p$ and noise $\mathbf{e}$ being $\mathcal{N}(0, \sigma^2 I)$. The entries in the exact output vector $\mathbf{q}$ are plotted in Figure 6.7 together with one realisation $\mathbf{q}^*$ with standard deviation (sdev) $\sigma = 0.001$. Although in this plot, $\mathbf{q}$ and $\mathbf{q}^*$ are undistinguishable, we shall see in the next section that use of the realisation $\mathbf{q}^*$ to estimate $\mathbf{u}_p$ via a direct approach leads to very large error magnifications.

Direct Approach

Given the above, (32) becomes

$$(B^T B)\hat{\mathbf{u}}_p = B^T D \mathbf{q}^o \qquad (57)$$

with the condition number $\kappa(B^T B)$ equal to 529. Note that the matrix D above is actually the discretisation of D without its last row. The last row was removed since it contains the flow boundary condition and so is not linked to the input $p(x) = 5 + 4x - 4x^2$.

Input estimates were first computed directly from the noisy data $\mathbf{q}^*$, i.e. without data smoothing. Average results of input estimation for 200 Gaussian noise realisations of $\mathbf{e}$ of varying sdev σ are in Table 6.2. In this table, estimates for $\mathbf{u}_p$ are the sample mean and sdev calculated over the sample of 200 noise realisations. Observe the high input estimate sdev even for noise levels that are very small relative to the output data values in the range 0 to 2, and this despite the favourable set-up of knowing exactly the form and number of basis functions.

Next the noisy data were smoothed using linear expansions $\Sigma_{n=0}^N s_n T_n^*(x)$ of shifted Chebyshev polynomials $T_n^*(x)$. The shifted Chebyshev polynomials are orthonormal over $[0, 1]$ and have useful properties in approximation theory (Abramowitz and Stegun 1970). Results for three Chebyshev expansions with $N = 4$, 6 and 8 are in Table 6.3. The absolute value of the largest approximation error between the true output $\mathbf{q}$ and the Chebyshev expansion at the discretisation nodes is 0.062 for $N = 4$, 0.046 for $N = 6$, and 0.036 for $N = 8$. In other words the approximations between true and smoothed data is good and certainly much better than can be achieved in practice. Despite this, observe the very poor accuracy of the input estimates. This is due to the fact that with K piecewise constant, the output surface has cusps at the points $x = 0.2, 0.4, 0.6, 0.8$ where K changes discontinuously (this feature is clearly evident in Figure

Table 6.2 Sample mean and standard deviation information for *direct approach* input estimates calculated over 200 output realisations $\mathbf{q}^*$. The true value of the unknown parameter is $\mathbf{u}_p = (5, 4, -4)$. As shown in Figure 6.7, the entries in the exact output data $\mathbf{q}$ varied between 0 and about 2. The values listed below are the realisation sample mean with the sample associated sdev in parentheses

Noise sdev σ	u_{p1}	u_{p2}	u_{p3}
10^{-5}	5 (0.04)	4 (0.31)	-4 (0.37)
10^{-4}	4.98 (0.45)	4.17 (3.4)	-4.23 (4.22)
10^{-3}	5.2 (4.3)	2.6 (32)	-2.5 (40)

Table 6.3 Stabilised *direct approach* input estimates based on one noisy data realisation $\mathbf{q}^*$. Stabilisation is provided by *smoothing noisy data* via expansions in terms of shifted Chebyshev polynomials of the first kind. The number N is the Chebyshev expansion order. The data realisation $\mathbf{q}^*$ were obtained with sdev $\sigma = 10^{-5}$. The true value of the unknown parameter is $\mathbf{u}_p = (5, 4, -4)$

N	Max. Chebyshev approx. error	u_{p1}	u_{p2}	u_{p3}
4	0.062	1.4 (0.0006)	49 (0.003)	-71 (0.003)
6	0.046	8.2 (0.39)	-18 (2.90)	24 (3.62)
8	0.036	8.6 (0.002)	-37 (0.02)	53 (0.03)

6.7). These cusps cannot be approximated with high accuracy unless the order N of the expansion is very large. This highlights the fact that to estimate the input p from the output p requires a good approximation to (the first) derivatives of q.

Although the table provides estimates based on a single realisation, the estimates remained essentially the same for other data realisations.

The shortcomings indicated above show that in practice, meaningful input estimation from output data via inversion of (56) is essentially not achievable from a direct approach.

Indirect Approach

Given the above setting, the indirect approach estimate (33) becomes

$$F\hat{\mathbf{u}}_p = B^T S^T L^T \mathbf{q}*$$

with $F = B^T S^T L^T LSB$. In this simulation exercise, the matrix L was first set to the identity, i.e. output data was assumed to be available at all discretisation nodes with thus $\mathbf{q}*$ of dimension $n = 100$. Next, L was set to sample over 25 equidistant locations in the interval $\Omega = [0, 1]$ with thus $\mathbf{q}*$ of dimension 25.

Average results of input estimation for 200 noise realisations of varying sdev are in Table 6.4. As was the case in Table 6.2, estimates for $\mathbf{u}_p$ in Table 6.4 are the sample mean and sdev calculated over the sample set. Observe that for the noise sdev $\sigma = 10^{-3}$ common to Table 6.2 and Table 6.4, estimation results are significantly better for the indirect approach estimates, even when only 25 output data are taken.

Stabilisation was implemented with $\xi(\mathbf{u}_p)$ in (47) set to $\xi(\mathbf{u}_p) = \mu \mathbf{u}_p^T \mathbf{u}_p$. Results are in Table 6.5. Note the decrease in variance achieved at the cost of introducing biases in input parameter estimates.

Table 6.4 Sample mean and standard deviation information for *indirect approach* input estimates calculated over 200 output realisations $\mathbf{q}*$. The simulation set-up is the same as that of Table 6.2 except that here results are based on 100 and 25 noisy output measurements. Note that output noise sdev values here are larger than in Table 6.2

Data #	Noise sdev σ	u_{p1}	u_{p2}	u_{p3}
100	10^{-3}	5.01 (0.03)	3.97 (0.17)	$-$ 3.98 (0.16)
100	10^{-2}	4.97 (0.36)	4.22 (1.75)	$-$ 4.25 (1.65)
25	10^{-3}	5.00 (0.07)	3.96 (0.32)	$-$ 3.96 (0.30)
25	10^{-2}	5.01 (0.67)	3.90 (3.20)	$-$ 3.80 (3.00)

Table 6.5 Stabilised *indirect approach* input estimates calculated over 200 output realisations $\mathbf{q}*$. Stabilisation was implemented by inclusion of the quadratic form $\mu \mathbf{u}_p^T \mathbf{u}_p$. The output noise sdev was fixed to $\sigma = 10^{-2}$ and the number of data points was 25. Stabilised estimates have thus to be compared with the last row of Table 6.4

μ	u_{p1}	u_{p2}	u_{p3}
10^{-6}	5.11 (0.58)	3.41 (2.8)	$-$ 3.43 (2.6)
10^{-5}	5.48 (0.23)	3.96 (1)	$-$ 1.63 (0.99)

6.8.4   Estimation of the parameter *K*

The impact of the ill-posed nature of estimating the parameter K from noisy output data $\mathbf{q}^*$ via an indirect approach such as (41) can be illustrated through structures of the associated gradient $\nabla_{\mathbf{u}_K} J(\mathbf{u}_K)$ given in (42). As indicated earlier, ill-conditioning will be reflected in the objective function having a very flat surface. Such flatness will be particularly severe in the direction of the eigenvectors of the Jacobian $G_{\mathbf{u}_K}$ associated with the smallest singular values. In our case, the singular values of $G_{\mathbf{u}_K}$ for $\mathbf{u}_K = (5, 3, 2, 1, 2, 4)$ and for 100 and 25 equispaced output data are respectively $(1.05, 0.29, 0.13, 0.054, 0.0026)$ and $(0.51, 0.15, 0.066, 0.026, 0.00088)$. This indicates that even when output data are available at each of the $n = 100$ discretisation node, the smallest singular value 0.0026 will cause the objective function $J(\mathbf{u}_K)$ to be very flat in the direction of the associated eigenvector.

6.9   CONCLUSIONS

Issues associated with the calibration of environmental systems have been discussed and illustrated. The salient points of the discussion are summarised as follows.

1. A major cause of ill-conditioning in inverse problems is due to the dependence of the solution on derivatives of the data. Such dependence is not continuous and so can cause error magnifications that corrupt estimates.
2. The dependence of the solution on derivatives of the data can be reinterpreted as the data being an indirect measurement of the unknown, with the instrument being an integrator. This indirect measurement perspective indicates that in the presence of data noise, high frequency information (or resolution) in the unknown is lost when passed to the data via integration. This is due to the smoothing property of integration that attenuates high frequency signals below noise levels. As a consequence, when solving an inverse problem, modellers should only reconstruct smooth, slowly varying components in the unknown.
3. Inverse problems can be solved in two ways: (i) by solving directly the governing equation for the unknown, using a data surface reconstructed smoothly over the whole simulation domain (*direct approach*), or (ii) by constructing an estimate that provides a best match between computed and historical data (*indirect approach*). While more demanding in set up and computer resources, indirect approaches have generally far better properties than direct approaches and should therefore be preferred in practice.
4. Stabilisation can be achieved in various ways. One method is to simply reduce the dimension of the unknown via linear combinations of m appropriate, smooth, basis functions. In this case, an important issue is to choose the functional form of the basis functions and the expansion size m so that a maximum amount of information about the unknown is reconstructed. Indeed, choosing basis functions that are too smooth or a value for m that is too small, may not do justice to the information content present in the data. Conversely, choosing high resolution basis functions or a value for m that is too large will cause estimates to be corrupted by error magnifications. For linear problems, Newsam (1982) showed that optimum basis functions and dimension m can be derived from a singular value decomposition of the linear operator at hand. However, for large discretisation sizes, singular value decompositions are computationally prohibitive. For non-linear inverse problems, a generic approach to derive optimum basis functions and dimension m is not available, as such optima are strongly problem dependent.
5. Stabilisation can also be achieved by injecting prior information about the unknown in the

form of constraints on its size, variability and/or distance to available data. Such constraints can in general be expressed as upper bounds on some quadratic forms and the resulting effect is to filter out problem sensitivity to data noise. This can be achieved without the need to reduce a priori the dimension of the unknown and hence without major concern about optimum choice of basis functions and dimension m.

6. Inverse problems can be solved in a stochastic framework where data and unknowns are viewed as random variates, or more generally random fields. The issue then becomes that of deriving probability density functions (pdfs) and computing (conditional) expectations. Since derivations of pdfs can be difficult, Monte Carlo simulations can be invoked to derive a posteriori pdfs numerically. Such approaches are receiving increased attention from the modelling community.

This chapter concludes with a word about validation. In any modelling exercise, validating results is an essential task. In an inverse problem, the following questions therefore arise: can a parameter and/or an input estimated from output data be validated? If the answer is yes, does the estimate need to be validated?

At first sight an answer to these questions is not obvious. The reason is that if an inverse problem is ill-conditioned because the unknown is insensitive to the data, then the unknown is not really relevant to the forward problem and estimate errors will not affect forward simulations in any significant way. Given that models are generally built to perform forward simulations, it may be concluded that validating parameters and/or input estimates is a minor task compared to validating predictions based on forward simulations. However, closer inspection of this issue shows the following: global insensitivity of a unknown to data is linked to non-uniqueness. At the limit, if the unknown is totally insensitive to the data then it is totally redundant and any estimate is as good as any other. On the other hand, partial insensitivity of the unknown to data is generally linked to non-continuous dependence between both. When the insensitivity is only partial, validating estimates of the components (such as low frequency components) sensitive to the data is a useful exercise, as it ensures that some true knowledge about the unknown which is relevant to the forward problem, has been recovered. In this light, prior to embarking on any calibration exercise it is essential to do some analysis to determine the structure and source of potential ill-conditioning.

APPENDIX

The solution to (17) in terms of K is best derived using the method of characteristics. For this let Γ denote the family of 3-D curves $\mathbf{x}(t) = (x_1(t), x_2(t), x_3(t))$ parameterised by t and solution to the characteristics ODEs

$$\frac{d\mathbf{x}(t)}{dt} = \nabla q \tag{A.1}$$

With this, over any characteristic curve $\mathbf{x}(t)$ in Γ, (17) becomes the ODE

$$\frac{dK}{dt} + (\nabla^2 q)K = -p \tag{A.2}$$

where it has to be understood that all variables in (A.2) are taken over $\mathbf{x}(t)$ i.e. at points $(x_1(t), x_2(t), x_3(t))$ solution to the characteristics ODEs (A.1). Multiplying (A.2) by the integrating factor

$$\psi(t) = e^{\int_0^t \nabla^2 q(\mathbf{x}(\tau))d\tau} \tag{A.3}$$

and integrating over $[0, t]$ on both sides yields the following solution to (A.2) (Zill and Cullen 1997)

$$K(t) = K(0)e^{-\int_0^t \nabla^2 q(\mathbf{x}(\tau))d\tau} - \int_0^t e^{-\int_\xi^t \nabla^2 q(\mathbf{x}(\tau))d\tau} p(\mathbf{x}(\xi))d\xi \qquad (A.4)$$

Observe that (A.4) has the same form as the solution (3) to the first ODE (2).

REFERENCES

Abramowitz, M. and Stegun, I.A. 1970. *Handbook of Mathematical Functions*. Dover, New York.

Anderssen, R. S. 1980. On the use of linear functionals for Abel-type integral equations in applications. In: Anderssen, R.S., de Hoog, F.R. and Lukas, M.A. (eds), *The Application and Numerical Solution of Integral Equations*. Alphen aan den Rijn, Netherlands and Sijthoff and Noordhoff, Rockville, MD, USA.

Barker, H.A. and Young, P. 1985. *Identification and System Parameter Estimation*. IFAC/IFORS Symposium on Identification and System Parameter Estimation (7th: York, UK). Published for the International Federation of Automatic Control by Pergamon Press.

Barnett, S. 1990. *Matrices, Methods and Applications*. Oxford Applied Mathematics and Computing Sciences Series. Clarendon Press, Oxford.

Bear, J. 1972. *Dynamics of Fluids in Porous Media*. American Elsevier, New York.

Box, G.E.P. and Jenkins, G.M. 1970. *Time Series Analysis, Forecasting and Control*, Holden Day, San Francisco.

Carrera, J. and Neuman, S. 1986. Estimation of aquifer parameters under transient and steady state conditions: No. 1, 2, 3. *Water Resources Research*, **22**(2), 199–242.

Chapman, T.G. 1985. Continuous convolution with hydrologic data. *Water Resources Research*, **21**(6), 847–852.

Cooley, R.L. 1985. A comparison of several methods for solving nonlinear regression groundwater flow problems. *Water Resources Research*, **21**(10), 1525–1538.

Craven, P. and Wahba, G. 1979. Smoothing noisy data with spline functions. *Numerische Mathematik*, **31**, 377–403.

Dagan, G. 1985. Stochastic modelling of groundwater flow by unconditional and conditional probabilities: the inverse problem. *Water Resources Research*, **21**(1), 65–72.

Dietrich, C.R. and Chapman, T.G. 1993. Unitgraph estimation and stabilisation using quadratic programming and difference norms. *Water Resources Research*, **29**(8), 2629–2635.

Dietrich, C.R. and Newsam, G.N. 1989. A stability analysis of the geostatistical approach to aquifer identification. *Stochastic Hydrology and Hydraulics*, **4**(3), 293–316.

Dietrich, C.R., Norton, J.P. and Jakeman, A.J. 1993. Ill-conditioning in environmental-system modelling. In: Jakeman, A., Beck, M. and McAleer, M. (eds), *Modelling Change in Environmental Systems*. John Wiley, Chichester, 37–57.

Dietrich, C.R., Green, T.R. and Jakeman, A.J. 1998. An analytical model for stream sediment transport: application to Murray and Murrumbidgee River reaches, Australia. *Hydrological Processes*, **13**(5), 763–776.

Elliott, D. 1980. Integral equations – ninety years on. In: Anderssen, R.S., de Hoog, F.R. and Lukas, M.A. (eds), *The Application and Numerical Solution of Integral Equations*. Alphen aan den Rijn, Netherlands and Sijthoff and Noordhoff, Rockville, MD, USA.

Goldstein, M. and Smith, A.F.M. 1974. Ridge type estimators for regression analysis. *Journal of the Royal Society*, **B36**, 284–291.

Golub, G.H. and Van Loan, C.F. 1989. *Matrix Computation*. Johns Hopkins University Press, Baltimore, MD.

Golub, G.H., Heath, M. and Wahba, G. 1979. Generalized cross-validation as a method for choosing a good ridge parameter. *Technometrics*, **21**(2), 215–223.

Griffel, D. H. 1981. *Applied Functional Analysis*. John Wiley, New York, 386.

Hadamard, J. 1932. *Le problème de Cauchy el les équations aux dérivées partielles linéaires hyperboliques.* Hermann, Paris.

Hemmerle, W.J. 1975. An explicit solution for generalized ridge regression. *Technometrics*, **17**, 309–313.

Hutchinson, M.F. 1984. *A Summary of Some Surface Fitting and Contouring Programs for Noisy Data.* Consulting Report No. ACT 84/6, CSIRO, Division of Mathematics and Statistics and Division of Water and Land Resources, Canberra.

Jakeman, A.J. and Hornberger, G.M. 1993. How much complexity is warranted in a rainfall–runoff model? *Water Resources Research*, **29**(8), 2637–2649.

Landau, L.D. and Lifshitz, E.M. 1989. *Fluid Mechanics.* Pergamon Press, New York.

Leopold, L.B. and Maddock, T. 1953. The hydraulic geometry of stream channels and some physiographic implications, *US Geological Survey Professional Paper* 252.

Marquardt, D.W. and Snee, R.D. 1975. Ridge regression in practice. *The American Statistician*, **29**, 3–20.

McLaughlin, D. and Townley, L.R. 1996. A reassessment of the groundwater problem. *Water Resources Research*, **32**(5), 1131–1161.

Newsam, G.N. 1982. Numerical reconstruction of partially known transforms. PhD thesis, Harvard University, Cambridge, MA.

Newsam, G. N. and Barakat, R. Essential dimension as a well-defined number of degrees of freedom of finite convolution operators appearing in optics. *Journal of the Optical Society of America*, **2**(11), 2040–2045.

Sen, M. and Stoffa, P.L. 1995. *Global Optimization in Geophysical Inversion.* Elsevier, Amsterdam.

Shannon, C. E. and Weaver, W. 1962. *The Mathematical Theory of Communication.* University of Illinois Press, Urbana.

Stakgold, I. 1967. *Boundary Value Problems of Mathematical Physics*, Vol. 1. Macmillan, New York.

Stakgold, I. 1968. *Boundary Value Problems of Mathematical Physics*, Vol. 2. Macmillan, New York.

Stepien, I. 1984. On the numerical solution of the Saint Venant equations. *Journal of Hydrology*, **67**, 1–11.

Tikhonov, A.N. and Arsenin, V.A. 1977. *Solutions of Ill-posed Problems.* V.H. Winston & Sons, Washington, DC.

Wolfram, S. 1991. *Mathematica: A System for Doing Mathematics by Computer.* Addison-Wesley, Redwood City, CA.

Yeh, W.W.H. 1986. Review of parameter identification procedures in groundwater hydrology: the inverse problem. *Water Resources Research*, **22**(2), 95–108.

Young, P.C. 1978. A general theory of modeling for badly defined systems. In: Vansteenkiste, G.C. (ed.), *Modelling Identification and Control in Environmental Systems.* Elsevier, Amsterdam.

Young, P. C. 1984. *Recursive Estimation and Times-Series Analysis.* Springer-Verlag, New York.

Young, P.C. and Pedregal, D. 1999. Recursive and en-block approaches to signal extraction. *Journal of Applied Statistics*, **26**(1), 103–128.

Zill, D.G. and Cullen, M.R. 1997. *Differential Equations with Boundary-Value Problems.* Brooks/Cole Publishing Company, NY, USA.

7 Data-Based Mechanistic Modelling and Validation of Rainfall–Flow Processes

PETER YOUNG[1]

Institute of Environmental and Natural Sciences, Lancaster University, UK

7.1 INTRODUCTION

One of the most interesting modelling problems in hydrology is the characterisation of the nonlinear dynamic relationship between rainfall and runoff. This has received considerable attention over the past 30 years, with mathematical and computer-based models ranging from simple 'black-box' representations to complex, physically based catchment models. Wheater et al. (1993) have categorised such models into the following four, broad types.

- *Metric models*, which are based primarily on observational data and seek to characterise the flow response largely on the basis of these data, using some form of statistical estimation or optimisation (e.g. Young 1986). These include purely black-box, time series models, such as the discrete-time transfer function or the neural network representations. Often, such models derive from, or can be related to, the earlier unit hydrograph theory, but this is not always recognised overtly.
- *Conceptual models*, which vary considerably in complexity but are normally based on the representation of internal storages, as in the original Stanford Watershed Model of the 1960s (Crawford and Linsley 1966), although the hypothesis of catchment-scale response is sometimes included, as in TOPMODEL (Beven and Kirkby 1979). The essential feature of all these models, however, is that the model structure is specified a priori, based on the hydrologist/modeller's perception of the relative importance of the component processes at work in the catchment, and then an attempt is made to 'optimise' the model parameters in some manner by 'calibration' against the available rainfall and flow data.
- *Physics-Based models*, in which the component processes within the models are represented in a more classical, mathematical–physics form, based on continuum mechanics solved in an approximate manner via finite difference or finite element spatio-temporal discretisation methods. A well-known example is the Système Hydrologique Européen (SHE) model (e.g. Abbott et al. 1986). The main problems with such models, which they share to some degree with the larger conceptual models, are two-fold: first, the inability to measure soil physical properties at the scale of the discretisation unit, particularly in relation to subsurface processes; and, second, their complexity and consequent high dimensional parameterisation. This latter problem makes objective optimisation and calibration virtually impossible, since the

[1] Also, Adjoint Professor, Centre for Resource and Environmental Studies, Australian National University, Canberra, Australia.

Model Validation: Perspectives in Hydrological Science. Edited by M.G. Anderson and P.D. Bates.
© 2001 John Wiley & Sons, Ltd.

model is normally so over-parameterised that the parameter values cannot be uniquely identified and estimated against the available data.

* *Hybrid Metric–Conceptual* (HMC) models, in which (normally quite simple) conceptual models are identified and estimated against the available data to test hypotheses about the structure of *catchment-scale* hydrological storages and processes. In a very real sense, these models are an attempt to combine the ability of *Metric models* to efficiently characterise the observational data in statistical terms (the principle of *parsimony*: Box and Jenkins 1970), with the advantages of *Conceptual models* that have a prescribed physical interpretation within the current scientific paradigm.

The models in the two middle categories, above, are often characterised by a large number of unknown parameters that need to be estimated ('optimised' or 'calibrated') in some manner against the observational rainfall-flow time series. Because the number of parameters is normally very large in relation to the information content of the data, however, such models are often *over-parameterised* and not normally *identifiable*, in the sense that it is impossible to estimate their parameters uniquely without imposing prior restrictions on a large subset of the parameter values prior to estimation (see e.g. Young et al. 1996). The author and his co-workers have addressed these problems of over-parameterisation and poor identifiability associated with large environmental models many times over the past quarter century (e.g. Young 1978, 1983, 1992, 1993, 1998a,b, 1999a; Young and Minchin 1991; Young and Lees 1993; Young and Beven 1994; Young et al. 1996; Shackley et al. 1998). And recently, Keith Beven and his co-workers (e.g. Franks et al. 1997; Beven, this volume, chapter 4) have used the term *equifinality* to describe the consequences of such over-parameterisation and non-identifiability: namely the existence of many different parameterisations and model structures that are all able to explain the observed data equally well, so that no unique representation of the data can be obtained within the prescribed model set.

There appear to be two main reasons for these identifiability problems. First, any limitations of the observational data can be important, since the available time series may not be sufficiently informative to allow for the estimation of a uniquely identifiable model form. In particular, the inputs to a system may not be *sufficiently exciting* (see e.g. Young 1984), in the sense that they do not perturb the system sufficiently to allow for unambiguous estimation of all the model parameters within an otherwise identifiable model structure. Second, even if the input does sufficiently excite the system, there are usually only a limited number of dynamic modes – the *dominant modes* of the system – that are excited to any significant extent, and the observed output of the system is dominated by their cumulative effect. The importance of this dominant mode concept in model identification and estimation is illustrated by the example in Appendix 1, which shows how the response of a 26th order simulation model can be duplicated with exceptional accuracy (0.001% error by variance) by a much simpler seventh order dominant mode model. This is typical of most high order linear systems and appears to carry over to nonlinear systems. For example, Young et al. (1996) and Young (1998b) have used similar analysis to show how the response of high dimensional, nonlinear global carbon cycle simulation models are accurately reproduced by differential equation models of much reduced order. This is also reflected in other recent work on the simplification of global climate models (Hasselman et al. 1997; Hasselman 1998).

In the above references, the author has stressed that dominant modal behaviour is a generic property of dynamic systems (see Young et al. 1996 and the prior references therein) and that it is probably the main reason for the limitation on the number of clearly identifiable parameters that can be estimated from observational data. The largest order identifiable system that the author has encountered from the analysis of real time series data, over the past 40 years, is a tenth order differential equation model for the vibrations in a cantilever beam (Young 1996, 1998a), where the very low damping of the system results in four dominant complex modes and the resulting model explains 99.74% of the experimental data. However, the identifiable order is normally much lower than this and many previous rainfall–runoff modelling studies (e.g. Kirkby 1976; Hornberger et al. 1985; Jakeman and Hornberger 1993; Young 1993, 1998b; Young and Beven 1994; Young et al. 1997, 1998; Ye et al. 1998) suggest that a typical set of rainfall–runoff observations contain only sufficient information to estimate up to a maximum of less than ten parameters within simple, nonlinear dynamic models of *dynamic order* three or less. In the rainfall–runoff example discussed later, for instance, there is clear evidence in the data of only three dominant modes between the effective rainfall input and the flow response (as described by a second order transfer function model with only five parameters): an instantaneous component, in which rainfall affects flow within the daily sampling interval; a 'quick' mode with a time constant of 1.02 days; and a 'slow' mode, with a time constant of 12.2 days. Other, less significant aspects of the flow response are a seasonal evapotranspiration component, modelled as the response of a first order transfer function to temperature variations, and a constant component. As a result, the total number of parameters, including those associated with the nonlinear effective rainfall generation, is only nine.

Given the above considerations, one has to seriously question whether it is possible to validate the large Conceptual and Physics-based models in any rigorous statistical sense, since non-uniqueness is endemic in such models and their over-parameterisation presents formidable obstacles to the application of the available statistical modelling and numerical optimisation procedures. Of course, this does not mean that model optimisation and limited forms of validation are impossible: there are many examples where large models have been successfully constructed and used to good effect in a variety of applications, from research studies to environmental management and planning. But it does mean that unconstrained, holistic optimisation of the model parameters in such large models is virtually impossible. It also limits the more rigorous application of statistical diagnostic and testing procedures, of the kind discussed in this chapter.

Both the Metric and HMC approaches avoid these 'large model' problems to some extent but the practical utility of the basic Metric model is limited by its lack of any clearly defined physical interpretation. In this chapter, therefore, attention is limited to the identification, estimation and validation of HMC-type models, where the problems of identifiability and optimisation are minimised (although, as we shall see, not eliminated) and the questions of validity can at least be approached in a statistically meaningful manner. Within the category of HMC models, however, two main approaches to modelling can be discerned; approaches which, not surprisingly, can be related to the more general *deductive* and *inductive*[1] approaches to scientific inference that have been identified by philosophers of science from Francis Bacon (1620), to Karl Popper (1959) and Thomas Kuhn (1962).

In the first hypothetico-deductive approach, the *a priori* conceptual model structure is effectively a theory of hydrological behaviour. This is normally based on the perception of the hydrologist/modeller and is strongly conditioned by assumptions that derive from current hydrological paradigms, based on past research and experience (e.g. the IHACRES model of Jakeman et al. 1990). The alternative *Data-Based Mechanistic* (DBM) approach is basically inductive, in the sense that it tries to avoid theoretical preconceptions as much as possible in the initial stages of the analysis. In particular, the model structure is not pre-specified by the

---

[1] The term 'induction' is open to different interpretations and inductive methods have, for example, tended to dominate many areas of social science. Here, we utilise the term more narrowly, within the context of statistical inference, in order to differentiate the DBM approach from the related but more conventional hypothetico-deductive approaches to HMC modelling.

modeller but, wherever possible, inferred directly from the observational data in relation to a more general class of models. Only then is the model interpreted in a physically meaningful manner, most often (but not always) within the context of the current hydrological paradigm: e.g. the models of rainfall-flow data in Young (1993, 1998b), Young and Beven (1994), Young et al. (1997, 1998) and Young and Tomlin (2000).

This physical interpretation is an essential element in all DBM modelling: no matter how well the DBM model explains the data, it is only considered truly credible if it can be interpreted in physically meaningful (albeit not always conventional – see below) terms. In this, the DBM approach harks back to the father of modern statistical inference, Karl Friedrich Gauss, who held that no hypothesis was satisfactory which rested on a formula and was not also a consequence of physical conjecture. For this reason, Gauss abandoned his work on the attraction between charged particles because he was unable to find a plausible physical interpretation of the formula he had obtained for the relationship between the relative motion and position of two particles. Of course, this lack of credibility does not preclude the purely data-based model from being used for certain applications, such as forecasting and control, where the physical interpretation may be relatively unimportant and where the success of 'black-box' models is well documented.

Since the DBM approach is inductive, it is not wedded as strongly to the current paradigms as the hypothetico-deductive approach: indeed, its intention is always to respect these paradigms but not allow them to dictate the structure of models (here rainfall–flow) if the data suggest otherwise. In other words and to use a Kuhnian (Kuhn 1962) interpretation of science, the DBM approach encourages the continual questioning of current paradigms and rejoices in its ability to promote paradigm change *if this is supported by observational data*. One example of this is the development of the *Aggregated Dead Zone* (ADZ) model for solute transport in rivers (e.g. Beer and Young 1983; Wallis et al. 1989; Young and Wallis 1994). Another is recent research on modelling and forecasting the relationship between government spending, private capital investment and unemployment in the USA during the last half century (Young and Pedregal 1998, 1999).

Another important aspect of HMC approaches to rainfall–flow modelling relates to the *objectives* of the modelling exercise. In the author's opinion, the search for a single, all encompassing model of any dynamic system is futile. Rather, the model-builder should be seeking *a model that suits the nature of the study objectives*. Of course, this objective orientation does not have to be precisely defined, since a model can simultaneously serve more than one purpose. But even more loosely defined objectives need to be considered carefully before the modelling exercise begins. In this sense, we will see that the IHACRES and DBM models of rainfall–flow processes, for instance, appear to have overlapping but not identical objectives. And these objectives affect, and have to be taken into account when evaluating, the overall utility and 'validity' of the models.

On the basis of the above observations, the present chapter discusses the IHACRES and DBM approaches to rainfall–flow modelling within the context of daily rainfall, flow and temperature time series from the well-known archive of data from the Coweeta Catchment (as collected over many years by scientists at the US Forest Service). Prior to this, however, some background to the statistical methods utilised in the chapter is presented. In particular, the basic aspects of statistical identification, estimation and validation are outlined and the DBM approach is described briefly. The historical development of the IHACRES and DBM rainfall–flow models and their interrelationships are then discussed briefly, preparatory to the presentation of the Coweeta example. Four HMC models are considered in this example: the standard DBM model; a modified version of the IHACRES model which is appropriate to the Coweeta data; the Bedford-Ouse model which was the progenitor of the IHACRES model;

and a 'simulation' version of the DBM model. All of these models are optimised using a similar numerical optimisation scheme based on a 2000 sample set of daily data (see Appendix 2). Following optimisation, the models are subjected to standard statistical diagnostic tests, prior to predictive validation on separate sets of data from the same catchment.

This detailed modelling analysis of the Coweeta data helps to highlight certain important questions about the wider definition of model validation within the HMC context; questions that relate back to the different approaches and objectives of modelling discussed above, and are considered briefly at the end of the chapter. It is concluded that none of the HMC models can be considered well validated at this time, although they can be deemed valid in the 'conditional' sense defined in this chapter (i.e. they have not been falsified by the analysis here: which is arguably all that can be achieved by models anyway). In particular, the nature of the nonlinearity in the models is not sufficiently well defined and requires more detailed scientific study. Despite this, all the models appear to have advantages over alternative rainfall–flow models and are undoubtedly useful in both scientific and predictive terms.

7.2 STATISTICAL IDENTIFICATION, ESTIMATION AND VALIDATION

The statistical approach to modelling assumes that the model is stochastic: in other words, no matter how good the model and how low the noise on the observational data happens to be, a certain level of uncertainty will remain after modelling has been completed. Consequently, full stochastic modelling requires that this uncertainty, which is associated with both the model parameters and the stochastic inputs, should be quantified in some manner as an inherent part of the modelling analysis. This statistical approach involves three main stages, as shown in Figure 7.1: (i) *identification* of an appropriate, identifiable model structure; (ii) *estimation* (optimisation, calibration) of the parameters that characterise this structure, using some form of estimation or optimisation; and (iii) predictive validation of the model on data sets different to those used in the model identification and estimation. In this section, we consider these three stages in order to set the scene for the later analysis. This discussion is intentionally brief, however, since the topic is so large that a comprehensive review is not possible in the present context.

7.2.1 Structure and Order Identification

In the hypothetico-deductive approach to model-building, the model constitutes a hypothesis or 'theory of behaviour' and it is normally selected beforehand, based on the current scientific paradigm. However, the subsequent processes of model estimation and validation are often considered as exercises in Popperian 'falsification' (Popper 1959) and so the initial perceived model structure may well be modified in the light of these. In the DBM approach, this questioning of the hypothetical model is more overt and the identification stage is considered as a most important and essential prelude to the later stages of model building. Nevertheless, in the case of HMC models, both approaches make use of statistical identification procedures to some extent. These usually involve the identification of the most appropriate *model order*, as defined in dynamic system terms, although the *model structure* itself can be the subject of the analysis if this is also considered to be ill-defined. In the DBM approach, for instance, the nature of linearity and nonlinearity in the model is not assumed a priori (unless there are good reasons for such assumptions based on previous data-based modelling studies) but is identified from the data using non-parametric and parametric statistical estimation methods. Once a suitable model structure has been defined, there are a variety of statistical methods for identifying model order,

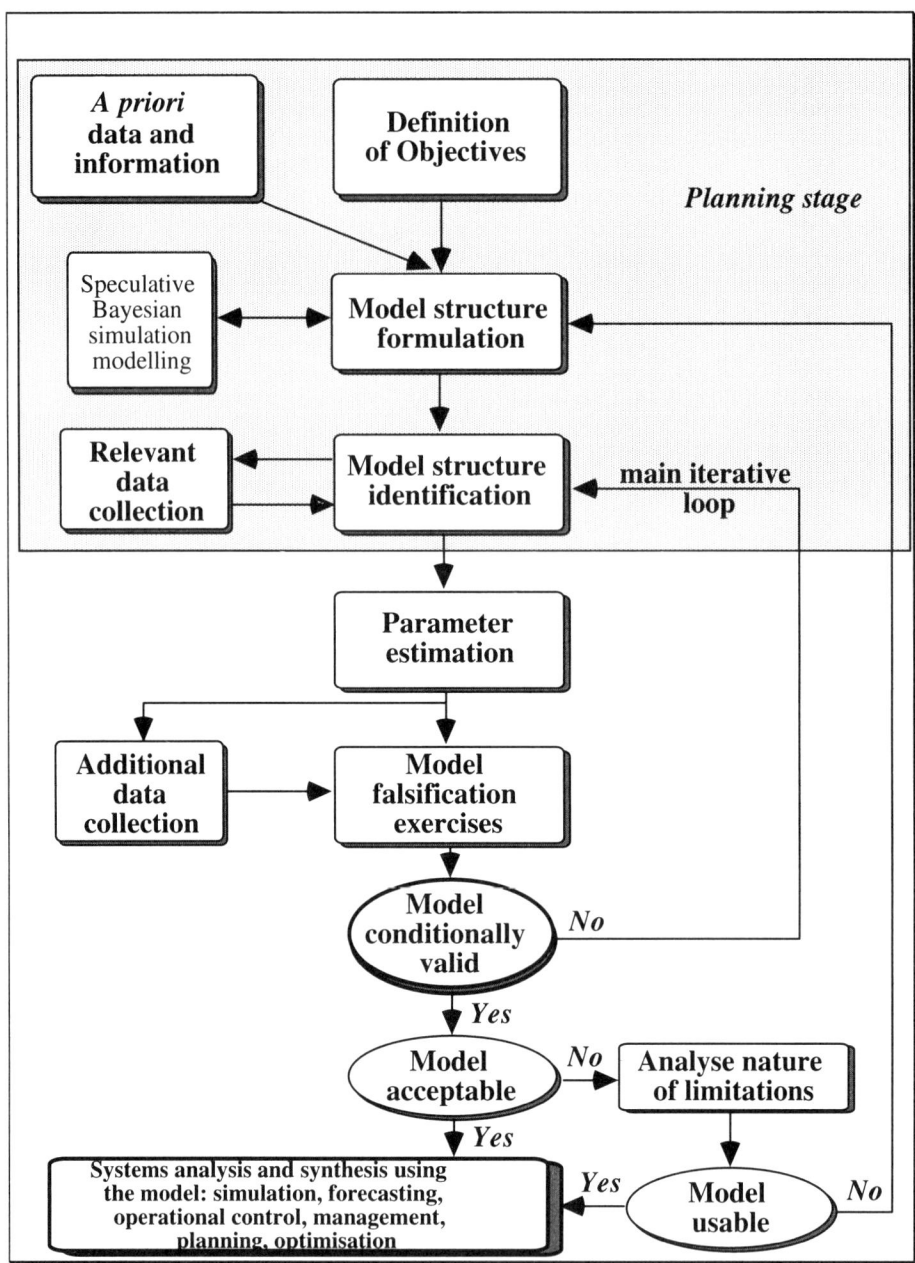

Figure 7.1 Stochastic dynamic modelling: a general statistical procedure for identification, estimation and conditional validation

some of which are discussed later in connection with the Coweeta example (see Appendix 3). In general, however, they exploit some *order identification statistics*, such as the correlation-based statistics popularised by Box and Jenkins (1970); the well-known Akaike Information Criterion (AIC: Akaike 1974) used for purely stochastic processes; and the YIC statistic defined in Appendix 3 for transfer function models.

7.2.2 Estimation (Optimisation)

Once the model structure and order have been identified, the parameters that characterise this structure need to be estimated in some manner. There are many automatic methods of estimation or optimisation available in this age of the digital computer, from the simplest, deterministic procedures, usually based on the minimisation of least squares cost functions; to more complex numerical optimisation methods based on statistical concepts, such as Maximum Likelihood (ML). In general, the latter are more restricted, because of their underlying statistical assumptions, but they provide a more thoughtful and reliable approach to statistical inference; an approach which, when used correctly, includes the associated statistical diagnostic tests that are considered so important in statistical inference. Moreover, the power of the modern computer is such that some of these restrictions are gradually being lifted, with the advent of stochastic approaches, such as Markov Chain Monte Carlo (MCMC) procedures (e.g. Gamerman 1997). In the present context, however, the estimation methods are based on nonlinear modifications of linear *Instrumental Variable* (IV) methods (e.g. Young 1984 and the references therein).

7.2.3 Conditional Validation

As the concluding section of this chapter will stress, validation is a complex process and even its definition is controversial. Some academics (e.g. Konikow and Brederhoeft 1992, within a groundwater context; Oreskes et al. 1994, in relation to the whole of the earth sciences) question even the possibility of validating models. To some degree, however, these latter arguments are rather philosophical and linked, in part, to questions of semantics: what is the 'truth'; what is meant by terms such as 'validation', 'verification' and 'confirmation'? etc. Nevertheless, one specific, quantitative aspect of validation is widely accepted; namely *predictive validation*, in which the predictive potential of the model is evaluated *on data other than that used in the identification and estimation stages of the analysis*. While Oreskes et al. (1994) dismiss this approach, which they term 'calibration and verification', their criticisms are rather weak and appear to be based on a perception that 'models almost invariably need additional tuning during the verification stage'. While some modellers may be unable to resist the temptation to carry out such additional tuning, so negating the objectivity of the validation exercise, it is a rather odd reason for calling the whole methodology into question. On the contrary, provided it proves practically feasible, there seems no doubt that predictive validation is an essential prerequisite for any definition of model efficacy, if not validity.

It appears normal these days to follow the Popperian view of validation (Popper 1959) and consider it as a continuing process of 'falsification'. Here, it is assumed that scientific theories (models in the present context) can never be proven universally 'true'; rather, they are 'not yet proven to be false'. In Figure 7.1, it is suggested that this yields a model that is considered 'conditionally valid', in the sense that it can be assumed to represent the best theory of behaviour currently available that has not yet been falsified. Thus, conditional predictive validation means that the model has proven valid in this more narrow predictive sense. In the rainfall–flow context, for example, it implies that, on the basis of the 'new' measurements of the model inputs (e.g. rainfall, temperature or evaporation) from the validation data set, the model produces flow predictions that are acceptable *within the uncertainty bounds associated with the model*. Note this stress on the question of the inherent uncertainty in the estimated model: one advantage of statistical estimation, of the kind considered in this chapter, is that the level of uncertainty associated with the model parameters and the stochastic inputs is quantified in the time series analysis. Consequently, the modeller should not be looking for perfect predictability (which no one expects anyway) but *predictability which is consistent with the quantified uncertainty associated with the model*.

It must be emphasised, of course, that conditional predictive validation is simply a useful statistical diagnostic which ensures that the model has certain desirable properties. It is not a panacea and it certainly does not prove the complete validity of the model if, by this term, we mean the 'establishment of the truth' (Oreskes et al. 1994). Models are, at best, approximations of reality designed for some specific objective, and conditional predictive validation merely shows that this approximation is satisfactory in this limited predictive sense. As we point out in the concluding section of this chapter, this may well not satisfy all of our requirements and the efficacy of the model will often need to be evaluated in much wider and practical terms.

7.3   DATA-BASED MECHANISTIC (DBM) MODELLING

Previous publications (Beck and Young 1975; Whitehead and Young 1975; Young 1978, 1983, 1992, 1993, 1998a,b; Young and Minchin 1991; Young and Lees 1993; Young and Beven 1994; Young et al. 1996; Young and Pedregal 1998, 1999) map the evolution of the DBM philosophy and its methodological underpinning in considerable detail, and so it will suffice here to merely outline the main aspects of the approach.

The main stages in DBM model building are as follows:

1. The important first step is to define the objectives of the modelling exercise and to consider the type of model that is most appropriate to meeting these objectives. The prior assumptions about the form and structure of this model are kept at a minimum in order to avoid the prejudicial imposition of untested perceptions about the nature and complexity of the model needed to meet the defined objectives.
2. An appropriate model *structure* is identified by a process of objective statistical inference applied directly to the time series data and based on a given *general class* of *linear* Transfer Function (TF) models *whose parameters are allowed to vary over time*, if this seems necessary to satisfactorily explain the data.
3. If the model is identified as predominantly linear or piecewise linear, then the *constant parameters* that characterise the identified model structure in step 2 are estimated using advanced methods of statistical estimation for dynamic systems. The methods used in the present chapter are the Refined Instrumental Variable (RIV) and Simplified RIV (SRIV) algorithms, which provide a robust approach to model identification and estimation that has been well tested in practical applications over many years. Full details of these methods are provided in Young and Jakeman (1979, 1980); Young (1984, 1985). They are also outlined in Young and Beven (1994) and Young et al. (1996).
4. If significant parameter variation is detected then the model parameters are estimated by the application of an approach to time- (or state-) dependent parameter estimation based on recursive Fixed Interval Smoothing (FIS): e.g. Young (1984, 1988, 1993, 1998a,b, 1999b), Young and Beven (1994). Such parameter variation will tend to reflect *nonstationary* and *nonlinear* aspects of the observed system behaviour. In effect, the FIS algorithm provides a method of non-parametric estimation, with the Time Variable Parameter (TVP) estimates

defining the non-parametric relationship, which then can often be interpreted in State-Dependent Parameter (SDP) terms (see Young 1993; Young 1998a,b; Young 2000a,b; Young and Beven 1994). These methods are outlined in Appendix 2.

5. If nonlinear phenomena have been detected and identified in stage 4, the non-parametric state-dependent relationships are normally parameterised in a finite form and the resulting nonlinear model is estimated using some form of numerical optimisation, such as nonlinear least squares or ML based on prediction error decomposition (Schweppe 1965). In the present chapter, this approach to nonlinear identification and estimation is required only to define the nature of the effective rainfall nonlinearity, which appears only at the input to the model, as described in subsequent sections. This approach is also outlined in Appendix 2.

6. Regardless of whether the model is identified and estimated in linear or nonlinear form, it is only accepted as a credible representation of the system if, in addition to explaining the data well, it also *provides a description that has direct relevance to the physical reality of the system under study.* This is a most important aspect of DBM modelling and differentiates it from more classical statistical modelling methodology.

7. Finally, the estimated model is tested in various ways to ensure that it is conditionally valid in the sense discussed above. This involves standard statistical diagnostic tests for stochastic, dynamic models, including analysis which ensures that the nonlinear effects have been modelled adequately (e.g. Billings and Voon 1986), as well as exercises in predictive validation and stochastic sensitivity analysis (see the later example).

Of course, while step 6 should ensure that the model equations have an acceptable physical interpretation, it does not guarantee that this interpretation will necessarily conform exactly with the current scientific paradigms. Indeed, one of the most exciting, albeit controversial, aspects of DBM models is that they can tend to question such paradigms. For example, DBM methods have been applied very successfully to the characterisation of imperfect mixing in fluid flow processes (e.g. Young et al. 2000) and, in the case of pollutant transport in rivers, have led to the development of the *Aggregated Dead Zone* (ADZ) model (Beer and Young 1983; Wallis et al. 1989). The practical success of this ADZ model, coupled with its formulation in terms of physically meaningful parameters, seriously questions certain aspects of the ubiquitous *Advection Dispersion Equation* (ADE) which preceded it as the most credible theory of pollutant transport in stream channels (Young and Wallis 1994).

7.4   THE BEDFORD-OUSE, IHACRES AND DBM MODELS

The IHACRES and DBM models have a common origin in the discrete-time rainfall–flow model developed by the author and his colleagues for the Bedford-Ouse Study in the early 1970s (Young 1974; Whitehead and Young 1975). This section introduces first this seminal model, and then discusses briefly the later development of the IHACRES and DBM models.

7.4.1   The Bedford-Ouse Model (BM)

This was probably the first HMC model to be suggested and it takes the following form:

$$y_t = \frac{B(z^{-1})}{A(z^{-1})} u_t + \xi_{t-\delta}; \quad t = 1, 2, \ldots, N \tag{1}$$

$$r_t^* = r_t \frac{T_r - T_m}{c} \qquad \text{(i)}$$

$$s_t = s_{t-1} + \frac{1}{\tau_s} \{r_t^* - s_{t-1}\} \qquad \text{(ii)} \qquad\qquad \text{(2)}$$

$$u_t = s_t \cdot r_t \qquad \text{(iii)}$$

where,

$$A(z^{-1}) = 1 + a_1 z^{-1} + \ldots + a_n z^{-n}; \quad B(z^{-1}) = b_0 + b_1 z^{-1} + \ldots + b_m z^{-m}$$

and z^{-i} is the backward shift operator: i.e. $z^{-i} y_t = y_{t-i}$. Equation (1) is a linear *Transfer Function* (TF) relationship between the effective rainfall (rainfall excess) input u_t and the measured flow y_t. This can be written in the following alternative equation form,

$$y_t = -a_1 y_{t-1} - \ldots - a_n y_{t-n} + b_0 u_{t-\delta} + b_1 u_{t-\delta-1} + \ldots + b_m u_{t-\delta-m} + \eta_t \qquad (3)$$

where δ represents the potential presence of any pure advective time delay in the system; $\eta_t = \xi_t + a_1 \xi_{t-1} + \ldots + a_n \xi_{t-n}$; and ξ_t represents uncertainty in the relationship arising from a combination of measurement noise, the effects of other unmeasured inputs and modelling error. Normally, ξ_t is modelled as an AutoRegressive (AR) or AutoRegressive-Moving Average (ARMA) stochastic process (see e.g. Box and Jenkins 1970; Young 1984).

The equations 2(i) to 2(iii) describe the assumed nonlinear relationship between the measured rainfall r_t and u_t, which takes the form of a multiplicative nonlinearity between r_t and a measure of soil moisture s_t. This soil moisture variable can be considered as an unobserved state of the system that has to be estimated from the data (see later discussion). The modified rainfall input r_t^* is a nonlinear function of the measured rainfall r_t and the mean monthly temperature T_m, as defined in equation 2(i). Like the linear TF parameters $a_i, i = 1, 2, \ldots, n$, and $b_j, j = 0, 1, 2, \ldots, m$, the nonlinear model parameters τ_s, T_r and c, are all unknown a priori and need to be estimated from the rainfall, flow and temperature measurements. In addition the TF polynomial orders n and m (which are normally ≤ 3), and any associated time delay δ, need to be identified as part of the identification and estimation procedure, which is discussed in the next section of the chapter.

The conceptual justification for the model defined by equations (1) and (2) is straightforward. First, the linear part of the model, as represented by the TF model in equation (1), can be considered in its alternative *infinite dimensional*, discrete-time 'convolution equation' form,

$$y_t = \sum_{i=1}^{\infty} g_i \cdot u_{t-i} + \xi_t \qquad (4)$$

which is obtained by simply dividing the numerator polynomial $B(z^{-1})$ by the denominator polynomial $A(z^{-1})$. This form of the model is interesting because the coefficients $g_i, i = 1, 2, \ldots, \infty$ are the ordinates of the impulse response function, or, in hydrological terms, the associated infinite dimensional unit hydrograph (IUH). The more conventional finite dimensional approximation of the IUH (the *Finite Impulse Response* (FIR) in systems terms) is obtained from the g_i by curtailing the number of these coefficients to only those that are numerically significant. This interpretation demonstrates that the TF model is a parametrically efficient ('parsimonious') representation of the unit hydrograph model which, given our comments in the previous section, has important implications on the statistical identification and estimation of the model.

Second, the nonlinear equations in (2) are an attempt to represent, in a simple manner, the soil-moisture, groundwater and seasonal evapotranspiration mechanisms in the system and modify the rainfall to account for these effects. In this regard, equation 2(i) multiplies the 'raw' rainfall measurements by a factor that is a function of the difference between a reference temperature T_r and the current monthly mean temperature T_m, thereby adjusting the rainfall to allow for seasonal factors. This initially modified rainfall r_t^* is then passed through the first order filter in equation 2(ii), which has a time constant (or residence time) τ_s. This can be conceptualised as a catchment storage equation that yields a constantly changing index of catchment storage s_t reflecting antecedent rainfall that has fallen on the catchment. In times of no rainfall, this index will gradually reduce in a manner defined by τ_s, while during high rainfall episodes it will increase sharply in value. The final effective rainfall measure u_t is obtained as the product of s_t and r_t, so modifying the measured rainfall to reflect the temperature and antecedent rainfall effects. Interpreted in these terms, the input nonlinearity can be considered as a refinement of the conventional 'antecedent precipitation index' approach (e.g. Weyman 1975) used in previous rainfall–runoff models (since the dynamic equation 2(ii) yields s_t as an exponentially weighted measure of the input r_t^* into the past, with the exponential decay defined by τ_s). The main difference lies in the overall framework of the model and the fact that the model in (1) and (2) is couched within a stochastic framework and so is identified and estimated using statistical methods.

7.4.2   The IHACRES Model

The IHACRES model (Jakeman et al. 1990; Jakeman and Hornberger 1993) is quite similar to the Bedford-Ouse model, from which it was derived, but its conceptual basis is perceived to be somewhat more acceptable in hydrological terms. The only differences lie in the nonlinear part of the model, where the equations in (2) are replaced by the following:

$$\tau_s(T_t) = \tau_s \cdot \exp[(20 - T_t)/f] \qquad \text{(i)}$$

$$s_t = c \cdot s_{t-1} + \frac{1}{\tau_s(T_t)} \{r_t - s_{t-1}\} \qquad \text{(ii)} \qquad\qquad (2a)$$

$$u_t = s_t \cdot r_t \qquad \text{(iii)}$$

Here, the main change relates to the time constant $\tau_s(T_t)$ in the storage equation (2a)(ii). In the BM model this is a time-invariant parameter estimated from the data, while in (2a)(ii), it is now a nonlinear function of the temperature T_t, as defined by equation (2a)(i). This temperature-dependent time constant $\tau_s(T_t)$, which now applies differential exponential weighting to the antecedent rainfall r_t, is inversely related to the rate at which catchment wetness declines (or potential evapotranspiration): this is arbitrarily defined as a constant τ_s at 20°C. The parameter f is a temperature modulation factor that accounts for the fluctuations in potential evapotranspiration and determines how $\tau_s(T_t)$ changes with temperature. Both parameters τ_s and f are estimated from the data, while the parameter c is set so that the volume of the effective rainfall (rainfall excess) is equal to the total stream flow volume over the estimation period of N samples.[1]

---

[1]   In the published IHACRES model, this parameter c is selected by the modeller and appears in equation (2a)(ii), but it performs the same function as here.

7.4.3 The DBM Model

The nonlinear part of the DBM model is much simpler than in either of the two previous models. Following the DBM approach discussed in the last section and the procedures outlined in Appendix 2, it is based on non-parametric estimation from the data and takes the form,

$$u_t = c \cdot s_t r_t; \quad s_t = f_s\{y_t\} \tag{2b}$$

In other words, the catchment storage term s_t in the IHACRES model equation (2a)(iii) is defined as a function $f_s\{y_t\}$ of the flow y_t. As pointed out in the above references, this relationship should *not* be interpreted as saying that effective rainfall is physically a function of flow, which is obviously impossible. Rather, the measured flow y_t is acting here as an objectively identified *surrogate* for the catchment storage s_t. This seems sensible from a hydrological standpoint, since flow is clearly a function of the catchment storage and its pattern of temporal change is likely to be similar. In most previous publications, $f_s\{y_t\}$ has been parameterised as a power law y_t^β identified objectively from the data during the non-parametric estimation phase of the analysis. In fact, this kind of relationship has been substantiated by subsequent research on the IHACRES model (e.g. Ye et al. 1998), which has shown the advantage of incorporating such a power law. Consequently, later versions of the IHACRES model replace (2a)(iii) by $u_t = c \cdot s_t^\beta r_t$, thus making the similarity between the two models even more transparent. However, later research (e.g. Young 2000b) suggests that the power law is not universally applicable and other parameterisations may well be preferable in some cases.

It will be noted that the DBM model (equations (1) and (2b)) does not include any temperature-dependent terms and so, in this basic form, it cannot account for seasonal evapotranspiration effects. Following the DBM approach, these should be identified and estimated from the data, rather than included a priori, as in the Bedford-Ouse and IHACRES models. As we shall see later, such analysis applied to the Coweeta data suggests strongly that the temperature dependency enters through a separate TF relationship, so that the more general DBM model, for the Coweeta data, takes the form:

$$y_t = b + \frac{B_1(z^{-1})}{A_1(z^{-1})} u_{t-\delta_1} + \frac{B_2(z^{-1})}{A_2(z^{-1})} T_{t-\delta_2} + \xi_t \tag{i}$$

$$u_t = c \cdot y_t^\beta \cdot r_t \tag{ii} \tag{2c}$$

where,

$$A_1(z^{-1}) = 1 + a_{11}z^{-1} + \ldots + a_{1n1}z^{-n1}; \quad B_1(z^{-1}) = b_{10} + b_{11}z^{-1} + \ldots + b_{1m1}z^{-m1}$$
$$A_2(z^{-1}) = 1 + a_{21}z^{-1} + \ldots + a_{2n2}z^{-n2}; \quad B_2(z^{-1}) = b_{20} + b_{21}z^{-1} + \ldots + b_{2m2}z^{-m2}$$

$T_{t-\delta_2}$ is the delayed temperature, and b is a constant or very slowly varying flow component, which seems to be present in the Coweeta flow data and is probably due to deep aquifer effects of some kind.

7.4.4 The DBMS Model

Unfortunately, the simplification of the nonlinear relationships in the above DBM models is obtained at a cost. Although the models so defined can function very well indeed in a forecasting context, they cannot be used directly for *simulation* purposes, where only the rainfall and temperature measurements are used to generate purely simulated flow outputs (e.g. in applications such as off-line management, planning and 'what-if' studies). This arises because the flow variable y_t, which is not available in such studies, is required to generate the effective rainfall via equations (2b) and (2c)(ii). Related DBM simulation models can be constructed, however, by introducing a catchment storage equation into the model that mirrors the identified DBM model rainfall–flow dynamics but does not utilise the flow measurement directly. The exact form of this storage equation cannot mimic the DBM model rainfall–flow dynamics exactly, of course, and so it has to be identified from the data by comparing the estimation results obtained with various possible approximations. In the case of the Coweeta catchment data discussed in the next section, for example, this has yielded the following DBM Simulation (DBMS) model:

$$y_t = b + \frac{B_1(z^{-1})}{A_1(z^{-1})} u_{t-\delta_1} + \frac{B_2(z^{-1})}{A_2(z^{-1})} T_{t-\delta_2} \qquad (i)$$

$$s_t = b + \frac{g}{1 + \alpha z^{-1}} r_t + \frac{B_2(z^{-1})}{A_2(z^{-1})} T_{t-\delta_2} \qquad (ii) \qquad\qquad (2d)$$

$$u_t = c \cdot s_t^{\beta} r_t \qquad (iii)$$

In equation (2d)(ii), $\alpha = 1 - (1/\tau_s)$; g and β are parameters optimised within the nonlinear storage part of the model; while b, $A_2(z^{-1})$ and $B_2(z^{-1})$ are assumed to be the same as those estimated in the linear TF part of the model (2d)(i). In other words, since the effective rainfall (a direct function of flow in the standard DBM model) is not available in this simulation mode of solution, the storage equation (2d)(ii) is an approximation of the rainfall–flow equation in (2d)(i) but with the measured rainfall entering linearly through the first order TF $g/(1 + \alpha z^{-1})$. Here, this simple, first order, TF form, with only two unknown parameters to be estimated (g and α), is necessary because the assumption of a higher order TF, such as $B_1(z^{-1})/A_1(z^{-1})$ in (2d)(i), leads to over-parameterisation and associated identifiability problems during optimisation.

   In relation to the nonlinear storage equation in the IHACRES model, note that the two-input model in (2d)(ii) effectively replaces equations (i) and (ii) of (2a). As a result, the overall DBMS model, excluding the model for the noise term ξ_t, has 11 parameters to be estimated, compared with only 9 for the IHACRES model and 9 for the standard DBM. This suggests that either the IHACRES is under-parameterised or the DBMS has too many parameters: this is discussed in the next section.

7.5   A PRACTICAL EXAMPLE: HMC MODELS OF RAINFALL–FLOW PROCESSES IN THE COWEETA CATCHMENT

Figure 7.2 is a plot of the rainfall, flow and temperature time series from the Coweeta catchment, as used for model identification and estimation in this example: there are 2000 daily measurements of each variable, from sample numbers 8310–10 309 out of a full data set of 17 532 daily samples. Independent one-year-long subsets, from samples 10 310–10 674 and 14 310–14 674, respectively, are used for the predictive validation tests reported below, although these are representative of the validation results obtained over all sections of the complete data set.

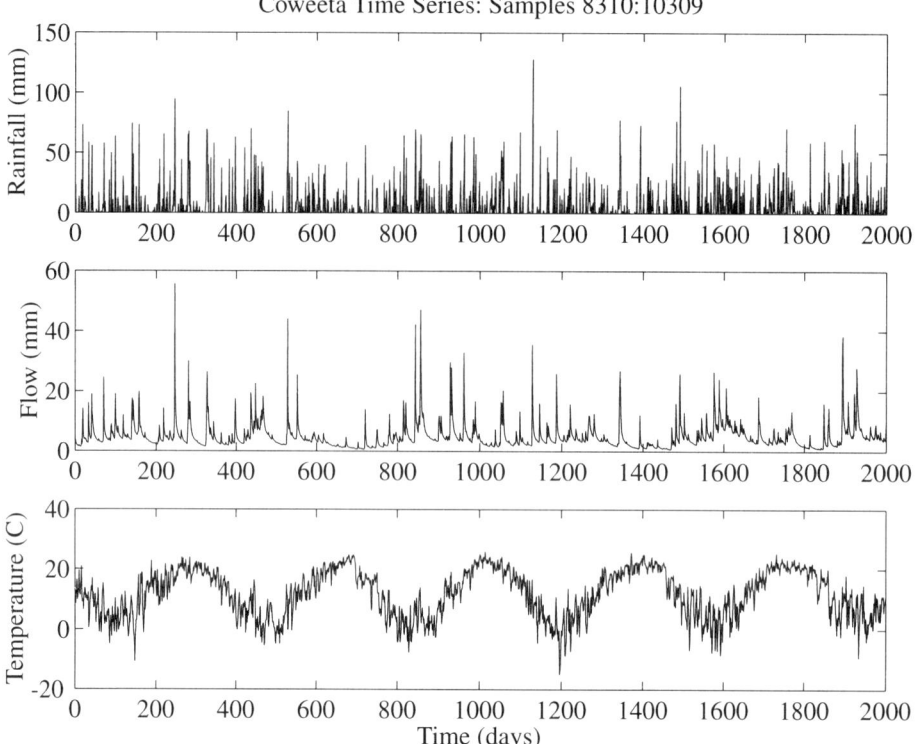

Figure 7.2 Daily rainfall, flow and temperature data from the Coweeta catchment used for model identification and estimation: samples 8310 to 10309

7.5.1   DBM Model Identification and Estimation

The first stage in DBM model identification confirms that the relationship between rainfall and flow is nonlinear and the non-parametric State-Dependent Parameter (SDP) analysis, as outlined in Appendix 2, yields the results shown in Figure 7.3. This is a plot of the estimated SDP against the flow y_t^* (where the star superscript indicates that the flow has been sorted in ascending magnitude) and the FIS non-parametric estimate $\hat{b}_{0,t|N}$ is shown as dots, with the standard errors shown dashed (see Appendix 2 for further explanation). The full line curve is the weighted least squares estimate of a power law relationship (see above discussion), which does not fit the data too badly and confirms, to some extent, the results of previous analysis of data from other catchments in the UK, Australia and the tropics (Young 1993, 1998b; Young and Beven 1994; Young et al. 1997, 1998; Ye et al. 1998; Chappell et al. 1999). It might be better, in this case, to consider a two-segment linear relationship, as used by Young (1993), or some other parameterisation (e.g. Young 2000b), rather than the power law. However, selection of the latter has the advantage of maintaining continuity with earlier work on both DBM and IHACRES models that have utilised such a relationship. Nevertheless, it must be emphasised that further research on the nature of the nonlinearity is required and it is clear that the power law relationship may not always be optimum.

FIS Non-Parametric Estimate of Rainfall-Flow Nonlinearity

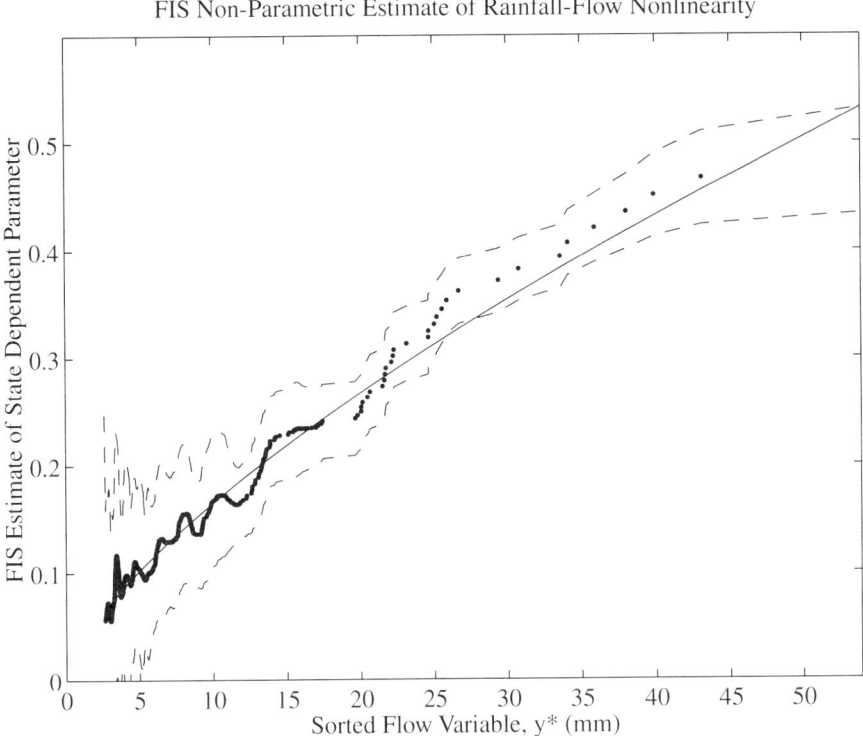

Figure 7.3 FIS non-parametric estimate of the rainfall-flow nonlinearity as a function of flow

Based on this power law parameterisation of the nonlinearity, the identification and estimation analysis outlined in Appendix 2 yields the following DBM model, which is simply a special example of equations (2c):

$$y_t = 2.05 + \frac{0.170 - 0.118z^{-1} - 0.022z^{-2}}{1 + 1.298z^{-1} - 0.347z^{-2}} u_t + \frac{-0.0042z^{-6}}{1 - 0.97z^{-1}} T_t + \xi_t$$

$$u_t = c \cdot y_t^{0.711} \cdot r_t; \quad c = \left\{ \sum_{t=1}^{t=N} y_t \Big/ \sum_{t=1}^{t=N} y_t^{0.711} \cdot r_t \right\} \tag{5a}$$

Here, the noise term ξ_t is modelled as an AR(5) process,

$$\xi_t = \frac{1}{1 - 0.269z^{-1} - 0.192z^{-2} - 0.062z^{-3} - 0.040z^{-4} - 0.056z^{-5}} e_t; \quad \text{var}(e_t) = 1.46 \tag{5b}$$

and the parameter c is selected so that total effective rainfall u_t over the observation interval is equal to the total flow Σy_t; and T_t is the temperature minus its mean value. The estimated covariance matrices associated with the three component model parameters and the noise model are given below:

Effective Rainfall–Flow TF

$$P(A_1, B_1) = 10^{-3} \begin{bmatrix} -0.9901 & -0.8596 & -0.0013 & 0.1983 & -0.1173 \\ -0.8596 & 0.7499 & 0.0016 & -0.1735 & 0.1046 \\ -0.0013 & 0.0016 & 0.0044 & -0.0067 & 0.0026 \\ 0.1983 & -0.1735 & -0.0067 & 0.0527 & -0.0308 \\ -0.1173 & 0.1046 & 0.0026 & -0.0308 & 0.0203 \end{bmatrix} \quad (5c)$$

Temperature–Flow TF

$$P(A_2, B_2) = 10^{-5} \begin{bmatrix} 0.6353 & -0.0665 \\ -0.0665 & 0.0093 \end{bmatrix} \quad (5d)$$

Non-linear Catchment Storage parameters

$$P(\beta, b) = 10^{-4} \begin{bmatrix} 1.8795 & 4.3312 \\ 4.3312 & 5.3894 \end{bmatrix} \quad (5e)$$

Noise Process

$$P(AR) = 10^{-3} \begin{bmatrix} 0.5025 & -0.1375 & -0.0983 & -0.0366 & -0.0278 \\ -0.1375 & 0.5393 & -0.1128 & -0.0940 & -0.0366 \\ -0.0983 & -0.1128 & 0.5559 & -0.1128 & -0.0985 \\ -0.0366 & -0.0940 & -0.1128 & 0.5394 & -0.1377 \\ -0.0278 & -0.0366 & -0.0985 & -0.1377 & 0.5032 \end{bmatrix}$$

Figure 7.4 compares the output of the deterministic part of the model (5a), $\hat{y}_t$, where,

$$\hat{y}_t = 2.05 + \frac{0.170 - 0.118z^{-1} - 0.022z^{-2}}{1 + 1.298z^{-1} - 0.347z^{-2}} u_t + \frac{-0.0042z^{-6}}{1 - 0.97z^{-1}} T_t$$

with the observed flow y_t: clearly the data are fitted well with the coefficient of determination $R_T^2 = 0.91$ (i.e. 91% of the flow data are explained by $\hat{y}_t$: see Appendix 3). After incorporating the AR(5) noise model, the coefficient of determination, based on the one-step-ahead prediction errors, $R^2 = 0.93$. Figure 7.5 is a plot of the deterministic output from the temperature–flow model, $\hat{y}_{2,t}$ (see below and Appendix 2) compared with the residual $y_t - b - \hat{y}_{1,t}$ (dotted), where $\hat{y}_{1,t}$ is the deterministic output of the effective rainfall–flow model. This shows how well the temperature–flow model explains the seasonal variations in flow due, presumably, to long-term evapotranspiration effects. Note also the 'spikes' in the residual $y_t - b - \hat{y}_{1,t}$, which suggests that there are outliers in the data, particularly during high flow events (see later).

Figures 7.6 to 7.9 present the results of correlation analysis applied to the model residuals. Figure 7.6 is the autocorrelation function (acf) of the deterministic model residuals $y_t - \hat{y}_t$: it is clear that there is considerable correlation in these residuals but, as shown in Figure 7.7, the final stochastic residuals $\hat{e}_t$ (the one-step-ahead prediction errors) are not correlated and subscribe to

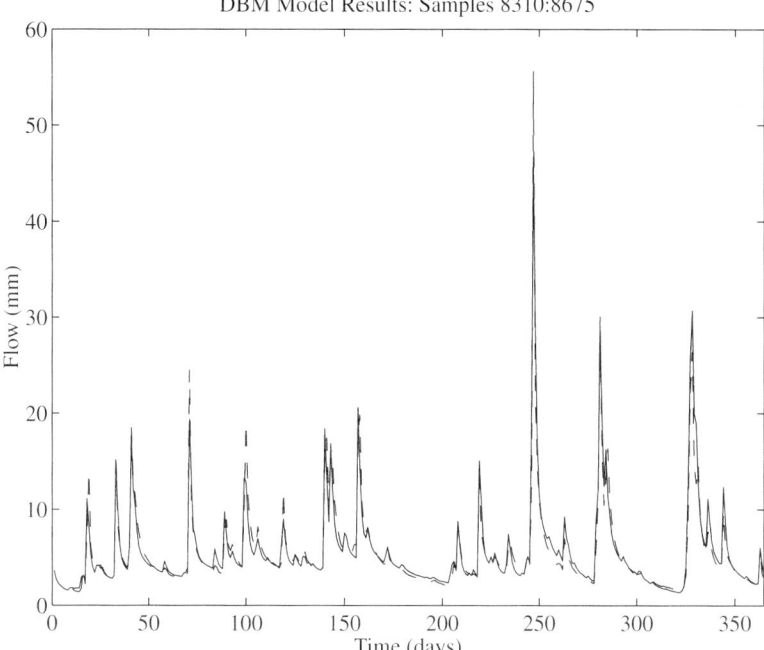

Figure 7.4   DBM modelling results: comparison of the deterministic model output $\hat{y}_t$ and the flow measurements y_t between samples 8310 and 8675

the assumption of white noise.[1] However, the 'spikes' mentioned above are clear on the plot of $\hat{e}_t$ in the top panel of the figure and a histogram of the $\hat{e}_t$, coupled with tests for normality and a normal probability plot, confirm that the amplitude distribution is far from Gaussian normality. Given this observation, it is necessary to be careful about the interpretation of the covariance matrices in (5c) to (5e), which are computed under the assumption of normally distributed residuals. However, the matrices still provide an indication of the uncertainty on the estimates and are useful in this regard for subsequent stochastic analysis, provided we bear in mind the distortions that will arise from the non-normality .

Figure 7.8 shows the cross-correlation function (ccf) between $\hat{e}_t$ and the temperature input T_t: although there is a larger correlation at lag zero, it falls only on the boundary of the significance bands (dotted) and so we can assume that $\hat{e}_t$ and T_t are reasonably uncorrelated, as required. Finally, Figure 7.9 is the ccf between $\hat{e}_t$ and the effective rainfall input u_t: here, there is evidence of just significant correlation at some lags, although the magnitude is always quite small. Apart from the clear deviation from normality, this is the only way in which the model (5a) fails its statistical diagnostic tests and these failures appear common in all rainfall–flow models that we have investigated using these statistical diagnostic methods.

One of the main reasons for the correlation between the stochastic residuals $\hat{e}_t$ and u_t, as well as the non-normality of $\hat{e}_t$, is the presence of the 'spikes' mentioned above. Figure 7.10 is a plot of $\hat{e}_t$ with the outliers (defined as values lying outside the ± 3 standard deviations bound) marked as circles. It is clear that there are numerous samples within the series $\hat{e}_t$ that appear to be outliers

---

[1] More precisely, the residual series constitutes 'weak white noise', in the sense that there is some autocorrelation when the squared or absolute values of the residuals are considered.

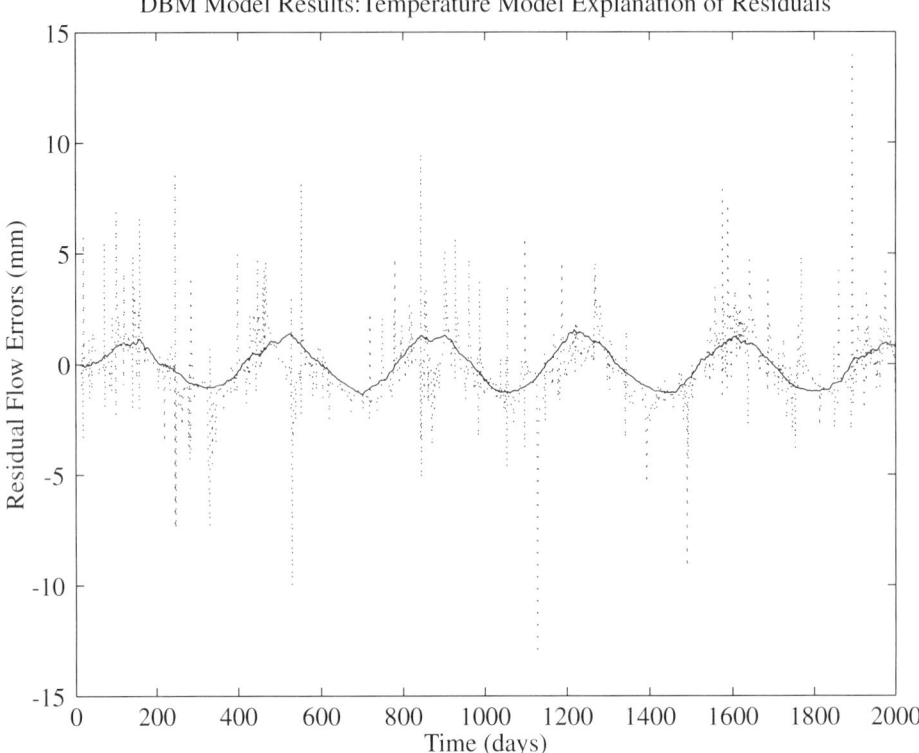

Figure 7.5 DBM modelling results: comparison of the deterministic output $\hat{y}_{2,t}$ of the temperature–flow model and the residual series $y_t - b - \hat{y}_{1,t}$

in this sense, and these most often occur at times of high rainfall and flow when one might expect measurement errors and extremes of behaviour to be present. One straightforward approach, in this situation, would be to repeat the identification and estimation analysis under the assumption that the outliers represent errors in the data and either should be given less weight in the estimation, or considered as missing samples. A related approach would be to consider that the $\hat{e}_t$ are heteroscedastic (i.e. their variance is not the same for all samples) and utilise some method of estimation which allows for this (e.g. Johnston and DiNardo 1997). These approaches have not been attempted in the present illustrative analysis, but it seems unlikely that it would make a great difference to the final estimated model's predictive performance (although it would affect the parameter estimates). Unfortunately, other approaches to the problem of non-normality, such as the use of the Markov Chain Monte Carlo (MCMC) methods mentioned previously, involve new, relatively untried methodology and can be computationally very intensive. Nevertheless, the non-normality of the residuals in rainfall–flow models is clearly significant and constitutes a suitable topic for future research.

7.5.2 Mechanistic Interpretation of the DBM Model

As discussed in Young (1992, 1998b) and following the procedure used in the previous publications mentioned above, the TF part of the model (5a) can be written in the following alternative

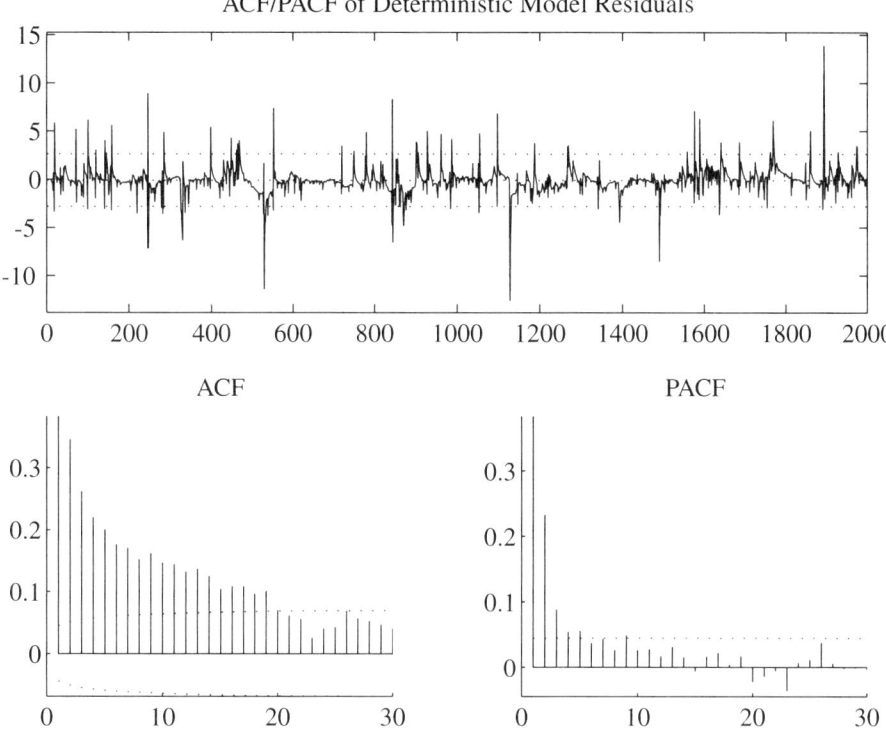

Figure 7.6 DBM modelling results: auto (ACF) and partial autocorrelation (PACF) functions of the deterministic model residuals $\xi_t = y_t - \hat{y}_t$

form:

$$y_t = 2.05 + \left[0.17 + \frac{0.077}{1 - 0.38z^{-1}} + \frac{0.026}{1 - 0.92z^{-1}} \right] u_t + \frac{-0.0042z^{-6}}{1 - 0.97z^{-1}} T_t + \xi_t \qquad (5f)$$

Considering only the deterministic parts of this model, the deterministic output $\hat{y}_t$ is given by,

$$\hat{y}_t = 2.05 + \hat{y}_{1,t} + \hat{y}_{2,t} \qquad (5g)$$

where:

$$\hat{y}_{1,t} = \left[0.17 + \frac{0.077}{1 - 0.38z^{-1}} + \frac{0.026}{1 - 0.92z^{-1}} \right] u_t; \quad \hat{y}_{2,t} = \frac{-0.0042z^{-6}}{1 - 0.97z^{-1}} T_t$$

Here, the variables $\hat{y}_{1,t}$ and $\hat{y}_{2,t}$ are, respectively, the modelled flow due to the effective rain and the temperature effects, with the former TF decomposed into a three-pathway, parallel flow form. A systems diagram of the model in this decomposed parallel form is shown in Figure 7.11 and the details of the decomposition are given below in Table 7.1.

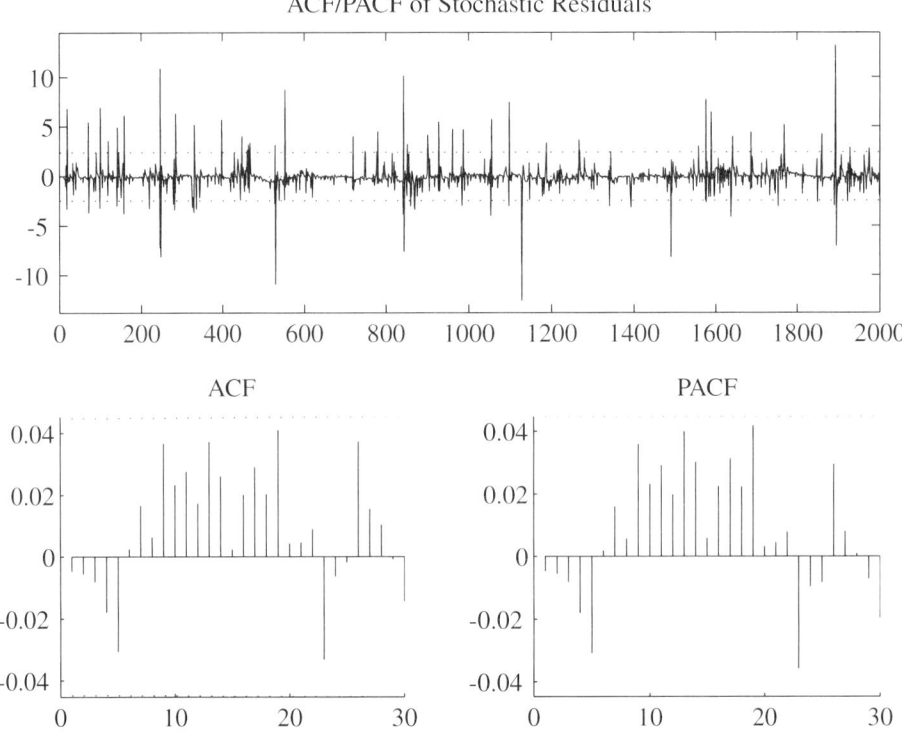

Figure 7.7   DBM modelling results: auto (ACF) and partial autocorrelation (PACF) functions of the stochastic model residuals $\hat{e}_t$ (one-step-ahead prediction errors)

Table 7.1   Parallel decomposition of the rainfall–flow model: Eig denotes the eigenvalue associated with the component; SSG denotes steady-state gain; TC denotes time constant (residence time); TD denotes pure time delay

Instantaneous TF (A)			
Eig	SSG	TC	% of $\hat{y}_{1,t}$ flow
0	0.17	0.00	27.3
Fast Flow TF (B)			
Eig	SSG	TC	% of $\hat{y}_{1,t}$ flow
0.38	0.12	1.02	19.9
Slow Flow TF (C)			
Eig	SSG	TC	% of $\hat{y}_{1,t}$ flow
0.92	0.33	12.21	52.8
Slow Temperature (Seasonal) Effect (D)			
Eig	SSG	TC	TD
0.97	− 0.14	32.9	6
	Constant Base flow		
	2.05 ($\forall t$)		

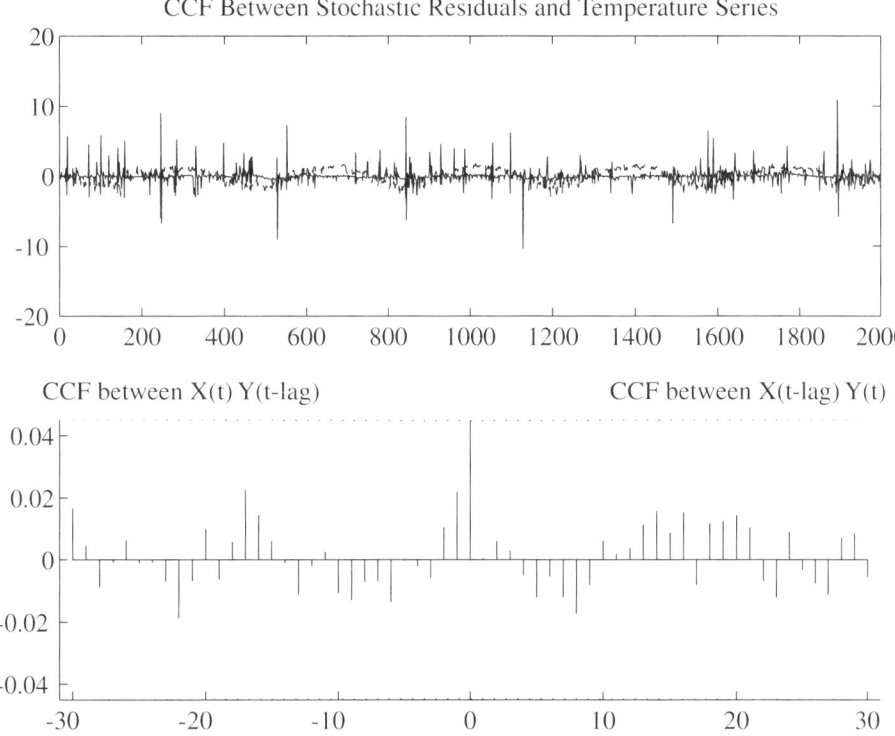

CCF Between Stochastic Residuals and Temperature Series

CCF between X(t) Y(t-lag)   CCF between X(t-lag) Y(t)

Figure 7.8 DBM modelling results: cross-correlation function (CCF) between the stochastic residuals $\hat{e}_t$ and the temperature input T_t

We now see that *one interpretation* of the model (5a)[1] is that the effective rainfall u_t reaches the river and affects the flow via three pathways, and that the flow is also affected by the prevailing temperature variations. This interpretation has the advantage that it makes reasonable physical sense. In particular, it suggests that the river flow is composed of five components: a very rapid instantaneous (i.e. within one day) effect; a 'fast flow' component with residence time 1.02 days, probably associated with surface processes; a 'slow flow' component with residence time 12.21 days, probably associated with subsurface and groundwater processes; and a very slow, 'base flow' component in the form of a 'constant flow' of 2.05 mm, probably due to deep aquifer effects and apparently constant over the 2000 days of data (but varying very slowly over the whole 17 532 data set). Note that the TFs associated with 'fast' and 'slow' flow parallel pathways can be interpreted directly as dynamic mass balance or storage equations (see Young and Beven 1994). The fifth, temperature-dependent component (the temperature variations about the mean passed through a TF with time constant 32.9 days) clearly accounts for long-term temperature-dependent effects, such as those arising from evapotranspiration processes. Here, the time delay (6 days) and long time constant of 32.9 days are required because of the considerable phase lag between the fluctuations in temperature and the apparent effect of this on the rainfall–flow dynamics.

---

[1] This decomposition is not unique: serial and feedback decompositions are equally feasible in mathematical system terms but the parallel decomposition seems most logical on physical grounds.

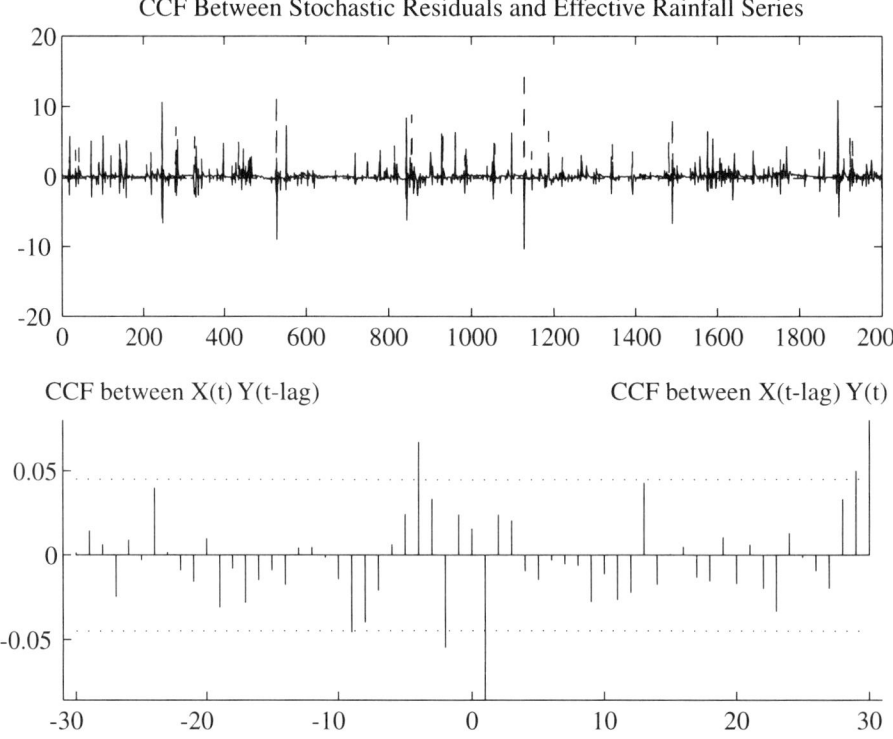

Figure 7.9 DBM modelling results: cross-correlation function (CCF) between the stochastic residuals $\hat{e}_t$ and the effective rainfall input u_t

Figure 7.12 is a plot of the flow estimates associated with each of the parallel pathways. The instantaneous effect is shown in the upper panel; the fast and slow flows below this; and the temperature-dependent effect in the lower panel. Figure 7.13 lumps these effects together: in the upper panel is the sum of the instantaneous and fast flow components, while the cumulative slow flows are plotted in the lower panel. The dashed line in this lower plot is the temperature-dependent component, showing how it dominates the slower flow components in the longer term.

7.5.3 Predictive Validation of the DBM Model

The model (5a) has been validated in a predictive sense, as discussed in earlier sections, by applying it to two other, one-year-long, segments of the data: from samples 10 310–10 674 and 14 310–14 674, respectively. Over the first segment, the coefficients of determination are $R_T^2 = 0.90$ and $R^2 = 0.91$, respectively, while over the second segment, they are $R_T^2 = 0.92$ and $R^2 = 0.95$. Since both of these results show that the model, without any re-estimation, continues to explain the data to a level commensurate with that obtained over the original estimation period between samples 8310 and 10 309, we can conclude that the model is 'conditionally valid' and can be used for forecasting applications. In such applications, the model would be best transformed into discrete-time, stochastic state space form embedded within a recursive estima-

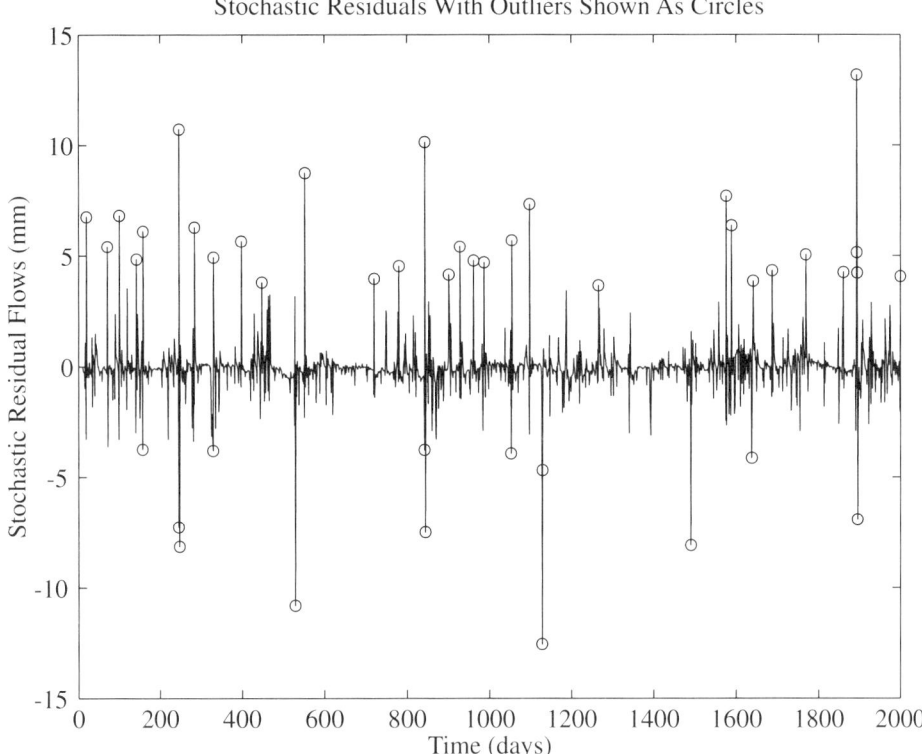

Figure 7.10 DBM modelling results: the stochastic residuals $\hat{e}_t$, with the outliers marked as circled points

tion mechanism of some sort (see e.g. Young 1984; Lees et al. 1994; Young et al. 1998; Young and Tomlin 2000). However, while the predictive validation results obtained above engender confidence in the model for such forecasting applications, it cannot be considered fully validated in a wider sense, as discussed later in Section 7.6.

7.5.4   The Effects of Uncertainty in the DBM Model

One final evaluation of the DBM model (5a) concerns the effects of the parametric uncertainty on the mechanistic interpretation of the model. The easiest approach to such analysis is to utilise Monte Carlo Simulation (MCS) methods (see Young et al. 1996; Parkinson and Young 1998; Young 1999a). Here, the model is simulated many times (in this case between 1000 and 5000 times) with the model parameters for each 'random realisation' of this kind generated by randomly sampling the parameters from their parent probability distribution functions (pdfs). In the present context, these pdfs are assumed to be multivariate Gaussian, with the mean values and covariance matrices shown in equations (5a) to (5e).[1]

---

[1] Note the earlier comment on the non-normality of the model residuals. In this case, it means that the covariance matrices will probably tend to underestimate the parametric uncertainty and this will naturally affect the propagation of this uncertainty in the MCS analysis.

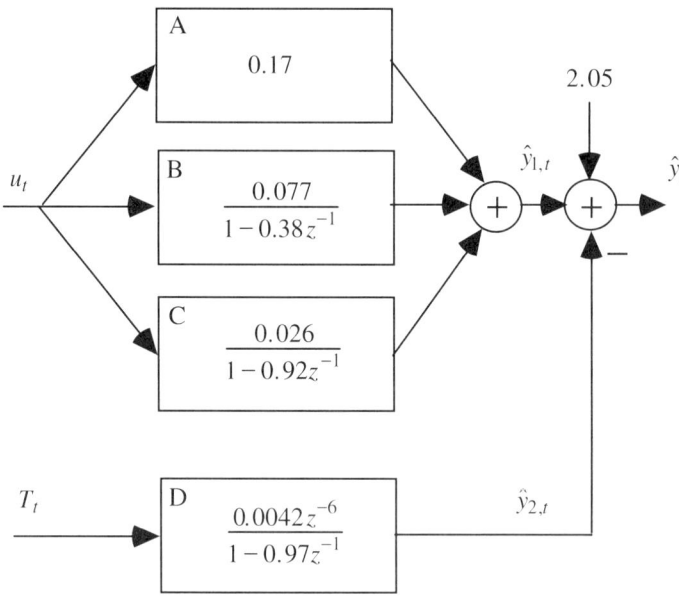

Figure 7.11 Systems block diagram of the DBM rainfall–flow model, showing the parallel flow decomposition inferred from the TF model

Two types of output are obtained from this analysis: first, a response analysis, where the model is perturbed by the actual measured inputs used during the estimation (i.e. y_t, r_t and T_t), with the stochastic noise ξ_t generated, for each realisation, from independent samples of white noise, via equation (5b), and, second, a parametric analysis, where each realisation is used to calculate only the 'derived', mechanistically meaningful parameters associated with the parallel flow decomposition. Typical MCS results in the present example are presented in Figures 7.14 to 7.16. These are not meant to be comprehensive in any way: they are simply illustrative of the kind of uncertainty analysis that is possible using stochastic simulation methods.

Figure 7.14 compares the mean of an ensemble of 1000 stochastic realisations (full line) with the actual data (circles) over the shorter period between samples 8550 and 8660, which is selected to avoid congestion in the graph. This mean response is virtually identical to the deterministic model output $\hat{y}_t$, as would be expected in this case. The dashed lines mark the envelope of the ensemble and thus show the limits of the uncertainty propagation arising from the various uncertain elements in the model. The measured response falls mainly within these bounds, showing that the stochastic model provides a reasonable description of the data. Figures 7.15 and 7.16 are typical of the results obtained from the parametric MCS analysis based on 5000 random realisations: Figure 7.15 shows the histograms of the residence time parameters for the slow and fast parallel TFs (see Table 7.1), while Figure 7.16 displays the histograms of the partition percentages associated with the parallel pathways, including the instantaneous flow effect. As required, the ensemble mean values agree almost exactly with the values of these parameters derived from the estimated TF parameters, as listed in Table 7.1.

The sample distributions in Figure 7.15 show that the estimated slow pathway residence time is much more uncertain than that of the fast pathway, and, while both show signs of minor

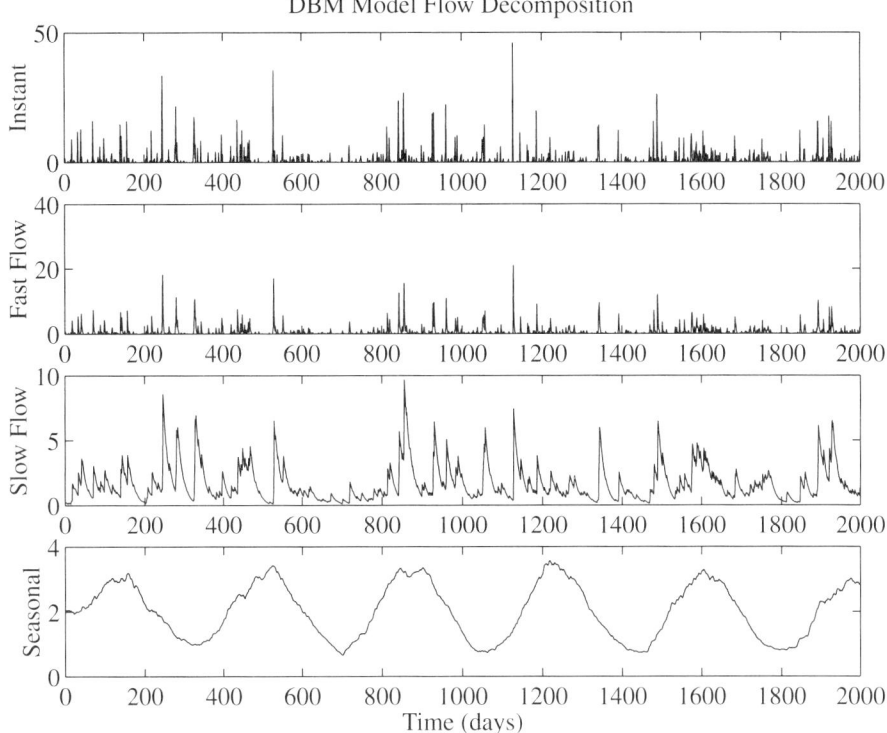

Figure 7.12 DBM modelling results: TF derived decomposition of flow showing the extracted instantaneous (within one day), fast flow, slow flow and seasonal (temperature-dependent) components

skewness towards larger values, that of the slow pathway is more pronounced. These results make physical sense, since there is less information in the data on the dynamics of the slow pathway over the sampling interval and it is also describing subsurface processes which are likely to be less well defined. In the case of Figure 7.16, there are some minor signs of skewness in the fast and slow pathway partition percentages, with the former skewed towards higher and the latter towards lower values. The instantaneous distribution seems quite symmetric.

7.5.5 Comparison of HMC Models

Similar analysis to that described in Sections 7.5.1 to 7.5.4 has been carried out on the other HMC models discussed in Section 7.4 and some of the estimation results are reported in Appendix 4. There are many ways in which the models can be compared and any one set of comparisons can be misleading. However, the results shown in Table 7.2, which compares the coefficients of determination R_T^2 and R^2 obtained for each model over the estimation and predictive validation periods, provide a reasonable impression of the comparative performance in this example.

These results are also reflected in the statistical diagnostics. The DBM and DBMS models are clearly superior in this regard: both satisfy most of the statistical requirements on the final stochastic residuals $\hat{e}_t$, failing only marginally in relation to the ccf between these residuals and

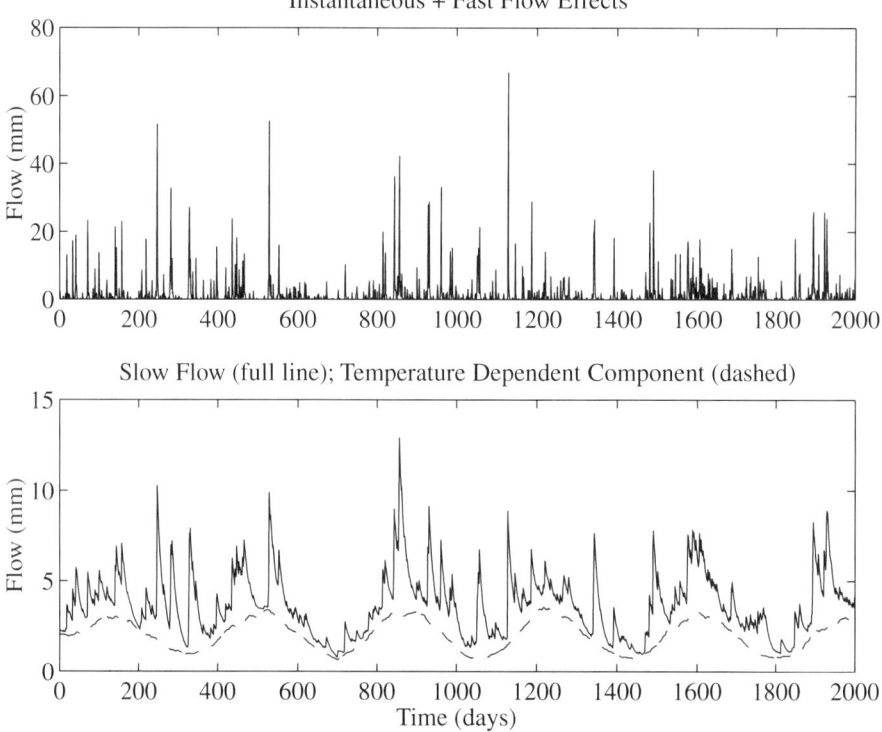

Figure 7.13 DBM modelling results: TF derived decomposition of flow showing the combined instantaneous and fast flow effects (upper); and the combined slow flow and temperature-dependent components (lower) Also, temperature-dependent component shown separately (dashed) in lower plot

Table 7.2 Comparison of coefficients of determination of the HMC models: the equals (=) sign shows that one or more model performed virtually the same

Model	Estimation (8310–10 309)		Validation 1 (10 310–10 674)		Validation 2 (14 310–14 674)		Validation 3: all 17 532 samples	
	R_T^2	R^2	R_T^2	R^2	R_T^2	R^2	R_T^2	R^2
DBM	0.91	0.93	0.90	0.91	0.92	0.95	0.85	0.90
DBMS	0.86	0.88	0.87	0.90	0.90	0.92	0.82	0.87
IHACRES	0.82	0.86	0.89	0.91	0.87	0.90	0.82	0.87
BM	0.84	0.87	0.88	0.91	0.91	0.92	0.82	0.86
Best	DBM	DBM	DBM	DBM =	DBM	DBM	DBM	DBM
2nd best	DBMS	DBMS	IHACRES	IHACRES =	BM	DBMS=	DBMS=	DBMS=

the effective rainfall input, but with clearly non-normal $\hat{e}_t$. On the other hand, the IHACRES and BM models only satisfy the acf requirement on their $\hat{e}_t$ series: they fail both of the ccf requirements, with many significant violations of confidence bounds at a range of lags. And, as expected, the distribution of the residuals is clearly non-normal with many outliers.

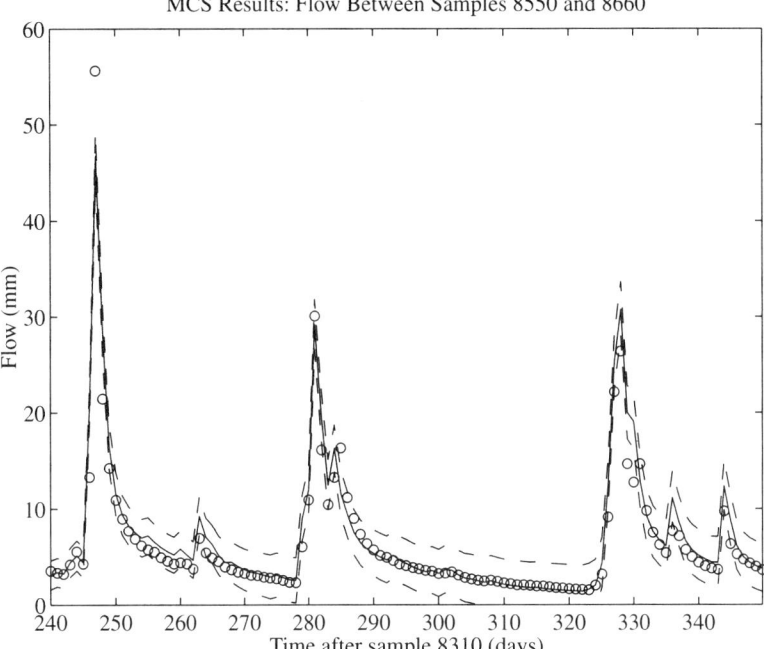

Figure 7.14 DBM modelling results. MCS flow response: mean (full line) and ensemble envelope (dashed), compared with the measured flow (circles)

An indication of how well or badly the model parameters are defined in any nonlinear least squares optimisation exercise is provided by how well the optimisation procedure converges, combined with the resulting covariance properties of the estimates. Here again, there are indications of some problems with the IHACRES and BM models. Whereas convergence is rapid and well defined in the case of the DBM and DBMS models, neither of the other two models converge under the same convergence criteria, and the Matlab algorithm *leastsq*, as used for optimisation in these studies, reports difficulties during convergence for both models. The covariance properties of the optimised estimates are also worse, with some '*t*' values (the estimated parameter divided by its standard error) noticeably lower in the case of the IHACRES and BM models (although it is well known that the covariance properties obtained during nonlinear numerical optimisation can be misleading because of the approximations involved).

One common aspect of all four models is the use of a TF model to describe the predominantly linear relationship between the effective rainfall and flow. But, as pointed out previously, the TF model is basically a parametrically efficient way of characterising the (hopefully) invariant, unit hydrograph characteristics of the catchment, which is quantified in the present context by the impulse response of the TF model. These TF-derived unit hydrographs for all four models are compared in Figure 7.17, where it is clear that they are all very similar. The initial responses, which are dominated by the 'quick flow' component, are virtually identical, while the somewhat longer tails of the IHACRES and BM hydrographs reflect the rather longer residence times of the 'slow flow' components in these cases (see Appendix 4). This is also confirmed by the TF-based flow decomposition results shown in Table 7.3, where the partition percentages for all the models are reasonably similar.

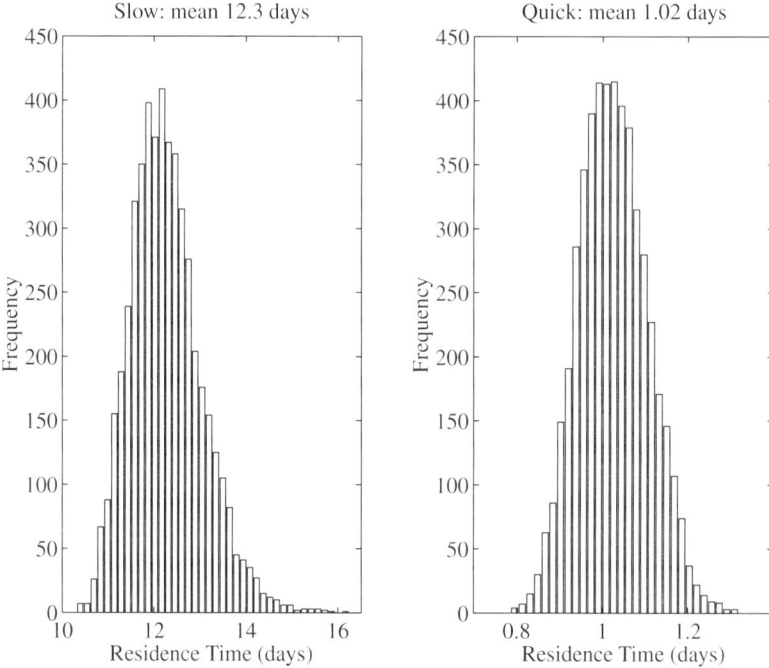

Figure 7.15 DBM modelling results. MCS generated histograms for the TF derived slow (left) and quick (right) residence time (time constant) parameters

These results are comforting in practical terms. They show that, although the four models are rather different in the manner that they characterise the nonlinear catchment storage characteristics, they are quite similar in their quantification of the underlying unit hydrograph aspects of the catchment dynamics. This is particularly important, of course, since the unit hydrograph is such a well accepted icon in both the theory and practice of hydrology.

Since the catchment storage part of all the four HMC models provides a nonlinear mechanism for generating the effective rainfall series, it is interesting to compare the relationship between the rainfall and effective rainfall series in each case. However, since this relationship is quite complex, a simple graph of effective rainfall against measured rainfall takes the form of a 'scatter plot' and it is difficult to compare these plots in any meaningful manner. One approach to this problem is to exploit the FIS approach to non-parametric estimation outlined in Appendix 2 and estimate a smooth relationship between effective rainfall and measured rainfall for each of the models. The results of this analysis, as shown in Figure 7.18, reveal that there are quite significant differences in the pattern of the transformation in each case. In particular, the DBM model nonlinear transform (full line) lies a little below the others (DBMS, dash–dot; IHACRES, dotted; BM, dashed) for rainfall less than 45 mm, and above them for rainfall greater than this. Moreover, this difference is very significant for rainfall greater than 70 mm, as we see in Figure 7.19. This shows plots of the difference between the effective and measured rainfall for all four models and the increased magnitude of the high rainfall events in the case of the DBM model (upper plot) is obvious. It seems, therefore, that since the linear parts of the models are reasonably well defined and quite similar, the superiority of the DBM model derives from its superior definition of effective rainfall during the higher rainfall episodes.

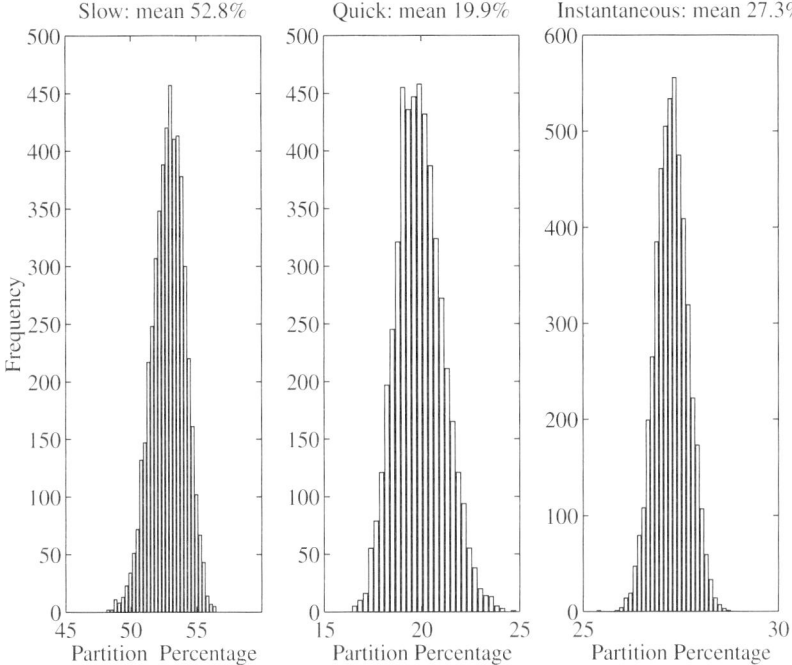

Figure 7.16   DBM modelling results. MCS generated histograms for the TF derived slow (left), quick (middle) and instantaneous (right) partition percentage parameters

Finally, it is necessary to ask why the DBM model performs significantly better than the other three models. The answer is fairly obvious: the DBM modelling strategy has uncovered the fact that the flow series itself provides a *surrogate measure* of catchment storage and so the model is able to exploit this in the definition of the effective rainfall nonlinearity. In the other models, on the other hand, the catchment storage is an entirely unobserved (sometimes termed 'latent') variable, an *estimate* of which is computed from the rainfall input using the conceptualised storage equations. The differences between these other models lies mainly, therefore, in the nature of these storage equations. The message from this result is also clear: a model should extract all of the useful information contained in the observational data. If measured variables are available which provide either direct (e.g. soil moisture or evapotranspiration measurements) or indirect (e.g. the flow measurement as a surrogate for catchment storage) quantification of important variables within the model, they should be incorporated into the model wherever this is possible.

This does not mean, however, that the concept of unobserved (latent) variables should be dismissed. On the contrary, it is essential in the present context for the construction of models that are useful in a simulation context. Moreover, systems research has shown how the definition and estimation of such latent state variables (state variable estimation and reconstruction, as in the Kalman filter) provides a very powerful, general approach to modelling stochastic dynamic systems. Indeed, one obvious development of the HMC models discussed in this chapter is to incorporate them into such a state estimation framework for both optimisation and on-line purposes. This is discussed later in the Conclusions (Section 7.6).

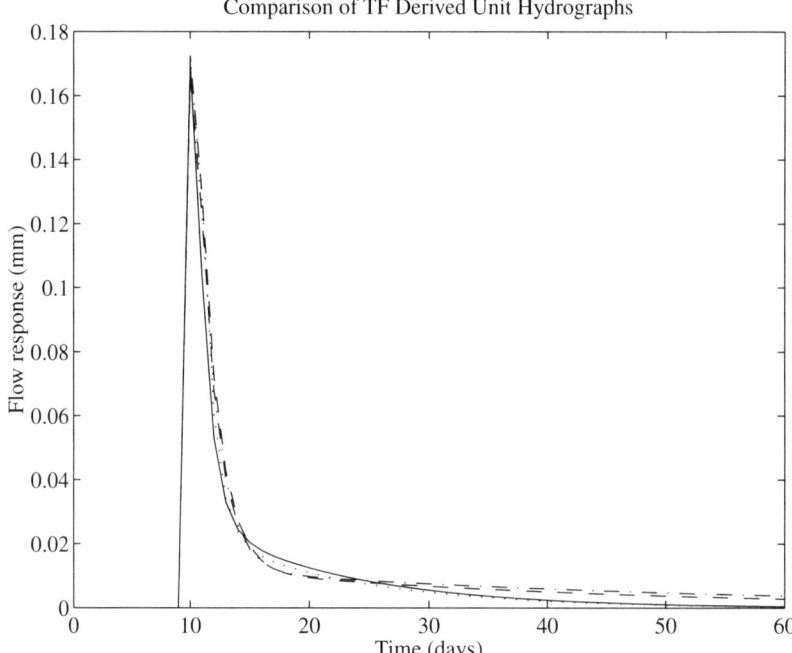

Figure 7.17 Comparison of the TF derived instantaneous unit hydrographs for the four HMC models: DBM (full line); DBMS (dotted); IHACRES (dashed); and BM (dash–dot)

Table 7.3 Comparison of TF-based flow decomposition partition percentages for the HMC models

Component	Instantaneous (%)	Fast (%)	Slow (%)
DBM	27.3	19.9	52.8
DBMS	26.1	25.6	48.3
IHACRES	21.0	29.7	49.3
BM	18.4	25.0	56.6
Average	23.2 ± 4.2	25.0 ± 4.0	51.8 ± 3.8

7.5.6   Discussion

The results presented in previous subsections show that all of the HMC models considered here perform reasonably well on the Coweeta catchment data. The DBM model (5a) is best in statistical diagnostic and predictive validation terms, with its simulation version, DBMS, having very similar diagnostics but not performing quite as well in the predictive validation tests. However, these DBMS validation results are not bad (see Table 7.2) and the model seems marginally superior to the IHACRES and BM models in this regard. Although the latter two models perform poorly in the diagnostic tests, their predictive validation results are quite acceptable. Indeed, they could be used with some confidence in forecasting applications, although they would probably be out-performed by the DBM model in most cases.

The only disadvantage of the DBM model is its inability to function directly in simulation terms. In such applications, its simulation version, DBMS, would have to be used and, as shown

FIS Non-Parametric Estimates of the Effective Rainfall Nonlinearities

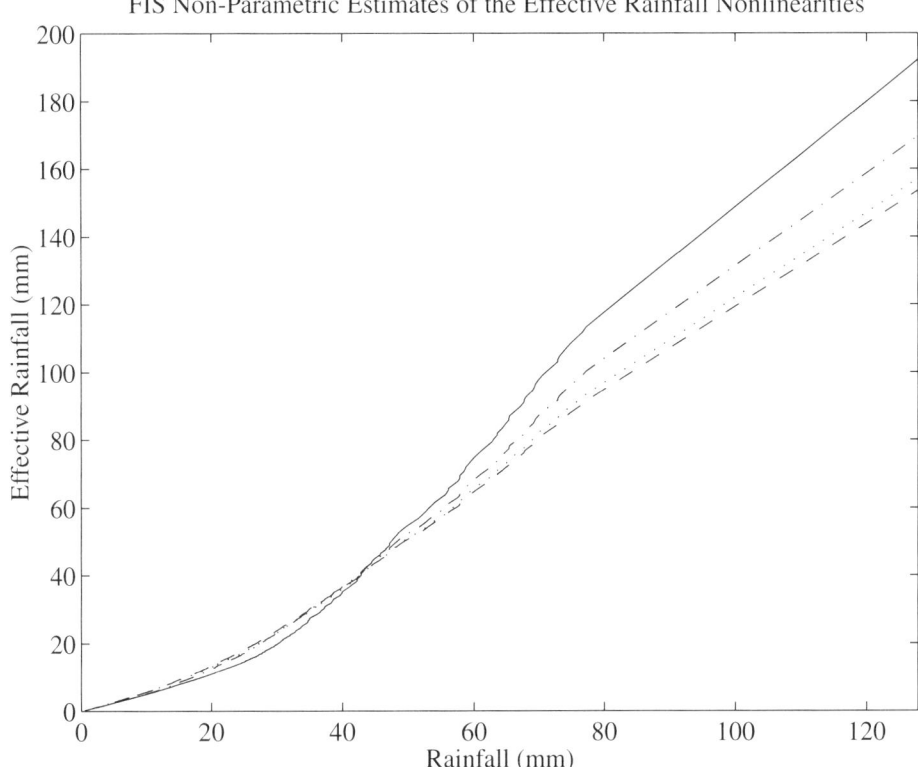

Figure 7.18 Comparison of the FIS estimated effective rainfall nonlinearities for the four HMC models: DBM (full line); DBMS (dash–dot); IHACRES (dotted); and BM (dashed)

above, the differences between this version and both the IHACRES and BM models is not nearly as significant. All three would provide reasonable models in simulation terms, with little to choose between them, and their predictive performance is quite similar, with only a marginal advantage to the DBMS model. Perhaps surprisingly, given its age, the BM model is still very competitive with both the IHACRES and DBMS models in all simulation and predictive respects. If we are to choose between the models, therefore, it is necessary to look further than these quantitative performance measures and consider other, more qualitative, properties of the models, such as their conceptual interpretations. This must be carried out very carefully, however, because it means moving away from objective measures of performance into much less trustworthy subjective areas, where the prior perceptions of the modeller can sometimes tend to bias rational judgement.

7.6 CONCLUSIONS

This chapter has been concerned with the process of constructing models for stochastic, dynamic systems. It has argued that there is no unique mathematical model of any real system and that, in consequence of this, the model has to be chosen on the basis of carefully defined objectives.

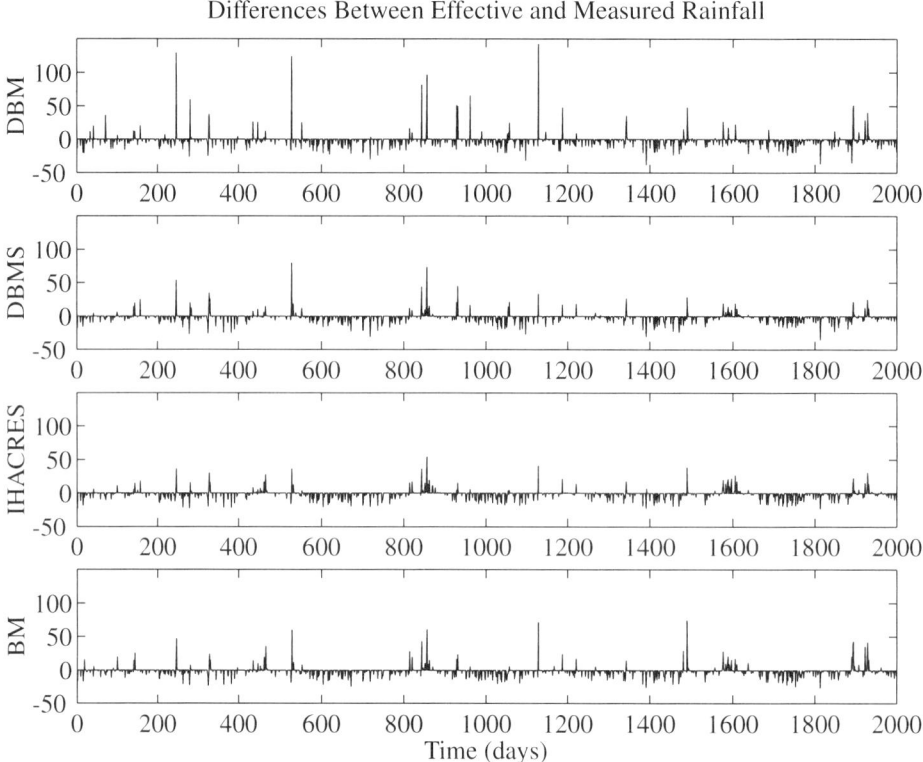

Figure 7.19 Comparison of the differences between the effective and measured rainfall for the four HMC models: DBM (top); DBMS (second); IHACRES (third); BM (lower)

Furthermore, the Data-Based Mechanistic (DBM) approach to modelling stochastic, dynamic systems appears to provide a suitably rigorous, statistical framework for such objective-oriented modelling exercises. In more specific terms, the chapter has been concerned with a detailed exploration of the Hybrid Metric–Conceptual (HMC) class of rainfall–flow models. And it has shown how such models may be statistically identified, estimated and validated using DBM methods, as demonstrated by the modelling of rainfall–flow processes in the Coweeta catchment. In relation to the topic of this book, the chapter has also raised questions about the meaning of the term 'validation' and to what extent quantitative procedures of predictive validation can engender confidence in a model. The Popperian concept of model falsification, combined with the idea of 'conditional validation', has been promoted as a reasonable approach to verifying the practical utility of DBM models, but the possible limitations of such quantitative methods have also been mentioned.

Clearly, predictive validation of a data-based rainfall–flow model, as defined in this chapter, is a *necessary* condition for its acceptance, regardless of the study objectives. If the model fails to perform adequately in this predictive sense then it is hard to see how it can be considered acceptable in any meaningful role, except as a vehicle for scientific debate. But, just as obviously, predictive validation is not a *sufficient* condition for the model's acceptance within the wider hydrological community. The hydrologist is rightly suspicious of the 'black-box' model that performs reasonably in forecasting terms but whose abstract form (e.g. in a model based on neural networks) makes its contact with hydrological reality very tenuous. In this sense, it has been argued that a model needs to have a physically interpretable conceptual basis that is reasonably acceptable within current hydrological paradigms. On the other hand, the model should not be overly constrained by these existing paradigms and the modeller should be willing, even anxious, to attack them if the DBM modelling procedure secures evidence from reliable experimental data that they may have serious limitations.

In relation to the specific rainfall–flow modelling exercise considered in this chapter, we have seen that a number of HMC models are competitive in their explanation and prediction of the time series data from the Coweeta catchment. If the modelling objective is *forecasting*, then the standard DBM model, particularly if it is made state adaptive (e.g. Young and Tomlin 2000), has definite theoretical and practical advantages. However, if the model is required for *simulation* purposes, in applications such as management, planning and certain aspects of scientific investigation, then the DBM model is not so useful and the alternative DBMS, IHACRES and BM models come into their own.

In regard to these latter three models, it would appear that the IHACRES and BM models are less satisfactory in statistical terms and the estimated parameters of the nonlinear parts of the models, in particular, are not very well defined. Equifinality plots (Franks et al. 1997; Young 1999a) tend to confirm this and suggest that there are some identifiability problems for these models, deriving mainly from the effective rainfall nonlinearity. Although the DBMS is marginally superior in statistical terms, its nonlinear catchment storage model is also rather ill-defined and it is more highly parametrised.

Overall, then, it would appear that there is little to choose between the different catchment storage equations used in the three simulation models and, indeed, that any reasonable conceptual storage model would yield similar predictive performance. This is a classic case of equifinality and it means that the models cannot be judged or 'validated' on the basis of their conceptual structure, even though hydrologists may hold great store by this aspect of the model. On the other hand, although all three simulation-type models are statistically inferior to the DBM model, they perform sufficiently well in a predictive sense to make them conditionally valid under our definition of this term. Moreover, they all have a conceptual basis that seems reasonably acceptable within current hydrological paradigms; and they can be employed easily within a simulation context, whether deterministic or stochastic.

It is their ability to function in this stochastic simulation sense that sets these latest HMC models apart from the deterministic simulation models that have so dominated rainfall–flow modelling in the past. The advent of very fast, desk-top (and portable[1]) computers, as well as developments in Monte Carlo simulation methods, mean that deterministic simulation should no longer be the norm. All systems are uncertain to some degree and even models constructed on a largely deterministic basis can now be subjected quite easily to powerful stochastic uncertainty and sensitivity analysis (see e.g. Young et al. 1996; Parkinson and Young 1998; Young 1999a). The author certainly believes that the era of deterministic reductionism should be terminated and that future hydrological research should be well grounded within a more realistic stochastic setting that accepts the inevitable uncertainty of nature and exploits statistical methods appropriate to this setting.

Another advantage of the HMC models is that they can be incorporated easily into a stochastic state space setting. This is useful for two main reasons. First, it can provide a framework for final *stochastic* optimisation, in which the associated Kalman filter is exploited to

---

[1] All of the analysis described in this chapter was carried out on a Macintosh Powerbook 3400c, within a Matlab™ software environment.

obtain maximum likelihood estimates of the model parameters (see Appendix 2). Second, the stochastic state space formulation is inherently recursive so that, when combined with the recursive estimation of the model parameters, which is straightforward in the case of the TF models considered here (Young 1984), it provides a vehicle for on-line adaptive modelling, forecasting and data assimilation. The DBM model appears to be particularly attractive in this context (e.g. Young and Tomlin 2000).

So what can we conclude at the end of this investigation of HMC rainfall–flow modelling? Unambiguously, there are good reasons for preferring the DBM model in forecasting applications. But which, of the other three HMC models (DBMS, IHACRES or BM), is preferable within the alternative simulation environment? This is almost an impossible question to answer since it requires us to reflect on what is meant by the term 'validation' in wider, semi-quantitative or qualitative terms, where personal subjective judgement becomes a factor. In particular, because all three models perform equally well, with only marginal statistical advantages to the DBMS model, it is necessary to decide which of them has the most attractive conceptual base. While the author has some preferences in this regard, only time will tell which of the models, or any future developments of them, will perform better in practical terms. And it is surely on such practical grounds that any preference should be given.

As regards future developments of the HMC models considered in this chapter, there remain two major problems that need attention. First, the poor identifiability and consequent equifinality of the nonlinear effective rainfall equations in the BM, IHACRES and DBMS models suggest strongly that more research is required on the nature and modelling of the rainfall–flow nonlinearity in such 'simulation' models. The DBM approach may help in such research, since the nonlinearity in the DBM model is much better defined and demonstrates the value of using measured rather than 'latent' variables in defining the nonlinear mechanisms. Second, in these models and most other rainfall–flow models the author and his colleagues have investigated, the model residuals are clearly non-normal and dominated by large outliers. The statistical analysis in this chapter has largely ignored this problem, but it is clear that superior estimation results and better quantification of the uncertainty in the stochastic models would be obtained if this aspect of rainfall–flow data is taken fully into account.

ACKNOWLEDGEMENTS

The author is grateful for the support of the UK Engineering and Physical Sciences Research Council under grant GR/G53866, and to Professor A.J. Jakeman of CRES, Australian National University, for his support during visits to CRES, which have strongly influenced the research reported in this chapter. He is also grateful to his colleagues Dr Laura Price, Dr Paul McKenna, Dr Nick Chappell and Professor Keith Beven for checking the chapter. Of course, the author remains responsible for any errors or omissions.

APPENDIX 1   DOMINANT MODE BEHAVIOUR

This appendix presents a simulation example that demonstrates the importance of dominant modal behaviour in model identification and estimation. The simulated transfer function model is of 26th order with one sample pure (advective) time delay and consists of 26 first order processes in series, each with unity gain and lag coefficient ranging from 0.1 in steps of 0.02 to 0.6 (e.g. time constant (residence times) from 0.43 to 1.96 days). In this sense, it could represent a 'lag and route' model of flow in a long river, or an ADZ model of solute transport, also in a long river. Since the model is an 'all pole' TF (i.e. the TF numerator is a scalar), the SRIV identification stage in the analysis is constrained to such models and a selection of the results are presented in Table 7.4 (even better fitting models can be identified if this restriction is removed). These were

Table 7.4   SRIV identification results

Model structure	YIC	R_T^2	σ_ξ^2	% error by variance	% error by stand. dev.
7 1 3	− 26.0291	0.999990	0.000139	0.0010	0.3131
6 1 4	− 22.8568	0.999936	0.000907	0.0064	0.7999
7 1 2	− 22.3942	0.999833	0.00235	0.0166	1.2876
6 1 3	− 22.5934	0.999819	0.00256	0.0181	1.3439
5 1 5	− 20.9560	0.999740	0.00367	0.0259	1.6091

obtained from the analysis of the noise-free impulse response data (the unit hydrograph in hydrological terms) generated by the 26th order simulation model. Although the best identified model is only seventh order, with a pure time delay of three sampling intervals, it is extremely well identified, with a very negative YIC value and a coefficient of determination R_T^2 of nearly unity (see Appendix 3). Neither the SRIV algorithm nor the PEM algorithm in Matlab converged at all for model orders greater than ten, indicating severe identifiability problems in these higher order cases. In other words, it is impossible to identify the true model from the simulation data, despite the fact that the data are noise free.

The best identified seventh order model takes the following form:

$$y_t = \frac{0.0001169}{1 - 5.65z^{-1} + 13.93z^{-2} - 19.39z^{-3} + 16.47z^{-4} - 8.54z^{-5} + 2.50z^{-6} - 0.32z^{-7}} u_{t-3}$$

where the denominator polynomial coefficients have been rounded to two decimal places. The standard errors on the estimated parameters are very small and so these are not reported. The steady-state gain of this reduced order model is insignificantly different from the unity value of the 26th order model. If the 26th order model is now perturbed by a *different* input, in the form of the Coweeta rainfall data from sample 8310 to 8810, then the above model performs in a very similar manner. The percentage error by variance is only 0.001 (0.32 by standard deviation) despite the fact that the input data is so different in this 'validation' test.

These results are confirmed by the power decomposition analysis of Liaw (1986), which shows that the percentage power contributed to the response of the 26th order model by the first seven modes is 99.6% of the total power in the model response, and that this power is distributed as shown in Table 7.5. In other words, practically all of the model response is explained by the effect of these dominant modes and, in fact, the first five modes explain approximately 97.5% of the response.

The message provided by this example is clear: a high order linear system can be approximated almost exactly by a much lower order linear system (often including an additional pure time delay) which reflects the dominant modes of the system's dynamic behaviour. Since this result is generic and certainly applies to most linear dynamic systems of the kind encountered in environmental and hydrological science, the consequences on model identification and parameter estimation are profound. In particular, the identifiable order of a dynamic system is clearly limited by its dominant mode characteristics, and it is impossible, in almost all practical situations, to unambiguously estimate the parameters from experimental data, even when the level of noise of the data is very low.

Table 7.5   Liaw analysis results

Dynamic mode	1	2	3	4	5	6	7
% power	48.30	28.47	11.67	6.37	2.60	1.69	0.49

APPENDIX 2   NON-STATIONARY AND NON-LINEAR MODEL ESTIMATION

This appendix outlines the approach to time variable and state-dependent parameter estimation employed in the DBM modelling of non-stationary and nonlinear systems. In the main text, this is used to identify the nature of the effective rainfall nonlinearity shown in Figure 7.3.

The TVP/SDP Model

In discrete-time, sampled data terms, assume that the behaviour of a typical measured variable y_t can be described by a general stochastic, dynamic equation of the form,

$$y_t = \mathcal{T}\{\boldsymbol{\chi}_t, \mu_t\} \tag{A2.1}$$

where $\mathcal{T}\{\cdot\}$ is a reasonably behaved, nonlinear function of the variables in an extended or *Non-Minimum State Space* (NMSS; see Young et al. 1987) defined by the following NMSS state vector,

$$\boldsymbol{\chi}_t = [y_{t-1} \ldots y_{t-n}\mathbf{u}_t^T \ldots \mathbf{u}_{t-m}^T\mathbf{U}_t^T \ldots \mathbf{U}_{t-q}^T]^T$$

We see that $\boldsymbol{\chi}_t$ is composed of the past values of y_t, as well as present and past values of a deterministic input (or exogenous) variable vector $\mathbf{u}_t$ with elements $u_{i,t}, i = 1, 2, \ldots, r$; and the present and past values of a vector $\mathbf{U}_t$ containing other exogenous variables $U_{j,t}, j = 1, 2, \ldots, p$. Finally, μ_t is an unobserved, zero mean, stochastic process with fairly general properties, which is the source of all stochasticity in the system and is assumed to be independent of the input variables $u_{i,t}$ and $U_{j,t}$.

This model assumes that y_t is causally related to the *primary* input variables $u_{i,t}$, while the vector $\mathbf{U}_t$ represents any other associated variables which *may* affect the system nonlinearly but whose relevance in this regard may not be clear prior to time series analysis. For example, in the rainfall–flow case, y_t is the flow measurement, $u_{1,t} = r_t$ is the rainfall, and $U_{1,t} = T_t$ is the temperature.

If, for simplicity, we consider the case of a single, primary, input variable, so that $r = 1$ then, following arguments similar to those presented in Young (1993) and Young and Beven (1994), the nonlinear model (A2.1) can be approximated by the following general TF model with time or state-dependent parameters,

$$y_t = \frac{B(t, z^{-1})}{A(t, z^{-1})}u_{t-\delta} = \frac{b_{0,t} + b_{1,t}z^{-1} + \ldots + b_{m,t}z^{-m}}{1 + a_{1,t}z^{-1} + \ldots + a_{n,t}z^{-n}}u_{t-\delta} + \xi_t \tag{A2.2a}$$

or,

$$y_t = x_t + \xi_t \tag{A2.2b}$$

where x_t, which can be considered as the noise-free output of the model, is defined as,

$$x_t = \frac{B(t, z^{-1})}{A(t, z^{-1})}u_{t-\delta}; \tag{A2.2c}$$

ξ_t is a stochastic noise term arising from the stochastic disturbance μ_t in (A2.1) which, like μ_t, is assumed to be statistically independent of the input variable u_t; and δ is any pure time delay affecting the relationship between u_t and y_t.

The model (A2.2) is a time/state-dependent parameter version of a standard TF model, such as equation (1) in the main text, where the polynomial parameters are assumed to be possible functions of the time index t. In the present context, these Time Variable (TVP) or State Dependent (SDP) parameters will reflect the nature of any non-stationary or nonlinear aspects of the system behaviour. And so their statistical estimates, based on the data $\{y_t, u_t\}$, should provide a potential source of information on the nature of the nonstationarity and/or nonlinearity in the system. The approximation of the nonlinear system by a TVP/SDP model such as (A2.2) is the key assumption in the first stage of DBM modelling. As shown in Young (1984, 1988, 1993, 1998a,b, 1999b) and Young and Beven (1994), TVP/SDP estimation is based on a powerful *Fixed Interval Smoothing* (FIS) method of recursive estimation. The details of FIS estimation are given in these prior publications and so it will suffice here to present only the latest improvements to the procedures described in these earlier publications.

FIS Estimation of TVPs

The FIS algorithm is applied to the following alternative vector form of the TF model (A2.2), where the pure time delay δ has been omitted to simplify the presentation:

$$y_t = \mathbf{z}_t^T \mathbf{a}_t + \eta_t; \quad t = 1, 2, \ldots, N \tag{A2.3}$$

where,

$$\mathbf{z}_t^T = [-y_{t-1} - y_{t-2} \cdots - y_{t-n} u_t \cdots u_{t-m}];$$
$$\mathbf{a}_t = [a_{1,t} a_{2,t} \cdots a_{n,t} b_{0,t} \cdots b_{m,t}]^T$$

and η_t is a stochastic noise term arising from the presence of ξ_t in (A2.2). We can say little about the statistical properties of η_t because of the general TVP nature of the model. Consequently, it will be assumed only that it is statistically independent of the input variable u_t.

Based on this model form, the FIS algorithm provides an estimate $\hat{\mathbf{a}}_{t|N}$ of the model parameter vector $\mathbf{a}_t$ at every sampling instant, conditional on the time series data $\{y_t, u_t\}$, over the whole observation interval $t = 1, 2, \ldots, N$. In addition, if the noise η_t is assumed to be a zero mean sequence of serially uncorrelated random variables (discrete white noise), it also yields a good estimate, at each sampling instant, of the covariance matrix $\mathbf{P}_t = E\{\tilde{\mathbf{a}}_{t|N} \tilde{\mathbf{a}}_{t|N}^T\}$, where $\tilde{\mathbf{a}}_{t|N} = \hat{\mathbf{a}}_{t|N} - \mathbf{a}_t$ is the estimation error. If η_t is not white noise, then the matrix $\mathbf{P}_t$, as derived from the FIS algorithm in the same manner, will not provide an accurate estimate of the covariance properties of the FIS estimates. Nevertheless, it still provides information on the *relative* accuracy of the parameter estimates which can prove useful in evaluating the detailed nature of the estimated parameter variations.

FIS Estimation of SDPs by Sorting and Backfitting

FIS estimation of the TVP vector $\mathbf{a}_t$ in (A2.3) is limited to slowly varying parameters: i.e. parameters that vary slowly in relation to the variations in the input and output variables y_t and u_t (see Young 1999b). In the case where $\mathbf{a}_t$ is a SDP vector, on the other hand, the estimation procedure normally needs to be modified. In particular, it becomes necessary to sort the data in a non-temporal order, so that the SDP variations are smoother and less rapid. In this manner, the FIS algorithm is able to estimate the parametric variations more easily and so obtain more accurate, lower variance estimates. For example, if the time series are sorted in some common 'ascending order' manner (the *sort* operation in Matlab), then the rapid natural variations in y_t and u_t are effectively eliminated from the data and replaced, in the sorted data space, by much smoother and less rapid variations. And if the SDPs are, indeed, related to these variables, then they will be similarly affected by the sorting. Following FIS estimation, however, these SDP estimates can be 'unsorted' (a trivial *unsort* operation to reverse Matlab's *sort*) and their true, rapid variation becomes apparent.

One obvious requirement with this new approach to SDP estimation is that the sorting of data, prior to FIS estimation, must be *common to all of the variables in the relationship* (A2.3). If an 'ascending order' strategy is selected, therefore, it is necessary to decide upon which variable in the model the sorting should be based. The simplest strategy is to sort according to the ascending order of the 'dependent' variable y_t (here flow). Depending upon the nature of each SDP in the vector $\mathbf{a}_t$, however, a single variable sorting strategy, such as this, may not produce entirely satisfactory results. If this is the case, then a more complicated, but still straightforward, 'backfitting' procedure is exploited. Here, each parameter is estimated *in turn*, based on the 'modified dependant variable (mdv)' series obtained by subtracting all the other terms on the right hand side of (A2.3) from y_t. In this situation, the sorting, at each such backfitting iteration, can be based on the single variable associated with the current SDP being estimated.

To clarify this backfitting procedure, let us consider the DBM rainfall–flow model discussed in the main text. The analysis starts by considering the simplest TVP/SDP relationship that might explain the dynamic relationship between the rainfall r_t, temperature T_t and flow y_t. Initial SRIV identification suggests that, in the case of the Coweeta data, this is the following first order model:

$$y_t = b + \frac{b_{0,t}}{1 + a_{1,t}z^{-1}}r_t + \frac{fz^{-6}}{1 + gz^{-1}}T_t + \xi_t \tag{A2.4}$$

where it will be noted that the second TF between T_t and flow is identified as being linear with *constant* parameters. If all terms except the noise ξ_t and TF between r_t and y_t are taken to the left hand side of the equation (since they involve only parameters that have been identified as constant over time), then equation (A2.4) can be written in the form,

$$y_t^* = \frac{b_{0,t}}{1 + a_{1,t}z^{-1}}r_t + \xi_t \tag{A2.5a}$$

where,

$$y_t^* = y_t - b - \frac{fz^{-6}}{1 + gz^{-1}}T_t \tag{A2.5b}$$

or, considering (A2.5a) in equation form,

$$y_t^* = -a_{1,t}y_{t-1}^* + b_{0,t}r_t + \eta_t \tag{A2.5c}$$

which is a first order example of the general model (A2.3) with y_t replaced by y_t^*.

In this case, sorting according to the ascending order of y_t^* yields sensible FIS estimation results (see Figure 7.3), in the sense that the smoothed SDP estimates are relatively smooth and the SDP model explains the data well with $R_T^2 = 0.9$. However, if this were not the case, backfitting would need to be invoked. Then, a typical algorithmic approach would be as follows:

- Assume that, without any sorting (i.e. as in the previous studies of Young 1993; Young and Beven 1994), FIS estimation has yielded prior TVP estimates $\hat{a}_{1,t|N}^1$ and $\hat{b}_{0,t|N}^1$ of $a_{1,t}$ and $b_{0,t}$, respectively.[1] An SDP estimation equation for $a_{1,t}$ can then be formulated as,

$$[y_t^* - \hat{b}_{0,t|N}^1 r_t]^{sy} = a_{1,t}^{sy}(y_{t-1}^*)^{sy} \tag{A2.6}$$

---

[1] These could be simply the SRIV *constant* parameter estimates, since the convergence of the backfitting procedure is not too sensitive to the prior estimates, provided they are reasonable. Here, for example, the SRIV constant parameter estimates yield an $R_T^2 = 0.74$ and SDP estimation raises this to $R_T^2 = 0.9$.

where the superscript sy denotes that all the variables are sorted in the ascending order of the right hand side variable y^*_{t-1} in (A2.5c), which is associated with the SDP $a_{1,t}$. This yields the FIS estimate $\hat{a}^{sy}_{1,t|N}$ of $a^{sy}_{1,t}$.

- $\hat{a}^{sy}_{1,t|N}$ is then 'unsorted' so that the SDP estimation equation for $b_{0,t}$ can be formulated as,

$$[y^*_t + \hat{a}_{1,t|N}y^*_{t-1}]^{su} = b^{su}_{0,t}r^{su}_t \tag{A2.7}$$

with the variables now sorted according to the ascending order of the rainfall variable r_t associated with $b_{0,t}$. This yields the FIS estimate $\hat{b}^{su}_{0,t|N}$ and the first iteration of the 'backfitting' algorithm is complete.

- This process is continued in an iterative manner (each time unsorting, forming the mdv, and sorting according to the current right hand side variable, prior to FIS estimation), until the FIS estimates of the SDP's $\hat{a}_{1,t|N}$ and $\hat{b}_{0,t|N}$ (which are each time series of length N) do not change significantly between iterations. The smoothing 'hyper-parameter' required for FIS estimation at each iteration is optimised by Maximum Likelihood (ML), based on prediction error decomposition (see Young 1999b).

It must be emphasised that the backfitting algorithm is described in the above simple manner for illustrative purposes. In the rainfall–flow case, as we have seen above, backfitting is not required at all. Also, the sorting at each stage does not have to be based on the variable with which the SDP is associated in the model. It general, it should be *associated with any measured variable on which the parameter variations may be dependent*. In the rainfall–flow case, for example, sorting of all variables in the model is based on the sorting order of the flow measurement y_t, since this is where the significant state dependency resides.

Since the SDP estimates resulting from the above backfitting algorithm are themselves time series, the algorithm constitutes a special form of non-parametric estimation and, as such, can be compared with the Generalised Additive Modelling (GAM) approach of Hastie and Tibshirani (1996). However, in both conceptual and algorithmic terms, the algorithm is significantly different from this earlier approach. Indeed, in the current dynamic modelling context, the GAM backfitting algorithm will only produce unbiased estimates of the SDPs in (A2.5c) if the noise η_t has certain special properties. In particular, it should constitute a zero mean, serially uncorrelated series that is independent of any noise on the model variables y^*_t and u_t (i.e. the model should represent the SDP version of the AutoRegressive eXogenous variable (ARX) model, rather than the more general TF model considered here). However, by exploiting a new instrumental variable form of the FIS algorithm (Young 2000a), the proposed backfitting algorithm is not limited in this manner and is able to produce unbiased non-parametric estimates of the TF model parameters.

Final Nonlinear Model Estimation

The primary aim of the non-parametric FIS analysis in previous sections of this appendix is to identify a sensible functional form for any nonlinearities in the model; *in other words, the analysis is aimed at nonlinear model structure identification*. In general, this identification phase is completed by fitting a *finite dimensional* parametric function to the non-parametric estimate of the nonlinearity using Weighted Least Squares (WLS) estimation, as discussed in Young (1993) and Young and Beven (1994). Figure 7.3 illustrates this procedure within the present rainfall–flow modelling context: the dots show the estimate $\hat{b}_{0,t|N}$ plotted against the sorted flow variable y^*_t and the full line is the WLS estimate of a power law relationship of the form $\hat{b}_{0,t|N} = \alpha(y_t)^\beta$ (see the comments about this in the main text). The TVP estimation procedure suggests that the lag parameter $a_{1,t}$ does not change too significantly over the observation interval and so it is assumed time-invariant in this case. As a result, the identified model, incorporating the power law, takes the form,

$$y^*_t = \frac{b_0}{1 - a_1 z^{-1}}u_t + \xi_t; \quad u_t = c(y_t)^\beta \tag{A2.8}$$

where $b_0 \cdot c = \alpha$. This constitutes the end of the first stage of nonlinear DBM model identification.

The model (A2.8) explains the Coweeta data reasonably well but further SRIV identification and estimation with u_t defined as in (A2.8) suggests that the dynamics are a little more complicated than this, with a second order relationship indicated between the modified (effective) rainfall u_t and y_t. As a result the finally identified model structure takes the form shown in equation (5a) of the main text. Once a plausible structural form for the nonlinear model, such as this, has been identified, it is necessary to finally re-estimate the model against the time series data. This is accomplished by some form of numerical optimisation (e.g. deterministic minimisation of the model residual variance; maximum likelihood estimation; prediction error minimisation etc.), the exact nature of which will tend to depend on the identified form of the model.

In the present rainfall–flow example, the nonlinearity resides only in the effective rainfall input, so that the linear and nonlinear parts of the model are relatively separable. As a result, the optimisation method used to obtain the model (5a) involves optimising the nonlinear component parameters with, at each step in the optimisation, the linear TF model parameters estimated by SRIV estimation. This can be considered as an approximation to a more elegant (but computationally more demanding) Maximum Likelihood (ML) procedure based on prediction error decomposition (see Schweppe 1965; Harvey 1989). However, recent research (Young (2000a), which considers a variety of nonlinear time series, including the modelling of blowfly population dynamics) has suggested that the recursive ML methodology is feasible for highly nonlinear stochastic systems and its application to rainfall–flow modelling is now being considered.

A Summary of the Complete DBM Procedure for a SISO System

In the case of a single input–single output system, the analysis discussed in the previous subsections can be summarised as follows:

1. Examine the available time-series data $\{y_t, u_t\}$ and use these to estimate the parameters in the best identified, constant parameter TF model using a reliable method of TF model identification and estimation (here the SRIV method). Apply standard statistical tests to the model residuals, including tests for nonlinearity (e.g. Billings and Voon 1986): if the results indicate linearity, then the linear model can be accepted and the analysis is complete. Alternatively, proceed to step (2).
2. Based on the analysis in (1) and any knowledge about the physical nature of the system, select: (a) the variables that could characterise the NMSS vector, and (b) the simplest TF model that appears capable of characterising the behaviour of the output variable y_t in relation to the observed input u_t (this will often be a first or second order model).
3. Use FIS estimation to obtain initial non-parametric TVP estimates of the parameters in the initial, simple TF model and define those parameters in $\hat{\mathbf{a}}_t$ which show significant variation over the observation interval. If any TVP/SDP parameters are identified, then obtain the FIS estimate of the TF model parameter vector $\hat{\mathbf{a}}_{t|N}$, if necessary (in the SDP case) using the backfitting algorithm. Note the relative accuracy of these estimates over time by reference to the appropriate diagonal elements of the covariance matrix $\mathbf{P}_{t|N}$.
4. Examine the nature of the FIS estimated time variation in the parameters in relation to all the variables in the defined NMSS vector, using devices such as scatter plots, correlation analysis etc., in all cases taking into account the relative accuracy of the FIS estimates identified in 3. On the basis of these results and a physically meaningful interpretation of the model, define nonlinear, parametric relationships that are able to approximate the FIS estimated non-parametric relationships.
5. On the basis of the results in 4, use WLS estimation to estimate the constant parameters that characterise these nonlinear laws, with the weighting defined by the diagonal elements of the covariance matrix $\mathbf{P}_{t|N}$.
6. Use the estimates of the parameters in 5 as starting values in a final model estimation stage where the (now hopefully constant) parameters in the identified nonlinear, stochastic model are estimated by some form of numerical optimisation based on the identified nonlinear model form and all the relevant data in the NMSS vector.
7. Analyse the residuals from the nonlinear estimation to ensure that there is no evidence of any residual nonlinearity not identified in steps 1–6. This should include standard statistical tests, such as correlation analysis and normality statistics, as used in the main text, as well as nonlinearity tests on the residuals (e.g. Billings and Voon 1986).

APPENDIX 3   TF MODEL ORDER IDENTIFICATION

In DBM and TF modelling, model order identification is based around the R_T^2, YIC and AIC criteria, which are defined as follows, where y_t is the measured output of the system:

(i) $R_T^2 = 1 - \dfrac{\hat{\sigma}^2}{\sigma_y^2};\quad \sigma_y^2 = \dfrac{1}{N}\sum_{t=1}^{t=N}[y_t - \bar{y}]^2;\quad \bar{y} = \dfrac{1}{N}\sum_{t=1}^{t=N}y_t$

(ii) $\text{YIC} = \log_e\dfrac{\hat{\sigma}^2}{\sigma_y^2} + \log_e\{\text{NEVN}\};\quad \text{NEVN} = \dfrac{1}{np}\sum_{i=1}^{i=np}\dfrac{\hat{\sigma}^2 \cdot \hat{p}_{ii}}{\hat{a}_i^2}$

(iii) $\text{AIC}(np) = N\log_e\hat{\sigma}^2 + 2 \cdot np$

Here $\hat{\sigma}^2$ is the variance of the model residuals; σ_y^2 is the variance of $y_t - \bar{y}$; $np = n + m + 1$ is the number of estimated parameters in the $\hat{\mathbf{a}}_N$ vector; $\hat{p}_{ii}$ is the ith diagonal element of the $\hat{\mathbf{P}}_t$ covariance matrix obtained from the estimation analysis (so that $\hat{\sigma}^2 \cdot \hat{p}_{ii}$ can be considered as an approximate estimate of the variance of the estimated uncertainty on the ith parameter estimate); and $\hat{a}_i^2$ is the square of the ith parameter estimate in the $\hat{\mathbf{a}}_N$ vector.

We see that the coefficient of determination R_T^2 is a statistical measure of how well the model explains the data: if the variance of the model residuals $\hat{\sigma}^2$ is low compared with the variance of the data σ_y^2, then R_T^2 tends towards unity, while if $\hat{\sigma}^2$ is of similar magnitude to σ_y^2 then it tends towards zero (and can even become negative). Note, however, that R_T^2 is based on the variance of the model errors $\hat{e}_t$ and it is *not* the more conventional coefficient of determination R^2 based on the variance of the *one-step-ahead prediction errors*: this is because R_T^2 is a more discerning measure than R^2 for TF model identification.

The YIC is a more complex, heuristic criterion. From the definition of R_T^2, we see that the first term is simply a relative measure of how well the model explains the data: the smaller the model residuals the more negative the term becomes. The second term, on the other hand, provides a measure of the conditioning of the Instrumental Product Matrix (IPM), which needs to be inverted when the IV normal equations are solved (see Young 1984). If the model is over-parameterised, then it can be shown that the IPM will tend to singularity and, because of its ill-conditioning, the elements of its inverse $\hat{\mathbf{P}}_t$ will increase in value, often by several orders of magnitude. When this happens, the second term in the YIC tends to dominate the criterion function, indicating over-parameterisation. An alternative justification of the YIC can be obtained from statistical considerations (see e.g. Young 1989). Although heuristic, the YIC has proven very useful in practical identification terms over the past ten years: it should not, however, be used as a sole arbiter of model order and improvements in its definition are being researched.

Finally, the Akaike Information Criterion AIC is a well-known identification criterion for AR processes (Akaike 1974) and is used here to identify the order of AR models for the noise process ξ_t, based on the analysis of the model residuals $\hat{e}_t$. Here, the first term is a measure of how well the model explains the data, while the second term is simply a penalty on the number of parameters in the model. Thus, as in the YIC, the AIC seeks a compromise between the degree of model fit and the complexity of the model.

APPENDIX 4   BM, IHACRES AND DBMS ESTIMATION RESULTS

The tables in this appendix (Tables 7.6 to 7.8) report the most significant of the estimation results for the BM, IHACRES and DBMS models. The complete results are not reported to conserve space.

Table 7.6   BM model

Instantaneous TF (A)			
Eig	SSG	TC	% of $\ddot{y}_{1,t}$ flow
0	0.17	0.00	18.4
Fast Flow TF (B)			
Eig	SSG	TC	% of $\ddot{y}_{1,t}$ flow
0.52	0.22	1.49	25.0
Slow Flow TF (C)			
Eig	SSG	TC	% of $\ddot{y}_{1,t}$ flow
0.98	0.51	43.0	56.6
	Constant Base flow		
	0.50 ($\forall t$)		

$\hat{\beta} = 0.95(0.02)$; $\hat{T}_r = 31.32(0.52)$; $\tau_s = 7.16(0.48)$.
(Note: to allow for comparison with the IHACRES model, T_m in equation (2)(i) was replaced by the daily temperature T_t.)

Table 7.7   IHACRES model

Instantaneous TF (A)			
Eig	SSG	TC	% of $\ddot{y}_{1,t}$ flow
0	0.17	0.00	21.0
Fast Flow TF (B)			
Eig	SSG	TC	% of $\ddot{y}_{1,t}$ flow
0.52	0.22	1.53	29.7
Slow Flow TF (C)			
Eig	SSG	TC	% of $\ddot{y}_{1,t}$ flow
0.97	0.40	33.9	49.3
	Constant Base flow		
	0.88 ($\forall t$)		

$\hat{\beta} = 0.92(0.04)$; $\hat{f} = 12.80(0.64)$; $\hat{\tau}_s = 4.80(0.41)$.

Table 7.8   DBMS model

Instantaneous TF (A)			
Eig	SSG	TC	% of $\ddot{y}_{1,t}$ flow
0	0.17	0.00	26.1
Fast Flow TF (B)			
Eig	SSG	TC	% of $\ddot{y}_{1,t}$ flow
0.38	0.17	1.03	25.6
Slow Flow TF (C)			
Eig	SSG	TC	% of $\ddot{y}_{1,t}$ flow
0.92	0.32	12.66	48.3
Slow Temperature (Seasonal) Effect (D)			
Eig	SSG	TC	TD
0.97	− 0.23	9.63	6
	Constant Base flow		
	1.82 ($\forall t$)		

$\hat{\beta} = 1.49(0.07)$; $\hat{c} = 0.87(0.007)$; $0.036(0.001)$.

REFERENCES

Abbott, M.B., Bathurst, J.C., Cunge, J.A., O'Connell, P.E. and Rasmussen, J.L. 1986. An introduction to the European Hydrology System (SHE). 2: Structure of a physically based distributed modelling system. *Journal of Hydrology*, **87**, 61–77.

Akaike, H. 1974. A new look at statistical model identification. *Institution of Electrical and Electronic Engineers, Transactions on Automatic Control*, **AC-19**, 716–723.

Bacon, F. 1620. *Novum Organum*.

Beck, M.B. and Young, P.C. 1975. A dynamic model for DO-BOD relationships in a non-tidal stream. *Water Research*, **9**, 769–776.

Beer, T. and Young, P.C. 1983. Longitudinal dispersion in natural streams. *American Society of Civil Engineering, Journal of Environmental Engineering*, **109**, 1049–1067.

Beven, K.J. and Kirkby, M.J. 1979. A physically-based variable contributing area model of basin hydrology. *Hydrological Sciences Journal*, **24**, 43–69.

Billings, S.A. and Voon, W.S.F. 1986. Correlation based model validity tests for nonlinear models. *International Journal of Control*, **44**, 235–244.

Box, G.E.P. and Jenkins, G.M. 1970. *Time Series Analysis Forecasting and Control*. Holden-Day, San Francisco.

Chappell, N.A., McKenna, P., Bidin, K., Douglas, I. and Walsh, R.P.D. 1999. Parsimonious modelling of water and suspended-sediment flux from nested-catchments affected by selective tropical forestry. *Philosophical Transactions, Royal Society of London B*, **354**, 1831–1846.

Crawford, N.H. and Linsley, R.K. 1966. *Digital Simulation in Hydrology: The Stanford Watershed Model IV*. Technical Report 39, Stanford University, California.

Franks, S., Beven, K.J., Quinn, P.F. and Wright, I. 1997. On the sensitivity of soil–vegetation–atmosphere transfer (SVAT) schemes: equifinality and the problem of robust calibration. Agricultural and Forest Meteorology, **86**, 63–75.

Gamerman, D. 1997. *Markov Chain Monte Carlo*. Chapman & Hall, London.

Hasselmann, K. 1998. Climate-change research after Kyoto. *Nature*, **390**, 225–226.

Hasselmann, K., Hasselmann, S., Giering, R., Ocana, V. and Storch, H. von 1997. Sensitivity study of optimal CO_2 emission paths using a simplified structural integrated assessment (SIAM). *Climate Change*, **37**, 345–386.

Harvey, A.C. 1989. *Forecasting Structural Time Series Models and the Kalman Filter*. Cambridge University Press, Cambridge.

Hastie, T.J. and Tibshirani, R.J. 1996. *Generalized Additive Models*. Chapman & Hall, London.

Hornberger, G.M., Beven, K.J., Cosby, B.J. and Sappington, D.E. 1985. Shenandoah watershed study: calibration of a topography-based variable contributing area hydrological model for a small forested catchment. *Water Resources Research*, **21**, 1841–1850.

Jakeman, A.J. and Hornberger, G.M. 1993. How much complexity is warranted in a rainfall–runoff model? *Water Resources Research*, **29**, 2637–2649.

Jakeman, A.J., Littlewood, I.G and Whitehead, P.G. 1990. Computation of the instantaneous unit hydrograph and identifiable component flows with application to two small upland catchments. *Journal of Hydrology*, **117**, 275–300.

Johnston, J. and DiNardo, J. 1997. *Econometric Methods*, 4th edition. McGraw-Hill, New York.

Kirkby, M.J. 1976. Hydrograph modelling strategies. In: Peel, R.J., Chisholm, M. and Haggett, P. (eds), *Processes in Physical and Human Geography*. Academic Press, London, 69–90.

Konikow, L.F. and Bredehoeft, J.D. 1992. Ground water models cannot be validated. *Advances in Water Resources*, **15**, 75–83.

Kuhn, T. 1962. *The Structure of Scientific Revolutions*. University of Chicago, Chicago (2nd edition 1980).

Lees, M., Young, P.C., Beven, K.J., Ferguson, S. and Burns, J. 1994. An adaptive flood warning system for the River Nith at Dumfries. In: White, W.R. and Watts, J. (eds), *River Flood Hydraulics*. Institute of Hydrology, Wallingford, UK.

Liaw, C.M. 1986. Model reduction of discrete systems using the power decomposition method. *Institution of Electrical Engineers*, **133**(D), 30–34.

Oreskes, N., Shrader-Frechette, K. and Belitz, K. 1994. Verification, validation, and confirmation of numerical models in the earth sciences. *Science*, **263**, 641–646.

Parkinson, S.D. and Young, P.C. 1998. Uncertainty and sensitivity in global carbon cycle modelling. *Climate Research*, **9**, 157–174.

Popper, K. 1959. *The Logic of Scientific Discovery*. Hutchinson, London.

Schweppe, F. 1965. Evaluation of likelihood functions for Gaussian signals. *Institution of Electrical and electronic Engineers, Transactions on Information Theory*, **11**, 61–70.

Shackley, S., Young, P.C., Parkinson, S.D. and Wynne, B. 1998. Uncertainty, complexity and concepts of good science in climate change modelling: are GCMs the best tools? *Climate Change*, **38**, 159–205.

Wallis, S.G., Young, P.C. and Beven, K.J. 1989. Experimental investigation of the aggregated dead zone model for longitudinal solute transport in stream channels. *Proceedings of the Institute of Civil Engineers, Part 2*, **87**, 1–22.

Weyman, D.R. 1975. *Runoff Processes and Streamflow Modelling*. Oxford University Press, Oxford.

Wheater, H.S., Jakeman, A.J. and Beven, K.J. 1993. Progress and directions in rainfall–runoff modelling. In: Jakeman, A.J., Beck, M.B. and McAleer, M.J. (eds), *Modelling Change in Environmental Systems*. John Wiley, Chichester, Chapter 5.

Whitehead, P.G. and Young, P.C. 1975. A recursive approach to time series analysis for multivariable systems. In: Vansteenkiste, G.C. (ed.), *Modeling and Simulation of Water Resource Systems*. North Holland, Amsterdam, 39–58.

Ye, S., Jakeman, A.J. and Young, P.C. 1998. Identification of improved rainfall–runoff models for an ephemeral low yielding Australian catchment. *Environmental Modelling and Software*, **13**, 59–74.

Young, P.C. 1974. Recursive approaches to time series analysis. *Bulletin of the Institute of Mathematics and its Applications*, **10**, 209–224.

Young, P.C. 1978. A general theory of modelling for badly defined dynamic systems. In: Vansteenkiste, G.C. (ed.), *Modeling, Identification and Control in Environmental Systems*. North Holland, Amsterdam, 103–135.

Young, P.C. 1983. The validity and credibility of models for badly defined systems. In: Beck, M.B. and van Straten, G. (eds), *Uncertainty and Forecasting of Water Quality*. Springer-Verlag, Berlin.

Young, P.C. 1984. *Recursive Estimation and Time-Series Analysis*. Springer-Verlag, Berlin, 69–98.

Young, P.C. 1985. The instrumental variable method: a practical approach to identification and system parameter estimation. In: Barker, H.A. and Young, P.C. (eds), *Identification and System Parameter Estimation*. Pergamon Press, Oxford, 1–16.

Young, P.C. 1986. Time-series methods and recursive estimation in hydrological systems analysis. In: Kraijenhoff, D.A. and Moll, J.R. (eds), *River Flow Modelling and Forecasting*. D. Reidel, Dordrecht, 129–180.

Young, P.C. 1988. Recursive extrapolation, interpolation and smoothing of non-stationary time-series. In: Chen, H.F. (ed.), *Identification and System Parameter Estimation 1988*. Pergamon Press, Oxford, 33–44.

Young, P.C. 1989. Recursive estimation, forecasting and adaptive control. In: Leondes, C.T. (ed.), *Control and Dynamic Systems, Vol. 30, Part 3*. Academic Press, San Diego, 119–165.

Young, P.C. 1992. Parallel processes in hydrology and water quality: a unified time series approach. *Journal of the Institute of Water and Environmental Management*, **6**, 598–612.

Young, P.C. 1993. Time variable and state dependent parameter modelling of nonstationary and nonlinear time series. In: Subba Rao, T. (ed.), *Developments in Time Series*. Chapman & Hall, London, 374–413.

Young, P.C. 1996. Identification, estimation and control of continuous-time and delta operator systems. In: Friswell, M.I. and Mottershead, J.E. (eds), *Identification in Engineering Systems*. University of Wales, Swansea, 1–17.

Young, P.C. 1998a. Data-based mechanistic modelling of engineering systems. *Journal of Vibration and Control*, **4**, 5–28.

Young, P.C. 1998b. Data-based mechanistic modelling of environmental, ecological, economic and engineering systems. *Environmental Modelling and Software*, **13**, 105–122.

Young, P.C. 1999a. Data-based mechanistic modelling, generalised sensitivity and dominant mode analysis. *Computer Physics Communications*, **117**, 113–129.

Young, P.C. 1999b. Nonstationary time series analysis and forecasting. *Progress in Environmental Science*, **1**, 3–48.

Young, P.C. 2000a. Stochastic, dynamic modelling and signal processing: time variable and state dependent parameter estimation. In: Fitzgerald, W.A., Walden, A., Smith, R. and Young, P.C. (eds), *Nonstationary and Nonlinear Signal Processing*. Cambridge University Press, Cambridge, 74–114.

Young, P.C. 2000b. The identification and estimation of nonlinear stochastic systems. In: Mees, A.I. (ed.), *Nonlinear Dynamics and Statistics*. Birkhauser, Boston, in press.

Young, P.C. and Beven, K.J. 1994. Data-Based Mechanistic (DBM) modelling and the rainfall–flow nonlinearity. *Environmetrics*, **5**, 335–363.

Young, P.C. and Jakeman, A.J. 1979. Refined instrumental variable methods of recursive time-series analysis: part I, single input, single output systems. *International Journal of Control*, **29**, 1–30.

Young, P.C. and Jakeman, A.J. 1980. Refined instrumental variable methods of recursive time-series analysis: part III, extensions. *International Journal of Control*, **31**, 741–764.

Young, P.C. and Lees, M.J. 1993. The Active Mixing Volume (AMV): a new concept in modelling environmental systems. In: Barnett, V. and Turkman, K.F. (eds), *Statistics for the Environment*. John Wiley, Chichester, 3–44.

Young, P.C. and Minchin, P. 1991. Environmetric time-series analysis: modelling natural systems from experimental time-series data. *International Journal of Biological Macromolecules*, **13**, 190–201.

Young, P.C. and Pedregal, D. 1998. Data-based mechanistic modelling. In: Heij, C. and Schumacher, H. (eds), *System Dynamics in Economic and Financial Models*. Wiley, Chichester, 169–213.

Young, P.C and Pedregal, D. 1999. Macro-economic relativity: government spending, private investment and unemployment in the USA 1948–1998. *Structural Change and Economic Dynamics*, **10**, 359–380.

Young, P.C. and Tomlin, C.M. 2000. Data-based mechanistic modelling and adaptive flow forecasting. In: Walshe, P. and Lees, M. (eds), *Flood Forecasting: What Does Current Research Offer the Practitioner?* British Hydrological Society Occasional Paper.

Young, P.C. and Wallis, S.G. 1994. Solute transport and dispersion in channels. In: Beven, K.J. and Kirkby, M.J. (eds), *Channel Networks*. Wiley, Chichester, 129–173.

Young, P.C., Behzhadi, M.A., Wang, C.L. and Chotai, A. 1987. Direct digital and adaptive control by input–output, state variable feedback. *International Journal of Control*, **46**, 1861–1881.

Young, P.C., Parkinson, S.D. and Lees, M. 1996. Simplicity out of complexity in environmental systems: Occam's Razor revisited. *Journal of Applied Statistics*, **23**(2,3), 165–210.

Young, P.C., Schreider, S. Yu. and Jakeman, A.J. 1997. A streamflow forecasting algorithm and results for the Upper Murray Basin. In: McDonald, A. D. and McAleer, M. (eds), *Proceedings MODSIM 97 International Congress on Modelling and Simulation*, Hobart, Tasmania, 1707–1712.

Young, P.C., Jakeman, A.J. and Post, D. 1998. Recent advances in the data-based modelling and analysis of hydrological systems. *Water Science and Technology*, **36**, 99–116.

Young, P.C., Price, L., Berckmans, D. and Janssens, K. 2000. Recent developments in the modelling of imperfectly mixed airspaces. *Computers and Electronics in Agriculture*, **26**, 239–254.

8 The Use of Remote Sensing to Validate Hydrological Models

D. PEARSON, M.S. HORRITT,[1] R.J. GURNEY AND D.C. MASON

University of Reading, UK

8.1 INTRODUCTION

Remote sensing has been held out to have great promise for advancing hydrology for many years. However, compared to meteorology or oceanography, the use of remote sensing is only widespread for estimating rainfall fields from ground- and space-based radars, and for estimating continental snow areas. Further advances in the use of the observations depend on the development of new models that need and can usefully assimilate new observations. This in turn depends on the development of hydrology into a science interested in water movement at a variety of scales, and not just locally, and on new quantitative observations. There are several candidate areas of hydrology. Passive microwave observations of snow volume are beginning to allow an understanding of the interannual variability of continental snow masses. Multi-angle radiometer and spectrometer data are beginning to allow much more accurate estimation of the radiation interception at the land surface. Spectrometer data of rivers and lakes are allowing the identification and mapping of eutrophication and other water quality problems. However, two particular areas are changing particularly rapidly as new models and new observations are brought together. The first is in passive microwave sensing of soil moisture, where new modelling approaches to time series of remote sensing observations are allowing soil moisture changes in the whole vadose zone to be inferred. The second is the use of different types of remotely sensed data to initialise and validate two- and three-dimensional models of river flooding. Both are difficult problems at the heart of hydrology, and both are now being made tractable by new observations and models. The approaches will be described in turn.

8.2 VALIDATION OF SOIL–VEGETATION–ATMOSPHERE MODELS USING OBSERVATIONS OF PASSIVE MICROWAVE EMISSION

8.2.1 Introduction

Section 8.3 explores some current issues in the hydrological modelling of floods. The evolution of such transient hydrological phenomena on land is largely governed by soil water content in the affected areas. Furthermore, the behaviour of groundwater is controlled by the soil's hydraulic properties, such as water content at saturation, the hydraulic conductivity, and others that will be discussed below. It is plain that a good knowledge of soil water content and

---

[1] Now at School of Geographical Sciences, University of Bristol, UK.

Model Validation: Perspectives in Hydrological Science. Edited by M.G. Anderson and P.D. Bates.
© 2001 John Wiley & Sons, Ltd.

hydraulic properties will enable us to improve our modelling and prediction of agricultural production, floods, aquifer recharge, transport of pollution, river flow, and of the effects of possible climate change.

Weather and climate are also subject to strong feedback with soil moisture. For example, Betts et al. (1996) showed that highly significant errors in predicted precipitation over North America occur in a weather model if the initial soil moisture content at the start of the simulation is wrong. Fast and McCorcle (1991), in a modelling study of the central USA, showed that the spatial distribution of soil moisture and also soil *type* on a grid of 104 km spacing are important. Phenomena like sea-breezes – referred to in the present context as non-classical mesoscale circulations (NCMCs) – arise due to spatial differences in evaporation and surface temperature induced by varying soil types and moisture contents. Significant changes in the humidity and temperature fields occur between the model and control runs in which soil type and soil moisture are spatially uniform. Furthermore, the heterogeneity-induced flows were found to interact with and modify synoptic-scale phenomena such as an incoming cold front and related moisture-convergence fields. Avissar and Pielke (1989) showed similar NCMC behaviour on a grid with an order of magnitude finer spacing than that used by Fast and McCorcle. Clearly, then, horizontal variations in soil type and soil moisture on different length scales are relevant to accurate modelling of atmospheric circulation. Njoku et al. (1999) go so far as to say: 'Soil moisture is one of the highest priority requirements of global change and operational forecasting not met by current or planned remote observing systems.'

Passive microwave remote sensing offers an approach to tackling all of these problems. Although at first sight it seems to give information only about a thin, surface layer of soil, use of models enables estimation of soil moisture content to be made through the vadose zone (i.e. from the surface down to the water table), and for soil hydraulic properties to be estimated.

This section continues with a brief introduction to the physical principles of passive microwave radiometry, then discusses applications. Applications naturally fall into four categories, each with a different modus operandi:

- point scale, no use of temporal variations;
- point scale, with use of temporal variations;
- heterogeneous area, no use of temporal variations;
- heterogeneous area, with use of temporal variations.

As we progress down this list, the science will become more speculative and forward-looking, and the models less well validated. A comprehensive literature review is not attempted – instead, selected publications are cited as illustrations.

8.2.2   Passive Microwave Remote Sensing

The fundamental radiometric quantity in this context is the *brightness temperature*, T_B. If a perfect black body could be made, it would emit thermal radiation with a radiance (or, loosely, 'brightness') given by Planck's law:

$$B_v = \frac{2hv^3}{c^2} \frac{1}{\exp(hv/kT) - 1} \quad [\mathrm{W\,m^{-2}\,sr^{-1}\,s^{-1}}] \tag{1}$$

where B_v is the spectral radiance at frequency v and temperature T, h is Planck's constant, c is the speed of light and k is Boltzmann's constant. At low frequencies, this expression can be expanded in a Taylor series about $v = 0$ to obtain the Rayleigh–Jeans Law:

$$B_v \simeq \frac{2kTv^2}{c^2} \qquad (2)$$

We can also consider the Rayleigh–Jeans approximation in terms of wavelength[1]:

$$B_\lambda \simeq \frac{2kTc}{\lambda^4} \qquad (3)$$

where λ is the wavelength. This approximation is correct to about one part in ten thousand at $T = 300\,\mathrm{K}$ and $\lambda = 0.21\,\mathrm{m}$. Thus, at a fixed value of the wavelength, the emitted radiance is very closely proportional to the temperature. A real body will emit thermal radiation more weakly, with an emissivity e:

$$B_{\lambda,\mathrm{real}} = eB_{\lambda,\mathrm{black\ body}} \qquad (4)$$

The Rayleigh–Jeans approximation enables us to define the brightness temperature:

$$T_B = eT \qquad (5)$$

The proportionality of radiance and brightness temperature enables us to replace the radiance with T_B in equations, provided that the Rayleigh–Jeans approximation is valid and a fixed wavelength (or a very narrow waveband) is considered.

The fundamental observation is that the brightness temperature of a soil decreases as its moisture content increases in the microwave part of the spectrum at constant thermodynamic temperature. The dry components of soil have a relative permittivity or dielectric constant, ε_r, typically around $\varepsilon_r \simeq 3.5$, while water has $\varepsilon_r \simeq 80$ at frequencies less than around 5 GHz (Njoku and Entekhabi 1996). Since air has $\varepsilon_r \simeq 1$, there is a discontinuity in the dielectric constant at the surface of soil, causing reflection of the upwelling thermal radiation and a consequent reduction in the brightness of the observable radiation that is transmitted through the surface. This is the reason for an emissivity less than unity. Increasing the water content increases the dielectric contrast, and so reduces the emissivity.

The behaviour of radiation at a smooth dielectric discontinuity is given by the Fresnel equations (e.g. Reitz et al. 1979; Ulaby et al. 1981). These give a different transmission coefficient for the horizontally and vertically resolved polarisations of the emitted radiation, in terms of the dielectric constants on each side of the interface, as a function of direction of observation. A deep homogeneous soil in thermodynamic equilibrium would emit radiation with an emissivity given directly by the Fresnel equations. Figure 8.1 illustrates this for a homogeneous sandy loam soil at two different volumetric water contents. The dielectric constants were taken from Figure 6(a) of Hallikainen et al. (1985). Observe that the emissivity is greater in the vertically polarised component of emitted radiation than in the horizontal – this is generally true for a given soil. Observe also that e decreases as water content increases at off-nadir observation angles of less than around 70° – this is always the case for small enough observation angles for a given soil.

The relation between a soil's dielectric constant and its volumetric water content depends on the soil's texture (i.e. the particle size distribution), salinity and, to a lesser extent, temperature

---

[1] Ulaby et al. (1981) get this wrong in their equation 4.39, and the error is propagated by Njoku and Entekhabi (1996) and others.

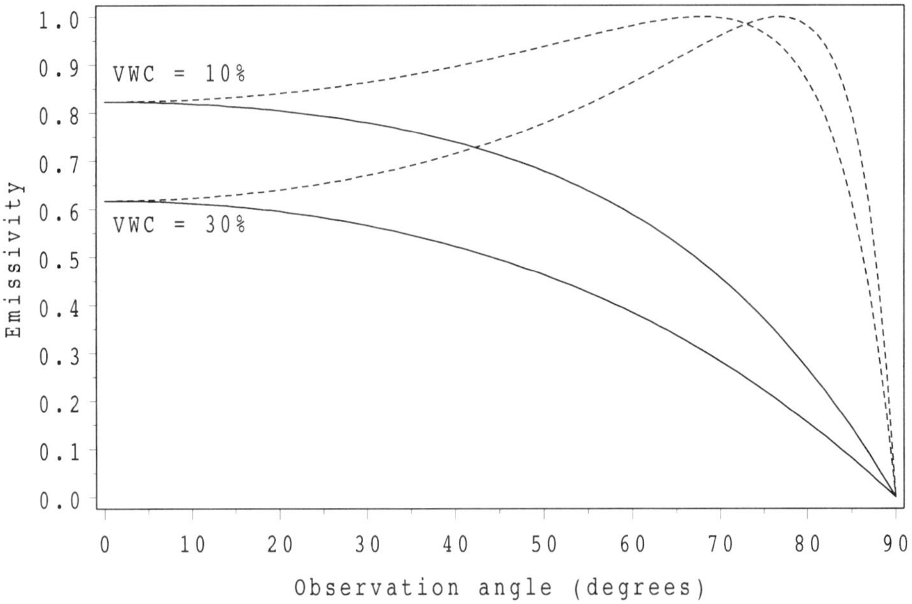

Figure 8.1   Emissivity of a sandy loam soil with two different values of the volumetric water content (W_s), as a function of the off-nadir angle of observation. Fresnel transmission at the surface of a vertically homogeneous soil is assumed. Solid lines – horizontal polarisation; broken lines – vertical

and the shapes of particles. Dependence on texture is a result of the strong attraction between the surface of soil particles and water molecules, which prevents the molecules from responding to external electromagnetic forces as freely as they do in the bulk liquid. The dielectric constant of moist soil has been modelled adequately, for example by Wang and Schmugge (1980) and Dobson et al. (1985).

Microwave radiation is exponentially attenuated by soil, so a radiometer will only receive radiation that originated from close to the surface – the exponential decay distance is known as the penetration depth, although 'emitting depth' would be a more accurate term. The attenuation coefficient is very largely a function of the imaginary part of the dielectric constant, ε_i (only the real part was discussed above), which depends on frequency (e.g. Wang and Schmugge 1980; Dobson et al. 1985). Thus the penetration depth depends on the radiometric frequency and on the soil moisture content. Many attempts have been made to estimate the penetration depth at different frequencies and conditions. Generally, it is reported to be between $\sim 0.06\lambda$ and λ (Owe and van de Griend 1998, and references therein).

A consequence of a non-zero penetration depth is that the radiation picked up by the radiometer originates from various depths in the soil profile, i.e. not just the surface. Thus the temperature T on the right hand side of equation (5) is a weighted average of temperatures down through the soil profile – an effective temperature, T_{eff}. Thus the emissivity is also an effective value, and equation (5) could more accurately be written as $T_B = e_{eff}T_{eff}$.

A similar attenuation occurs in vegetation, so any measurement of soil moisture taken above vegetation is adversely affected.

All of the above factors must be taken into account when choosing which frequency or frequencies at which to operate a microwave radiometer. A low frequency is preferred to minimise the attenuation by vegetation, and it also seems advantageous to have a soil penetration depth that is not too shallow, to prevent unwanted transient effects. The frequency of choice is 1.4 GHz, as this has a penetration depth up to about 20 cm, and can see through vegetation to some extent, while being in a protected astronomy band so electromagnetic interference should ideally be absent. Furthermore, atmospheric effects and extraterrestrial sources are rather small at this frequency. Therefore the rest of this part of the chapter will mainly be concerned with L-band radiometry.

8.2.3   Point-scale Radiometric Measurements

Here, we discuss experiments and models relating to relatively small, ground-based radiometers that observe a patch of ground with a field of view covering no more than around 10 m^2. Such an instrument, SWaMP-L (Pearson et al. 2000) is shown in Figure 8.2. As the horizontal scale is small, it is assumed that the problem is one-dimensional, i.e. depth in the soil is the only non-ignorable spatial coordinate.

Figure 8.2 The SWaMP-L radiometer deployed in the field. The circular parabolic dish antenna is clearly visible at the top of a steerable gantry. The white box immediately behind the antenna contains control and data acquisition equipment

Bare Soil

Although the Fresnel equations have pedagogical value, and in some cases provide a reasonable model of microwave emission, often a more sophisticated approach is required to adequately model experimental results. There are two major causes of failure of the Fresnel equation approach over bare soil: vertical heterogeneity and surface roughness.

Soil water and heat evolve in a dynamic and coupled way (Daamen and Simmonds 1996), obeying nonlinear coupled diffusion equations. As a result, it is extremely unlikely that a given soil column will have homogeneous vertical profiles of water content and temperature. If there are abrupt changes in the dielectric constant in the profile, for example due to a wetting front propagating downwards after rainfall or irrigation, strong reflection of microwave radiation will occur at the interface. Even if gradients are gentle, some reflection will take place within the soil. Therefore the upwelling radiance just below the surface is a combination of contributions from all depths in the profile, modified by interaction with the soil between them and the surface. Once the subsurface upwelling radiance is calculated, the Fresnel equations can *then* be used to estimate the emissivity, if the surface is smooth. Fortunately, volume scattering is negligible at L-band, so it is not necessary to solve the full equations of radiative transfer.

Many subsurface radiation transport models have been proposed, and Schmugge and Choudhury (1981) provide a useful review. The models fall naturally into two categories – coherent and incoherent. Incoherent models consider only flows of energy, quantified in the time-averaged Poynting vector, without considering the phase of the thermal radiation. Coherent models do take account of the phase of the electromagnetic waves, and are thus potentially more accurate – especially in the presence of discontinuities in the dielectric constant in the soil profile. A very common coherent approach is that of Wilheit (1978), who considered the soil to be composed of many thin layers, each of which is itself homogeneous, as an approximation to a profile that may be smoothly varying.

Roughness of the soil's surface confounds the situation further. Surface scattering both alters the directional distribution of radiation, and partially scrambles the state of polarisation.[1] Qualitatively, increasing roughness increases the brightness temperature, due to scattering between the roughness elements. As Mudaliar (1999) says, 'The literature on scattering from rough surfaces and random media is vast.' Models range from the simple and *ad hoc*, through semi-empirical to rigorous calculations of scattering of waves at a rough surface (e.g. Sarabandi and Chiu 1997). Rigorous calculation requires a priori knowledge of either the exact description of the surface roughness or a statistical description. Since neither of these are likely to be available, a simpler approach is required in practical applications. A promising method is one of the oldest, due to Wang and Choudhury (1981). They proposed that the emissivities be expressed as follows:

$$e_{eff,h}(\theta) = 1 - [(1 - Q)R_h(\theta) + QR_v(\theta)]\exp(- h\cos^2\theta) \tag{6}$$

$$e_{eff,v}(\theta) = 1 - [(1 - Q)R_v(\theta) + QR_h(\theta)]\exp(- h\cos^2\theta) \tag{7}$$

where $e_{eff,h}$ and $e_{eff,v}$ are the horizontal and vertical effective emissivities, Q is a polarisation mixing factor due to the surface roughness, R_h and R_h are the Fresnel reflectances and h is a parameter characterising height. We will refer to this as an h–Q model. This model was shown to work well. However, Wang et al. (1983) re-wrote equations (6) and (7), replacing the $\cos^2\theta$ dependence with a general function $G(\theta)$. It was found that setting G to a constant value of unity

---

[1] Although the thermal emission at any point in the soil is incoherent and unpolarised, polarisation arises through Fresnel-like effects.

gave the best fit between the model and their data. Furthermore, they noticed an anomalous reduction in T_B at 5 GHz over a smooth, wet field, concluding that the Wilheit (1978) radiation transport model was inadequate in this case. However, this problem does not appear to have been addressed in the literature since then.

Figure 8.3 illustrates that the simple, physically based models discussed above can be remarkably accurate. This shows measurements of T_B over a bare loamy sand soil, at 5 GHz and 1.4 GHz. Superimposed are results from a model incorporating the Wilheit (1978) model, the Wang and Choudhury (1981) roughness model and the Wang and Schmugge (1980) semi-empirical dielectric model. The ability of the model to simulate the data is remarkably good.

Vegetated Soil

Vegetation has a marked effect on microwave radiometric observations, through two mechanisms: attenuation and perturbation of the signal from the soil; and emission from the vegetation itself. Vegetation has a depolarising effect on the signal, because the soil signal, which is attenuated, is much more strongly polarised than the vegetation's own thermal emission (Wigneron et al. 1995). This latter phenomenon is due to the lack of a sharp boundary to a canopy, hence a lack of a sharp discontinuity in dielectric properties.

Thus estimation of soil moisture beneath a canopy requires correction for these effects. As in the case of soil radiation models, there is great variation in the complexity and realism of vegetation radiation models. Perhaps the simplest useful model is the τ–ω model, illustrated in Figure 8.4. Here, the soil's brightness temperature, $T_{B,s}$ is attenuated by passing through the vegetation, a fraction t_v passing through without interception. The self-emission brightness

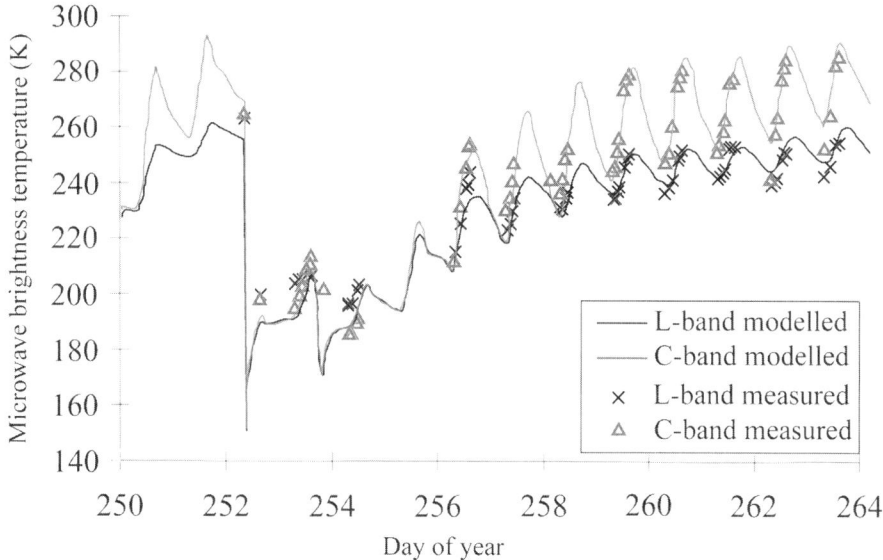

Figure 8.3 Observed and modelled time series of brightness temperature over bare soil, at L-band and C-band. The model was a combination of Wilheit (1978) radiative transfer and an h–Q roughness modifier. This diagram was supplied by L.P. Simmonds of the University of Reading, in connection with the work of Burke et al. (1998), but has not been published

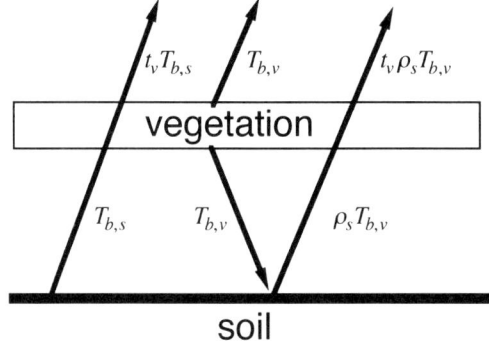

Figure 8.4 Schematic illustration of the brightness temperatures that are taken into account in the τ–ω model

temperature of the vegetation, $T_{B,v}$, is directly sensed by the overhead radiometer, while the same quantity is reflected from the soil by a reflectance ρ_s, and then attenuated by the vegetation by the same factor t_v. Thus the total brightness temperature observed by the radiometer is given by the sum of these three components:

$$T_B = t_v T_{B,s} + (1 + t_v \rho_s) T_{B,v} \tag{8}$$

By the definition of brightness temperature, this expression becomes

$$T_B = \varepsilon_s t_v T_s + (1 + t_v \rho_s) \varepsilon_v T_v \tag{9}$$

where T_s and T_v are the effective temperatures of the soil and vegetation. Now ε_v can be expressed as $(1 - t_v)(1 - \omega)$, according to van de Griend et al. (1996), where ω is the single scattering albedo.[1] Thus we obtain:

$$T_B = \varepsilon_s t_v T_s + (1 + t_v \rho_s)(1 - t_v)(1 - \omega) T_v \tag{10}$$

Equations (8)–(10) depend on the angle of observation, and each can be repeated once for each polarisation. It is usually assumed that Beer's law holds, i.e. that

$$t_v = e^{-\tau \sec \theta} \tag{11}$$

where τ is the extinction optical depth of the canopy and θ is the angle of observation. Thus this type of model is generally referred to as a τ–ω model. When coupled to an h–Q model, we will refer to it as a τ–ω–h–Q model. Furthermore, following Jackson and O'Neill (1990), it is often assumed that the optical depth is linearly related to the areal water content of the vegetation, W_v:

$$\tau = b W_v \tag{12}$$

The soil's emissivity can be related to its water content and temperature by any of the methods discussed under 'Bare soil' above – typically, the roughness model of Wang and Choudhury

---

[1] The single scattering albedo is defined as the fraction of intercepted radiation that is scattered. The rest of the radiation that is intercepted is absorbed. Historically (Chandrasekhar 1960) the symbol ϖ_0 was used, but this seems to have become ω over time.

(1981) is used, along with Wilheit's model (1978) or the Fresnel equations. Equation (10) is sometimes simplified by assuming that $T_v \simeq T_s$, e.g. O'Neill et al. (1996). Thus:

$$T_B \simeq T_s[\varepsilon_s t_v + (1 + t_v \rho_s)(1 - t_v)(1 - \omega)] \tag{13}$$

At 1.4 GHz, the single scattering albedo is small. Setting it to zero in equation (13) yields the simplest useful form of equation (10):

$$T_B \simeq T_s(1 - \rho_s t_v^2) \tag{14}$$

Njoku and Entekhabi (1996) provide a useful illustration of the effect of vegetation on above-canopy T_B. Assuming equation (14) and that the emissivity of soil is linearly related to W_s, equation (14) leads to:

$$\frac{\partial T_B}{\partial W_s} = \left(\frac{\partial T_B}{\partial W_s}\right)_{bare} e^{-2bW_v \sec\theta} \tag{15}$$

where $(\partial T_B/W_s)_{bare}$ is the sensitivity to W_s that would occur if the soil were bare. They assume values of $b = 0.12$ and $b = 0.24$ for 1.4 GHz and 5 GHz, respectively (taken from Jackson and Schmugge 1991), to obtain the results shown in Figure 8.5. This must be examined in parallel with real data. Wigneron et al. (1995) measured wheat water contents up to almost 3.5 kg m^{-2}; Jackson and O'Neill (1990) measured values of W_v up to ~ 6.5 kg m^{-2} in a soybean crop. In reference to Figure 8.5, one should not be too surprised if remote sensing estimation of W_s is less accurate under such canopies than over bare soil.

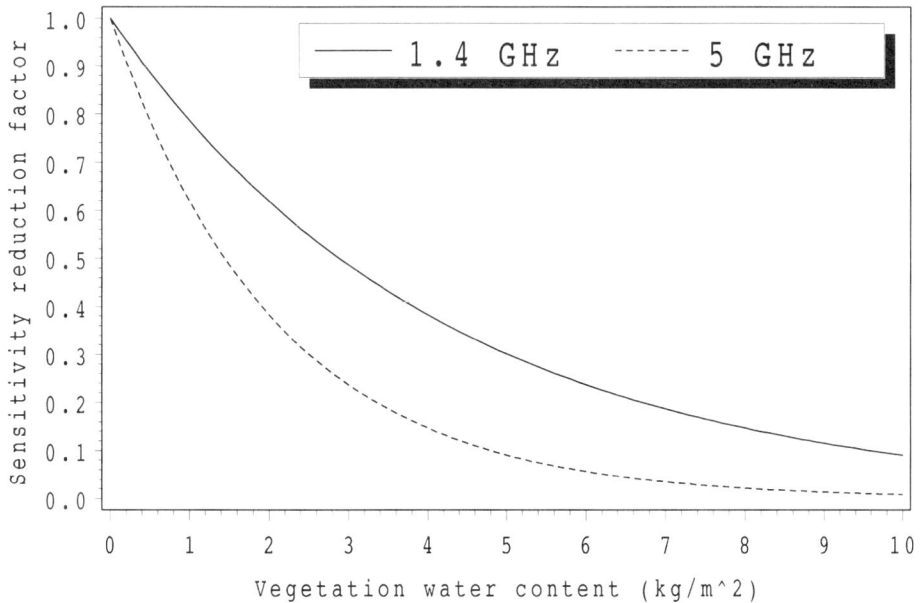

Figure 8.5 Calculated values of the sensitivity reduction factor for estimation of W_s, as a function of vegetation water content. The simplified $\tau-\omega$ model of equation (14) is used

According to equations (10) and (11), estimation of ε_s also requires ω and τ – hence the name τ–ω model. Thus at least three pieces of information are required. The careful analysis of Wigneron et al. (1995) exemplifies the use of the τ–ω model – this procedure is known as the τ–ω *method*. In this paper, various ways of acquiring the necessary data are tried, such as observing the scene from different angles, or at two frequencies (1.4 GHz and 5 GHz). They achieved a relative accuracy in the estimation of independently measured values of W_s and W_v of 15% over cropped fields. However, they had to insert values of Q and h into the analysis, values which were not derived as part of the remote sensing problem per se.

More complex models of radiation transport in vegetation are found in the literature. These fall naturally into two groups: discrete models, which assume a random arrangement of discrete scatterers such as leaves and stalks (e.g. Lang and Sidhu 1983 – applied practically by O'Neill et al. 1996); and continuous methods, which assume that the vegetation can be represented by a continuous dielectric medium that has fluctuations in its density (e.g. Jin 1989), quantified by mean quantities and autocorrelation functions. The former are useful in understanding and modelling the physical processes involved, but usually contain too many parameters to be of much use in the estimation of ground-level properties from remotely sensed data. The latter are less physically realistic than discrete models, but are of more use in estimation, as they contain fewer parameters that are a priori unknown.

Le Vine and Karam (1996) investigate the validity of equation (12) using a discrete model. They conclude that linearity between τ and W_v is 'not unreasonable' for canopies whose components are small compared with the wavelength. But in general, the relationship is a complicated one, and future work may replace the linear model of equation (12) with a more accurate one based on a priori information on canopy architecture, when available.

Wigneron et al. (1993) replace equation (10) with a full radiative transfer calculation, assuming a continuous canopy with strong spatial fluctuations in the dielectric constant. The statistical distribution of the canopy is fully specified by only three parameters: the variance of the fluctuating dielectric constant, and its vertical and horizontal correlation lengths. Thus, estimation of the canopy's properties from remotely sensed data is feasible. From these parameters, the single scattering albedo, extinction coefficient and scattering phase matrix can be derived for use in the equation of radiative transfer. This model was coupled to an h–Q soil roughness model and optimised to radiometric data over a growing soybean crop, at 1.4 GHz, 5.05 GHz and 36.5 GHz. The results are complex and perturbed by certain experimental difficulties, but clearly point to further use of this technique.

Point-scale Radiometric Measurements and Time Series Modelling

The estimation problems addressed in the preceding section did not make use of the information that is hidden in time series of data. Each estimate was essentially temporally isolated from the others, even in a series of measurements, and any point in a given series of measurements could generally be considered as if the rest did not exist.

A drawback of passive microwave remote sensing is that it only samples through the shallow penetration depth. However, interactions between vadose zone water and the atmosphere are affected by the moisture content deeper than the penetration depth – the vadose zone acts as a 'memory' of past rainfall.

A more ambitious approach to the use of radiometric data is to use the passive microwave observations to drive physically based models of water and energy fluxes in the soil–plant–atmosphere column. Thus, data-constrained modelling has the potential to estimate soil water contents below the penetration depth. This could be done by using the estimated near-surface soil water content as a boundary condition on a soil water model; or a more sophisticated data-assimilation method can be adopted, as will be demonstrated below.

Li and Islam (1999) used a four-layer model of heat and water movement in soil down to a depth of 1.89 m. Latent and sensible heat fluxes into the atmosphere were predicted, including (for latent heat) evaporation from the soil surface, transpiration and evaporation from rain intercepted by the canopy. Full details are found in papers cited by Li and Islam. Although this is a relatively simple model compared with some (e.g. SWEAT: Daamen and Simmonds 1996), it is claimed that it 'can capture land surface dynamics from the diurnal cycle to seasonal time scales'. A data set from FIFE (Sellers et al. 1992) was used, the relevant parts of which included meteorological data, and near-surface soil moisture measurements, determined by direct sampling. The model was first run forced by meteorological data such as measured precipitation, radiation etc., at a site in the USA, and the soil moisture content and heat fluxes were compared with those independently measured. As simulated time passed, differences arose between the modelled and measured values of W_s, due to inescapable imperfections in the data and model. A second run was then performed, in which the model's near-surface water content was changed to the measured value whenever this measurement was taken, i.e. every day. This was done as a surrogate passive microwave radiometry measurement, as microwave data were not available. Assimilating the soil water measurements into the model in this way resulted in a slight improvement in the prediction of soil water content from the surface down to 0.21 m depth, and in the predicted heat fluxes. A further simulation ran the model with continuous rain falling at a rate equal to the climatological average, without assimilation of soil moisture measurements. Not surprisingly, the model then did a rather poor job of predicting soil moisture contents. Repeating the run with assimilated daily measurements of θ_s resulted in a considerable improvement in predictive ability. However, much work remains to be done to establish the value of this method.

Galantowicz et al. (1999) were considerably more successful in their use of a more sophisticated soil model and assimilation scheme. The bare soil model solved the coupled nonlinear diffusion equations for heat and liquid water from the surface down to 1 m depth, using a discretisation of 31 layers. This was coupled to a simplified radiative transfer model that simulated measurements made by an L-band radiometer. The soil model was run for a period of eight days, starting with a deliberately poor representation of the initial profile of soil moisture (as compared with measurements made by buried instruments). The upper boundary condition was provided by measurements of meteorological variables every 15 minutes. The model was updated by assimilating passive microwave observations of the brightness temperature roughly every hour during the day, by a Kalman filter method. This adjusted the values of volumetric water content in the modelled soil layers in such a way as to bring the modelled brightness temperature closer to the observed value, at each measurement. The Kalman filter is, under certain conditions, an optimal way of adjusting models to fit observations, and embodies a considerably more sophisticated approach than the direct method of Li and Islam. The result of this was that the modelled soil moisture profile gradually approached the measured profile, while the error bars gradually shrank in a way that bracketed the measured profile.

A second study in the same paper attempted to update the model using Kalman filtering again, but assimilating passive microwave measurements once every three days, to simulate a satellite-borne radiometer. First, a simulated four-month data set of soil moisture and temperature was generated by the soil model described in the previous paragraph, driven by real measurements of precipitation and other meteorological measurements. The radiative transfer model was used to generate simulated T_B values every three days, which were degraded by Gaussian noise with a standard deviation of 2 K. Then the model was again run with an initially wrong profile of soil moisture, assimilating the degraded simulated T_B measurements. Furthermore, the model was not driven by precipitation measurements, so the only information on W_s

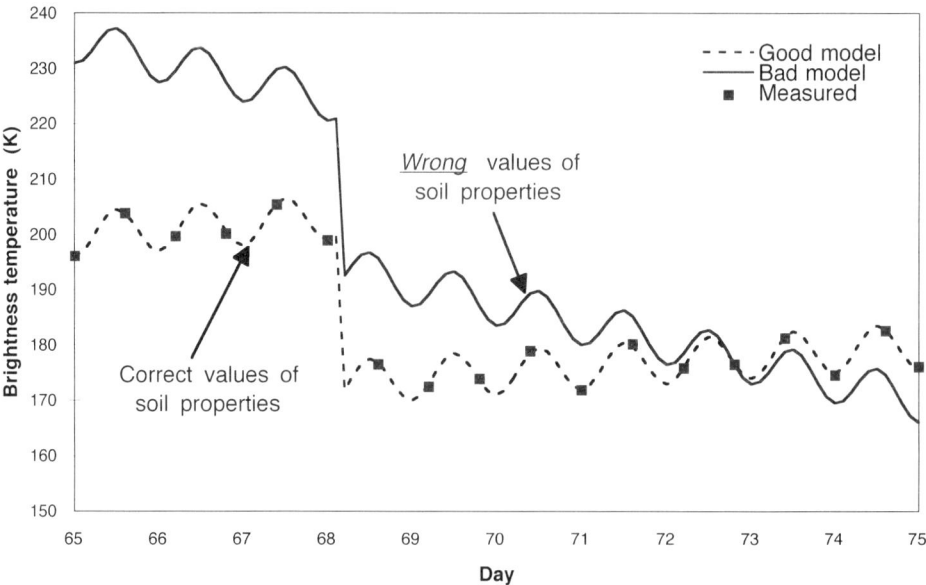

Figure 8.6 Cartoon showing how data-constrained modelling can be used to estimate soil hydraulic properties, given a time series of T_B measurements

that was available to the Kalman filter was by the proxy of brightness temperature. The resulting attempt to reconstruct the simulated time series of depth-profiles of W_s was remarkably good, with a depth-averaged RMS error of typically 2%. However, sometimes the 'retrieved' value of W_s lagged that of the simulated measured data set, when it rained at the beginning of an inter-update period, rather than towards the end of it – this is an unavoidable consequence of the finite repeat time of the putative satellite.

A further advance in the use of time series data is in the estimation of soil hydraulic properties. Consider a model of water flow in a soil, controlled by parameters such as the saturated hydraulic conductivity, the air-entry potential, and others. If the model accurately reflected the physical processes in a real soil, then the parameters in the model would have to be close to those descriptive of the real soil, in order to provide a good simulation of a time series of measurable phenomena, such as W_s or T_B. This is the approach taken by Burke et al. (1998): adjust the parameters in a soil hydrology model (Daamen 1996) until a time series of T_B measurements is simulated as well as possible, by minimising the RMS error in the simulated time series, as illustrated in Figure 8.6. This technique was remarkably successful in estimating the hydraulic properties of several bare soil types, as compared with lab-based destructive sampling measurements of the same parameters, as shown in Table 8.1. This was extended to vegetated soil by simulating the T_B measurements by use of a τ–ω model for the vegetation. Although preliminary, this also seems a promising line of research.

The method of Galantowicz et al. assumed independent knowledge of the soil hydraulic properties. It now appears that these can also be estimated by use of time series. Indeed, Galantowicz et al. suggest that this can be done in an optimal way by extending the Kalman filter, which is an established analysis method, but has not yet been applied in this context.

Table 8.1 Three parameters describing hydraulic properties, for three different soils, are shown. The air entry potential is the matric potential at which air begins to enter the soil, from an initially saturated state. The water release exponent describes the curvature of the graph of W_s against matric potential, as described in Daamen and Simmonds (1996). The rows labelled 'Lab.' show the values obtained by analysing soil samples in a laboratory. The rows labelled 'Est.' give the results of fitting the SWEAT model to time series of brightness temperatures. Values typeset in bold are the best estimates, bracketed left and right by rough indicators of uncertainty. The Wilheit (1978) model was used with an h–Q roughness modifier

Soil parameter		Loam	Sandy loam	Loamy sand
Air entry potential	Lab.	-0.41	-1.14	-0.82
ψ_e (J kg^{-1})	Est.	$(-0.25, \mathbf{-0.35}, -0.50)$	$(-1.0, \mathbf{-1.2}, -1.45)$	$(-0.55, \mathbf{-0.80}, -1.05)$
Water release	Lab.	5.7	3.5	3.4
exponent b	Est.	$(3.6, \mathbf{4.2}, 4.6)$	$(3.0, \mathbf{3.2}, 3.4)$	$(3.1, \mathbf{3.4}, 3.7)$
Bulk density	Lab.	1.31	1.35	1.27
ρ_b (g cm^{-3})	Est.	$(1.21, \mathbf{1.32}, 1.42)$	$(1.32, \mathbf{1.40}, 1.50)$	$(1.27, \mathbf{1.33}, 1.39)$

Discussion

We have demonstrated that physically based models are required to estimate near-surface soil moisture from passive microwave measurements at the point scale. There are many models of the interaction of microwave radiation with soil and vegetation that can be used to do this. In fact, many more models than have been summarised here are found in the literature, and many more ground situations can also occur (e.g. forests have not been discussed here, nor snow, nor row crops). Thus the field may appear to be in a state of confusion, with a lack of validated models. However, the understanding of the physics of the radiative processes has reached some level of maturity, and the proliferation of models reflects the confidence with which that understanding is held. We can, in fact, state that most of the modern models that are used in this context have been well validated by theory and experiment. Njoku and Entekhabi (1996) express a similar opinion in the conclusion to their review. The chief difficulties in applying these models lie in estimating the correct values of the parameters within the models (such as Q or ω etc.). However, data-constrained modelling now appears to be able to address this problem also. A remaining problem that has not been addressed in any detail is subsurface vertical heterogeneity in soil properties – e.g. how deep is the soil?

8.2.4 Remote Sensing of Large Heterogeneous Areas

No low frequency (e.g. L-band at 1.4 GHz) imaging instruments with a spatial resolution useful for hydrological studies have yet been placed in orbit – SMOS (Kerr 1998) may be the first, a few years from the time of writing (2000). However, several airborne imaging missions have been flown using the Push-Broom Microwave Radiometer (PBMR) (Wang et al. 1990) and the Electronically Steered Thinned Array Radiometer (ESTAR) (Le Vine et al. 1990).

Airborne Remote Sensing

O'Neill et al. (1996) present analyses of data from PBMR and ESTAR, flown over corn canopies of different maturities before and after rainfall. The brightness temperatures were modelled by a

τ–ω–h–Q model in which the vegetation parameters were inserted after having been carefully modelled by a fluctuating continuum dielectric model. This, in turn, required detailed information on canopy architecture, which was gathered manually on the ground. The estimated soil water contents compared remarkably well with measurements made at ground level. Although this method requires the input of information on canopy architecture that cannot be feasibly estimated by remote sensing, it does show that airborne passive microwave remote sensing has the potential to estimate W_s accurately.

O'Neill et al. (1996) essentially treated fields as points, similar to the point-scale measurements described above. Mapping of W_s has also been done by use of airborne imaging radiometers. Wang et al. (1990) describe PBMR overflights of the FIFE research site, covering an area of approximately 7×14 km in Oklahoma. The raw data are spatial maps of T_B, from which soil moisture maps had to be estimated. First, an estimated emissivity $\hat{e}$ was obtained through use of measured surface temperatures from an infrared thermometer that was mounted with the PBMR. Then the emissivity map was converted to a W_s map. This was done by deriving a linear regression line relating the emissivity to the near-surface W_s, for each point in the area. Two parameters are needed to describe a straight line. One of these was estimated by assuming that the soil was saturated to a constant W_s of 43% immediately after rainfall. The other was estimated by use of results from two ground-instrumented watersheds in the study area, with different biophysical properties. It was noticed that graphs of observed (by PBMR) $\hat{e}$ vs. measured (by ground-based instruments) W_s for these areas could be described rather well by straight lines, and that the lines for the two watersheds intersected at a certain point in the graph. It was then assumed that *all* parts of the site could be described by straight lines passing through that point on the graph. In summary, the analysis process was

$$\left.\begin{cases} T_B(x, y) \\ T_{surf}(x, y) \end{cases}\right\} \rightarrow \hat{e}(x, y) \rightarrow W_s(x, y) \tag{16}$$

where x, y are spatial coordinates. A scatter plot of PBMR-estimated W_s against measured values on the ground, for all available ground data, gave an encouragingly tight cluster around a straight line. A simple τ–ω–h–Q model containing only τ and h was fit to the data from the two watersheds described above. It was shown that this is an adequate representation of the effects of vegetation. However, the vegetation water content was less than 0.5 kg m^{-2}, so little reduction in sensitivity to ground moisture was expected. The above method was found to fail in certain areas of thick, dead vegetation that had a high emissivity, thus obscuring the signal from the ground.

Another airborne mission used data from PBMR and ESTAR (Jackson et al. 1993) over the Walnut Gulch watershed in Arizona. It was found that maps of rainfall, estimated by interpolation from a network of rain gauges, correlated well with maps of the *change* in brightness temperature that occurred due to that rainfall.

Mattikalli et al. (1998a,b) describe ESTAR overflights of the Little Washita watershed in Oklahoma. A map of T_B obtained from ESTAR data acquired on 10 June 1992 shows that T_B is very strongly correlated with an independently acquired map of soil type. June 10 was dry, following a rainy period that ended on 9 June. The June 10 T_B map shows low values of T_B at the east and west ends of the area, where the soil was relatively fine in texture, and higher values of T_B in the central region of relatively coarser soil, which allowed the preceding rain to infiltrate and drain away relatively quickly. Soil moisture maps were estimated from T_B maps for 10–18 June, using a τ–ω–h–Q model. The saturated hydraulic conductivities of the soils in the area, k_{sat}, were also estimated by optimising a soil hydrology model to measurements made at ground level, much as described by Burke et al. (1998) and discussed above. Then a simple scatter plot of

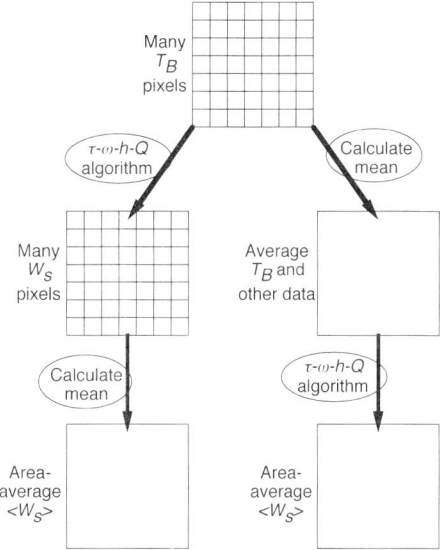

Figure 8.7   The aggregation paradigm: 'analyse then average; or average then analyse'

k_{sat} vs. the change in W_s over two days showed a fairly good correlation. Mattikalli et al. (1998a) suggest that this enables passive microwave data to be used to roughly and quickly provide maps of k_{sat}. This work extends the earlier, less sophisticated method of Hollenbeck et al. (1996).

Spaceborne Remote Sensing

A fundamental feature of planned spaceborne passive microwave remote sensing at L-band is the large size of the ground resolution elements (loosely speaking, 'pixels'). This is a function of the inverse relation between the angular size of the field of view of the instrument and the diameter of the antenna. A small field of view and small pixels would require an immense antenna that could not feasibly by launched.

Although no imaging low frequency orbiting radiometers have been launched, a small number of simulations have been undertaken using airborne instruments. Jackson et al. (1995) started with the ESTAR data taken over the Little Washita watershed (as discussed above), and averaged the 200 m pixels to make larger ones with edges 1.6 km long, then again at 9.0 km. This is an *aggregation* problem. The aggregation investigation paradigm is illustrated in Figure 8.7. A pixelated image or data set is treated pixel-by-pixel, then the final result (in this case, a map $W_s(x, y)$) averaged over the area of interest. Alternatively, the areal average of the data set is calculated, then the analysis model is applied to the single averaged value, again obtaining (in this case) an area-averaged estimate $< W_s >$. More succinctly: '*average then analyse; or analyse then average*'. Many variations are possible, for example the dominant vegetation type may be taken to cover the whole area, rather than the area-averaged properties of the vegetation. The two resulting averages are likely to be different because of the nonlinear nature of nearly all the processes involved (e.g. equation (10)). Jackson et al. applied a simplified τ–ω–h–Q algorithm to the averaged pixels, to obtain a relation between area-averaged T_B and W_s for eight synthesised pixels. The degree of correlation was encouraging, but not conclusive.

A similar analysis was undertaken by Drusch et al. (1999), using ESTAR data from the SGP'97 experiment in Oklahoma. Here, the ESTAR data were resampled onto a grid of 800 m resolution, then synthetic pixels were created by averaging, with sides of lengths 2.4 km, 10.4 km and 31.4 km. Thousands of large pixels were synthesised by allowing them to overlap – thus there was a very high degree of correlation between synthetic pixels. A τ–ω–h–Q-type algorithm was used again at each spatial scale, and also with the unaveraged 800 m data. The mean $< W_s >$ was again formed according to the two recipes of Figure 8.7 for each synthesised big pixel, and these values are compared for different sizes of pixel in Figure 8.8. The results are encouraging, but it should be borne in mind that the large pixels are very highly correlated, so the effective number of points in each graph would have to be greatly reduced to analyse the statistical significance of the correlations that are obtained.

Discussion

It has been demonstrated that a good map of near-surface soil water content can be inferred, given a map of L-band brightness temperature, if the vegetation is not too thick. However, in each case ancillary data such as information on soil texture, vegetation properties and so forth, are required. In some cases, such as the Oklahoma experiments described above, soil and vegetation maps are available, but this is not the case over much of the land surface of the world. Elsewhere, possibilities lie in multi-observation methods such as the τ–ω method for estimating vegetation properties, and the h–Q method for soil roughness properties. There is also scope for mapping vegetation type 'on the fly' by simultaneous use of an imaging spectrometer and imaging microwave radiometer. It should be noted that as the density of vegetation increases, land–atmosphere interactions become increasingly dominated by the vegetation, while the radiometric sensitivity to soil moisture decreases. Thus if the object of mapping $T_B(x, y)$ is to investigate these interactions, rather than obtain $W_s(x, y)$, the loss of sensitivity due to vegetation may not be of great importance – this is an open question which needs to be addressed by modelling and measurement. It is possible that repeated overflights may give sufficient information to determine soil hydraulic properties, extending and combining the approaches of Burke et al. (1998) and Galantowicz et al. (1999). Such an approach relies on the validity of SVATs at the scale of a pixel, i.e. usually $\gtrsim 100$ m. This an another open question (e.g. Kabat et al. 1997) that needs to be addressed by measurement and modelling.

While the utility of airborne radiometric mapping over small watersheds is established

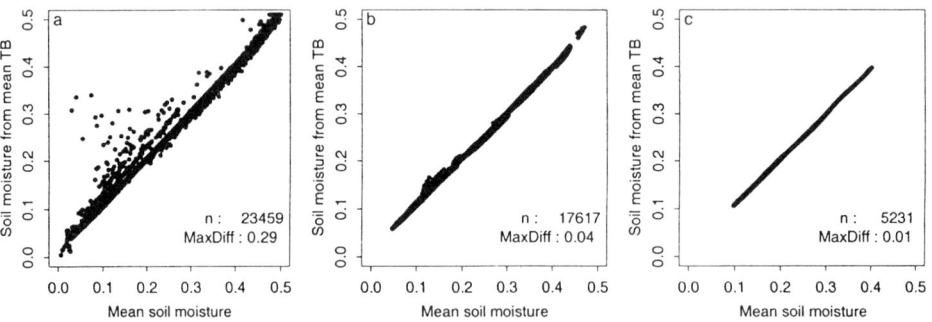

Figure 8.8 Aggregation results, taken from Drusch et al. (1999). Each graph shows results from aggregation over all the available synthetic pixels, of sides 2.4 km, 10.4 km and 31.4 km, respectively. In each graph, a single point represents the two area-averaged values of W_s over a single pixel, following the two analysis paths of Figure 8.7

(although not fully validated), it is plain that continental or global mapping by aircraft is not feasible. However, spaceborne L-band radiometry has its own set of problems, chiefly arising from poor spatial resolution. The operational configuration of SMOS is not yet finally defined, but it is likely that a pixel will have linear dimensions of typically 30 km (Kerr 1998). In view of the nonlinear processes involved, one wonders if the signal from a heterogeneous pixel that big has a useful relation with the ground-level properties of the scene. For example, if the pixel is 50% vegetation and 50% bare soil (to choose an easy example), attenuation of the signal by the vegetation will result in a biased estimate of $< W_s >$ if this is not corrected. As in the case of airborne radiometry, land use classification will help this problem in some areas, but in general, ancillary high resolution data will be required, for example from Landsat TM images. Leaf area index and W_v can be estimated from such multispectral data (e.g. Senay et al. 2000), at least for vegetation that is not too dense – corresponding with the range of validity of microwave remote sensing. Clearly, estimation of W_s under dense vegetation is not going to be achieved with a useful degree of accuracy.

SMOS is planned to be a two-dimensional synthetic aperture radiometer, essentially taking a 'snapshot' of a large area. As the observation footprint is swept forward by the orbital motion of the platform, each point in the footprint will be observed from different angles. Thus there exists the possibility of using a τ–ω algorithm to compensate for vegetation effects in the estimation of W_s – indeed, Kerr (1998) says that temporal variation of the vegetation optical depth, mapped globally, will be a useful data set. However, as in the case of direct estimation of W_s, the validity of the τ–ω algorithm over heterogeneous pixels is questionable, and needs to be addressed carefully.

We can go beyond the estimation of soil moisture maps and consider the mapping of deep vadose soil moisture status, in the manner of Galantowicz et al. (1999), and/or soil hydraulic properties according to Burke et al. (1998). In heterogeneous pixels, this will undoubtedly require the disaggregation of this heterogeneity by use of ancillary high resolution data, as discussed above. This begs the question of how much information can be extracted from a time series of T_B measurements. For example, can soil hydraulic properties be estimated in a pixel that contains two contrasting types of landscape (e.g. crop and bare soil, or forest and crop)? And so on for higher degrees of heterogeneity? Another speculative question is, can we assimilate spaceborne radiometric observations to predict heat fluxes into the atmosphere over homogeneous and heterogeneous pixels? The answers to these questions are that we do not know, although point-scale and airborne studies have pointed the way to go towards answering them.

8.2.5   Validation of Advanced Methods

The value of a spaceborne passive microwave radiometer such as SMOS can be greatly increased if data products more advanced than soil moisture maps can be produced. However, as indicated in the previous section, even the estimation of soil moisture is not without problems over heterogeneous areas. Understanding of this and more ambitious analyses will necessarily rely heavily on theory and modelling, as the data sets required for full validation cannot be acquired. Nevertheless, it is essential to validate the models *as fully as possible*. The University of Reading, UK, will shortly begin a campaign of experiments aimed at partially verifying modelling and estimation over heterogeneous areas, using the SWaMP-L radiometer shown in Figure 8.2, augmented by a new C-band radiometer mounted on the same platform. This will be done in two ways. Firstly, radiometric measurements will be done at the 'point' scale of a few square metres, over managed small, heterogeneous plots of differing soil and vegetation types, with heterogeneous irrigation schemes. Thus no spatial resolution will occur, as the whole plot will fit within the field of view of the radiometer; but observations will be made at different angles. Such small plots will interact with an atmospheric boundary layer that is essentially homogeneous in the horizontal plane, and thus will not simulate large-scale heterogeneity in a fully realistic way. However, there is a corresponding advantage, which is that boundary layer heterogeneity will be decoupled from the measurement process, thus making the latter easier to understand.

The next set of validation experiments will be undertaken over a more natural size of plot, such as field of the order of 1 ha in size. This will be densely instrumented with biophysical diagnostics. Spatially resolving radiometric measurements will be made by a combination of linear motion of the radiometer and angular scanning transverse to the direction of linear motion. These experiments will be complementary to the point-scale measurements, in that the components of the heterogeneous plot will be spatially resolved, but a single integrated measurement of the whole plot will not be possible – although it will be simulated by composition of the individual point-scale observations that cover it.

Thus we expect to contribute to the validation of horizontally heterogeneous models that will be required in the analysis of future global radiometry missions, and to guide the development of improved models and analyses.

8.3   VALIDATION OF RIVER FLOOD MODELS USING SYNTHETIC APERTURE RADAR AND LIDAR DATA

8.3.1   Introduction

The second approach illustrated in this chapter explicitly uses the spatial properties of remotely sensed data.

Two-dimensional numerical hydraulic models are at present the most sophisticated tools available for the prediction of fluvial flood extent, depth and flow velocities, but their use is hampered by the lack of distributed validation and calibration data. Recent research has represented a move away from one-dimensional models of compound channel flow towards models of higher dimensionality, but validation and calibration data and techniques have failed to keep pace with this advancement in hydraulic modelling. We are thus placed in the situation where two- (or higher) dimensional models are being validated against point (effectively-zero dimensional) hydrometric data (Bates et al. 1998a), and this disparity between the dimensionality of hydraulic model structure and validation data must be addressed. These models also require input data (topography, friction parameterisation) of appropriate dimensionality and spatio-temporal resolution. Such parameters are often left to be defined by the calibration process, which effectively reduces the utility of the models, and may lead to an ill-posed problem (especially given the current dearth of distributed data). Indeed, a poorly posed calibration problem can be used to mask a multitude of evils in the model formulation, which may only become apparent when adequate distributed parameterisation and validation data becomes available.

This section addresses one approach to solving some of these problems: the use of remotely sensed data to provide input and validation data for two-dimensional models of fluvial flood flow. First, satellite remote sensing of the flood boundary is discussed as a means of providing synoptic validation data for the hydraulic model. Synthetic aperture radar is identified as the most suitable sensor, and appropriate image processing tools are developed for the extraction of the flood boundary. This is then compared with model predictions for a reach of the river Thames, UK. There then follows a discussion of the use of airborne laser altimetry for topographic and vegetation mapping, with the aim of providing bathymetry and friction parameters for the model. Thus the problems of data provision and model validation discussed above are both addressed.

8.3.2 Remote Sensing of Flood Extent: a Review

Ground observations of flood events can provide valuable data such as stage and discharge readings from gauging stations and inundation extent from high water marks or surveys of flood deposits. However, such measurements are time and resource intensive and may be difficult to obtain during flood events when installations may be destroyed and sampling may become difficult or dangerous. Remote sensing, from both space and airborne platforms, offers the opportunity to collect spatially distributed data rapidly and over large areas (Schultz 1988; Bates et al. 1997; Smith 1997) without the need for costly ground surveys, thus satisfying the requirements outlined in the introduction to this section.

The chief way in which remote sensing will enhance our understanding of the flood modelling process is from the provision of distributed validation data from maps of flood extent. Low floodplain gradients mean that the position of the shoreline is very sensitive to changes in water-level predictions, and should provide a stern test of model performance. There is the potential to make other measurements from space, such as water levels, through radar altimetry (Koblinsky et al. 1993), and information on flow velocities by using suspended sediments as natural tracers (Currey 1977). In this chapter, however, we concentrate on the mapping of flood extent as being the easiest way of obtaining spatial data while providing most useful information for model validation.

Three types of differences between land and water can be detected by remote sensing: differences in emissivity (passive), differences in the reflection of natural radiation (also passive) and differences in the reflection of satellite generated radiation (active). In terms of emissivity, field experiments (Schmugge 1988) using passive microwave radiometry have shown a correlation with soil water content. This has enabled the difference in brightness temperature, as measured by AVHRR (Advanced Very High Resolution Radiometer) thermal infra-red channels, to produce flood inundation extent maps (Barton and Bathols 1988). Passive techniques based on reflectance have been employed: Rango and Salomonson (1974) found that near infra-red (0.8–1.1 μm) images from the Landsat MSS could be used to distinguish between dry soil and vegetation (good infra-red reflectors) and wet soil, stressed, damaged or wet vegetation and standing water (poor infra-red reflectors) with a resolution of $\sim 80\,m$. The main problem with these techniques using visible and infra-red wavelengths is that they are incapable of penetrating the cloud cover associated with many flood events.

Active microwave reflectance techniques such as Synthetic Aperture Radar (SAR) have considerable potential to overcome the limitations of the above methods. SAR is an imaging radar system (Ulaby et al. 1982) that gives high ground resolution (25 m for ERS-1 SAR) and is capable of both cloud cover penetration and 24 hour operation. In the simplest model of SAR imaging of flooded areas, smooth water acts as a specular reflector (i.e. it has a low radar back scattering coefficient) and returns little energy back to the satellite or aircraft. Land areas are rougher and have higher back scattering coefficients. Thus water appears dark in the image with land brighter. The situation is, however, complicated as the water surface may be roughened by wind or broken by protruding vegetation, the image is degraded by speckle and the side-looking radar system can produce complex returns from topography which introduce distortion into the images.

To date much of the literature on delineating flood boundaries using SAR has adopted a qualitative approach. Tholey (1995) used ERS-1 images combined with a Geographic Information System (GIS) to carry out land use impact studies. Bonansea (1995) compared ERS-1 imagery to SPOT data, also using a GIS, while others (Badji and Dautrebande 1995; Kannen 1995; Noyelle et al. 1995) have used multi-temporal SAR scenes, one of which observes a flood, to contribute the red, green and blue channels of a colour image in order to aid visual interpretation. The use of SAR imagery compares favourably with other remote sensing systems: Imhoff et al. (1987) compared Landsat MSS and SIR-B (shuttle imaging radar) images with colour and infra-red aerial photography of monsoon flooding in Bangladesh, SAR imagery providing a better identification of the flooded area (85% correct) compared to Landsat (64% correct) when compared to the air photograph control. Biggin and Blyth (1996) compared a SAR scene of a flood of the River Thames, UK, with simultaneous air photographs. The images were analysed by eye with the SAR image correctly identifying approximately 80% of the flooded area.

Of the complications noted above, the effects of flooded vegetation are well explored. Multiple reflections from the water surface and upright vegetation enhance backscattering giving flooded vegetation a bright return on the image, the magnitude of this effect being a function of radar look angle, wavelength and polarisation. Inundated forests have been identified (Richards et al. 1987) with high returns in the L-band (~ 20 cm), but this effect is reduced at shorter wavelengths (~ 6 cm) due to increased volume scattering in the canopy. Ormsby et al. (1985) found that the X-band (~ 3 cm) gave bright returns for flooded marshland but no backscatter enhancement in forests. Henderson (1995) found that flooded forest and swamp gave high returns and Ramsey (1995) found an inverse relationship between SAR returns and the depth of marsh flooding. Solomon (1993) found that the appearance of rivers in tropical forest areas varied depending on the orientation of the river with respect to the radar look direction, with the water acting as a specular reflector when looking along the river but giving higher returns when looking across due to multiple water/vegetation reflection. Wang et al. (1995) have modelled these interactions mathematically, their results predicting that for C-band (wavelength ~ 6 cm) SAR at an incidence angle of approximately 20° (the characteristics of ERS-1 and 2 SARs) flooding should increase backscatter by approximately 2.6 dB.

The coherent nature of the SAR imaging system means phase information can also be used for surface mapping. Wegmuller et al. (1995) found that water could be identified as regions of low backscatter and low interferometric correlation between 2 SAR scenes. Corr et al. (1995) used correlation between ERS-1 images taken three days apart to distinguish between fields in winter and deciduous and coniferous forest, but achieved poor results for 35-day repeat cycles. Crop growth and farming activities tend to reduce correlation, making classification more difficult. Coherence mapping also tends to generate results at a lower spatial resolution than the SAR image itself.

A major drawback of SAR imagery is that it is degraded by a high level of noise (also known as speckle). This noise is generally non-Gaussian and multiplicative in nature (regions of high intensity are subject to more noise). This speckle means that often traditional image processing techniques (as surveyed in Sonka et al. 1993) are inappropriate and new approaches have to be developed. For example, White (1994) developed a simulated annealing algorithm which adopted a probabilistic approach for speckle reduction, results showing images that are smoothed (de-speckled) without loss of spatial resolution. Oliver et al. (1994) use local image texture to map clearings in rain forest, the measure being smoothed using a simulated annealing algorithm. This demonstrates that speckle, rather than simply being noise that degrades SAR image quality, can sometimes be a property useful in image classification.

Maps of flood extent derived from remotely sensed data have so far only been used to calibrate and assess simple models and empirical relationships for flood events. Imhoff et al. (1987) derived an approximate linear relationship between inundated area (as measured from SAR imagery) and water levels measured by hydrographs, and Ramamoorthi (1988) adopted a similar approach. These empirical schemes must be calibrated by a number of flood events, and their validity is restricted to the range of floods observed, and therefore extension to extreme and as yet unobserved floods may not be viable. Smith et al. (1995) used SAR imagery to calibrate a simple power law model relating width of glacial braided streams to discharge. Moll and Overmars (1990) outlined a scheme for the calibration of a one-dimensional flow model using Landsat TM data. River levels predicted by the model were converted to inundation extent via empirical relationships, and compared with remotely sensed flood areas. Bates et al. (1997) compared the results from two-dimensional modelling of a 60 km reach of the Missouri river, USA, with Landsat TM data. The study only modelled steady state flow within the channel, so the validation data was limited to the inundation extent over a few isolated islands and sand bars.

8.3.3 Flood Model Validation from SAR Imagery

Given that its all weather capability makes synthetic aperture radar the most appropriate sensor for remote flood mapping, we begin to address some of the problems associated with its use. The problem central to this and many other SAR processing problems is that of making best use of the sensor's high ground resolution (which is commensurate with that of current hydraulic models) while combating the high level of noise in the imagery. Traditional noise reduction techniques typically rely on spatial averaging, but this tends to reduce the spatial resolution of the imagery (blurring occurs). Simulated annealing methods (White 1994) can smooth noisy images while maintaining sharply defined edge features, but the technique is computationally expensive and still requires a post-processing stage (edge detection, linking and vectorisation) to define the flood shoreline. Instead, we get around the problem by developing a statistical active contour technique, which is capable of segmenting noisy SAR images to an accuracy of ~ 1 pixel.

Active contour models (snakes) (Cohen 1991; Williams and Shah 1992) have recently gained favour in image segmentation applications because of their ability to link detected edges to produce smooth region boundaries. The snake method uses a dynamic curvilinear contour to search the edge image space until it settles upon image region boundaries. The approach adopted here (see Horritt 1999 for a full description) follows the statistical snake concept of Ivins and Porrill (1994), where the snake operates directly on the image statistics that are calculated along the length of the snake. Thus, by looking at a number of pixels, the effects of noise can be reduced without a loss of resolution perpendicular to the shoreline, and texture can also be incorporated as extra information in the segmentation procedure. Local mean and texture measures are used in a statistical framework to decide whether a group of pixels should be included in the region inside the snake, which aims to identify the flood as a region of homogeneous speckle statistics. The snake then expands or contracts accordingly, being driven by an energy functional that it aims to minimise. A curvature measure is also included in the functional, so that the final smoothness of the contour can be specified, which enables small-scale features to be ignored and again reduces the effects of noise.

The algorithm has been tested on two ~ 15 km reaches of the upper Thames, where oblique colour aerial photography was acquired within ~ 2 hours of an ERS-1 overpass during a flood in 1992. After registration and analysis, the photography can be used to define a flood shoreline to an accuracy of approximately 15 m (just over 1 SAR pixel of 12.5 m), and compared with the shoreline found from the SAR imagery. Figure 8.9 (in plate section) compares the two shorelines for a 3×3 km area. Wind roughening of the water surface is evident from the significant returns from the flooded area. Protruding vegetation (a complex of hedgerows to the top right) have also produced higher returns as expected, and caused the algorithm to misclassify as unflooded a part of that region. The figure also demonstrates another (unexpected) problem: the long island in the flood (as identified from the airphoto data) has given very similar backscatter to that from the flood, and has therefore been misclassified. Analysis of further unflooded SAR imagery and the landcover map of Great Britain (Fuller et al. 1994) indicates that rough meadow (a common landcover type in floodplain environments) often gives returns similar to those from a wind-roughened open water surface, even when in an unflooded state, so this overestimation of the flooded region is likely to be a common occurrence. Comparing the shorelines over the whole of the two reaches, 75% of the shoreline region (defined as a strip 300 m either side of the shoreline) is identified correctly. Figure 8.10 shows the distribution of distance errors between the two shorelines. The major part of the SAR shoreline is seen to lie within 20 m of the true boundary (as delineated from the airphoto data), but with significant portions > 50 m away. These generally correspond to large misclassification errors (the algorithm defining whole fields, for example, incorrectly) rather than random errors in the shoreline location.

Given that the SAR shoreline is subject to a degree of uncertainty, it is better to include it into the modelling process in an uncertain fashion, rather than treating it as the definite flood boundary. A Bayesian technique is used (Horritt et al. 2000) to give a probabilistic flood classification, based on pre-flood imagery. The rationale behind this is that one of the main sources of error is unflooded areas being classified as flooded because of their low radar backscatter. Figure 8.11 (in plate section) shows a probability map for a ~15 km reach of the Thames, developed using this Bayesian approach, again covering a strip 300 m either side of the shoreline. Assigning a probability rather than a shoreline has two advantages: the raster data will be easier to deal with when model results are assessed, and it allows the validation/

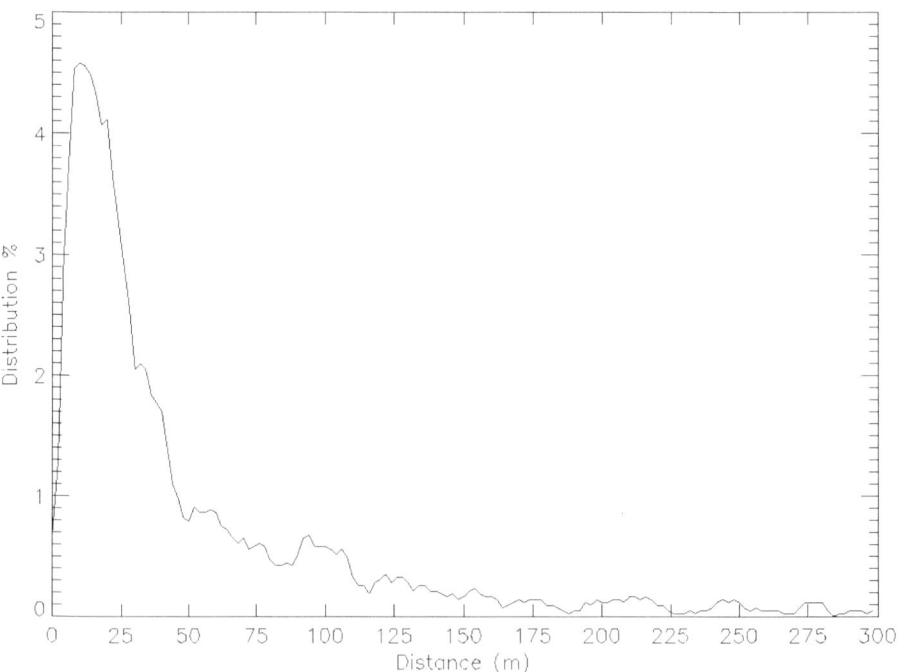

Figure 8.10   Distribution of distance errors between SAR and airphoto data

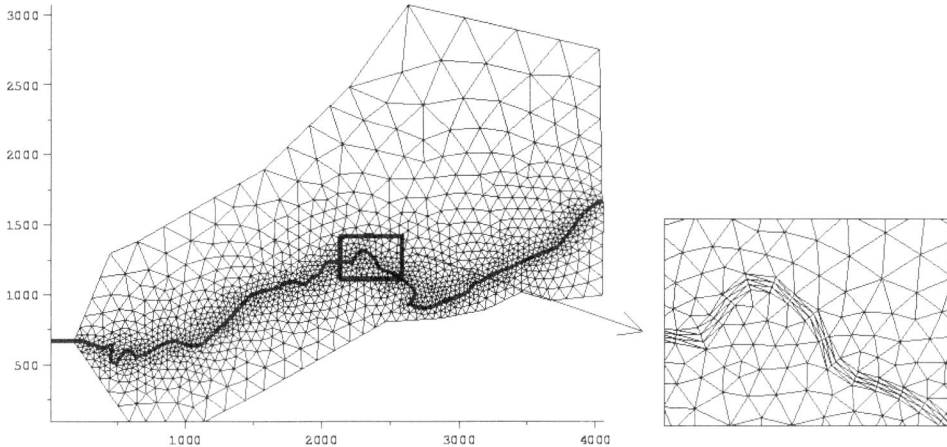

Figure 8.12   Finite element mesh for the Thames reach

calibration process to pay more heed to regions of high and low flood probabilities (areas of high classification certainty) and ignore regions of intermediate probability (uncertain areas).

With the flood probability map of Figure 8.11, we can now begin to assess the performance of a two-dimensional numerical hydraulic model. The TELEMAC-2D model (Bates et al. 1996; Hervouet and Van Haren 1996) aims to solve the St Venant equations for shallow water flow:

$$\frac{\partial \mathbf{v}}{\partial t} + (\mathbf{v} \cdot \nabla)\mathbf{v} + g\nabla(z_0 + h) - \frac{\varepsilon}{\rho}\nabla^2\mathbf{v} + \frac{g\mathbf{v}|\mathbf{v}|}{C^2 h} = 0 \qquad (17)$$

$$\frac{\partial h}{\partial t} + \nabla(h\mathbf{v}) = 0 \qquad (18)$$

$\mathbf{v}$ is a two-dimensional depth-averaged velocity, h is the flow depth, g the acceleration due to gravity, z_0 the ground elevation, C the Chezy coefficient of bed friction ρ the water density, ε the turbulent eddy viscosity (assuming the simplest model of turbulence). The two equations represent the conservation of momentum and mass respectively. The model solves these equations using a finite element technique over a mesh of triangular elements. This model has been extensively validated for fluvial flow studies against point hydrometric data (Bates et al. 1998a) and in a limited fashion against analytical solutions of the governing equations (Horritt 2000). The model requires input data in the form of a DEM, channel cross-sections and bed friction parameters, and boundary conditions, such as upstream inflow. In this case, a very simple friction parameterisation is used, based on using only two friction values, one for the channel and one for the floodplain areas. The turbulent eddy viscosity, ε, is also used as a calibration parameter. Topographic data was provided by a 50 m stereophotogrammetric DEM for the floodplain and Environment Agency cross-sections of the channel.

This preliminary test of model performance uses a short (~ 5 km) reach for which the mesh shown in Figure 8.12 has been developed. Nine simulations were performed, covering a variety of floodplain and channel frictions (Table 8.2) and two eddy viscosities. The values of channel friction and eddy viscosity were constrained by attempting to reproduce the estimate of bankful discharge of $40 \, m^3 \, s^{-1}$. The area correctly predicted by the model can be defined as:

$$A_{correct} = \underbrace{\sum P(\text{flooded})}_{\substack{\text{Area predicted as} \\ \text{wet by model}}} + \underbrace{\sum 1 - P(\text{flooded})}_{\substack{\text{Area predicted as} \\ \text{dry by model}}} \qquad (19)$$

with the sums being calculated on a pixel by pixel basis from the flood probability map of Figure 8.11. Table 8.2 summarises the performance of the nine model runs, along with some other crude predictions of the flood state, and the best and worst possible predictions. The results show that the model predictions are more accurate than the crude (totally flooded or totally dry) predictors, and also an improvement on using a simple planar water free surface height interpolator (Figure 8.13, in plate section). It is noticeable, however, that the two predictions coincide very closely for much of the reach, especially to the south where the floodplain is bounded by steep slopes.

Varying the bed friction and turbulent eddy viscosity parameters has appeared to make little difference in the accuracy of the model predictions, the difference between the best and worse results being shown in Figure 8.14 (in plate section). This apparent insensitivity (at least within this range of parameters) can be viewed in two ways. On the one hand, the modelling process is quite robust with respect to friction parameters, and these values do not need to be defined too closely to achieve reasonable results. On the other hand, the rationale behind the calibration process is that we can vary the friction parameters and thus mask poor process representation, but this has proved ineffectual in improving model predictions in this case. The fact that the simple planar water surface gives very similar results to that from the full two-dimensional model seems to indicate that the flow (and hence the shoreline) is topographically driven, rather than by friction, and this would go some way to explaining the lack of sensitivity evident in this study. Given this insensitivity, the only way to improve model results in this case is to improve model process representation. For example, the largest discrepancy between the predicted and observed shorelines is towards the downstream end of the reach. The flooding here may actually

Table 8.2 Summary of the performance of the nine model runs, along with some other crude predictions of the flood state, and the worst and best possible predictions

Simulation number/predictor	Channel friction $(m^{\frac{1}{2}} \, s^{-1})$	Floodplain friction $(m^{\frac{1}{2}} \, s^{-1})$	Turbulent eddy viscosity $(m^2 \, s^{-1})$	% correct
1	60	20	0.05	64.0
2	60	10	0.05	64.2
3	60	30	0.05	62.8
4	75	20	0.05	63.3
5	90	20	0.05	62.7
6	75	20	0.1	63.4
7	90	20	0.1	63.2
8	75	30	0.1	62.7
9	75	10	0.1	64.0
Zero flooding	Domain completely dry			55.0
Complete flooding	Domain completely flooded			45.0
Best prediction	Area within SAR shoreline flooded			72.0
Worst prediction	Area outside SAR shoreline flooded			28.0
Planar interpolation	Water surface approximated as planar			56.7

be due to flow back up the reach from beyond the downstream end, so the predictions may be improved by extending the model domain, and only ending it where the flow is more tightly constrained.

This study has served to confirm the feasibility of validating flood flow models using remotely sensed imagery, and shed some light on the nature of the distributed calibration process, demonstrating that we cannot always rely on friction calibration to improve model results. Of course, in this simple study, various assumptions have been made that need to be justified and their validity assessed. Uniform channel and floodplain friction have been assumed, for the sake of simplicity, but friction is likely to vary throughout the domain. This will probably reaffirm the calibration dichotomy: we will be able to improve model results without including more physical processes in the model, but at the expense of having to search through a parameter space of higher dimensionality for the optimum fit, and thus perhaps formulating an ill-posed problem. On the remote sensing side, it has been assumed here that the flood boundary detected from the SAR imagery is the shoreline (a contour of zero depth). The real shoreline may be obscured by vegetation, or register as a region of higher radar backscatter due to the multiple reflection mechanisms discussed in the review section of this chapter. This would mean that the detected shoreline would in fact be a pseudo-contour with water depths anywhere between 0 and ~ 50 cm, depending on typical vegetation depths. Topographic representation will also have a large effect on flood prediction (Bates et al. 1998b), as errors in the DEM will affect shoreline location, and some flow processes may be dominated by small-scale features (such as ditches and dykes) which are absent from the airphoto DEM. A possible resolution to these issues is proposed in the next section, where the use of airborne laser altimetry for vegetation and topographic mapping is discussed. There are, of course, other modelling issues (choice of numerical techniques, mesh resolution, process representation), which remain to be addressed, but the use of remote sensing validation data creates a rigorous framework within which this can be carried out.

8.3.4   Extraction of Surface Topography and Vegetation Information from LiDAR and CASI Data

The current crude representation of topography, friction and the wetting/drying process in two-dimensional flood flow models prevents the generation of accurate inundation predictions. Uncertainty over these factors leads to a heavy reliance on parameter calibration, particularly of friction, to optimise the model, and as a result the distributed predictions of water depth and velocity that the model produces cannot be relied upon (Beven 1989).

Airborne scanning laser altimetry (*Light Detection And Ranging* (LiDAR)) has recently gained favour for accurate dense topographic mapping, as it can penetrate the vegetation canopy and give actual ground elevations (Flood and Gutelius 1997). Topographic information can be generated over large areas at a horizontal resolution of 1–3 m and a vertical accuracy of ± 15 cm. This allows the production of accurate DEMs at a spatial resolution which exceeds that of the model grid. LiDAR can also produce information on vegetation properties, enabling the development of physically based models for frictional effects on vegetation floodplains.

Previous studies (e.g. Lin 1997) have successfully demonstrated the feasibility of extracting surface topography and vegetation height from LiDAR data using one-dimensional profiling LiDARs which capture the whole of the LiDAR range waveform. Scanning laser altimeters return only the range of the last and/or first significant return on the waveform, and effectively sacrifice range waveform information for two-dimensional spatial information.

To investigate the feasibility of using airborne scanning laser altimetry a small study was carried out over part of the Thames floodplain near Faringdon, Oxfordshire, UK, using an

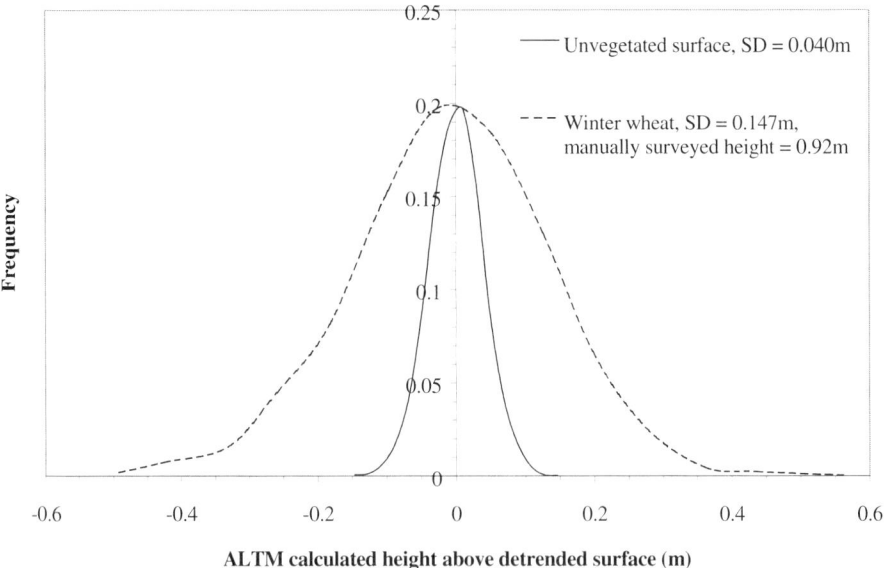

Figure 8.15 Detrended LiDAR height variation for a featureless surface and a field of winter wheat

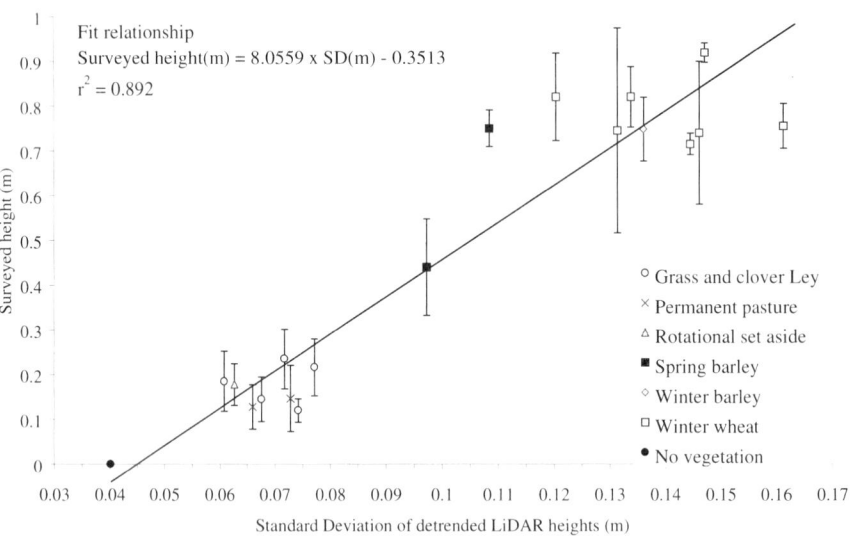

Figure 8.16 Regression of manually surveyed crop heights against standard deviations of LiDAR heights

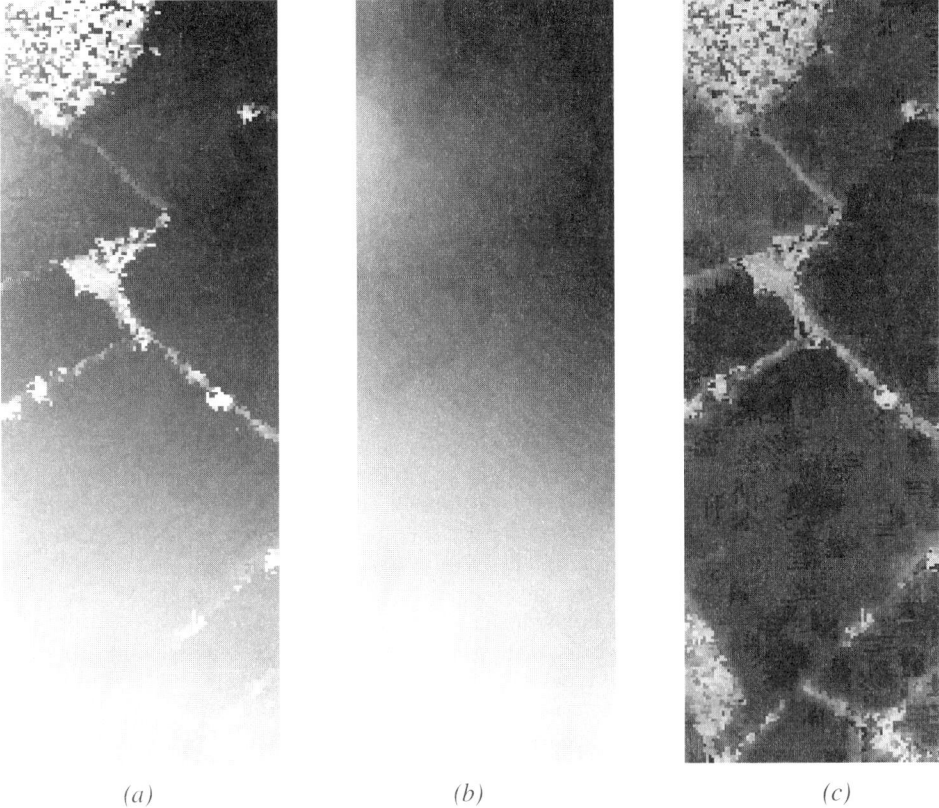

(a)   *(b)*   *(c)*

Figure 8.17   (a) LiDAR data of Oxfordshire test area. (b) Derived surface topography map. (c) Derived vegetation height map

Optech ALTM LiDAR, which measured only the time of the last significant return (Davenport et al. 2000). Each spot height in the ALTM data represents the height of incidence of the narrow (20 cm-diameter) laser pulse with either the ground, the top of the vegetation canopy, or some point in between. Vegetation height was measured by analysing high frequency height changes over small local areas, in this case windows of size 10×10 m. The height data were detrended prior to analysis to avoid exaggerating estimated vegetation height in regions of high slope.

Figure 8.15 shows a histogram of the differences in height of the points in each window from the average height in the window, constructed from many windows sampled over two regions with different vegetation heights. For a flat unvegetated surface the narrow spread of heights ($\sigma = 4$ cm) is due mainly to uncertainty in aircraft position and altitude, and return signal timing. The broader spread of returns from a 92 cm-height crop of winter wheat ($\sigma = 14.7$ cm) is also shown.

As these distributions (in common with those of other crop heights) are normally distributed, their variances (a robust measure) may be used in a regression analysis to estimate an empirical relationship with surveyed vegetation height. Figure 8.16 shows the result of such a regression from 18 fields containing a variety of crop types and heights. Vegetation height may be estimated to about 10 cm accuracy using this method.

A simple range image segmentation procedure is used to extract maps of surface and vegetation heights semi-automatically from a LiDAR height image. The segmenter clusters similar adjacent height differences into regions such as fields or hedges. A texture image is formed by measuring the local height standard deviation around each image pixel. Discontinuities between regions of different texture are determined using a texture edge detector, and strong edges are thresholded and linked to form closed regions. Within regions of short vegetation, local vegetation height is determined using the regression method above, while local surface height is average local height less an empirically determined fraction of the vegetation height. A more elaborate technique is used to determine surface and vegetation heights in regions of tall vegetation such as woods and hedges. Figure 8.17a shows a LiDAR subimage from the Oxfordshire test site, and Figures 8.17b and 8.17c show the surface topography and vegetation height maps extracted from it. The segmentation may also be used to improve the model mesh generation process, e.g. allowing long thin triangles to be generated within hedges, and more equilateral triangles in fields.

In the future, multi-spectral CASI (Compact Airborne Spectrometer Instrument) data will be obtained simultaneously with the LiDAR data, and this will allow vegetation type and density to be obtained as well as giving improved vegetation height estimates. From these data simple biomechanical models of floodplain friction parameters based on vegetation height, density and stiffnesss will be developed (Kouwen and Li 1980).

Future use of LiDAR data may also involve improving the interpretation of SAR returns for inundation mapping. The effects of floodplain vegetation on radar returns (Ramsey 1995) are not currently taken into account in the SAR segmentation algorithm. As a vegetation canopy is progressively flooded, the radar plant/water scattering mechanism will pass through three regimes. At first, the flood will be obscured by scattering from the canopy (if thick enough). When the flood level approaches the top of the canopy, multiple reflections between the water surface and the upright vegetation may result in higher backscatter than for the unflooded case. Finally, the canopy is overtopped and open-water scattering occurs. The depths at which these regimes are valid will depend principally on the vegetation depth and density, retrievable from the LiDAR/CASI data, as well as on sensor parameters such as incidence angle, wavelength and polarisation. LiDAR data and unflooded SAR imagery will be used in a quasi-physical modelling scheme that will aim to predict SAR returns as a function of flooding depth.

8.4   CONCLUSION

These two apparently different examples have a methodological approach in common. The approach now being adopted in both cases is to start with a physically based hydrological model of the process being studied. A good understanding of the physical basis of the remote sensing observation is also used. This understanding allows the model output to be compared with the new observations so as to minimise errors and reduce biases. The hydrological model can then be adjusted either in its parameter values or its parameterisation, and the results compared again with observations. This approach to validation allows new understanding to develop as we realise the limitations in the numerical models that encapsulate our understanding. In both examples, time series of observations have been shown to be essential to understand the changing hydrological fluxes. This is particularly true as with remote sensing we only ever measure a state variable, and not a flux directly, except in the case of radiant fluxes themselves. The approach to validation recommended in this chapter lends itself to the methods of assimilation used by meteorologists in weather prediction. Such methods are only just starting to be used in hydrology with remotely sensed data, but will surely be a key to more operational adoption of remote sensing. After years of promises, it now appears that remote sensing is starting to change our understanding of hydrology in unique ways.

REFERENCES

Avissar, R. and Pielke, R.A. 1989. A parameterization of heterogeneous land surfaces for atmospheric numerical models and its impact on regional meteorology. *Monthly Weather Review*, **117**, 2113–2136.

Badji, M. and Dautrebande, S. 1995. Characterisation of flood inundated areas and delineation of poor internal drainage soil using ERS-1 SAR imagery. *Proceedings of the 1st ERS Thematic Working Group Meeting on Flood Monitoring*, ESA-ESRIN, Frascati, Italy.

Barton, I.J. and Bathols, J.M. 1989. Monitoring floods with AVHRR. *Remote Sensing of Environment*, **30**, 89–94.

Bates, P.D., Anderson, M.G., Price, D., Hardy, R. and Smith, C. 1996. Analysis and development of hydraulic models for floodplain flows. In: Anderson, M.G., Walling, D.E. and Bates, P.D. (eds), *Floodplain Processes*. John Wiley, Chichester, 215–254.

Bates, P.D., Horritt, M.S., Smith, C.N. and Mason, D.C. 1997. Integrating remote sensing observations of flood hydrology and hydraulic modelling. *Hydrological Processes*, **11**, 1777–1795.

Bates, P.D., Stewart, M.D., Siggers, G.B., Smith, C.N., Hervouet, J-M. and Sellin, R.J.H. 1998a. Internal and external validation of a two-dimensional finite element code for river flood simulations. *Proceedings Institute Civil Engineers Water, Maritime and Energy*, **130**, 127–141.

Bates P.D., Horritt M.S. and Hervouet J-M. 1998b. Investigating two-dimensional finite element predictions of floodplain inundation using fractal generated topography. *Hydrological Processes*, **12**, 1257–1277.

Betts, A.K., Ball, J.H., Beljaars, A.C.M., Miller, M.J. and Viterbo, P.A. 1996. The land surface–atmosphere interaction: a review based on observational and global modeling perspectives. *Journal of Geophysical Research*, **101**, 7209–7225.

Beven, K. 1989. Changing ideas in hydrology – the case of physically-based models. *Journal of Hydrology*, **105**(1–2), 157–172.

Biggin, D.S. and Blyth, K. 1996. A comparison of ERS-1 satellite radar and aerial photography for river flood mapping. *Journal of the Chartered Institute of Water Engineers and Managers*, **10**(1), 59–64.

Bonansea, E. 1995. Mapping 1994 floods in Piedmonte Region, Italy: an example of remote sensing and GIS application. *Proceedings of the 1st ERS Thematic Working Group Meeting on Flood Monitoring*, ESA-ESRIN, Frascati, Italy.

Burke, E.J., Gurney, R.J., Simmonds, L.P. and O'Neill, P.E. 1998. Using a modeling approach to predict soil hydraulic properties from passive microwave measurements. *IEEE Transactions on Geoscience and Remote Sensing*, **36**, 454–462.

Chandrasekhar, S. 1960. *Radiative Transfer*. Dover Publications, New York.

Cohen, L.D. 1991. On active contour models and balloons. *CVGIP: Image Understanding*, **53**(2), 211–218.

Corr, D.G., Keyte, G.E. and Whitehouse, S. 1995. Studies of decorrelation in multi-temporal SAR imagery. *IGARSS '95 – 1995 International Geoscience and Remote Sensing Symposium*, vols 1–3, 1026–1028.

Currey, B. 1977. Identifying water flood movement. *Remote Sensing of Environment*, **6**, 51–61.

Daamen, C.C. and Simmonds, L.P. 1996. Measurement of evaporation from bare soil and its estimation using surface resistance. *Water Resources Research*, **32**, 1393–1402.

Davenport, I.J., Bradbury, R.B., Anderson, G.Q.A., Hayman, G.R.F., Krebs, J.R., Mason, D.C., Wilson, J.D. and Veck, N.J. 2000. Improving bird population models using airborne remote sensing. *International Journal of Remote Sensing*, **21**, 2705–2717.

Dobson, M.C., Ulaby, F.T., Hallikainen, M.T. and El-Rayes, M.A. 1985. Microwave dielectric behaviour of wet soil – Part II: dielectric mixing models. *IEEE Transactions on Geoscience and Remote Sensing*, **23**, 35–46.

Drusch, M., Wood, E.F. and Simmer, C. 1999. Up-scaling effects in passive microwave remote sensing: ESTAR 1.4 GHz measurements during SGP'97. *Geophysical Research Letters*, **26**, 879–882.

Fast, J.D. and McCorcle, M.D. 1991. The effect of heterogeneous soil moisture on a summer baroclinic circulation in the central United States. *Monthly Weather Review*, **119**, 2140–2167.

Flood, M. and Gutelius B. 1997. Commercial implications of topographic terrain mapping using scanning airborne laser radar. *Photogrammetric Engineering and Remote Sensing*, **63**(7), 780.

Fuller, R.M., Groom, G.B. and Wallis, S.M. 1994. The landcover map of Great Britain: an automated classification of Landsat thematic mapper data. *Photogrammetric Engineering and Remote Sensing*, **15**, 1357–1362.

Galantowicz, J.F., Entekhabi, D. and Njoku, E.G. 1999. Tests of sequential data assimilation for retrieving profile soil moisture and temperature from observed L-band radiobrightness. *IEEE Transactions on Geoscience and Remote Sensing*, **37**, 1860–1870.

Hallikainen, M.T., Ulaby, F.T., Dobson, M.C., El-Rayes, M.A. and Wu, L-K. 1985. Microwave dielectric behaviour of wet soil – Part I: empirical models and experimental observations. *IEEE Transactions on Geoscience and Remote Sensing*, **23**, 25–34.

Henderson, F.M. 1995. Environmental factors and the detection of open surface water using X-band radar imagery. *International Journal of Remote Sensing*, **16**, 2423–2437.

Hervouet, J-M. and Van Haren, L. 1996. Recent advances in numerical methods for fluid flows. In: Anderson, M.G., Walling, D.E. and Bates, P.D. (eds), *Floodplain Processes*. John Wiley, Chichester, 183–214.

Hollenbeck, K.J., Schmugge, T.J., Hornberger, G.M. and Wang, J.R. 1996. Identifying soil hydraulic heterogeneity by detection of relative change in passive microwave remote sensing observations. *Water Resources Research*, **32**, 139–148.

Horritt, M.S. 1999. A statistical active contour model for SAR image segmentation. *Image and vision computing*, **17**, 213–224.

Horritt, M.S. 2000, Development of physically based meshes for two-dimensional models of meandering channel flow. *International Journal for Numerical Methods in Engineering*, **47**(12), 2019–2037.

Horritt, M.S., Mason, D.C. and Luckman, A.J. 2000. Flood boundary delineation from synthetic aperture radar imagery using a statistical active contour model. *International Journal of Remote Sensing*, in press.

Imhoff, M.L., Vermillion, C., Story, M.H., Choudhury, A.M., Gafoor, A. and Polcyn, F. 1987. Monsoon flood boundary delineation and damage assessment using space borne imaging radar and Landsat data. *Photogrammetric Engineering and Remote Sensing*, **53**(4), 405–413.

Ivins, J. and Porrill, J. 1994. Statistical snakes: active region models. *Proceedings of the 5th British Machine Vision Conference*, 2 vols, 377–386.

Jackson, T.J. and O'Neill, P.E. 1990. Attenuation of soil microwave emission by corn and soybeans at 1.4 and 5 GHz. *IEEE Transactions on Geoscience and Remote Sensing*, **28**, 978–980.

Jackson, T.J. and Schmugge, T.J. 1991. Vegetation effects on the microwave emission of soils. *Remote Sensing of Environment*, **36**, 203–212.

Jackson, T.J., Le Vine, D.M., Griffis, A.J., Goodrich, D.C., Schmugge, T.J., Swift, C.T. and O'Neill, P.E. 1993. Soil moisture and rainfall estimation over a semiarid environment with the ESTAR microwave radiometer. *IEEE Transactions on Geoscience and Remote Sensing*, **31**, 836–841.

Jackson, T.J., Le Vine, D.M., Swift, C.T., Schmugge, T.J. and Schiebe, F.R. 1995. Large area mapping of soil moisture using the ESTAR passive microwave radiometer in Washita'92. *Remote Sensing of Environment*, **53**, 27–37.

Jin, Y-Q. 1989. The radiative transfer equation for strongly fluctuating, continuous random media. *Journal of Quantitative Spectroscopy and Radiative Transfer*, **42**, 529–537.

Kabat, P., Hutjes, R.W.A. and Feddes, R.A. 1997. The scaling characteristics of soil parameters: from plot scale heterogeneity to subgrid parameterization. *Journal of Hydrology*, **190**, 363–396.

Kannen, A. 1995. Use of SAR data for flood mapping and monitoring in Thuringen, Germany. *Proceedings of the 1st ERS Thematic Working Group Meeting on Flood Monitoring*, ESA-ESRIN, Frascati, Italy.

Kerr, Y.H. 1998. SMOS: Soil moisture and ocean salinity. *http://www-sv.cict.fr/cesbio/smos/*

Koblinsky, C.J., Clarke, R.T, Brenner, A.C. and Frey, H. 1993. Measurement of river level variations with satellite altimetry. *Water Resources Research*, **29**(6), 1839–1848.

Kouwen, N. and Li, R-H. 1980. Biomechanics of vegetative channel linings. *ASCE Journal of the Hydrology Division*, **106**, 1085–1103.

Lang, R.H. and Sidhu, J.S. 1983. Electromagnetic backscattering from a layer of vegetation: a discrete approach. *IEEE Transactions on Geoscience and Remote Sensing*, **21**, 62–71.

Le Vine, D.M. and Karam, M.A. 1996. Dependence of attenuation in a vegetation canopy on frequency and plant water content. *IEEE Transactions on Geoscience and Remote Sensing*, **34**, 1090–1096.

Le Vine, D.M., Kao, M., Tanner, A.B., Swift, C.T. and Griffis, A. 1990. Initial results in the development of a synthetic aperture microwave radiometer. *IEEE Transactions on Geoscience and Remote Sensing*, **28**, 614–619.

Li, J. and Islam, S. 1999. On the estimation of soil moisture profile and surface fluxes partitioning from sequential assimilation of surface layer soil moisture. *Journal of Hydrology*, **220**, 86–103.

Lin, C.S. 1997. Waveform sampling lidar applications in complex terrain. *International Journal of Remote Sensing*, **18**(10), 2087–2104.

Mattikalli, N.M., Engman, E.T., Jackson, T.J. and Ahuja, L.R. 1998a. Microwave remote sensing of temporal variations of brightness temperature and near-surface soil water content during a watershed-scale field experiment, and its application to the estimation of soil physical properties. *Water Resources Research*, **34**, 2289–2299.

Mattikalli, N.M., Engman, E.T., Ahuja, L.R. and Jackson, T.J. 1998b. Microwave remote sensing of soil moisture for estimation of profile soil property. *International Journal of Remote Sensing*, **19**, 1751–1767.

Moll, J.R. and Overmars, J.F.M. 1990. River hydrometry by pattern recognition in remote sensing. *Proceedings of an International Symposium on Remote Sensing in Water Resources*. International Association of Hydrogeologists Publication, the Netherlands, 443–452.

Mudaliar, S. 1999. Scattering from a rough layer of a random medium. *Waves in Random Media*, **9**, 521–536.

Njoku, E.G. and Entekhabi, D. 1996. Passive microwave remote sensing of soil moisture. *Journal of Hydrology*, **184**, 101–129.

Njoku, E.G., Rahmat-Samii, Y., Sercel, J., Wilson, W.J. and Moghaddam, M. 1999. Evaluation of an inflatable antenna concept for microwave sensing of soil moisture and ocean salinity. *IEEE Transactions on Geoscience and Remote Sensing*, **37**, 63–78.

Noyelle, J., Delimiere, S. and Marinelli, L. 1995. Identification of flooded areas in the Rhone. *Proceedings of the 1st ERS Thematic Working Group Meeting on Flood Monitoring*, ESA-ESRIN, Frascati, Italy.

Oliver, C.J., Blake, A. and White, R.G. 1994. Optimum texture analysis of synthetic aperture radar images. *Algorithms for Synthetic Aperture Radar Imagery*, **2230**(34), 389–398.

O'Neill, P.E., Chauhan, N.S. and Jackson, T.J. 1996. Use of active and passive microwave remote sensing for soil moisture estimation through corn. *International Journal of Remote Sensing*, **17**, 1851–1865.

Ormsby, J.P., Blanchard, B.J. and Blanchard, A.J. 1985. Detection of lowland flooding using active microwave systems. *International Journal of Remote Sensing*, **5**, 317–328.

Owe, M. and van de Griend, A.A. 1998. Comparison of soil moisture penetration depths for several bare soils at two microwave frequencies and implications for remote sensing. *Water Resources Research*, **34**, 2319–2327.

Pearson, D., Burke, E.J., Cartmell, I., Gurney, R.J., Jarrett, M.J. and Simmonds, L.P. 2001. First results and analyses from the SWaMP-L Radiometer. *Remote Sensing and Hydrology*. Ed. Owe, M., Brubaker, K., Ritchie, J., and Rango, A. IAHS, 2001. ISBN 1-901502-46-5.

Ramamoorthi, A.S. 1988. Flood damage forecasting by satellite remote sensing. *Proceedings of an International Symposium on Remote Sensing in Water Resources*. International Association of Hydrogeologists Publication, the Netherlands, 433–442.

Ramsey, E.W. 1995. Monitoring flooding in coastal wetlands by using radar imagery and ground based measurements. *International Journal of Remote Sensing*, **16**, 2495–2502.

Rango, A. and Salomonson, V.V. 1974. Regional flood mapping from space. *Water Resources Research*, **10**, 473–484.

Reitz, J.R., Milford, F.J. and Christy, R.W. 1979. *Foundations of Electromagnetic Theory*. Addison-Wesley, Reading, MA.

Richards, J.A., Woodgate, P.W. and Skidmore, A.K. 1987. An explanation of enhanced radar backscattering from flooded forests. *International Journal of Remote Sensing*, **8**, 1093–1100.

Sarabandi, K. and Chiu, T. 1997. Electromagnetic scattering from slightly rough surfaces with in-homogeneous dielectric profiles. *IEEE Transactions on Antennas and Propagation*, **45**, 1419–1430.

Schmugge, T.J. 1988. Microwave approaches in hydrology. *Photogrammetric Engineering and Remote Sensing*, **46**, 495–507.

Schmugge, T.J. and Choudhury, B.J. 1981. A comparison of radiative transfer models for predicting the microwave emission from soils. *Radio Science*, **16**, 927–938.

Sellers, P.J., Hall, F.G., Asrar, G., Strebel, D.E. and Murphy, R.E. 1992. An overview of the First International Satellite Land Surface Climatology Project (ISLSCP) Field Experiment (FIFE). *Journal of Geophysical Research*, **97**, 18 345–18 371.

Senay, G.B., Ward, A.D., Lyon, J.G., Fausey, N.R., Nokes, S.E. and Brown, L.C. 2000. The relations between spectral data and water in a crop production environment. *International Journal of Remote Sensing*, **21**, 1897–1910.

Shultz, G.A. 1988. Remote Sensing in Hydrology. *Journal of Hydrology*, **100**, 239–265.

Smith, L.C. 1997. Satellite remote sensing of river inundation area, stage and discharge: a review. *Hydrological Processes*, **11**(10), 1427–1439.

Smith, L.C., Isacks, B.L., Forster, R.R., Bloom, A.L. and Preuss, I. 1995. Estimation of discharge from braided glacial rivers using ERS-1 SAR: first results. *Water Resources Research*, **31**, 1325–1329.

Solomon, S.I. 1993. Methodological considerations for the use of ERS-1 SAR imagery for the delineation of river networks in tropical forest areas. *Proceedings of the 1st ERS-1 Symposium*, ESA-ESRIN, Frascati, Italy, SP-359, 595–600.

Sonka, M., Vaclav, H. and Boyle, R. 1993. *Image Processing, Analysis and Machine Vision*. Chapman & Hall, London.

Tholey, N. 1995. Monitoring flood events with remote sensing data: an example of ERS-1's contribution to flood events in northern and southern France regions. *Proceedings of the 1st ERS Thematic Working Group Meeting on Flood Monitoring*, ESA-ESRIN, Frascati, Italy.

Ulaby, F.T., Moore, R.K. and Fung, A.K. 1981. *Microwave Remote Sensing: Active and Passive. Vol. 1, Microwave Remote Sensing Fundamentals and Radiometry*. Addison-Wesley, Reading, MA.

Ulaby, F.T., Moore, R.K. and Fung, A.K. 1982. *Microwave Remote Sensing: Active and Passive. Vol. 2.* Addison-Wesley, London.

Van de Griend, A.A., Owe, M., de Ruiter, J. and Gouweleeuw, B.T. 1996. Measurement and behaviour of dual-polarization vegetation optical depth and single scattering albedo at 1.4 and 5 GHz frequencies. *IEEE Transactions on Geoscience and Remote Sensing*, **34**, 957–965.

Wang, J.R. and Choudhury, B.J. 1981. Remote sensing of soil moisture content over bare field at 1.4 GHz frequency. *Journal of Geophysical Research*, **86**, 5277–5282.

Wang, J.R. and Schmugge, T.J. 1980. An empirical model for the complex dielectric permittivity of soils as a function of water content. *IEEE Transactions on Geoscience and Remote Sensing*, **18**, 288–295.

Wang, J.R., O'Neill, P.E., Jackson, T.J. and Engman, E.T. 1983. Multifrequency measurements of the effects of soil moisture, soil texture and surface roughness. *IEEE Transactions on Geoscience and Remote Sensing*, **21**, 44–51.

Wang, J.R., Shiue, J.C., Schmugge, T.J. and Engman, E.T. 1990. The L-band PBMR measurements of surface soil moisture in FIFE. *IEEE Transactions on Geoscience and Remote Sensing*, **28**, 906–914.

Wang, Y., Hess, L.L., Filoso, S. and Melack, J.M. 1995. Understanding the radar backscattering from flooded and non-flooded Amazonian forests: results from canopy backscatter modelling. *Remote Sensing of Environment*, **54**, 324–332.

Wegmuller, U., Werner, C.L., Nuesch, D. and Borgeaud, M. 1995. Forest mapping using ERS repeat-pass SAR interferometry. *Earth Observation Quarterly*, **49**, 4–7.

White, R.G. 1994. A simulated annealing algorithm for radar cross-section estimation and segmentation. *Applications of Artificial Neural Networks*, **2243**(60), 231–240.

Wigneron, J-P., Kerr, Y., Chanzy, A. and Jin, Y-Q. 1993. Inversion of surface parameters from passive microwave measurements over a soybean field. *Remote Sensing of Environment*, **46**, 61–72.

Wigneron, J-P., Chanzy, A., Calvet, J-C. and Bruguier, N. 1995. A simple algorithm to retrieve soil moisture and vegetation biomass using passive microwave measurements over crop fields. *Remote Sensing of Environment*, **51**, 331–341.

Wilheit, T.T. 1978. Radiative transfer in a plane stratified dielectric. *IEEE Transactions on Geoscience Electronics*, **16**, 138–143.

Williams, D.J. and Shah, M. 1992. A fast algorithm for active contours and curvature estimation. *CVGIP: Image Understanding*, **55**(1), 14–26.

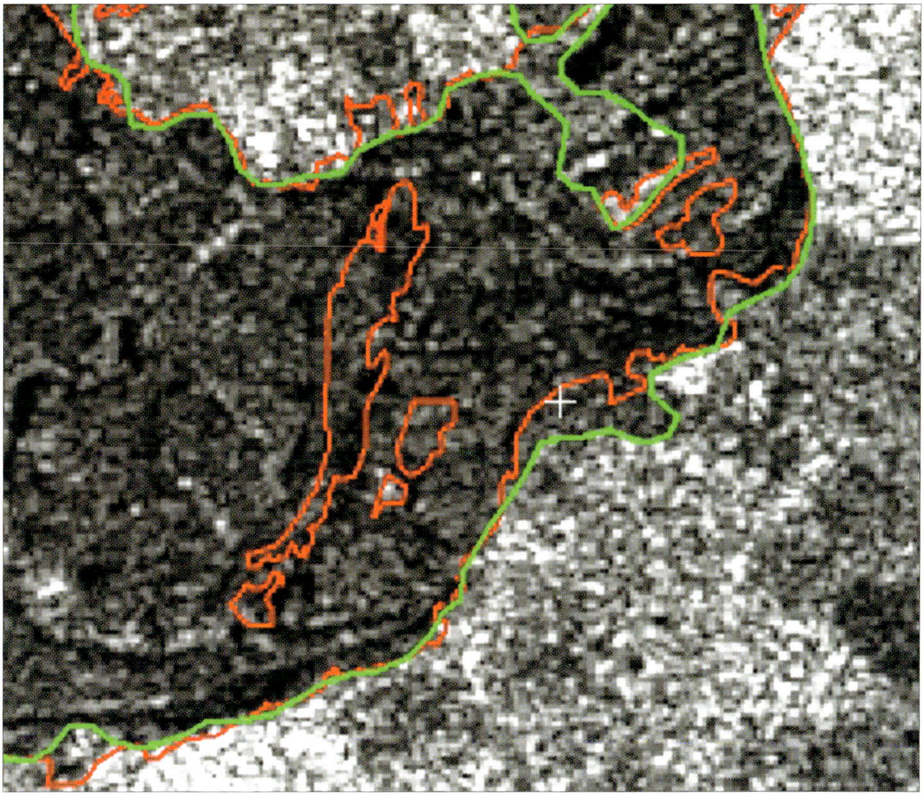

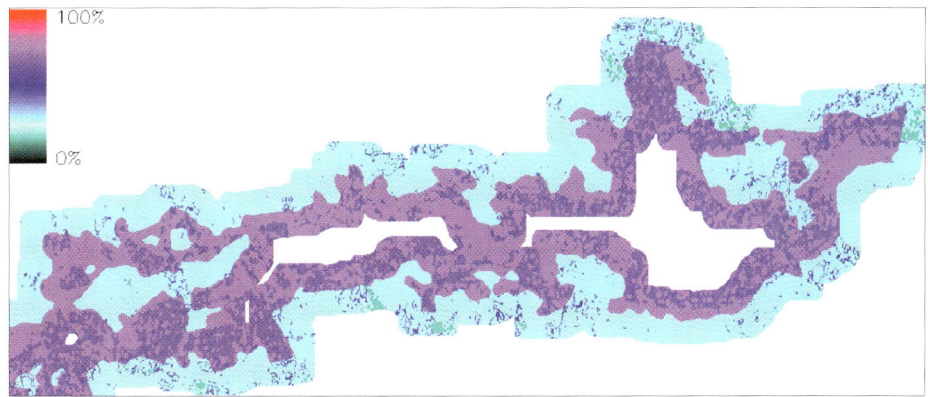

Figure 8.9 Comparison of SAR derived shoreline (green) and aerial photography (red)

100%

0%

Figure 8.11 Flood probability map derived from SAR imagery

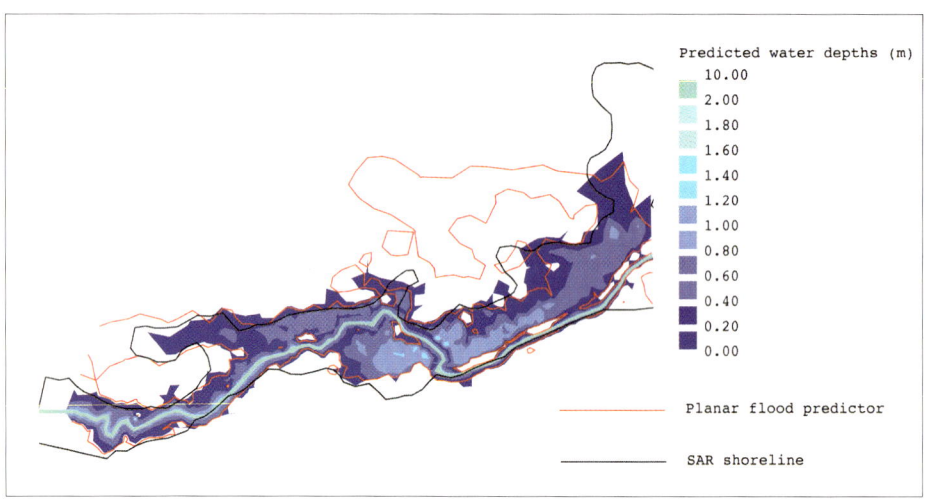

Figure 8.13 Comparison of best model prediction (run 6) and a simple shoreline predictor, formed by taking the intersection of a planar water surface with the DEM

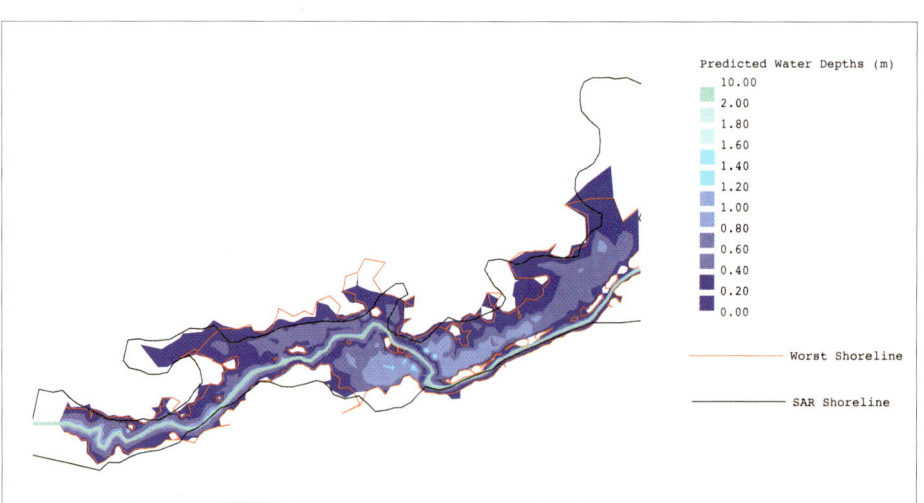

Figure 8.14 Comparison of best and worst model predictions, along with the flood shoreline derived from SAR imagery

9 Soil Water Relations

H.W.G. BOOLTINK
Laboratory of Soil Science and Geology, Wageningen University, The Netherlands

9.1 INTRODUCTION

Simulation models are useful tools for examining many problems in agricultural, environmental and land use studies. They can serve as tools to gain understanding of the behaviour of systems or they can be used to evaluate management options by using Decision Support Systems (DSS). The first type of model is generally of a complex nature and it requires many input parameters, whereas the second type of model is more robust, less descriptive in terms of modelled process concepts and less data demanding.

Hoosbeek and Bryant (1992) and Bouma and Hoosbeek (1996) described models in a diagram with two perpendicular axes (Figure 9.1). The first axis describes the conceptual modelling detail, ranging from strictly empirical to completely mechanistic. The other axis illustrates the type of answers a model is giving varying from quantitative to qualitative. So models can range from empirical–qualitative to mechanistic–quantitative. Models and modelling approaches are always subordinate to the aim for which they are used. Modelling is, therefore, hardly ever a purpose in itself. The specific aim of a study determines: (i) the hierarchical level(s) at which the study should focus (spatial scale), (ii) its time horizon (temporal scale), and (iii) its conceptual structure, i.e. the components that are endogenous and exogenous to the model and the relationships between components that are considered in the model.

Using Figure 9.1 a soil characteristic can be determined using various methods and techniques, depending on the question raised, the accuracy required or data available. In Figure 9.1 these knowledge levels are expressed as K-levels ranging from $K1$ (user), $K2$ (expert), $K3$ (specialist: simple comprehensive model), $K4$ (specialist: complex comprehensive model) to $K5$ (specialist: complex models of aspects). Data can be transferred from one K-level to the other on different scales (research chain). The example, indicated in Figure 9.1, depicts an example of a research chain exploring management options to prevent salinisation. A local manager ($K2$) of the water distribution board wants to know where and how much water he can apply on different soil units within his region. To do so, he has to differentiate between the various soil units. Information collected on this level is used to run a simple simulation model ($K3$). In this simple model some chemical interactions are not adequately described and a more complex model ($K5$) is necessary to solve these specific questions. Using this $K5$ knowledge in the $K3$ model, aggregation to the region can be performed and the manager is able to answer his research question on the regional scale in a empirical–quantitative way. This type of research chain can be made for almost any research question and contributes to a high extent to the transparency of the methodology used.

Model Validation: Perspectives in Hydrological Science. Edited by M.G. Anderson and P.D. Bates.
© 2001 John Wiley & Sons, Ltd.

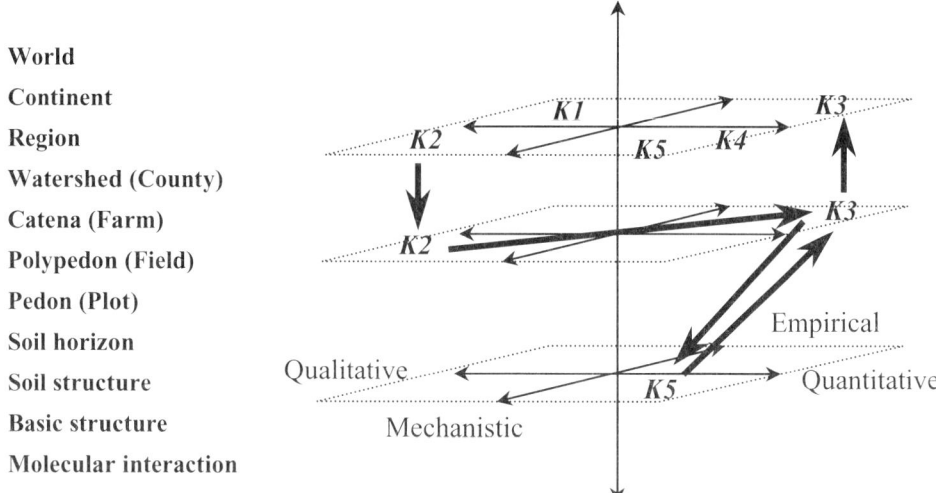

Figure 9.1   Scale hierarchy and knowledge diagram according to Hoosbeek and Bryant (1992) and Bouma and Hoosbeek (1996)

The aim of simulation modelling in agricultural, environmental and land use studies can be categorised in explorative studies and predictive studies. Explorative studies try to answer '*what-if*' questions and make explicit assumptions about behaviour of decision-makers; the actual decision-making is, however, exogenous to the model. Predictive studies try to predict the future behaviour of the simulated characteristic. In the latter type of study, decision-making is endogenous to the model.

As indicated in Figure 9.1 soil water physics can be applied on almost any scale ranging from molecular interaction to world. Especially when scaling up, the interaction between different processes can strongly influence the soil's physical behaviour. The ideal conditions of laboratory measurements can no longer be maintained on the field scale. The use of, for example, a desorption retentivity curve, measured in the laboratory, is of limited use at the field where it is influenced by continuous wetting and drying, resulting in hysteresis and most likely strong deviation of the laboratory measured desorption curve.

In macroporous soils water flow can be turbulent, or water is moving as film-flow along macropore walls (Bouma and Dekker 1978; Germann 1987). Darcy's law is no longer a valid concept here, other methods should be used in this type of soil. Interaction of soil organic matter and soil particles can result in hydrophobicity (Steenhuis et al. 1994; Dekker 1998; Ritsema 1998). Application of isotropic and homogeneous infiltration in such a case can lead to underestimation of agro-chemical breakthrough (Ritsema 1998).

Adequate monitoring and experimentation should be carried out in all these cases. However, to obtain insight into aspects of such a complex system and to perform virtual (digital) experiments, simulation models are indispensable. Simulation models used in this context can help us to gain understanding of processes and allow better, narrowly focused experimentation.

In general, our knowledge and information on these problems is limited and data collected uncertain. To enable a reliable development, application and interpretation of these models, it is therefore necessary to perform a thorough analysis of data and models. Data and models need to be analysed simultaneously in a coherent way, relevant to the purpose they are used for. Highly uncertain data in combination with an excellent model will not result in reliable results, nor will the combination of qualitative good data with an inaccurate model. For this reason model analysis should always include a sensitivity and uncertainty analysis.

In this chapter a sensitivity analysis is defined as 'the study of the influence of variations in model parameters, initial conditions etc. on model outputs'; and an uncertainty analysis as: 'the study of the uncertain aspects of a model, and of their influence on the model' (Janssen et al. 1993).

For sensitivity analysis it comes down to specifying the nominal values of model inputs and appropriate variations around these nominal values. This can be achieved through measurements, estimates through pedotransfer functions (Bouma and Van Lanen 1987; Tietje and Tapkenhinrichs 1993) or (spatial) interpolation (Finke 1993). Using Monte Carlo simulation modelling procedures the sensitivity of the simulation model and the accuracy of the predicted values can be quantified (Booltink 1994).

For uncertainty analysis it will depend on the nature of the uncertainty and on the information available. If the simulation model has been programmed correctly, the uncertainties (errors) in the model will usually be minor, compared to the other uncertainties. Uncertainties in the model structure are usually difficult to quantify. The uncertainty in the other sources is usually described in a probabilistic setting by specifying their probability distributions and their mutual correlations (Janssen et al. 1993).

The results of a sensitivity or uncertainty analysis are established under conditions and assumptions concerning, for example, the validity of the model and the correctness of the distributions. The extrapolation of the obtained conclusions to other applications is therefore only warranted if these assumptions are adequate and if the model describes the real phenomena in a correct way. As a consequence, the possible arbitrariness and subjectiveness in the specification of the model inputs should be an important point of concern since these facts limit the value of the results obtained. Janssen (1995a) advocates the use of a robustness study here. A robustness analysis should be applied to obtain an impression of the consequences of possible mis-specifications in the employed choices and assumptions (e.g. try different distributions and correlations for uncertainty analysis; try different nominal points for sensitivity analysis). The item of robustness analysis is closely related to model validation.

At present, the only way to perform a robustness analysis will be to repeat the model analysis using different specifications of the model inputs. For computationally intensive models such a second study will generally not be feasible. Less time consuming robustness analyses can be performed using so called meta-models. Using the simulation results of the original sensitivity or uncertainty analysis (linear) regression models can be derived between model inputs and simulated outputs. These regression models (meta-models) require little computational effort and can therefore be used to replace the original model in the robustness analysis. The derivation and interpretation of these meta-models will be discussed further in Sections 9.3.2 and 9.3.4.

Good quality data are crucial for the application of simulation models and statistical analysis techniques, to feed models and to validate modelling results. Every specific data value has its quality label in terms of reliability of determination (what methods were used), for what area of land is it representative, and what external and internal factors have determined its values. With the availability of electronic information systems such as GIS, data can be derived from other sources, i.e. through the use of pedotransfer functions, through neural networks, or through dynamic simulation modelling. Quality assessments of these types of data are hardly ever considered but are truly important when conclusions, based on these data, are derived. Despite all developments in modelling, data acquisition and data analysis techniques, basic data are the cornerstone for further development, validation, running and extrapolating models and their applications.

The use of a simulation model has to be discussed in coherence with (i) its application purpose, (ii) its spatial and temporal applicability, and (iii) its specific sensitivity for model input parameters. Data quality determines to a large extent the accuracy of the simulation results. Data quality assessment should therefore be an indispensable part of simulation studies, especially in cases where model calibration and model validation are important issues.

In this chapter theoretical aspects of such an approach are discussed and illustrated with different case studies. In Section 9.2 data necessary for simulating soil water processes are discussed. Data or model input parameters could be collected through various measuring methods (9.2.1) or estimated through expert knowledge or derived using pedotransfer functions (9.2.2). The effects of both procedures and a combination of them are presented in two case studies in Sections 9.2.3 and 9.2.4.

Uncertainty of model input data and the sensitivity of simulation models to input parameters are an important source of modelling inaccuracies. Section 9.3 deals with Monte Carlo simulation procedures. The technique of Monte Carlo modelling using Latin Hypercube sampling techniques is discussed in 9.3.1. Section 9.3.2 discusses the methodology of sensitivity or uncertainty analysis through Monte Carlo techniques. Section 9.3.3 explains the rotated random scan procedure for calibrating simulation models. Section 9.3.4 discusses model validation aspects. A case study, in which these theoretical concepts are applied, is shown in Section 9.3.5.

Main conclusions and discussion on the methodology applied and its coherency is finally discussed in Section 9.4.

9.2   DATA FOR MODELLING SOIL WATER RELATIONS

Methods to determine[1] soil physical characteristics are generally described in terms of the underlying physical theory or the equipment used. Little attention is usually given to operational aspects such as applicability, complexity, costs involved or obtained accuracy. Practice has shown that neither can all methods be applied to all soils nor can all measurements be done at all times. The choice of a method should be an indispensable part of research study. The questions such as the use of the data in a study, and the required accuracy given the tools and questions being asked should determine the choice of a method. Here we will discuss operational and accuracy aspects of methods to determine soil physical characteristics for simulation purposes, with special emphasis on the hydraulic conductivity and moisture retention characteristic and on the occurrence of bypass flow of water in soils.

9.2.1   Measuring Soil Physical Properties

Soil Moisture Retention

The moisture retention curve describes the relation between water content (θ) and the soil moisture potential (ψ). Various methods exist to determine the $\theta(\psi)$ relation, generally referred to as the moisture retention characteristic. The function relates a capacity factor (water content) to an intensity factor (water potential); it is, therefore, the ability of a soil to store water. A distinction is generally made in laboratory and field methods. Klute (1986) gives an overview of

---

[1] Directly through measurements (Section 9.2.1) as well as indirectly through estimation procedures such as pedotransfer functions (Section 9.2.2).

the theory and operational aspects of a broad listing of various methods widely used by many researches all over the world. Here we will focus on the applicability, complexity, costs involved, and the obtained accuracy.

In Table 9.1 operational aspects of five common procedures to measure soil moisture retention characteristics are presented. The suction method is still widely used. It uses suction plates or sand beds to desorp saturated core samples to standard suction heads after which the water content is gravimetrically determined. The method is relatively simple to apply, does not need complex calculations and time per sample is relatively low. The accuracy, however, is unpredictable and the reliability therefore low, since the method is susceptible to all kinds of errors. Poor contact between sample and suction plate or sand bed may lead to inadequate desorption, leading to an overestimation of the water content. There is not a control whether the obtained potential has been established.

The overpressure method uses undisturbed samples in the low pressure range of the curve and disturbed samples at high suctions. This transition can easily lead to discontinuities in the curves, especially when the bulk density of the sample is not determined with great precision.

A more general error source in the determination of the retention characteristic is dealing with the effect of hysteresis. Desorption curves differ from adsorption curves as a result of irregular pore sizes that have different 'emptying' and 'filling' characteristics during drying and wetting. Adsorption methods are rarely used, if ever, due to long equilibrium times at lower pressure ranges.

In the field, both desorption and adsorption occurs. The use of 'standard' desorption curves in simulation models will overestimate the moisture content of a soil. For reliable estimates of dynamic behaviour of the moisture content desorption curves can be combined with field measured retentivity points, e.g. use TDR (Topp et al. 1980) in combination with measurement of the matrix potential by transducer tensiometry. Models describing the hysteric behaviour of soils have been developed (Hogarth et al. 1988); however, the parameterisation of these models involves many uncertainties and does not automatically lead to reliable results.

Iterative inverse simulation modelling techniques such as the one-step outflow (Parker et al. 1985) and multi-step outflow method (Van Dam et al. 1990) have been successfully used in many studies. These methods use an iterative simulation procedure from well-controlled outflow experiments to inversely calculate the moisture retention and conductivity curve simultaneously. Since both characteristics are determined from the same dynamic flow experiment they closely copy flow conditions as occurring under field conditions. Disadvantages of this method are its relatively high costs in terms of equipment and the high complexity of the data analysis.

Table 9.1 Operational aspects of five common methods to measure soil moisture retention characteristics. H refers to high, A to average, L to low; F to field method and L to laboratory method

Method	Field/Lab.	Preparation Time	Costs	Complexity	Accuracy
Suction method[a]	L	A	A	L	L	L
Pressure extractor[a]	L	A	A	A	A	A
Vapour equilibrium[a]	L	L	H	L	A	L
In situ determination[a]	F	A	H	H	A	A
Column method[b]	L	H	H	H	A	A

[a]Klute (1986).
[b]Bouma (1980).

Hydraulic Conductivity

The hydraulic conductivity K of a soil is a measure of the soil's ability to conduct or transmit water. More precisely it is defined as 'the flux at unit gradient of the hydraulic head at a particular pressure h'. At zero or positive head K is referred to as the saturated hydraulic conductivity (K_{sat}). At negative pressure heads K is referred to as the unsaturated hydraulic conductivity (K_{unsat}). Methods to measure (K_{sat}) and (K_{unsat}) have been published in various manuals (e.g. Klute 1986); their technical aspects will not be discussed here. Attention will be focused on the application and relevance of the results obtained. Ten methods for measuring (K_{sat}) and (K_{unsat}) are listed in Table 9.2.

K_{sat} can be measured relatively easily by many methods with acceptable accuracy. However, in structured soils containing macropores K_{sat} is not uniquely described. These types of soils generally have a bi-modal pore size distribution with one peak in the macropore domain and one in the micropore domain. The large macropores dominate K_{sat} to a large extent. In cases where macropores are emptied and the soil matrix is still saturated, K_{sat} values decrease dramatically. Booltink et al. (1991) showed a drop of K_{sat} from 5500 cm day^{-1} to 10 cm day^{-1} when reducing the effects of macropores in an experiment where micropores were still saturated and measured matrix potentials were equal to zero. Bouma (1982) and Booltink et al. (1991) replaced K_{sat} and introduced the term $K_{(sat)}$ for this condition. $K_{(sat)}$ is defined as the saturated hydraulic conductivity of a soil at a matrix potential that is just saturated, i.e. the transition point between saturated and unsaturated conditions. Measurement of $K_{(sat)}$ is less easy than K_{sat}. The suction crust infiltrometer (Booltink et al. 1991) has proved to give reliable measurements of $K_{(sat)}$. By lowering the suction applied to the top of the sample, the boundary pressure

Table 9.2 Operational aspects of ten common methods to measure soil saturated and unsaturated hydraulic conductivity. H refers to high, A to average, L to low; F to field method and L to laboratory method

Method	Field/Lab.	Preparation	Time	Costs	Complexity	Accuracy
Saturated						
Column method[a]	F L	A	A	A	A	A
Auger hole method[b]	F	L	L	L	L	L
Inverse auger hole[b]	F	L	L	L	L	L
Drain outflow[b]	F	L	H	A	L	L
Unsaturated						
Improved evaporation[c]	L	L	A	A	H	A
Hot-air method[d]	L	L	L	L	A	L
Sorptivity method[e]	L	L	L	A	H	L
One/multi-step[f]	L	L	A	A	H	A
Crust method[g]	F L	A	H	L	L	A
Instantaneous profile[b]	F	A	H	L	L	A

[a]Bouma (1980), Booltink and Bouma (2000).
[b]Klute (1986).
[c]Halbertsma and Veerman (1994).
[d]Arya et al. (1975).
[e]Dirksen (1979).
[f]Van Dam et al. (1990).
[g]Booltink et al. (1991).

head between saturated and unsaturated (zero vs. negative) can be determined. The hydraulic conductivity at that point can be regarded as $K_{(sat)}$. Especially for modelling studies, using Darcian types of flow models, this $K_{(sat)}$ value will give more reliable results since the effects of bypass flow cannot be adequately modelled with this type of model.

For K_{sat}, $K_{(sat)}$ and (K_{unsat}) it is of great importance that measurements are done on samples that meet or exceed the Representative Elementary Volume (REV)[1] of a soil. Lauren et al. (1988) advocate the use of at least 20 structure elements in a sample to measure hydraulic conductivity. Especially for laboratory measurements, using relatively small samples, this can be a problem. Increasing the number of samples does not automatically compensate this shortcoming since the 'average' curve does not necessarily represents field conditions in the best possible way. If these large-scale spatial correlations exist, a well-designed spatial sampling design either based on geostatistics or on a classical stochastical approach is needed.

Bypass Flow

The occurrence of macropores in soils has an important impact on water and solute movement. Beven and Germann (1982) and White (1985) reviewed the experimental evidence that water flow in soils containing macropores cannot be described accurately by models based on Darcian flow theory only.

Bouma (1984) defined macropore or bypass flow as: 'The flow of free water along macropores into and through an unsaturated soil matrix', and Beven and Germann (1982) defined it as: 'the flow of water through a system of large pores that allows fast velocities and bypasses the soil matrix'.

Increased awareness of the need to improve water-use efficiency and of the risks of rapid transport of pesticides and fertilisers to deeper soil layers leads to the necessity to describe bypass flow in quantitative terms. Methods to measure bypass flow and related processes in a standardised way are, however, still lacking (see e.g. Klute 1986). Booltink et al. (1991; Booltink and Bouma 1991) developed a method that allows simultaneous characterisation of soil morphological features of a structured soil in combination with bypass flow processes such as infiltration, bypass flow, internal catchment and adsorption of water along macropore walls and in dead-end macropores. The schematic set-up of this method is depicted in Figure 9.2.

The set up allows different rain intensities and measures matrix pressure head in combination with outflow characteristics. A typical example is presented in Figure 9.3. Data obtained through this method can be directly used in a modelling approach as described in Section 9.3.5. Reliability and accuracy of the method are relatively high, since phenomena necessary to model bypass flow are directly measured under boundary conditions comparable to field circumstances. Execution of the measurements, however, is time consuming in terms of collecting samples and measurement. Additional calculations are limited.

9.2.2   Estimating Soil Physical Properties

Existing soil survey data are a valuable source of information to generate soil physical input data for simulation models describing water and solute movement (Wösten 1987; Finke and Bosma 1993; Wösten and Van der Zee 1993). Bouma and Van Lanen (1987) propose the use of pedotransfer functions to translate basic soil properties obtained in soil surveys to parameters describing soil physical properties. They distinguish two types of pedotransfer functions: (i)

---

[1] The REV denotes the minimum volume of soil needed to permit a representative measurement (Lauren et al. 1988).

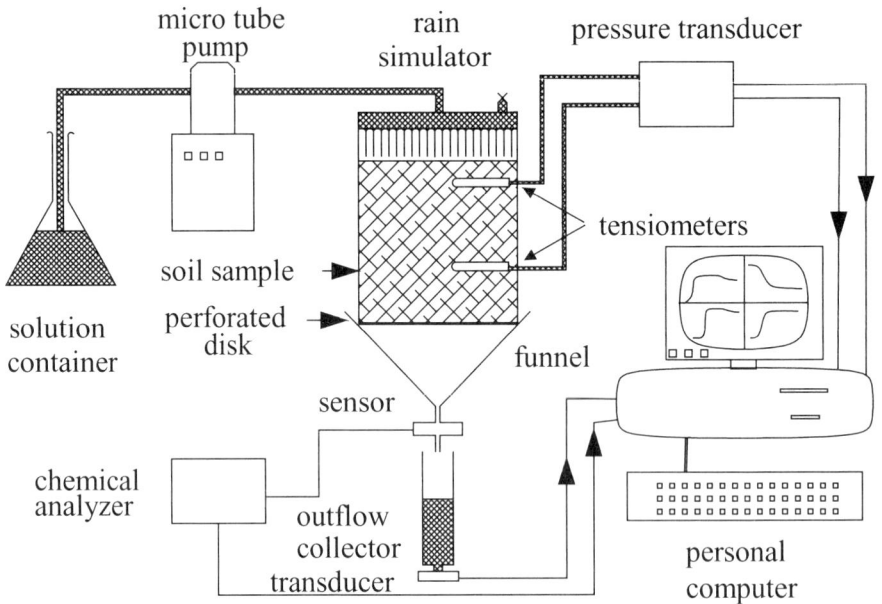

Figure 9.2   Schematic set-up of the computer-controlled device for measuring bypass flow

continuous pedotransfer functions, relating physical characteristics to discrete soil data such as clay percentage, organic matter content or bulk density; (ii) class pedotransfer functions relating soil physical properties to soil data defined in terms of classes (e.g. 'silt loam' texture or 'medium' organic matter content).

Numerous pedotransfer functions have been developed over the past years. Vereecken (1988), Vereecken et al. (1992) and Tietje and Tapkenhinrichs (1993) presented reviews on a wide range of pedotransfer functions in terms of applicability and accuracy. The effects of uncertainty in pedotransfer functions on simulated soil behaviour have been studied by Finke et al. (1996). In their study they compare the use of continuous and class pedotransfer functions in break-through calculations of absorbing and non-absorbing agro-chemicals. They conclude that the effect of uncertainty in pedotransfer functions plays an important role in the aspects of soil related to moisture movement (non-absorbing chemicals). On the other hand, uncertainty in pedotransfer functions did not show a significant effect on simulation studies dealing with transport of absorbing agro-chemicals. In the first type of study the more precise continuous pedotransfer functions are the best to apply; in the second, class pedotransfer functions will perform with satisfying accuracy.

Bouma et al. (1996) used soil survey data for modelling solute transport in the vadose zone on two soil types with a different texture. Error propagation from collecting field data through pedotransfer functions and simulation modelling were quantified in probabilistic terms. The results of this study are briefly described in Case study I, Section 9.2.3.

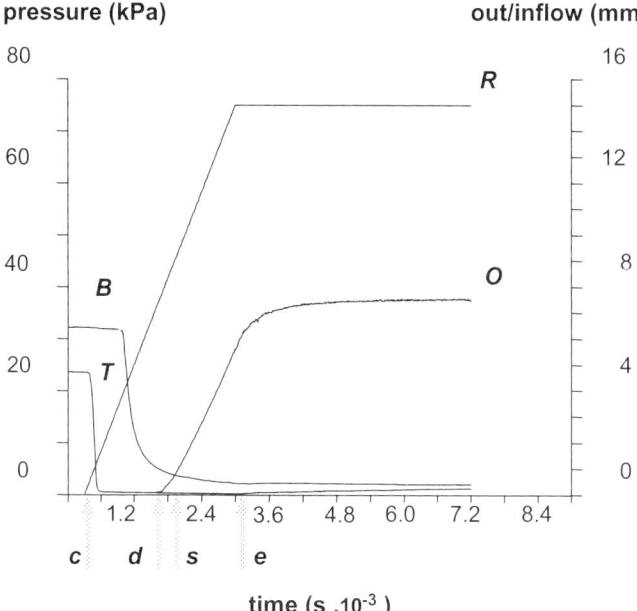

pressure (kPa)   **out/inflow (mm)**

time (s .10⁻³)

Figure 9.3   Results of a typical bypass flow measurement. Arrows indicate characteristic points in time: c refers to start of water entering the macropores ($T_{(c)}$), d to the point of initial breakthrough at the bottom of the sample ($T_{(d)}$), s to the starting point of the linear outflow ($T_{(cs)}$), and e to the end point of linear outflow ($T_{(ce)}$). B and T refer to the bottom and top tensiometer, respectively. R refers to the cumulative rain applied and O the cumulative outflow

9.2.3   Case Study I: Reliability and Uncertainty in Soil Survey Data for Modelling Solute Transport

Introduction

Modelling of non-point-source pollutants in the vadose zone is crucial in the development of innovative land management schemes, which allow multiple forms of land use while maintaining acceptable levels of soil quality. Because human and animal health is involved, modelling has major implications and should therefore include expressions of accuracy and precision for both input parameters and results. Many questions deal with regional land use, and soil survey information is therefore necessary. The case study to be discussed here focuses on the determination of chemical fertilisation of agricultural land to an extent which is acceptable from both a production and environmental point of view. Balancing these two requirements represents a crucial political question at this time in the Netherlands. Considering the above discussion on using soil survey data, the study will address the issue as to how results of modelling are affected by the accuracy of basic input. All procedures require massive data handling and manipulation, making use of GIS systems indispensable. However, GIS systems only facilitate and smooth procedures. Measurements and simulation models provide the keys to solutions being pursued.

   The field site being characterised measured 7 ha and occurred in the Wieringermeer polder of the Netherlands. Soils were classified as loamy, mixed, mesic Typic Udifluvents (Soil Survey Staff 1975). A geostatistical analysis of spatial variability resulted in a recommendation to make

Table 9.3 Main soil characteristics of the functional layers used

| Functional layer | Texture | Organic matter (%) | Water content | | Alpha (cm^{-1}) | n (–) | Gamma (–) | Saturated conductivity (cm day^{-1}) |
			Residual (cm^3 cm^{-3})	Saturated (cm^3 cm^{-3})				
B1	Loam to clay loam	2–4	0.000	0.450	0.1387	1.1205	− 2.8550	243.0
B2	Sandy loam to loam	0–1	0.000	0.470	0.0469	1.1292	− 4.1497	104.1
B3	Loam, clay loam	0–1	0.037	0.587	0.0267	1.2193	11.155	201.6

Table 9.4 Descriptions of the soil profiles used in the simulation study

Profile	Depth (m)	Functional layer
Profile 1	0.00–0.40	B1
	0.40–1.20	B2
Profile 2	0.00–0.40	B1
	0.40–0.45	B2
	0.45–1.20	B3

402 point observations (borings) in a triangular grid, which were each characterised in terms of texture, structure, organic matter content and colour. Different genetic soil horizons were distinguished (Soil Survey Staff 1975). In this study, however, attention was focused on *functional* horizons, which consist of combinations of adjacent genetic horizons with properties that do not differ significantly. Emphasis is here on physical properties, specifically on the hydraulic conductivity and moisture retention. Distinction of functional horizons is based on comparison of multiple measurements in genetic horizons, combining horizons that have comparable properties (e.g. Finke and Bosma 1993). Only three functional horizons, also referred to as 'building blocks', were distinguished. These three functional horizons can be used to represent physical properties at any depth in the 402 point observations. Soil physical characteristics for the three functional horizons are shown in Table 9.3.

Simulation Procedure

The simulation model LEACHM (Hutson and Wagenet 1991), which was extended with a potato growth submodel (Feddes et al. 1988), was used to simulate the water regime, nitrate leaching and potato growth in two soils. Two soils, belonging to the database of 402 observations, were selected for further analysis. The first had a dominantly sandy loam texture (Profile 1) and the other had a silty clay loam texture (Profile 2). Table 9.4 shows the depths at which the functional horizons occur in both soils. Simulations were made for the subsequent growing seasons of 1989 and 1990, for which field-monitoring data were available.

Nitrogen fertilisation rates applied in profiles 1 and 2 were based on official farmer guidelines. The necessary hydraulic characteristics were measured with the one-step outflow method (K_{unsat}, ψ-θ) (Doering 1965; Parker et al. 1985) and the crust test (K_{sat} and K_{unsat} near saturation) (Hillel and Gardner 1969; Booltink et al. 1991). A soil survey was made of the study area on which a preliminary selection of functional horizons was based. The definite selection was based on 43 measured hydraulic characteristics, following procedures defined by Finke (1993). According to this procedure measured hydraulic characteristics were used in a simultaneous fit of Van Genuchten's parameters (Van Genuchten 1980) using the software package RETC (Van Genuchten et al. 1991). This program provides estimates and probability intervals and mutual parameter correlations for Van Genuchten's parameters. Figure 9.4 depicts the measured hydraulic characteristics for the three functional layers and their 32% and 68% confidence levels.

To adequately express the effect of the variability of the basic hydraulic parameters on modelling results, Monte Carlo techniques were applied using the internal variability within the three functional horizons. The Monte Carlo analysis was carried out using UNCSAM 1.1 (Janssen et al. 1993; Booltink 1994). Sampling of the distributions was done using the Latin Hypercube sampling technique (McKay et al. 1979; Iman and Conover 1980). When Latin

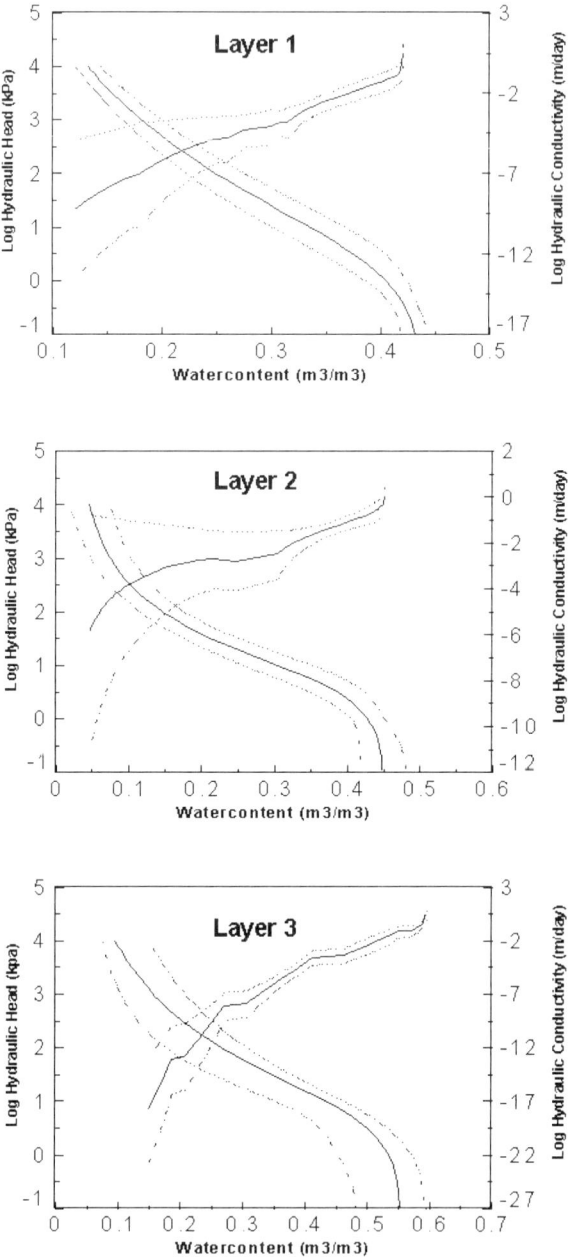

Figure 9.4 Retentivity and conductivity characteristics of functional layers 1, 2, and 3. The solid line refers to the average of the measurements, the dotted lines depict the average ± 1 standard deviation

Hypercube sampling is used the parameter space is adequately represented when using $2p < N < 5p$ samplings, where N is the number of samplings and p is the number of parameters considered. For the 16 parameters in this study 80 samplings were carried out. For each profile, therefore, 80 simulations were performed.

To compare the Monte Carlo simulation results with some more general parameter estimation approaches additional simulations were carried out for both profiles. First the average parameter set, as obtained with RETC, was used as input in LEACHM. In the second approach, for each profile, soil textural data were used to derive Van Genuchten's parameters from the Dutch national soil-physical database the 'Staring series' (Wösten 1987; Wösten and Van Genuchten 1988). The 'Staring Series' are based on measurements made all over the Netherlands and provide $K(\theta)$ and $h(\theta)$ data for broad standard textural classes as distinguished in soil survey. These classes function as more generalised class-pedotransfer functions, as compared with the ones for the functional layers in this study which were based on measurements within the area of study. Testing and comparing both functions will show whether it pays to make the latter type of investment.

Simulation Results

Results of the RETC fitting procedure for the hydraulic Van Genuchten parameters are presented in Table 9.5. The results of the 80 simulations with the extended LEACHM model for profiles 1 and 2 are presented in Table 9.6.

The production of potatoes is significantly higher in profile 1, which is due to the higher upward flux from the water table to the root zone. This holds for the total water-limited production, the water- and N-limited production and for tuber yields (water and N-limited). Significant differences are found too when comparing N-leaching during the growing season.

Table 9.5 Results of RETC fitting procedure for the three functional layers B1, B2 and B3. WCS refers to the saturated water content; Alpha, N and L are fitting parameters from the RETC procedure; and CONDS is the saturated hydraulic conductivity. WCR, finally, is the water content residual (if not specified: equal to 0)

Variable	Value	Std. dev	95% confidence limits	
			Lower	Upper
B1-WCS	0.450	0.0096	0.431	0.470
B1-ALPHA	0.1387	0.0244	0.0884	0.1890
B1-N	1.1204	0.0052	1.1096	1.1314
B1-L	− 2.8549	0.3560	− 3.5883	− 2.1217
B1-CONDS	243.0	71.02	96.6	389.2
B2-WCS	0.469	0.0211	0.426	0.513
B2-ALPHA	0.0469	0.0121	0.0218	0.0720
B2-N	1.2911	0.0238	1.2421	1.3402
B2-L	− 4.1496	0.3100	− 4.7884	− 3.5110
B2-CONDS	104.1	38.64	24.5	183.7
B3-WCR	0.037	0.0169	0.002	0.073
B3-WCS	0.5872	0.0038	0.579	0.595
B3-ALPHA	0.0269	0.0010	0.0248	0.0292
B3-N	1.2129	0.0106	1.1906	1.2353
B3-L	11.1553	0.7490	9.5874	12.7233
B3-CONDS	201.6	18.4	163.0	240.2

Table 9.6 Results of the Monte Carlo simulations for profiles 1 and 2. The mean and standard deviation are based on 80 model realisations per profile. All units are in kg ha^{-1}. Production is expressed as dry matter; the NO_3-N load is the annual cumulative amount of nitrate N leached from the profile divided by the annual cumulative amount of water leached from the profile

	Mean	Standard deviation
Potential biomass production	14 594	
Profile 1		
Actual biomass production		
Water limited	13 920	131.5
Water and nitrogen limited	11 239	226.7
Actual tuber yield	6 219	237.4
NO_3-N load	190	1876
Profile 2		
Actual biomass production		
Water limited	11 154	160.1
Water and nitrogen limited	9 267	313.7
Actual tuber yield (W + N limited)	4 337	319.5
NO_3-N load	6.7	0.745

Relatively low leaching was observed in profile 2 and values had a low standard deviation. Not only was total leaching in profile 1 about 30 times as high, but standard deviations were very high as well. Fluxes are lower in the clayey profile 2. While heavy showers induce rapid flow through in the more sandy profile 1, hydraulic characteristics of profile 2 tend to strongly reduce this flow through, causing not only lower breakthrough of N but lower standard deviations as well. The probability of high breakthrough values in profile 1 is, therefore, significantly higher. Also, the amount of N leaching in profile 1 is unacceptable from an environmental point of view as it far exceeds the 45 kg ha^{-1} yr^{-1} limit, which is being used in environmental regulations in the Netherlands. The official guidelines for fertilisation were used for both soils when making these simulations. They are acceptable for profile 2, but certainly not for profile 1. Exclusive emphasis on yield, ignoring effects of the associated leaching, has long been the guideline for setting up fertilisation recommendations and application schemes. Criteria for fertiliser application should be better adjusted to the soil type to prevent leaching, including the effect of soil variability on leaching probabilities. Here, simulations of yields and leaching rates, as presented, can play an important role in which the calculated variation is as important as the average value.

The results of the additional two simulation procedures using the average parameter set from the RETC procedure and the 'Staring' class-pedotransfer function are summarised in Table 9.7. For both profiles the results obtained using the RETC-averages coincide well with the averages of the Monte Carlo procedure. The use of the 'Staring' class-pedotransfer function in the more sandy profile 1 shows a large difference with the average of the Monte Carlo simulation. A negative load would refer to a net uptake from the groundwater, which is highly unlikely. For profile 2 the 'Staring' class-pedotransfer function gives results comparable with the RETC and Monte Carlo procedure.

Results indicate that use of generalised national databases, such as the 'Staring series', should be used with care because major errors can result, particularly in the rapid responding, sandier soils with relatively high fluxes.

Table 9.7 Simulation results for profiles 1 and 2. Average refers to the simulation using the best fit of the RETC procedure, Staring refers to the results obtained using the class-pedotransfer function. All units are in kg ha^{-1}. Production is expressed as dry matter, the NO_3-N load is the annual cumulative amount of nitrate N leached from the profile divided by the annual cumulative amount of water leached from the profile

Simulation	NO_3-N load	Actual biomass production		Tuber yield (W + N limited)
		Water limited	Water and nitrogen limited	
Profile 1				
Average	229	13 645	10 657	5603
Staring	− 3.4	14 270	12 543	7671
Profile 2				
Average	6.8	11 190	9 172	4285
Staring	5.1	11 141	8 936	4634

9.2.4   Case Study II: Collecting Farm-scale Soil Physical Data for Simulation Modelling in Precision Agriculture

Introduction

Advanced information technology has allowed the development of various techniques that make it possible to realise precision agriculture. Precision agriculture indicates that differences in soil and crop conditions within a field are the basis for 'precise' forms of management which differ as the agricultural equipment moves over a field. Important here are Global Positioning Systems (GPS) and Geographic Information Systems (GIS). Proceedings of International Conferences summarise current developments (Robert et al. 1993, 1995, 1997; Stafford 1997).

Even though the concept of precision agriculture is relatively simple and attractive, the practical realisation of operational field systems, integrating all aspects of precision farming, has not yet been achieved. The possible exceptions are systems where chemical fertilisation rates are based on multiple fertility samples taken within a field. However, such systems represent only a fraction of what is conceptually possible. The main problem is a need to predict soil and weather conditions for the next growing season, as weather conditions are very important in determining plant growth and solute fluxes. Climate data, obtained by averaging weather data for a large number of years, is unsuitable for use because of weather variation among the years. Use of computer simulation modelling is attractive to predict crop growth and solute fluxes for a series of many years (e.g. Verhagen et al. 1995). Results can be stored in a database and data, which corresponds with actual conditions in a given year, can be retrieved as time moves on during a particular growing season. Such simulation modelling can also be used to identify areas within a field which consistently show different growth patterns and solute fluxes as compared with other areas (e.g. Verhagen and Bouma 1997).

Use of simulation models should, however, proceed with care because their use is only justified when they are properly calibrated and validated. Calibration and validation can be achieved by comparing measured and calculated moisture contents in the soil (e.g. Verhagen et al. 1995; Booltink and Verhagen 1996). This is costly and can only be done to a limited extent.

In this study we will demonstrate the use of soil physical characteristics obtained through various procedures. Pedotransfer functions (class as well as continuous) will be combined with a limited number of retentivity and conductivity measurements to obtain maximum accuracy and reliability of simulated soil water responses.

Data Collection

The study was carried out on the Van Bergeijk farm in the Netherlands, a commercial farm of approximately 100 ha in the southwest of the Netherlands. Soils consist of marine deposits and are generally calcareous, and texture ranges from fine loam to heavy clay loam. With the excellent drainage induced by a dense system of tile-drains, these soils are considered to be prime agricultural soils. A detailed soil survey was carried out at the Van Bergeijk farm in the spring of 1997. Crop rotation on this farm is mainly winter wheat, consumption potatoes and sugar beet.

On geo-referenced sites augurings were taken (612 in total) and layer characteristics such as texture, organic matter, soil colour, thickness of the layer etc. were determined and stored in the field computer. Using data from a GPS the data was geo-referenced and stored in a soil database, which is used to calculate soil physical parameters. Using a simple simulation model (Finke and Bosma 1993) functional characteristics were calculated and functional layers were determined. The 612 augurings with a total of about 3000 horizons could be characterised using only 18 different functional layers. These 18 functional layers were sampled in two folds for soil physical analysis. Retentivity and conductivity measurements were done in the laboratory using the suction crust infiltrometer (Booltink et al. 1991), the multi-step outflow method (Van Dam et al. 1990) and additional retentivity measurements using pressure extractors (Klute 1986). For every functional layer one set of Van Genuchten's (Van Genuchten 1980) was calculated (data set I).

The second set of soil physical characteristics was determined using the Staring series (see Section 9.2.2). Soil texture and organic matter data were used to classify each of the 3000 soil horizons in this class pedotransfer function and related soil physical characteristics as given in the Staring series were selected (data set II).

Thirdly, soil texture, organic matter and estimated bulk density information from the soil database was used to estimate soil physical characteristics for each of the 3000 individual soil horizons (data set III).

A fourth set of physical characteristics was obtained by combining the measurements of the functional layer (data set I) with soil physical characteristics from the continuous pedotransfer function (data set III).

An intensive field-monitoring scheme has been followed. Monitoring, implying incorporation of the time dimension, provides time series required to feed and calibrate simulation models. Two monitoring schemes will be discussed here, concerning weather conditions and soil water contents:

- The Van Bergeijk farm is equipped with a modern weather station, registering rainfall, humidity, temperature and global radiation. Data are initially stored for 12-minute intervals and converted to daily values. Potential evapotranspiration rates were calculated.
- A monitoring plot has been installed in every field to measure groundwater levels and soil water contents at different depths. During the growing season measurements are taken at least once every week; measurements are continued at a lower frequency during winter. Groundwater levels are essential input parameters for the simulation model. Water contents, measured using Time Domain Reflectometry (TDR), provide calibration series to assess model performance.

Simulation Procedure

In this case study the integrated simulation model WAVE has been used. WAVE (Water and Agro-chemicals in soil and Vadose Environment, Vanclooster et al. 1994) integrates four existing models, including dynamic simulation of water flow based on the SWATRER model (Dierckx et al. 1986), a nitrogen model based on the SOILN-model (Bergström et al. 1991), a heat and solute transport model based on the LEACHM-model (Wagenet and Hutson 1989) and the SUCROS crop growth model (Spitters et al. 1989). Two conceptual changes in the original WAVE model were made. First, water uptake by plant roots was originally modelled assuming preferential uptake at the top compartments excluding roots in deeper layers. In the revised version the total water uptake is the integral over the root zone. Second, nitrogen uptake in the original version was modelled using the nitrogen concentration in the leaves as driving variable for total nitrogen uptake. The revised version links nitrogen uptake to biomass production as described below. No revisions were made in the four core modules.

Water stress is calculated according to Feddes et al. (1978), in which the maximum water uptake is defined by a sink term as a function of depth. Water uptake is reduced at characteristic high and low pressure head values.

Stress resulting from nitrogen deficiencies is calculated using the 'critical nitrogen concentrations' as defined by Greenwood and co-workers (Greenwood et al. 1985; Greenwood et al. 1990). They describe the decrease in N-percentage with increasing plant mass. When the actual uptake is insufficient to sustain the necessary concentration as defined by biomass, production is proportionally reduced to the ratio of the actual and required uptake.

The WAVE model was executed for each of the four data sets and compared to measured soil water contents on two depths on three different fields.

Simulation Results

In Figures 9.5 and Figure 9.6 soil moisture retentivity and conductivity curves are presented for soil layers B12, O13, O8 and O3. As can be seen from these figures the best fit between measured retentivity c.q. conductivity data was obtained by the multi-step optimisation procedure. These results, however, did not obtain the best fit with the time series of water contents. In Figure 9.7 the results of using the four different soil physical data sets is shown for the three fields. Modelling accuracy for the fourth data set, i.e. combination of measured water content saturated and shape parameters of the Van Genuchten's closed form equation was far superior to the other options. This result corresponds with the work of Vereecken (1988), who found in a sensitivity analysis of various pedotransfer functions that the saturated water content in the retentivity curve explained about 80% of the variability in his data. The saturated hydraulic conductivity appeared to be the most important variable in the conductivity curve.

From this study it can be concluded that measurements of soil physical characteristics can be successfully combined with estimates using pedotransfer functions. Strong aspects of the two approaches (accurate values for direct measurements and detailed spatial information for the continuous pedotransfer functions) can be combined and will have a synergetic effect on the simulation results.

9.3   MONTE CARLO SIMULATION PROCEDURES

9.3.1   Monte Carlo Techniques

Monte Carlo based techniques rely on the fact that variations in model inputs can generally be

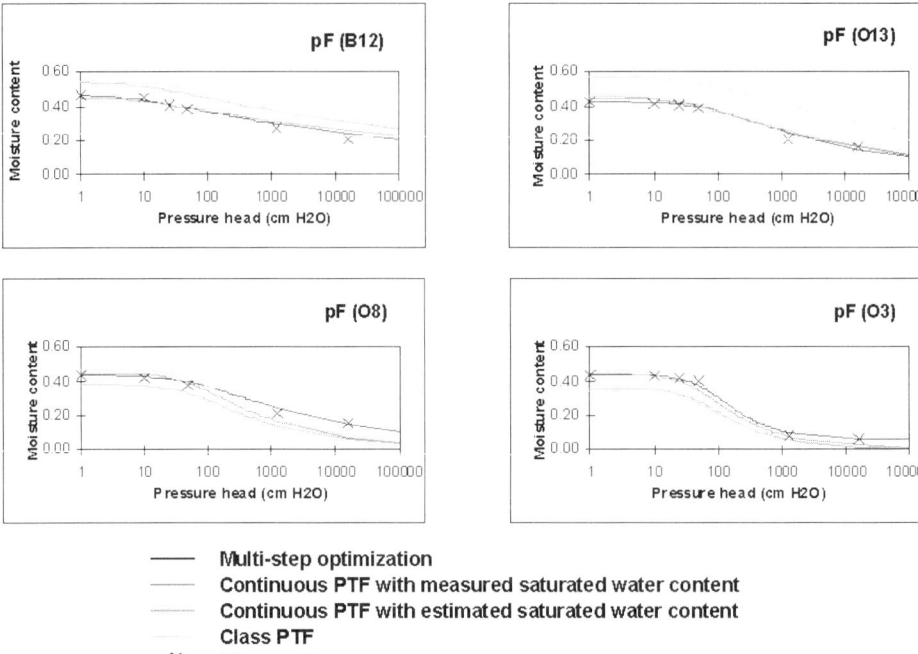

———   Multi-step optimization
————   Continuous PTF with measured saturated water content
········   Continuous PTF with estimated saturated water content
             Class PTF
X   Measured

Figure 9.5   Retentivity curves for a typical profile on the Bergeijk farm. B indicates topsoil, O subsoil, the number indicates the diffferent types of top and subsoils. The multi-step optimisation uses time series of outflow in combination with measured retentivity data. This method, therefore, produces the best fit between retentivity measurements and estimated values

described by probability density functions. After specifying these distributions sampling from these distributions is performed, resulting in a set of model inputs. All these inputs are used to simulate the model output. Further analysis of the distributed modelling results, in terms of means, variances of the simulated outputs and correlation between model input and output, provides information on sensitivity and uncertainty of the model and its inputs.

This Monte Carlo approach is simple and can be applied to almost any simulation model. However, the computational effort on large models with many input parameters is becoming extremely large in terms of time and data management. In case of random sampling from the distributions the number of samples to be taken should be larger than ten times the number of parameters included in the Monte Carlo analysis (Janssen et al. 1993).

Latin Hypercube sampling was developed by McKay et al. (1979) and Iman and Conover (1980) and is a more efficient sampling technique. This technique uses a stratified way of sampling from separate inputs, on the basis of a subdivision of the range of each input in N disjunct equipropable intervals. Sampling one value in each interval according to the associated distribution, one obtains thus N sampled values for each model input. The sampled values for the first input are subsequently randomly paired to the sampled values of the second input, third etc., which finally results in N combinations of p parameters. When using Latin Hypercube sampling the input parameter space is adequately represented with a number of samples (N) between $2p$ and $5p$. Recalling for random sampling $N > 10p$.

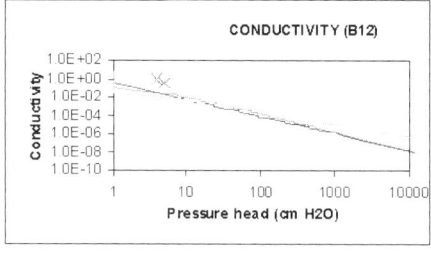

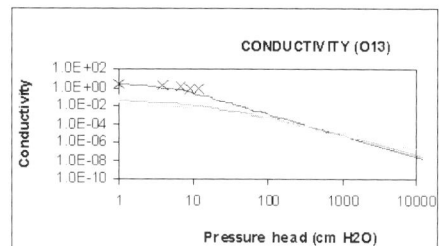

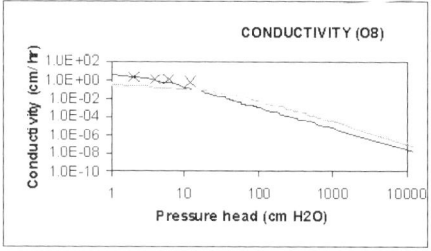

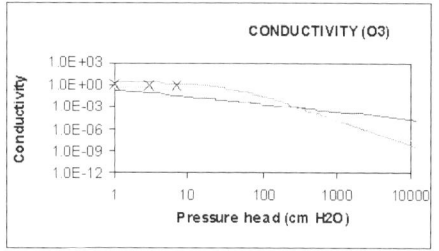

——— **Multi-step optimization**

——— **Continuous PTF with measured saturated hydraulic conductivity**

········ **Class PTF**

X   **Measured**

Figure 9.6 Conductivity curves for a typical profile on the Bergeijk farm. B indicates topsoil, O subsoil, the number indicates the different types of top and subsoils. Measured conductivity data (except for the saturated hydraulic conductivity) were not included in the optimisation procedure

In many environmental and agricultural modelling studies, model input parameters are correlated. In Figure 9.8, two hypothetically correlated, normally distributed model inputs (*a* and *b*) are presented. The shaded area represents the correlation between the two parameters. When ignoring this correlation in the sampling procedure, any combination within the dashed square is considered to be a valid combination of *a* and *b*, thus resulting in many unrealistic parameter combinations and therefore model outcomes. Distributions of model outputs will therefore be unnecessarily wide and not reflect reality. Iman and Conover (1982) developed a method using the distribution independent restricted pairing technique for incorporating correlations in the sampling procedure and to avoid spurious correlations due to the sampling process.

9.3.2 Sensitivity and Uncertainty Analysis

Janssen et al. (1993) described various statistics to quantify the sensitivity and uncertainty of model inputs to model outputs. Most of them are based on regression and correlation between model inputs and simulated outputs obtained through a Monte Carlo simulation procedure. Here we will briefly present a few measures, based on linear regression analysis. For full detail on these methods see Janssen et al. (1993).

The basis of this sensitivity/uncertainty analysis is the linear regression model as depicted in

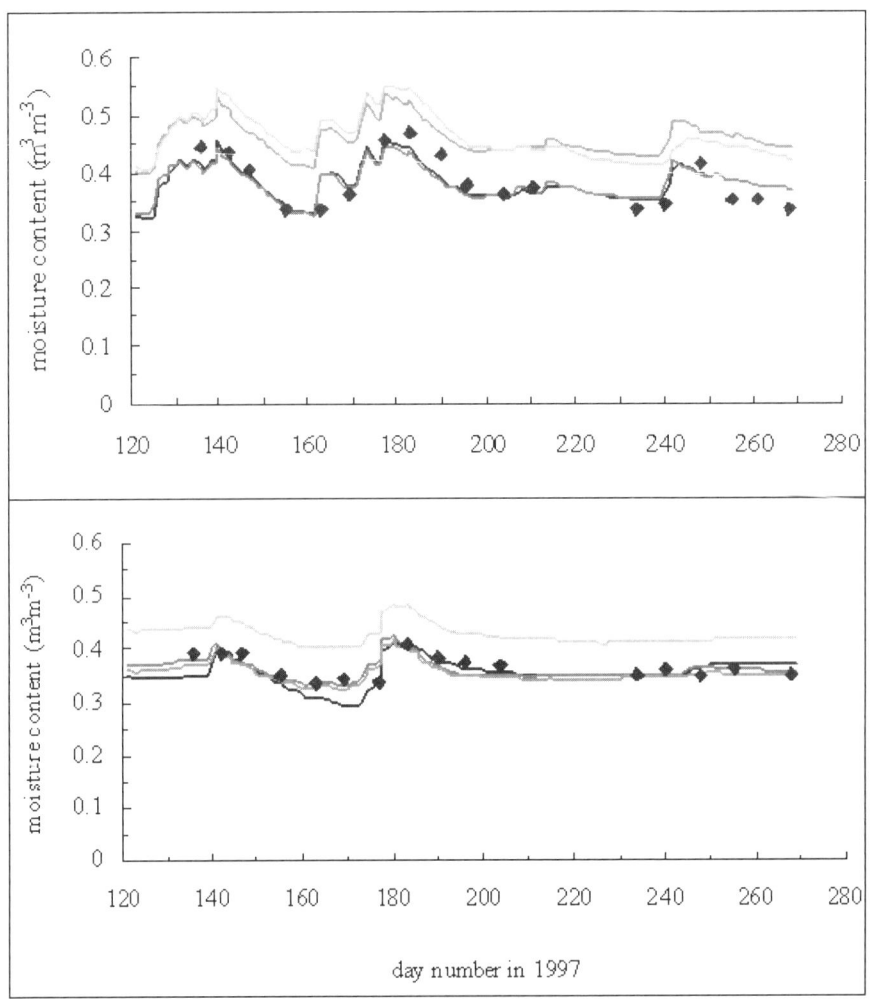

_____ **Multi-step optimization**

_____ **Continuous PTF with measured saturated water content
and estimated hydraulic conductivity**

_____ **Continuous PTF with estimated saturated water content
and estimated hydraulic conductivity**

_____ **Class PTF**

♦   **Measured**

Figure 9.7 Comparison of simulation results using the four different sets of soil hydraulic properties for three different plots (A–C–D) on two depths for 1997. Black dots indicate TDR-measured water contents

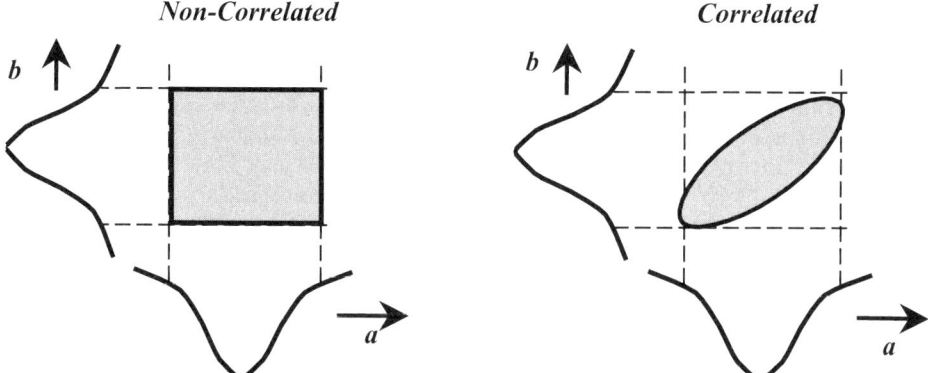

Figure 9.8 Sampling from uncorrelated and correlated distributions

equation (1). The regression model output is represented by $y(k)$ in the kth simulation run $(k = 1, \ldots, N)$ and the associated model inputs by $x_i(k)$.

$$y(k) = \beta_0 + \sum_{i=1}^{p} \beta_i x_i(k) + e(k) \qquad k = 1, \ldots, N \tag{1}$$

Further β_0, β_i denote the ordinary regression coefficient (ORC) and $e(k)$ denotes the regression residual. The ORC (β_i) can be considered as the absolute sensitivity measure, measuring the absolute change Δy of y if x_i changes with amount Δx_I.

The regression coefficient (R^2) (equation (2)) is a number between 0 and 1, expressing the validity of the linear regression model $y(k)$ to approximate the original model output. $R^2 \approx 1$ indicates a good approximation.

$$R^2 = 1 - \frac{S_e^2}{S_y^2} \tag{2}$$

where S^2 denotes the variance of the associated quantity.

In the case where x_I is uncorrelated with the other model inputs, $\beta_i S_{xi}$ measures the linear uncertainty contribution of source x_I. The related standardised regression coefficient (SRC) (β_i^s) is given in equation (3):

$$\beta_i^s = \beta_i \frac{S_{xi}}{S_y} \qquad i = 1, \ldots, p \tag{3}$$

The SRC measures the fraction of uncertainty contributed by x_I. It will only give a valid impression of the relative uncertainty contribution if model inputs show no substantial correlation.

A different way of standardising the quantities in equation (1) is obtained by dividing the data by their average values. This coefficient is called the Normalised Regression Coefficient (NRC) (β_i^n) and measures the relative change of $\dfrac{\Delta y}{\bar{y}}$ of y due to a relative change of of $\dfrac{\Delta x}{\bar{x}}$ of x_i:

$$\beta_i^n = \beta_i \frac{\bar{X}_i}{\bar{y}} \tag{4}$$

The above-listed measure for sensitivity/uncertainty analysis is not complete and does not deal with non-linearities; however, it provides useful information in many simulation studies. When encountering important non-linearities one should therefore apply other methods, e.g. consider non-linear regression models or apply data transformations on y and x_i. Useful information can be obtained by a thorough study of the characteristics and relations of the sampled parameters x_i and the simulated model outputs y, e.g. by making scatter plots of these quantities.

9.3.3   Rotated Random Scanning for Model Calibration

In environmental agricultural applications of models one often encounters parameters that cannot be measured directly (e.g. the diffusion distance of a molecule from the soil to a plant root) or are difficult to measure (e.g. rate constants for the turnover of organic matter in the soil). Especially when using complex simulation models, only seldom have all model inputs have been measured or determined uniquely. In terms of model applications these categories of parameters are called 'ill-defined' (Keesman 1989). Model calibration is a means to make the simulation model applicable for the case study involved. A viable alternative, for example, trial and error parameter adjustment is to use the set-theoretic approach developed by Janssen et al. (1993) and Janssen (1995b). This approach aims at finding a set of parameter vectors leading to model results, which are deemed 'acceptable'. This requires in the first place an adequate definition of when to judge the model results as acceptable (against the background of the available data, prior information and potential application), and in the second place adequate methods to determine the requested parameter set. For this latter problem Janssen (1995b) proposed the use of the Rotated Random Scan (RORASC) procedure, an iterative Monte Carlo search of the parameter space inspired by Keesman and Van Straaten (1988, 1989), Keesman (1989, 1990). This method tries to efficiently update the current available set of acceptable parameter vectors by iteratively applying rotations or transformations of the parameter space, followed by a uniform random scan in the transformed space. In this way one hopes to gradually and efficiently zoom in on the final acceptable parameter set.

The RORASC procedure consists of three basic steps (Janssen 1993):

1. *Initialisation.* Before starting the iterative procedure, an initial set of parameters has to be determined, which serves as a starting point for the subsequent iterative simulations. A straightforward way to obtain such a set is as follows:
   - Sample the originally specified parameter space, e.g. by a Monte Carlo sampling procedure.
   - Simulate the model for these sampled parameter combinations and determine which of them give acceptable simulation results. According to predefined criteria, one can manually or automatically perform this step, depending on the availability of accurate data of the potential application.

   The above sketched approach yields two sets of parameter combinations: a set leading to acceptable simulation results and a set leading to non-acceptable simulation results.
2. *Generating new candidate samples.* First all available acceptable parameter combinations are collected and checked as to whether or not the predefined total number of acceptable combinations is reached. If this is not the case, the generation of a new set of parameters is

achieved by transforming or rotating the original parameter space to focus on the current set of parameters leading to acceptable simulation results. This transformation is based on the decomposition of the covariance matrix of the accepted parameters, as described in detail by Janssen (1993). Subsequently a random-scan on the basis of a uniform random sampling in this transformed space is performed which serves as a new set of candidate samples.

3. *Simulation and acceptance.* Model simulations are performed for the newly generated candidate samples, and it is decided whether the results are acceptable or non-acceptable. One then returns to the next iteration.

The exact progress of the above iterative process depends on a number of choices and options which have to be specified by the user in the initialisation phase and when generating new candidate samples; these are discussed in detail by Janssen (1993). To obtain a reliable covariance matrix when using Latin Hypercube sampling in combination with ROASC, a number of samples $N > p$ is recommended, where p is the number of parameters to be sampled.

9.3.4   Model Validation

The meaning of the word validation is not uniquely defined. Hassanizadeh and Carrera (1992) described the multi-faceted aspects of validation in a special issue of *Advances in Water Resources*. Here we will only touch the aspects of conceptual validation and validation of performance.

Model validation can be regarded as the last of three steps in the application of simulation models. The first step is the determination of the parameters to which the model is sensitive. The most sensitive and the ill-defined parameters are subsequently used to fine-tune or calibrate the model. The last step concerns the analysis of the model validity. Two types of validity can be distinguished: (i) conceptual validation and (ii) validation of performance.

Conceptual validation represents the validity of the aggregated model structure, i.e. it indicates whether or not the model is capable of simulating the processes according to a user-predefined accuracy. A conceptual validation can be performed using the results of a Monte Carlo based sensitivity, uncertainty or robustness analysis. When equation (1) has been derived, the ordinary regression coefficient R^2 (equation (2) represents the explained variance of the calibrated model of the total variance of all model realisations. An R^2 close to 1 is an indication for a well-performing model, whereas a low R^2 indicates that the sensitive parameters do not control model outputs. The latter case can be an indication of an inadequate model structure, i.e. too many subprocesses and/or too many parameters with mutual influence that lead to unbalanced simulation results.

A validation of performance is generally performed through a simulation study on an independent data set. Comparing simulated values with measured values (preferably time series) yields insight into the model performance on independent data. Quality criteria for this comparison have to be defined by the user and cannot be generalised. Required modelling accuracy or quality of the independent data set are a few of the many criteria that can influence this choice.

9.3.5   Case Study III: Simulation of Bypass Flow on Field Scale

Data Collection

At the experimental farm 'De Kandelaar' in Eastern Flevoland in the Netherlands a tile-drained research site of approximately 100×300 m was used for this experiment. Tile-drains in this site are 150 m long and 48 m apart. The soil was classified as a mixed, mesic Hydric Fluvaquent (Soil

Survey Staff 1975). A soil profile is presented in Table 9.8. Many small macropores were present in the plough layer (0.0–0.3 m), and a few very large in the subsoil. These large cracks originate from structural soil shrinkage after reclamation of the polder and are present permanently throughout the year. Average contact areas for vertically and horizontally oriented macropores (Vercon and Horcon, respectively, in Table 9.9) for top- and subsoil (respective suffixes __up and __lo) showed a decline of 85 to 90% for top- compared to subsoil.

A soil survey was carried out to quantify spatial variability. However, no spatial correlation could be detected. Therefore, all variability was regarded as variation of a stochastic nature. Continuity of flow patterns and contact areas between bypass flow water and macropore walls were quantified using tracer experiments with methylene blue and iodine tracers in combined field–laboratory experiments. These experiments are described elsewhere as reported by Booltink et al. (1993), Booltink (1994).

In six research plots tensiometer profiles were installed at 0.05, 0.25, 0.35, 0.50, 0.80 and 1.00 m depth. Groundwater piezometers were installed on each of the six plots and on a line perpendicular to the tile-drains. Deep groundwater (4.5–5.0 m) was measured on two plots. A fortnightly monitoring scheme was maintained as much as possible during the year.

A CR10-datalogger (Campbell) was used for continuous monitoring drain discharge from one tile-drain in the centre of the study site. Measuring resolution of the drain outflow meter was 0.03×10^{-3} m. The CR10-datalogger also recorded rain intensity and amount, soil surface temperature and groundwater level at one plot within the experimental field with a 15-minute time resolution.

The parameters Ctd_1, Ctd_2 and Ctd_3 in the third order polynomial (equation (10)) for drain discharge, were fitted on simultaneously measured groundwater levels and drain discharge rates in the winter period 1989–1990 using multiple regression techniques.

On each of the six plots in every distinguishable soil horizon, samples were taken in large steel cylinders, 0.20 m long and 0.20 m diameter. These samples were analysed in the laboratory using the following methods:

- *Suction crust infiltrometer*: determination of the saturated conductivity and unsaturated conductivity near saturation (Booltink et al. 1991).
- *One-step outflow method*: for unsaturated conductivity and water retention curve (Parker et al. 1985).
- *Pressure extractors*: used to measure water retention data at high suctions (Klute 1986).

Van Genuchten parameters, for describing the hydraulic functions (Van Genuchten 1980), were determined on both retentivity and outflow data, as described by Booltink (1994). Basic water retention and hydraulic conductivity curves, to be used as input in the simulation model, were established by averaging individual curves for each soil layer.

Table 9.8 Soil profile description of the Kandelaar experimental farm

A_p	0.00–0.30 m:	Clay (42% clay) with a moderate, medium angular blocky structure and an abrupt wavy boundary
2C	0.30–0.70 m:	Clay loam (40% clay), strong, very coarse prismatic structure
3C	0.70–1.03 m:	Silty clay loam (30% clay), strong, very coarse prismatic structure
$4BC_b$	1.03–1.20 m:	Sand, single grain

Model Description

The numerical simulation model LEACHM (Leaching Estimation and Chemistry Model) was used as the base model (Wagenet and Hutson 1989). LEACHM considers different processes in a variable soil profile, with or without plant growth. Processes included are: transient fluxes of water; fluxes and transformation of nitrogen, pesticides and salts; evapotranspiration; and rainfall. In this study, the water flow submodel LEACHW was used (Wagenet and Hutson 1989; Wagenet et al. 1989).

In LEACHW, water flow is calculated using a finite-difference solution of the Richards' equation:

$$\frac{\delta \theta}{\delta t} = \frac{\delta}{\delta z}\left[K(\theta)\frac{\delta H}{\delta z} \right] - U(z,t) \tag{5}$$

where θ is the volumetric water content ($m^3\,m^{-3}$), t is time (day), K is the hydraulic conductivity ($m\,day^{-1}$), H is hydraulic head (m), composed of matrix potential ψ and profile depth z, and U is a sink term representing water uptake by plants (day^{-1}).

The original Campbell K–θ–ψ equations to describe soil hydraulic properties (Campbell 1974; Hutson and Cass 1987), were replaced by Van Genuchten's closed form equations (Van Genuchten 1980).

Since LEACHW originally does not consider bypass flow, the model was extended with the following modifications:

1. For every rain or irrigation event (30 minutes interval), the amount of water that could not infiltrate through the soil surface during the time period was determined. This surplus is stored on the soil surface and when a certain threshold value for surface storage ($MinSS$) is exceeded, water starts to flow into the macropores. Not all macropores present are equally accessible, due to small differences in microrelief (Booltink et al. 1993). A maximum surface storage ($MaxSS$) was therefore defined. Above this level, water flows directly into the macropores. Between $MinSS$ and $MaxSS$ excess water is divided proportionally between surface storage and bypass flow. Water remaining on the surface continues to infiltrate after the rain has stopped.
2. In the macropore domain, water transport is simulated using a tipping-bucket approach. Propagation of the water front in the macropores (Bp_{prop} ($mm\,day^{-1}$)), i.e. tipping-bucket switching times, is calculated using a regression model equation (6) based on rain intensities (R_{inten}) and measured morphological properties (Booltink et al. 1993)

$$Bp_{prop} = a_{trv}*R_{inten} + d_{trv}*G_{fac} + b_{trv} \tag{6}$$

Parameters a_{trv}, b_{trv} and d_{trv} are empirical constants. The dimensionless geometry factor (G_{fac}) is defined as (equation (7)):

$$\frac{1}{G_{fac}} = \left[\frac{Vs_l}{100} \right](D_s3_l - 1) \tag{7}$$

where Vs_l is the volume of methylene blue stains (% vol.) in the limiting soil layer and $D_s3_l - 1$ is the depth-weighted fractal dimension of the stains. High values of Vs_l, which are an indication of either a large number of small pores or few big water conducting macropores, will lead to a reduction of Bp_{prop}. The fractal dimension, on the other hand, gives

Table 9.9 Overview of parameters used in Monte Carlo analysis

Parameter	Distribution	Mean	Variance	Minimum	Maximum	Description		Dimension
$Vercon\_up$	Normal	0.174	0.004225	0.044	0.304	Vertical contact area	0.00–0.30 m depth	$m^2\,m^{-2}$
$Vercon\_lo$	Log-normal	0.0211	0.0145	0.001	0.07	Vertical contact depth	0.30–1.03 m depth	$m^2\,m^{-2}$
$Horcon\_up$	Log-normal	0.0115	0.0017	0.001	0.10	Horizontal contact area	0.00–0.30 m depth	$m^2\,m^{-2}$
$Horcon\_lo$	Log-normal	0.0016	0.0019	0.001	0.04	Horizontal contact area	0.30–1.03 m depth	$m^2\,m^{-2}$
Vsl	Log-normal	0.964	1.133	0.10	8.0	Volume of soil stained		%
Ds_{3l}	Normal	2.011	0.0177	1.75	2.26	Fractal dimension of stained area		—
b_{trr}	Normal	$1.718E-04$	$3.650E-10$	$1.302E-04$	$2.133E-04$	Constant		—
a_{trr}	Normal	$-1.456E-07$	$1.399E-15$	$-2.271E-07$	$-6.410E-08$	Constant		—
d_{trr}	Normal	$-7.661E-04$	$7.938E-07$	$-2.707E-03$	$1.175E-03$	Constant		—
$MinSS$	Uniform			$5.000E-05$	$5.000E-04$	Minimum surface storage		m
$MaxSS$	Uniform			$1.000E-05$	$3.000E-04$	Maximum surface storage (to be added to $MinSS$)		m
WCS_1	Normal	0.484	$1.000E-04$	0.46	0.50	Water content saturated	0.00–0.30 m depth	$m^3\,m^{-3}$
WCS_2	Normal	0.623	$1.000E-04$	0.60	0.64	Water content saturated	0.30–0.70 m depth	$m^3\,m^{-3}$
WCS_3	Normal	0.620	$2.500E-03$	0.50	0.72	Water content saturated	0.70–1.03 m depth	$m^3\,m^{-3}$
WCS_4	Normal	0.320	$1.000E-04$	0.30	0.34	Water content saturated	1.03–1.50 m depth	$m^3\,m^{-3}$

Parameter	Distribution					Description		Units
CON_0	Uniform			$1.000E-04$	0.02	Saturated hydraulic conductivity	0.00–0.02 m depth	m day^{-1}
CON_1	Uniform			0.1	0.20	Saturated hydraulic conductivity	0.02–0.30 m depth	m day^{-1}
CON_2	Normal	0.59	0.0160	0.050	1.25	Saturated hydraulic conductivity	0.30–0.70 m depth	m day^{-1}
CON_3	Log-normal	0.22	0.0049	0.010	0.48	Saturated hydraulic conductivity	0.70–1.03 m depth	m day^{-1}
CON_4	Normal	0.25	0.0125	0.050	0.40	Saturated hydraulic conductivity	1.03–1.50 m depth	m day^{-1}
Ctd_1	Normal	$1.637E-01$	$6.204E-04$	0.114	0.213	Constant in drainage discharge function		—
Ctd_2	Normal	$-2.150E-03$	$1.887E-06$	$-4.870E-03$	$5.740E-04$	Constant in drainage discharge function		—
Ctd_3	Normal	$2.971E-05$	$3.232E-10$	$-5.932E-06$	$6.540E-05$	Constant in drainage discharge function		—
Cd_1	Normal	$-3.100E-03$	$4.706E-06$	$-7.778E-03$	$1.586E-03$	Constant in deep drainage discharge function		—
Cd_2	Normal	$1.618E-05$	$4.165E-11$	$2.234E-06$	$3.012E-05$	Constant in deep drainage discharge function		—
Cd_3	Normal	$-9.194E-09$	$2.465E-17$	$-1.992E-08$	$1.532E-09$	Constant in deep drainage discharge function		—
Cd_0	Normal	$-9.223E-02$	$3.043E-02$	-0.469	0.285	Constant in deep drainage discharge function		—
x_{struc_0}	Uniform			0.010	0.120	Structure element size	0.00–0.02 m depth	m
x_{struc_1}	Uniform			0.040	0.120	Structure element size	0.02–0.30 m depth	m
x_{struc_2}	Uniform			0.200	0.400	Structure element size	0.30–0.70 m depth	m
x_{struc_3}	Uniform			0.150	0.300	Structure element size	0.70–1.03 m depth	m

For a uniform distribution only minimum and maximum values are specified.

information on the geometry of water conducting macropores. A macropore system with a $D_s 3_l - 1$ value of 1 consists mainly of vertically oriented cracks; a value of 2, on the other hand, indicates that horizontal cracks dominate the system.

3. When water flows into macropores it is absorbed into macropore walls. Lateral absorption of water from vertically continuous macropores is based on absorption described by a diffusivity equation (equation (8)). Gravity forces, described by the Darcy equation (equation (9)) dominate vertical absorption of water on horizontal pedfaces of structural elements.

$$Qp_{hor} = -D(\theta)\frac{\Delta\theta}{\Delta\left[\dfrac{x_{struc}}{2}\right]} \tag{8}$$

$$Qp_{ver} = -K(\psi)\left[\frac{\Delta\psi}{\Delta\left[\dfrac{x_{struc}}{2}\right]} + 1\right] \tag{9}$$

where Qp_{hor} and Qp_{ver} represent the potential horizontal and vertical absorption fluxes (m day^{-1}), respectively, D is soil water diffusivity in m^2 day^{-1} and X_{struc} is the structure element diameter (m). $\Delta\psi$ and $\Delta\theta$ were calculated as a geometrical mean between saturation (macropore wall) and the actual value, respectively, of ψ and θ in the adjacent soil matrix as simulated by the Richards' equation. The potential absorption fluxes were reduced for the area of soil in contact with bypass water as indicated by the occurrence of stains. Contact areas for horizontal and vertical absorption (Horcon and Vercon, respectively) were determined using dye tracers in laboratory experiments. This procedure was described and is based on stratification of methylene blue stained macropores into sets of horizontally and vertically oriented macropores using the ratio between area and perimeter (Ar/Pe^2 < 0.015) (Bouma et al. 1977).

The reduced, actual absorption fluxes were added as an additional sink term to Richards' equation. The surplus of water not absorbed during the residence time in a given compartment is added to the next compartment and finally to the groundwater.

4. Originally LEACHW did not provide a special bottom boundary condition which can simulate dynamic groundwater levels. Simulation of such a groundwater level was realised by combining a freely draining profile with an impermeable soil layer at the bottom of the profile. Deep drainage towards ditches (Qd) was simulated using a third order polynomial (equation (10)):

$$Qd(h) = Cd_0 + Cd_1 h + Cd_2 h^2 + Cd_3 h^3 \tag{10}$$

where h is the pressure potential in the bottom compartment and the constants Cd_0, Cd_1, Cd_2 and Cd_3 are fitting parameters. A similar procedure was followed to simulate discharge through tile-drains (Qtd). Parameter h, in that case, represents the groundwater level above a specified drain depth and constants are indicated as: Ctd_0, Ctd_1, Ctd_2 and Ctd_3, respectively.

Sensitivity Analysis

The modified LEACHW model was tested with data from the Kandelaar experimental farm. The objective of this test was: (i) to obtain information on the performance of the model and (ii) to calibrate the model on measured time series of groundwater levels. For this purpose, a Monte

Carlo analysis was used as described in Sections 9.3.1 and 9.3.2. In this Monte Carlo analysis, the uncertainty of model inputs is characterised in terms of distributions with or without correlations among the various parameters.

From exploratory simulation runs the maximum compartment thickness was determined to be 0.02 m. The total profile depth was set at 1.5 m and consists, therefore, of 75 compartments. During the exploratory runs, using the soil layers as determined in the soil survey, bypass flow occurred only in a limited number of cases. An additional topsoil layer of 0.02 m, which reflected surface resistance, caused by surface smearing during ploughing and small loose structure elements, was, therefore, added.

Model simulation outputs were evaluated in a sensitivity analysis. In the sensitivity analysis, the variation of the various parameters was tested against simulated water balance terms by means of a linear regression model (equation (1)).

The sensitivity analysis was carried out on the simulated mass balance terms for the winter season 1989–1990. Since not all parameter combinations show linear relations with the various mass balance terms, ordinary regression coefficients were replaced by their rankings, where the smallest value gets ranking 1, the next one 2 etc. This technique is therefore not sensitive to small deviations from linearity and can be used when relations among parameters are slightly non-linear (Iman and Conover 1982).

Results are summarised in Table 9.10. The saturated hydraulic conductivity of topsoil (Con_0) dominated mostly. Parameters regulating tortuous water transport in macropores, a_{trv}, b_{trv} and d_{trv} also have significant impact on cumulative absorption into macropores walls. Leachate at the bottom of the soil profile is influenced mainly by the regression parameters of the deep drainage characteristic (Cd_0, Cd_1, Cd_2 and Cd_3).

Most significant model parameters (underlined in Table 9.10) were selected for further model calibration. Selection was based on their importance in the sensitivity analysis as well as on their contribution to one of the mass balance terms.

Model Calibration

Although model input parameters and their distributions reflect individual variability, not all combinations of these input parameters necessarily result in realistic simulation results. Further calibration is therefore required. In this study the Rotated random Scan (RORASC) procedure, described in Section 9.3.3 was used.

Since the number of selected parameters was eight, in total 80 simulation runs ($N > 10p$) were executed in every RORASC cycle. Parameters not varied were fixed at values obtained from the best simulation run in the first Monte Carlo session.

Calibration was performed on measured cumulative drain during the winter season 1989–1990 and the measured time series of groundwater levels in the same period. In the RORASC optimisation procedure only three iterative steps had to be performed. The third step did not lead to significantly better simulation results, indicating that the parameters had reached their limits. It also indicates that the initial parameter space was close to the calibrated one.

Ill-defined parameters slowly concentrate around their most likely values. The uniform distribution of the ill-defined parameter $MinSS$ becomes log normal in Figure 9.9. In Figure 9.10 measured groundwater levels are presented with the 99% confidence intervals of the simulation results. The smaller band between the upper and lower 1% confidence intervals demonstrates the effect of the RORASC procedure, which is especially obvious from days 60 to 110. When interpreting these data one has to consider the difference in sensitivity for the groundwater above and below the depth of the tile-drains of 0.82 m. Whereas a simulated deviation of a few centimetres below the tile-drain level will have little effect on the water balance, if the

Table 9.10 Parameters that contributed with 95% significance to the simulated cumulative mass balance terms using a linear regression model (equation (1)). The fraction of the total explained variance (R^2) and the average simulated amount (μ) in the mass balance is indicated, together with the standard deviation (σ)

Infiltration parameter	t-stat	Evaporation parameter	t-stat	Absorption parameter	t-stat	Bypass parameter	t-stat	Drain discharge parameter	t-stat	Leachate parameter	t-stat
CON_1	16.82	CON_0	18.43	CON_0	−10.36	CON_1	−18.04	CON_0	−8.71	Cd_1	12.41
CON_0	13.76	CON_1	14.35	CON_1	−9.62	CON_0	−13.78	Cd_1	−7.28	Cd_2	9.91
CON_2	2.04	$MinSS$	2.59	$Vercon\_lo$	9.11	CON_2	−2.02	CON_1	−6.79	Cd_0	9.51
		CON_2	2.05	$Horcon\_lo$	5.66			Cd_2	−5.66	Cd_3	6.79
				d_{trv}	2.76			Cd_0	−5.36	CON_4	−5.33
				b_{trv}	2.69			WCS_3	−4.35	WCS_3	5.30
				a_{trv}	2.14			CON_4	3.99	CON_3	4.76
				Cd_0	2.04			Cd_3	3.73	$Vercon\_lo$	2.52
								WCS_4	−2.01	CON_2	2.51
R^2	0.76		0.79		0.67		0.78		0.62		0.63
μ (mm)	195.0		76.4		8.3		64.6		125.8		70.8
σ (mm)	32.7		4.8		11.2		25.2		11.6		10.4

Number of simulations

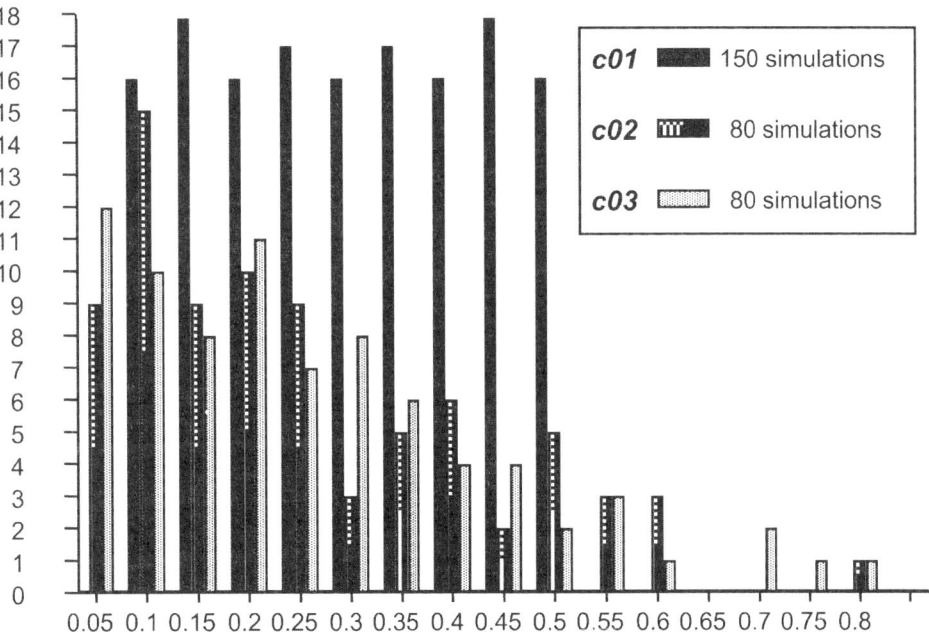

Minimum Surface Storage [MinSS] (mm)

Figure 9.9 The effect of the RORASC procedure on surface storage parameter *MinSS*. c01, c02 and c03, respectively, depict steps 1, 2 and 3 in the RORASC calibration procedure. The original uniform distribution (c01) is transformed into a log-normal distribution

groundwater level is above tile-drain depth an equivalent deviation will have a large effect on the drain discharge because of high drain discharge.

Model Validation

The ordinary regression coefficient R^2 (equation (2)) represents the explained variance of the calibrated model of the total variance of all model realisations (Section 9.3.4). This characteristic is used to validate the model concept. The results are depicted in Table 9.10. All terms in the water balance are simulated with adequate precision with R^2 ranging from 0.78 to 0.62. Coefficients of determination $\left(\dfrac{\sigma}{\mu} \times 100\%\right)$ are low, around 15% except for absorption (135%) and bypass (39%). The first, however, is only of minor importance in the total water balance (8.3 mm for the entire winter season).

Using the input parameters as determined in the third RORASC step, a Monte Carlo simulation of 80 model runs was used to validate the model on an independently collected data set in the winter season 1991–1992 from the same experimental field.

In Figure 9.11, measured groundwater levels in the winter season 1991–1992 are compared to the simulation results for that same period. Except for the bypass flow event around day 65, all

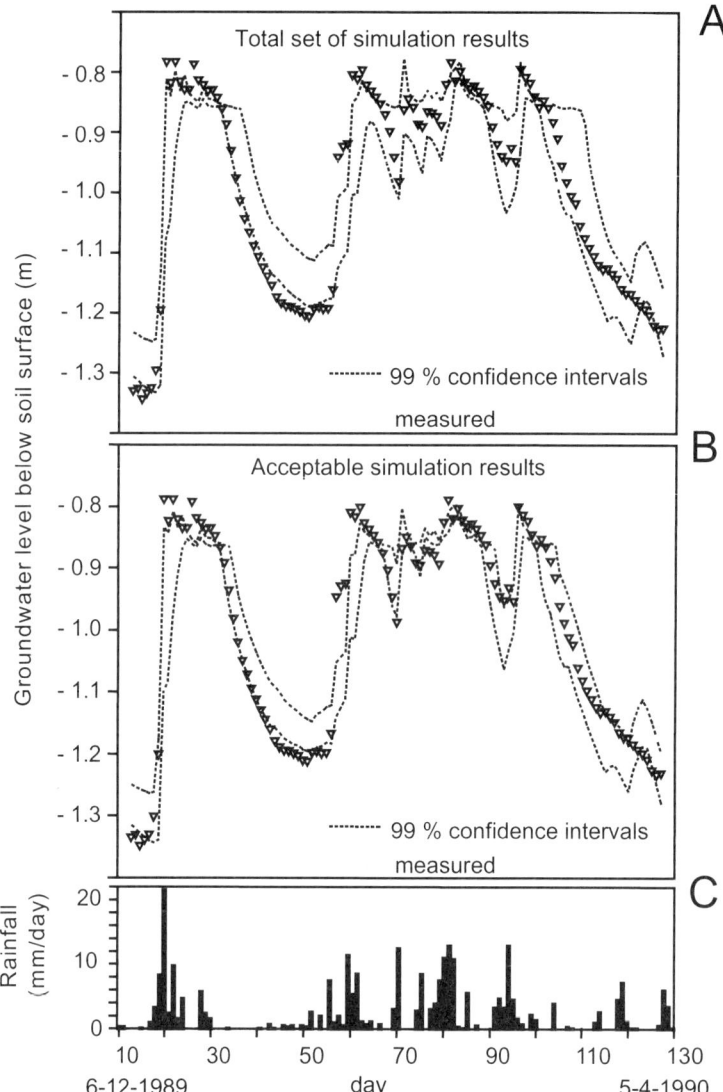

Figure 9.10 Comparison of measured groundwater levels below soil surface for the winter season 1989–1990 with upper and lower 99% confidence intervals. The total of all simulation results are presented in A, the acceptable simulation results only are shown in B. C represents the daily precipitation during the simulation period

events were well simulated by the model. The model overestimates the groundwater level from days 45 to 60, although this does not lead to large mass balance errors, as discussed in the model calibration section (9.3.5). The generally small difference between the 95 and 99% confidence intervals in Figure 9.11 indicates that the calibrated parameter space was within the physical limits and does, therefore, not lead to large simulation errors or outliers. Only around day 95 can a distinct difference between the 95 and 99% levels be seen. In this case study the model was

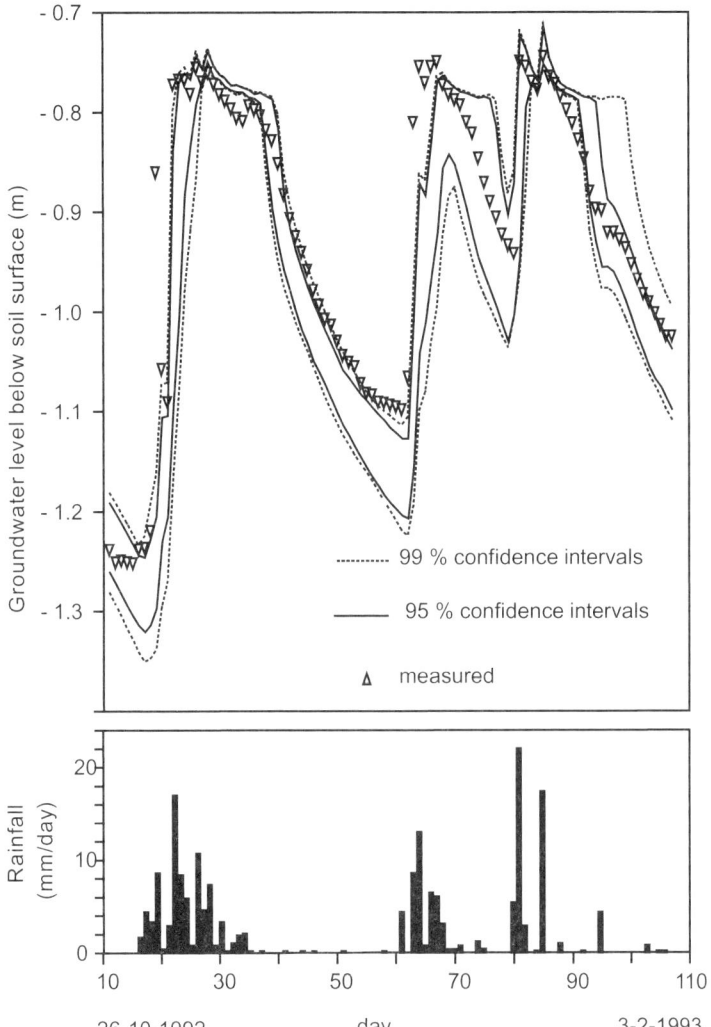

Figure 9.11   Top: validation results of simulated groundwater levels below soil surface for the winter season 1991–1992. Bottom: the daily precipitation during the simulation period.

validated against data obtained from a single piezometer. However, when dealing with trends, slopes or other sources of spatial variability this can easily lead to erroneous results and more information should be included.

The average simulated cumulative drain discharge at the end of the period of 79.9 mm was close to the measured cumulative drain discharge of 86.9 mm. Cumulative surface infiltration is the most important term in the water balance. Bypass flow is also substantial and was calculated to be 48.1 mm. Absorption along macropores in this winter period was relatively small at 8.7 mm. The mass balance error was small: for most simulations the value was 0.6 mm with only a few outliers.

9.4 CONCLUDING REMARKS

Model validation has not been defined uniquely and various procedures are being used to validate simulation models. However, the lack of such a unique system is not the most fundamental problem in modelling studies. Risk assessment through uncertainty and sensitivity analysis of simulation models and the results produced by them, is more likely to increase reliability of simulation models than a well-defined validation method. In cases where basic assumptions and data are doubtful, or where calibration of the model was performed in a irreproducible way, the predictable capabilities of the model will be limited or sometimes even erroneous.

Transparent modelling procedures are necessary to avoid these kind of problems. Studies in which simulation models are used should follow a strategy in which the choice of a model is the logical consequence of a procedure in which first the modelling necessity and scale of the problem has been established. When the problem and the model have been selected, data requirements should be determined based on sensitivity and uncertainty of the simulation model, the input data and the desired accuracy of the model outputs. Such a research chain should be documented in a transparent reproducible way for others than the modellers, and model results should preferably be presented in probabilistic terms. The information that a critical threshold value such as the $50\,\mathrm{mg\,l^{-1}}$ nitrate level is exceeded once per year or once per ten years is of much higher value than the fact that it is exceeded.

Such a modelling approach is only feasible if data and models are well documented and easily applicable by others than the developers themselves. The international consortium for agricultural systems approaches (ICASA) promotes the use of standard data sets to store experimental data to be used for modelling purposes in a coherent unique format (Van Kraalingen and Hunt 1997). Initiatives such as CAMASE (1995, 1996), which contains a register of agro-ecosystem models, all well documented and described, can play a important role here. Also the development of accessible databases that contain data necessary for modelling studies need to be encouraged. The UNSODA database (Leij et al. 1996) for the USA, as well as the HYPRES database (Wösten et al. 1998) for Europe, are initiatives that need full support. Financing of this type of methodological studies is, however, still problematic. Funding possibilities as provided by ISNAR (International Service for National Agricultural Research) through their Ecoregional Initiative are a good example of a framework that concentrates on methodology development instead of direct modelling results.

Modelling accuracy strongly depends on model quality but also on the input data quality. Excessive field and laboratory measurement campaigns do not necessarily lead to more reliable data. As shown in case study II the combined use of pedotransfer functions and a limited number of field measurements resulted in the best results during validation. This 'local adjustment' of 'global' pedotransfer functions seems an attractive option to collect laborious and, therefore, costly soil characteristics, especially in studies where large amounts of these soil characteristics are required.

ACKNOWLEDGEMENTS

The following persons are acknowledged for their contribution to this chapter. Professor J. Bouma, Ir.B.J. van Alphen, P.D Peters, A. and J. van Bergeijk, Ing. J. Oosterhuis, Dr P.A. Finke. Finally the BCRS is acknowledged for financing one of the case studies discussed.

REFERENCES

Arya, L.M., Farrell, D.A. and Blake, G.R. 1975. A field study of soil water depletion patterns in presence of growing soybean roots: 1 Determination of hydraulic properties of the soil. *Soil Science Society of America Journal*, **40**, 424–430.

Bergström L., Johnsson, H. and Tortensson, G. 1991. Simulation of nitrogen dynamics using the SOILN model. *Fertiliser Research*, **27**, 181–188.

Beven, K. and Germann, P. 1982. Macropores and water flow in soils. *Water Resources Research*, **18**, 1311–1325.

Booltink, H.W.G. 1994. Field scale distributed modelling of bypass flow in a heavily structured clay soil. *Journal of Hydrology*, **163**, 65–84.

Booltink, H.W.G. and Bouma, J. 1991. Physical and morphological characterization of bypass flow in a well-structured clay soil. *Soil Science Society of America Journal*, **55**, 1249–1254.

Booltink, H.W.G. and Bouma, J. 2000. Column method. In: Elrick, D. et al. (eds), *Methods of Soil Analysis. Physical and Mineralogical Methods*, 3rd edition, in press.

Booltink, H.W.G. and Verhagen, J. 1996. Using decision support systems to optimize barley management on spatial variable yields. In: Kropff, M.J. et al. (eds), *Applications of Systems Approaches at the Field Level*. Kluwer, Dordrecht, 219–213.

Booltink, H.W.G., Bouma, J. and Giménez, D. 1991. A suction crust infiltrometer for measuring hydraulic conductivity of unsaturated soil near saturation. *Soil Science Society of America Journal*, **55**, 566–568.

Booltink, H.G.W., Hatano, R. and Bouma, J. 1993. Measurement and simulation of bypass flow in a structured clay soil: a physico-morphological approach. *Journal of Hydrology*, **148**, 149–168.

Bouma, J. 1980. Field measurement of soil hydraulic properties characterizing water movement through swelling clay soils. *Journal of Hydrology*, **45**, 149–158.

Bouma, J. 1982. Measuring the hydraulic conductivity of soil horizons with continuous macropores. *Soil Science Society of America Journal*, **46**, 438–441.

Bouma, J. 1984. Using soil morphology to develop measurement methods and simulation techniques for water movement in heavy clay soils. In: Bouma, J. and Raats, P.A.C. (eds), *Proceeding ISSS Symposium on Water and Solute Movement in Heavy Clay Soils*, Wageningen, Netherlands, 27–31 August 1984. Publication 37, Institute for Land Reclamation and Improvement, Wageningen, 298–316.

Bouma, J. and Dekker, L.W. 1978. A case study on infiltration into dry clay soil. I. Morphological observations. *Geoderma*, **20**, 27–40.

Bouma, J. and Hoosbeek, M.R. 1996. The contribution and importance of soil scientists in interdisciplinary studies dealing with land. In: Wagenet, R.J. and Bouma, J. (eds), *The Role of Soil Science in Interdisciplinary Research*. Soil Science Society of America, Special Publication, 45.

Bouma, J. and Van Lanen, J.A.J. 1987. Transfer functions and threshold values: from soil characteristics to land qualities. In: Beek, K.J., Burrough, P.A. and McCormack, D.E. (eds), *Proceedings of the International Workshop on Quantified Land Evaluation*. ISSS and SSSA, Washington, DC, 27 April–2 May 1986. International Institute Aerospace Survey, Earth Science Publication No 6. ITC Publ. Enschede, The Netherlands, 106–111.

Bouma, J., Jongerius, A., Boersma, O., de Jager, A. and Schoonderbeek, D. 1977. The function of different types of macropores during saturated flow through four swelling soil horizons. *Soil Science Society of America Journal*, **41**, 945–950.

Bouma, J., Booltink, H.W.G. and Finke, P.A. 1996. Use of soil survey data for modeling solute transport in the vadose zone. *Journal of Environmental Quality*, **25**, 519–526.

CAMASE, 1995. *Register of agro-ecosystems models version I*. Plentiger, M.C. and de Vries, F.W.T. (eds), Research Institute for Agrobiology and Soil Fertility (AB-DLO), Wageningen, The Netherlands.

CAMASE, 1996. *Register of agro-ecosystems models version II*. Plentiger, M.C. and de Vries, F.W.T. (eds), Research Institute for Agrobiology and Soil Fertility (AB-DLO), Wageningen, The Netherlands.

Dekker, L.W. 1998. Moisture variability resulting from water repellency in Dutch soils. PhD Thesis, Wageningen Agricultural University.

Dierckx, J., Belmans, C. and Pauwels, P. 1986. *SWATRER, A Computer Package for Modeling the Field Water Balance. Reference Manual*. Catholic University, Leuven.

Dirksen, C. 1979. Flux-controlled sorptivity measurements to determine soil hydraulic property functions. *Soil Science Society of America Journal*, **43**, 827–834

Doering, E.J. 1965. Soil water diffusivity by the one-step method. *Soil Science*, **99**, 322–326.

Feddes, R.A., Kowalik, P.J. and Zarandny, H. 1978. *Simulation of Field Water Use and Crop Yield*. Simulation Monographs, PUDOC, Wageningen.

Feddes, R.A., de Graaf, M. and Bouma, J. 1988. Simulation of water use and production of potatoes as affected by soil compaction. *Potato Research*, **31**, 225–239.

Finke, P.A. 1993. Field scale variability of soil structure and its impact on crop growth and nitrate leaching in the analysis of fertilizer scenarios. *Geoderma*, **60**, 89–107.

Finke, P.A. and Bosma, W.J.P. 1993. Obtaining basic simulation data for a heterogeneous field with stratified marine soils. *Hydrological Processes*, **7**, 63–75

Finke, P.A., Wösten, J.H.M. and Jansen, M.J.W. 1996. Effects of uncertainty in major input variables on simulated functional behaviour. *Hydrological Processes*, **10**, 661–669.

Germann, P.F. 1987. The three modes of water flow through a vertical pipe. *Soil Science*, **144**, 153–154.

Greenwood, D.J., Neeteson, J.J. and Draycott, A. 1985. Response of potatoes to N-fertilizer: quantitative relations for components of growth. *Plant and Soil*, **85**, 163–183.

Greenwood, D.J., Lemair, G., Gosse, G., Cruz, P., Draycott, A. and Neeteson, J.J. 1990. Decline in percentage N of C3 and C4 crops with increasing plant mass. *Annals of Botany*, **66**, 425–436.

Halbertsma, J.M. and Veerman, G.J. 1994. *A New Calculation Procedure and Simple Set-up for the Evaporation Method to Determine Soil Hydraulic Functions*. DLO-Staring Centre, Wageningen, The Netherlands.

Hassanizadeh, S.M. and Carrera, J. 1992. Editorial: introduction to special issue on validation. *Advances in Water Resources*, **15**, 75–83.

Hogarth, W.L. Hopmans, J., Parlange, J-Y. and Haverkamp, R. 1988. Application of a simple soil-water hysteresis model. *Journal of Hydrology*, **98**, 21–29.

Hoosbeek, M.R. and Bryant, R. 1992. Towards quantitative modeling of pedogenesis – a review. *Geoderma*, **55**, 183–210.

Hutson, J.L. and Cass, A. 1987. A retentivity function for use in soil water simulation models. *Journal of Soil Science*, **38**, 487–498.

Hutson, J.L. and Wagenet, R.J. 1991. Simulating nitrogen dynamics in soils using a deterministic model. *Soil Use and Management*, **7**(2), 74–78.

Iman, R.L. and Conover, W.J. 1980. Small sample sensitivity analysis techniques for computer model, with an application to risk assessment. *Communications in Statistics*, **A9**, 1749–1842. Rejoinder to comments, pp. 1863–1974.

Iman, R.L. and Conover, W.J. 1982. A distribution free approach to inducing rank correlations among input variables. *Communications in Statistics*, **B11**, 311–334.

Janssen, P.H.M. 1993. *RORASC and SELACC: Software for Performing the Rotated-Random-Scanning Calibration Procedure*. CWM Memorandum. RIVM, Bilthoven, The Netherlands.

Janssen, P.H.M. 1995a. *GENSEN: A Program for Generalized Sensitivity Analysis*. RIVM Report No. 733001006, RIVM, Bilthoven, The Netherlands.

Janssen, P.H.M. 1995b. *RORASC: Software for the Rotated-Random Scan Calibration*. RIVM Report No. 733001005, RIVM, Bilthoven, The Netherlands.

Janssen, P.H.M., Heuberger, P.S.C. and Sanders, R. 1993. *UNCSAM 1.1: A Software Package for Sensitivity and Uncertainty Analysis; Manual*. RIVM Report No. 959101004, RIVM, Bilthoven, The Netherlands.

Keesman, K.J. 1989. A set-membership approach to the identification and prediction of ill-defined systems: application to a water quality system. PhD thesis, University of Twente, Enschede, The Netherlands.

Keesman, K.J. 1990. Membership-set estimation using random scanning and principal component analysis. *Mathematics and Computers in Simulation*, **32**, 535–543.

Keesman. K.J. and Van Straaten, G. 1988. Embedding of random scanning and principal component analysis in set-theoretic approach to parameter estimation. *Proceedings 12th IMACS World Congress*, Paris.

Keesman, K.J. and Van Straaten, G. 1989. Identification and prediction propagation of uncertainty in models with bounded noise. *International Journal of Control*, **49**, 2259–2269.

Klute, A. 1986. *Methods of Soil Analysis Part 1. Physical and Mineralogical Methods*, 2nd edition. American Society of Agronomy, Monograph 9, Madison, WI.

Lauren, J.G., Wagenet, R.J., Bouma, J. and Wösten, J.H.M. 1988. Variability of saturated hydraulic conductivity in a Glossaquic Hapludalf with macropores. *Soil Science*, **145**, 20–28.

Leij, F.J., Alves, W.J., van Genuchten, M.Th. and Williams, J.R. 1996. *Unsaturated Soil Hydraulic Database, UNSODA 1.0 User's Manual*. Report EPA/600/R–96/095, US Environmental Protection Agency, Ada, OK.

McKay, M.D., Beckman R.J. and Conover, W.J. 1979. A comparison of three methods for selecting values of input variables in the analysis of output from a computer code. *Technometrics*, **28**, 211–217.

Parker, J.C., Kool, J.B. and van Genuchten, M.Th. 1985. Determining soil hydraulic properties from one-step outflow experiments by parameter estimation: II Experimental studies. *Soil Science Society of America Journal*, **49**, 1354–1359.

Ritsema, C.J. 1998. Flow and transport in water repellent sandy soils. PhD Thesis, Wageningen Agricultural University, the Netherlands.

Robert, P.C., Rust, R.H. and Larsen, W.L. (eds). 1993. *Soil Specific Crop Management*. ASA–CSSA–SSSA, Madison, WI.

Robert, P.C., Rust, R.H. and Larsen, W.L. (eds). 1995. *Site Specific Management for Agricultural Systems*. ASA–CSSA–SSSA. Madison, WI.

Robert, P.C., Rust, R.H. and Larsen, W.L. (eds). 1997. *Site Specific Management for Agricultural Systems*. ASA–CSSA–SSSA. Madison, WI.

Soil Survey Staff, 1975. *Soil Taxonomy: A Basic System of Soil Classification for Making and Interpreting Soil Surveys*. USDA–SCS Agricultural Handbook 436. US Government Printing Office, Washington, DC.

Stafford, J. 1997. *Precision Agriculture 1997. Vol I: Spatial Variability in Soil and Crop. Vol II: Technology, IT and Management*. SCI-Bios Scientific Publishers, Oxford.

Steenhuis, T.S., Ritsema, C.K., Dekker, L.W. and Parlange, J.Y. 1994. Fast and early appearance of solutes in groundwater by rapid and far-reaching flows. *15th International World Congress of Soil Science Transactions 2a*. Acapulco, Mexico, 10–16, July 1994, 184–203.

Tietje, O. and Tapkenhinrichs, M. 1993. Evaluation of pedo-transfer functions. *Soil Science Society of America Journal*, **57**, 1088–1095.

Topp, G.C., Davis, J.L. and Annan, A.P. 1980. Electromagnetic determination of soil water content: measurement in coaxial transmission lines. *Water Resources Research*, **16**, 574–582.

Van Dam, J.C., Stricker, J.N.M. and Droogers, P. 1990. *From One-Step to Multi-Step: Determination of Soil Hydraulic Functions by Outflow Measurements*. Department of Water Resources, Report No. 7. Wageningen Agricultural University, Wageningen.

Van Genuchten, M.Th. 1980. A closed-form equation for predicting the hydraulic conductivity of un-saturated soils. *Soil Science Society of America Journal*, **44**, 892–898.

Van Genuchten, M.Th., Leij, F.J. and Yates, F.R. 1991. *The RETC Codes for Quantifying the Hydraulic Functions of Unsaturated Soils*. US Salinity Laboratory, Riverside, CA.

Van Kraalingen, D.W.G. and Hunt, L.A. 1997. *ICUTIL v. 1.0, Fortran Software Interface for Icasa v.1 Data Standard. User Guide*. AB-DLO, Wageningen.

Vanclooster, M., Viane, P., Diels, J. and Christiaens, K. 1994. *Wave: a Mathematical Model for Simulating Water and Agrochemicals in the Soil and Vadose Environment – Reference and User's Manual (Release 2.0)*. Institute for Land and Water Management, Leuven.

Vereecken, H. 1988. Pedotransfer functions for the generation of hydraulic properties for Belgian soils. PhD thesis. Catholic University, Leuven.

Vereecken, H., Diels, J., Orshoven, J., Feyen, J. and Bouma, J. 1992. Functional evaluation of pedotransfer functions for the estimation of soil hydraulic properties. *Soil Science Society of America Journal*, **56**, 1371–1378.

Verhagen, J. and Bouma, J. 1997. Modelling soil variability. In: Pierce, F.J. and Sadler, J. (eds), *The State of Site Specific Management for Agriculture*. ASA–Misc. Publ. ASA, CSSA, SSSA, Madison, WI, pp. 55–67.

Verhagen, A., Booltink, H.W.G. and Bouma, J. 1995. Site specific management: balancing production and environmental requirements at farm level. In: Ritchie, J.T. and Bouma, J. (eds), *ICASA: Role of Agronomic Models in Interdisciplinary Research. Agricultural Systems*, **49**, 369–384.

Wagenet, R.J. and Hutson, J. 1989. *LEACHM, A Process-Based Model of Water and Solute Movement, Transformations, Plant Uptake and Chemical Reactions in the Unsaturated Zone.* Cornell University, Ithaca, NY.

Wagenet, R.J., Hutson, J.L. and Biggar, J.W. 1989. Simulating the fate of a volatile pesticide in unsaturated soil: a case study with DBCP. *Journal of Environmental Quality,* **18**, 78–84.

White, R.E. 1985. The influence of macropores on the transport of dissolved and suspended matter through soil. *Advances in Soil Science,* **3**, 95–121.

Wösten, J.H.M. 1987. *Description of the Water Retention and Hydraulic Conductivity Characteristics of Top- and Sub-soils in the Netherlands: The Staring Database.* Report 1932, Soil Survey Institute, Wageningen (in Dutch).

Wösten, J.H.M. and Van Genuchten, M.Th. 1988. Using texture and other soil properties to predict the unsaturated soil hydraulic functions. *Soil Science Society of America Journal,* **52**, 1762–1770.

Wösten, J.H.M. and van der Zee, S.E.A.T.M. 1993. Vulnerability of three spatially variable soil groups for solute leaching. *Hydrological Processes,* **7**, 235–247

Wösten, J.H.M., Lilly, A., Nemes, A. and Le Bas, C. 1998. *Using Existing Soil Data to Derive Hydraulic Parameters for Simulation Models in Environmetal Studies and Landuse Planning; Final Report on the European Funded Project.* SC-DLO Report 156, Wageningen.

10 A Hydromechanical Approach to Preferential Flow

P. GERMANN
Geographisches Institut, Universitat Bern, Switzerland

10.1 INTRODUCTION

Preferential flow in soils is usually viewed as flow along macropores. The latter are seen as structural voids in soils extending at least in one dimension over considerable depth. They allow for much faster wetting front advancements than does flow in smaller pores. The smaller pores are thought to exert capillary forces onto the water, and flow in them will be referred to as ordinary. Preferential flow along macropores implies a well-defined relation between their geometry and the flow process. Numerous studies actually try to establish this relationship.

A more functional approach will be presented here. Preferential flow is considered to follow a fluid mechanical principle which differs from the one applicable to ordinary flow. The latter flow is dominated by the reversible diffusion of capillary potential, whereas preferential flow is governed by the irreversible dissipation of momentum at a length scale extending considerably beyond the one of potential diffusion. The flowpath geometries *contact length of mobile water per unit cross-sectional area of soil*, $[\ell/A][\text{m}^{-1}]$ and *average film thickness of mobile water*, $F[\text{m}]$, are linked with a dimensionless *scaling parameter a*. Combined with gravity and the water's viscosity, the three parameters quantify preferential flow. They follow from parameter interpretation when the concept of momentum dissipation is applied to measurements. Because dissipative and diffusive flow frequently concur at the scales of soil horizons and profiles, a range of mixed flow behaviour is to be expected.

In this chapter, the theoretical base of momentum dissipation will be briefly introduced. The approach will be validated against more than 100 measurements of preferential flow, and the limitations of its applicability are discussed. Some examples of applying the approach to flow in soil profiles, and a discussion about its implication for the hydrology of hillslopes will conclude the chapter.

10.2 THEORY

10.2.1 Force Balance During Porous Media Flow

Germann and Di Pietro (1999) presented the macroscopic force balance of water in porous media from the macroscopic linear momentum balance in terms of the volume flux density $q_b = \theta\mathbf{v}$ as

Model Validation: Perspectives in Hydrological Science. Edited by M.G. Anderson and P.D. Bates.
© 2001 John Wiley & Sons, Ltd.

$$\theta \frac{d(q_b/\theta)}{dt} = -g\mathbf{T}\nabla(\psi + z) + \eta\nabla^2 q_b + \eta\mathbf{B}q_b \qquad (1)$$

where θ [m³ m⁻³] is the volumetric soil moisture, $\mathbf{v}$ [m s⁻¹] is its macroscopic velocity vector, g [m s⁻²] is acceleration due to gravity, ψ [m] is the matric potential, η [m² s⁻¹] is the kinematic viscosity of water, $\mathbf{T}$ [m m⁻¹] is a tensor related to the tortuosity of the volume occupied by the liquid, $\mathbf{B}$ [m⁻²] is a second order tensor related to the liquid–solid contact area, z [m] is depth, t [s], [h], [d] represents time, and the index b refers to the bulk of volume flux density. Thus $\mathbf{T}$ and $\mathbf{B}$ are functions of θ, and $\eta \approx 10^{-6}$ [m² s⁻¹] is the coefficient of momentum dissipation (Dingman 1984). The expression on the left hand side of equation (1) is the total rate of momentum change per time of the moisture's unit mass per unit volume of the REV. The first term on the right hand side expresses the supply of momentum arising from gravity and capillarity, the second term represents the irreversible dissipation of momentum due to viscosity within fluid layers, and the last term states the irreversible dissipation of momentum due to friction at the liquid–solid interface in proportion to $\mathbf{B}$. (The force balance per unit volume of water evolves when both sides of equation (1) are multiplied by the density of water, ρ [kg m⁻³].)

Friction forces dominate flow when a porous medium consists of highly tortuous, narrow and heterogeneous paths along those the momentum carried by flow completely dissipates within an REV. The mean velocity of flow is slow and the regime is quasi-static. The total rate of momentum change within the liquid layers is negligible. Flow is due to the gradient of total potential within a continuously defined potential field. The resistance to flow is expressed by Darcy's hydraulic conductivity, which is defined as

$$K = \frac{\theta g \mathbf{T}}{\eta \mathbf{B}} \qquad (2)$$

[m s⁻¹]. It implies momentum dissipation towards the solid at the REV scale. Flow in unsaturated porous media (i.e. $\theta < E$, where E [m³ m⁻³] is the porosity) can be treated as reversible diffusion of capillary potential, where the soil hydraulic diffusivity is given by $D = \frac{K d\psi}{d\theta}$ [m² s⁻¹].

There might be flowpath geometries showing very modest tortuosities and very small liquid–solid contact areas, eventually approaching the extremes of $\mathbf{T} \approx 1$, $\mathbf{B} \to \mathbf{B}_{min}$, and thus $K \to K_{max}$. Germann and Di Pietro (1996) demonstrated that $\mathbf{B} = \mathbf{B}_{min}$ during fully developed laminar flow in cylindrical pores. In mechanically stable soils, only a small portion of pores may carry flow over considerable distances under the extreme conditions. These pores will be referred to as macropores (i.e. paths of preferred flow, prefential flowpaths) in which resistance to flow is minimal and momentum dissipates mainly according to the second and third terms on the right hand side of equation (1). Germann and Di Pietro (1999) postulated that the average velocity, and thus the momentum of flow within macropores, may be so dominant at times that water films detach from the remaining bulk of soil moisture, leading to bimodal flow. As a consequence, the force balance, equation (1), no longer applies to the entire REV when the wetting front of preferred flow moves considerably ahead of the front of the soil moisture's bulk. Measurements of $^{18}O/^{16}O$-ratios (Beven and Germann 1982) in the drainage water from, and bromide tracer experiments (Germann and Di Pietro 1999) in the Rietholzbach weighing lysimeter (with depth and diameter of 2.2 m and 2.0 m, respectively), revealed that water resides approximately 6 months in the soil. However, about 20 wetting fronts per year which originate from heavy rains move within 1 to 2 days to the drainage outlet. Thus, fronts from preferential flow move about 100 to 200 times faster than those from ordinary flow, supporting the notion of water films being torn away.

Capillarity may dominate flow in cylindrical tubes with diameters ϕ_{max} of at least 10 mm, as Germann (1987) reported. However, minor disturbances may suddenly alter the flow regime to the point of being dominated by momentum dissipation. Germann et al. (1997) set the condition $D/\eta \geq 1$ to define the lower diameter limit of pores able to dissipate momentum, leading to minimal pore diameters of $\phi_{min} \approx 10\,\mu m$. This leaves a range of pore diameters of about 1000:1 in which flow may be dominated by either the reversible diffusion of capillary potential or the irreversible dissipation of momentum. However, the closer $D/\eta \to 1$, the smaller will be the probability that momentum dissipation dominates flow.

The generation of a minimal momentum of flow is related to sufficiently high volume fluxes along individual flowpaths. This requires path geometries allowing for unrestricted flow over a considerable distance and high enough overall volume flux densities. In addition, the water demand for approaching capillary equilibrium diverts flow from preferential to ordinary, thus reducing momentum. Therefore, high enough antecedent soil moisture and related capillary potential are additional requirements for maintaining momentum of flow.

Preferential flow may be triggered by infiltration at the soil surface as well as by any abrupt changes of hydraulic conditions within soil profiles. Little is known at present about the initial and boundary conditions which may lead to preferential flow as perceived in this contribution. Therefore, the following considerations constitute preferential flow, implicitly assigning slower moving soil moisture to ordinary flow, which is assigned to the realm of Richards' (1931) equation.

10.2.2 Momentum of Flow in Soils

Consider the slab $\Delta\ell$–F–Δz of water in Figure 10.1 flowing downwards along a wall of a vertical macropore, where $\Delta\ell$–Δz is the contact area between the solid and stagnant water on one side, and the moving water (i.e. preferential flow) on the other side. The orientations of $\Delta\ell$ [m] and Δz [m] are horizontal and vertical, respectively, and F [m] is the thickness of the water film. Mobile volumetric soil moisture is defined as

$$w = \frac{\ell F}{A} \tag{3}$$

[$m^3\,m^{-3}$], where A [m^2] is the cross-sectional area of the soil and $\ell = \Sigma\Delta\ell$ [m] is the entire length of contact in A between w and the parts of the soil–water system at rest.

Germann and Di Pietro (1999) applied Newton's law of shear to derive the momentum balance for steady film flow as

$$I_{v,h}(F) = \pm\frac{1}{6\eta}F^3\rho g\Delta z T\ell \tag{4}$$

[$kg\,m\,s^{-1}$], where tortuosity $T \geq 1$ is the scalar of $\mathbf{T}$ in the z-direction and the indices v and h refer to the momentum's vertical and horizontal components, respectively. Gravity continuously adds to the vertical component of the film's momentum whereas viscosity reduces its horizontal component due to dissipation.

The volume flux density was established as

$$q = bw^a \tag{5}$$

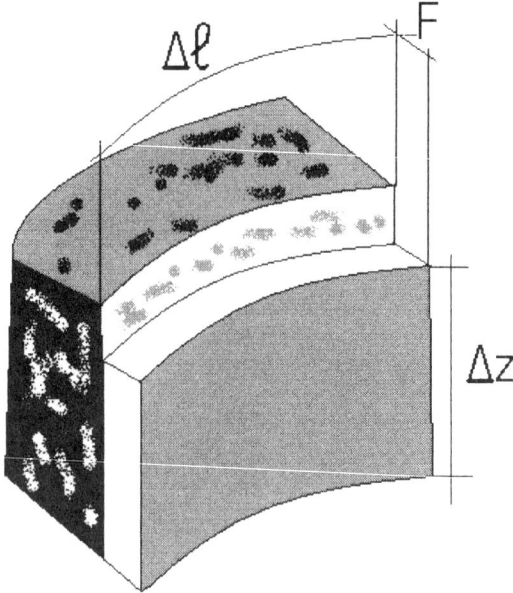

Figure 10.1 Schematic representation of a water film flowing vertically down along the wall of a macropore. $\Delta\ell$ [m] is the contact length between mobile water and the parts at rest of a soil–water–air system, F [m] is the thickness of the water film, and Δz is a depth section

$[\mathrm{m\,s^{-1}}]$, where $b\,[\mathrm{m\,s^{-1}}]$ is conductance and $a\,[-]$ is an exponent. From previous considerations it follows that $a = 3$. Because Di Pietro and LaFolie (1991) found $4 < a < 8$ from drainage experiments, the exponent will be treated as a variable that needs to be estimated from data.

The conductance parameter is

$$b = \frac{g}{2a\eta}\frac{A^2}{\ell^2} \tag{6}$$

The specific momentum components of the entire mobile water film (i.e. momentum per A) are

$$i_{v,h} = \pm\,\rho T q z_W(t) = \pm\,\rho T b w^a z_W(t) \tag{7a,b}$$

$[\mathrm{kg\,m^{-1}\,s^{-1}}]$, where the film extends from the surface to the time-dependent depth of the wetting front, $z_W(t)$. The water film accelerates when $i_v > i_h$ and reduces its velocity when $i_v < i_h$. Because $\dfrac{\partial i}{\partial w} \cong aw^{a-1}$, any reduction of w, for instance due to capillarity, reduces velocity and momentum of the water film, eventually leading to its complete cessation.

Water is assumed to infiltrate as a rectangular pulse with volume flux density $q_S\,[\mathrm{m\,s^{-1}}]$ and duration t_S [s], leading to the initial and boundary conditions of

$$t \leq 0 \text{ and } t \geq t_S\text{:} \qquad q(0,t) = w(0,t) = 0 \tag{8a}$$

$$0 \leq t \leq t_S\text{:} \qquad q(0,t) = q_S; \quad w(0,t) = w_S \tag{8b}$$

$$0 \leq z \leq \infty\text{:} \qquad q(z,0) = w(z,0) = 0 \tag{8c}$$

Initiation and cessation of water input to the soil surface represent physical and mathematical flow discontinuities. They release wetting and draining shock fronts which are designated with indices W and D, respectively. Their velocities are defined as

$$c_W = \frac{q}{w} = bw^{a-1} = b^{1/a}q^{(a-1)/a} \tag{9a}$$

$$c_D = \frac{dq}{dw} = abw^{a-1} = ab^{1/a}q^{(a-1)/a} \tag{9b}$$

Thus, the position of the wetting-shock front as function of time is

$$z_W(t) = tbw^{(a-1)} = tb^{1/a}q^{(a-1)/a} \tag{10a}$$

and the time lapse for the wetting front to move to depth Z is

$$t_W(Z) = \frac{Z}{bw^{a-1}} = \frac{Z}{b^{1/a}q^{(a-1)/a}} \tag{10b}$$

The position of the draining front is

$$z_D(t) = (t - t_S)abw^{(a-1)} = (t - t_S)ab^{1/a}q^{(a-1)/a} \tag{11a}$$

and the time lapse for the draining front to move to depth Z is

$$t_D(Z) = t_S + \frac{Z}{abw_S^{(a-1)}} = t_S + \frac{Z}{ab^{1/a}q_S^{(a-1)/a}} \tag{11b}$$

The time–depth relations of the wetting and draining fronts are called their characteristics. They are illustrated in Figure 10.2.

A trailing wave evolves in the depth range of $0 \le z \le z_D(t)$ over which mobile moisture increases from $w(0) = 0$ to $w(z_D) = w_S$. It is defined within the limits of $0 \le z \le z_D(t)$ and $t_S \le t < \infty$ as

$$w(z, t) = \left[\frac{z}{ab(t - t_S)} \right]^{1/(a-1)} \tag{12}$$

(Germann 1985). The draining front intercepts the wetting front at time

$$t_I = t_S \frac{a}{a - 1} \tag{13a}$$

[s] and at depth

$$z_I = t_S \frac{a}{a - 1} b^{1/a}q^{(a-1)/a} = t_S \frac{a}{a - 1} bw^{(a-1)} \tag{13b}$$

[m]. Thus,

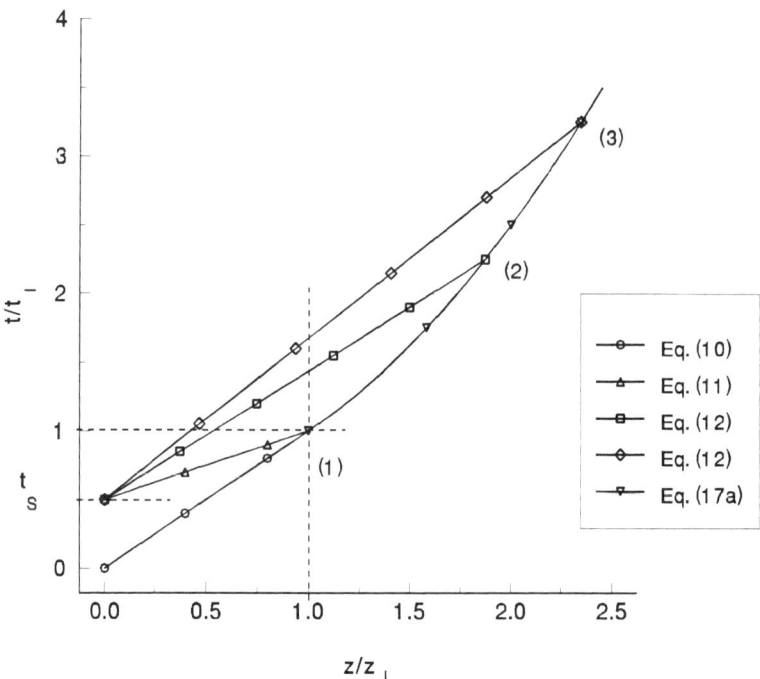

Figure 10.2 Spatio-temporal evolution of the wetting and draining fronts. The lines $(0, 0)$-(1) and $(0, t_S)$-(1) are the characteristics of the wetting and draining fronts, respectively; the lines $(0, t_S)$-(2) and $(0, t_S)$-(3) are the characteristics of two examples of an infinite number of water films constituting the trailing wave; the curve (1)-(2)-(3) represents the position in time and space of the wetting front for $z \geq z_I$ and $t \geq t_I$

$$0 \leq t \leq t_S; \qquad 0 \leq z \leq z_W(t) \qquad i_{v,h} = \pm \rho T b w^a z_W(t) = \pm \rho T q z_W(t) \qquad (14a)$$

$$w(z,t) = w_S \qquad (14b)$$

$$q(z,t) = b w_S^a \qquad (14c)$$

$$t_S \leq t \leq t_I; \qquad 0 \leq z \leq z_D(t) \qquad i_{v,h} = \pm \frac{z_D(t)^{(2a-1)/(a-1)} T \rho (a-1)}{(2a-1) b^{1/(a-1)} [a(t-t_S)]^{a/(a-1)}} \qquad (15a)$$

$$w(z,t) = \left[\frac{z}{ab(t-t_S)} \right]^{1/(a-1)} = w_S \left[\frac{t_D(z) - t_S}{t - t_S} \right]^{1/(a-1)} \qquad (15b)$$

$$q(z,t) = \left[\frac{z}{ab^{1/a}(t-t_S)} \right]^{a/(a-1)} = q_S \left[\frac{t_D(z) - t_S}{t - t_S} \right]^{a/(a-1)} \qquad (15c)$$

$$t_S \leq t \leq t_I; \qquad z_D(t) \leq z \leq z_W(t) \quad i_{v,h} = \pm \rho T b w^a [z_W(t) - z_D(t)] = \pm \rho T q [z_W(t) - z_D(t)] \qquad (16)$$

$$\text{where} \qquad [z_W(t) - z_D(t)] = b w^{a-1}[at_S - (a-1)t]$$

$$\text{or} \qquad\qquad\qquad = b^{1/a} q^{(a-1)/a}[at_S - (a-1)t]$$

$$w(z,t) = w_S \qquad (14b)$$

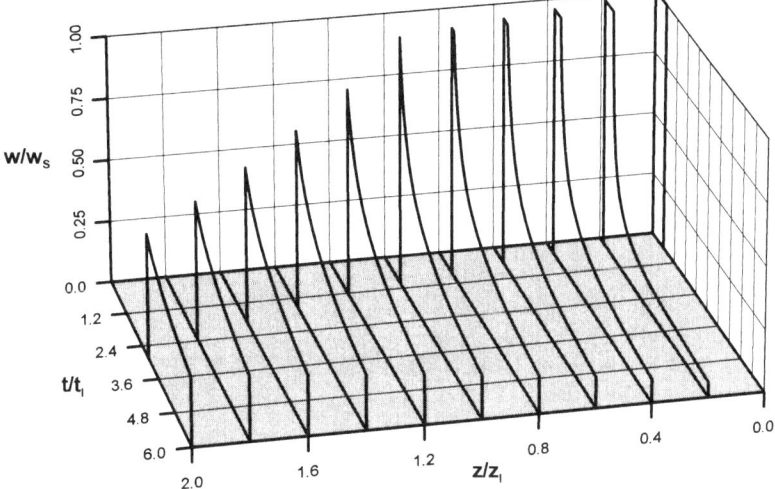

Figure 10.3 Kinematic wave, $w(z, t)/w_S$. Time and depth scales are normalised with t_I and z_I, (equations (13a,b), respectively

$$q(z, t) = bw_S^a \tag{14c}$$

Germann (1985) presented the advancement of the wetting front after $t \geq t_I$ as

$$z_W(t \geq t_I) = z_I \left[\frac{(t - t_S)}{t_S}(a - 1) \right]^{1/a} \tag{17a}$$

Volume flux density and mobile soil moisture at the wetting front are

$$q(z_W, t \geq t_I) = q_S(a - 1)^a \left[\frac{t_S}{t - t_S} \right] \tag{17b}$$

$$w(z_W, t \geq t_I) = w_S(a - 1) \left[\frac{t_S}{t - t_S} \right]^{1/a} \tag{17c}$$

Equations (8) to (17) are elements of a kinematic wave approach to flow in structured soils (Beven and Germann 1981; Germann 1985, 1990a) in which the propagation of fronts is dealt with according to the method of characteristics. A complete kinematic wave is shown in Figure 10.3.

The generalisation of the flux law of preferential flow, equation (5), thus becomes

$$q = \frac{g}{2\eta} \frac{F^a}{a} \left[\frac{\ell}{A} \right]^{a - 2} \tag{18}$$

10.2.3   Momentum of Infiltration

Infiltrating water carries momentum, and its time rate of change produces a force which needs to be considered in the balance of momentum. According to Germann et al. (1997), average momentum carried by infiltration is

$$\bar{i}_b = \frac{\rho}{\Delta t} q_b^2(t) \int_t^{t+\Delta t} dt \tag{19}$$

Equation (19) indicates the sensitivity of momentum with respect to variations in infiltration rates.

10.3   VALIDATION OF MOMENTUM DISSIPATION APPROACH TO FLOW IN STRUCTURED SOILS

10.3.1   Principle

Validation refers to making or being made valid, and *valid* means (i) to be logically sound or (ii) seen to be in agreement with the facts (Webster 1992). Thus, validation requires at least two steps: (1) *demonstration of the logical soundness* of approaching preferential flow from first fluid mechanical principles and (2) *demonstration of the agreement between the approach and the facts*. In principle, both steps have previously been presented (Germann et al. 1997; Germann and Di Pietro 1999), and only key expressions and illustrative examples will be validated here.

However, the validation procedure bears the risk of data acquisition methods being inherently geared towards the theoretical requirements of the approach. Therefore, proper validation needs step (3) *stating the limitations of an approach's applicability*.

10.3.2   Validation Step (1): Logic of Momentum Dissipation Approach

The momentum dissipation approach to flow in structured soils is based on Newton's law of shear, which led to the basic equation (4). The entire approach is based on a set of mathematical expressions, thus allowing for the most rigorous application of mathematical logic. It is particularly worth noting that Newton developed the concept of shear and provided the mathematical tools for its proper application.

10.3.3   Validation Step (2): Parameter Estimation from Measurements

Procedure

The model parameters a and b are estimated from linear regression analysis applied to the falling limb of either $w(Z, t)$ or $q(Z, t)$ for $t \geq t_p(Z)$. Equations (15b) and (15c) are linearised to the type of

$$Y_i = U(w, q)X_i + V(w, q) \tag{20}$$

according to

$$\ln\left(\frac{w(Z, t_i)}{w_{\max}(Z)}\right) = \frac{-1}{a-1}\ln(t_i - t_S) + \frac{1}{a-1}\ln(t_p(Z) - t_S) \tag{21}$$

and

$$\ln\left(\frac{q(Z,t_i)}{q_{max}(Z)}\right) = \frac{-a}{a-1}\ln(t_i - t_S) + \frac{a}{a-1}\ln(t_D(Z) - t_S) \tag{22}$$

where the index i refers to the data pairs. In contrast to equations (14b,c), the local amplitudes of the kinematic wave, $w_{max}(Z)$ and $q_{max}(Z)$, respectively, replace $w_S \geq w_{max}(Z)$ and $q_S \geq q_{max}(Z)$, recognising that some of the input water may get stuck between the surface and Z, the depth of observation. It follows that

$$a = \frac{U(w) - 1}{U(w)} = \frac{U(q)}{U(q) + 1} \tag{23}$$

and

$$t_D(Z) - t_S = e^{-V(w,q)/U(w,q)} \tag{24}$$

Conductance b is estimated with equation (10b) or (11b):

$$b = w_{max}^{(1-a)} \frac{Z}{a.(t_D - t_S)} = q_{max}^{(1-a)} \left(\frac{Z}{a.(t_D - t_S).}\right)^a \tag{25}$$

The contact length per cross-sectional area, $[//A]$ follows from equation (6) and the film thickness F from equation (3).

Figure 10.4 illustrates the procedure and the definitions, using the data from Site 7033, depth $Z = 0.53$ m, run 3 (see also Figure 10.10).

Soil moisture variations due to sprinkling were measured with TDR equipment. The wave guides were horizontally installed at depths 0.16, 0.33, 0.53, 0.93 and 1.33 m. Each wave guide consisted of a pair of rods of stainless steel, 50 mm apart, with diameters and lengths of 6 mm and 0.25 m, respectively. An RF Pulse transformer (500 kHz to 1 GHz, and 50 to 200 Ω), was used to stabilise the signal on the Tektronix 1502B cabletester. The wave guides were multiplexed with a SDMX50 50W Coax Multiplexer, which was controlled by a 21X Campbell Micrologger. The TDR system was calibrated according to Roth et al. (1990), who separated the impact of the wave-guide geometry from the soil properties, such as bulk density and the contents of clay and organic matter, on the dielectric constant. The frequency of data recording was set to $1/300\,s^{-1}$.

The results of the regression analysis are $U(w) = -0.61$ and $V(w) = 4.117$ with $N = 35$ and $r^2 = 0.97$. Further, $t_D(Z) - t_S = 853$ s; $b = 0.11\,m\,s^{-1}$; $[//A] = 4108\,m^{-1}$; $F = 8.0\,\mu m$, and $q_{max} = bw_{max}^a = 1.3 \times 10^{-5}\,m\,s^{-1}$. Specific momentum related to q_{max} was $\frac{i_{v,h}}{TZ} = 1.35 \times 10^{-2}\,kg\,s^{-1}\,m^{-2}$; momentum of the wave from the surface to $Z(= 0.53\,m)$ was $i_{v,h,max} = 7.16 \times 10^{-3}\,kg\,s^{-1}\,m^{-1}$ (equation 14a)), and average momentum of input was $\bar{i}_b = 1.92 \times 10^{-7}\,kg\,s^{-1}\,m^{-1}$ (equation (19)). It is interesting to note how average momentum of input increased in the soil due to its structure. However, momentum of input is considered important to the initiation of preferential flow.

For further details see Germann (1985), Mdaghri Alaoui et al. (1997), and Germann and Di Pietro (1999). Usually, the coefficients of determination between the momentum dissipation approach and the data are $r^2 > 0.9$.

Data Sets Used for Validation

Three sets of data will be referred to in the following discussion. *Set (1)* comprises the parameters

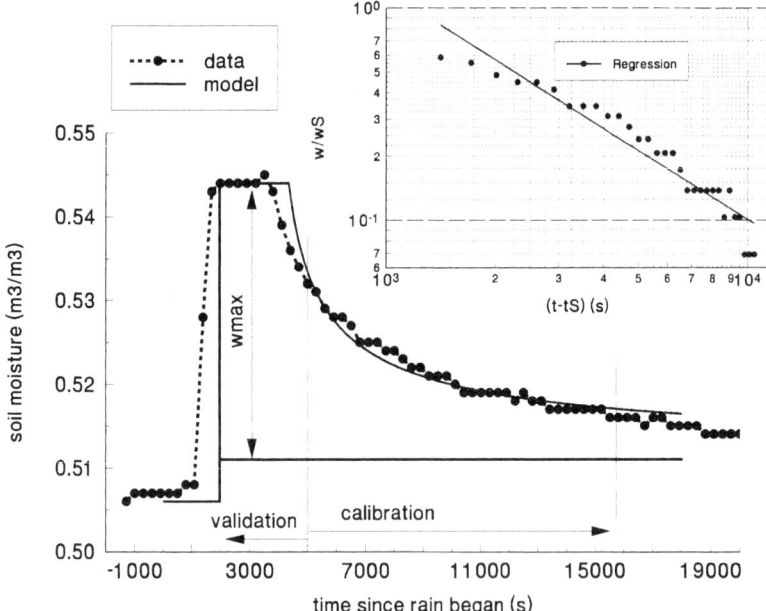

Figure 10.4 Illustration of estimating the model parameters from soil moisture variations. Moisture cut-off at 0.51 m³ · m⁻³ corresponds to terminal soil moisture after the kinematic wave has ceased at about 50 000 s after sprinkling. The inset shows the regression, equation (20), applied to the transformed data of the calibration section. Model performance in the validation section is purely based on the data in the calibration section and on w_{max} which was estimated from the data (Site 7033, depth 0.53 m, run 3, see also Figure 10.10)

derived from 112 measurements of $w(Z, t)$ at various depths in the range of $0.1 \leq Z \leq 1.2$ m in 18 different soils and with varying initial and boundary condtions (Germann and Bürgi 1996; Mdaghri et al. 1997; Germann and Di Pietro 1999). *Set (2)* consists of 13 $q(Z, t)$ time series which were derived from drainage experiments in four soil columns at a depth range of $0.2 \leq Z \leq 0.7$ m. One column was of an undisturbed soil (Mdaghri Alaoui 1998), one of a reconstituted soil and the remaining two columns were artificial porous media containing macropores (Germann 1985; Di Pietro and LaFolie 1991; Germann and Di Pietro 1996). *Set (3)* includes 16 drainage events from the weighing lysimeter Rietholzbach, whose diameter and depth are 2 m and 2.2 m, respectively, and input was natural precipitation (Germann and Di Pietro 1999). The data sets are characterised in Table 10.1.

Concluding Remarks

The many successfull applications of the momentum dissipation approach to data suggests its general validity according to Step (2). Moreover, the use of data from the trailing wave to back-calculate the arrival time of the wetting front at the depth of observation adds to the credibility of the approach.

Table 10.1   Characterisation of the three data sets used for validation

Parameter		w-version	q-version	Lysimeter drainage
$a_{maximum}$	—	5.99	5.60	2.21
$a_{minimum}$	—	2.10	2.85	2.00
$b_{maximum}$	$m\,s^{-1}$	5.96×10^{-3}	230	8.50×10^{-2}
$b_{minimum}$	$m\,s^{-1}$	2.61×10^{-4}	0.754	3.00×10^{-3}
$w_{maximum}$	$m^3\,m^{-3}$	0.1070	0.0787	0.0091
$w_{minimum}$	$m^3\,m^{-3}$	0.0039	0.0082	0.0024
$q_{maximum}$	$m\,s^{-1}$	2.48×10^{-5}	2.14×10^{-4}	9.66×10^{-7}
$q_{minimum}$	$m\,s^{-1}$	1.51×10^{-8}	4.10×10^{-6}	9.76×10^{-8}
$[\ell/A]_{maximum}$	m^{-1}	331 130	1122	18 620
$[\ell/A]_{minimum}$	m^{-1}	12	63	5012
$F_{maximum}$	m	1.45×10^{-3}	5.89×10^{-4}	1.15×10^{-3}
$F_{minimum}$	m	1.58×10^{-7}	1.00×10^{-5}	2.34×10^{-4}
Number of events	–	112	13	16

Table 10.2   Coefficients and their 95%-confidence limits of the regression $a = u_1 \cdot \log[\ell/A] + u_2 \cdot \log[F] + u_3$ (equation (26)), applied to the basic parameters of the momentum dissipation approach as derived from the w- and the q-version, and lysimeter drainage

	w-version	q-version	Lysimeter drainage
u_1	$-0.19 \leq \mathbf{0.07} \leq 0.33$	$-1.76 \leq -\mathbf{0.24} \leq 1.28$	$-0.45 \leq -\mathbf{0.30} \leq 0.15$
u_2	$0.79 \leq \mathbf{1.07} \leq 1.35$	$0.15 \leq \mathbf{1.15} \leq 2.15$	$-0.36 \leq -\mathbf{0.12} \leq 0.12$
u_3	$7.79 \leq \mathbf{8.38} \leq 8.97$	$6.64 \leq \mathbf{9.49} \leq 12.34$	$2.41 \leq \mathbf{2.89} \leq 3.37$
N	112	13	16
r^2	0.81	0.66	0.72

10.3.4   Validation Step (3): Limitations of the Momentum Dissipation Approach

q-Version vs. w-Version

The momentum dissipation approach to flow in structured soils is entirely based on the constants g and η, and the flow parameters a, F and $[\ell/A]$, equation (18). Further, free surface flow is assumed i.e. only gravity drives flow and no outside macroscopic pressure is acting on the mobile water. The application of the q- and the w-versions of the approach to corresponding data should therefore reproduce two overlapping frequency distributions of the basic parameters a, F and $[\ell/A]$ if the approach represents momentum dissipation due only to gravity. The statistical confidence ranges of the coefficients u_1, u_2 and u_3 of the regression

$$a = u_1.\log[\ell/A] + u_2.\log[F] + u_3 \tag{26}$$

for the w-version are expected to overlap with the corresponding ranges of the q-version. Table 10.2 compiles the coefficients and their 95%-confidence limits. From the overlapping confidence intervals it is concluded that the parameters u_1, u_2 and u_3 representing the 112 w-version time series also represent the 13 q-version time series, and that momentum dissipation during flow in soils is equally well representable by either version. Figure 10.5 illustrates and generally supports the notion.

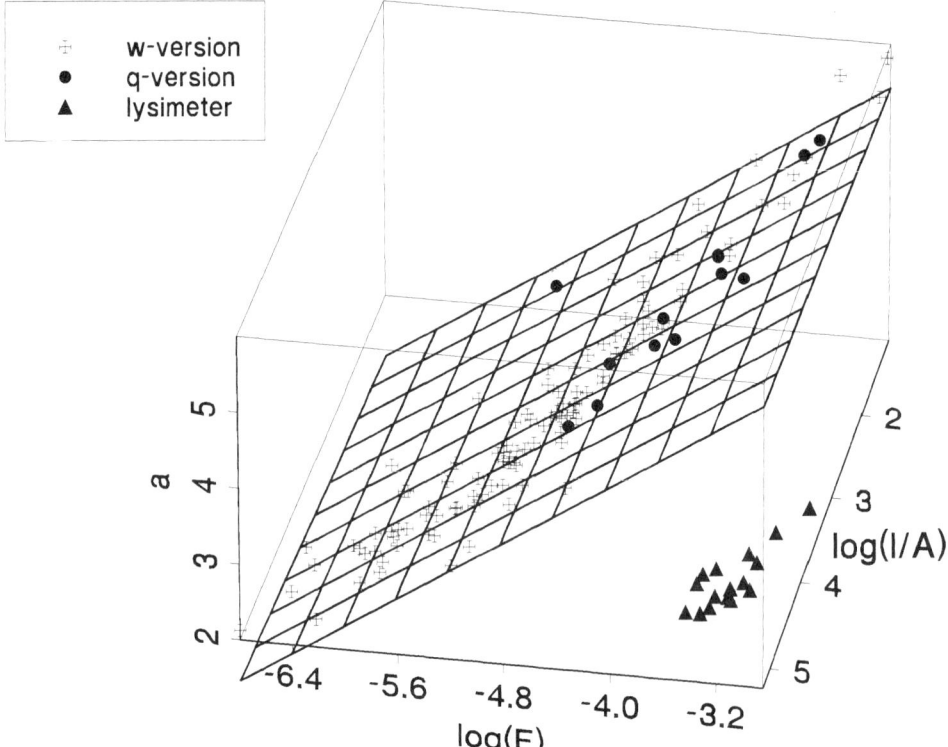

Figure 10.5   Graphic representation of the regression $a = u_1 \log[\ell/A] + u_2 \log[F] + u_3$ for the w- and q- version of the data which are represenented by crosses and circles, respectively. The solid triangles summarise the parameters from lysimeter drainage data which are located distinctly outside the range of the other two data sets

However, the 16 data triplets a, F and $[\ell/A]$ derived from the Rietholzbach lysimeter drainage data are located distinctly outside the ranges of the other two data sets, as the triangles in Figure 10.5 clearly demonstrate. The extraordinary film thicknesses F and their modest variation suggest flow in saturated porous media. The ranges of coefficients in Table 10.2 support the notion of film thickness variation being insignificant during lysimeter drainage because the 95%-confidence interval of u_2 includes zero.

Alternative Approach to Lysimeter Drainage

If time variations of lysimeter drainage are assumed to depend on time-variable hydraulic heads rather than on time-variable film thicknesses then drainage flow should follow from a linear reservoir approach. The lysimeter's impermeable bottom and hull support the notion.

From Poiseuille's law it follows that

$$Q(t) = -P\left[\frac{h(t)}{x}\right] \tag{27}$$

where $Q[\mathrm{m}^3\,\mathrm{s}^{-1}]$ is the volume flux of drainage, the permeability $P = \dfrac{\pi g}{8\eta} \mathrm{f}(\Sigma R^4)[\mathrm{m}^3\,\mathrm{s}^{-1}]$ is a

fourth power function of the flowpaths' radii R [m], and h [m] is the hydraulic head acting along a not well known but presumably constant flow restricting distance x [m]. The linear reservoir function is

$$Q(t) = A_{Lys}\Delta\theta\frac{dh}{dt} \tag{28}$$

where A_{Lys} is the horizontal cross-sectional area of the lysimeter and $\Delta\theta[\mathrm{m}^3\,\mathrm{m}^{-3}]$ is the average moisture release from the soil due to $\dfrac{dh}{dt}$. Integration of the combined equations (27) and (28) leads to the exponential function of $h(t) = h_0 \exp[-tC_R]$. With $h(t)$ the only time variable in equation (27), we get

$$Q(t) = Q_0 \exp[-tC_R] \tag{29}$$

where the reservoir constant is $C_R = \dfrac{P}{\Delta\theta A_{Lys}x}\,[\mathrm{s}^{-1}]$. It can be estimated from two points of the recession limb of drainage flow, say $q(t)/q_{max} = (1.0$ and $0.5)$. Figure 10.6 shows the momentum dissipation (equation (15c)) and the linear reservoir (equation (29)) approaches fitted to drainage data from the Rietholzbach lysimeter.

It is interesting to note that the data follow the linear reservoir approach during the early part

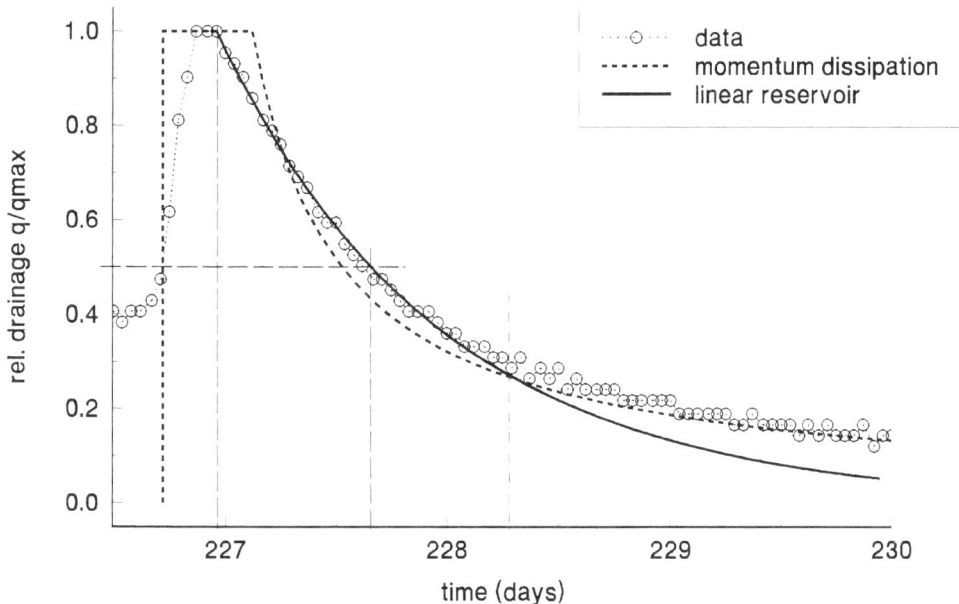

Figure 10.6 Relative drainage flow from the Rietholzbach lysimeter. The linear reservoir function and momentum dissipation represent the data well during early and later drainage, respectively

of the recession limb, whereas momentum dissipation represents the data well at later times. The transition is indicated by the crossing of the two approaches. The behaviour is considered reasonable because positive pressure may have acted at the beginning of drainage recession which was followed by free surface flow later on. However, independent measures are required to cleanse the discussion from speculation. Film thickness variations may eventually provide for a criteria to separate pressure-controlled flow from free-surface flow, but no well-defined threshold can be offered at the present.

Momentum dissipation still acts during lysimeter drainage, but it is most likely driven by pressure gradients in addition to gravity. The film thickness cannot freely react on variations of flow rates because of the pores' complete saturation.

As will be shown in Section 10.4.2, soil moisture variations during infiltration may reveal pressure build-up in some instances.

Concluding Remarks

Parameter estimation from soil moisture variations according to equations (20) to (25) is not permissible if additional pressures are acting on mobile soil moisture. Hence the differences between the distributions of the basic parameters derived from lysimeter drainage and the parameter ranges derived from the q- and w-versions of the approach.

Exponential and power functions i.e. equations (29) and (15c), may produce fairly similar results; however, the underlying physical processes may differ considerably. A parameter range of misinterpreting the momentum dissipation approach still exists at the present, and extreme caution is recommended not to abuse it.

10.3.5   Interpretation of Exponent *a*

The exponent a may indicate the degree of dissipative flow. Germann and Di Pietro (1996) demonstrated that a cylindrical flowpath hampers flow the least, and $a = 2$. The ratio $[\ell/A]$ then disappears in equation (18) because the contact length in this special case becomes a variable of the cylinder radius. On the other hand, $a > 3$ may macroscopically express the increased mechanical interactions between the stagnant parts of the flow system and the mobile water during free-surface flow. Thus, Germann and Di Pietro (1996) classified flow in soils accordingly:

$2 \leq a \leq 3$ momentum dissipation dominates flow
$3 \leq a \leq \approx 6$ mixed flow behaviour
$\approx 6 \leq a < \infty$ potential diffusion governs flow.

There is phenomenological support to this classification scheme. When antecedent soil moisture is low, infiltration in macroporous soils is frequently classified as matrix flow, and the parameters a and b are either indeterminable or a assumes high values (see the examples in Section 10.4.2). As initial soil moisture (i.e. water content in the matrix) increases, the shape of the observed time series $w(Z, t)$ increasingly approaches the one expected from dissipation theory, and the related exponent a decreases towards 2.

The classification scheme is also supported from the momentum dissipation approach itself. Depth of front interception z_I is viewed as a measure for the limitation of the penetration depth of preferential flow, i.e. it indicates the strength of preferential flow. Equation (13b) shows that $z_I \rightarrow z_{I,\max}$ for $a \rightarrow 2$ when the same conditions of t_S and q_S prevail.

In addition, Brooks and Corey (1964) determined the hydraulic conductivity function simultaneously with the air conductivity by matching the permeabilities $k = (K/\mathbf{f})\,[\mathrm{m}^2]$ of water and air at comparable degrees of saturation, where $\mathbf{f} = (g/\eta)\,[\mathrm{m}^{-1}\,\mathrm{s}^{-1}]$ is known as the fluidity of the respective fluid. They found that the relative conductivity of the wetting fluid is

$$K_{rw} = S_e^{(2+3\lambda)/\lambda} \tag{30}$$

where $S_e\,[\,-\,]$ is the degree of water saturation, and $\lambda\,[\,-\,]$ describes the pore size distribution. A wide range of pore sizes involved in the flow process produces small λ, and the smaller the range of pore sizes, the larger becomes λ. It follows from Brooks and Corey's (1964) experiments that $1.82 \leq \lambda \leq 7.30$, and the exponent $\varepsilon = (2 + 3\lambda)/\lambda$ is thus in the range of $3.27 \leq \varepsilon \leq 4.10$, which coincides with the proposed range of $2.0 \leq a \leq \approx 6$.

The upper limit of $a \approx 6$ at the transition from mixed to mainly capillarity dominated flow is empirically based. We found that one or both of the basic parameters F and $[\ell/A]$ may assume unrealistic values, and curve fitting according to equations (20) to (25) leads often to low r^2 values. However, no clear upper threshold can be defined because the macroscopic approach and the real process allow for mixed behaviour.

Low exponents a are generally accompanied by thinner film thicknesses F. The classification scheme thus suggests that, at comparable mobile soil moisture w, preferential flow is increasingly facilitated by increasing contact lengths, i.e. larger volume portions of the soil participate in preferential flow and not necessarily just thicker films. It is also worth noting that, according to the confidence limits of factor u_1 in Table 10.2, $[\ell/A]$ is not significantly correlated with the exponent a. However, we hope to clarify more rigorously the relations among the model parameters once larger data sets are available.

10.3.6 Flow Dimensions Compared with Morphometric Data

Daniel et al. (1997), for instance, counted the burrows produced by the earthworm *Aporrectodea nocturna*. They found about $N_P = 1600$ pores m^{-2} from earthworms in the diameter range of $1.6 \leq \phi \leq 6\,\mathrm{mm}$, covering a total area of about $0.01\,\mathrm{m}^2\,\mathrm{m}^{-2}$ per cross-sectional area of soil. This would lead to a maximum observed water content of $w_O \approx 0.01$, when full-pipe flow is assumed, an average pore radius of

$$\overline{\phi_O} = \sqrt{\frac{4w_O}{\pi N_P}} = 2.8\,\mathrm{mm} \tag{31}$$

and a total contact length of

$$[\ell/A]_O = N_P\overline{\phi_O}\pi = 2\sqrt{\pi w_O N_P} = 14.2\,\mathrm{m}^{-1} \tag{32}$$

Although the assumption of full-pipe flow and the estimated w_O and $\overline{\phi_O}$ represent extremes, they belong to the ensemble of values calculated with the momentum dissipation model. Pores need not be completely filled during flow, and frequently $w \geq w_O$ and $\overline{\phi} \leq \overline{\phi_O}$. The range of $63 \leq [\ell/A] \leq 1122\,\mathrm{m}^{-1}$ from the q-version, as reported in Table 10.1, thus seems plausible. The upper limit of the w-version range of $12 \leq [\ell/A] \leq 331\,130\,\mathrm{m}^{-1}$ needs further evaluation. Limited data suggest that extremely high $[\ell/A]$ values are related to small exponents, conductances, and film thicknesses in the ranges of $2.0 \leq a \leq 2.2$, $8 \times 10^{-5} \leq b \leq 2.6 \times 10^{-4}\,\mathrm{ms}$ and $10^{-7} \leq F \leq 10^{-6}\,\mathrm{m}$, respectively, perhaps pushing momentum dissipation towards the limit of $D/\eta \to 1$.

10.3.7   Concluding Remarks

Validation of the momentum dissipation approach led to two types of hydromechanical processes of preferential flow:

1. *Free surface flow in unsaturated soils*, which may well relate to preferential flow as observed, for instance, by Lawes et al. (1882): 'The drainage water of a soil may thus be of two kinds: it may consist (1) of rainwater that has passed with but little change in composition down the open channels of the soil; or (2) of the water discharged from the pores of a saturated soil.'
2. *Pressure dominated flow* when the macropores are water saturated and which may lead to pushing out old water as described, for instance, by McDonnell (1991).
3. *Transitions from one type to the other one*, as indicated in Figure 10.6, are considered highly dynamic at the pore scale and somewhat smoother at the scale of soil horizons or profiles.

Further experimental experience under conditions as close as possible to natural flow is required to assess the importance of transitional behaviour for runoff production, in particular in producing subsurface storm flow, in hillslopes and catchments.

10.4   APPLICATION OF MOMENTUM DISSIPATION TO THE HYDROLOGY OF SOIL PROFILES

10.4.1   Extension of the Approach to Variable Input Rates

Principles

Any input function $q_S(t)$ can be approached to the desired accuracy with a series of n square pulses $P_j, j = 1 \ldots n$, each of intensity $\overline{q}_j$ and duration Δt_j, thus

$$q_S(t) = \overline{q}_j \left(t = \tau \Big|_0^{\Delta t_j} + \sum_{m=1}^{j-1} \Delta t_m \right) \tag{33}$$

In analogy to the input pulse, equations (8a–c), each pulse P_j has to be routed separately according to the velocities of the shock fronts $c_{W,j}$ and $c_{D,j}$, and the corresponding trailing wave needs to be considered. This procedure establishes a Lagrangean approach to flow, and the spatio-temporal positions of flow properties are used to route the discontinuities. Once a pulse P_j has been overtaken by pulse P_{j+1}, the former loses its identity and the latter dominates flow. The propagation velocities are illustrated in Figure 10.7 as slopes of the chords and tangents of $q(w)$. The slopes of chords (1) and (3) represent the wetting front velocities $c_{W,j}$ and $c_{W,j+1}$ of two isolated pulses according to equation (9a). The slopes of the tangents (2) and (4) represent the draining front velocities $c_{D,j}$ and $c_{D,j+1}$ according to equation (9b). The trailing wave, equations (15b,c), evolves because each lamina at soil moisture w in the range of $w_j \geq w \geq 0$ moves with the corresponding velocity of $c_{D,j} \geq c(w) \geq 0$, which is represented by the ensemble of tangents to the function $q(w)$ in the appropriate range.

Case 1: $\overline{q}_{j+1} > \overline{q}_j$

If the volume flux density of pulse P_{j+1} is superior to the one of pulse P_j, the resulting shock front moves with the velocity

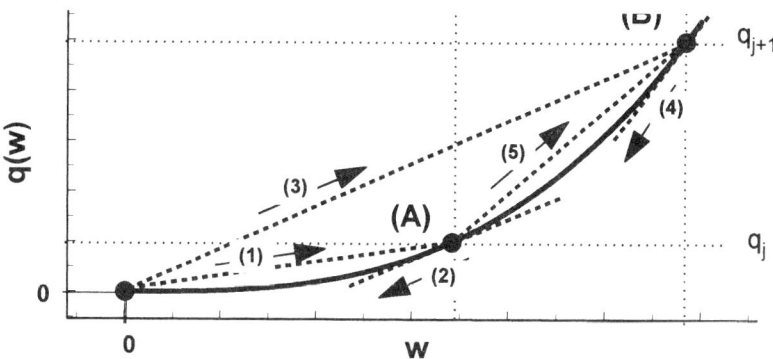

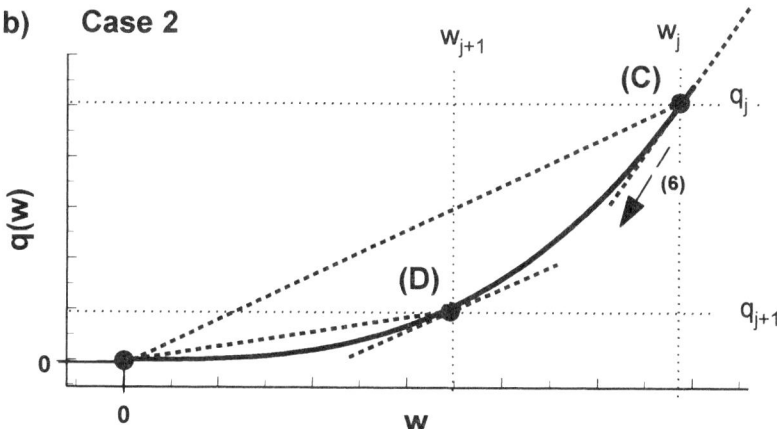

Figure 10.7 Relation $q = bw^a$, equation (5), and the front velocities. (a) Case 1: Sudden increase of volume flux density from 0 to q_j leads to the wetting front velocity $c_{W,j}$ which is equal to the slope of chord (1) i.e. the line (0, 0-A). The draining front velocity $c_{D,j}$ equals the slope of the tangents (2) through point (A). Similarly, the front velocites $c_{W,j+1}$ and $c_{D,j+1}$ correspond to slopes of the chord (3), i.e. the line (0, 0-B) and tangent (4) in point (B), when volume flux density suddenly increases from 0 to q_{j+1}. The front from suddenly increased volume flux density from q_j to q_{j+1} moves with a velocity which corresponds to the slope of chord (5), line (A)-(B). Case 2: Sudden decrease of input volume flux density from q_j to q_{j+1} releases an infinite number of laminae, each moving with a velocity which corresponds to the slope of a tangent between points (C) and (D). The slope of tangent (6) expresses the velocity of $c_{D,j}$

$$c_{W,j+1} = \frac{\overline{q}_{j+1} - \overline{q}_j}{w_{j+1} - w_j}$$

(34)

which corresponds with the slope of chord (5) in Figure 10.7a.

Case 2: $\bar{q}_{j+1} < \bar{q}_j$

If the volume flux density of pulse P_{j+1} is inferior to the one of pulse P_j, the resulting draining shock front moves with the velocity $c_{D,j}$, and the trailing wave, equations (15b,c), reduces its moisture and volume flux density along $q(w)$ to w_{j+1} and $\bar{q}_{j+1}$, respectively, as explained in Figure 10.7b.

Case 3: $\bar{q}_{j-1}, \bar{q}_{j+1} > 0$ and $\bar{q}_j = 0$

If the volume flux density of the intermittent pulse P_j drops to $\bar{q}_j = 0$, pulse P_{j+1} will glide on the trailing wave of pulse P_{j-1}, and equation (34) has to be modified to

$$c_{W,j+1} = b\frac{[w_{j+1}^a - w_{j-1}^a(z,t)]}{[w_{j+1} - w_{j-1}(z,t)]} \tag{35}$$

If the exponent a is an integer, equation (35) expands to the following series

$$c_{W,j+1} = b[w_{j+1}^{a-1} + w_{j+1}^{a-2}w_{j-1}(z,t) + \ldots + w_{j+1}w_{j-1}(z,t)^{a-2} + w_{j-1}(z,t)^{a-1}] \tag{36}$$

The acceleration effect of P_{j-1} on the propagation of P_{j+1} can be demonstrated for $a = 2$ as

$$c_{W,j+1} = b[w_{j+1} + w_{j-1}(z,t)] \tag{37}$$

From equation (15b) it follows that

$$w_{j-1}(z,t) = \left[\frac{z}{ab(t - t_{j-1})}\right]^{1/(a-1)} \tag{38}$$

where

$$t_{j-1} = \sum_{m=1}^{j-1} \Delta t_m \tag{39}$$

thus, due to P_{j-1}, the front velocity of pulse P_{j+1} increases with depth and decreases with time lapsed since the cessation of P_{j-1} at t_{j-1}. The wetting front depth of P_{j+1} follows from numerical integration of

$$c_{W,j+1}(z,t) = b\frac{\left[w_{j+1}^a - \left[\dfrac{z}{ab(t - t_{j-1})}\right]^{a/(a-1)}\right]}{\left[w_{j+1} - \left[\dfrac{z}{ab(t - t_{j-1})}\right]^{1/(a-1)}\right]} = \frac{dz}{dt}\bigg|_{P_{j+1}} \tag{40}$$

over time since the release of P_{j+1} at the soil surface. The parameters $a = 3$, $b = 0.15\,\mathrm{m\,s^{-1}}$, and $w_{j-1} = w_{j+1} = 0.02\,\mathrm{m^3\,m^{-3}}$ were applied when solving equation (40). The results are offered in Figure 10.8, which shows the effect of P_{j-1} on P_{j+1} at various release times after the cessation of P_{j-1}. An increased front velocity is still recognisable when P_{j+1} is released 7200 s after the cessation of input which led to P_{j-1}.

Case 3

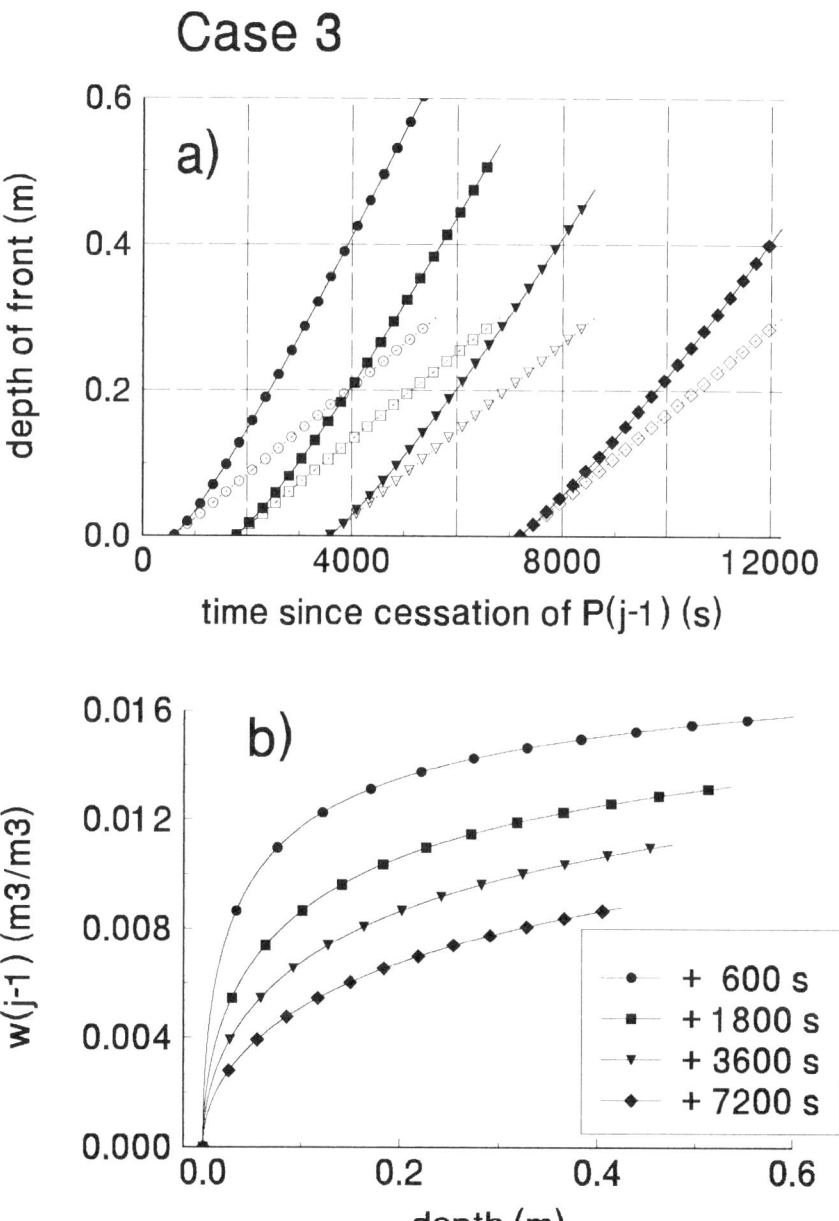

Figure 10.8 Case 3: Numerical integration of equation (40) using the parameters: $a = 3$, $b = 0.15\,\text{m s}^{-1}$, and $w_{j-1} = w_{j+1} = 0.02\,\text{m}^3\,\text{m}^{-3}$. (a) The solid symbols represent the front depths of pulses P_{j+1} which were released 600, 1800, 3600 and 7200 s after the cessation of pulse P_{j-1}. The corresponding open symbols indicate the front depths of the same pulses P_{j+1} if pulse P_{j-1} were absent. The differences between the lines with solid and corresponding open symbols indicate the accelerating effect of P_{j-1} on P_{j+1}. (b) Mobile soil moisture w_{j-1} vs. depth at the moment it is encountered by the front of pulse P_{j+1}

In conclusion, this theoretical example demonstrates the need to look more closely at the sequences of and time lags between runoff generating precipitation pulses as they do not necessarily propagate independently.

10.4.2   Characterisation of Observed Rapid Soil Moisture Variations

Five patterns of soil moisture reactions on sprinkling or rain were repeatedly established from TDR-moisture readings. The patterns are shown in Figures 10.9 to 10.11 as indicated with boxed-in numbers.

Pattern 1: No variations (see Figure 10.9). Steady flow prevails, including the possibility of zero flow. The front of infiltration has not penetrated to the depth of observation because (a) infiltrating water was adsorbed by capillarity between the surface and the depth of observation or (b) preferential flow was intercepted by a layer without macropores.
Interpretation: Preferential flow is not discernible; however, it may occur under alternate initial and boundary conditions.
Pattern 2: Gradual increase of soil moisture to a final content which is distinctly higher than antecedent soil moisture and which remains for a prolonged period of time (see Figure 10.10). Soil moisture has moved to the depth of observation mainly as diffusive flow in the matrix, i.e. the diffusion of capillary potential dominates.
Interpretation: Preferential flow is not discernible; however, it may occur under alternate initial and boundary conditions, as shown in Figure 10.10, depth 1.33 m, runs 1 to 3.
Pattern 3: Rapid increase of soil moisture, followed by a concave decrease shortly after cessation of input. Final soil moisture remains between antecedent and peak moisture (see Figure 10.10).
Interpretation: Typical behaviour of preferential flow according to momentum dissipation.
Pattern 4: Rapid increase of soil moisture to, and remaining for some time at a maximum, followed by a more or less convex decrease after the cessation of input (see Figure 10.9).
Interpretation: Preferential flow to a layer without macropores somewhere beneath the level of observation (as indicated by pattern 1 flow at the 0.22 m depth of the Leissigen 3 soil profile) with subsequent lateral drainage. Profile Leissigen 3 most likely produces subsurface storm flow between the 0.12 and 0.22 m depths. Retarded moisture decrease along an impermeable layer may only occur under sprinkling of limited horizontal extent, and the reaction pattern under conditions of natural rainfall may look different. Pattern 4 is most likely related with positive pressures which may occur temporarily above layers of low permeability during infiltration similar to drainage from the Rietholzbach lysimeter (see Section 3.4.2 and Figures 10.5 and 10.6).
Pattern 5: Combined profile. Almost simultaneous and rapid moisture increases at several depths (patterns 3 or 4) with an impermeable layer underneath (pattern 1) (see Figure 10.11).
Interpretation: Indicative of preferential flow in a soil whose matrix is of limited hydraulic conductivity, and with an impermeable layer at some depth (at 0.8 m in the example of Figure 10.11). Typical flow in a stagnogley.

The examples in Figure 10.10 demonstrate that flow pattern alterations due to a series of infiltration experiments always followed the order of $1 \rightarrow 2 \rightarrow 3$, and no reversed order has been observed so far, increasing antecedent soil moisture being the reason. Flow pattern alterations indicate the dynamics involved during infiltration, particularly switching from capillarity diffusion to momentum dissipation dominated flow.

Soil moisture variations at various depths due to a series of three infiltraton experiments, as shown in Figures 10.9 and 10.10, also show the reproducibility of the experiments (note the varying moisture scales when moving among graphs.) The examples in Figures 10.9 to 10.11

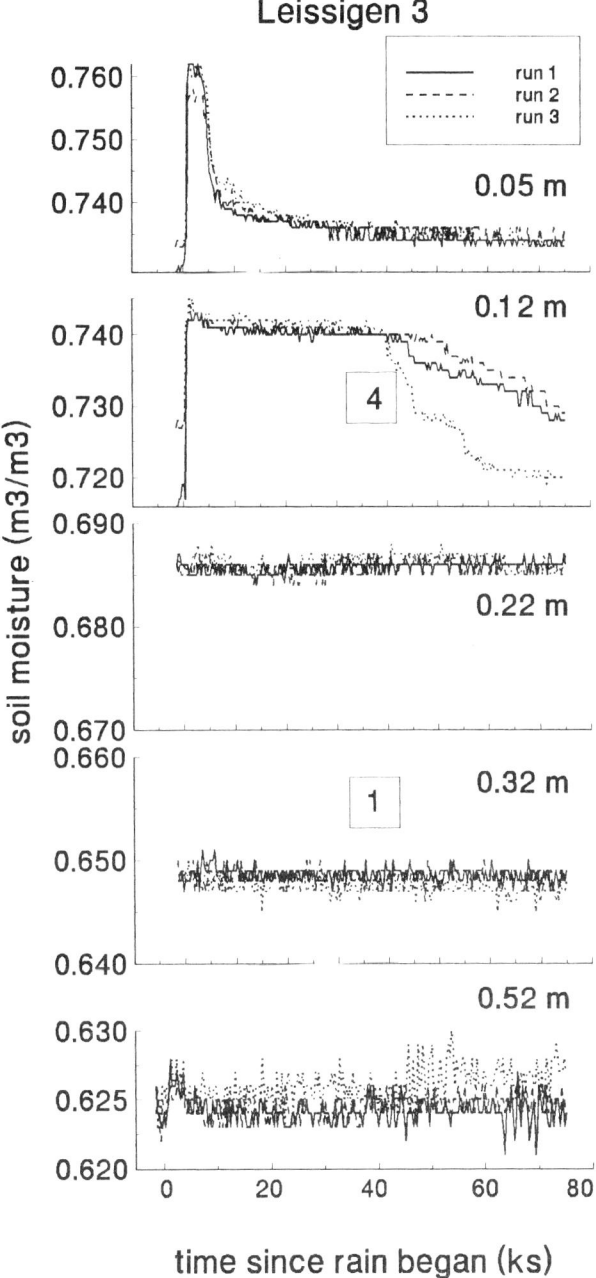

Figure 10.9   Soil moisture variations in the soil profile Leissigen 3 as the result of three infiltration experiments. Water application was by evenly sprinkling 50 mm during 1 h, i.e. $q_b = 1.4 \times 10^{-5}$ m s^{-1}. About 1 day lapsed between runs 1, 2 and 3. (Note the variation of scales when moving among the figures)

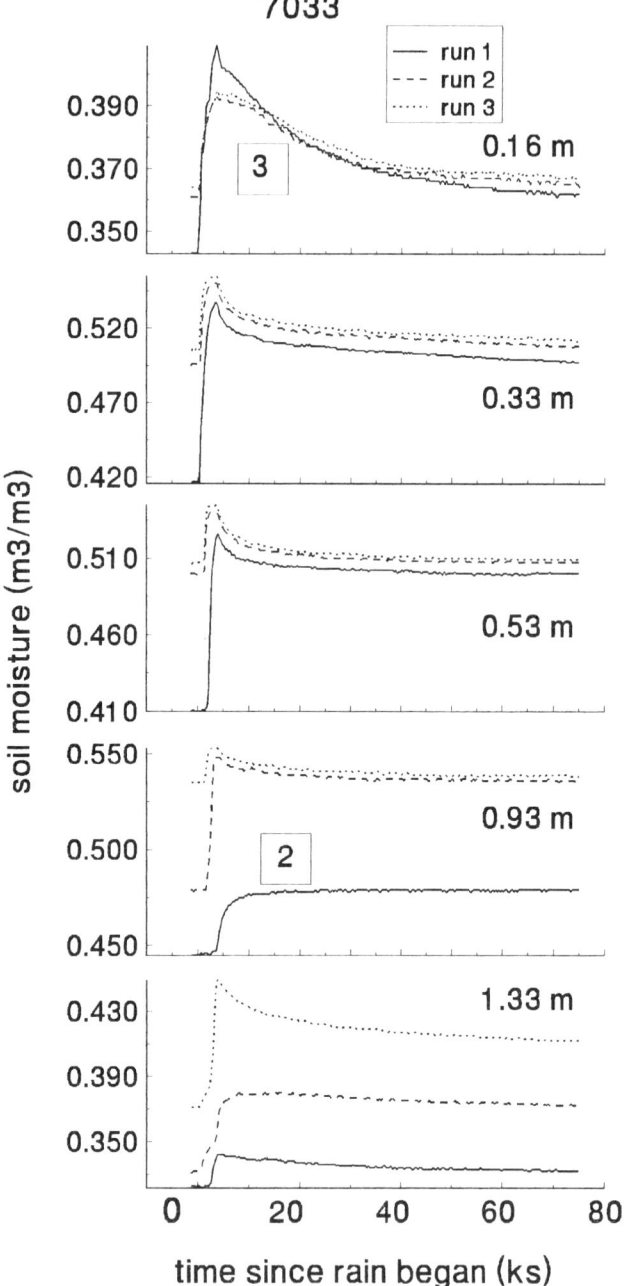

Figure 10.10   Soil moisture variations in the soil profile site 7033, due to the same experimental procedure followed at Leissigen 3. Depth 0.93 m shows clearly the transition from pattern 2 to 3 from run 1 to run 2 due to increased antecedent soil moisture. (Note the variation of scales when moving among the figures)

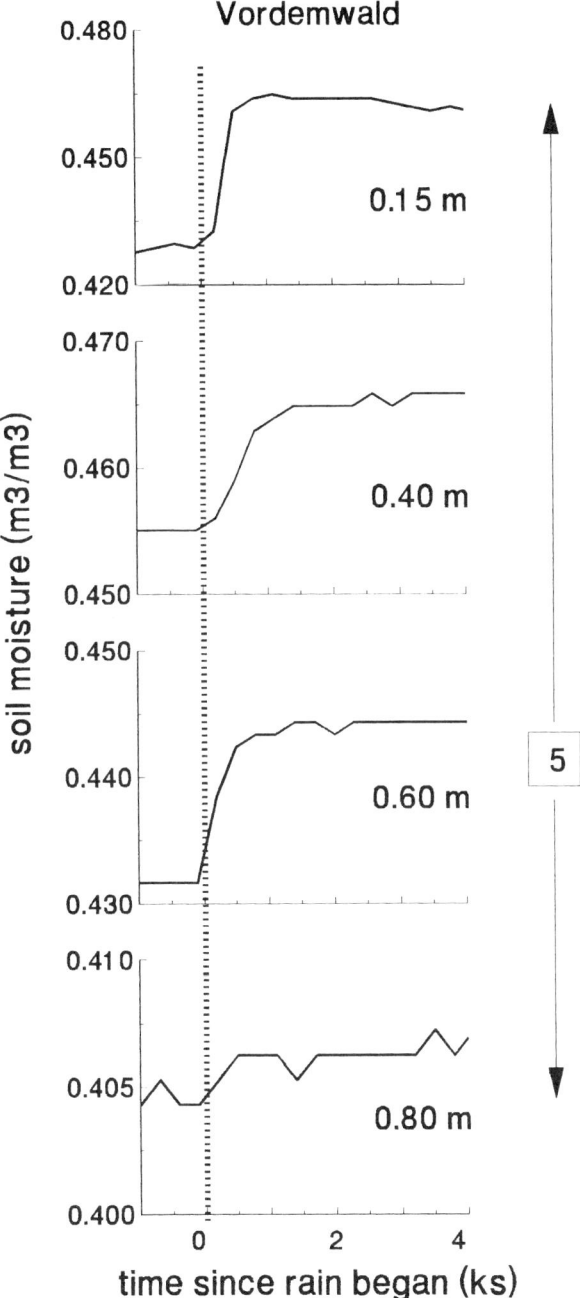

Figure 10.11   Soil moisture variations in the soil profile Vordemwald. Pattern 5 relates to the behaviour of an entire stagnogley profile. (Note the variation of scales when moving among the figures)

clearly indicate the short durations and rapid advancements of preferential flow, and how little moisture variations are actually involved.

10.4.3   Dimensions of Functional Macropores

The dimensions of preferential flow (i.e. hydraulically active macropores) are assessed with the momentum dissipation approach. Soil moisture variations in Figure 10.9 at depths 0.93 and 1.33 m demonstrate that increasing antecedent soil moisture may transform initial pattern 2 to pattern 3. Thus, a series of infiltration experiments is required to ultimately distinguish preferential from ordinary flow. The vertical extent of patterns 3, 4 and 5 (excluding the impermeable layer of the latter) indicate preferential flow to possible depths of at least 1 m. The patterns of drainage from the Rietholzbach lysimeter suggest preferential flow over depths exceeding 2 m, although a lower range might be completely waterlogged at times. There is evidence from deposits of garbage incineration slags that preferential flow may penetrate under natural rainfall conditions as deep as 12 m into the distinct dual-porosity medium (Hartmann 2000). However, the same restrictions most certainly apply as they were stated for lysimeter drainage.

The volume fraction of pores participating in preferential flow i.e. *minimum macroporosity*, is set equal to the amplitude of the kinematic wave, w_{max}. The *minimum widths* of macropores are equal to the maximum film thickness F [m], and specific contact length $[\ell/A]$ [m^{-1}] presents an additional geometric parameter.

10.4.4   Concluding Remarks

Momentum dissipation offers a hydromechanical concept to assess dynamic macropore flow at the scale of soil profiles. The concept allows for experimental and theoretical assessments of the evolution of varying input pulses, including breaks between them. Soil moisture may react to infiltration over a depth range of more than 1 m and within less than 1 h as the examples in Figures 10.9 to 10.11 demonstrate. The dynamic aspect is pronounced when the rapid reactions lead to short-term positive pressures.

The examples have shown that only small portions of soil moisture, mostly far less than 0.1 m^3 m^{-3}, need to participate in the flow process. As a consequence, soil moisture variations have to be recorded at short time intervals with high resolutions. Serious concern is raised to whether the piezometers with diameters of usually more than about 10 mm react fast enough to detect the hydrodynamics of macropore flow and its consequences for runoff production.

Application of momentum dissipation to flow in field soils led to five reaction patterns. With progressing experience, they may eventually provide for links between the morphology and the rapid hydrology of soil profiles. Pattern distinction is viewed as a gross screening procedure of dissipation against diffusion dominated flow. Refined screening relies on the interpretation of the basic parameters a, F and $[\ell/A]$ as demonstrated in Figures 10.5 and 10.6.

The approach of momentum dissipation is based on local momentum balance between the vectors parallel and perpendicular to flow (equations (7a,b) and Figure 10.1), respectively. It represents local quasi-steady flow and is thus not practical to completely characterise transient macropore flow. However, abrupt detaching of mobile water from the surrounding stagnant counterparts during the initial phase (i.e. the formation of a wetting front) and asymptotic attaching during the gradual die-off phase towards the end of the kinematic wave (i.e. the trailing wave) seem reasonably well modelled, as Figure 10.4 and similar examples show.

Momentum differences between points along macroscopic flowpaths are thought to bear more information on flow because the momentum differences reveal acceleration or deceleration of macropore flow between points, eventually leading to the prediction of whether preferential flow will cease, lead to pressure dominated flow or continue. Germann and Di Pietro (1999) presented experimental evidence of macropore flow acceleration and deceleration.

10.5   APPLICATION OF MOMENTUM DISSIPATION TO THE HYDROLOGY OF HILLSLOPE SOILS

Spatio-temporal momentum variations may link the hydrologic processes in soil profiles with those in hillslopes by improved understanding of lateral subsurface stormflow generation. The degree of spatial and temporal resolutions of the soil processes necessary to the better understanding of hillslope process is debatable. The instrumentation required to produce data for complete experimental assessments of momentum dissipation in an entire hillslope may soon turn to insurmountable problems, given the spatial variations among soil properties, such as the thickness of horizons, and extents and densities of macropore systems. However, because inertia of flow processes generally increases with increasing scales of the flow systems, it may suffice to experimentally study infiltration in a few selected profiles in detail, and to use the results as indices for larger hydrologic units. For instance, transitions from moisture pattern 1 to 2 to 3, or from 2 to 3 in a soil profile due to increased antecedent soil moisture may indicate the transition from diffusion to dissipation dominated flow for a wider area which, in turn, is most likely related to the initiation of rapid subsurface storm flow. Likewise, the occurrence of patterns 1, 4 and 5 may indicate readiness for producing overland flow or the acceleration of subsurface storm flow, when positive pressures build up. The readiness of process transitions depends on the depth to the impermeable horizons, antecedent soil moisture, and intensity and duration of input. As demonstrated in Section 10.4.1, the sequence of input pulses, including the durations of breaks between them, also need to be considered.

   Lateral subsurface flow is based on the same principles of momentum dissipation as vertical macropore flow as, for instance, Germann (1990a,b) showed, and the sequences of input pulses, including breaks between them, are considered important hydrologic features. The resulting kinematic waves tend to diffuse and intercept with one another, thus smoothing input variations along the flowpaths. In general, shock fronts will be damped and flow will vary more gradually.

   However, it is conceivable that the opposite behaviour may occur. Incidentally, kinematic waves may accumulate. Pressures may build up almost simultaneously in larger areas of a hillslope or catchment, thus leading to extraordinary shock waves along the flowpaths, similar to the evolution of avalanches. Actually, this scenario is thought to initiate mudslides. Faeh (1997) and Scherrer (1997) experimentally and computationally simulated the effects of extreme precipitations.

   The gross distinction at the profile scale between capillary diffusive and momentum dissipative flow, and the finer separation of the latter into free surface and pressure dominated flow, may lead to distinguishable subsurface storm flow and drainage behaviours. As a consequence, the various separated sections of hydrographs may not necessarily be due to drainage from various water storages such as shallow and deep groundwaters (i.e. spatial distributions of water storage), but due to different flow processes within the same soils and hillslopes (i.e. temporal distributions of water flow).

10.6   SUMMARY AND CONCLUSIONS

Preferential flow is viewed as dominated by irreversible momentum dissipation, whereas ordinary flow (i.e. matrix flow) is governed by reversible diffusion of capillary potential. Both flow types are based on a hydromechanical balance of linear momentum. Ordinary flow is related to complete linear momentum dissipation within an REV of capillary potential diffusion, whereas momentum is considered to dissipate over much larger spatial units during preferential flow. The concept allows for theoretical and experimental assessment of transitional behaviours between the two flow types.

From actual soil moisture θ only small volume fractions in the range of $0.005 \leq w \leq 0.1$ participate in preferential flow. Its initiation is viewed as detaching w from θ, the condition considered is $D/\eta > 1$. However, the critical ratio $(w/(\theta - w))$ is thought to depend strongly on θ and the momentum of flow, in addition to the presence of adequate soil structures (i.e. macropores). Likewise, preferential flow lasts only short periods of time after the cessation of its initiating water input. Both accelerated and decelerated preferential flow occur (Germann and Di Pietro 1999).

The momentum dissipation approach relates preferential flow to the basic parameters contact length of mobile water per unit cross-sectional area of soil, $[\ell/A]$ $[\text{m}^{-1}]$, average film thickness of mobile water, F [m], and the exponent a, whose physical meaning is obscure at the present and which may be viewed as a non-linear matching factor more likely related to the flow process rather than the macropore morphology.

The experimental back-ups of the momentum dissipation approach to preferential flow presented here and elsewhere indicate that wetting fronts may advance within minutes to depths beyond 1 m and that preferential flow is hardly discernible some hours after cessation of input to the soil surface. Further, momentum during flow in soils is by orders of magnitude higher than average momentum of input flow, suggesting that actual local momentum of drops and other flow concentrations are triggering preferential flow. As a consequence, experimental procedures geared towards improving our understanding of preferential flow have to stress high time resolutions.

Validation of the momentum dissipation approach led to two subcategories. Free surface flow corresponds to the theory presented in Section 10.2, and pressure dominated flow occurs when macropores are saturated. The theoretical approach is easily expandable to include pressure effects. The approach was applied to assess the hydrologic behaviour of a series of input pulses, including the accelerating effect of a persisting pulse on a following one, hinting at substantial effects even after hour-long breaks between input pulses.

Based on the approach, five patterns of soil moisture variations due to infiltration were identified. They are cautiously related to no-flow (pattern 1), matrix flow (pattern 2), free surface and saturated dissipative flow (patterns 3 and 4), and flow in a stagnogley profile (pattern 5). The fact that the same rate and duration of input led to different moisture variation patterns in the same soil profile, Figure 10.10, indicates the importance of antecedent soil moisture on the type of hydrologic process.

Once a larger database on soil moisture variations due to infiltrations is available, we hope to relate reaction patterns to soil morphologic features. For instance, readily recognisable reddish oxidation speckles indicate temporary water saturation, whereas dense horizons are frequently accompanied by oxidation and pale to greenish-bluish reduction spots indicating low permeability (i.e. confinement of macropore penetration depth).

The theoretical and experimental approach in its current version is suited best to the a posteriori flow classification. Transitions among moisture variation patterns occur spontaneously. They are reasonably well characterised by the shock front advancements. However, their occurrence is difficult to predict. With increasing numbers of experiments, we feel confident of narrowing down the range of hydrologic conditions which lead to the transitions.

Bronstert's (1999) thorough review on detailed hillslope hydrologic modelling lists the assessments of depth and storage capacity of macroporosity, and initiation of lateral flow as the most crucial issues of model improvement. Along the same line of thought, this contribution suggests that concentrating research on flow behaviours, and transitions among them, may prove more promising than the efforts put on detailed descriptions of flow-active macropores.

ACKNOWLEDGEMENTS

Data collection was made possible through several grants from the Swiss National Science Foundation and the Federal Office of Forest Management, and through the relentless efforts of Abdallah Mdaghri Alaoui, Therese Bürgi and Thomas Niggli.

REFERENCES

Beven, K. and Germann, P. 1981. Water flow in soil macropores. II. A combined flow model. *Journal of Soil Science*, **32**, 15–29.

Beven, K. and Germann, P. 1982. Macropores and water flow in soils. *Water Resources Research*, **8**(5), 1311–1325.

Bronstert, A. 1999. Capabilities and limitations of detailed hillslope hydrological modelling. *Hydrological Processes*, **13**(1), 21–48.

Brooks, R.H. and Corey, A.T. 1964. *Hydraulic Properties of Porous Media*. Hydrology Papers Number 3, Colorado State University, Fort Collins, CO.

Daniel, O., Kretzschmar, A., Capowiez, Y., Kohl, L. and Zeyer, J. 1997. Computer assissted tomography of macroporosity and its application to study the activity of the earthworm *Aporrectodea nocturna*. *European Journal of Soil Science*, **48**(4), 727–738.

Di Pietro, L. and LaFolie, F. 1991. Water flow characterization and test of a kinematic wave model for macropore flow in a highly contrasted and irregular double-porosity medium. *Journal of Soil Science*, **42**, 551–563.

Dingman, S.L. 1984. *Fluvial Hydrology*. W.H. Freeman, New York.

Faeh, A. 1997. *Understanding the Process of Discharge Formation under Extreme Precipitation. A Study Based on the Numerical Simulation of Hillslope Experiments*. Mitteilungen der Versuchsanstalt für Wasserbau, Hydrologie und Glaziologie No. 150, ETH, Zurich.

Germann, P.F. 1985. Kinematic wave approach to infiltration and drainage into and from soil macropores. *Transactions ASAE*, **28**, 745–749.

Germann, P.F. 1987. The three modes of water flow through a vertical pipe. *Soil Science*, **144**(2), 153–154.

Germann, P.F. 1990a. Preferential flow and the generation of runoff. 1. Boundary-layer flow theory. *Water Resources Research*, **26**(12), 3055–3063.

Germann, P.F. 1990b. Macropores and hydrologic hillslope processes. In: Anderson, M.G. and Burt, T.P. (eds), *Process Studies in Hillslope Hydrology*. John Wiley, Chichester, 327–364.

Germann, P. and Bürgi, Th. 1996. Kinematischer Ansatz zur in-situ Erfassung des Makroporenflusses während Infiltrationen. *Zeitschrift für Kulturtechnik und Landentwicklung*, **37**, 221–226.

Germann, P.F. and Di Pietro, L. 1996. When is porous media flow preferential? A hydromechanical perspective. *Geoderma*, **74**, 1–21.

Germann, P.F. and Di Pietro, L. 1999. Scales and dimensions of momentum dissipation during flow in soils. *Water Resources Research*, **35**(5), 1443–1454.

Germann, P. F., Di Pietro, L. and Singh, V.P. 1997. Momentum of flow in soils assessed with TDR-moisture readings. *Geoderma*, **80**, 153–168.

Hartmann, F. 2000. Modellrechnungen zur Beschreibung der Wasserbewegung durch eine Müllschlacken-deponie unter besonderer Berücksichtigung der Porenstruktur. Diss. ETH-Zürich No. 13732.

Lawes, J.B., Gilbert, J.H. and Warrington, R. 1882. *On the Amount and Composition of Rain and Drainage Water Collected at Rothamsted*. William Clowes and Sons, London.

McDonnell, J.J. 1991. Preferential flow as a control of stormflow response and water chemistry in a small forested watershed. In: Gish, T.J. and Shirmohammadi, A. (eds), *Preferential Flow – Proceedings of the*

National Symposium, December 16–17, Chicago, IL, published by the American Society of Agricultural Engineers, 50–58.

Mdaghri Alaoui, A. 1998. Transfert d'eau et de substances (bromure, chlorure et bactériophages) dans des milieux non saturés à porosité bimodale: Expérimentation et modélisation. PhD dissertation, Faculty of Sciences, University of Bern. *Geographica Bernensia* G 55.

Mdaghri Alaoui, A., Germann, P., Lichner, L. and Novak, V. 1997. Preferential transport of water and [131]iodide in a clay loam assessed with TDR-techniques and boundary-layer flow theory. *Hydrology and Earth System Sciences,* 1(4), 813–822.

Richards, L.A. 1931. Capillary conduction of liquids in porous mediums. *Physics,* **1**, 318–333.

Roth, K., Schulin, R., Flühler, H. and Attinger, W. 1990. Calibration of Time Domaine Reflectometry for water content measurement using a composite dielectric approach. *Water Resources Research,* **26**(10), 2267–2274.

Scherrer, S. 1997. *Abflussbildung bei Starkniederschlägen. Identifikation von Abflussprozessen mittels künstlicher Niederschläge.* Mitteilungen der Versuchsanstalt für Wasserbau, Hydrologie und Glaziologie No. 147, ETH, Zürich.

Webster's *Encyclopedic Dictionary of the English language,* 1992. Lexicon Publications, Inc. Danbury, CT (USA).

11   Validation of Snow Models

ROBERT E. DAVIS, RACHEL JORDAN, STEVEN DALY AND GEORGE KOENIG
CRREL, Hanover, New Hampshire, USA

11.1   INTRODUCTION

Snow model means different things to different developers or users, from providing lower boundary conditions and hydrology for atmospheric circulation models, to predicting details on snow processes for hydrological investigations, to estimating fluxes of melt water and chemical species for biological studies. Snow models may include calculations that help users specify the boundary conditions at the atmosphere–snow interface, or boundary conditions at the interface between snow and terrestrial materials or ice. In this chapter we define a snow model as the set of equations and algorithms that describe the processes and properties of snow cover, above the soil and below the snow–atmosphere interface at a plot scale ($\approx 100\,\mathrm{m}^2$). We examine the issues with validating snow models that have explicit spatial distribution and that use spatially distributed inputs for the sake of obtaining spatially distributed outputs. Hence we consider a spatially distributed snow model as an ensemble of snow model runs, initialised and driven by locally estimated variables that explicitly represent spatially contiguous areas. Maps or images of snow extent or other properties can be produced by this approach, as compared with spatially indexed models that run on tabular data lumping area characteristics. Spatially distributed models may also use tabular data to specify controls on model input and initial conditions but each model run has a pointer to an area or set of areas on a map.

Spatially distributed snow models potentially have wide application. Recently they have shown potential to improve operational hydrology, and we expect that this application will see intense development. The feeling in the operational community seems to suggest that lumped parameter models have reached an asymptote with regard to value added to forecasting when traded off against calibration issues. Research and operational efforts to implement spatially distributed models of snow have some common interrelated challenges. These include: (1) assessing appropriate model complexity for particular development and application; (2) segmenting and/or parameterising landscape and terrain data into model cells or polygons at suitable scales; (3) separating and estimating error due to forcing variables versus snow model performance; and (4) validating and/or updating spatial model predictions. This chapter aims to review one-dimensional snow process models and discuss their validation. We also review recent progress in spatially distributed modelling of snow and discuss the significantly greater problems of validation associated with the third and fourth challenges listed above.

We do not mean to provide an exhaustive review of progress in snow modelling to date. Other papers have done rather comprehensive jobs of this, through reviewing snow model applications (Dozier 1987; Leavesley 1989), discussing uses of remote sensing (Dozier 1992; Rango 1993), surveying spatial modelling approaches (Bales and Harrington 1995; Kirnbauer et al. 1994) and

Model Validation: Perspectives in Hydrological Science. Edited by M.G. Anderson and P.D. Bates.
© 2001 John Wiley & Sons, Ltd.

recently assessing the performance of a variety of snow models currently incorporated in land surface schemes of climate models (Slater et al. 2000). Bergström (1991) discusses model complexity and validation with respect to hydrologic applications and some related philosophical considerations. Braun et al. (1994) described the use of snow models driven with different meteorological variables, and different basin discretisation for hydrological modelling. Of the most important issues including those listed above, nearly all have been described since the late 1980s and early 1990s. Specifically, this chapter presents our views on validating snow models, from the plot to regional scales, and how these depend on model complexity and the variables employed in validation. We provide examples of actual validation tests using the simple temperature index model described by Daly et al. (1999) and the detailed model SNTHERM (Jordan 1991; Jordan et al. 1999) to illustrate the kinds of tests one can use to validate snow models.

11.2   One-Dimensional Snow Process Models

Snow science has seen many advances in predicting snow processes at a site using one-dimensional models (Marsh 1999). Figure 11.1 schematically shows the processes considered by some of the more complex, physics-based approaches (e.g. SNTHERM – Jordan 1991; Jordan et al. 1999). Model validation efforts at a study-plot or local-area scale use manual or remote

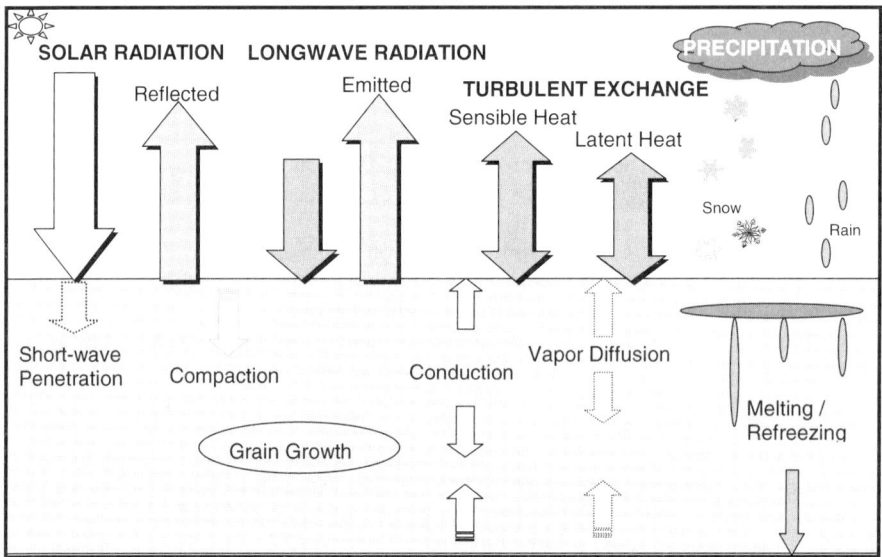

Figure 11.1   Schematic diagram of surface and snow cover energy and mass transfer processes. Net radiation consists of the balance between incoming and upwelling solar and longwave radiation. Gradients of temperature and humidity drive turbulent exchange according to wind speed. Precipitation dominates mass inputs to snow cover, but sublimation and condensation can prove important in some areas and during some events. Within a snow cover the important processes include penetration of solar radiation, grain growth, compaction, heat flow by conduction and vapour diffusion, and heat and mass flow with movement of liquid water. Hydrological applications have primary interest in liquid water flux at the bottom of the pack

measurements and have become the standard approach to building model credibility. Different hydrological applications suggest inclusion of different processes or parameterisations that lump processes. For example, many applications employing snow models in hydrological forecasting tend to use temperature index approaches to physically lump the components of surface energy exchange. This primarily stems from these models having more easily met requirements for input data, as well as lower computational expense. Moreover, it proves less complex for forecasting purposes to generate air temperature and precipitation based on climatology than the individual components of the surface energy exchange over snow. On the other hand, research investigations of snow processes and application of snow models for avalanche forecasting tend to use the most process-detailed models available (Brun et al. 1989). Melloh (1999) and Slater et al. (2000) recently reviewed the use of snow models with a range of physical detail and input data requirements. Table 11.1 shows the input requirements and state or output variables of two models we use for examples in this chapter that lie near opposite extremes of model characterisation of snow processes. The snow model SSARR__grid (Daly et al. 1999) represents a good example of the simplest snow models used by hydrologists. It requires air temperature and precipitation, used with a calibrated melt factor to estimate the mass balance of snow.

Snow model complexity controls the type of measurements that can be used to validate model predictions. In the next sections we focus on validation of one-dimensional models operating at the plot scale. We consider different variables that one may choose in validation exercises, first starting with bulk properties of a snow cover. Since presumably one can accurately specify the boundary conditions in such cases there should exist little ambiguity in distinguishing sources of error between the snow model itself and the handling of upper and lower boundary conditions. We next consider spatially distributed snow models and the difficulties in validating maps of snow properties.

11.2.1   Validation of Snow Models using Measurements of Bulk Snow Properties

Physically lumped or conceptually simple snow models may only allow testing the presence or absence of snow, snow depth and/or snow water equivalent and basal outflow (e.g. Martinec 1975; Kondo and Yamasaki 1990; Sommerfeld et al. 1991; Walland and Simmonds 1996;

Table 11.1   Snow input, state and output variables for two models near the ends in the range of complexity of snow models

	Input variables	State/output variables
SSARR__grid	Air temperature Precipitation (snow, rain) Melt factor (calibrated) Base temperature (calibrated)	Cold content Snow water equivalent Basal outflow
SNTHERM	Incoming solar radiation Incoming longwave radiation Air temperature Wind speed Humidity Precipitation (snow, rain)	Temperature profile Density profile Liquid water content profile Grain size profile Leaf litter on surface Albedo Snow water equivalent Snow depth Basal outflow

Douville 1997; Essery 1997; Daly et al. 1999). Variables for validating conceptually simple models may also include the bulk properties density and average temperature (e.g. Wigmosta et al. 1994; Tarboton et al. 1995; Yang et al. 1997; Albert and Krajeski 1998; Marks et al. 1998). Simple models that do not explicitly represent many of the surface or internal processes generally require some calibration of state or energy transfer parameters (e.g. degree-day factor, critical melting temperature), which can limit their applicability to areas or meteorological conditions governing the calibration procedure (Blöschl and Kirnbauer 1991). Calibration also introduces an element of subjectivity in validation tests. During the high melt season, we can consider three state variables: depth of the snow pack, snow water equivalent (SWE), and basal water outflow for validating snow models. These variables depend on boundary fluxes and internal processes as:

$\Delta SWE = f(\text{Precipitation} - \text{sublimation} - \text{basal outflow})$
$\Delta \text{Snow depth} = f(\text{Precipitation} - \text{sublimation} - \text{melt} - \text{compaction})$
$\text{Basal outflow} = f(\text{Rainfall} + \text{melting/refreezing} + \text{internal flow field})$

In periods of overnight freeze-up or during rain-on-cold snow events, we could consider snow temperature as a state variable, but one finds it difficult to determine an appropriate average temperature of the snow cover that reflects the mean thermal state. In terms of the seasonal and annual water balance of snow-dominated basins, the SWE of a snow cover represents the most important hydrological variable.

Case Studies

This section uses case studies of snow modelling at the plot scale (Jordan et al. 2000) to illustrate the issues in validation using bulk properties. We chose three snow study plots that represent a large range in snow accumulation and ablation rates and processes:

- *Mammoth Mountain, California, USA.* The cooperative snow study plot at Mammoth Mountain Ski Area (MMSA) lies in the Sierra Nevada at about 2926 m elevation (37 39'N, 119 02'W). This subalpine Sierra site lies at the transitional timberline and occupies a terrain bench that has extensive fetch from south, through east to west-northwest. The site receives continual sun on clear days with no shading from terrain and only slight shading from nearby trees. Average maximum snow accumulation reaches slightly over 3 m, representing about 1 m of water equivalent by April. However, the maximum snow depth at the site can range to over 5 m, which requires substantial structures to support meteorological instrumentation.
- *Sleepers River Research Watershed (SRRW), Danville, Vermont, USA.* The study plot at SRRW lies in a clearing in a forest of mixed hardwoods at 560 m elevation (44 29'N, 72 09'W). Average annual precipitation reaches 125 cm, 25% falling as snow, with snow cover persisting from early December to mid-April. The snow melt model development work of Anderson (1968, 1976) and the recent model work of Albert and Krajeski (1998) used measurements from this site. This site also produced one of six data sets chosen for the project on Intercomparison of Models of Snowmelt Runoff (WMO 1986).
- *CRREL Energy Balance Site, Hanover, New Hampshire, USA.* The relatively sheltered site lies on a west-facing slope above the Connecticut River on the grounds of the Cold Regions Research and Engineering Laboratory at 151 m (43 38'N, 72 08'W). Prevailing winds come down the river valley from the north. Surrounding hills, nearby trees or buildings shade the tower location early or late in the day and limit the fetch to 100 m. The relatively shallow snow cover rarely reaches 60 cm, and may melt completely during mid-season thaws.

All three sites monitored the necessary meteorological and snow cover parameters for running and validating a detailed snow cover model. The validation tests used SNTHERM (Jordan 1991; Jordan et al. 1999), a physically based one-dimensional mass and energy balance model that simulates details in changing snow cover properties (Figure 11.1) over time in response to meteorological forcing (Table 11.1). All three simulations use the same standard SNTHERM default parameters as shown in Table 11.2, not tuned specifically to the sites.

For comparison, a simple temperature index approach, SSARR_grid, predicted SWE and the basal outflow of melt water at the MMSA study plot. Table 11.1 compares the relative complexity of variables used by each model, SNTHERM and SSARR_grid. The SSARR_grid snow model (Daly et al. 1999) evolved from the 'Snow-Band' snow melt computation, part of the Streamflow Synthesis and Reservoir Regulation (SSARR) model (US Army Corps of Engineers 1991). The routine estimates the residual SWE and liquid water available at the soil surface by applying a melt factor to a temperature index. At least three parameters in this model require calibration, the base air temperature below which the model does not allow melt, the threshold temperature separating rain from snowfall and the snow melt factor. Model simulations for MMSA used a linear function to specify the snow melt factor based on the accumulated temperature index (Daly et al. 2000). The calibration procedure used hourly temperature and precipitation measurements during the winters 1985–1999 from 22 observation stations in the Sierra Nevada to calibrate the SSARR_grid melt factor. A sum-of-squares minimisation process exploited the difference between the SWE values calculated by SSARR_grid and the measured SWE. The optimisation followed the downhill simplex method of Nelder and Mead (1965) as presented by Press et al. (1992). This method belongs to the class of multidimensional minimisation procedures. We decided to pool the stations and all the years of data to produce one set of optimised results for the entire Sierra test region. This resulted in a base temperature of $0.36\,°C$, a temperature to separate rain and snow events of $1.65\,°C$ and a melt factor that ranged from $6.56 \times 10^{-5}\,mm/°C$ at an accumulated temperature index of 0, to $6.47 \times 10^{-3}\,mm/°C$ at an accumulated temperature index of 50. The melt factor reached an asymptote for higher values of the accumulated temperature index.

Snow Depth and Snow Water Equivalent

Snow depth constitutes one of the most fundamental variables that one can directly and easily measure. Manual probing, manual observations from an in situ stake and various ranging devices can yield reliable, unambiguous data. This generally holds true for a study plot, but not

Table 11.2   Default parameters used in SNTHERM simulations

Parameter	Value
Roughness length for momentum	5 mm
Roughness lengths for heat and moisture	Andreas (1987)
Stability correction for stable atmosphere	Beljaars and Holtslag (1991)
Windless exchange coefficient for sensible heat	$1.0\,W\,m^{-2}\,K^{-1}$
Snow emissivity	0.98
Fraction of incoming solar radiation with wavelengths $< 1.12\,\mu m$	0.41
Irreducible liquid water saturation	0.04
Snow viscosity	$1.0 \times 10^{6}\,kg\,s\,m^{-2}$
Density cutoff for metamorphic compaction	$min(100\,kg\,m^{-3}, 1.15 \times$ new snow density$)$

necessarily for a study area, which might exhibit a large variation of snow depth (or *SWE*) over short distances, such as in a forest. Freshly fallen snow compacts at about 1% per hour, but the rate varies with snow type and slows rapidly with time. In the accumulation and pre-melt periods, the precipitation rate, snowfall density and snow compaction rate control the snow depth – making it difficult to attribute sources of model error. During the melt period, with low compaction rates, snow depth can prove a useful indicator of snow ablation. Spatial variations in snow depth at the plot scale, however, can easily account for timing discrepancies of a day or more.

As examples of comparing SNTHERM model output to snow depth for validation, Figure 11.2 shows observed (triangles or lines) and modelled time series (dotted lines) of snow depth at the three case study plots. The figures show close agreement between observed and modelled snow depth. Simulations at MMSA and SRRW, Figures 11.2a and 11.2b, respectively, cover the melt period only from just before the maximum accumulation, whereas the simulation at Hanover (CRREL), Figure 11.2c, includes the entire winter season. Note the difference of range in the *y* axes of the plots. The snow depths at Mammoth Mountain and some from Hanover came from observations in periodic snow pits and reconstruction from precipitation measurements made on snow boards (triangles). An acoustic snow depth sensor measured snow at the other two sites (solid line). Mammoth has the deepest snow cover of the three sites, exceeding 2

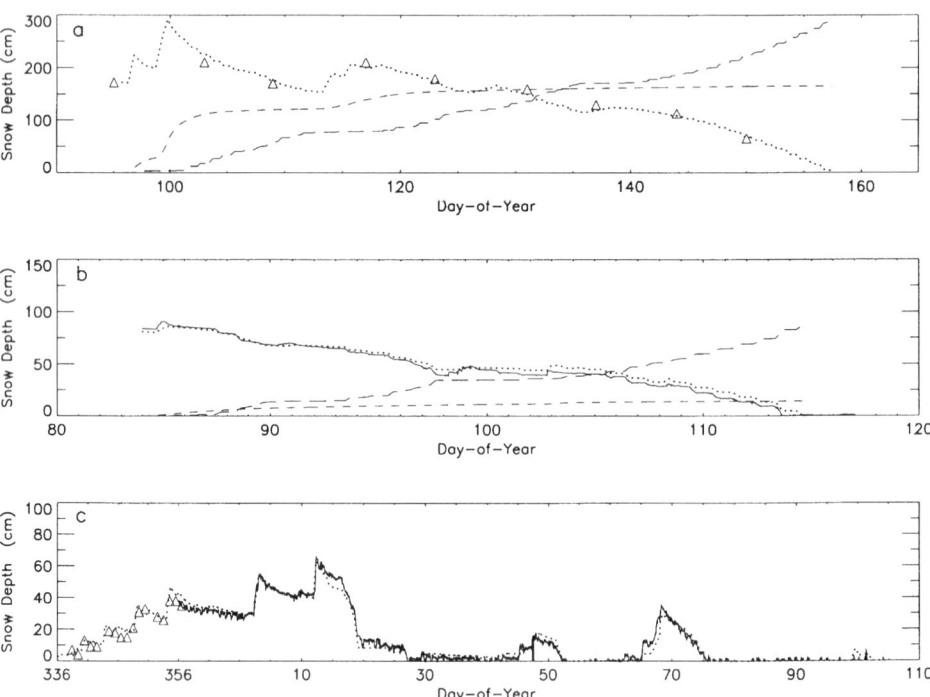

Figure 11.2 Predicted and observed snow depth: (a) between DOY 95 and 163, 1994, at Mammoth Mountain, CA; (b) between DOY 84 and 118, 1997, at Sleepers River Research Watershed, VT; and (c) between DOY 335, 1995, and DOY 103, 1996 at CRREL, Hanover, NH. The dotted line represents the SNTHERM simulation and the triangles and solid line show measurements. Plots (a) and (b) also show simulated contributions to depth reduction from densification (short-dashed line) and melting (long-dashed line)

metres at peak accumulation, while Sleepers River had an intermediate amount between MMSA and CRREL, which had the shallowest cover. At MMSA we see large variation in snow depth, particularly in the early part of the test period, while daily snow depth losses averaged 3 to 4 cm per day over the simulation. The variation of the trend in snow depth at SSRW showed much less change, with depth decreasing at a steady rate during the test period. At CRREL on DOY 17–19, advection of warm air followed by an intense rainstorm (Anderson and Larson 1996; Hogan 1996) melted almost the entire snow pack within 32 hours. The simulated loss of snow depth during this event agrees closely with these observations.

Ignoring snow redistribution by wind, the evolution of snow depth results from the interplay between two processes, snow densification and melting. Many simple models do not include densification, instead assuming a uniform density. This has the consequence that one must consider uncertainty due to variance of actual density around the specified constant value when using snow depth as a surrogate to validate simple snow models. In Figure 11.2a and b the plots of snow depth for the Mammoth and Sleepers River also show the depth reduction from snow densification (short-dashed line) and from melting (long-dashed line). Snow depth reduction due to densification at Mammoth (Figure 11.2a) occurs primarily in response to the loading induced by large storms. For example on DOY 100 the plot received nearly 100 cm of new snow. A period of rapid densification followed. Compaction losses at Mammoth diminished as the pack densified and after DOY 120 they accounted for less than 0.5 mm per day. Thus, during this period agreement between the calculated and measured snow depth at this site proved a good indicator that the model simulated the energy balance and melt rate correctly. At the Sleepers River site the change in depths reflected the melting conditions since densification had a small effect on the depth signal (Figure 11.2b). The peak snow depth reached 87 cm, typical for this site, and roughly 40% of that observed at MMSA in 1994. The saw-tooth nature of the snow depth trace during the accumulation period at CRREL (Figure 11.2c) reflects the rapid compaction rate of new snow.

Snow water equivalent represents the potential water input to the terrestrial hydrological system. Snow depth and SWE relate through the average snow density as $SWE =$ snow depth $\times$ snow density$/1000$, where the units follow the SI convention. Changes in SWE relate directly to changes in the total mass of the snow pack. Like snow depth, one can directly measure SWE, in this case by obtaining the mass of a snow core obtained from snow boards or the whole pack (e.g. Church 1935 and 1937; Goodison 1981; Goodison et al. 1987). One can approximate SWE by reconstructing measurements from snow pit data, or by using devices that sense the mass of snow overlying a pillow or other weighing gauge. Although both snow depth and SWE relate to snow melt, neither provides direct measures of the melt rate of the ice matrix. Nevertheless, if the ablated SWE has the correct timing then it represents a reasonable surrogate of simulated melting.

Figure 11.3a compares measured SWE reconstructed from snow pits at Mammoth Mountain with the SNTHERM simulation. Comparison between Figures 11.2a and 11.3a shows the differences between the snow depth and SWE signal. Reduction in snow depth between DOY 101 and DOY 113 corresponded primarily with simulated melting at the top of the snow pack. Melt water infiltrated the dry snow pack at 20 to 40 cm per day. Loss of SWE (Figure 11.3a) beginning on DOY 108 corresponded with arrival of the wetting front at the bottom of the snow pack. At MMSA we found the melting rate sensitive to estimated albedo. For instance, final loss of snow cover varied by ± 5 days when we reran the model using albedo computed with full cloud cover or totally clear skies. We estimate an uncertainty of ± 1 day in the simulated timing of snow loss resulting from uncertainty in albedo computations. The Sleepers River Research Watershed had the most northerly and coldest location of the three sites. The 1996–97 season had particularly calm winds, rarely reaching $3 \, \mathrm{m \, s^{-1}}$, and thus the lowest of the three simula-

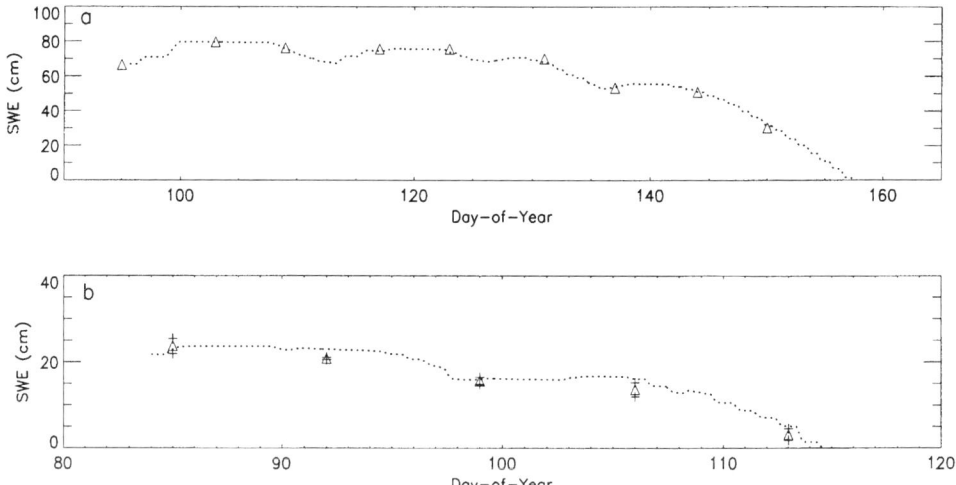

Figure 11.3 Predicted and observed snow water equivalent (*SWE*): panel (a) between DOY 95 and 163, 1994, at Mammoth Mountain, CA; and panel (b) between DOY 84 and 118, 1997, at Sleepers River Research Watershed (SRRW), VT. At SRRW, we plot the minimum, maximum and mean observed values (b). The dotted line represents the SNTHERM simulation and the triangles and solid line show measurements

tions. Figure 11.3b shows close agreement between observed and simulated *SWE* using SNTHERM. The ablation for the 34-day melting period averaged about 0.8 cm per day, half that at Mammoth Mountain, due both to lower insolation and lower winds.

The temperature index model SSARR—grid underpredicted the melting rates in the early part of the melting season at MMSA. The model began to catch up later in the season, predicting total melt by DOY 162 five days later than the predicted end of snow by SNTHERM, but three days later than the last observed basal outflow. Figure 11.4 shows the comparison between measured and simulated *SWE*. Divergence between the SSARR—grid predictions and the observations begins about DOY 115 and gradually increases almost to the end of snow. In general the model performed adequately during periods when sensible heat transfer to the snow represented a significant component of the surface energy exchange, but underpredicted melt during periods when the radiation balance between solar and longwave increased to become comparable to, or greater than, the magnitude of the sensible flux. Figure 11.5 shows the components of surface energy flux during the simulation period at MMSA as represented by the SNTHERM simulation using the measurements shown in Table 11.1. After DOY 135 the radiation balance became increasingly positive and the average sensible heat flux did not increase until later, as shown in Figure 11.5. This model behaviour reflects the calibration of the melt factor. The majority of the 22 sites used in calibrating SSARR—grid lie below the regional timberline, so that sensible heat transfer and longwave radiation play a relatively more important role than in the subalpine and alpine zones.

Melt Water Outflow

With comparisons of model results and measurements of snow depth, *SWE* and water outflow one can thoroughly assess the bulk thermal performance of snow models. Lysimeters allow the

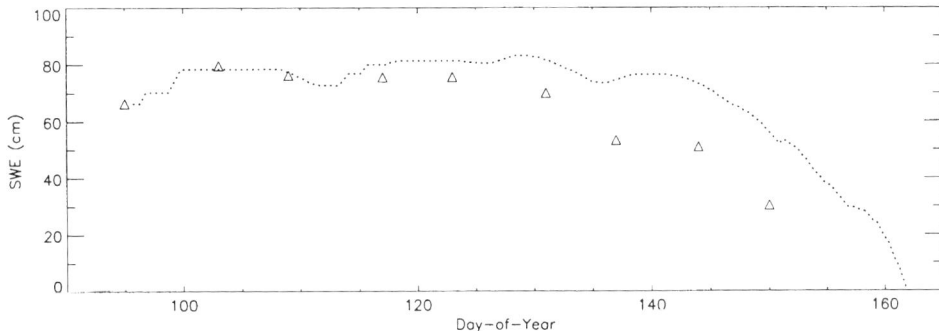

Figure 11.4 Predicted and observed snow water equivalent (*SWE*) between DOY 95 and 163, 1994, at Mammoth Mountain, CA, using the SSARR__grid snow model. The triangles show measurements and the dotted line shows the simulation

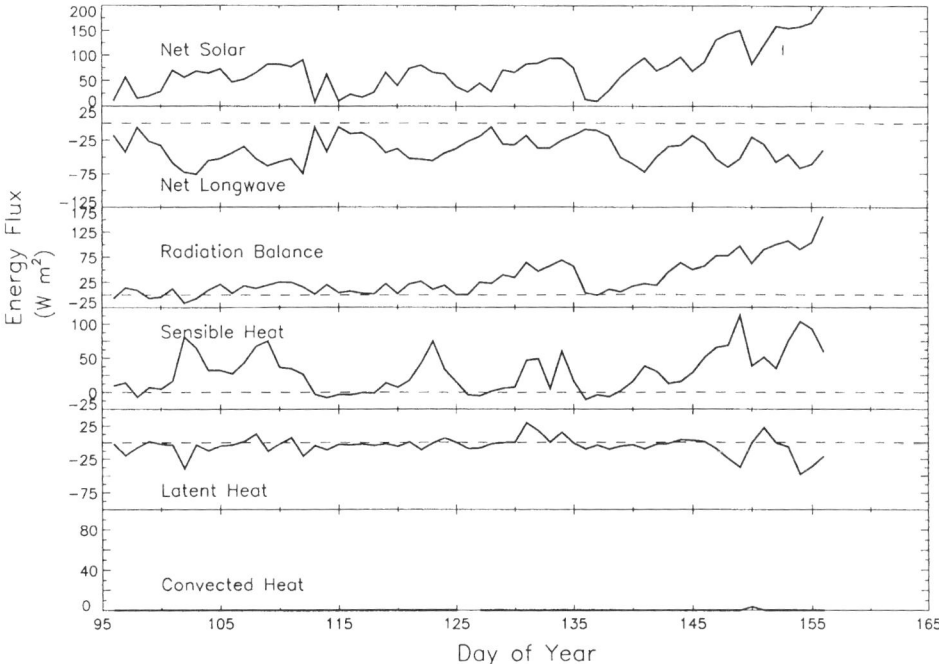

Figure 11.5 Fluxes of the surface energy exchange predicted by SNTHERM at Mammoth Mountain, CA, during the test period 1994. Convected heat refers to the heat transfer from rain entering the snow cover

direct measurement of the melt water arriving at the base of a snow cover. However, cold early season snow packs tend to consist of heterogeneous stratigraphy and microstructures compared with ripe snow cover and this character strongly affects the water flow patterns (Kattelmann 1989), which in turn can make it difficult to interpret observations from lysimeters. Stratigraphic inhomogeneities in permeability and capillary tension impede and horizontally divert the water flow and enhance the development of flow fingers (Marsh and Woo 1984a,b; Kattelmann and Dozier 1999). Figure 11.6 shows the snow pit wall cut into an early season snow pack after dye, originally applied to the surface, had found its way towards the base of the snow cover. Deeper and sloped snow packs can result in substantial flow of melt water in lateral directions. In cold snow packs or during overnight cooling, retained water often freezes to form series of ice crusts or ice layers. Thus, there is potential both for slowing of the melt water wavefront due to retention and freezing and of rapid penetration of the snow pack in flow channels or fingers. These processes can fully control water flow in the early melt season of alpine snow covers, where conventional understanding of water retention and transmission can prove of little use (Kattelmann and Dozier 1999). Snow behaves as a filter for rain-on-snow events, delaying influx of water to the ground surface layer, and as a source of water during the melt season. Because snow ripens to a relatively coarse material, water generally passes more rapidly through late-season snow than through soil provided that the snow has reached a thermal state to

Figure 11.6   Photograph of dye flow (grey). Photo: R. Kattelmann

prevent freezing of transmitted water. In older, well-metamorphosed snow (typical of the melt season) water infiltrates the pack with speeds of the order of 50 to 70 cm/h and thus, for snow packs shallower than about 5 m, the diurnal melt wave will exit the bottom of the pack within the same day, assuming a ripe snow pack.

Figure 11.7 shows how one can validate snow model predictions of pack outflow by comparing predicted and observed outflow. This figure shows the results from MMSA and SRRW. The water flow algorithm in SNTHERM assumes gravitational flow and a uniform wetting front, and does not address known complexities of water flow, such as fingering and retardation of the wetting front by capillary barriers (Illangasekare *et al.* 1990). At MMSA the observations represent the adjusted mean of three lysimeters, and adjustments accounted for the mass balance observed in nearby snow pits. At SRRW the observations came from only one lysimeter. In both simulations, SNTHERM predicted early outflow events not observed by the lysimeter systems. As described above, heterogeneous flow at the beginning of the ablation season causes melt water to follow routes that do not lead directly to the base of the pack. Thus, the lysimeters can exhibit large variance in flow, with some showing effective collection areas of several square metres and others showing no flow (Kattelmann 1989; Sommerfeld et al. 1994; Harrington and Bales 1998). The coefficient of variation between the three MMSA lysimeters decreased from around 1.50 at the start of outflow to around 0.40 towards the end of the melt. Final snow cover loss at MMSA gauged by the three lysimeters ranged over a four-day period, DOY 156–159.

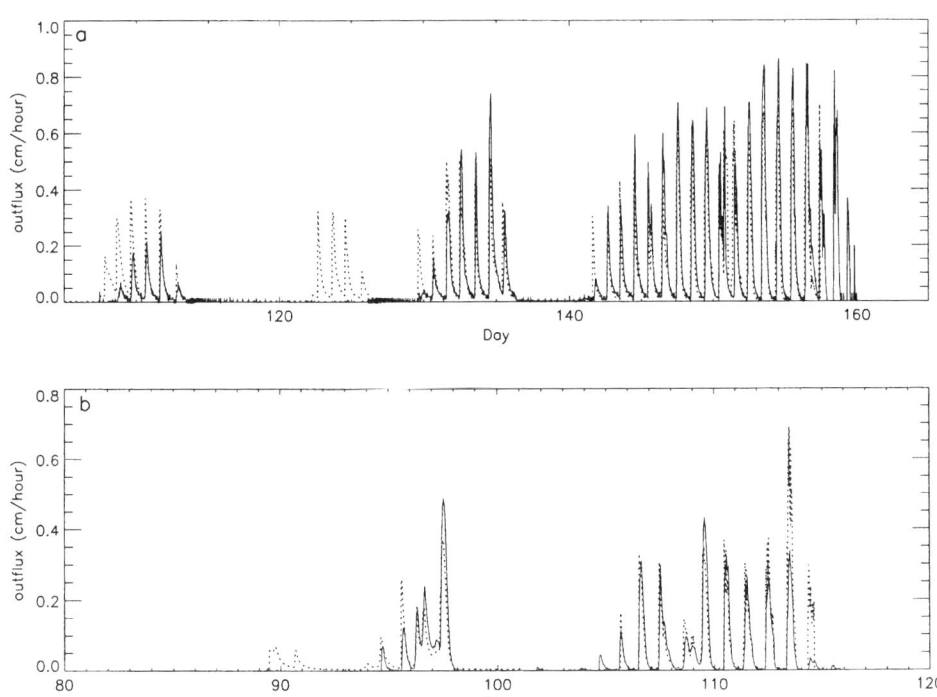

Figure 11.7 Observed outflow from snow cover and predictions from SNTHERM: (a) between DOY 95 and 163, 1994, at Mammoth Mountain, CA and (b) between DOY 84 and 118, 1997, at Sleepers River Research Watershed, VT. The solid line shows the measurements from lysimeters, while the dotted line shows the simulation results

Because SNTHERM predicted more outflow in the earlier part of the season, the final outflow occurred on DOY 157, two days ahead of the longest flowing lysimeter.

Figure 11.8 shows the comparison between outflow predicted at MMSA by SSARR_grid and the observations from the lysimeters. The daily modelled outflow did not reach as low levels or as high as observed or modelled by SNTHERM. On a 24-hour basis the daily volumes compared more favourably than the peaks indicate. On the other hand, SSARR_grid did not predict much flow during DOY 124–126, which SNTHERM did (Figure 11.3a). Overall, and consistent with the *SWE* trends shown in Figure 11.4, SSARR_grid underpredicted outflow during the periods of significant melt, with one exception, until the last days of melt when it overpredicts. The exception occurred over DOY 150–152 during a rain-on-snow event at MMSA. During this event SSARR_grid overpredicted melt/rain outflow in terms of total daily flow. Since SSARR_grid treats the snow conceptually as a bucket, with bulk parameters describing water retention and melting, the snow does not retain water proportional to the effects of different snow layers at their individual properties (e.g. Kattelmann 1986). Moreover, the model melts snow with a proportionality constant calibrated from sites that represented a range of situations, from forest to subalpine. The model cannot adapt easily to changes in the relative contributions of surface energy fluxes (Figure 11.5).

11.2.2   Validating Snow Models using Detailed Measurements of Snow Properties

As models become more complex in process detail, so does the opportunity to use more types of measurements for model testing. The more complex, process-detailed, energy balance approaches simulate surface temperature and albedo, as well as vertical profiles of the properties density, temperature, wetness and grain size (e.g. Brun et al. 1989; Jordan 1991; Morris et al. 1997; Jordan et al. 1999). These models present the most extensive opportunity for testing and validation. Moreover, these models best represent the actual processes in snow cover because they characterise the snow as a layered medium (Blöschl and Kirnbauer 1991; Colbeck 1991; Loth et al. 1998a,b). One can use measurements of snow stratigraphy, individual layer properties and/or measurements of albedo and surface optical grain size. Testing of these models with measurements provides a more clear way to differentiate between errors in model representation of snow processes and errors associated with calculating energy and mass fluxes at the upper and

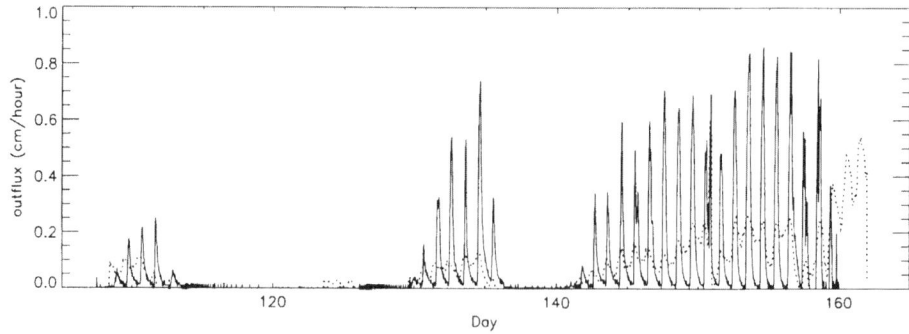

Figure 11.8   Observed outflow from snow cover and predictions from SSARR_grid between DOY 95 to 163, 1994, at Mammoth Mountain, CA. The solid line shows the measurements from lysimeters, while the dotted line shows the simulation results

lower boundaries of the snow cover. Moreover, they can and should provide reference standards for simpler approaches (Blöschl and Kirnbauer 1991; Koivusalo and Heikinheimo 1999; Yang et al. 1999).

Snow Density Profiles

Like snow water equivalent, one measures snow density by directly obtaining the mass of a known volume of snow. Snow density and metamorphic state control the thermal conductivity of the snow (Arons and Colbeck 1995, 1998; Sturm et al. 1997) as well as the heat capacity of individual layers. Thus prediction of the vertical profile of density in snow will affect the prediction of the temperature profile (Morris et al. 1997) and the flux of meltwater in terms of satisfying any thermal constraints to water transmission (Colbeck 1978). In Figure 11.9 we show an example comparing simulated and measured snow density profiles at the CRREL energy balance site. Small snow cutters with $100\,cm^3$ nominal volume and $33\,cm^3$ volume for thin layers provided the samples for weighing the mass. The first sustainable snow cover of the 1995–96 winter fell on DOY 335, built to a maximum of around 70 mm SWE by DOY 15 and then rapidly melted during a heavy rainfall over DOY 18–19 (see Figure 11.2). SNTHERM built the snow pack using observed precipitation and snow density. One can still observe differences in initial snow density from the DOY 335 stratigraphy of the measured and modelled snow cover in Figure 11.9a, over three weeks into the simulation on DOY 13, as shown by Figure 11.9b. However, the simulation underpredicted snow density in the middle of the snow pack. Figure 11.9b for DOY 13 shows a persistence and amplification of this trend. Note that the density profile for this older snow pack appears smoother than that in Figure 11.9a. The overdependence of thermal conductivity on snow density and the underprediction of snow compaction acted in a compensatory manner, so that temperatures in this model run agreed well with observations.

Snow Temperature Profiles

Researchers and field practitioners have directly measured snow temperatures for decades and with few biases, except near the surface of the snow cover (i.e. within about 10–15 cm) where

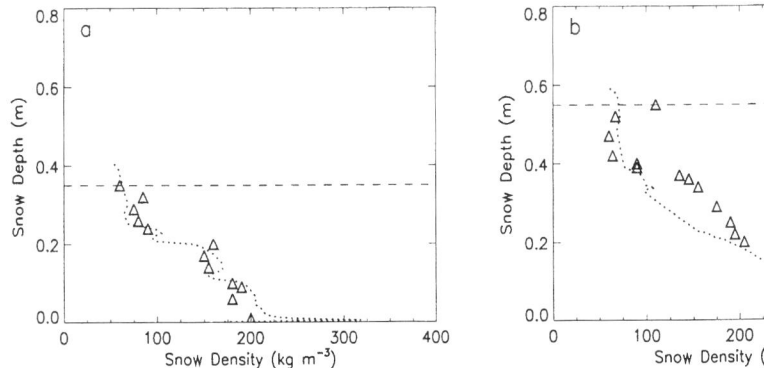

Figure 11.9   Predicted (dotted line) and measured (triangles) snow density profiles for (a) DOY 355, 1995, and (b) DOY 13, 1996, at CRREL, Hanover, NH

penetrating solar radiation can cause errors. Figure 11.10 (Jordan et al. 1989) compares simulated and measured temperatures profiles from the CRREL energy balance site. This DOY 37 example from a 1987 winter field test shows typical diurnal temperature variations in the snow cover for a clear day. In-pack thermocouples spaced at 5 cm intervals provided the profile of measurements, while the surface measurement used an adjustable probe, reset to the snow surface at least once daily. Though the two surface thermocouples had shielding from direct sunlight, measurements made in this manner generally showed temperatures 0.5 to 1.0°C higher than those computed from pyrgeometer observations.

The general trends in all of the panels in Figure 11.10 show that SNTHERM does a reasonable job of estimating temperature profiles and, by extension, the heat flow processes within a snow cover. However, higher measured temperatures well within the pack and the more pronounced curvature of the measured profile suggest that the simulation predicted thermal conductivity too high. Because thermal conductivity shows strong correlation with snow density (e.g. Sturm et al. 1997), such an error could result from either overprediction of snow compaction or from inaccuracies in the functional dependency on snow density. This example came from the second day of the simulation, so little snow compaction has occurred and the error most likely relates to the second type. In an intercomparison of algorithms, Glendinning (1998) corroborates that SNTHERM may overpredict thermal conductivity, particularly for medium density snow.

Snow Wetness and Melt Transmission

The precise timing of melt outflow from snow models concerns hydrologists, particularly in considering streams that show a pronounced diurnal cycle of runoff. Liquid water transmission through snow in the early part of the melting season follows irregular patterns of finger flow and background wetting (see Figure 11.6). At present, there exist no reasonable theoretical descriptions, developed into model parameterisations, of these complex flow patterns, and despite some recent progress (Illangasekare et al. 1990) an adequate modelling description of multipath flow in snow appears a long way off. Thus it is customary to model infiltration with a theory for uniform wetting fronts using Darcy's Law, such as a kinematic wave approximation (Colbeck 1972; Singh et al. 1997) or some other approach (e.g. Albert and Krajeski 1998). Because mature snow packs have relatively homogeneous structure, this approach can provide reasonable estimates of basal outflow to compare with lysimeter data (e.g. Tuteja and Cuname 1997; Albert and Krajeski 1998). Colbeck (1978) suggested that the volumetric water content of freely draining ripe snow rarely exceeds 10%. While retained or residual water amounts to around 5% during overnight drainage of a ripe snow pack, infiltration into dry snow can have a much lower immobile deficit (see Denoth et al., 1979; Kattelmann 1986; Dullien 1992). For example, Kattelmann and Dozier (1999) recently showed that retained water may amount to less than 4% in volumetric content during the early runoff season. SNTHERM predictions, however, consistently predicted early arrival of basal outflow in cold snow at MMSA and SRRW. This, in conjunction with underestimation of snow density suggests an immobile water content higher than 4%, at least in terms of how SNTHERM handles water flow. This poorly parameterised model variable thus remains a significant source of error for validating snow model predictions of early-season melting and rain-on-snow events. Immobile water retained within the interstices of the snow pack by capillary tension only becomes available for outflow when the snow melts. Although various methods can measure snow wetness (Denoth et al. 1979; Davis et al. 1985), due to flow heterogeneity one cannot readily compare snow wetness measurements in the pack with predictions for snow model validation. The exceptions include right at the surface of snow, with its relatively uniform wetness field, or profiles late in the melting season. Observations in snow

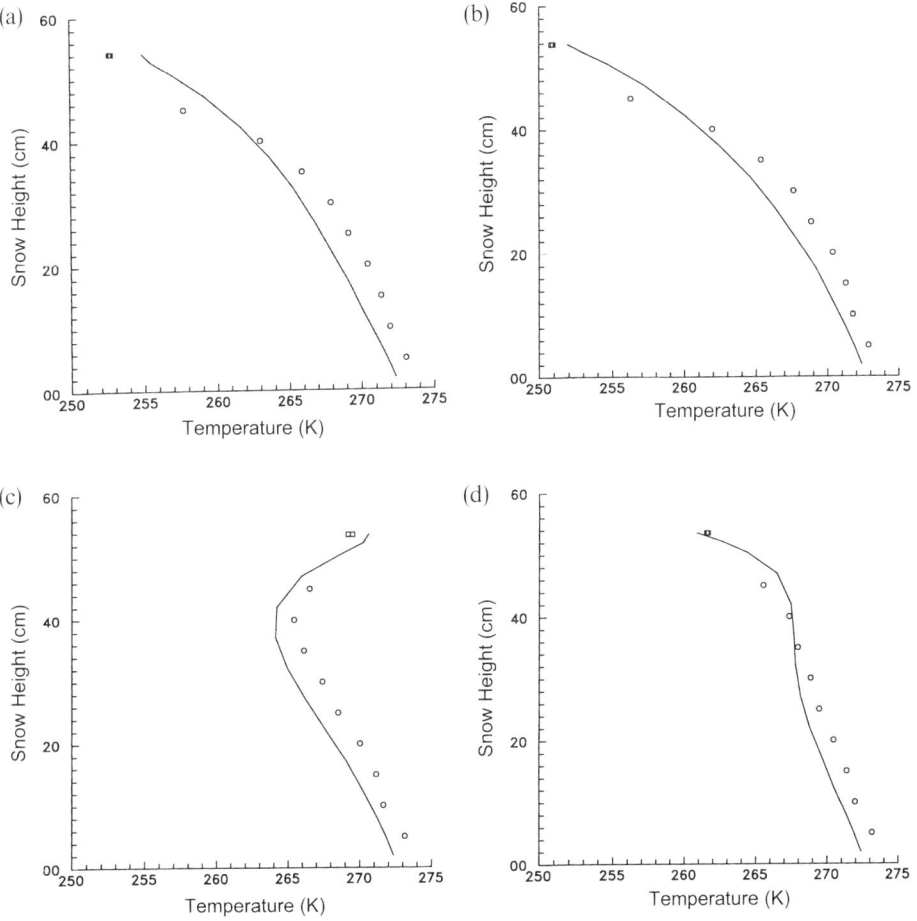

Figure 11.10 Predicted and measured snow temperature profiles for DOY 37, 1987, at CRREL, Hanover, NH: (a) 0000 hours, (b) 0600 hours, (c) 1200 hours and (d) 1800 hours. Circles show the measurements and the solid line shows the SNTHERM simulation

pits can measure densification from retained or refrozen melt water, which can provide important clues about water movement history.

Snow Grain Size and Spectral Reflectance

Snow models predict a characteristic mean grain size (Brun et al. 1989; Jordan 1991; Jordan et al. 1999). On the other hand, Colbeck et al. (1990) pragmatically define snow grain size as the largest dimension of a grain or particle. The terms grain and particle in turn have the definition: the smallest characteristic subunit of snow texture recognisable with a hand lens (Colbeck et al. 1990). These definitions leave much about measuring grain size unresolved in terms of how one gets these observations. Moreover, they create some unresolved issues concerning the theory of metamorphism of snow. Specifically, one finds that the maximum dimension of snow grains decreases with time as atmospheric crystal forms evolve towards snow cover forms (e.g. Bader et al. 1954; Colbeck 1983). Then mean particle sizes tend to increase, sometimes dramatically as in the case of the formation of surface hoar (Colbeck 1988) or grain clusters (Colbeck 1979). On the other hand, snow models predict that grain sizes increase monotonically from some small size, useful for predicting snow albedo and permeability, to larger forms. We have no complete theory that describes the evolution of new snow grains through the various crystal forms possible before melting. Therefore, in validating snow model predictions of grain size one should probably ignore Colbeck et al.'s (1990) definitions, and instead make careful statements about the textural variables predicted by the model and make even more careful measurements of the variables in the field.

Snow research does address measurements of snow texture in ways that can relate to snow model variables. Perla (1982) described methods to prepare section cuts on undisturbed samples of snow for the purpose of making stereological measurements of the ice matrix. Perla (1985) and Perla and Ommanney (1985) showed how measurements from sections could help characterise snow microstructure during metamorphism. Though tedious and time consuming, these techniques can yield several equivalent grain sizes, such as the size of the sphere with equal surface area-to-volume ratio as the bulk snow (Davis et al. 1987). Davis et al. (1993) used SNTHERM, with a grain size initialised using this particular equivalent size to predict the dynamics of the spectral reflectance of snow at MMSA during a period of rapid warming. The trends in grain size and spectral reflectance followed the measurements except during a phase of grain cluster formation as the snow began to melt. SNTHERM does not predict the size of grain clusters in snow, although cluster size probably represents a suitable scatterer size for albedo and permeability calculations. Nonetheless, this work pointed to the possibility of using spectral reflectance measurements for validating snow models. Glenndinning and Morris (1999) explicitly incorporated algorithms for spectral reflectance into SNTHERM and found reasonable agreements with reflectance measurements in Antarctica.

As an example of comparing model predictions of grain sizes with observations, we show in Figure 11.11 a comparison of snow grain sizes, predicted by SNTHERM and measured from section planes. The sphere of equal surface area-to-volume ratio, adjusted for snow density, initiated the SNTHERM model run and also provided the validating data based on section measurements (Shih et al. 1997). As the modelled and measured grain size equivalent increased, the scatter also increased (Figure 11.11). SNTHERM underestimated the grain radii of the larger sizes. Like the results of Davis et al. (1993) a period of melting and refreezing preceded the final day of the simulation and this accounted for most of the largest grain sizes. This shows the difficulty in predicting grain sizes when the snow exhibits heterogeneous processes, such as gain clustering, ice lens formation and fingering during early melting events.

In terms of working with individual snow grains, disaggregating the snow in a repeatable fashion and acquiring a representative set of observations represent current issues unresolved across the snow research community. Sturm and Benson (1997) revived interest in sieving the snow to produce a measure of grain size distribution. This method produces repeatable results, with the possible exception of wet snow, which must be refrozen before processing (Bergen 1975). On the other hand, this technique may not produce results from new snow valuable for modelling. We know of no complete snow models explicitly predicting the grain size distribution that could be compared to the results of sieving. CROCUS, the snow model of Brun et al. (1989), has two parameters regarding snow texture, the mean convex particle size and a dendricity

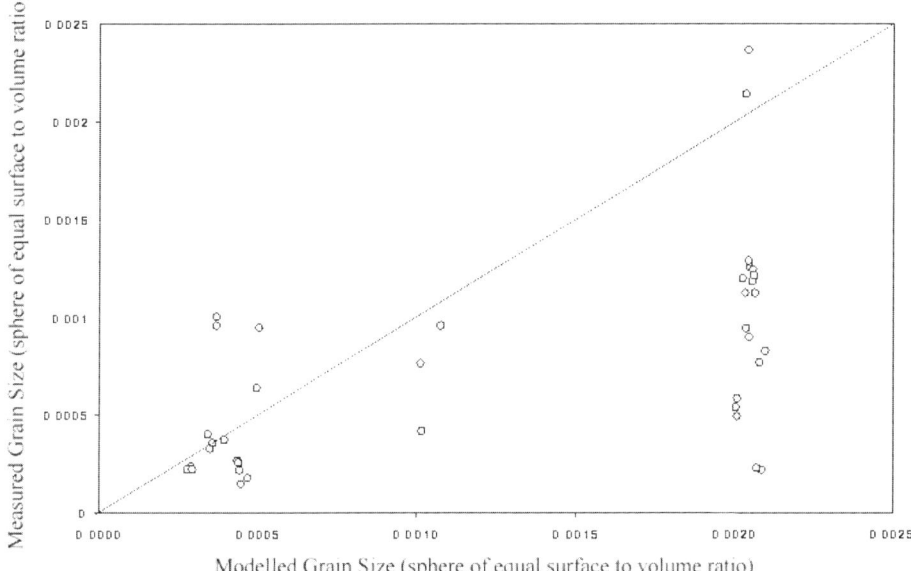

Figure 11.11 Comparison of measured and modelled grain size. Measurements from snow sections provided a grain size index from microstructure, the size equivalent to the sphere of equal surface area-to-volume ratio as the bulk snow of a layer. SNTHERM had difficulty predicting the larger sizes primarily due to inability to handle the complexities of liquid water flow

factor, which provides an index of the metamorphic state. Brun and Pahaut (1991) described a method to retrieve snow samples from the field for analysis of disaggregated grains. Lesaffre et al. (1998) extended this work by developing computer-aided techniques for estimating grain properties using image analysis and this snow retrieval method.

11.3 SPATIALLY DISTRIBUTED SNOW MODELS

With a few exceptions, spatially distributed snow models consist of multiple runs of one-dimensional snow models applied to many individual spatial domains. Within a spatial domain most distributed snow models assume plane-parallel, laterally isotropic conditions in the boundary conditions as well as in the snow. For validation purposes, one would like to make the assumption that spatial discretisation handles both local non-vertical processes and spatial heterogeneity. Modellers select the scale over which they explicitly treat spatial heterogeneity due to differences in land cover change, terrain and other factors controlling net energy exchange or snow accumulation. They can do this by segmenting the support data, land and terrain attributes, or by formulating parameterisations that describe spatial depletion patterns. The modeller must face the trade-off between the number (size) of spatial domains (i.e. the spatial resolution) and the complexity of the one- or two-dimensional snow model. One can investigate land cover and terrain heterogeneity to guide selection of suitable domains (Blöschl 1999). The scale affects the outcome (e.g. Brun et al. 1994; Kite 1995; Cline et al. 1998a), and inevitably the application will influence the degree to which the modelling approach must describe spatial variance in snow processes.

As long as the vertical and lateral distances over which the important snow processes operate remain much less than the dimension of the discrete spatial unit, modellers have tended to ignore subresolution processes and their effects. Exceptions to this trend include distributed models of heat advection over patchy snow and blowing snow (e.g. Liston et al. 1993; Liston and Sturm 1998), and water flow (Illangasekare et al. 1990). In the discussion below we review different approaches to spatially distributed snow models in the explicit sense; the results produce a map of snow properties. Exercises developing and validating spatially distributed models have used two types of spatial modelling elements, grid cells at some regular spacing and irregular domains that arise from specification of areas with relatively uniform hydrological, energy or mass flux response. This section reviews different approaches and identifies factors that have immediate impact of the amount and kind of effort required to perform model validation. The following section describes snow property mapping techniques that offer possibilities to compare with predicted maps of snow properties for model validation.

Traditionally, atmospheric modelling at the regional and global scales has used the regular grid approach (e.g. Essery 1997; Sellers et al. 1996; Yang et al. 1999). These models obtain feedback from the land surface modelling schemes through surface properties such as albedo, surface temperature and snow moisture, but tend to use simple treatments of snow processes (Slater et al. 2000). Regional and global scale modellers have begun to test and integrate more complex snow models that also fully couple with atmospheric processes. Loth et al. (1993) and Yang et al. (1997) have developed and tested snow models to use in assessing the optimal complexity of snow models for use in general circulation models (Loth and Graf 1998a,b; Yang et al. 1999). Few of the regional and global modelling efforts involving snow models have made a thorough effort in comparing predictions with snow maps. Modelling at these relatively coarse spatial scales has just begun to address the effects of subgrid heterogeneity, such as snow cover fraction, terrain and energy fluxes (e.g. Avissar 1992; Marshall and Oglesby 1994; Arola and Lettenmaier 1996; Walland and Simmonds 1996). This becomes important when one considers the disparity between the spatial resolution of the typical support data, the modelling elements and the validation data. Luce and co-workers used the model by Tarboton *et al.* (1995) to examine the effects of spatial distribution and aggregation of meteorological input (Luce *et al.* 1997) and assess basin-wide influences of snow distribution (Luce *et al.* 1998). This effort has conceptual similarity with the validating snow models with large cell size and internal heterogeneity. These efforts begin to provide frameworks for using ground-based observations as validating data in modelling exercised using large spatial domains.

Regional and local scale applications have also used regular discretisation of land cover and terrain data. At fine spatial resolutions (e.g. $\leq 1000\,\mathrm{m}$) most of these investigations have assumed explicit handling of heterogeneity in snow cover processes. Blöschl et al. (1991a) distributed a snow model over an alpine site in the Austrian Alps and used maps of snow cover extent at $25\,\mathrm{m}$ resolution to compare with predicted depletion patterns. They observed partial snow cover of model cells in the validation data, while the model predicted snow or no snow (binary) patterns. But overall the predicted patterns of snow extent compared well with the patterns derived from aerial photography. Extending this work, Blöschl et al. (1991b) compared the distributed model with simpler approaches including distribution by elevation band and a parametric approach. They found the distributed model more realistic than the other approaches, that is, the simpler models did not reproduce as reasonable spatial results with similar meteorological inputs. Turpin et al. (1999) also found similar results when using depletion curves to predict snow extent for comparison to maps. Wigmosta et al. (1994) distributed a coupled vegetation, snow and soil model at a spatial resolution of 1000 m over a catchment with heterogeneous land cover. Binary snow depletion patterns showed overall similarity with maps derived from satellite imagery (NOAA AVHRR), but uncertainty in the relationship between mapped binary snow cover and model snow depth may have thwarted more detailed comparison.

An alternative to using regular cell geometry for modelling domains comes from combining cells with similar land cover and terrain attributes, which can reduce snow cover variability within cells while potentially reducing the number of model runs. As with regular grid spacing, the spatial scale of response units in relation to heterogeneity in net energy flux, snow cover properties or hydrologic response control the degree to which subresolution processes affect the results over each domain. The model domains, patches or polygons, can also have similar hydrologic response (Leavesley and Striffler 1979). Leavesley and Stannard (1990) described categorical combination of land cover and terrain classes, which produced units that they assumed to have uniform hydrological response. Model simulations incorporated snow extent maps to update model progress in each response unit. Davis et al. (1995) used categorical combination of snow classes, slope, aspect and soil type in a distributed model simulation over a small area. The scale of the modelling domains had a similar characteristic dimension to the spatial resolution of the snow extent maps used for validation. Melloh et al. (1997) also used this approach to apply SNTHERM (Jordan 1991) to operational modelling. Snow extent maps and surface observations provided periodic updating of snow distribution to reduce sensitivity to error in initialisation data and input meteorology. Davis et al. (1997) combined categories of forest cover attributes, species, tree height and canopy density to spatially distribute an empirical snow model over a test area of the boreal forest. Link and Marks (1999) applied a physically based snow model over the same test area using domains that also combined forest attributes. Both modelling approaches compared well with site data, but the efforts did not compare model predictions against measured snow maps.

11.4 VALIDATING SPATIALLY DISTRIBUTED SNOW MODELS

Many tests of snow model predictions distributed over large areas have suffered from the difficulty in quantitative evaluation due to variability in snow extent patterns and snow physical properties. Many studies have compared estimated snow melt volume against runoff from a watershed. We view this as suspicious at best since calculating a reasonable water balance for medium to large basins basically constitutes a many-to-one mapping. Moreover, such comparisons contain no information about the validity of spatial model results. Some studies have made the effort to measure snow properties over time at a wide variety of sites in the modelling domain. While this type of validation has great importance to the modelling community, it becomes increasingly cost prohibitive when testing snow model predictions over larger and larger areas. Some combination of ground measurements and remote sensing products may produce the most robust validation exercises of spatially distributed snow models.

11.4.1 Snow Extent

Maps of snow cover area (SCA) reveal the presence or absence of snow. Government agencies and commercial enterprises map snow extent operationally with multispectral sensors, and many areas have sufficient cloud-free periods to compile time series of snow cover patterns. A time series of snow extent maps produces a record of the duration of snow cover on different areas of a study basin, quantised into the periodicity of the images used to make the maps. Various research efforts have made use of this concept to estimate the total SWE at different points in time, or over different areas of a watershed useful for model initiation, validation or update (e.g. Martinec and Rango 1981; Sommerfeld *et al.* 1991; Cline *et al.* 1998b). Many researchers have carried out tests comparing measured SCA with snow model results (e.g. Blöschl et al. 1991a,b; Wigmosta et al. 1994; Davis et al. 1995; Harrington et al. 1995; Hartman et al. 1999; Turpin et al. 1999).

Users of snow extent maps for validating and/or updating snow models must face two choices in validating predictions for individual model domains. While most modellers to date assume SCA goes to zero as model SWE goes to zero, one should explicitly specify a threshold snow water equivalence separating area considered as 'snow' from 'no snow' (Davis et al. 1995). Alternatively, one can develop a functional relation that expresses snow cover extent as a function of snow water equivalent (Ferguson 1986; Shook et al. 1993; Luce et al. 1999). Either approach will require different thresholds or different functions depending on land cover and terrain. Validating snow models with SCA also depends on the spatial resolution of the maps and model, and the spatial variance of snow accumulation and ablation. Confidence in maps of snow extent generally increases as the pixel size decreases relative to the model domain element. When both have fine spatial scale, then binary maps (snow or no snow) function suitably since the amount of snow water equivalent present at low snow may prove negligible, except for instances of large contrasts in subresolution ablation patterns. When the SCA maps have spatial scale finer than the model resolution then binary maps yield snow cover fraction in the model elements and the user must again decide between a thresholding approach and determining the depth–area function. Unfortunately, the relevant literature contains little information on the minimum detectability of snow using different remote sensors.

Figure 11.12 illustrates some of these concepts. The rows of SCA maps indicate the time domain, increasing downward, while the columns distinguish between three techniques used to map the snow. The four rows show SCA, mapped every other day over a test area in central Michigan during a snow melting period. The first column of Figure 11.12 (a–d) shows all detectable snow mapped in a binary fashion, indicated by white pixels. This case represents the sort of data recovered from aerial photographs and some binary algorithms used on digital image data. The National Operational Hydrologic Remote Sensing Center, run by NOAA in the US (Cline and Carroll 1999) produces this type of map. The second column in Figure 11.12 (e–h) also shows SCA represented by a binary pattern, but with a threshold of 50%: a pixel categorised as snow when more than about 50% of the pixel area has snow and no snow otherwise (black). We chose this example to represent the type of product proposed for the NASA multispectral sensor MODIS (Hall et al. 1995). In the third column (Figure 11.12 i–l) we have shown snow extent depicted as a grey level corresponding to the fractional coverage per pixel, the type of product produced by spectral mixture modelling (Nolin et al. 1993; Rosenthal and Dozier 1996; Painter et al. 1998). If one considers snow depletion as indicated by partial coverage in the smallest elements, then this representation contains the most information.

In Figure 11.13 we show comparison of predicted and measured SCA for the same test area. The SNTHERM simulations started with complete snow cover on the central Michigan test area about 10 days prior to the end date. Over the test area two categories of SWE initialised the model runs based on field surveys: 7 cm in the forested areas of the upper left and lower right corners of the maps and 4 cm SWE along the valley running from lower left to upper right. The observed SCA map in Figure 11.13 portrays all detected snow as white. The model-derived SCA map used a threshold of 0.4 cm SWE as the cutoff below which a pixel had 'no snow'. We based the selection of this threshold on previous experience with this area with snow maps derived

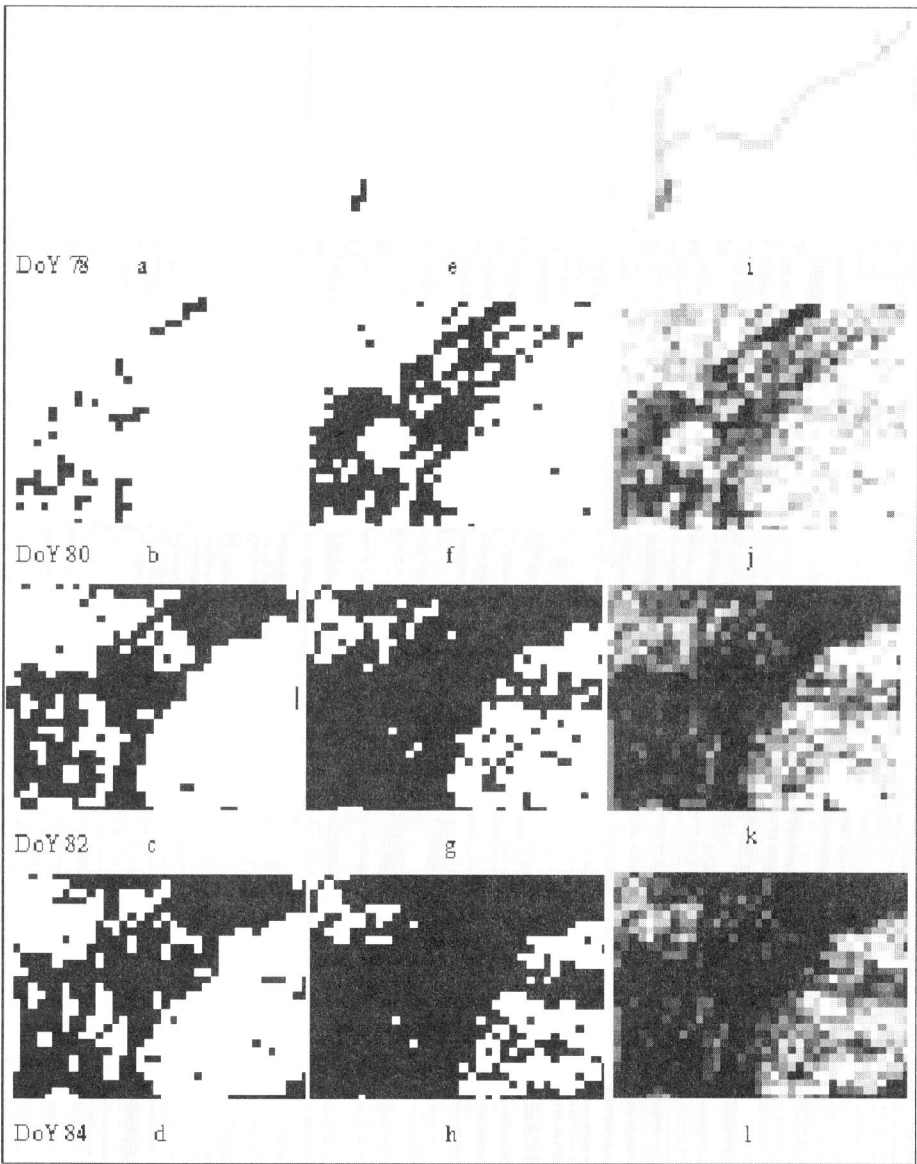

Figure 11.12 Time series snow maps of test area near Grayling, Michigan, winter 1994. Columns show snow extent represented with different methods. The first column (a–d) shows pixels with any detected snow (white) verses no snow (black). The second column (e–h) shows pixels with more than 50% snow extent (white). The third column shows snow extent represented as fractional cover (0–100%) as a grey scale between black and white

from aerial photography (Davis et al. 1995). This example shows that although the total *SCA* over the test area compared favourably, 63% measured versus 64% modelled, significant differences in the spatial distribution of residuals (grey) occurred mostly where snow had exposure to wind in the valley. Over a larger test area these residual differences could account for important hydrologic sources. Stehman (1999) describes other quantitative techniques for the comparison of spatial data sets.

11.4.2 Snow Water Equivalent

Researchers have pursued the recovery of the spatial distribution of *SWE* using airborne and spaceborne measurements since the 1970s. For example, operational products in Canada estimate the spatial distribution of *SWE* over large areas using passive microwave measurements with resolutions of about 35 km (Goodison et al. 1987). Forest cover, other mixed pixel

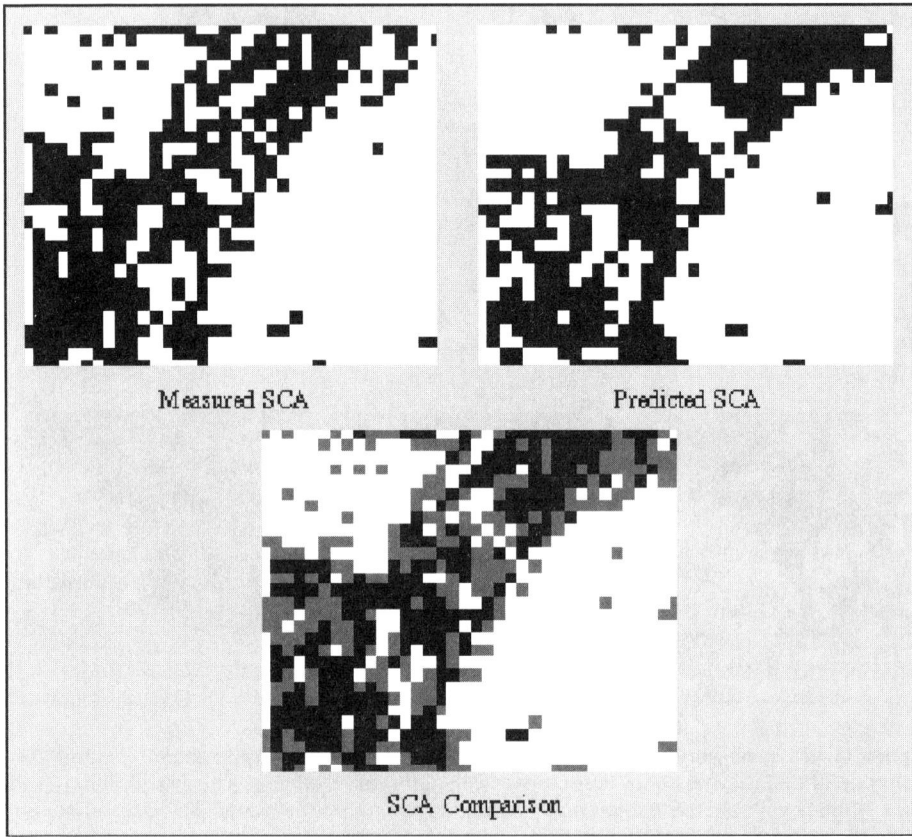

Figure 11.13 Measured, predicted and compared snow extent of test area near Grayling, Michigan, winter 1994 for DOY 81. Measured total *SCA* of the test area was 63% (upper left), while the predicted *SCA* was 64% (upper right). However, 22% of the test area had incorrect predictions (bottom – grey)

effects and the natural variation of snow properties affecting its apparent emissivity control the recovery of SWE from passive sensors (Chang *et al.* 1997). Thus we can state that this product does not appear generally suited for validating snow models for all but the largest most homogeneous areas using large modelling elements. Operational products in the US provide estimates of SWE along aircraft flight lines that exploit the attenuation of natural terrestrial gamma radiation by the snow water equivalent (Peck *et al.* 1980; Carroll 1987). These products provide the average spatial snow water equivalent along transects 250–300 m wide by several kilometres long. Again, using these measurements for validating snow models will prove a challenge to approaches with model domain sizes less than several kilometres. Recent investigations have demonstrated methods to estimate SWE using measurements from synthetic aperture radar (SAR) (Shi and Dozier 2000a). Shi and Dozier (2000b) also show how one can infer snow depth and particle size from SAR. However, the lack of wide variation of snow cover properties, land cover and terrain in testing these algorithms has prevented a thorough determination of the efficacy of different approaches and has prevented a reasonable determination of the conditions limiting recovery of SWE from radar. These represent experimental methods that need further testing. Thus we may not see an instance soon where researchers use passive microwave or radar estimates of SWE to quantitatively validate snow models at most spatial scales of hydrological interest. Ironically, we may soon see the use of high-resolution, process-detailed snow models to aid in interpolating ground measurements for validating remote sensing algorithms to recover SWE.

Other possibilities exist to validate spatially distributed snow models using maps of SWE. For example, developments to map snow water equivalent by interpolation of point and transect measurements have made much progress in recent years, in some cases using remote sensing to provide information in data-sparse areas. The following discussion reviews recent methods to model the spatial distribution of SWE more or less independently from the predictions of spatial snow modelling as described above. However, anyone who validates snow models using SWE maps derived from the techniques reviewed below faces the additional task of demonstrating their accuracy. In terms of using SWE maps to validate models we can consider this as comparing models with other models, which does not fit the criteria one normally assumes for validation. Nonetheless we review these additional works that use innovative or hybrid techniques to estimate the spatial distributions of SWE.

Using an empirical approach, Burkard *et al.* (1991) reported on efforts to assign weights to potential snow accumulation in different land cover units that made use of relational models developed with extensive measurements. Schroeter *et al.* (1991) and Schroeter and Whiteley, (1992) applied these methods to assessing whether accurate distributions of snow cover could improve operational runoff forecasts. Their snow model ran on domains consisting of the landscape units defining the weights for potential water equivalent. Elder *et al.* (1991, 1998) used binary regression trees with extensive, detailed snow surveys to interpolate snow water equivalent over alpine terrain at high spatial resolutions where the independent variables, elevation, slope, aspect etc. had a broad dynamic range. McManamon *et al.* (1993) developed an operational method for mountainous areas based on measurement statistics at snow sites and melt-factor class from a conceptual snow model. Snow water equivalent estimates were distributed to areas determined from categorical combination of land and melt classes. Lapen and Martz (1996) reported on the difficulties of attributing snow depth to terrain variables in a prairie setting, where snow accumulation is complicated by wind redistribution, which depends on a variety of weather variables. Ling *et al.* (1995) showed some utility of a method of kriging to distribute snow water equivalent between measurements at much larger scales in the Upper Colorado River Basin for mesoscale analysis. Carroll *et al.* (1995) used this type of approach to investigate improvement of operational products of the National Weather Service (Carroll 1995;

Carroll and Cressie 1996, 1997). Binary snow cover maps provided information on proximity of areas classified as having no snow, but within snow-covered areas a geostatistical model relying on standardised observational data distributed SWE.

Using physically based models, Liston and Sturm (1998) and Pomeroy *et al.* (1997) modelled the distribution of snow in arctic settings using land cover classes. The models perform mass budgets as snow blows from source areas to sink areas based on energy balance principles. Both of these efforts showed accumulation patterns that reconciled with binary snow maps. Also using a physically based approach, Cline *et al.* (1998a) extended previous attempts to model the spatial distribution of snow water equivalent through the use of snow maps (Martinec and Rango 1981; Sommerfeld *et al.* 1991). These studies used snow maps to regulate the amount of net energy accumulated in a model domain. The time series of snow extent patterns provided estimates of snow duration, which when combined with the available energy input converted to SWE. Cline *et al.* (1998a) used physically based calculations of net radiation and turbulent fluxes, rather than the degree-day approaches used by others. With net energy weighted by the observed fractional snow extent per pixel, the predicted distribution of snow water equivalent showed remarkable agreement with detailed snow surveys (Cline *et al.* 1998a).

11.4.3   Snow Surface Temperature

Energy balance approaches to modelling snow solve for the surface temperature. At first there may seem great potential in using remotely sensed land surface temperature for validating spatially distributed snow models. However, there remain two basic challenges to this concept. First, one cannot currently make the statement that the retrieval of snow surface temperatures from spaceborne sensors has matured as a technology. Algorithms require some empirical calibration (Qin and Karnieli 1999) and terrain corrections (Lipton and Ward 1997). This severely limits the confidence in surface temperature fields one would use for model validation. Second, retrieved surface temperatures have a characteristic variation due to high-frequency temporal forcing by meteorological variables, while model output represents relatively filtered results, considering the time steps used in modelling exercises and the smoothing of spatial discretisation. Nonetheless, with new sensors and products representing surface temperature we may find modellers more willing to develop techniques for comparing measured and modelled surface temperature fields (Silberstein et al. 1999).

11.4.4   Snow Surface Wetness

Various methods can measure the presence of liquid water in the surface layer of the snow pack at the plot scale and recent research has shown this possible with microwave remote sensors as well. While observations of snow using passive microwave sensing has long shown difficulty in measuring the SCA or SWE of wet snow, time series of observations combined with air temperature measurements can provide detection of wet snow (Walker and Goodison 1993). Koskinen et al. (1997) have shown that with time series of single-frequency, single-polarisation SAR, one can also detect the onset of snow wetness. With polarimetric SAR, Shi and Dozier (1995a,b) showed that snow wetness recovered from imagery agreed to within a few per cent of ground observations carried out at the time of the overpass. In these cases, the SAR measurements had spatial resolutions reasonable for validating distributed snow models using various approaches.

11.4.5   Snow Surface Texture

In the previous discussion on validating one-dimensional snow models with measurements of grain size, we cited examples where researchers developed links between model-predicted grain size and spectral reflectance of snow. Multispectral and hyperspectral imagery offers an important potential for validating process-detailed snow models, through spectral reflectance and/or optical grain size. Nolin and Dozier (1993) showed how to retrieve optical grain size from spectral imagery and later Painter et al. (1998) showed how consideration of snow grain size affected estimation of the fractional coverage of snow, on a pixel basis, and suggested methods for determined both fractional SCA and optical grain size from one algorithm. Fily et al. (1999) used the snow model CROCUS to estimate grain sizes over the Alps in a semi-distributed approach and compared the results with a grain size index from the Landsat Thematic Mapper, with reasonable agreement. This type of validation must still demonstrate, through plot studies, that the snow model can predict grain sizes across the full range found in seasonal snow packs, and that one can determine how profiles of grain sizes combine to contribute to the average optical grain size integrated vertically in the snow surface layers and over the size of the image pixel. Further, we do not know the degree to which the bi-directional reflectance distribution function of snow plays a role in the retrieved spectral reflectance.

11.5   SUMMARY AND CONCLUSIONS

Most hydrological applications of snow models require spatial distribution. Spatially distributed modelling of snow has begun to reach a mature state, at least in terms of showing its feasibility in (1) providing accurate lower boundary conditions for atmospheric models, (2) predicting spatial relationships of some of the details of snow processes and (3) showing useful input to runoff routing models for hydrological forecasts. One-dimensional models of snow processes explain much of the detail of snow dynamics at the local scale. One can use a variety of measurements to validate snow models at the plot scale to unambiguously define the performance. This chapter has reviewed various snow measurements and issues with their use. Model detail in many ways limits the type of measurements used for validation. However, some measurements, such as grain size and wetness, seem easy to make but do not currently provide useful data for validating snow models except under special circumstances. Spatially distributed snow models can present huge challenges regarding their validation. Researcher must take more care than in the past in validating distributed snow models. Spatial models require spatial measurements for validation, or at least independent spatial models of snow properties. Currently the single most useful and cost effective type of validation data consists of time series of snow extent. This chapter has summarised the problems with using snow extent as they relate to the type of algorithm and product. Remote sensing of snow shows some promise to provide maps of other types of snow properties, but the uncertainty associated with these products currently exceeds by far the uncertainty involved in the spatial distribution of a reasonably detailed snow model. Relational or hybrid techniques that distribute snow water equivalence across a test area based on snow extent and point measurements currently appear to offer the best means for using this variable in validating distributed snow models.

ACKNOWLEDGEMENTS

Many thanks to anonymous reviewers who helped improve the manuscript. Helpful discussions with Don Cline, Kelly Elder, Janet Hardy, Bob Harrington, J.C. McKenzie, Rae Melloh and Curtis Woodcock led up to this review. The US Army Project 4A762784AT42 and NASA Grant NRA-92-OSSA-1 helped fund the analytical work and measurements at SRRW and CRREL. We thank the ERDC–CRREL technical and research staff for providing measurements at SRRW and CRREL The US Army Engineer Sacramento District, Water Management Section (CESPK-ED-D) and the Corps Civil Works Remote Sensing Research Program, Work Unit CWIS 32039, also provided resources supporting preparation of this chapter. We also gratefully acknowledge Mammoth Mountain Ski Area and the Valentine Eastern Sierra Reserve for their support in acquiring field data. The NASA Southwest Earth Science Applications Center at the University of Arizona supported continued research to develop methods to map snow water equivalent in montane environments and the development of methods to use mesoscale climate analyses to drive the snow models.

REFERENCES

Albert, M.R. and Krajeski, G. 1998. A fast, physically-based point snow melt model for use in distributed applications. *Hydrological Processes*, **12**(10–11), 1809–1821.

Anderson, E.A. 1968. Development and testing of snowpack energy balance equations. *Water Resources Research*, **4**(1), 19–37.

Anderson, E.A. 1976. *A Point Energy and Mass Balance Model of a Snow Cover*. NOAA Technical Report NWS 19, Office of Hydrology, National Weather Service, Silver Spring, MD.

Anderson, E. and Larson, L. 1996. The role of snowmelt in the January 1996 floods in the Northeastern United States. *Proceedings, 53rd Eastern Snow Conference*, Williamsburg, VA, 141–149.

Andreas, E.L. 1987. A theory for the scalar roughness and the scalar transfer coefficients over snow and sea ice. *Boundary-Layer Meteorology*, **38**, 159–184.

Arola, A. and Lettenmaier, D. 1996. Effects of subgrid spatial heterogeneity on GCM-scale land surface energy and moisture fluxes. *Journal of Climate*, **9**, 1339–1349.

Arons, E.M. and Colbeck, S.C. 1995. Geometry of heat and mass transfer in dry snow: a review of theory and experiment. *Reviews of Geophysics*, **33**(4), 463–493.

Arons, E.M. and Colbeck, S.C. 1998. Effective medium approximation for the conductivity of sensible heat 3in dry snow. *International Journal of Heat and Mass Transfer*, **41**(17), 2653–2666.

Avissar, R. 1992. Conceptual aspects of a statistical–dynamical approach to represent landscape subgrid-scale heterogeneities in atmospheric models. *Journal of Geophysical Research*, **97**, 2729–2742.

Bader, H., Haefeli, R., Bucher, E., Neher, J., Eckel, O. and Thams, C. 1954. Snow and its metamorphism. US Army Cold Regions Research and Engineering Laboratory (SIPRE), AD-030 965. Translation of 'Der Schnee und seine Metamorphose'. *Geologie der Schweiz. Geotechnische Serie, Hydrologie* **3**, 340. 3 p.

Bales, R.C. and Harrington, R.F. 1995. Recent progress in snow hydrology. *Reviews of Geophysics*, **33**, 1011–1020.

Bergen, J.D. 1975. Possible relation of albedo to the density and grain size of natural snow cover. *Water Resources Research*, **11**(5), 745–746.

Beljaars, A.C.M. and Holtslag, A.A.M. 1991. Flux parameterisation over land surfaces for atmospheric models. *Journal of Applied Meteorology*, **30**, 327–341.

Bergström, S. 1996. Principles and confidence in hydrological modelling. *Nordic Hydrology*, **22**, 123–136.

Blöschl, G. 1999. Scaling issues in snow hydrology. In: Hardy, J.P., Albert, M.R. and Marsh P. (eds), *Advances in Hydrological Processes: Snow Hydrology: The Integration of Physical, Chemical and Biological Systems*. Wiley Interscience, Chichester, 2149–2178.

Blöschl, G. and Kirnbauer, R. 1991. Point snowmelt models with different degrees of complexity – internal processes. *Journal of Hydrology*, **129**, 127–147.

Blöschl, G., Kirnbauer, R. and Gutknecht, D. 1991a. Distributed snowmelt simulations in an alpine catchment 1. Model evaluation on the basis of snow cover patterns. *Water Resources Research*, **27**(12), 3181–3188.

Blöschl, G., Kirnbauer, R. and Gutknecht, D. 1991b. Distributed snowmelt simulations in an alpine catchment 2. Parameter study and model predictions. *Water Resources Research*, **27**(12), 3171–3179.

Braun, L.N., Brun, E., Durnad, Y., Martin, E. and Tourase, P. 1994. Simulation of discharge using different methods of meteorological data distribution, basin discretization and snow modelling. *Nordic*

Hydrology, **25**(1–2), 129–144.

Brun, E. and Pahaut, E. 1991. An efficient method for a delayed and accurate characterization of snow grains from natural snowpacks. *Journal of Glaciology*, **37**(1127), 420–422.

Brun, E., Martin, E., Simon, V., Gendre, C. and Coleou, C. 1989. Energy and mass model of snow cover suitable for operational avalanche forecasting. *Journal of Glaciology*, **35**(121), 333–342.

Brun, E., Durand, Y., Martin, E. and Braun, L. 1994. Snow modelling as an efficient tool to simulate snow cover evolution at different spatial scales. In: Jones, H.G., Davies, T.D., Ohmura, A. and Morris, E.M. (eds), *Snow and Ice Covers: Interactions with the Atmosphere and Ecosystems*. IAHS Publication, 163–174.

Burkard, M.B., Whiteley, H.R., Schroeter, H.O. and Donald, J.R. 1991. Snow depth/area relationships for various landscape units in southwestern Ontario. *Proceedings, 48th Eastern Snow Conference*, 51–65.

Carroll, S.S. 1995. Modelling measurement errors when estimating snow water equivalent. *Journal of Hydrology*, **172**, 247–260.

Carroll, S.S. and Cressie, N. 1996. A comparison of geostatistical methodologies used to estimate snow water equivalent. *Water Resources Bulletin*, **32**(2), 267–278.

Carroll, S.S. and Cressie, N. 1997 Spatial modelling of snow water equivalent using covariances estimated from spatial geomorphic attributes *Journal of Hydrology*, **190**, 42–59.

Carroll, S.S., Day, G.N., Cressie, N. and Carroll, T.R. 1995. Spatial modelling of snow water equivalent using airborne and ground-based snow data. *Environmetrics*, **6**, 127–139.

Carroll, T. 1987. Operational airborne measurements of snow water equivalent and soil moisture using terrestrial gamma radiation in the United States. In: Goodison, R.G., Barry, R.G. and Dozier, J. (eds), *Large Scale Effects of Seasonal Snow Cover*. IAHS Publication No. 166, 213–223.

Chang, A.T.C., Foster, J.L., Hall, D.K., Goodison, B.E., Walker, A.E. and Metcalfe, J.R. 1997. Snow parameters derived from microwave measurements during the BOREAS winter field campaign. *Journal of Geophysical Research*, **102**(D24), 29 429–29 444.

Church, J.E. 1935. Principles of snow surveying as applied to forecasting stream flow. *Journal of Agricultural Research*, **51**, 97–130.

Church, J.E. and Marr, J.C. 1937. Further improvements of snow-survey apparatus. *American Geophysical Union Transactions*, **18**, 607–617.

Cline, D.W. and Carroll, T.R. 1999. Influence of snow cover beneath obscuring clouds using optical remote sensing and a distributed snow energy and mass balance model. *Journal of Geophysical Research* **104**(D16), 19631–19644.

Cline, D.W., Elder, K. and Bales, R.C. 1998a. Scale effects in a distributed SWE and snowmelt model for mountain basins. *Hydrological Processes*, **12**, 1527–1536.

Cline, D.W., Bales, R.C. and Dozier, J. 1998b. Estimating the spatial distribution of snow in mountain basins using remote sensing and energy balance modelling. *Water Resources Research*, **34**(5), 1275–1285.

Colbeck, S.C. 1972. A theory of water percolation in snow. *Journal of Glaciology*, **11**(64), 369–385.

Colbeck, S.C. 1978. The physical aspects of water flow through snow. In: Chow, V.T. (ed.), *Advances in Hydroscience*, vol. 11. Academic Press, New York, 165–206.

Colbeck, S.C. 1979. Grain clusters in wet snow. *Journal of Colloid and Interface Science*, **72**(3), 371–384.

Colbeck, S.C. 1983. Theory of metamorphism of dry snow. *Journal of Geophysical Research*, **88**(C9), 5475–5482.

Colbeck, S.C. 1988. Ideas on the micrometeorology of surface hoar growth on snow. *Boundary-Layer Meteorology*, **44**, 1–12.

Colbeck, S.C. 1991. Layered character of snow covers. *Reviews of Geophysics*, **29**(1), 81–96.

Colbeck, S.C., Akitaya, E., Armstrong, R., Gubler, H., Lafeulle, J., Lied, K., McClung, D. and Morris, E. 1990. *The International Classification for Seasonal Snow on the Ground*. The International Commission on Snow and Ice of the International Association of Scientific Hydrology and International Glaciology Society.

Daly, S.F., Ochs, E.S., Brooks, P.F., Pangburn, T. and Davis, E.M. 1999. Distributed snow process model for use with HEC-HMS. *Proceedings, ASCE Conference on Cold Regions Engineering*, 16–19 August 1999. American Society of Civil Engineers, Lebanon, NH, 538–549.

Daly, S.F., Davis, R.E., Ochs, E. and Pangburn, T. 2000. An approach to spatially distributed snow modelling of the Sacramento and San Joaquin basins, *California Hydrological Processes*, **14**, 3257–3271.

Davis, R.E., Dozier, J., Perla, R. and LaChapelle, E.R. 1985. Field and laboratory measurements of snow liquid water by dilution. *Water Resources Research*, **21**(9), 1415–1420.

Davis, R. E., Dozier, J. and Perla, R. 1987. Measurement of snow grain properties. In: Orville-Thomas, W.J. and Jones, H.G. (eds), *Seasonal Snowcover: Physics, Chemistry, Hydrology*. NATO ASI Series C: Mathematical and Physical Sciences Vol. 211. Reidel, Dordrecht, 53–74.

Davis, R.E., Nolin, A.W., Jordan, R. and Dozier, J. 1993. Towards predicting temporal changes of the spectral signature of snow in visible and near-infrared wavelengths. *Annals of Glaciology*, **17**, 143–148.

Davis, R.E., McKenzie, J.C. and Jordan, R. 1995. Distributed snow process modelling: an image segmentation approach. *Hydrological Processes*, **9**, 865–875.

Davis, R.E., Woodcock, C.E., Hardy, J.P., Ni, W.G., Jordan, R. and McKenzie, J.C. 1997. Spatially-distributed modelling of snow in the boreal forest: a simple approach. *Eastern Snow Conference and Western Snow Conference, Proceedings, Joint 54th and 65th*, 20–28.

Denoth, A., Seidenbush, W., Blumthaler, M. and Kirchlechner, P. 1979. Study of water drainage from columns of snow. *CRREL Special Report 79-1*. US Army Cold Regions Research and Engineering Laboratory.

Douville, H. 1997. Local and global stand-alone tests of the Météo-France snow parameterisation. *Annals of Glaciology*, **25**, 165–169.

Dozier, J. 1987. Recent research in snow hydrology. *Reviews of Geophysics*, **25**(2), 153–161.

Dozier, J. 1992. Opportunities to improve hydrologic data. *Reviews of Geophysics*, **30**(4), 315.

Dullien, F.A.L. 1992. *Porous Media. Fluid Transport and Pore Structure*, 2nd edition. Academic Press, San Diego.

Elder, K., Dozier, J. and Michaelsen, J. 1991. Snow accumulation and distribution in an alpine watershed. *Water Resources Research*, **27**(7), 1541–1552.

Elder, K., Rosenthal, W. and Davis, R.E. 1998. Estimating the spatial distribution of snow water equivalence in a montane watershed. *Hydrological Processes*, **12**(10–11), 1793–1808.

Essery, R. 1997. Seasonal snow cover and climate change in the Hadley Centre GCM. *Annals of Glaciology*, **25**, 362–366.

Fily, M., Dedieu, J.P. and Durand, Y. 1999. Comparison between results of a snow metamorphism model and remote sensing derived snow parameters in the Alps. *Remote Sensing of Environment*, **68**, 254–263.

Ferguson, R. 1984. Magnitude and modelling of snowmelt runoff in the Cairngorm mountains Scotland. *Hydrologic Sciences Journal*, **29**, 49–62.

Ferguson, R. 1986. Parametric modelling of daily and seasonal snowmelt using snow pack water equivalent as well as snow covered area. Proceedings 2nd Scientific Assembly IAHS, Budapest. *IAHS Publication No. 155*, 151–161.

Glendinning, J.H.G. 1998. The modelling of radiative transfer in snow at visible and near infrared wavelengths. PhD Thesis, University of Reading.

Glendinning, J.H.G. and Morris, E.M. 1999. Incorporation of spectral and directional radiative transfer in a snow model. *Hydrological Processes*, **13**(12/13), 1761–1772.

Goodison, B.E. 1981. Compatibility of Canadian snowfall and snow cover data. *Water Resources Research*, **17**(4), 893–900.

Goodison, B.E., Glunn, J.E., Harvey, K.D. and Slater, J.E. 1987. Snow surveying in Canada: a perspective. *Canadian Water Resources Journal*, **12**(2), 27–42.

Hall, D.K., Riggs, G.A. and Salomonson, V.V. 1995. Development of methods for mapping global snow cover using Moderate Resolution Imaging Spectroradiometer data. *Remote Sensing of Environment*, **54**(2), 127–140.

Harrington, R. and Bales, R.C. 1998. Interannual, seasonal, and spatial patterns of meltwater and solute fluxes in a seasonal snow pack. *Water Resources Research*, **34**(4), 823–831.

Harrington, R.F., Elder, K. and Bales, R.C. 1995. Distributed snowmelt modelling using a clustering algorithm. In: Tonnessen et al. (eds), *International Symposium on Biogeochemistry of Seasonally Snow-covered Catchments*, Boulder, CO, 1–14 July 1995. IAHS Publication No. 228, 167–174.

Hartman, M.D., Baron, J.S., Lammers, R.B., Cline, D.W., Band, L.E., Liston, G.E. and Tague, C. 1999. Simulations of snow distribution and hydrology in a mountain basin. *Water Resources Research*, **35**(5), 1587–1603.

Hogan, A.W. 1996. Inferring dynamic winter variables. *Proceedings, 53rd Eastern Snow Conference*,

Williamsburg, VA, 205–212.

Illangasekare, T.H., Walter, R.J. Jr, Meier, M.F. and Pfeffer, W.T. 1990. Modelling of meltwater infiltration in subfreezing snow. *Water Resources Research*, **26**(5), 1001–1012.

Jordan, R. 1991. One-Dimensional Temperature Model for a Snow Cover: Technical Documentation for SNTHERM 89, *CRREL Special Report*. SR91–16 *US Army Cold Regions Research and Engineering Laboratory*.

Jordan, R.E., Andreas, E.L. and Makshtas, A.P. 1999. Heat budget of snow-covered sea ice at North Pole 4. *Journal of Geophysical Research*, **104**(C4), 7785–7806.

Jordan, R., O'Brien, H. and Albert, M.R. 1999. Snow as a thermal background: preliminary results from the 1987 field test. *US Army Cold Regions Research and Engineering Laboratory* Report SR 89–07.

Kattelmann, R.C. 1986. Measurements of snow layer water retention. In: Kane, D.L. (ed.), *Proceedings, Symposium: Cold Regions Hydrology*, Fairbanks, Alaska, 1986. American Water Resources Association, Bethesda, MD, 377–386.

Kattelmann, R.C. 1989. Spatial variability of snow-pack outflow at a site in Sierra Nevada, USA. *Annals of Glaciology*, **13**, 124–128.

Kattelmann, R.C. and Dozier, J. 1999. Observations of snowpack ripening in the Sierra Nevada, California, USA. *Journal of Glaciology*, **45**(151), 409–416.

Kirnbauer, R., Blöschl, G. and Gutknecht, D. 1994. Entering the era of distributed snow models. *Nordic Hydrology*, **25**, 1–24.

Kite, G.W. 1995. Scaling of input data for macroscale hydrologic modelling. *Water Resources Research*, **31**(11), 2769–2781.

Koivusalo, H. and Heikinheimo, M. 1999. Surface energy exchange over an arctic snow pack: comparison of two snow energy balance models. *Hydrological Processes*, **13**, 2395–2408.

Kondo, J. and Yamazaki, T. 1990. Prediction model for snowmelt, snow surface temperature and freezing depth using a heat balance method. *Journal of Applied Meteorology*, **29**(5), 375–384.

Koskinen, J.T., Pulliainen, J.T. and Hallikainen, M.T. 1997. Use of ERS-1 SAR data in snow melt monitoring. *IEEE Transactions on Geoscience and Remote Sensing*, **35**(3), 601–610.

Lapen, D.R. and Martz, L.W. 1996. An investigation of the spatial association between snow depth and topography in a prairie agricultural landscape using digital terrain analysis. *Journal of Hydrology*, **184**, 277–298.

Leavesley, G.H. 1989. Problems of snowmelt runoff modelling for a variety of physiographic and climatic conditions. *Hydrologic Science Journal*, **105**, 205–223.

Leavesley, G.H. and Stannard, L.G. 1990. Application of remotely sensed data in a distributed-parameter watershed model. In: Kite, G.W. and Wankiewicz, A. (eds), *Proceedings Workshop on Applications of Remote Sensing in Hydrology*, Saskatoon, Saskatchewan, 13–14 February 1990. Environment Canada, 47–68.

Leavesley, G.H. and Striffler, W.D. 1979. Mountain watershed simulation model. In: Colbeck, S.C. and Roy, M. (eds), *Proceedings Meeting on Modelling of Snow Cover Runoff*, 26–28 September 1978, Hanover, NH. CRREL Special Report 79-36, US Army Cold Regions Research and Engineering Laboratory, 379–386.

Lesaffre, B., Pougatch, E. and Martin, E. 1998. Objective determination of snow grain characteristics from images. *Annals of Glaciology*, **26**, 112–118.

Ling, C.-H., Josberger, E.G. and Thorndike, A.S. 1995. Mesoscale variability of the Upper Colorado River snow pack. *Nordic Hydrology*, **27**, 313–322.

Link, T. and Marks, D. 1999. Distributed simulation of snowcover mass- and energy-balance in the boreal forest. *Hydrological Processes*, **13**(iii), 2439–2451.

Lipton, A.E. and Ward, J.M. 1997. Satellite-view biases in retrieved surface temperatures in mountain areas. *Remote Sensing of Environment*, **60**, 92–100.

Liston, G.E. 1995. Local advection of momentum, heat, and moisture during the melt of patchy snow covers. *Journal of Applied Meteorology*, **34**(7), 1705–1715.

Liston, G.E. and Sturm, M. 1998. Simulating arctic Alaska snowdrifts using a numerical snow-transport model. *Journal of Glaciology*, **44**(148), 498–516.

Liston, G.E., Brown, R.L. and Dent, J.D. 1993. Two-dimensional computational model of turbulent atmospheric surface flows with drifting snow. *Annals of Glaciology*, **18**, 281–286.

Loth, B. and Graf, H-F. 1998a. Modelling the snow cover in climate studies. 1. Long-term integrations under different climatic conditions using a multilayered snow-cover model. *Journal of Geophysical Research, Atmospheres*, **103**(10), 11 313–11 327.

Loth, B. and Graf, H-F. 1998b. Modelling the snow cover in climate studies. 2. The sensitivity to internal snow parameters and interface processes. *Journal of Geophysical Research, Atmospheres*, **103**(10), 11 329–11 340.

Loth, B., Graf, H-F. and Oberhuber, J.M. 1993. Snow cover model for global climate simulations. *Journal of Geophysical Research*, *20*, **98**(D6), 10 451–10 464.

Luce, C.H., Tarboton, D.G. and Cooley, K.R. 1997. Spatially distributed snowmelt inputs to a semi-arid mountain watershed. *Eastern Snow Conference and Western Snow Conference, Proceedings*, Joint 54th and 65th, 344–353.

Luce, C.H., Tarboton, D.G. and Cooley, K.R. 1998. The influence of the spatial distribution of snow on basin-averaged snowmelt. *Hydrological Processes*, **12**, 1671–1685.

Luce, C.H., Tarboton, D.G. and Cooley, K.R. 1999. Subgrid parameterisation of snow distribution for an energy and mass balance snow-cover model. *Hydrological Processes*, **13**(12/13), 1921–1933.

Marks, D., Kimball, J., Tingey, D. and Link, T. 1998. The sensitivity of snowmelt processes to climate conditions and forest cover during rain-on-snow: a case study of the 1996 Pacific Northwest flood. *Hydrological Processes*, **12**(10–11), 1569–1587.

Marsh, P. 1999. Snow cover formation and melt: recent advances and future prospects. In: Hardy, J.P., Albert, M.R. and Marsh P. (eds), *Advances in Hydrological Processes: Snow Hydrology: The Integration of Physical, Chemical and Biological Systems*. Wiley Interscience, Chichester, 2117–2134.

Marsh, P. and Woo, M.K. 1984a. Wetting front advance and freezing of meltwater within a snow cover 1. Observations in the Canadian Arctic. *Water Resources Research*, **10**(12), 1853–1864.

Marsh, P. and Woo, M.K. 1984b. Wetting front advance and freezing of meltwater within a snow cover 2. A simulation model. *Water Resources Research*, **10**(12), 1865–1874.

Marshall, S. and Oglesby, R.J. 1994. Improved snow hydrology for GCMs. Part 1: snow cover fraction, albedo, grain size, and age. *Climate Dynamics*, **10**(1–2), 21–37.

Martin, E., Brun, E. and Durand, Y. 1997. Snow-cover simulations in mountainous regions based on general circulation model outputs. *Annals of Glaciology*, **25**, 42–45.

Martinec, J. 1975. Snowmelt runoff model for stream flow forecasts. *Nordic Hydrology*, **6**, 145–154.

Martinec, J. and Rango, A. 1981. Areal distribution of snow water equivalent evaluated by snow cover monitoring. *Water Resources Research*, **17**(5), 1480–1488.

McManamon, A., Szeliga, T.L., Hartman, R.K., Day, G.N. and Carroll, T.R. 1993. Gridded snow water equivalent estimation using ground-based and airborne snow data. *Eastern Snow Conference and Western Snow Conference, Proceedings*, Joint 50th and 61st, 75–81.

Melloh, R.A. 1999. A synopsis and comparison of selected snow algorithms. *CRREL Report* 99-8. US Army Cold Regions Research and Engineering Laboratory.

Melloh, R.A., Daly, S.F., Davis, R.E., Jordan, R. and Koenig, G.G. 1997. Operational distributed snow dynamics model for the Sava River, Bosnia. *Eastern Snow Conference and Western Snow Conference, Proceedings*, Joint 54th and 65th, 152–162.

Morris, E.M., Bader, H.P. and Weilenmann, P. 1997. Modelling temperature variations in polar snow using DAISY. *Journal of Glaciology*, **43**(143), 180–191.

Nelder, J.A. and Mead, R. 1965. A simplex method for function minimization. *Computer Journal*, **7**, 308–313.

Nolin, A.W. and Dozier, J. 1993. Estimating snow grain size using AVIRIS data. *Remote Sensing of Environment*, **44**, 231–238.

Nolin, A.W., Dozier, J. and Mertes, L.A.K. 1993. Mapping alpine snow using a spectral mixture modelling technique. *Annals of Glaciology*, **17**, 121–124.

Painter, T.H., Roberts, D.A., Green, R.O. and Dozier, J. 1998. The effect of grain size on spectral mixture analysis of snow-covered area from AVIRIS data. *Remote Sensing of Environment*, **65**, 320–332.

Peck, E., Carroll, T. and Vandermark, S.C. 1980. Operational aerial snow surveying in the United States. *Hydrological Sciences Bulletin*, **25**, 51–62.

Perla, R. 1982. Preparation of section planes in snow specimens. *Journal of Glaciology*, **28**(98), 199–204.

Perla, R. 1985. Snow in strong or weak temperature gradients. Part II: section-plane analysis. *Cold Regions Science and Technology*, **11**(2), 181–186.

Perla, R. and Ommanney, C.S.L. 1985. Snow in strong or weak temperature gradients. Part 1: experiments and qualitative observations. *Cold Regions Science and Technology*, **11**(1), 23–35.

Pomeroy, J.W., Marsh, P. and Gray, D.M. 1997. Application of a distributed blowing snow model to the Arctic. *Hydrological Processes*, **11**(11), 1451–1464.

Press, W.H., Teukolsky, S.A., Vetterling, W.T. and Flannery, B.P. 1992. *Numerical Recipes*. Cambridge University Press, New York.

Qin, Z. and Karnieli, A. 1999. Progress in the remote sensing of land surface temperature and ground emissivity using NOAA–AVHRR data. *International Journal of Remote Sensing*, **20**(12), 2367–2393.

Rango, A. 1993. Snow hydrology processes and remote-sensing. *Hydrological Processes*, **7**(2), 121–138.

Rosenthal, W. and Dozier, J. 1996. Automated mapping of montane snow cover at subpixel resolution from the Landsat Thematic Mapper. *Water Resources Research*, **32**(1), 115–130.

Schroeter, H.O. and Whiteley, H.R. 1992. Does detailed areal distribution of snow cover improve the utility of a hydrological model for watershed management. *Proceedings, 49th Eastern Snow Conference*, 181–192.

Schroeter, H.O., Boyd, D.K. and Whiteley, H.R. 1991. Area snow accumulation–ablation model (ASAAM): experience of real-time use in southwestern Ontario. *Proceedings, 48th Eastern Snow Conference*, 25–38.

Sellers, P.J., Randall, D.A., Collatz, G.J., Berry, J.A., Field, C.B., Dazlich, D.A., Zhang, C., Collelo, G.D. and Bounoua, L. 1996. A revised land surface parameterisation (SiB2) of atmospheric GCMs. Part I: Model formulation. *Journal of Climate*, **9**, 676–705.

Shi, J. and Dozier, J. 1995a. Inferring snow wetness using C-band data from SIR-C's polarimetric synthetic aperture radar. *IEEE Transactions on Geoscience and Remote Sensing*, **33**(4), 905–914.

Shi, J. and Dozier, J. 1995b. Corrections to 'Inferring snow wetness using C-band data from SIR-C's polarimetric synthetic aperture radar'. *IEEE Transactions on Geoscience and Remote Sensing*, **33**(6), 1340.

Shi, J. and Dozier, J. 2000a. Estimation of snow water equivalence using SIR-C/X-SAR, Part I: Inferring snow density and subsurface properties. *IEEE Transactions on Geoscience and Remote Sensing*, **36**(6), 2465–2474.

Shi, J. and Dozier, J. 2000b. Estimation of snow water equivalence using SIR-C/X-SAR, Part II: Inferring snow depth and particle size. *IEEE Transactions on Geoscience and Remote Sensing*, **38**(6), 2475–2487.

Shih, S-E., Ding, K-H., Kong, J.A., Yang, Y.E., Davis, R.E., Hardy, J.P. and Jordan, R. 1997. Modelling of millimeter wave backscatter of time-varying snowcover. *Journal of Electromagnetic Waves and Applications*, **11**, 1289–1298.

Shook, K., Gray, D.M. & Pomeroy, J.W. 1993. Temporal variation in snowcover area during melt in prairie and alpine environments. *Nordic Hydrology*, **24**(2–3), 183–198. Presented at the *9th Northern Research Basins Symposium*, Whitehorse, Yukon, Canada, 11–15 August 1992.

Silberstein, R.P., Sivapalan, M. and Wyllie, A. 1999. On the validation of a coupled water and energy balance model at small catchment scales. *Journal of Hydrology*, **220**, 149–168.

Singh, V.P., Bengtsson, L. and Westerstrom, G. 1997. Kinematic wave modelling of vertical movement of snowmelt water through a snowpack. *Hydrological Processes*, **11**, 149–167.

Slater, A.G., Schlosser, C.A., Desborough, C.E., Pitman, A.J., Henderson-Sellers, A., Robok, A., Vinnikov, K.Ya., Mitchell, K., Boone, A., Branden, H., Chen, F., Cox, P.M., de Rosnay, P., Dickinson, R.E., Dai, Y-J., Duan, Q., Entin, J., Etchevers, P., Gedney, N., Gusev, Ye.M., Habets, F., Kim, J., Koren, V., Kowalczyk, E., Nasonova, O.N., Schaake, J., Shmakin, A.B., Smirnova, T.G., Verseghy, D., Wetzel, P., Xue, Y., Yang, Z-L. and Zeng, Q. 2000. The representation of snow in land-surface schemes: results from PILPS 2(d). *Journal of Hydrometeorology*, **2**(1), 7–25.

Sommerfeld, R.A., Musselman, R.C., Wooldridge, G.L. and Conrad, M.A. 1991. Performance of a simple degree-day estimate of snow accumulation to an alpine watershed. In: Bergman et al. (eds.), *Snow, Hydrology and Forests in High Alpine Areas*. IAHS Publication No. 205, 221–228.

Sommerfeld, R.A., Bales, R.C. and Mast, A. 1994. Spatial statistics of snowmelt flow: data from lysimeters and aerial photos. *Geophysical Research Letters*, **21**(25), 2821–2824.

Stehman, S.V. 1999. Basic probability sampling designs for thematic maps accuracy assessment. *International Journal of Remote Sensing*, **20**(12), 2423–2441.

Sturm, M. and Benson, C.S. 1997. Vapor transport, grain growth and depth-hoar development in the subarctic snow. *Journal of Glaciology*, **43**(143), 42–59.

Sturm, M., Holmgren, J., König, M. and Morris, K. 1997. The thermal conductivity of seasonal snow. *Journal of Glaciology*, **43**(143), 26–41.

Tarboton, D.G., Chowdhury, T.G. and Jackson, T.H. 1995. Spatially distributed energy balance snowmelt model. In: Tonnessen, K.A. et al. (eds), *Proceedings, International Symposium on Biogeochemistry of Seasonally Snow-covered Catchments*, Boulder, CO, 1–14 July 1995. IAHS Publication No. 228, 141–155.

Turpin, O., Ferguson, R. and Johansonn, B. 1999. Use of remote sensing to test and update simulated snow cover in hydrological models. *Hydrological Process*, **13**(12/13), 2067–2078.

Tuteja, N.K. and Cunname, C. 1997. Modelling coupled transport of mass and energy into the snowpack model development, validation and sensitivity analysis. *Journal of Hydrology*, **195**, 232–255.

US Army Corps of Engineers, 1956. *Snow Hydrology: Summary Report of the Snow Investigations*. North Pacific Division, Portland, OR.

US Army Corps of Engineers, 1991. *User Manual: SSARR Model Streamflow Synthesis and Reservoir Regulation*. North Pacific Division, Portland, OR.

WMO, 1986. *Intercomparison of Models of Snowmelt Runoff*. Operational Hydrology Report No. 23, WMO-No. 646, WMO, Geneva.

Walker, A.E. and Goodison, B.E. 1993. Discrimination of a wet snow cover using passive microwave satellite data. *Annals of Glaciology*, **17**, 307–311.

Walland, D.J. and Simmonds, I. 1996. Sub-grid-scale topography and the simulation of Northern Hemisphere snow cover. *International Journal of Climatology*, **16**, 961–982.

Wigmosta, M.S., Lettenmaier, D.P. and Vail, L.W. 1994. A distributed hydrology–vegetation model for complex terrain. *Water Resources Research*, **30**(6), 1665–1679.

Yang, Z.L., Dickinson, R.E., Robock, A. and Vinnikov, K.I.A. 1997. Validation of the snow submodel of the biosphere–atmosphere transfer scheme with Russian snow cover and meteorological observational data. *Journal of Climate*, **10**(2), 353–373.

Yang, Z.L., Dickenson, R.E., Shaikh, M., Gao, X., Bales, R.C., Sorooshian, S. and Jin, J. 1999. Simulation of snow mass and extent in general circulation models. *Hydrological Processes*, **13**(12/13), 2097–2113.

12 Groundwater

HUBERT J. MOREL-SEYTOUX
Atherton, California, USA

12.1 INTRODUCTION

This chapter starts with the premise that the reader either knows by now, from the preceding chapters, or already knew before, what is meant by a 'model'. Thus, to the reader's relief, no further explanation or definition will be attempted. At the end of the chapter the reader may wonder about this writer's wisdom not to have applied the same reasoning to the term 'validation' . . . To his defence it may be said that he was foolish enough to accept the task of writing this chapter, which, probably wisely, many more qualified experts would not, . . . which of course is no defence at all.

Not being a recognised expert on the subject, I have relied on the views of others to develop and/or support my own . . ., noting that no consensus exists. To keep the debate lively I have attempted to present the subject, at least in some parts, in the form of a dialogue of quotes '. . .'. However, beyond a semantic debate on the appropriate or inappropriate, honest or misleading, use of terms such as 'validation' or 'verification', there remain important questions regarding the basic tasks that must be undertaken to convince oneself and especially others, that a particular model has a useful practical value. It is not possible to define authoritatively all the steps to be carried out in model development and evaluation in an exhaustive and detailed fashion, but it is good to be aware of the many possible pitfalls. Several sections of this chapter discuss the sources of errors in models for the benefit of the uninitiated model developer and/or user. It is wise to guard against these errors by following a reasonable protocol of model use and development. In California, the Bay-Delta Modeling Forum (for the San Francisco bay and the Sacramento–San Joaquin delta) felt the need for such a protocol and has produced a recent version (19 November 1999).

I have attempted to assess how close the practitioners are to being able to 'validate' a groundwater model in a meaningful way. I have chosen to answer that question by following chronologically, over a couple of decades, the work of a few scientists whose careers have been dedicated to that objective. The choice of the individuals was clearly subjective as many other names come readily to mind. However, their competency is clear and their scientific quest is representative.

The emphasis is on groundwater quantity, not quality. Probably at least 95% of a problem of contamination would be solved if one knew where the water is really going. There can be no valid transport model if the water flow model is not valid. Is it possible to validate a groundwater model in the first place? That is the question.

In this chapter I have referred to myself indifferently as: I, we ('*pluriel de majesté*!') or the 'writer'. When referring to documents by others, I have used the term: 'author(s)'. Although I

Model Validation: Perspectives in Hydrological Science. Edited by M.G. Anderson and P.D. Bates.
© 2001 John Wiley & Sons, Ltd.

have followed the wording of quoted documents, to avoid the distraction of variations of spellings, hyphenation etc. I have standardised these to match the rest of the text. I have also added italic for emphasis.

12.2   IMPORTANT QUESTIONS

12.2.1   What is Model Validation?

A variety of terms are used in the literature to describe the *level of confidence* one has in a particular model. Invariably, models (even if they are perfectly correct in all other ways) involve some *parameters* whose values cannot be determined directly by simple field measurements. The values of the parameters are 'inferred' from observations by selecting them to, in some sense, 'best' enable the model to reproduce observations. That preliminary phase, before a model can be applied for a study, is referred to as 'identification' by some (e.g. Yeh 1986), 'calibration' by others (e.g. Cooley 1979; Hill et al. 1998) and in some quarters as 'history matching' (e.g. Morel-Seytoux 1966; Bredehoeft and Konikow 1993) or even as the 'inverse' problem (e.g. Hill 1998). The outcome of the procedure (whatever name it is given) is a set of 'estimates' of these parameters. Given that these estimates are 'uncertain' (approximate) one often tries to provide 'confidence limits' (e.g. Cooley 1997) for them.

Added to these parameter uncertainties are a number of *errors* that can be present in the structure and conception of the model. Will the model perform well for conditions different from the period of time (or the spatial pattern) for which it was calibrated? Typically one tries to answer that question by 'checking the performance' of the model on a period of time (or a spatial pattern) different from the one for which it was calibrated. That step is often referred to as 'verification' but a 'purist' might prefer the term 'performance check' or possibly 'performance assessment' though that expression has been used in a somewhat different context (e.g. Ewing et al. 1999) and for that reason will not be used here. Thus so far we have defined calibration and performance check (PC). Note that the usage of these terms employed here has not received ('*tant s'en faut*'!) universal acceptance.

What is model validation? As the name would imply in *lay* language it means a procedure, the outcome of which is to declare the model 'valid'. It is easiest to define what a valid model is through the attributes of such model. It accurately can predict the behaviour of the system that is modelled, under any circumstances, any scenario of excitations (forcing functions, driving mechanisms), boundary and initial conditions. Which begs the next question.

12.2.2   Is Model Validation Possible?

Clearly absolute validation would require an infinite number of tests. Validation in an absolute sense is therefore impossible. Validation of a model can only be done in a 'relative' sense, limited (1) circumstantially by the conditions prevalent within the data sets used for calibration and PC and (2) intrinsically by the conceptual structure of the model. The adjective 'relative' by its nature is subjective and one may ask the question: relative to what? To other models, and with what comparison scale?

12.2.3   Should the Terms Validation and Verification be Used At All?

According to Bredehoeft and Konikow (1993, p. 178)

to the general public, proclaiming that a ground-water model is validated carries with it an aura of correctness that we do not believe many of us who model would claim. *We can place all the caveats we wish*, but the public has its *own understanding* of what the word implies. Using the word *valid* with respect to models misleads the public; *verification* carries with it similar connotations as far as the public is concerned.

Indeed, should these terms be used? It is of course a matter of opinion. For instance, in a comment on an article (Oreskes et al. 1994) in *Science*, Rykiel (1994, p. 330) states:

'Semantically there is little to choose among the terms "verify," "validate," "confirm," . . . and "substantiate," because they are synonymous in ordinary language. . . . Thus, "verification" and "validation" acquire special disciplinary meanings for testing simulation models.' Nevertheless Rykiel continues: 'Modellers themselves . . . should draw some important lessons from Oreskes et al.: (i) make clear that "verification" and "validation" are used in a technical sense; (ii) if necessary, don't use the terms if they are likely to be misunderstood and create a false sense of truth rather than consensus'.

12.2.4   Why is it Important?

Here is one illustration. In the process of disposing of spent nuclear fuel (in the United States) the Environmental Protection Agency (EPA) promulgated a standard in the 'Containment Requirements'. These are 'among the first federal standards and regulations that *mandate* use of numerical *models* of a proposed technological system in a formal determination of system *acceptability*' (Ewing et al. 1999, p. 936). If use of numerical models is indeed mandated, the question of their 'validity' is certainly an important one and an issue that cannot be ignored.

12.3   GROUNDWATER VERSUS OTHER HYDROLOGIC COMPONENTS

The distinction between surface and ground waters is often quite artificial. That (trivial) realisation did not, however, become fully explicit in the literature till around 1970 (e.g. Morel-Seytoux et al. 1980). Yet it is only recently that a description of river dynamics has been incorporated into the standard USGS groundwater MODFLOW (Jobson and Harbaugh 1999). This writer, primarily between 1970 and 1985 (e.g. Morel-Seytoux et al. 1973; Morel-Seytoux 1985; Morel-Seytoux and Restrepo 1987a,b,c) developed specialised models to describe the behaviour of both types of water and their *dynamic* interactions, accurately and efficiently. For this reason, the following sections will include examples for both surface and ground waters (including the vadose zone). Because it is possible to see surface water, to sample it readily, to measure its discharge etc., it is easier to test surface water models and to develop a trust in them, or eventually replace them, at least partially, by relatively simple campaigns of measurements. On the other hand, because it is not possible to see subsurface water, to have access to it, to measure its velocity etc., groundwater models are even more necessary but at the same time users are not inclined to trust them. Groundwater models (modellers) are needed but not trusted.

12.4   TYPES OF MODELS

We find it useful to distinguish three types of models.

12.4.1 Generic Models

A generic model is a model that performs a particular function, *given* known values of the parameters. An example would be a model that is 'supposed' to solve the unsaturated flow equation in one dimension under specified upper and lower boundary conditions of the first (Dirichlet), second (Neumann) and third (Cauchy or Fourier or 'general') type for a homogeneous porous medium. For a generic model only the technique of numerical solution (and its software implementation) may be in doubt and it is necessary to check by comparison with known solutions (e.g. analytic solutions) that the numerical procedure is adequate. Generally it is possible to quantify the numerical errors of such generic models. On the other hand, it is possible that in a generic model the equation solved does not really represent the behaviour of the system. Checking that a particular model algorithm solves correctly Richards' equation does not guarantee that the equation itself represents correctly unsaturated water flow, for example if air viscous resistance, counterflow and compression effects are important (e.g. Morel-Seytoux and Khanji 1975; Morel-Seytoux 1983a,b; Schultze et al. 1999).

12.4.2 Site-specific Models

One example would be the use of the generic model just discussed for interpretation of results of column experiments in the laboratory. In this case the site of concern is the column. Calibration will be necessary. Even if 'successfully' calibrated for a set of data and verified for another, there is no simple way to analyse how well the calibrated model would represent the behaviour of the column if totally different boundary conditions were present in a simulation. There is simply no known way to extrapolate the quantified error, under performance check, to other conditions. In this case the errors come from several sources, such as: (1) the conceptual error of assuming that flow at a Darcy scale (cross-section of a few square centimetres) can be represented by a mathematically one-dimensional model, (2) the use of conceptual theoretical curves for retention and relative permeabilities (Morel-Seytoux and Nimmo 1999) and (3) the inferred values of saturated hydraulic conductivity etc.

12.4.3 Future-Scenarios Models

Models are put to use for management purposes. Typically if a groundwater model has been calibrated in a certain region, it is because a problem has arisen in the region concerning the use or exploitation of the aquifer(s). Unhappiness with the current situation generally motivates the study. Thus the model will be used to investigate *other* management strategies for the *future*. Immediately one can see two causes for the difficulty of validation of the results of application of the model under these circumstances. Statistical inference (Mood and Graybill 1963) is possible only if made for the same statistical population from which the sample was drawn. Quoting these authors (p. 142): one 'can draw valid probabilistic conclusions about the sampled population, but one must use *personal judgement* to extrapolate to the target population, and the *reliability* of the extrapolation *cannot* be *measured* in relative frequency–probability terms'. If circumstances change (different management of the system) then clearly the inferences are not rigorously valid. Second, who can predict the various contingencies and different conjunctures to exist into the future with reasonable certainty? The future-scenarios model is the combination of the site-specific model of the physical system with a forecasting of the (demographic, social, regulatory, economic, legal) future situation. In this chapter we do not wish to confound the question of *validation of the scenarios* with that of the site-specific hydrologic model. In this chapter we only consider the generic and site-specific hydrologic (groundwater) models.

12.5   SOURCES OF MODEL ERRORS

12.5.1   Errors in the Comprehension of the System to be Modelled

In attempting to solve problems associated with water management, one has to deal with systems which include two major, and quite distinct, types of components: the *natural* ones and the *man-made* ones. The latter ones are usually fairly well known because they were designed with specific criteria (one knows, for example, the dimensions of a spillway and its rating curve for discharge versus elevation or those of a concrete canal that conveys water from one part of a State to the other). On the other hand, nature does not tell us what Manning's roughness coefficient is for the innumerable heterogeneous segments of the rivers which crisscross the plains or valleys of the system, especially under extreme conditions of flood with overflowing banks etc. In the context of groundwater, 'the natural parts of a Performance Assessment (PA) system are usually heterogeneous, physically inaccessible, and therefore hard to map and characterise in the details required for effective study of the system via mathematical models' (Ewing et al. 1999, p. 936). Thus one must accept the fact that the system being modelled will always be described in a less than perfect way as to its physical static or dynamic characteristics. This introduces a fundamental error that *cannot* be circumvented and will always be present. The question that must be addressed is, nevertheless: how to reach decisions in spite of the uncertainties associated with our comprehension of the system? 'By stating that validation should be abandoned, we' (Bredehoeft and Konikow 1993, p. 178) 'do not mean to imply that model testing and evaluation should be abandoned or that models are not useful. Far from it; both of us have spent much of our careers developing and using models for analysis'.

12.5.2   Conceptual Errors in the Description of the System to be Modelled

Having understood that reality is too complex to be modelled perfectly, one must develop a schematic view of the system and its behaviour. One needs to superimpose on reality our view of it and, naturally, a highly simplified one at that. In this process of conceptualisation and simplification, an excellent understanding of the behaviour of the natural system is needed in order to separate what is important and essential from what is secondary or tertiary. Conceptualisation is required regarding both the static description of the system and its dynamic characteristics.

In the static characteristics category one finds the geometrical and topographic description of say a watershed. How long is the main river? Can we approximate it by a succession of highly straightened segments with sharp turns at the junctions or must we subdivide the river into a very large number of reaches to accommodate changes in width, slope, roughness, direction etc.? Comes the hard question: how do we know the error committed by not carrying out the greatest level of refinement in the description? How do we test the model conception error resulting from a given level of coarseness in the description of such factors?

In the dynamic category one finds the description of the physical laws that govern the processes. It is fair to say that usually we know these laws at the molecular, microscopic or at the punctual scale in the continuum mechanics sense. At the molecular scale we know Henry's law; at the microscopic scale we know Fick's law; at the punctual scale (or column scale, say cross-section of size 1 to 40 square centimetres, 20 cm deep, the Darcy scale) we know Darcy's law. In water management we are not interested, usually, in these scales. We have to deal with thousands of square kilometres. For example, to calculate infiltration in a basin as a result of rainfall events, shall we subdivide the top soil layer into billions of 100 cm^2 cross-section, 20 cm deep, soil columns because we know Darcy's law at that scale? Hardly practical though some merchants (either peddling computers or watershed models) would like us to believe so. At this point the infiltration process has to be conceptualised. There are several ways to proceed. A common way is to simply resort to analogy and to schematically assimilate the top soil to a lumped 'reservoir' (or a few such reservoirs) with a variety of spillways and conduits to represent the transmission and retention characteristics of the soil at the area scale of 1 to 100 square kilometres and for depths from 1 to 10 metres. Another approach is to extend to a much larger scale a satisfactory column law of infiltration (say Horton's or Greeen and Ampt's) which depends on physically based parameters such as saturated hydraulic conductivity of the soil, and use in the formulas *effective* values of the parameters. Finally, a third approach starts from the punctual laws and derives, more or less exactly, by multiple integration in space, time, expectation and process sense, a law valid at the larger scale of a parcel (1 to 10 square metres), hillslope (10 to 1000 square metres) or watershed (a few square kilometres). The problem with that approach is that to carry the integration one must conceptualise the laws of chance that underlie the spatial distribution of the soil parameters and the type and degree of connectivity between the columns or parcels that make up the hillslope or the watershed (e.g. Morel-Seytoux 1999). We shall not here discuss the pros and cons of these distinct approaches; we want only to point out that all may potentially introduce serious model errors by failing to represent adequately the processes involved at the *practical* decision scale. Finally it is quite conceivable that some conceptual models commit the 'sin of omission', i.e. do not account for a significant primary phenomenon. For example, when studying infiltration into a layered soil, one must account for the fact that some layers may be unsaturated and transmit by far less water than if they were assumed to be saturated. In this case the influence of capillarity would have been omitted.

12.5.3   Model Equations Errors

Starting from an analysis of the system and a conceptualisation of its static and dynamic characteristics, one proceeds to express that knowledge in mathematical symbolism, leading to a system of equations and logical statements. The equations can be written in differential, integral and/or algebraic forms. Naturally if the conceptualisation was severely in error or omitted significant processes, the mathematical formalism will not correct the analysis. Thus we shall here only discuss the *additional* errors that might be introduced at this level. If the form of the equations was differential in nature, again they only apply at the punctual scale. They have to be integrated. If it is done analytically then no error is introduced save for the possible wrong choice of boundary conditions. For example treating a river in hydraulic connection with an alluvial aquifer as a constant head boundary amounts to providing an inexhaustible source of recharge for the aquifer. This may be acceptable if the river is the Mississippi near New Orleans but certainly not for the South Platte, Arkansas and Rio Grande rivers in the state of Colorado.

12.5.4   Parameter Estimation Errors

In rare occasions the physical parameters appearing in the equations can be directly measured. For example, there is no major problem in measuring the width of a river. However, in almost all situations the parameters have to be estimated indirectly through a calibration procedure. The difficulty is that the estimation is always circumstantial and conditional. Pumping test procedures do not lead directly to transmissivity. From the measured drawdowns and the pumping rates in the pumped well one infers through an 'inference model' (typically an analytical solution, which is based on a lot of assumptions) what the value of a uniform transmissivity would have to be so that the observations and the calculations by the inference model match in some 'best' sense. In the context of investigating the one-dimensional exchange flux between a fully penetrating stream and an aquifer, Nachabe and Morel-Seytoux have shown (1995a, p. 353) that *no effective uniform* value of transmissivity could reproduce the temporal pattern of the aquifer drawdown over both the short-term and the long-term if the transmissivity instead of being perfectly uniform varied randomly in space about a stationary mean. Figure 12.1 shows the evolution of drawdown at a distance of 600 m from the stream bank for a reference run (geometric mean of transmissivity of 150 m^2/day and a value of standard deviation of the natural logarithm of transmissivity of one) and for two deterministic runs, with respectively effective uniform values of transmissivity equal to the arithmetic mean (247 m^2/day) and the harmonic mean (91 m^2/day). The aquifer, initially in equilibrium with the river is responding to a sudden permanent step river stage drop of 10 m.

Errors enter in any model because the assumptions of the inference model may be unreasonable for the given aquifer or because the selected criterion for the best match is not appropriate or because the match is fortuitous and only applicable under the conditions of the (pumping or river drawdown) test. Such estimations are, again, usually conditional and circumstantial. For

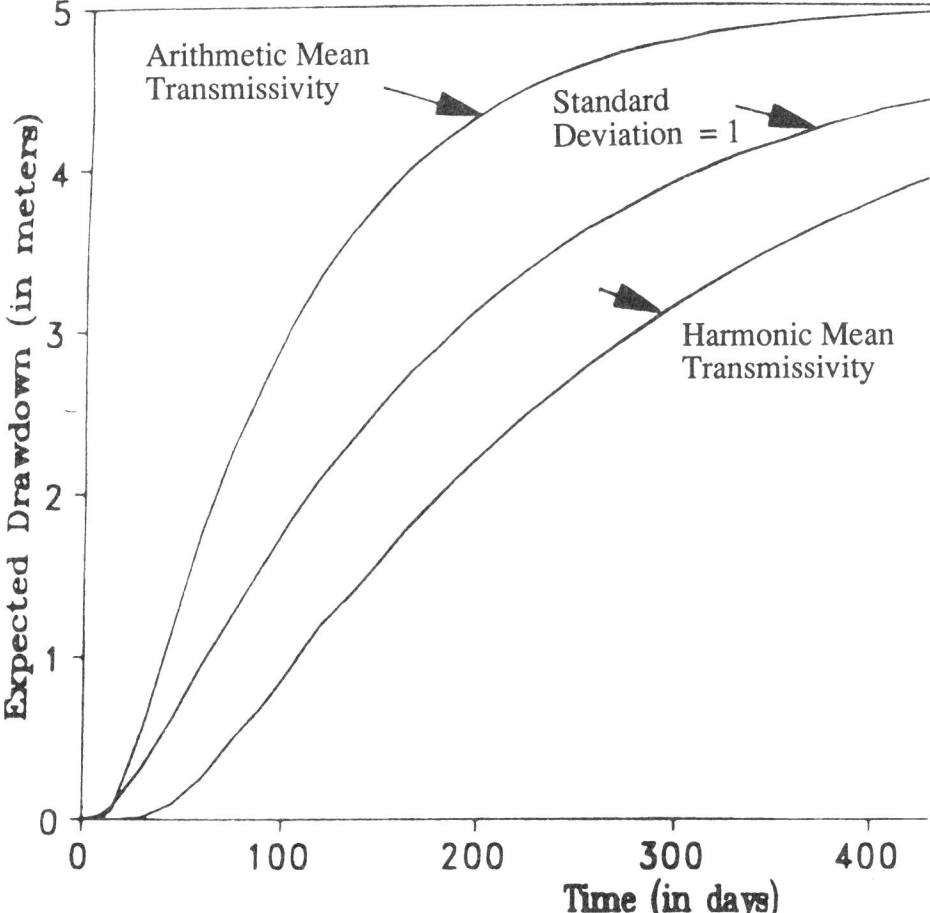

Figure 12.1 Drawdown 600 m from river bank simulated by three different models in response to a permanent step river stage drawdown of 10 m

example the estimated parameters of an infiltration model may give a good match in the prediction of runoff given rainfall because all the events used for calibration displayed a single clearly marked peak, but would be totally inapplicable if the event displayed several peaks. How does one quantify the error induced by parameters estimated for a certain type of condition when one wishes to simulate different conditions? Typically the reason why models are built is precisely because one is not satisfied with current management. Thus future operations will often lead to events drawn from a different population than the one used in the calibration.

12.5.5   Input and Data Errors

Precipitation is measured at only a few points in space and what is needed is a continuous description of it. This is again something one has to live with as measurements, even if done without errors, will never cover the entire domain under investigation. Sometimes the temporal distribution of rainfall will have a significant impact on the amount of runoff generated in a watershed. Again how much error is induced by using large time averages versus knowing the precise distribution of the rainfall rates in time?

12.5.6   Numerical Solution Errors

Whether in differential or integral form, the system of equations will have to be solved. If the system cannot be solved analytically then approximations are needed, be they of the finite difference or element type. Thus what is solved is not the original system but another one, hopefully close to the original one. The literature is full of discussions of the truncation errors associated with various schemes but for most systems these analyses do not apply. For example, it was concluded from a proof given in many textbooks (e.g. Remson et al. 1971, pp. 81 and 86), and thus believed, that implicit schemes are unconditionally stable. Experienced modellers have empirically discovered, probably to their sorrow at least once, that the validity of the statement was limited to homogeneous systems. Not that they were not warned. . . . if they read further (Remson et al., p. 96): 'It is unfortunate that knowledge of the applicability of such methods to problems with variable coefficients or problems on nonrectangular domains is fragmentary.'

    A measure of the finite differences approximation error is obtained by comparing the approximate solution with an exact analytical one. In groundwater the Theis solution is a favourite benchmark. When the comparisons are made, time steps and grid spacing are extremely small and the test results look good, but in practice the time steps and grid sizes used are orders of magnitude larger than the ones used in these tests. What is the real truncation error for these large time and grid increments?

12.5.7   Confounding of Errors

The most frustrating problem in trying to assess the magnitude of one sort of errors versus another kind is that they end up being confounded. For example if a finite difference model with coarse space and time increments is used to calibrate transmissivities based on observations, the parameters are conditional on the spacing. The same groundwater model used with a very fine grid (ignoring for now the problem of interpolation) may end up giving bad predictions because a wrong model with the wrong type of parameters when calibrated on good data will perform reasonably well, but it cannot be used reliably under different conditions than those for the calibration.

12.5.8   Interpretation Errors

It is very easy, especially when using automatic calibration procedures, to come to a decent match but for the wrong reasons. Different models describing different mechanisms may lead to similar decent matches with proper adjustment of their parameters. However, it is possible that with the calibration events the *trigger* for a completely different path in the computational sequence may never have been pulled because the necessary threshold is never reached either for the calibration or the PC period (or spatial pattern). Thus a good understanding of the detailed workings of a model is needed for the interpretation of the results.

12.6   SOURCES OF MODELLING ERRORS

We use the term 'modelling' to refer to the action of using a model for the purpose of conducting studies of water management.

12.6.1   'Wrong Choice of Model' Error

The first task for a water manager is to choose a model that is appropriate for the purpose of the investigation to be performed. A wise user should be somewhat leery of using a model for rainfall–runoff that was developed in England, France or Germany, and tested only there, for applications in Saudi Arabia. Either the basis for the model is thoroughly scrutinised and deemed applicable or a priori another model is selected. Naturally the necessity to scrutinise the model requires that a thorough documentation for the model be available. A poorly documented model should not be used even if it claims it can do the job, or you may suffer from a 'trusting' error.

When it comes to the particular choice of a boundary condition, *trust* should not be allowed to obscure judgement and *impede scrutiny*. It may be worthwhile to share an experience of reviewing the results of three different studies of a (controversial) plan for extraction of water from an aquifer. However, not to lose the reader at this point in too many details, the discussion is reported in the Appendix.

12.6.2   Wrong Choice of Increments Error

This has been discussed in Sections 12.5.4 and 12.5.6.

12.6.3   Wrong Calibration and Performance Check Procedures

At some stage in the study process it will be necessary to calibrate a lot of parameters. First one must understand clearly the structure of the model to proceed with calibration. Let us say one has 20 years of *daily* data of rainfall and runoff for a catchment. Some parameters affect mass balance, some affect the medium-term dynamics and some the short-term dynamics. There will be evapotranspiration parameters, aquifer parameters, soil parameters and river parameters to calibrate. One should not attempt immediately to calibrate all these parameters *jointly* on the 20 years of daily runoff. The long-term mass balance for runoff will be affected by rainfall and evapotranspiration. Except for obvious transcription errors on rainfall these data will be taken essentially at face value. Thus to establish a 20 years mass balance for the system one needs to adjust the coefficients involved in the calculation of evapotranspiration. If data of potential evapotranspiration were available, derived from pan evaporation, typically a coefficient is adjusted to account for the variation between theoretical potential evaporation and the actual potential evaporation for the various watersheds in the basin. This coefficient is adjusted to guarantee a perfect mass balance for the 20 years of record, i.e. cumulative volume of runoff as calculated is the same as the observed one. The time scale for that calibration is 20 years. Next one can look at dry weather seasons. During these periods, over several months usually, runoff is driven by the parameters that condition aquifer recharge. One thus calibrates the recharge parameters on the characteristic shape of the recession curves of runoff. The time scale for calibration is now one or a few months. Next one looks at volumes of flood events for well-characterised 'single rainfall – single discharge peak events' to calibrate the parameters that control infiltration and thus excess rainfall and runoff. The time scale now is several days to a couple of weeks. Next one looks at the volumes under, and to a lesser degree the shape of, the discharge hydrographs under conditions of 'double rainfall – double discharge peak events' with half a day to a couple of days separation between the rainfall events in order to calibrate the parameters that affect redistribution of moisture in the soil top two layers. The time scale is now a couple of days. Finally one looks at daily values during the flood events to estimate the parameters that control propagation and attenuation in the flood hydrographs. By now all parameters have been calibrated but unfortunately there is some dependence between the parameters so that the calibration steps must be repeated again from the large time scale down to the smallest scale at which the runoff data are known. This discussion illustrates that it is not possible to calibrate intelligently and reliably a model without both a good understanding of the phenomena and of the specific model structure used to represent them. An error in modelling, which is not easily quantifiable, is the 'lack-of-knowledge-induced' error.

Now one would not use all the data to calibrate the parameters. One would select some of the years for calibration and some for performance check. It is wise to select for calibration the years that did not exhibit extreme behaviours. The purpose of modelling studies is often to extrapolate to situations that have not been encountered through the historical record. Thus this partition in the record will demonstrate or not the ability of the model to extrapolate and will provide a quantification of the errors that are likely to be encountered for the more extreme situations. Naturally for actual use of the model in the future one now recalibrates the model with all the years of record, but one has a (hopefully) conservative basis for estimation of errors for the future.

12.7   DELIMITING SIGNIFICANCE OF ERRORS (SENSITIVITY ANALYSIS)

One needs to quantify the *intrinsic* model errors, i.e. the typical errors say in predicting runoff (if the model's basic function is to predict runoff) as a function of seasons (dormant versus growing season) or flow conditions (e.g. during floods or recessions). One needs to study the impact of an error in estimation of certain parameters on certain quantities of interest such as instantaneous discharges or seasonal cumulative values. Is it important to know the exact value of transmissivity in the aquifer to predict peak discharge during a flood? In that case the answer would be no (practically) all the time and one could dispense with the exercise. Next one needs to study the impact of an error in certain parameters on a management decision. It is quite conceivable that a model that is not very accurate in predicting hydrograph shapes during flood events could be perfectly acceptable to size a new reservoir with interannual capacity. Whereas the intrinsic error can be secured once and for all by running the model, the 'derived' errors for a particular study will depend on the objective of the study and an investigation is required for each individual study.

12.8   CHARACTERISTICS OF USEFUL MODELS

A model is useful, despite being wrong, when its quantified *intrinsic* error is compatible with the acceptable accuracy of a management decision made on the basis of use of the model. For example, if a relatively large error in prediction of runoff during dry weather seasons leads nevertheless to a relatively narrow distribution in the needed size of a low flow augmentation dam, then it is useful. If the reverse holds the model in its present form is not useful. The intrinsic error must be reduced by a better calibration or by a change in structure of the model or a combination.

Even if the intrinsic error results in an unacceptable level of error for the management decisions deduced from the use of the model, the model may have utility in a *relative* sense if one wishes to compare the effectiveness of different strategies. However, one must be sure that the model incorporates properly the factors that condition the different responses between strategies. For example, if one is concerned about the effect of a pumping well near a river on a downstream surface water right holder, one must be sure that the groundwater model does not treat the river as a constant head boundary and that the model accounts for the dynamic flow propagation and for the associated fluctuating river stage.

12.9   A PROTOCOL FOR WATER AND ENVIRONMENTAL MODELLING: EXECUTIVE SUMMARY (Bay Delta Modeling Forum (BDMF), California)

Except for a few words, here and there, Section 12.9 is the executive summary of a document created for BDMF by a small group of writers under the leadership of Richard Satkowski (State Water Resources Conservation Board), Chair, and including Jay Lund (University of California at Davis), Ted Roefs (US Bureau of Reclamation), Austin Nelson (consultant) and Morel-Seytoux. The full document is available from BDMF (see the end of the section for full information). The executive summary was prepared by Jay Lund and Richard Satkowski.

12.9.1   Introduction

Mathematical computer models have become indispensable for the planning and management of California's complex water systems. However, models can generate controversy in water management, particularly when their bases and assumptions are unclear. This document presents general protocols and guidelines to better support the development and use of models in water and environmental planning and management. These protocols and guidelines are based on a substantial and broad academic and professional consensus on how computer models should be developed and used.

Adherence to modelling protocols and guidelines will result in better models and modelling studies by:

- Improving the development of models.
- Providing better documentation of models and modelling studies.
- Providing easier professional and public access to models and modelling studies.
- Making models and modelling studies more easily understood and amenable to examination.
- Increasing stakeholder, decision-maker, and technical staff confidence in models and modelling studies.

Technical staff must obtain support from managers and supervisors of modelling activities to adequately implement these protocols and guidelines. This support comes in the forms of adequate budget and time for proper model development as well as management discipline to ensure that protocols and guidelines are followed in an efficient and effective way. While adherence to these protocols may increase the budget and time requirements for an individual modelling study, these efforts should enhance the credibility and effectiveness of modelling work and reduce the effort needed to respond to technical controversies. The 'bottom line' is that adherence to these protocols will reduce the overall costs for modelling, decision-making and water management.

12.9.2   Solving Water Problems

Computer models do not resolve water conflicts; people do. However, modelling can assist in that role by:

• Furthering understanding of the problem.
• Defining solution objectives.
• Developing promising alternatives.
• Evaluating alternatives.
• Providing confidence in solutions.
• Providing a forum for negotiations.

12.9.3   Model Development Process

Most modelling professionals agree that model development should proceed along the standardised steps summarised in Table 12.1 (ES-1). These steps are intended to ensure that the model (1) addresses the intended problem, (2) reasonably represents the system, and (3) results are reasonably tested. In addition, the process helps ensure that the entire model development

Table 12.1 (ES-1)   Major steps in model development

Step	Name	Purpose
1.	Problem Identification	Solving the right problem
2.	Define Modelling Objectives	Define use for model and standard of success
3.	Formulation of Model	Mathematical similarity to the described system
4.	Selection and study of numerical solution	Numerical similarity to the mathematical formulation of the problem
5.	Algorithm Check	Test model based on model behaviour
6.	Model Calibration	Set constants to represent system behaviour and characteristics using a given (calibration) data set
7.	Performance Check	Test model by comparison with other field data
8.	Documentation of Model	Make model understandable to users
9.	Update and Support of Model	Maintain and improve the model's usefulness

Table 12.2 (ES-2)   Steps in model application for planning and policy studies

1. Define study objectives
2. Define how model outputs relate to the performance of alternatives
3. Define a base case
4. Define alternatives
5. Identify model version and input data
6. Model results
7. Summarise and discuss the performance of each alternative
8. Discuss study limitations

process is documented so that others will know what has been done and are clearly informed about the model's limitations for use.

12.9.4   Use of Models in Planning Studies

Aside from model development, the use of models for particular planning or policy problems should also follow a logical pattern, integrated with a logical planning process. This process is summarised in Table 12.2 (ES-2).

12.9.5   Regulatory Aspects of Model Use in Planning Studies

Environmental impact analysis is an integral part of planning. Models are commonly used to evaluate project alternatives and the environmental impacts associated with those alternatives. The intent of the California Environmental Quality Act (CEQA) and the federal National Environmental Policy Act (NEPA) process is to make environmental documents a decision-aiding document rather than the primary decision-making report. Often, the environmental document and the decision-making report can be developed together. A CEQA environmental impact report must describe the existing environmental setting from local and regional perspectives. When a proposed project is evaluated under CEQA, the analysis must use existing physical conditions as the baseline, not future projections.

12.9.6   Stakeholder and Public Review of Modelling Efforts

Providing stakeholders and the public with an early acquaintance of the model or modelling study can often reduce the technical controversies involved in modelling. A variety of methods are available to reduce the technical controversies involved in modelling and better integrate modelling activities into larger study activities, including the following:

- *Public Participation.* Proper planning requires adequate review and consultation with interested and affected stakeholders, agencies, organisations, and individuals. Efforts to secure public participation should be pursued through public workshop, meetings, and technical advisory and citizens committees.
- *Technical Advisory Committees.* Technical Advisory Committees are a common way to provide ongoing review for modelling and planning studies and typically come in two forms, as a committee of technical people representing stakeholders or a committee of recognised independent technical experts.
- *Shared Vision Modelling.* Shared Vision Modelling is the common development and use of a model, or set of models, by a group of diverse stakeholders and/or decision-makers. Its

purpose is to remove as many technical disagreements as possible from the conflict so that efforts can focus on interpretation of the result, rather than arguments about the model. The development or use of shared vision modelling is usually a prelude to development and evaluation of alternatives as well as meaningful negotiations among stakeholders.

- *Peer Review.* Peer review is a method for reviewing models in a timely, open, fair, and helpful manner. All models require some level of peer review to assure that they are properly used. The Bay-Delta Modeling Forum has developed a peer review process that is intended to inform stakeholders and decision-makers of (1) whether or not a given model is a suitable tool, and (2) the limits on the use of the model. Recent legislation requires the California Environmental Protection Agency to conduct an external scientific peer review of the scientific basis for any water quality rule.

12.9.7   Public Access to Models and Data

Models and data used in public decision-making should be available for public scrutiny, like any other calculation or analysis presented in a public forum. For modelling studies used in public arenas, the model, model documentation and data sets used should be made available either through the services of agency staff, a consultant, or a web site on the Internet.

12.9.8   Implementation of Modelling Protocols

The Forum 'accepted' this report on 19 November 1999 and is in the process of assisting Forum members and other interested parties in implementing the modelling protocols contained in this report. As specified in the Forum bylaws, it should be noted that this report does not necessarily represent the views of the governing bodies of the represented organisations or the individual members of the Forum.

12.9.9   Mission of the Bay-Delta Modeling Forum

The Forum is a statewide, non-profit, non-partisan, 'consensus' organisation whose mission is to increase the usefulness of models for analysing water-related problems in the San Francisco Bay, Sacramento–San Joaquin Delta and Central Valley system. For more information, visit the Forum's webpage at *www.sfei.org/modelingforum/* or contact: Bay-Delta Modeling Forum, c/o San Francisco Estuary Institute, 1325 South 46th Street, Richmond, CA 94804, USA; e-mail: modelingforum@sfei.org; phone: +1-510-2319539; John Williams, Executive Director; phone: +1-530-7537081.

12.10   VARIOUS VIEWS ON VALIDATION, VERIFICATION, CALIBRATION AND PERFORMANCE CHECK

12.10.1   Konikow

Reading attentatively the publications of Konikow, one strongly gets the message that the *concept* that hydrologic models in general, and groundwater models in particular, can be validated is a *fallacious* one. Together with Bredehoeft (1993, p. 178) he makes the point that calibration, better called 'history matching' (when using time series for the process of adjustment of the parameters of a model) is just that, and 'to claim more for that process, using words like validate and verify, is to delude ourselves, mislead the public, and make us look foolish to our

307

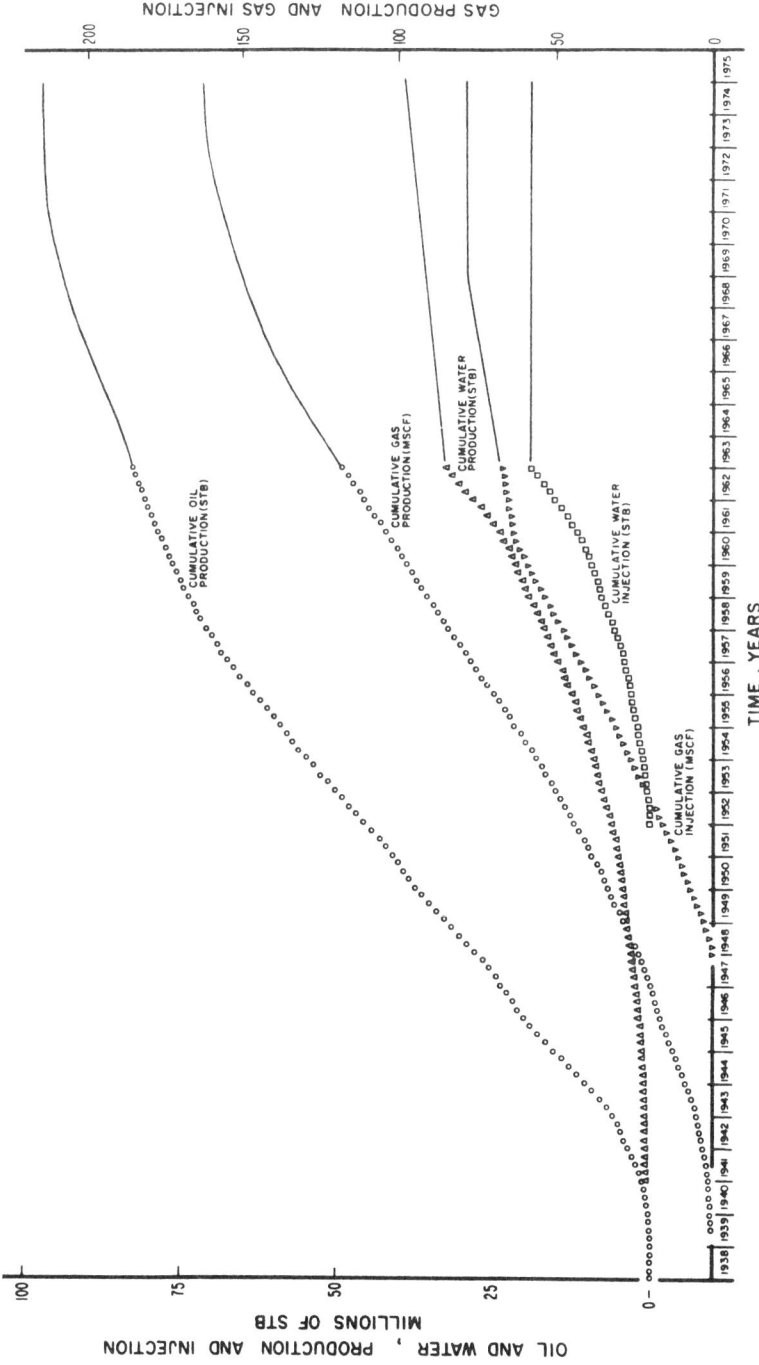

Figure 12.2 Production–injection schedules for typical forecast

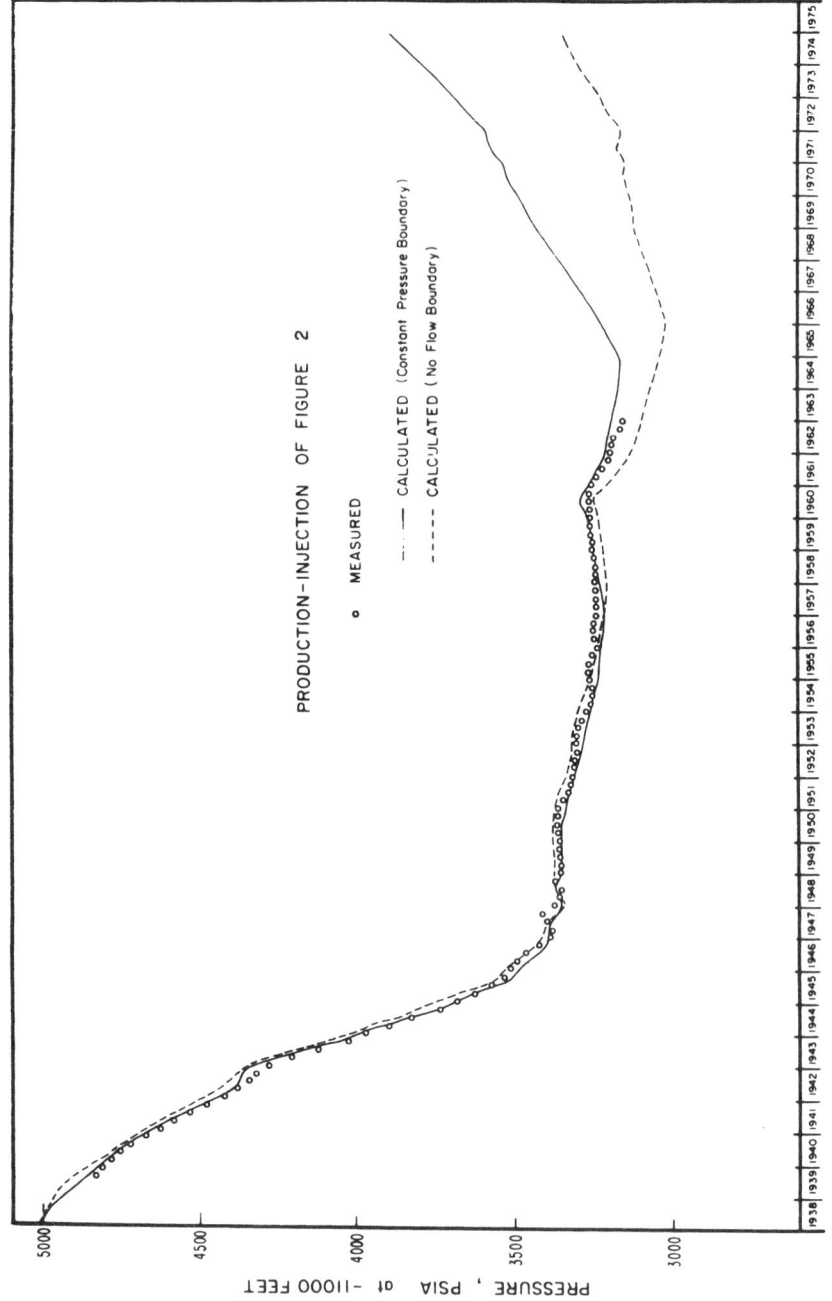

Figure 12.3 Evolution of reservoir pressure simulated by two different models during and after the calibration period, with schedules shown in Figure 12.2

scientific colleagues'. No history matching, however apparently good it may seem, can by its nature contain information about the long-term behaviour. Figures 12.2 and 12.3 illustrate the point. Two equally good history matchings of the data in Figure 12.3, differing only in the choice of the boundary condition, lead to widely diverging future predictions of reservoir pressure with the injection schedule of Figure 12.2, for the North Virden Scallion Unit #1 (Morel-Seytoux 1966).

'Geologic systems are extremely difficult to model over extended periods, and uncertainties are so large that careful attention has to be paid to possibilities of *alternative* conceptual models, incomplete descriptions of natural systems, and difficulty of describing future configurational patterns correctly' (Ewing et al. 1999, p. 942). 'Failure to establish firmly the basis of the conceptual models . . . and failure to *challenge* these conceptual models can have a profound effect on the results' (Ewing et al., p. 943). For over two decades, until recently, the literature was inundated with papers using stochastic groundwater and solute transport equations derived under the assumption that the structure of a spatially varying parameter (such as hydraulic conductivity) could be viewed as one of small random pertubations about a *stationary* mean. This totally ignored the fact that such an approach is incompatible with the realities of lens and layers, the rule rather than the exception in geologic systems. However, this school of thought was so entrenched that it was difficult to publish material that challenged its premises (Morel-Seytoux and Rathnayake 1988; Rathnayake and Morel-Seytoux 1989).

> One way to assess the predictive accuracy of groundwater models is by comparing the actual response of a groundwater system with that predicted by the model, and performing such a comparison a *sufficiently* long time after the prediction was made so that the state of the system at the time of evaluation will not be dominated by its 'memory' of conditions during the calibration period . . . This type of assessment of model reliability has been called a 'postaudit' (Konikow 1986).
>
> (Konikow 1995, p. 61)

One postaudit involves the alluvial aquifer system of the Salt River Valley and the lower Santa Cruz River basin, near Phoenix, Arizona. Much of the deviation between what was predicted and what happened was due to the inacurracy concerning the future pumping rates. As stated previously in Section 4.3 we are not concerned with the validation of the scenarios. However, independently of the errors from the incorrect forecast of pumping amounts and thus the 'bias in the model predictions . . ., the data show a relatively wide scatter, indicating that the model prediction is imprecise. . . . Hence there are *other* significant sources of error' (Konikow 1995, p. 64). Most likely some significant processes were omitted in the description of the system. Naturally no one will omit a significant process knowingly, so that the very modeller, not aware of the omission, is not in a position to realise that an error is introduced, the magnitude of which is totally unknown. It is always in post-auditing by others that the omissions are discovered, though in this particular case it was not clear even in the post-audit, what exactly may have caused the imprecision of the model.

In another paper (Konikow et al. 1997) the authors point out a conceptual defect in the analysis of a HYDROCOIN (Hydrologic Code Intercomparison) variable-density groundwater benchmark problem. Here is an example of a *conceptual* error in a model, as well as lack of judgement on the part of some of the code users.

> The previous numerical implementations by HYDROCOIN teams of a constant-concentration boundary to represent salt release by lateral dispersion only . . . were flawed because this boundary condition allows the release of salt into the flow field by both dispersion and advection. . . . The

importance and sensitivity to the *manner* of *specification* of this boundary does not appear to have been recognised previously in the analysis of the problem.

(Konikow et al. 1997, p. 2253)

What does this interesting case boil down to? It reveals the lack of availability on the part of some models to deal with the fact that a boundary condition must be applied '*on* the boundary' and not '*in* the adjacent cell domain'. The boundary is a line (in two dimensions) not a cell surface. It is a subject that most groundwater textbooks ignore. In fact few groundwater models in use today are probably constructed correctly, including (apparently) MOCDENSE. Most groundwater model-builders seem unwilling to include a boundary node in the block-centred finite difference scheme because it destroys some of the symmetry of the matrices and a great deal of attention to *details* is required to construct the matrix of coefficients. Ewing et al. rightly stated (1999, p. 948) 'As with many complex modelling procedures, *the devil is in the details*'.

What were the consequences of this improper implementation of the constant concentration boundary condition in the models used in the HYDROCOIN project?

The HYDROCOIN project was set up specifically to 'verify' complex numerical models. The particular problem (case 5), discussed by Konikow et al., has no simple analytical solution to serve as a ground truth and basis of comparison. There was no way to 'verify' the model results except by comparison with other 'unverified' models. This checks consistency, not accuracy. All of the original applications for HYDROCOIN were *implemented* in a manner that yielded significant salt release into the flow field by a longitudinal dispersive flux and advective flux of salt from the numerical 'boundary condition', in addition to the stipulated release by transverse dispersion. They were not solving the same problem that was the conceptual basis of the test case. Konikow et al. (1997, p. 2258) were able to make their case by using the code MOCDENSE, with the following block-centred finite difference expression for the flux at the boundary:

$$q_c = - \frac{V_x \alpha_T}{n}(C_s - C) \tag{1}$$

where q_c is the solute flux, V_x is the velocity in a tangential direction to the boundary, α_T is the transverse dispersivity, n is the distance normal to the boundary (to be precise the distance from the boundary to the centre of the adjacent cell, i.e. half the grid cell width), C_s is the specified concentration and C is the average concentration in the cell adjacent to the boundary. They referred to this as an 'equivalent' third-type boundary condition (BC), but it is in reality a *correct* consequence of a constant concentration boundary condition (CCBC) *if* properly implemented. For equation (1) to represent a 'real' third-type BC, C_s would have to be a 'representative value of concentration' outside the domain enclosed by the boundary and C would be the concentration *at* the boundary. However, because the CCBC is apparently not implemented correctly in MOCDENSE, it was necessary to use a trick, consisting of placing an extra grid outside the domain of interest (Konikow et al., 1997, p. 2255, Figure 3c) and thus bypass the MOCDENSE implementation of the CCBC in favour of the equivalent third-type BC. Because a portion of the salt dome is included in the grid of the groundwater model it is necessary to treat it as a porous medium with a hydraulic conductivity, 'several orders of magnitude lower than in the active flow field' (p. 2259). Still if the standard MOCDENSE code was used, the n in equation (1) would be incorrect because in the 'standard' code it would most probably be assigned the value of the distance between the cell centres when in fact it should be half of that. Ignoring this point of detail for now, the following statement of Konikow et al. (1997, p. 2260) remains correct. 'Although having an accurate numerical model is important and must be assured, the differences in accuracy in this problem among the solvers are minimal (or even trivial) relative to the magnitude of the difference in the final solution that is generated by different treatments of the constant concentration boundary condition.' If several specialists do not apply correctly the *simplest* boundary condition in a groundwater model, then clearly in more complicated cases it is very probable that conceptual errors creep rather frequently in models, thus making model comparison an unrewarding and possibly somewhat acrimonious task.

So far the discussion in this section dealt primarily with the problem of conceptual errors for which no quantitative measure can be asserted. Given a model (correct or not) involving a certain number of parameters, is it possible to quantify through calibration on observations the accuracy of the estimation of the parameters and the accuracy of the model itself? Again in reference to confidence limits placed on parameters, Bredehoeft and Konikow (1993, p. 178) stated: 'these confidence limits do not bound errors arising from the selection of a *wrong* conceptual model, or from problems arising with numerical solution algorithms'. Naturally, because textbook statistical procedures exist for linear models, it is easier to limit one's investigations to linear systems and to deal with parameter uncertainty in that context rather than to try to assess the error caused, for example, by 'omission' of certain phenomena. On the other hand, textbook procedures for computing confidence intervals for parameters of general nonlinear models do not exist. Now it appears that Cooley has tried to address this question not only of quantifying the uncertainty on the parameters for a nonlinear model but on the model itself. Were he and his co-workers successful? Before moving to this topic let us recall a few more points from Konikow.

In a discussion of Tsang's paper (1991) perhaps the most important statement, not emphasised in later papers, is: 'Consistency is not equivalent to either accuracy or validity, and a *lack of invalidation* is not equivalent to *validation*' (Konikow 1992, p. 622).

Finally, evaluation of model capabilities for prediction is hampered by the typical lack of complete documentation of previous uses of models. An article attempting a post-audit (Goode and Konikow 1990) illustrates that model 'checking' is difficult because of poor documentation of what previous investigators really did. Is it intentional 'shredding' to leave no trace? The difficulty in finding out what the earlier investigators did exactly is manifest by use of words such as '*Apparently*, the calibrated steady-flow model with no Big Lost River recharge was used' (Konikow 1992, p. 423) and '*Although not described in detail* in his report, we *assume* that this recharge had the same spatial distribution' (p. 419).

12.10.2 Cooley

Over a quarter of a century, Cooley has done research (1) to find efficient numerical ways to estimate model parameters, (2) to provide a measure of their reliability and of the results obtained by using the calibrated models for prediction.

Much could be said (but will not be said) concerning the contribution of Cooley to the development of efficient numerical techniques for parameter estimation and confidence measures. Our discussion here has a narrower focus. To what extent does the information developed in the process of calibration contribute to confidence in the calibrated model?

Because it is not easy to measure in the field the parameters appearing in *physically* based models,

> consequently, such models are usually adjusted by a trial-and-error process until calculated and observed water levels match in 'satisfactory' fashion. . . . These problems have stimulated considerable research into optimisation methods of estimating the hydrogeologic parameters, such as hydraulic conductivity or transmissivity and storativity, although much less research has been directed toward computing sources, sinks and *boundary fluxes.*
>
> (Cooley 1977, p. 318)

[We have seen in the previous section that inaccurate description of such fluxes could introduce significant errors in the computed results; we also discuss in the Appendix how the 'clever' choice of boundary condition can bias the results in favour of a desired outcome.] This 1977 paper describes the necessary steps to convert the original estimation problem formulation to one of an equivalent regression analysis (incorporating the 'ridge regression' technique).

> The method described in this report is, in essence, an elaborate regression procedure where the regression equation is given by (2) or in linearised form by (11). Thus . . . some measures of reliability and goodness of fit are given. . . . It is important to note that these measures are *strictly* correct only for the linearised solution and only apply approximately to the nonlinear model.
>
> (Cooley 1977, p. 322)

The contribution resulting from this conversion is the estimation of a measure of reliability of the parameters and, apparently of the model, since it is said further that 'the presence of normally distributed residuals . . . indicates that . . . there is no reason, based on the observed head data, to reject the notion that the model is correct. It appears that this is a *valid* partial indication of the adequacy of a particular basic model [e.g. the model given by (1)] and its attendant zonation scheme.' (Cooley 1977, p. 323). We note that the indication is 'valid' but the model is only 'adequate'. It is not possible to reject the model as incorrect at some confidence level but no measure of the error of type II, i.e. to accept when it is in fact incorrect (or equivalently no power of the test; Graybill 1961, p. 41) is provided. To calculate the power of the test one would need to have an alternative hypothesis. No such alternative hypothesis is presented (for very practical reasons). It would be very difficult to calculate the power of the test (e.g. Cooley and Naff 1990, p. 42). Finally it must be said that the model for which parameters were estimated assumes that the system is in *steady-state*.

The methodology was used to calibrate parameters on two different aquifers (Cooley 1979). One of the conclusions was:

> The models for the two case studies are characterised by large standard errors for the parameters. This indicates that many widely different sets of parameters would yield similar head distributions. However, in both cases, fit of the model to the data was very good. Thus even in the presence of a moderate amount of noise in the observed head data, the parameters used to define the model are *highly nonunique*.
>
> (Cooley 1979, p. 617)

This undesirable outcome may have led Cooley in two later papers (1982, 1983) to use prior information to alleviate the difficulty, while still maintaining the assumption of a steady-state groundwater. The papers indicate that proper scaling of the parameters and prior information are important to reduce the uncertainty in the parameter estimation.

> Regression analysis using (1) no prior information and (2) prior information, yield sets of parameters that are very similar, but use of prior information results in much smaller standard errors than disregarding it. The interpretation is that the set of parameters computed using no prior information is much better than indicated by the parameter standard errors.
>
> (Cooley 1983, p. 676)

One has to assume that this statement applies only if there *really* exists prior information with some degree of reliability even though it is not used in the parameter estimation. That statement only refers to the case studies and cannot be accepted generally. On page 675 of this paper, one reads: 'Indeed, it is standard modelling practice to assume that agreement between an independently estimated set of parameters and a set resulting from the calibrated model indicates that the model is a good approximation of the real system.' This statement has to be taken with a (large) grain of salt.

Cooley et al. (1986) make a convincing case that it is possible to discriminate between models and rank them *quantitatively*. Using statistical tools (Graybill 1976) and their own variations (Cooley et al. 1986, p. 1769), they test hypotheses between various models and can conclude from the discrimination analysis for example that 'model run 6 or model run 16 can be regarded as the best model' (Cooley et al. 1986, p. 1771). What differentiates these various models (or model runs; 21 of them) is the inclusion or exclusion of some parameters as part of the set to estimate, or different types of zonation. The basic model is the same: a finite element numerical approximation to the 'classical' groundwater equation for an isotropic porous medium in a steady-state condition. The steps in the analysis are somewhat sophisticated (for a clear tutorial description of the subject of hypothesis testing, see Cooley and Naff 1990). However, comparing the numbers in their Table 3 (the value of the test statistics for various hypotheses testing, between model run 1 and 2, then 3 etc. to 10) and those in Table 2, it is quite clear that two model runs with very close estimates of the error variance (corrected for degrees of freedom) will be declared as not significantly different. In the same paragraph the authors continue: 'It should be noted that *best* does not indicate the model that incorporates all known physical features, but means the model that incorporates all significant features, which are those that have a significant effect on model results.'

In several following papers (Cooley and Vecchia 1987; Vecchia and Cooley 1987; Cooley and Hill 1992) Cooley and co-workers have perfected their approach to set confidence and prediction intervals for nonlinear regression models (issued from the numerical formulation of the groundwater equation). The cases tested were hypothetical and led, among others, to the following conclusions. 'Considerations of reasonable values for random errors in the dependent variable in addition to parameter uncertainty can have the effect of considerably widening prediction intervals' (Cooley and Vecchia 1987, p. 581). 'The third data set . . . points to the problems that can arise from extreme nonlinear behaviour of the regression model' (Vecchia and Cooley 1987, p. 1237). In a 1993 paper Cooley sets the stage from the beginning by remarking:

> Application of the methods reviewed by these authors in computing variances and covariances of model output quantities for practical problems again appears to be confined to the linearisation and Monte Carlo methods. Both of these methods have serious drawbacks. Use of linearisation methods is restricted to systems characterised by small parameter variances, and Monte Carlo methods are computationally intensive.
>
> (Cooley 1993a, p. 17)

In the companion paper (Cooley 1993b) procedures are proposed to avoid the Monte Carlo method and the linearisation method, and yet lead to reasonably narrow confidence intervals, as opposed to the method presented in part I of this two paper set. We were inclined to think, at that stage of our study of Cooley's work, that in spite of remarkable technical accomplishments in methodology since Cooley set his mind to the task in the 1970s, by 1993 the goal to obtain both a rigorous and practical quantitative assessment of model uncertainty remained somewhat distant. (However, closer inspection of the work has led the writer to soften that statement as we shall see later on.) We say this in spite of the author's more optimistic (and possibly correct view): 'The method of model construction by hypothesis testing and sequential adding data to update the posterior distribution "*should*" allow efficient construction of a model of known and acceptable uncertainty' (Cooley 1993b, p. 44) and

> In conclusion, this two-part study indicates that model uncertainty can be quantitatively investigated

by using efficient and robust techniques to compute nonlinear prior and posterior confidence intervals for model output. The new methods allow quantitative assessment of improvement in model uncertainty resulting from using observed calibration data in addition to prior information on model parameters.

(Cooley 1993b, p. 45)

The reservations on the part of the writer stem from the fact that careful study of the methodology points to a certain number of assumptions made in the derivation of the otherwise rigorous approach followed, so that a subjective element remains. This is illustrated by statements of the kind: 'Furthermore this writer does not "*believe*" that the condition . . . will be common for most functions . . . of interest in modeling studies. If this condition were to exist, then the stationary point still "*might*" not represent a value more extreme than occurs in the contour.' (Cooley 1993a, p. 27). Our suspicion is confirmed by a statement by Cooley, four years later (which now, I realise, has to be interpreted as applying to other work, not his own!):

However, most standard methods of calculating confidence intervals are either approximate or can be applied only to restricted types of models. . . . Probably for these reasons, uncertainty analysis techniques proposed for ground-water models have often involved approximating means and variances only, with probabilistic inferences such as those associated with confidence intervals being omitted.

(Cooley 1997, p. 869)

In this paper four methods of establishing confidence limits are discussed. Two of them were used by Cooley for the first time: the original bootstrap method and a newer regression-based bootstrap one. Ultimately for practical and accuracy reasons the latter two are not considered useful alternatives. 'In contrast, the likelihood method appears to be very promising for what may be a wide range of problems'. Here comes the writer's change of opinion. Further study of the crucial paper (Cooley and Vecchia 1997) confirms the optimism of the authors. A parameter domain is defined by the equation:

$$\underline{q}^T(\underline{b})\underline{\underline{\omega}}\,\underline{q}(\underline{b}) \leq d^2_{1-\alpha} \tag{2}$$

where the superscript T indicates the transpose of a vector, the underbar a column vector, the double underbar a matrix, $\underline{b}$ is a set of parameters, $q(.)$ a function of the parameters, ω a known problem-dependent symmetric positive definite matrix, $1 - \alpha$ a confidence level (say 95%) and $d^2_{1-\alpha}$ a parameter defined such that the probability that (2) be true is $1 - \alpha$, given that random $\underline{B}$ comes from a prior distribution for the reasonable domain of the true parameters. Given such probability density function, say $f_{\underline{B}}(\underline{b})$, then it is possible to obtain confidence limits on the parameters and a function of these parameters by an optimisation problem. One difficulty in the approach is finding the value of $d^2_{1-\alpha}$ but the authors point out that it can be done relatively easily using a Monte Carlo technique to generate a sufficient number of random vectors $\underline{B}$ from the probability density function (pdf) $f_{\underline{B}}(\underline{b})$ (for details see Cooley and Vecchia 1997, top of right column of p. 585). Thus the approach is exact and feasible. Elements of subjectivity remain in the choice of the functions $\underline{q}(\underline{b})$ and $f_{\underline{B}}(\underline{b})$. The authors point out that 'for the method to be of practical use, the information required to determine the pdf for $\underline{B}$ must be kept to a minimum. Hence, it is assumed that at most only three types of information on the parameters are available: calibrated values, extreme values and ordering' (Cooley and Vecchia 1997, p. 583). The continued work in later papers on the determination of confidence and prediction intervals has to be understood as efforts not to improve on the exactness of the method, since it is exact, but in finding *other exact* methods leading to *more constrained* and *smaller* confidence regions and/or *approximate* yet good methods that are less computationally demanding. Cooley (1993a) was successful in the former attempt by finding a more judicious function $q(\underline{b})$ as a linear combination of the original quadratic form and a log(pdf) function. There is, however, a subjective element that remains in finding confidence limits for an arbitrary function of the parameters, $g(\underline{b})$: 'The sizes of confidence intervals are controlled by the interaction of the variability and degree of nonlinearity of $g(\underline{b})$ over parameter space with the size and shape of the confidence region for β (the true parameter vector)' (Cooley 1993a, p. 30) because the $f_{\underline{B}}(\underline{b})$ prior pdf function is subjective.

The next paper (Hill et al. 1998) provides a very comprehensive test of a nonlinear regression calibration procedure. It is in part a rebuttal to those who doubted its merits!

> Nonlinear regression was introduced to groundwater modeling in the 1970s, but has been used little to calibrate numerical models of complicated groundwater systems. Apparently, nonlinear regression is *thought* by many to be incapable of addressing such complex problems. With what we believe to be the *most complicated synthetic test case* used for such a study, this work investigates using nonlinear regression in groundwater model calibration.
>
> (Hill et al. 1998, p. 520)

The goals of the paper are clearly established: '(1) to present an approach that makes nonlinear regression methods more useful for the types of problems typical of groundwater; (2) to use a synthetic test case to evaluate the method and some general issues of model calibration.' (Hill et al. 1998, p. 521). There is no doubt that the authors have developed an effective procedure to estimate model parameters. Let us remember our focus. Is the model thus calibrated, validated? In Mood and Graybill terms (1963, p. 3): 'The use of statistical tools is not merely a matter of picking out the wrench that fits the bolt; it is more a matter of selecting the correct one of several wrenches which *appear* to fit the bolt *about equally well but none of which fit it exactly*'. The authors describe convincingly a procedure that allows them to select the *correct* model in the sense of Mood and Graybill. Some results of this comprehensive paper are quite curious. 'Thus, for this problem, using prior information to force optimal parameters values to be realistic reduced model accuracy; using prior information to include complexity not directly supportable by the other data improved model accuracy' (Hill et al. 1998, p. 527). This statement prepares the conclusion: 'Taken with conclusion 2, conclusion 5 produces a dilemma, because unrealistic optimal parameter values (as determined by comparing optimal and measured values) are said to indicate a less accurate model, but parameter values cannot be expected to equal measured values because parameter values are accommodating model errors' (Hill et al. 1998, p. 533). Possibly what that says is that if prior information is made more constraining the model will be less accurate, but on the other hand it may lead to narrower confidence and prediction regions. Unfortunately the paper did not calculate these. In other words if Figure 10 of the paper showed confidence intervals in addition to the best estimates for *different* models, the conclusion as to which was the best model may have been altered. Finally, the case study, to be fully authoritative, suffers from the fact that it deals only with a steady-state case and the treatment of the river is unrealistic in not including the fluctuations in stage that accompany fluctuations in discharges typical of most rivers with the seasons and years and the resulting change in 'reach transmissivity' (Morel-Seytoux et al. 1973, alias stream conductance).

12.10.3   Workshop Proceedings: 'Characterization and Measurement of the Hydraulic Properties of Unsaturated Porous Media' (M.Th. van Genuchten et al. 1999)

Little has been said of the validation of models for the unsaturated zone in this chapter. Yet an entire workshop was recently devoted to the matter of securing information about the parameters that would be present in models of water flow and solute transport in the vadose zone. From the views expressed by the participants can one feel comfortable regarding the ability to estimate the parameters appearing in the equations? My impression is not at this moment. Let us thus review some statements coming from various authors in these Proceedings.

> Since frequently *no* measurements of the unsaturated hydraulic conductivity are available, it is *popular* to derive the shape of the unsaturated conductivity function from the water retention curve.
>
> (Durner et al. 1999, p. 817)

> There is a significant deviation between observed and simulated outflow during step 2 and 3, i.e. at the pressure range near the air entry point, which prevails for any model. Even the B6 model that fits all other stages of the outflow perfectly, shows this deviation. We attribute this to dynamic effects, which cannot be described by a time-invariant model of hydraulic properties.
>
> (Durner et al. 1999, p. 825)

We note that in this paper no confidence limits for the parameters, nor prediction intervals for the simulation variables, nor performance check are provided.

'Dual porosity models have recently been developed . . . but application of these models is hampered by a lack of information concerning the nature of hydraulic conductivity near saturation.' (Jarvis et al. 1999, p. 839.) 'Therefore one important topic for future research is the incorporation of description of time-varying near-saturated hydraulic conductivity of the surface soil' (Jarvis et al. 1999, p. 846).

'Since fracture and matrix are not in equilibrium during transient flow events, it is important to determine whether the tensiometer gauge pressure reflects the fracture or matrix components or how the two are combined. Moreover, the tensiometer itself may lead to local redistribution of water between the fracture and the matrix, affecting the system to be studied' (Finsterle and Faybishenko 1999, p. 867).

> A comparison of simulation with the classical Richards' model and a two-phase flow model show that dynamic effects are to a considerable part caused by non-negligible resistance of air flow. . . . If dynamic nonequilibrium occurs between the water content and the water potential during transient water flows the presently used standard simulation technique for describing saturated/unsaturated flow cannot be reliably applied under transient flow conditions.
>
> (Schultze et al. 1999, p. 877)

> Furthermore, the phenomenon is interfering with the description of hysteresis in the hydraulic properties; without separating the two processes it is not possible to investigate hysteresis in a reliable manner . . . The importance of nonequilibrium effects in simulation of field scale water flow under natural conditions is not known presently, and should be investigated in the future.
>
> (Schultze et al. 1999, p. 890)

> Our results show that: 1) Performance of the tested Pedotransfer functions, PTF, was quite variable. Results ranged from acceptable to very poor . . . 2) the validity of a given PTF should not be considered as general, which in turn implies that further work is required to define more precisely the range of soil types over which each PTF performs correctly.
>
> (Bastet et al. 1999, p. 990)

> Many methods have been proposed for measuring or estimating these hydraulic properties, varying considerably in their time requirement. There have been relatively few attempts to evaluate their performance. . . . Direct comparison of the properties showed varying degrees of agreement, with more than an order of magnitude difference between parameters in some cases. . . . Results show that the differences between the simulations using different hydraulic property sets were much smaller

than might be expected from the large differences in the properties themselves. . . . These conclusions are formally restricted to the scenarios evaluated, and we recommend that further scenarios be investigated.

(Bond et al. 1999, p. 1161)

Part of the problem with current investigations in the vadose zone is that it is performed mostly with agronomic and agricultural applications in mind. Thus much of the interest is in column or plot hydrology, not in watershed hydrology. The earlier cited comment by Hill et al., namely that 'parameter values cannot be expected to equal measured values because parameter values are accommodating model errors' leads us to expect that model errors in the vadose zone will be more significant that in aquifers and the measured punctual field values will have even less relevance for watershed vadose zone models. Grayson et al. (1992) said so much:

> Measured parameter values do not integrate the response of the 'elemental' area, and there is an inconsistency in scale between that used in measurement of field variables and the way in which they are applied in models. This conflict caused Klemes (1986, p. 187S) to remark, 'It also seems obvious that search for new measurement methods that would yield areal distributions, or at least reliable areal totals or averages of hydrologic variables . . . would be a much better investment for hydrology than the continuous pursuit of a perfect massage that would squeeze the nonexistent information out of the few poor anaemic point measurements.' This sentiment could be extended to include infiltration and hydraulic resistance parameters. The result of these approximations is that the parameters lose their physical significance and the model becomes limited in terms of predictive capability because parameter values cannot be determined a priori. In addition, there are complex interactions between parameter values and the fundamental process descriptions.
>
> (Grayson et al. 1992, p. 2660)

Possibly one paper (Morel-Seytoux 1999, pp. 1425–1437) in the Proceedings addressed this issue of scale and the matter of these complex interactions.

12.11 CONCLUSIONS

'One can draw valid probabilistic conclusions about the sampled population, but one must use personal judgement to extrapolate to the target population, and the reliability of the extrapolation cannot be measured in relative-frequency–probability terms' (Mood and Graybill 1963, p. 142). Models typically will be used to make forecasts of what may happen in the future and as a result it is impossible to tell if the sampled population will remain representative of the target population. A model can be used for scientific prediction, that is prediction within the bounds of the theory, processes and conditions used to construct the model. It generally cannot safely be used for prediction into the future without regard for possible changes in the underlying system characteristics that were used to construct the model. In other words 'validation' in the layman sense is not attainable. For unsaturated flow and transient systems, it may not be possible currently to do it even in the technical sense as the tools for statistical analysis are not well developed.

   Calibration methodology, on the other hand, at least for groundwater flow under steady-state conditions, has made considerable progress and scientific prediction is possible. Cooley and his co-workers, as we have seen in this chapter, have developed appropriate procedures to assess uncertainty of parameters (confidence intervals), values of model-computed variables (confidence intervals), and predictions of these same variables (prediction intervals). Prediction intervals, as defined in the statistical literature, assume that the same model as used for calibration also applies for prediction. As an extension, anything new that is added must be known and have known prediction uncertainty. Proper use of prediction intervals must be based on these caveats. Thus, prediction intervals must be very carefully interpreted and used.

The writer wonders why Cooley has limited his efforts to 'steady-state' conditions. Is it because in order to use the statistical methods it is necessary to make the assumption that the model is correct and Cooley is of the opinion that the theory of transient groundwater flow is more approximate than its steady-state counterpart? Testing the calibration and statistical methods on actual field data would be further complicated by the presence of serious lack of model-fit problems. One can only hope that Cooley (and many others) now will bite the bullet and address the calibration problem for transient situations and on actual field data.

The problem is even more difficult when we include the possibility of substantial model error from sources such as heterogeneity of a smaller scale than included in the model. In this case could not the formulation developed for stochastic groundwater flow by Nachabe and Morel-Seytoux (1995a,b), which is not limited by assumptions either of smallness of the variance of the model parameters or of stationarity of the parameter means, be used in a nonlinear regression formulation? This would open the possibilities of determination of confidence and prediction intervals, following Cooley's well seasoned approach, which are fundamental to assess the reliability of a groundwater model.

APPENDIX

Background

The Hydrologic Engineering Center (HEC) was somewhat puzzled by the fact that three different studies of a proposed pumping plan led to considerably different results, even though they used the same data and the same model. A review of the three studies was requested by the US Army Corps of Engineers and the review was submitted in a report (McLaughlin 1984). The purpose of the report was not primarily to review the studies conducted by the consulting firms, but rather to discuss the 'technical choices which influence the predictions made by groundwater models' (McLaughlin 1984, p. 1). The investigations were viewed merely as case studies to support the primary purpose of the report. At the suggestion of Arlen Fedman, chief of the Research Branch, HEC, the writer read the report and sent comments in a letter. As opposed to McLaughlin's, the writer's concern, as quoted in the letter below, was primarily now focused on the reason for the differences in the results.

Letter, Dated 5 November 1985

The report is a good introduction to modelling (McLaughlin 1984, pp. 18–53) with quite a few useful tips to avoid pitfalls. There is no doubt that, in an aquifer with little previous development, a large concentration of wells with high pumping rates compared to previous pumpage will induce a cone of depression which can be well described by the Theis equation for an average transmissivity representative of the pumping zone (in this case the Ciniza well field). As McLaughlin states (1984, p. 69): 'Jacob's solution . . . is consistent with the results of the case study – the predicted drawdown was greater when T and S were assumed to be small (Model 2) and less when T and S were assumed to be large (Model 3). This is a physically plausible result since head gradients must become steeper as conductivity and compressibility decrease if a given pumping rate is to be sustained' (McLaughlin 1984, p. 71). The Theis solution (*because* it is a solution of the groundwater equation) *has to* reflect that factor, and so will a numerical groundwater model. In GENERAT (documented in AQUISIM), an important component of SAMSON, we have incorporated the Theis solution to generate the unit hydrograph ordinates (known also as discrete kernels) of drawdowns due to pumping to describe more accurately the cone of depression in the vicinity of large capacity wells, which tends to be dramatically smoothed by the finite difference modelling procedure when the grid size is large. As stated in the report 'it should be remembered that the spatial resolution of a discretised model is limited by the size of its elements or grid cells. Pumpage from a point located within a given element is *effectively spread* over the element' (McLaughlin 1984, p. 47).

'It follows from the brief review presented that the dominant inputs in the San Andres–Glorieta case study must be the aquifer parameters, particularly the transmissivity and storage coefficient values' (McLaughlin 1984, p. 69). In some respect I disagree with this statement. These parameters will affect considerably (re: Theis solution) the local drawdowns near a concentrated pumpage centre. However, the selected boundary conditions and the selection of the boundary locations where the boundary conditions are applied, have a far more important influence on the impact of the pumping wells at the *regional* scale. It is well known that if a straight boundary near a pumping well is considered a no-flow boundary (i.e. as done in Model 2) the drawdown will be as much as twice the drawdown calculated when the boundary is considered a constant head one (i.e. as done in Model 3). In addition, in one case (no flow boundary) no steady-state can be reached, whereas in the second case (constant head) steady-state is reached *regardless* of the *magnitude* of the pumping rate. I believe that this difference in boundary conditions between Models 2 and 3 is as important as the differences in transmissivities and has more influence on the solution away from the Ciniza well field. In my classes I tend to exaggerate (to make a point) and I will tell my students that in groundwater the *only* acceptable boundary condition is a *no flow boundary* condition so that in any study, even fairly local, one should include a sufficiently large area in order to reach the limits of the aquifer. The problem with a constant head boundary is that it *creates* water, a limitless supply of it at that (perpetual machine). Thus the assumption of a constant head boundary is only justifiable at the sea shore, by the great lakes and along the mighty Mississippi. I have seen people treat a South Platte or a Rio Grande as a constant head boundary and I can bet that the rivers provided more inflow to the aquifer than the USGS gauged upstream flow (these models of course *do not keep track* of surface flows). Unless the aquifer is a mighty one it is not reasonable to assume that heavy pumpage will not affect the heads at the boundaries. It is like assuming a hydraulic jump at the boundary. The mighty surrounding aquifer is super critical. Short of a no flow boundary, then a third type boundary condition should be used. That boundary condition is one that specifies neither the head nor the flux but a relation between them. It is the boundary condition that applies to describe a stream–aquifer interaction. For this reason I often call it the stream–aquifer boundary condition. The relation involves a parameter (boundary conductance) which requires careful identification. However, I am convinced that a 'sensitivity analysis' would show that the solution is less sensitive to that value than to the selected value of head at the boundary. In addition a 'desk-top' analysis of the surrounding aquifer can provide an estimate of that boundary conductance which depends on the flow convergence and values of transmissivity outside the study area. By the way the 'line of discharge' (McLaughlin 1984, p. 65) used in Model 2 is an indirect procedure to change the no-flow condition to a 3rd type but without attempting to estimate the boundary conductance.

In summary, given the scant information about Models 1, 2, 3 results (that was a disappointment, very little was given in terms of transient behaviour of system, calculated fluxes along the constant head boundaries etc.) it is nevertheless clear that by their very assumptions (aquifer parameters, types of boundary conditions) Models 1 and 3 were optimistic (they wanted to *limit* drawdowns and to *create* water) and Model 2 was pessimistic (they wanted to predict *scary* drawdowns and they were not about to *concede* that there was an available groundwater supply). 'What is less obvious is why qualified hydrologists should derive such different input estimates from the same database. This issue is clearly worth further investigation' (McLauglin 1984, p. 59). I have no doubt the hydrologists were competent. In fact they knew *very well* what parameters to choose and what assumptions to make in order to obtain results that would meet their clients' desire. Obviously Models 1 and 3 were developed for a client that wanted the developments to proceed and Model 2 was carried out for a client that did not favour the development. What *is* needed is an *independent* study, from a party that has no axe to grind (hint) . . .

Changing subject somewhat, one of my concerns with SAMSON is that to a large degree its use would curb the subjectivity that this report uncovers. I suspect that various parties (consulting engineers, water lawyers etc.) that thrive on those ambiguities are not interested in the application of such a tool that would restrict considerably their latitude for playing the game. There is no room in SAMSON to 'create' or for that matter to 'embezzle' water. Recently I read an article commenting on the Navy spy scandal. The writer was concerned with the permeating attitude in society, condoned by the current administration, the 'go-get it' attitude. The last three US attorney generals have ended in jail. So why should not a petty officer be adept at free enterprise? And the writer to conclude with this gem: 'Moral Majority, where are you when we need you?'

ACKNOWLEDGEMENTS

I want to thank Lenny Konikow and Dick Cooley for many useful exchanges on the subject by regular or e-mail. Their prompt, open and kind cooperation was appreciated. Naturally I am solely responsible for the opinions expressed in this chapter regarding their work. This chapter was prepared while the writer participated in the volunteer programme of the US Geological Survey with the team of John R. Nimmo in Menlo Park, California. I extend my thanks to John for providing me with a convenient access to the USGS facilities, besides doing research together.

I regret that time did not permit me to review other works on the subject, in particular that of Bill Yeh (and truly many others). I hope that I shall be able to do so in later contributions on the subject.

REFERENCES

Bastet G., Bruand, A., Voltz, M., Bornand, M. and Quétin, P. 1999. Performance of available pedotransfer functions for predicting the water retention properties of French soils. In: van Genuchten, M.Th., Leij, F.J. and Wu, L. (eds), *Proceedings of International Workshop on Characterization and Measurement of the Hydraulic Properties of Unsaturated Porous Media*, Riverside, California, 22–24 October 1997. Published by University of California, Riverside, 981–991.

Bond W.J., Cresswell, H.P., Verburg, K. and McKenzie, N.J. 1999. Functional comparison of methods for obtaining soil hydraulic properties. In: van Genuchten, M.Th., Leij, F.J. and Wu, L. (eds), *Proceedings of International Workshop on Characterization and Measurement of the Hydraulic Properties of Unsaturated Porous Media*, Riverside, California, 22–24 October 1997. Published by University of California, Riverside, 1161–1172.

Bredehoeft, J.D. and Konikow, L.F. 1993. Ground-water models: validate or invalidate? *Ground Water*, **31**(2), 178–179.

Cooley, R.L. 1977. A method of estimating parameters and assessing reliability for models of steady-state groundwater flow, 1, theory and numerical properties. *Water Resources Research*, **13**(2), 318–324.

Cooley, R.L. 1979. A method of estimating parameters and assessing reliability for models of steady-state groundwater flow, 2, application of statistical analysis. *Water Resources Research*, **15**(3), 603–617.

Cooley, R.L. 1982. Incorporation of prior information on parameters into nonlinear regression ground-water flow models, 1, theory. *Water Resources Research*, **18**(4), 965–976.

Cooley, R.L. 1983. Incorporation of prior information on parameters into nonlinear regression ground-water flow models, 2, applications. *Water Resources Research*, **19**(3), 662–676.

Cooley, R.L. 1993a. Exact Scheffe-type confidence intervals for output from groundwater flow models, 1, use of hydrogeologic information. *Water Resources Research*, **29**(1), 17–33.

Cooley, R.L. 1993b. Exact Scheffe-type confidence intervals for output from groundwater flow models, 2, combined use of hydrogeologic information and calibration data. *Water Resources Research*, **29**(1), 35–50.

Cooley, R.L. 1997. Confidence intervals for ground water models using linearization, likelihood and bootstrap methods. *Ground Water*, **35**(5), 869–880.

Cooley, R.L. and Hill, M.C. 1992. A comparison of three Newton-like nonlinear least-squares methods for estimating parameters of ground-water flow models. In: Russell, T.F., Ewing, R.E., Brebbia, C.A., Gray, W.G. and Pinder, G.F. (eds), *Computational Methods in Water Resources IX, Vol. 1, Numerical Methods in Water Resources – Proceedings of the 9th International Conference on Computational Methods in Water Resources*, Denver, CO, 1992. Elsevier, New York, 379–386.

Cooley, R.L. and Naff, R.L. 1990. Regression modeling of ground-water flow. In: *Techniques of Water-Resources Investigations of the US Geological Survey*, book 3, chapter B4. US Geological Survey Publication, Reston.

Coley, R.L. and Vecchia, A.V. 1987. Calculation of nonlinear confidence and prediction intervals for ground-water flow models. *Water Resources Bulletin*, **23**(4), 581–599.

Cooley, R.L., Konikow, L.F. and Naff, R.L. 1986. Nonlinear-regression groundwater flow modeling of a deep regional aquifer system. *Water Resources Research*, **22**(13), 1759–1778.

Durner W., Priesack, E., Vogel, H.J. and Zurmühl, T. 1999. Determination of parameters for flexible hydraulic functions by inverse modeling. In: van Genuchten, M.Th., Leij, F.J. and Wu, L. (eds), *Proceedings of International Workshop on Characterization and Measurement of the Hydraulic Properties of Unsaturated Porous Media*, Riverside, California, 22–24 October 1997. Published by University of California, Riverside, 817–829.

Ewing, R.C., Tierney, M.S., Konikow, L.F. and Rechard, R.P. 1999. Performance assessments of nuclear waste repositories: a dialogue on their value and limitations. *Journal of Risk Analysis*, **19**(5) 933–958.

Finsterle S. and Faybishenko, B. 1999. What does a tensiometer measure in fractured rock? In: van Genuchten, M.Th., Leij, F.J. and Wu, L. (eds), *Proceedings of International Workshop on Characterization and Measurement of the Hydraulic Properties of Unsaturated Porous Media*, Riverside, California, 22–24 October 1997. Published by University of California, Riverside, 867–875.

Goode, D.J. and Konikow, L.F. 1990. Re-evaluation of large-scale dispersivities for a waste chloride plume: effects of transient flow. In: Kovar, K. (ed.), *ModelCARE 90: Calibration and Reliability in Groundwater Modeling*. IAHS Publication No. 195, 417–426.

Graybill F.A. 1961. *An Introduction to the Linear Statistical Models*. McGraw-Hill, New York.

Graybill F.A. 1976. *Theory and Application of the Linear Model*. Duxbury, North Scituate, MA.

Grayson, R.B., Moore, I.D. and McMahon, T.A. 1992. Physically based hydrologic modeling. 2. Is the concept realistic? *Water Resources Research*, **26**(10), 2659–2666.

Hill, M.C. 1998. *Methods and Guidelines for Effective Model Calibration, with Application to UCODE, a Computer Code for Universal Inverse Modeling*. US Geological Survey, Water Resources Investigations Report 98–4005.

Hill, M.C., Cooley, R.L. and Pollock, D.W. 1998. A controlled experiment in ground water flow model calibration. *Ground Water*, **36**(3), 520–535.

Jarvis N., Messing, I., Larsson, M.H. and Zavattaro, L. 1999. Measurement and prediction of near-saturated hydraulic conductivity. In: van Genuchten, M.Th., Leij, F.J. and Wu, L. (eds), *Proceedings of International Workshop on Characterization and Measurement of the Hydraulic Properties of Unsaturated Porous Media*, Riverside, California, 22–24 October 1997. Published by University of California, Riverside, 839–850.

Jobson, H.E. and Harbaugh, A.W. 1999. Modifications to the diffusion analogy surface-water flow model (DAFLOW) for coupling to the modular finite difference ground-water flow model (MODFLOW). *US Geological Survey*, Open-File Report 99-217.

Klemes, V. 1986. Dilettantism in hydrology: transition or destiny? *Water Resources Research*, **22**(9), 177S–188S.

Konikow, L.F. 1986. Predictive accuracy of a ground-water model – lessons from a postaudit. *Ground Water*, **24**(2), 173–184.

Konikow, L.F. 1992. Discussion of 'The modeling process and model validation' by Chin-Fu Tsang. *Ground Water*, **30**(4), 622–623.

Konikow, L.F. 1995. The value of postaudits in groundwater model applications. In: El-Kadi, A.I. (ed.), *Groundwater Models for Resources Analysis and Management*. Lewis Publishers, Boca Raton, 59–78.

Konikow, L.F. and Bredehoeft, J.D. 1992. Ground-water models cannot be validated. *Advances in Water Resources*, **15**(1), 75–83.

Konikow, L.F., Campbell, P.J. and Sanford, W.E. 1996a. Modeling brine transport in a porous medium: a re-evaluation of the HYDROCOIN Level 1, Case 5 problem. In: Kovar, K. and Van Der Heijde, P. (eds), *Calibration and Reliability in Groundwater Modeling. Proceeding ModelCARE '96 Conference*. IAHS Publication No. 237, 363–372.

Konikow, L.F., Goode, D.J. and Hornberger, G.Z. 1996b. A three-dimensional method-of-characteristics solute-transport model (MOC3D). *US Geological Survey Water-Resources Investigations Report 96-4267*.

Konikow, L.F., Sanford, W.E. and Campbell, P.J. 1997. Constant-concentration boundary condition: lessons from the HYDROCOIN variable-density benchmark problem. *Water Resources Research*, **33**(10), 2253–2261.

McLaughlin, D.B. 1984. *A Comparative Analysis of Groundwater Model Formulation – The San Andres–Glorieta Case Study*. Hydrologic Engineering Center, US Army Corps of Engineers, June 1984.

Mood, A.M. and Graybill, F.A. 1963. *Introduction to the Theory of Statistics*. McGraw-Hill, New York.

Morel-Seytoux, H.J. 1966. *Flow of Immiscible Fluids in Porous Media*. Chevron Research Company, Technical Memorandum, September 1966.

Morel-Seytoux, H.J. 1983a. Infiltration affected by air, seal, crust, ice and various sources of heterogeneity (special problems). *Proceedings of ASAE National Conference on Advances in Infiltration*, Chicago, 12–13 December 1983, 132–146.

Morel-Seytoux, H.J. 1983b. Superiority of two-phase formulation for infiltration. *Proceedings of ASAE National Conference on Advances in Infiltration*, Chicago, 12–13 December 1983, 34–47.

Morel-Seytoux, H.J. 1985. Conjunctive use of surface and ground waters. In: Asano, T. (ed.), *Artificial Recharge of Groundwater*. Butterworth Publishers, Boston, 35–67.

Morel-Seytoux, H.J. 1999. Infiltration characteristics: from column to parcel, to hillslope. A physical and stochastic theory for spatial, temporal and process integration. In: van Genuchten, M.Th., Leij, F.J. and Wu, L. (eds), *Proceedings of International Workshop on Characterization and Measurement of the Hydraulic Properties of Unsaturated Porous Media*, Riverside, California, 22–24 October 1997. Published by University of California, Riverside, 1425–1437.

Morel-Seytoux, H.J. and Khanji, J. 1975. Equation of infiltration with compression and counterflow effects. *Hydrological Sciences Bulletin*, **XX**(4), 505–517.

Morel-Seytoux, H.J. and Restrepo, J.I. 1987a. *SAMSON. Vol. I: User's Manual*. Colorado Water Resources Research Institute, Colorado State University, Fort Collins.

Morel-Seytoux, H.J. and Restrepo, J.I. 1987b. *SAMSON. Vol. II: Calibration to the South Platte River Basin*. Colorado Water Resources Research Institute, Colorado State University, Fort Collins.

Morel-Seytoux, H.J. and Restrepo, J.I. 1987c. *SAMSON. Vol. III: Introduction to Cyber 205*. Colorado Water Resources Research Institute, Colorado State University, Fort Collins.

Morel-Seytoux, H.J. and Rathnayake, R.M.D. 1988. A radical departure from traditional description of solute transport in a large aquifer system. Proceedings 8th Annual Hydrology Days, 19–21 April 1988. HYDROLOGY DAYS Publications, Fort Collins, CO, 294–313.

Morel-Seytoux, H.J. and Nimmo, J.R. 1999. Soil water retention and maximum capillary drive from saturation to oven dryness. *Water Resources Research*, **35**(7), 2031–2041.

Morel-Seytoux, H.J., Young, R.A. and Radosevich, G. 1973. Systematic design of legal regulations for optimal surface–groundwater usage. Final Report to OWRR for first year of study, Environmental Resources Center, Colorado State University, Completion Report Series No. 53, August 1973.

Morel-Seytoux, H.J., Illangasekare, T., Bittinger, M.W. and Evans, N.A. 1980. Potential use of a stream–aquifer model for management of a river basin: case of the South Platte River in Colorado. In: IAWR (ed.), *Proceedings of International Association on Water Pollution Research Specialized Conference* on 'New Developments in River Basin Management', Cincinnati, OH, 29 June–3 July 1980, 1975–1987.

Nachabe, M.H. and Morel-Seytoux, H.J. 1995a. Scaling the ground water flow equation. *Journal of Hydrology*, **164**, 345–361.

Nachabe, M.H. and Morel-Seytoux, H.J. 1995b. Perturbation and Gaussian methods for stochastic flow models. *Advances in Water Resources*, **18**(1), 1–8.

Rathnayake, D.H. and Morel-Seytoux, H.J. 1989. Nondispersive description of dispersion for simulation of solute transport in a large system. Proceedings Symposium on 'Groundwater Contamination'. IAHS Third Scientific Assembly, 10–19 May 1989. IAHS Publication No. 185, 3–10.

Remson, I., Hornberger, G.M. and Moltz, F.J. 1971. *Numerical Methods in Subsurface Hydrology with an Introduction to the Finite Element Method*. Wiley Interscience, New York.

Rykiel, E.J. 1994. Letters, *Science*, 264, 329.

Schultze, B., Ippisch, O., Huwe, B. and Durner, W. 1999. Dynamic nonequilibrium during unsaturated water flow. In: van Genuchten, M.Th., Leij, F.J. and Wu, L. (eds), *Proceedings of International Workshop on Characterization and Measurement of the Hydraulic Properties of Unsaturated Porous Media*, Riverside, California, 22–24 October 1997. Published by University of California, Riverside, 877–892.

Tsang, C-F. 1991. The modelling process and model validation. *Groundwater* **29**(6), 825–831.

Van Genuchten, M.Th, Leij, F.J. and Wu, L. 1999. Characterisation and measurement of the hydraulic properties of unsaturated porous media. University of California, Riverside, CA.

Vecchia, A.V. and Cooley, R.L. 1987. Simultaneous confidence and prediction intervals for nonlinear regression models with application to a groundwater flow model. *Water Resources Research*, **23**(7), 1237–1250.

Yeh, W.W-G. 1986. Review of parameter identification procedures in groundwater hydrology: the inverse problem. *Water Resources Research*, **22**(2), 95–107.

13   Validation of Hydraulic Models

PAUL D. BATES AND MALCOLM G. ANDERSON
School of Geographical Sciences, University of Bristol, UK

13.1   MODELS IN THE EARTH SCIENCES

> Propositions concerning empirical matters of fact . . . I hold to be hypotheses, which can be probable
> but never certain. And in giving an account of their method of validation I claim also to have
> explained the nature of truth.
>
> (Ayer 1936, p. 31)

A cursory examination of any major earth science journal will demonstrate the current fascination of the scientific community with numerical models. There are very good reasons for this:
models allow processes and outcomes to be connected and unlike statistical approaches may
demonstrate causality. The insight models provide may be viewed as holistic compared to the
limited data sets that can be collected by field investigations and they have at least the potential
to make extrapolative prediction in a way that regression-based models find difficult. Models
often have an elegance that appeals to their practitioners but more importantly they allow
significant science on a small budget. Which is easier: the development of a numerical model of
ocean hydrodynamics or the implementation of an extensive oceanic monitoring programme to
collect a similar range of data? With more powerful computing facilities the sophistication of the
science that can be conducted is increasing rapidly and reinforces this position.

In moving to more impressive applications we should not lose sight of the fact that models are
just another tool for conducting science and should not be dissociated from basic scientific
methods and philosophy. As the above quote from A.J. Ayer emphatically suggests, model
parameters and predictions should be considered as propositions concerning empirical matters
of fact that can be verified (see Ayer 1936) or falsified (see Popper 1959). A model is thus both an
embodiment of a hypothesis and the numerical experiment to facilitate hypothesis testing, and
hence formally links epistemology and methodology (see Hanfling 1981). According to Harvey
(1969, p. 35) such a posteriori models serve to express the notions contained in a theory in
different form such as mathematical notation or on into computer code. Braithwaite (1953, pp.
88–91) clarifies the definition of models by saying that they are the end point of a deductive
calculus that commences with theory and points out:

> Thus there are great advantages in thinking about a scientific theory through the medium of thinking
> about a model; to do this avoids the complications and difficulties involved in having to think
> explicitly about the language or other form of symbolism by which the theory is represented. The use
> of models allows of a philosophically unsophisticated approach to an understanding of the structure
> of a scientific deductive system.
>
> (Brathwaite 1953, pp. 92–93)

Model Validation: Perspectives in Hydrological Science. Edited by M.G. Anderson and P.D. Bates.
© 2001 John Wiley & Sons, Ltd.

Further, a model is only a legitimate hypothesis if we can state what observations would lead us to verify or falsify it. In deciding the criteria under which we either verify or falsify a model–hypothesis we are making a judgement concerning what constitutes proof, and therefore our current best interpretation of reality. Conclusive verifiability may be impossible for many scientific problems and has led Ayer to distinguish between 'strong' verification, where the truth of a proposition (in this case a model) can be conclusively established, and 'weak' verification, where the likelihood of a proposition's truth must be expressed in terms of probability. Most environmental modelling problems belong to the latter class. While model methodological flaws (over-calibration and uncontrolled equifinality, indeterminate effective parameters, the difficulty of measuring what a model predicts, the difficulty of obtaining observations of particular processes etc.) may limit the scope for verification/falsification, these may be problems which can ultimately have a technical solution. Rather, the major dilemma for modellers will continue to be epistemological debate over what constitutes sufficient verification/falsification evidence to confidently utilise a particular model for a given application. Such debates over what constitutes proof should be ongoing and form a significant part of 'normal science'. However, it is a reflection of the scientific health of the discipline, and perhaps the contentious nature of the tasks to which models are put, that a book devoted to a single such debate can exist at all.

Models are now an accepted part of the scientist's methodological toolkit and can play a critical role at all stages of a scientific investigation. This plurality of function is actually a problem in debates on model validation as what constitutes a 'valid' scheme will vary with application. Many of the chapters (Oreskes and Belitz, Beven, Morel-Seytoux, chapters 3, 4 and 12) in this book focus on the use of models as predictive tools, and while this is perhaps an ultimate goal, many scientific investigations are conducted with models that do not have this aim in mind. Models can be used to examine contemporary environments by exploring mechanisms, testing for process dominance or quantifying flux rates. They can be used to simulate past conditions to aid environmental reconstruction, they can be used forensically to investigate engineering failures (see Hervouet 2000) and they can be used for design of both field instrumentation and construction work (see Kemp et al. 1989). All these may require a different approach to validation to that demanded by predictive studies and they can be more or less tolerant of parameter and conceptual uncertainties than the development of a predictive tool might allow. For example, in reconstructing past fluvial environments a useful result (in the sense that knowledge is advanced) may be obtained from a model that contains a level of uncertainty that would be unacceptable in a code used to aid construction of an underground nuclear waste repository with a 10 000 year design life.

In considering validation we should also note that prediction and explanation are not necessarily as synonymous as the empiricist position would lead us to suppose. Lumped conceptual schemes in hydrology such as the Stanford Watershed Model (see Gregory and Walling 1973, pp. 228–230) replicate catchment behaviour in turning rainfall to runoff but give no insight into the environmental processes by which this transformation occurs. Model parameters are fitted by optimisation and act, in signal processing terms, as series of gain settings to translate an input time series into an output time series. There is no particular magic in this, and four to five such gain parameters are usually sufficient to turn any given rainfall signal into a runoff signal that is a reasonable match to some observed data. Thus, the model predicts but does not explain and is, scientifically, a dead end. Primarily this is because the empirical hypothesis that such a model embodies is not verifable/falsifiable as the model parameters and internal fluxes have no physical meaning. Using Ayer's (1936) terminology such a model is not a legitimate hypothesis and any proposition that fails to satisfy this principle, that is not a tautology, is metaphysical and thus meaningless. While numerous authors (following Beven 1989) have argued that physically based distributed schemes are no better than the lumped conceptual models they replaced, they do embody a physical process hypothesis that can, at least in principle, be internally tested and used to explain observed behaviour. While certain grid-square effective parameters may currently be impossible to measure, we can specify what is required and the possibility exists that in the future these measurements will become technically feasible. This will never be true of a lumped conceptual scheme where the model parameters have no real basis. A similar distinction exists in hydraulics between simple non-storage flood routing methods (see Shaw 1983, pp. 378–380) and hydrodynamic models which solve some derivation of the physically based Navier–Stokes flow equation (see for example Bates et al. 1998a).

Clearly, many environmental models, for example groundwater codes, concern processes for which we can obtain no direct observations, yet these processes still have physical significance and are a legitimate subject for scientific investigation. In the case of the validation of a physically based groundwater code one can at least conceive the data collection programme (given a very generous research funding organisation) that would generate sufficient data to confidently verify/not falsify the model to allow its use for a given task. For a distributed hydraulic model this might in the limit comprise simultaneous collection of flow velocity data at each computational point in the model at each time step. In reality not all this data will be necessary, and techniques such as the GLUE procedure of Beven and his co-workers (see for example Freer et al. 1996) can quantify the value of individual data items in the evaluation process. Hence we can determine how much data we require to achieve given levels of uncertainty reduction. The possibility, at least in principle, of verification/falsification is a significant advantage, both in terms of formulating models as true empirical hypotheses in the sense advocated by Ayer and in more formally linking explanation and prediction. Most scientists would now accept a quasi-realist position that explanation and advancement of knowledge should be our primary aim (see Richards 1990) and that the ultimate expression of this will be an ability to predict system behaviour. A science that does not explain is ultimately unsatisfactory.

Predictive validation is therefore only one particular facet of the model evaluation process and accordingly we should consider a hierarchy of validation aims. We can ask a series of questions about any given model:

- Does the model provide a good solution of its constituent equations?
- Have we captured the variability and sensitivity of the real world system?
- Can we match available contemporary data and explain processes?
- Can we match an extended run of historical data?
- Can we use the scheme as a predictive tool?
- Can we export the technique to other catchments or climatic regimes?

These provide a series of increasingly stringent tests for hypothesis verification/falsification and increasingly more general science. We move from models used as interpolators to models as extrapolators of data, and each stage is conditional on further data not falsifying the original model. The question of acceptance/rejection at each stage should be based on whether the tests conducted allow the model to be confidently used for a particular purpose. Numerical models should accomplish interpolation tasks relatively well if boundary conditions are known and the model equation base is appropriate to the task in hand. In moving to history matching and predictive validation a good model can rapidly become a bad one if boundary conditions or processes change. This can often be beyond the scope of the modeller to deal with. For example, contemporary boundary condition data for hydraulic models are relatively easy to measure using stage recorders and rated sections. However, prediction of the hydrodynamics of the 100 year recurrence interval flood is subject to all the additional uncertainties contained in the flood frequency statistics that would typically be needed to assign flow rate boundary conditions to the model. There is a contrast here between general and site-specific validation. The former involves evaluating whether a model can perform the task that its developers claim, the latter broadens this to include the full range of uncertainties present in a given application data set. Of course, creating a model that cannot, in reality, be parameterised leads quickly to reductionism, yet inappropriate use of particular boundary conditions or calibrations may undermine an otherwise adequate code. A degree of common sense is necessary to disentangle these objectives.

In keeping with our view that model validation is more a practical rather than a philosophical problem, we seek to establish in this chapter the complexity of hydraulic model and calibration that can be supported by typically available validation data sets. These include analytical solutions, laboratory experiments and real world data which range from point measurements of bulk flow to distributed fields of model state variables. In doing so we attempt to highlight the model structural constraints that may limit validation and speculate on the existence of a 'glass ceiling' to the validation of hydraulic models.

13.2   A BRIEF OVERVIEW OF HYDRAULIC MODELS

This chapter is concerned with physically based hydraulic models as these represent the only schemes that are legitimate hypotheses and that can potentially be internally validated. Lane (1998) provides an excellent review of these approaches and here it is merely sufficient to give a brief overview of the main families of models used in hydraulics to enable their validation to be sensibly discussed.

All physically based hydraulic models are derived from the three-dimensional Navier–Stokes momentum equation, which for an incompressible fluid of constant density can be expressed in cartesian vector notation as:

$$\rho \frac{D\mathbf{u}}{Dt} = -\nabla p + \mu \nabla^2 \mathbf{u} + F \tag{1}$$

where ρ is the fluid density [with dimensions ML^{-3}]; $\mathbf{u}$ is the velocity $[LT^{-1}]$; t is the time [T], p is the pressure $[MLT^{-2}]$; μ is the viscosity $[MLT^{-2}]$ and F is the set of terms (for example gravity, Coriolis and friction) to be included in the specification of a particular problem.

Which when combined with the equation of continuity:

$$\nabla \cdot \mathbf{u} = 0 \tag{2}$$

gives a system of equations that can be solved to yield the three-dimensional velocity vector $\mathbf{u} = (u\ v\ w)$, where u, v and w are the three components of $\mathbf{u}$ in the x, y and z directions, respectively, and pressure, p, for a given point in time and space. In free surface models the pressure is typically replaced with the flow depth, h [L].

Theoretically, equations (1) and (2) can be used, with appropriate boundary conditions, to fully describe any open channel flow. However, direct application of these equations to turbulent flows requires a sufficiently resolved model discretisation capable of capturing the smallest time and length scales of turbulent motion. The smallest scale turbulent motion in a given flow is described by the Kolmogorov length scale, which for a typical fluvial flow with $\mathbf{u}$ equal to $1\,\mathrm{ms}^{-1}$, the largest length scales of the order of $1\,\mathrm{m}$ and kinematic viscosity, v, of $1.0 \times 10^{-6}\,\mathrm{m}^2\,\mathrm{s}^{-1}$ is approximately $0.03\,\mathrm{mm}$ (Hervouet and Van Haren 1996). Thus, the grid spacing needs to be much less than this to capture all turbulent motions and the computation will need to be unsteady with a time step small enough to capture the evolution of the most transient eddies. The number of grid cells required will therefore depend on the flow Reynolds number (R_e) and for practical applications with R_e in excess of 10^6 Direct Numerical Simulation of turbulence will be impossible. Such DNS models are, however, becoming feasible for flow problems of simple geometry at low Reynolds number (see, for example, Akhavan et al. 2000; Bastiaans et al. 2000; Evangelinos et al. 2000) and these can be used as benchmark data sets in the early stages of model validation.

For most hydraulic problems of practical interest we are not interested in the details of the instantaneous velocity field but, rather, the overall flow properties. In this case the most common approach is to use some statistical analysis of the turbulence structure of the flow to determine its impact on the mean flow field. This method, typically termed Reynolds (1895) averaging, assumes that each variable (e.g. $\mathbf{u}$) can be split into two components: a mean value ($\bar{\mathbf{u}}$) and a random variation about it ($\mathbf{u}'$). We assume that the random variations about the mean are normally distributed and that over a sufficient time period these cancel to zero:

$$\bar{\mathbf{u}} = \frac{1}{\Delta t} \int_0^{\Delta t} \mathbf{u}\, dt \quad \text{and} \quad \bar{\mathbf{u}}' = \frac{1}{\Delta t} \int_0^{\Delta t} \mathbf{u}'\, dt = 0 \tag{3}$$

We can then replace $\mathbf{u}$ in equations (1) and (2) with ($\bar{\mathbf{u}}$) and obtain the Reynolds Averaged Navier–Stokes equations or RANS. This does however lead to the introduction of new terms in equation (1) representing the shear stress on the mean flow due to turbulence. These so-called Reynolds stresses may be given, for example, as:

$$\bar{\tau}_{xz} = -\rho(\bar{u}'\bar{v}') \tag{4}$$

where $\bar{\tau}_{xz}$ is the Reynolds stress in the x–z plane.

As these depend on the instantaneous velocity fluctuations their existence is a problem for a time-averaged model as these values are unknown. To close the RANS equations we need to provide values for the Reynolds stresses by introducing some model of turbulence. Here a variety of schemes are available but the most simple are based on the Boussinesq approximation. This method treats turbulent momentum transfer in a similar manner to the approach adopted for viscous forces, as both can be thought of as shear stresses. Here a turbulent viscosity coefficient, μ_t, is introduced with dimensions similar to the molecular viscosity, μ. The overall fluid shear stress, τ [$ML^{-1}T^{-2}$] including molecular and turbulent components then becomes:

$$\tau = (\mu + \mu_t)d\bar{u}/dy \tag{5}$$

As with molecular viscosity, for convenience μ_t is often expressed in kinematic form, v_t, with dimensions [L^2T^{-1}]. Simple models of turbulence, such as the zero equation, mixing length, one equation and k–ε (two equation) schemes, merely attempt to provide a value for v_t. For the zero equation model v_t is either specified by the user or expressed as proportional to the water depth and friction velocity with the proportionality coefficient parameterised on the basis of experimental data (Rastogi and Rodi 1978). For higher order approaches v_t is obtained via differential equations of increasing complexity that express v_t as a function of the turbulent kinetic energy (k) and its dissipation rate (ε). More complex approaches, such as algebraic and Reynolds stress models, operate by formulating differential transport equations for the Reynolds stresses themselves (Younis 1996). A final approach worth mentioning is Large-Eddy Simulation or LES. This lies between direct numerical simulation of turbulence and classical turbulence modelling. It exploits the fact that at small eddy scales the turbulence is, statistically, relatively homogeneous. Thus the time evolution of large-scale heterogeneous eddies are explicitly represented on the numerical grid and only the small-scale, more isotropic eddies are statistically treated. As the turbulence model only concerns the sub-grid-scale motions the development of the overall flow field should be less dependent on the choice of turbulence model, and in general relatively simple sub-grid closure schemes can be applied. For an overview of turbulence modelling approaches the reader is referred to Rodi (1980) and Rodi et al. (1997).

While the discussion so far has been entirely based on three-dimensional approaches, many hydraulic problems do not require this level of complexity in order to be adequately modelled. In many circumstances a modeller may assume that significant flow field variations only occur in a reduced number of dimensions. Two-dimensional approaches typically involve integration of the flow field over the depth to produce depth-averaged values of the velocity. This results in a number of different equation sets such as the St Venant equations, which assume a hydrostatic pressure distribution, or the Boussinesq equations, which do not. For example, the St Venant equations are given in non-conservative form as:

Continuity equation

$$\frac{\partial h}{\partial t} + \vec{u}_d \cdot \overrightarrow{\text{grad}}(h) + h \, div(\vec{u}_d) = 0 \tag{6}$$

Momentum equations

$$\frac{\partial u_d}{\partial t} + \vec{u}_d \cdot \overrightarrow{grad}(u_d) + g\frac{\partial h}{\partial x} - div(v_t \cdot \overrightarrow{grad}(u_d)) = S_x - g\frac{\partial Z_f}{\partial x} \tag{7}$$

$$\frac{\partial v_d}{\partial t} + \vec{u}_d \cdot \overrightarrow{grad}(v_d) + g\frac{\partial h}{\partial y} - div(v_t \cdot \overrightarrow{grad}(v_d)) = S_y - g\frac{\partial Z_f}{\partial y} \tag{8}$$

where u_d, v_d are the depth-averaged velocity components [with dimensions LT^{-1}] in the x and y Cartesian directions [L]; Z_f is the bed elevation [L]; v_t is the kinematic turbulent viscosity [L^2T^{-1}]; S_x, S_y are the source terms (friction, Coriolis force and wind stress) and g is the gravitational acceleration [LT^{-2}].

These can then be solved using some appropriate numerical procedure to obtain predictions of the water depth, h, and the two components of the depth-averaged velocity, u_d and v_d. The Shallow Water equations are most often applied to flows that have a large areal extent compared to their depth and where there are large lateral variations in the velocity field. They are thus well suited to the computation of overbank flood flows, tides, tsunamis or even dam breaks (see Hervouet and Van Haren 1996).

For even simpler flow situations the two-dimensional St Venant equations can be reduced further into one-dimensional form (see, for example, Fread 1984, 1993; Samuels 1990; Singh 1996; Ervine and MacLeod 1999). This considers conservation of momentum between two flow cross-sections Δx apart and yields a first order partial differential which in conservative form may be expressed as:

$$\frac{\partial Q}{\partial t} + \frac{\partial (Q^2/A)}{\partial x} + gA\left(\frac{\partial h}{\partial x} + S_f\right) = 0 \tag{9}$$

where Q is the flow discharge [L^3T^{-1}]; A is the flow cross-section area [L^2] and S_f is the friction slope [$-$].

Similarly, the continuity equation can be expressed as:

$$\frac{\partial Q}{\partial x} + \frac{\partial A}{\partial t} = 0 \qquad (10)$$

which with appropriate boundary conditions can be solved to yield estimates of Q and h in both space and time. Equations (9) and (10) represent a fully dynamic wave; however, further simplification can be employed to yield diffusion and kinematic wave analogies to the fully dynamic wave model. One should note that turbulent viscosity terms appear in both two- and three-dimensional hydraulic models but not in the simpler one-dimensional forms. One-dimensional methods tend to be used for wave routing problems where lateral and vertical velocity variations can be assumed negligible. Knight and Shiono (1996) suggest that in-bank flows in river channels are a good example of such a situation, and for this reason the one-dimensional St Venant equations form the basis of most standard hydraulic river modelling codes such as MIKE11, ISIS and HEC-RAS.

Typically, for all the equation systems discussed no general solution exists for boundary conditions of practical interest and a numerical approximation therefore needs to be employed which requires some finite discretisation of both space and time. Although we need to discretise the time step, all the above hydraulic models can be used in either dynamic or steady-state modes through manipulation of the boundary condition specification. Before discussing numerical approximation procedures it is worth noting that for many hydraulic situations (floods, tides, dam breaks etc.) the flow field boundary changes with time. A choice therefore exists for two- and three-dimensional approaches between Lagrangian approaches, where the computational grid deforms to follow the flow field boundary (Lynch and Gray 1980; Kawahara and Umetsu 1986; Benkhaldoun and Monthe 1994), and Eularian approaches which utilise a fixed numerical grid. The latter methods must then introduce some additional correction to account for the impact of wetting and drying processes on mass and momentum conservation. While Lagrangian approaches would seem well suited to the problem, their use other than as research tools is quite problematic both in terms of stability and computational cost. For this reason most recent approaches to wetting and drying have focused on the use of fixed numerical grids (Leclerc et al. 1990; Defina et al. 1994; Ip et al. 1998; Tchamen and Kawahita 1998; Bates and Hervouet 1999). The reduced computational cost of the numerical procedure also allows use of a more resolved spatial discretisation and thus a greater degree of topographic complexity to be included in the model. This is clearly critical as topography is a major control on the inundation process. As one-dimensional schemes ignore lateral flow field variation such considerations do not arise. Thus to determine inundation extent from a one-dimensional model the water surface elevations predicted at each cross-section need to be projected back on to some Digital Elevation Model of the terrain.

Numerical solutions of the above equations generally use either finite difference, finite volume or finite element methods. It is beyond the scope of this chapter to discuss these methods in detail; however, a brief description of those characteristics which relate to model validation is appropriate. All numerical approximations require some discretisation of space, typically as a structured or unstructured numerical mesh. We assume that as the grid resolution tends to an infinitely small size the approximate solution tends to the true solution. Ideally we need a sufficiently resolved grid such that those flow features relevant to the development of the overall velocity field are explicitly modelled and that the solution obtained is independent of the mesh resolution chosen. The same is true of the time step used in the model. The choice of space and time discretisation is thus application specific. Lastly, most hydraulic problems are of the initial value-boundary value type. Initialisation of any model therefore requires specification of initial values for each dependent variable at each computational node at the first time step and the time evolution (if any) of the dependent variables at each boundary node throughout the duration of the simulation. It is common to assume an impermeability condition on all boundaries that are not subject to free surface flow such that interactions with the surrounding catchment are ignored. There are a number of studies that attempt to overcome this limitation (Kohane and Welz 1994; Stewart et al. 1999), but these are the exception rather than the rule.

13.3   VALIDATION OF HYDRAULIC MODELS

The above discussion demonstrates that compared to most hydrologic codes, hydraulic models have relatively few parameters that can or need to be calibrated by a user and need relatively few data sets in order to develop a simulation. Typically, parameterisation consists of finding values for the boundary friction and the turbulent viscosity coefficient, v_t, although the requirement for the latter disappears in all but the simplest turbulence models. Necessary data sets consist of a description of the domain topography at an appropriate scale, and boundary and initial condition information. Boundary conditions largely consist of quantities (flow rates, stage, tidal range) that are relatively easy to measure, while approximations to the initial conditions can often be developed by the model itself. For example, initial conditions for a flood simulation in a compound channel are often taken as the water depths and flow velocities predicted by a steady-state simulation with inflow and outflow boundary conditions at the same value as those used to commence the dynamic run. While most natural system are rarely in steady state, careful selection of simulation periods to coincide their start with near steady-state conditions can minimise the impact of this assumption. As we deal with surface flows it should be relatively easy to gain access in order to make boundary condition or validation measurements, although, as we will see, this is all too infrequently undertaken. Finally, one often has an extended run of historical data from which to develop extreme value frequency distributions for estimating higher return interval flows.

This is not to say that uncertainties do not exist, merely that the scope and nature of the parameterisation/calibration/validation problem is different to that for other hydrology models. For example, in terms of the problem physics, flow processes are much more transient. Flow is turbulent and varies over scales many orders of magnitude in size from that of the flow geometry (i.e. the water depth in an open channel flow or ~ 1–10 m) to the Kolmogorov length scale ($\sim 1 \times 10^{-5}$ m in open channel flow, Hervouet and Van Haren 1996). Friction parameters are also a major source of uncertainty and these may vary both spatially and temporally through the simulation. Measuring values for the resistance coefficient is extremely difficult, yet correct estimation of frictional losses is a critical aspect of any hydraulic modelling study. A final potential problem in the validation of hydraulic models is the possible significance of processes not included in either the boundary condition specification or the model equation base. These might include two-phase (gas/fluid or sediment/fluid) interactions, hydrological exchanges with the surrounding catchment and atmosphere, or energy dissipation by wave breaking.

Topographic, boundary condition and initial condition data are subject to uncertainty even when measured values are used. Topography has, until recently, been a major source of error in hydraulic models. One-dimensional codes have traditionally been parameterised through ground survey of topography and their cross-sectional basis integrates well with such data sources. Such data are highly accurate but time consuming to collect and ignore potentially important topographic features occurring between cross-sections. For this reason much recent interest has been shown in more automated survey techniques such as stereo photogrammetry, laser altimetry (Li 1997) or interferometric radar systems on airborne or spaceborne platforms. Such techniques produce Digital Elevation Models of varying accuracy and resolution but certain of these systems have ready potential for integration with hydraulic models. For example, airborne scanning laser altimetry systems (LiDAR) can sample terrain at rates of up to $\sim 50\,km^2$ per hour with horizontal resolution of up to 1 point per $10\,m^2$ and with vertical accuracy of ± 10–$15\,cm$ at one standard deviation for the WGS84 geoid model (Marks and Bates 2000). Such data is produced in raster or TIN data formats and integrates well with two- and three-dimensional hydraulic codes. For example, Marks and Bates (2000) describe a typical application of a two-dimensional finite element hydraulic model to a $12\,km$ reach of the River Stour in Dorset, UK, for which LiDAR data were available. Here the maximum resolution finite element mesh that could reasonably be computed on a medium power workstation consisted of $\sim 11\,000$ elements and ~ 6500 computational nodes. By contrast, within the model domain $\sim 300\,000$ x, y, and z topographic coordinates were available from LiDAR survey and resulted in high levels of data redundancy. The only other published topographic data source for this reach, and the one previously used to parameterise the model, was the UK National series $1:10\,000$ scale maps. These contained only two contours with 10 m spacing and accuracy of $\pm 1.25\,m$ and ~ 15 spot heights for the whole $\sim 4\,km^2$ floodplain area. In terms of topography we are therefore in the process of moving rapidly, at least in the developed world, from a data-poor to a data-rich modelling environment with many attendant possibilities for model development. For example, Bates and Hervouet (1999) use the redundant data from a LiDAR survey and a method independently proposed by King and Roig (1988) and Defina et al. (1994) to develop a sub-grid scaling for equations (6), (7) and (8) that better accounts for wetting and drying processes when using fixed numerical grids.

The same argument cannot be so readily made in terms of boundary and initial conditions. Here, hydraulic modellers are limited by the density of the national river/tidal gauging network in the country in which they work (see Bates et al. 1998a). Otherwise, dedicated gauging structures need to be installed or inflow and outflow velocity profiles across the channel need to be measured (see Lane and Richards 1998). This constraint effectively limits the amount of data that can be collected. One also has to be aware that using some kind of forcing function (such as an extreme value distribution) to estimate boundary conditions for flows for which no field measurements exist can constitute a large source of additional uncertainty. Potentially, this may swamp other uncertainties in the model and, while the model may still be a good representation of the problem physics, the forecast obtained may be substantially in error. Such forcing functions therefore need to be used in conjunction with studies of error propagation through the model structure.

While hydraulic modellers may have access to excellent topographic data sets and modest amounts of boundary condition data compared to other areas of the hydrological sciences, field validation data is rather poorly available given the relative ease with which one can gain access to the flow field. Depending on the dimensionality of the code, we require flow rates, water depths, inundation extents and velocities that are independent of the model boundary conditions. A distinction is typically been made (Bates et al. 1998a) between external validation, where the comparison takes place at outflow points from the modelled domain, and internal valida- tion, which uses flow measurements taken in the interior of the model. Clearly, it is much more difficult to ensure that the former type of validation is fully independent of the prescribed boundary conditions, although each model/flow configuration needs to be considered separ- ately. For example, in the case of critical flow a well-posed model can be obtained without specifying a downstream boundary condition as the Froude number is greater than one and, hence, information from the downstream boundary does not propagate upstream. Measurements of flow and stage at the downstream external boundary of the reach being considered are thus fully independent of the model. For sub-critical flow this is not the case and it is usual to prescribe the water surface elevation as the downstream boundary condition to obtain a well-posed problem. Even here, the measured flow rate will still be weakly independent of the water surface elevation and can be used, with caution, for model validation.

Thus, in terms of routine, nationally and globally collected measurements, the spacing of river and tidal gauging stations rarely permits the use of large numbers of internal validation points. Bates et al. (1998a) found that for a number of rivers in the UK and the USA, the typical distance between river gauges was about 15 km. For example, in a model of 60 km of the Missouri River in Nebraska only two internal gauges were available for model validation. This is to be expected as the spacing reflects the flood warning priorities of the Agencies that maintain river gauging networks. There are very few exceptions to this rule, and the situation deteriorates for less industrialised countries. Compared to other earth sciences there is no basic programme of field data collection for scientific purposes in hydraulics. Elements of the World Climate Research Programme (WCRP), such as the World Ocean Circulation Experiment in oceanography (WOCE, see, for example, Didden and Schott 1992) and the Global Energy and Water Cycle Experiment (GEWEX, see, for example, Berbery et al. 1999) in hydrology, provide such an impetus in other fields. Hydraulics does not even have the equivalent of the experimental catchment programme that has contributed much basic process data for model validation in hydrology (e.g. Bathurst 1986). The lack of such a fundamental data collection programme in hydraulics has clearly influenced model development within the discipline and will ultimately limit what we are able to achieve. This is becoming more acute as we continue to develop higher dimensionality models with more sophisticated representations of flow turbulence and attempt to transfer these to practical applications. By comparison, the most dense hydraulic field data sets seem rather modest in light of the large budget experiments conducted in some of the other hydrological sciences. The most comprehensive field installation the authors of this paper have so far found is maintained by the UK Environment Agency on a 500 m reach of the River Blackwater in southern England and was designed to monitor the hydraulic performance of a river restoration scheme. The monitoring equipment consists of a rated section to measure discharge entering the reach and five approximately equally spaced water stage recorders. No velocities are recorded and maximum flood discharge in the compound section is only $\sim 10 \, \text{m}^3 \, \text{s}^{-1}$. Individual researchers are beginning to collect some high temporal resolution velocity measurements using instruments such Acoustic Doppler Velocimeters (ADV) and electromagnetic current meters (see, for example, Ferguson et al. 1996) and comparisons to model output are now being conducted (e.g. Lane et al. 1995; Nicholas and Smith 1999). Yet we are still a long way from the availability of benchmark data sets.

A reason for this lack of emphasis on field studies has been the availability of a number of analytical solutions for the Navier–Stokes equations and its derivatives (see, for example, Horritt 2000; Panchang et al. 2000) and the ease of scale model construction for surface flow problems (see, for example, Knight and Sellin 1987; Sellin and Willetts 1996). While analytical solutions are an important and essential starting point for the model validation process, they only apply to simplified physical situations and hence cannot fully capture the complexity of real world flows. Scale models have played an important role in hydraulics and in the past provided a means of simulating fully three-dimensional turbulent flow when computer power did not exist to conduct an equivalent numerical calculation. The availability of such data, and the engineering focus of much of hydraulics has tended to mitigate against large-scale field data collection exercises in natural channels.

13.4   APPROACHES TO MODEL VALIDATION

In Section 13.1 we argued that a hierarchy of validation aims should be considered, commencing with comparison to analytical solutions and progressing to simulation of more realistic laboratory and field data sets to provide an increasingly stringent set of model tests. We also put forward the view that the decision as to whether a model has been sufficiently validated/falsified should be taken on practical grounds. Here one should determine the extent to which each particular validation test increases our confidence that the model can be used to fulfil a given application. Here we review how typically available data in hydraulics can be used in the validation process. In particular, we wish to determine the complexity of model that can be discriminated using each data set and explore the linkage between calibration and validation with reference to a number of specific examples.

13.4.1   Analytical Solutions

These consist of algebraic equations that can be directly solved to yield a solution of one of the fundamental hydraulic equation sets discussed in Section 13.2. While for many of the fundamental equations no general solution exists, for particular boundary conditions or simple geometries a solution of a reduced form of the equations may be directly obtained. Such analytical solutions yield data sets that can then be used for model validation. For example, Horritt (2000) used an analytical solution of the two-dimensional St Venant equations in plane polar coordinates to develop optimal mesh generation strategies for meandering channel flow problems. Similar approaches to model validation have been followed in hydraulics by numerous authors including Lynch and Gray (1980), Moramarco et al. (1999), Moramarco and Singh (2000) and Panchang et al. (2000).

   Pseudo-analytical solutions can also be obtained merely through careful manipulation of model boundary conditions. Bates and Hervouet (1999), for example, generated an analytical solution for tidal run up on a 1 km × 1 km planar beach (Figure 13.1) and used this in the validation of a new moving boundary approach for two-dimensional models. Here, manipulation of the water surface elevation at the seaward boundary of the model allowed a sinusoidal tide to be simulated entering the domain. By simplifying the beach topography to a planar slope and assuming a horizontal water surface, the position of the inundation front as it propagated across the domain and the water depth at any given point could be exactly known. Moreover as the new moving boundary algorithm was based on a scaling of the continuity equation (equation (6)) derived from the sub-grid topography, specification of a planar slope allowed exact specification of the scaling parameters. As an example, Figure 13.2 shows the comparison between a standard finite element solution of the St Venant equations with the new method proposed by Bates and Hervouet (1999) for water depths of 0.1, 0.05, 0.02 and 0.01 m at 3000 s into the simulation during the wetting phase. These data are sufficiently detailed to discriminate between the two approaches as we can be sure that the analytical data are exact. By employing their new technique Bates and Hervouet were able to improve the prediction of shallow water depth contours for moving boundary problems in hydrodynamics and demonstrate that this had a significant impact on the velocities predicted by the model over the whole domain. While the pseudo-analytical solution did not provide estimates of these velocities, those predicted using the Bates and Hervouet algorithm were less subject to unstable oscillations than those predicted by standard methods.

   Using the validation criteria put forward in Section 13.1 we should ask whether this constitutes sufficient proof to confidently use the model for simulation or prediction of moving

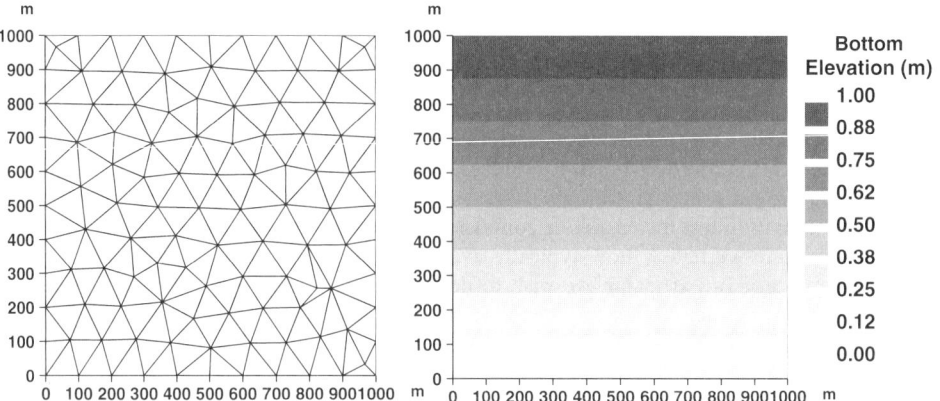

Figure 13.1   Finite element mesh and topography representing a planar beach generated as a simple analytical test case for wetting and drying problems. The cell resolution is approximately 100 m

boundary problems in shallow water. The answer here is clearly 'no', yet the analytical solution has certainly demonstrated the feasibility of the new approach and provides a rigorous metric for a first evaluation of alternative methods. Analytical solutions have the advantage that they provide exact data that is often both spatially and temporally distributed in a way that field data can rarely match, yet ultimately the simplicity of the physical problem limits the conclusions that can be drawn. We therefore require some real flow data to be sure that the method can be transferred to field situations.

13.4.2   Scale Model Comparison

Prior to a move to field data sets with measurement uncertainty and relatively sparse data density, comparison of numerical codes to laboratory scale models is a key step in hydraulic model development (see, for example, Moulin and Ben Slama 1998; Bates et al. 1999; Olsen and Kjellesvig 1999; Rajendran et al. 1999; Hervouet 2000; Sauvaget et al. 2000). This is particularly so as a large number of data sets are available covering a variety of flow situations and high accuracy data can be generated by scale models at reasonable spatial and temporal resolution. In this category one could also include validation data sets generated by other models of higher dimensionality and resolution. In this way quite substantial progress in model validation can be achieved without recourse to expensive field programmes and in this sense at least hydraulic modellers are data-rich. For example, Stoesser (1999) gives details of a series of laboratory and numerical test cases used in the validation of a new Reynolds Averaged Navier–Stokes (RANS) model, HYDRO, being developed to simulate fully turbulent flows in natural, vegetated open compound channels. For this code, analytical solutions and objective tests are used to check the integrity of the coding before proceeding to the series of comparisons to the scale model data detailed in Table 13.1. Table 13.1 demonstrates the range of experimental data available to hydraulic modellers even when the specific aim of the code, in this case the simulation of flow with submerged and emergent natural vegetation, is quite focused. This is achieved using a newly developed roughness closure scheme for calculating vegetational friction losses described by Fischer-Antze et al. (in press) and initially implemented in the SSIIM model (Olsen and

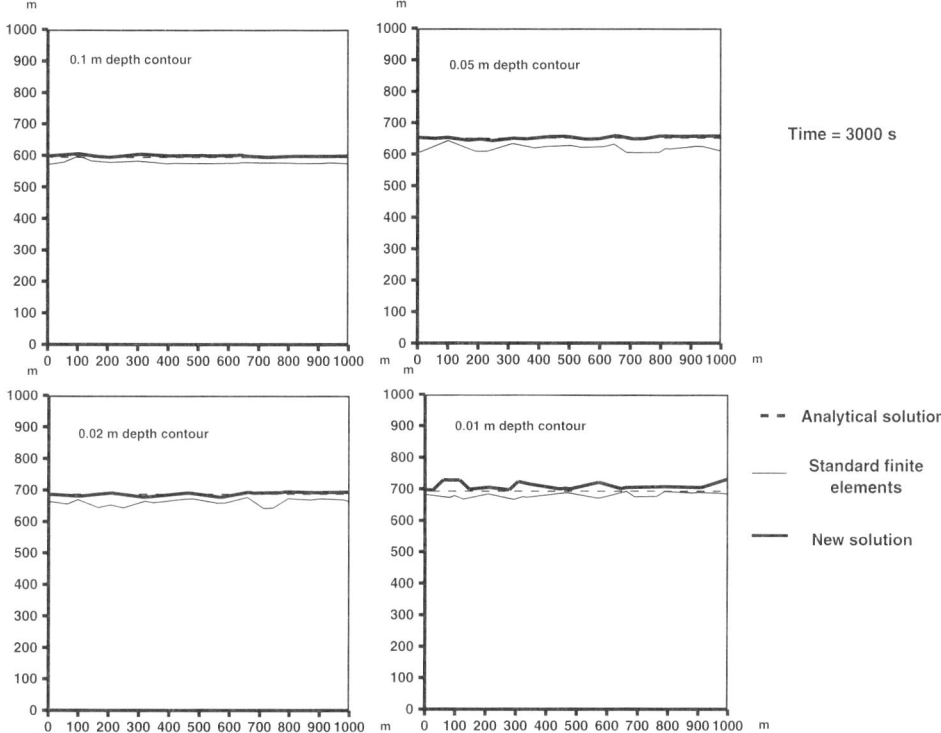

Figure 13.2 Comparison of wetting phase shallow water depth contours predicted by standard finite elements and the new technique developed by Bates and Hervouet (1999). Each method is validated against the pseudo-analytical solution for the run up of a sinusoidal tide on the planar beach shown in Figure 13.1. At 3000 s into the simulation the analytical solution predicts that the 0.1, 0.05, 0.02 and 0.01 m contours should be positioned, respectively, at $y = 600$, 650, 680 and 690 m

Stokseth 1995). This is undoubtedly a comprehensive programme with the particular elements chosen to test in a structured way specific elements of the model's behaviour. Nevertheless, certain of the cases under consideration, such as lid driven cavity flow and flow over a backward facing step, could be considered standard test cases that any hydraulic code should be able to simulate. Successful completion of these tests will substantially increase our confidence in the utility of the model. As an example of the results that may be obtained Figure 13.3 shows a comparison of the velocity fields simulated by both the RANS model and a Direct Numerical Simulation for the problem of lid driven cavity flow (Ghia et al. 1982). Here flow at $R_e = 1000$ and $R_e = 10\,000$ has been used as a first step in the validation of both laminar and turbulent flow simulations developed using **HYDRO**. Similarly, Figure 13.4 shows a comparison of measured depth-averaged velocity profiles from the experiments conducted by Pasche (1984) into flow with emergent rigid vegetation to computational results from the SSIIM model using the roughness closure model of Fischer-Antze et al. (in press). The model with new roughness closure is clearly able to replicate the observed flow velocities on the vegetated floodplain and within the unvegetated main channel for a variety of plant densities.

Table 13.1 Scale model and DNS experiments used in the validation of the Reynolds Averaged Navier–Stokes (RANS) model HYDRO described by Stoesser (1999). Specifically HYDRO aims to simulate compound channel flow with submerged and emergent natural vegetation and the specific validation research design has been chosen with this in mind

No.	Experiment	Type	Description
1.	Ghia et al. (1982)	DNS	Lid driven cavity flow at $R_e = 1000$ (laminar flow) and $R_e = 10\,000$ (turbulent flow)
2.	Le et al. (1997)	DNS	Turbulent flow over a backward-facing step at $R_e = 667$ and $R_e = 5100$
3.	Tominaga and Nezu (1991)	Scale model	Overbank flow in a straight compound channel
4.	Martinuzzi and Tropea (1993)	Scale model	Turbulent flow over a submerged article at $R_e = 1.2 \times 10^5$
5.	Wood and Tong (1989)	Scale model	Overbank flow in a straight compound channel with tracer transport
6.	DeVriend and Geldof (1983)	Scale model	180° turn-around bend to simulate secondary motions
7.	Hicks et al. (1990)	Scale model	270° turn-around bend to simulate secondary motions
8.	EPSRC Flood Channel Facility Series B experiments – smooth surface (Knight and Sellin 1987)	Large scale model	Overbank flow of smooth, meandering compound channel flow with constant roughness
9.	EPSRC Flood Channel Facility Series B experiments – rough surface (Knight and Sellin 1987)	Large scale model	Overbank flow in a smooth, meandering compound channel flow with high roughness elements on the floodplain
10.	Tsujimoto et al. (1991) and Lopez and Garcia (1997)	Scale model	Straight channel with submerged rigid vegetation
11.	Pasche (1984)	Scale model	Overbank flow in a smooth, straight compound channel with emergent rigid vegetation on the floodplain

The example taken from Stoesser (1999) shows that the use of data from laboratory models and DNS does allow a degree of experimental control to be exercised and measurements of fully three-dimensional turbulent flows to be obtained relatively easily. However, often the flow geometries considered are rather simplified. Thus the above discussion of the HYDRO model only considers rigid rather than flexible vegetation and simple channel shapes. Whilst there is some potential for scale model development that overcomes these constraints this, combined with the difficulty of scaling the flow Reynolds number, means that scale model comparison can only form part of a model validation strategy, albeit a very valuable and potentially comprehensive one.

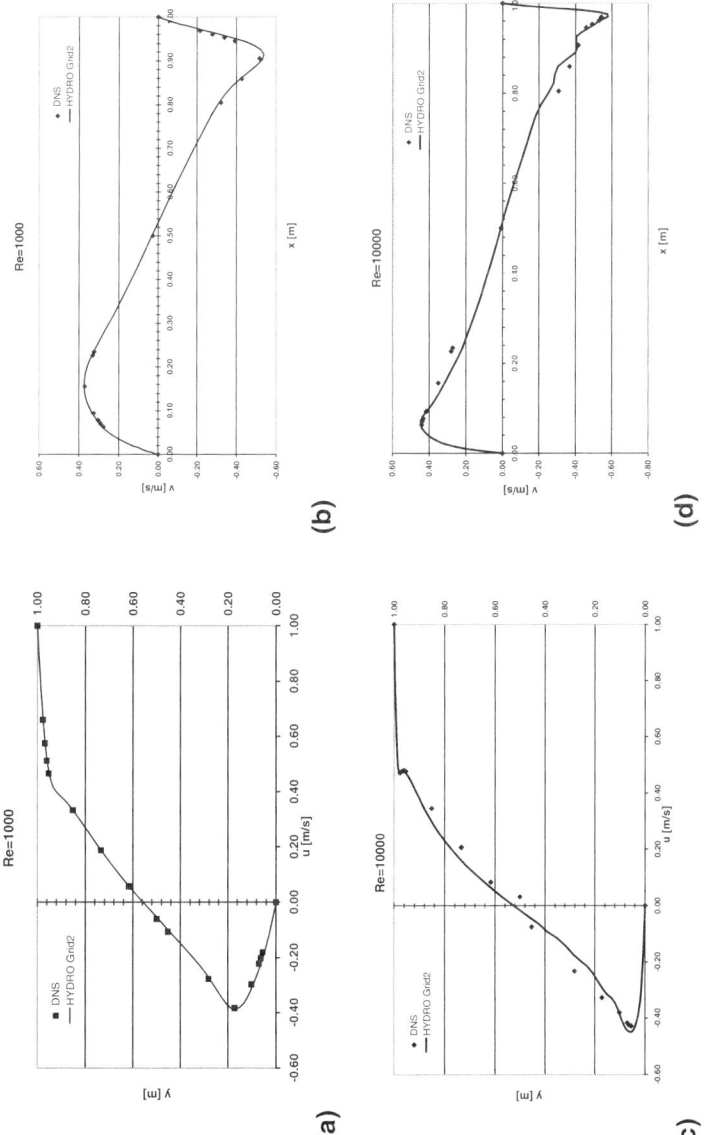

Figure 13.3 Comparison of a DNS for lid driven cavity flow developed by Ghia et al. (1982) with results from the HYDRO finite volume RANS solver for: (a) u velocity at $R_e = 1000$; (b) v velocity at $R_e = 1000$; (c) u velocity at $R_e = 10000$ and (d) v velocity at $R_e = 10000$

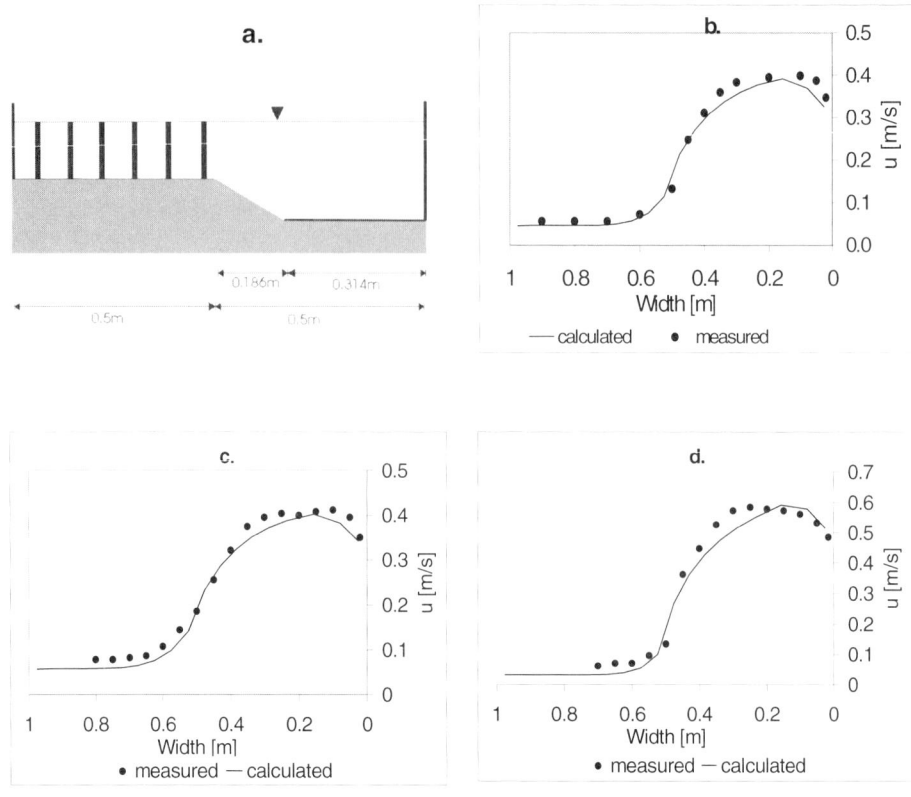

Figure 13.4 Comparison of measured depth-averaged velocity profiles from the experiments conducted by Pasche (1984) to computational results from the SSIIM code developed by Olsen and Stokseth (1995) enhanced with the new roughness closure algorithm developed by Fischer-Antze et al. (in press). (a) shows the cross-sectional geometry and vegetation arrangement used in the scale model, while (b) shows a comparison of experimental and computational results for a plant density 2.69 m² m⁻³ (c) for a plant density of 1.34 m² m⁻³ and (d) for a plant density of 10.76 m² m⁻³

13.4.3 Benchmarking

Construction of more than one model of a system can reveal biases and errors in a single model. Hence the benchmarking of codes can play an important role in model validation. We should perhaps distinguish between comparison of codes that are conceptually similar and those that are conceptually different. In the former case we are interested in determining how numerical solution and choice of discretisation impacts on model results, while for the latter we can ask questions regarding the level of physical processes representation necessary to replicate some observed data set. For example, Stoesser (1999) in validating the HYDRO code also simulated tests 1–11 in Table 13.1 with another three-dimensional finite volume solver for the Navier–Stokes equations, SSIIM (Olsen and Stokseth 1995) to provide a check on the quality of results obtained. Similar studies have been reported by Lynch and Officer (1985) and Harpin et al. (1995).

13.4.4 Field Data Sources

In moving away from analytical and laboratory data to field observations for model validation we (hopefully) increase the realism of the simulations performed, and hence the generality and significance of scientific evidence obtained. However, we also increase substantially the uncertainties inherent in particular data sets that may significantly constrain our ability to make unequivocal statements. For example, S.N. Lane, et al. (1999) validated a three-dimensional numerical model for the case of flow in a parallel confluence against velocities measured in a laboratory model and in the field. Correlations between observed and predicted velocities were of the order of 0.9 for the laboratory data and 0.7 for the field data. Lane et al. argued that this was due to the difficulty of determining adequately boundary condition, topography and roughness parameters in the case of the model based on field data. Of these Lane et al. concluded that an inability to specify complex bed topography was the fundamental limitation on model performance in the field application (see Lane and Richards, this volume, chapter 16 for a more detailed discussion).

Nevertheless, for many hydrological modellers comparison to field data is the key test a model must undergo in order to prove its utility, despite the relatively data-poor environments in which we work and the consequent loss of experimental control. To compensate for unmodelled or unmeasured processes and incomplete parameterisation and validation data, a degree of calibration is typically introduced for field applications. Calibration implies a further loss of experimental control as a unique solution to a problem may not exist. According to the terminology of Dietrich (this volume, chapter 6) the problem is thus ill-conditioned. Validation evidence from field tests of hydraulic models should therefore be interpreted with caution and, preferably, be part of a wider series of model tests.

Broadly, three classes of field data are available: external and internal bulk flow measures, vector data and point scale data. As mentioned previously only the very first of these is routinely available, despite the fact that it may be relatively easy to replicate such data with a variety of models. The key question to answer when considering the utility for model validation of each data type is what complexity of model does it allow one to conclusively discriminate. Essentially, we wish to identify those combinations of input data, model and calibration procedure that can explain all the available data. As it is likely that more than one data/model/calibration structure will be capable of this, we should favour the simplest admissible combination.

External and Internal Bulk Flow Measures

As noted in Section 13.3, global and national hydraulic data sets consist almost solely of bulk flow measures. These are usually recorded as time series of water levels and discharges, with the latter typically being obtained via a rating curve rather than by direct measurement. Given the typically spacing of river and tidal gauging stations, external validation is far more common than internal, but this may be less discriminatory in terms of model identification. As Dietrich (this volume, chapter 6) notes 'if the goal of a modelling exercise is to predict solely the time evolution of some bulk property of a complex system then spatial lumping (or spatial integration) can be invoked to filter out spatial variability'. In other words, lumped models may be able to fit bulk flow data equally as well as more complex spatially distributed schemes. External and internal bulk flow measures allow the flow routing behaviour of hydraulic models to be validated. However, the complexity of scheme needed to lag and attenuate an input waveform into a suitable output can be very simple indeed, even if intermediate measures of the waveform are available. In addition, more complex flow models may require information to be specified at the downstream boundary that may affect the independence of external validation tests.

A complicating factor in the validation of flood routing models is the additional process complexity that occurs during out-of-bank flow. Here floodplain storage and release of water, momentum exchange between main channel and floodplain (Knight and Shiono 1996) and the vigorous mixing that results from spillage of water from the downstream apex of meander loops (Sellin and Willetts 1996) considerably complicates the hydraulic environment. Not all these processes need operate and this will condition choice of model. For example, one can envisage flood events where the dominant process is the overbank storage of channel water and its subsequent infiltration or return to the channel. Choice of model also depends on the information required by the user. Thus, if a distributed velocity field is required a consideration of storage, momentum exchange effects and river meandering in a spatially distributed model (e.g. Bates et al. 1992) will be necessary even if bulk flow data are all that is available for validation. However, if storage volumes only are required a more spatially lumped representation of the floodplain may suffice. While a consideration of the flow physics may indicate the need for a complex flow model, lack of distributed validation data may again allow simpler models to be calibrated to compensate for unmodelled processes and replicate available bulk flow data equally well. For example, Bates et al. (1998a) develop a number of two-dimensional finite element models for reach-scale out-of-bank flood routing prediction (see Figure 13.5) and identify the minimum calibration complexity that allows such schemes to replicate external and internal bulk flow data (see Figure 13.6). Although these models include the additional hydraulic processes described above and enable prediction of the velocity vector field, there is little doubt that simpler one-dimensional models could be calibrated to replicate the available data equally well. Bates et al. (1998a) also consider the utility of internal validation data through simulation of the densely instrumented 500 m reach of the River Blackwater, UK, mentioned in Section 13.3. This extended the confidence that could be placed in the model predictions but was only a moderately more stringent evaluation of the model's abilities as it represented a rather similar type of test.

In summary, bulk flow measures provide only a limited test of model ability and are not sufficient to conclusively identify anything but the most simple of hydraulic schemes. This is largely because lack of parameter data means that all hydraulic schemes applied to field data sets require some degree of calibration. While more physically realistic schemes eliminate the need for certain hydraulic processes to be subsumed within the calibration, it is impossible to quantify this effect as the physically realistic range for typical calibration parameters is either large (in the case of boundary friction) or largely unknown (in the case of turbulent eddy viscosity). Hence, a calibrated parameter set for a complex model may be indistinguishable from a similar set for a simpler scheme even though the latter compensates for a greater number of unmodelled processes. Clearly, discrete bulk flow data, while being an important component of hydraulic model validation, has only limited strength as a stand-alone piece of evidence. One thus needs to consider either vector data or spatially distributed fields of model state variables to conclusively discriminate between one-dimensional and higher order models.

Vector Data Sources

Vector data sources, specifically synoptic images of flood inundation extent, appear to have considerable potential for improving the field validation of hydraulic models (Bates et al. 1997; Pearson et al. this volume, chapter 8). While such data is only available at the moment for research purposes there is good chance that it may become routinely available in the near future.

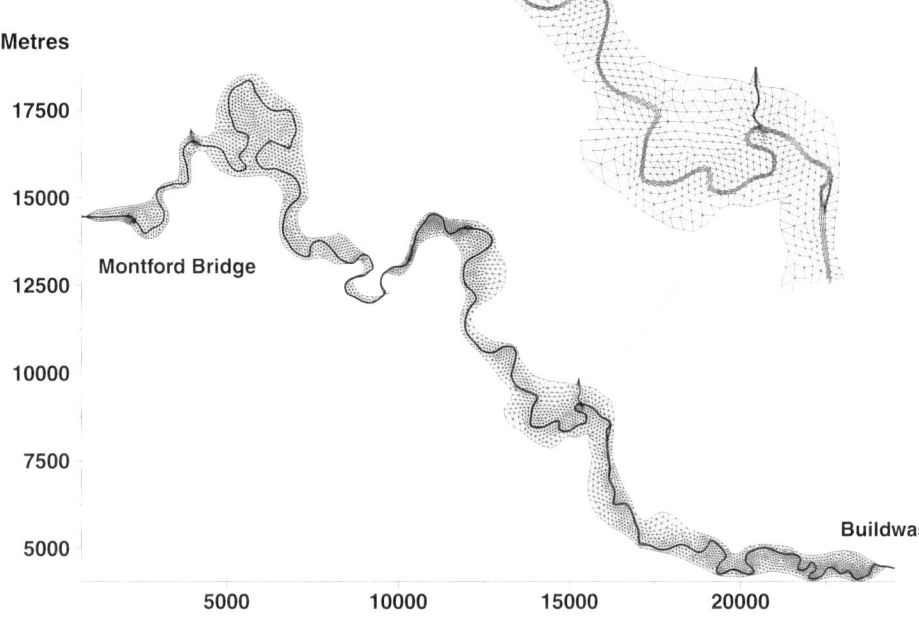

Figure 13.5   Example of a two-dimensional finite element mesh developed by Bates et al. (1998a) for a 40 km reach of the River Severn, UK

Use of airborne oblique photography or satellite data sources such as Synthetic Aperture Radar (SAR) can allow accurate determination of flood inundation at a specific point in time. For example, Horritt (1999) presents an algorithm capable of segmenting a SAR image into wet and dry areas to an accuracy of ~1 pixel (about 25 m for ERS-1 SAR data). This provides an approximation to the zero water depth contour that can be used to validate the ability of hydraulic models to simulate the complex problem of flood propagation over previously dry areas.

The advantage of such data is that it is continuous rather than discrete and for relatively flat floodplain and tidal marsh topography, small changes in water surface elevation can result in large changes in the shoreline position. Hence, in order to replicate a floodplain shoreline adequately a model must include topographic data at the relevant scale and generate an accurate distributed field of the flow depth. Inundation extent is potentially therefore a powerful test of a hydraulic model's capabilities as it is both a sensitive measure and its simulation requires a model capable of dealing with dynamic wetting and drying over complex topography.

Recent tests have, however, shown that inundation extent perhaps does have the discriminatory power we previously hoped (Bates and De Roo 2000). Here three different flood inundation models of varying complexity were used to simulate major flooding that occurred over a 20 day period in 1995 on a 35 km reach of the River Meuse on the Belgium/Netherlands border. These were, in order of increasing complexity, a planar approximation to the water surface, a specifically developed raster-based inundation model and a two-dimensional finite element scheme. The planar approximation treated the water surface as a non-dynamic 'lid' derived from the maximum stages recorded at gauges at the upstream and downstream ends of the reach. This

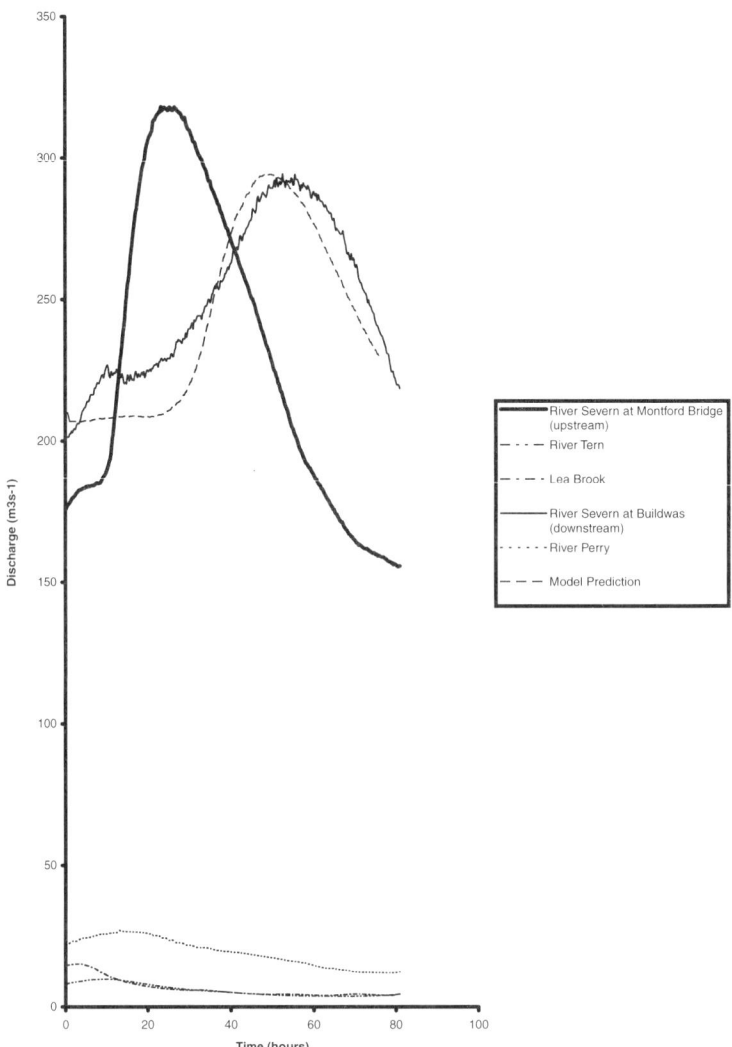

Figure 13.6 Validation of the two-dimensional finite element model shown in Figure 13.5 against external discharge data measured at the downstream boundary of the reach (Buildwas). Also shown is the upstream inflow discharge at Montford Bridge, and inflow discharges from three gauged tributary rivers: the Perry, the Tern and the Lea Brook. Even with multiple inputs it is relatively easy to calibrate a variety of models to fit such data and it is likely that a number of model calibrations will fit the data (within some reasonable error band) equally well

was then intersected with a high resolution Digital Elevation Model and can be considered as the simplest possible inundation extent predictor. The raster model used a finite difference kinematic model to route flood water downstream with floodplain storage routed across the floodplain using a two-dimensional diffusion wave approach. This was developed to incorporate

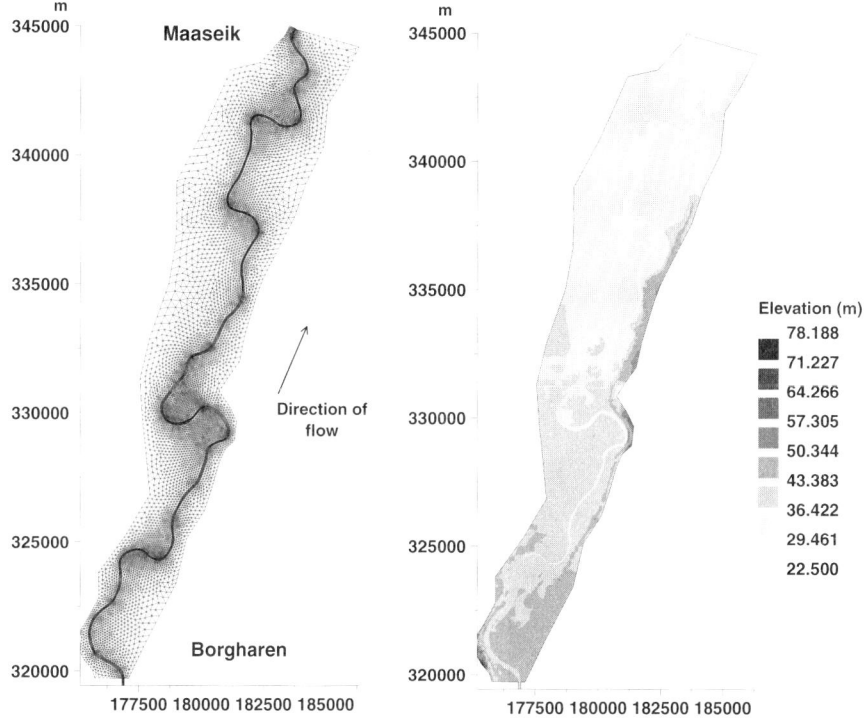

Figure 13.7   Finite element mesh and topography developed for the River Meuse between Borgharen and Maaseik. The mesh consists of 9639 nodes and 18 939 elements

the simplest process representation capable of simulating wave propagation and dynamic flooding. The finite element model utilised a full solution of equations (6), (7) and (8) and can be thought of as the standard model for this class of problem (see Figure 13.7). The model predictions of inundation extent were validated against shorelines derived from oblique air-photo data and ERS-1 SAR imagery of the flood. Results showed that the raster model was most successful in capturing the inundation extent field over the full reach (see Figure 13.8) and was able to correctly classify a maximum of 81.9% of the airphoto-derived inundated area. The planar approximation achieved a classification accuracy of 69.5%, while the two-dimensional finite element model managed only 56.2%. Assuming complete inundation of the floodplain would give a default classification accuracy of 67%, so this figure provides a benchmark that an inundation model should exceed. Thus, contrary to expectations, the most complex model failed to capture the inundation field, largely because the computational cost involved in the two-dimensional finite element simulation meant that element sizes of up to 250 m had to be used, severely limiting the topographic complexity that could be incorporated. The planar approximation failed to account for dynamic effects over the whole reach, but improved in accuracy close to the gauging stations which it used as control points. For example, for the 7 km reach below the upstream gauging station at Borgharen the raster model achieved a classification accuracy of 85.5% against the airphoto data, while the planar approximation was only margin-ally worse at 83.7%. However, with the planar approximation one cannot define a priori when it

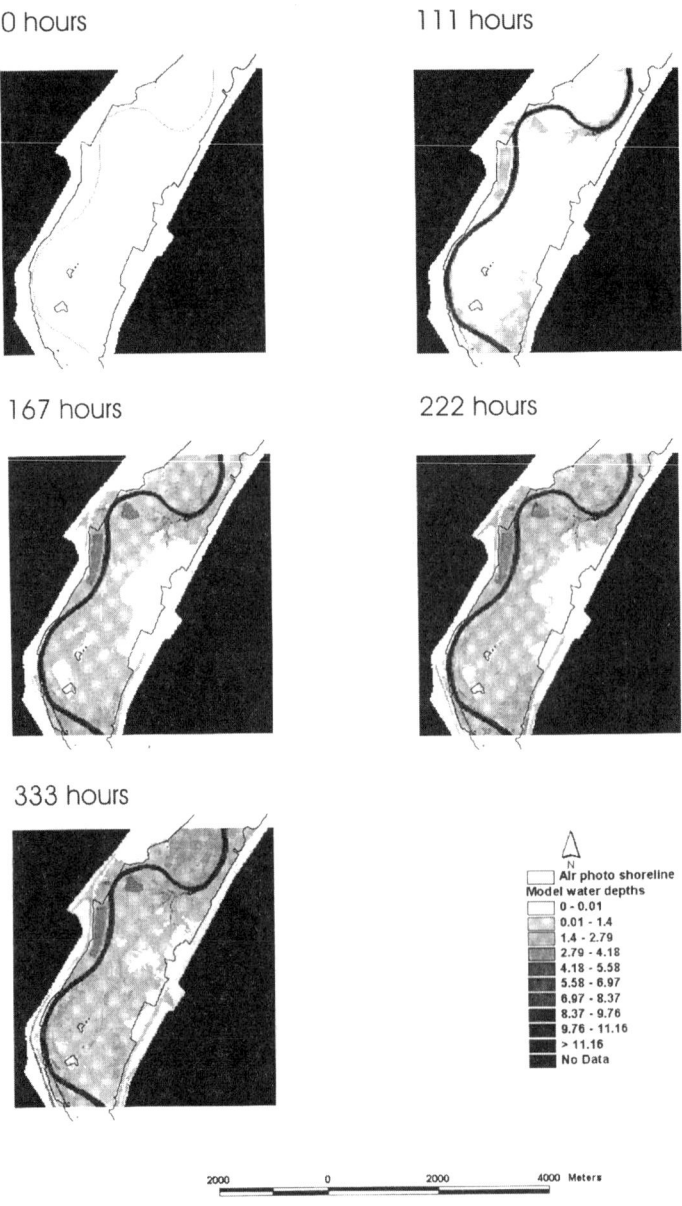

Figure 13.8 Time series of inundation extent predicted by a dynamic simulation of a raster-based flood inundation model using a 25 m resolution DEM for the 7 km reach downstream of the River Meuse, Belgium, below the gauging station at Borgharen. This represents the upper portion of the full 35 km reach simulated. The model water depths are compared to the airphoto derived shoreline sampled at approximately 160 hours into the simulation

will be appropriate and its success is dependent on an accurate delineation of the floodplain area.

Given errors in derivation of the shoreline data it seems likely that the best inundation models identified in this study are approaching the prediction limit for this class of problem, although the data available did not allow adequate checking of the inundation dynamics simulated by each code. Even airphoto data, which one may assume is more accurate than satellite sources, may only be able to correctly classify of the order of 90% of the true inundated area. Thus, while more complex codes applied at a sufficiently high resolution might show an apparent improvement in predictive ability it would be difficult to tell with current data sources whether or not this was justified. It follows that current inundation extent data sets can be replicated with rather simple models. Reasons for this lie in the fact that the steepest process gradients in river floods occur in channel areas that can considered as the near-field of the hydraulic problem. However, despite the known complexity of channel/floodplain flow (e.g. Sellin and Willetts 1996) these processes do not, in Young's (this volume, chapter 7) terminology, sufficiently excite the system to be apparent in inundation extent data taken in the hydrodynamic far-field. In terms of inundation extent the dominant mode of the system would appear to be rather simple: a kinematic wave and simple storage cell routing on the floodplain. In a similar fashion to bulk flow data sources, therefore, currently available inundation extent data may not permit the identification of anything but simple models.

Point Scale Data

In moving to point scale (primarily velocity) data for model validation we move our concern from far-field reach scale and regional data to near-field local data. Such data is never routinely available but would, in the light of the previous discussion, seem to be the only field data capable of conclusively discriminating between one-dimensional and higher order schemes and between higher order schemes themselves. Very few comparisons between two- and three-dimensional numerical models and field velocity data have been undertaken, and the scale of hydraulic problem that has typically been investigated has necessarily been small (domain sizes of typically < 40 m in length). Lane et al. (1995) compared a two-dimensional depth averaged model to velocities measured in a proglacial stream and found correlations between observed and measured data of 0.72 for downstream velocities and 0.57 for cross-stream velocities. Using an improved velocity data set measured for a gravel bed river confluence using Acoustic Doppler Velocimetry (ADV) and including comparison to a three- as well as a two-dimensional model S.N. Lane et al. (1999) found improved correlations between observations and model predictions. For the two-dimensional model these were found to be 0.9 for downstream velocity and 0.65 for cross-stream velocity. In the case of the three-dimensional model results averaged over the vertical (to give a two-dimensional field), the respective correlations were 0.85 and 0.64. Lane et al. did, however, note that the vertical velocities predicted by the three-dimensional code were significantly less accurate (correlation coefficient of 0.498 when compared to appropriate field data) as a result of the difficulty of representing the complex bed topography present in gravel rivers (see Lane and Richards, this volume, chapter 16, for a more detailed discussion). Similarly, Nicholas and Smith (1999) attempted to validate a three-dimensional model against two-dimensional velocities measured using an Electromagnetic Current Meter (ECM) in a braided proglacial stream and found a correlation of 0.88 for scalar velocity (the vector product of the downstream and cross-stream components measured by the ECM). Field data collection for overbank flow velocities in compound channels is currently being undertaken by the Universities of Glasgow and Lancaster in the UK in order to test three-dimensional models of meandering compound channel flow for domains approximately 1 km in length (Ervine, personal communication). Despite the infrequent nature of overbank flow making this a logistically difficult project, initial results are encouraging (see Morvan et al. 2000) and will provide a significant extension of the in-channel data collection already performed by S.N. Lane et al. (1995, 1999) and Nicholas and Smith (1999). An exception to the small-scale validation of point velocity predictions is provided by Kodama et al. (1996) who attempted validation of a three-dimensional multi-level tidal current model for Tokyo Bay. Here the domain was approximately 10×60 km in size and was meshed with 1216 elements at the surface level with progressively fewer elements as the depth decreased. The model was compared to mean and tidal currents determined for ten measurement sites over a two month period. Hence, to overcome the constraints imposed by the scale of the domain and the difficulty of obtaining synchronous velocity measurements at a number of points, temporal averaging of the velocity field was necessary. The coarse mesh resolution required to render large-scale problems computationally feasible also hampers direct comparison with field velocity measurements as the velocities computed by the model are grid square effective values rather than point scale data.

Lastly, one can use measures of sediment erosion and deposition as surrogates for hydraulic variables. The advantage of this is that quantities such as sediment deposition may be easier to measure in the field and can provide a time-integrated measure of the impact of hydraulic processes at a given point in the model. Thus, one may use sediment trap measurements of contemporary floodplain deposition (Nicholas and Walling 1997) and Cs^{137} radionuclide estimates of medium-term deposition rates (Nicholas and Walling 1997; Hardy et al. 2000) to validate linked hydraulic/sediment transport models. The disadvantage here is that an additional level of model and data uncertainty is introduced to the validation process; however, if these can be overcome then this may be a powerful validation technique.

A number of points emerge from these studies. First, S.N. Lane et al. (1999) note the difficulty of measuring exactly what the model predicts. Turbulent velocities are represented in a RANS-based numerical model as a time- and space-averaged ensemble value. Lane et al. give the example of potential discrepancies between the measurement volume over which velocity is sampled (typically of the order of cubic centimetres) and the volume of a computational cell in a typical three-dimensional numerical model (typically much greater than the measurement volume). Model predictions may thus not be representative of the volume over which measurements were obtained. Similar problems occur with time averaging of the instantaneous turbulent velocities (see equation (3)) and, with two-dimensional models, depth averaging the velocity field. Second, S.N. Lane et al. (1999) note that despite the relatively high correlations obtained the level of unexplained variance in these analyses remains high (40–50%, and up to 75% for vertical velocities). Both S.N. Lane et al. (1999) and Nicholas and Smith (1999) point out that high levels of unexplained variance may be inevitable in gravel bed rivers with high relative roughness where it may be difficult to supply sufficiently detailed topographic information to the model. This may limit the model's ability to resolve flow around individual grains and grain clusters that can have a significant effect on the flow velocity distribution, particularly if the roughness length approaches the model cell size. Lastly, studies of the response of higher order numerical models to parameter variation have demonstrated complex spatial sensitivity (Lane et al. 1994; Bates et al. 1996; Bates et al. 1998b; Lane and Richards 1998), even when uniform parameter changes are applied. For example, Bates et al. (1998b) looked at a hypothetical river–floodplain reach with realistic topography generated through fractal analysis (see Figure 13.9) to assess model sensitivity to uniform parameter change. The impact of a spatially uniform 20% increase and decrease in boundary friction coefficients was analysed in terms of bulk outflow (Figure 13.10), inundation extent (Figure 13.11) and the spatially distributed velocity field (Figure 13.12). Model response in terms of bulk flows was simple and linear but for local hydraulics complex spatial sensitivity was indicated. The sensitivity of inundation extent to

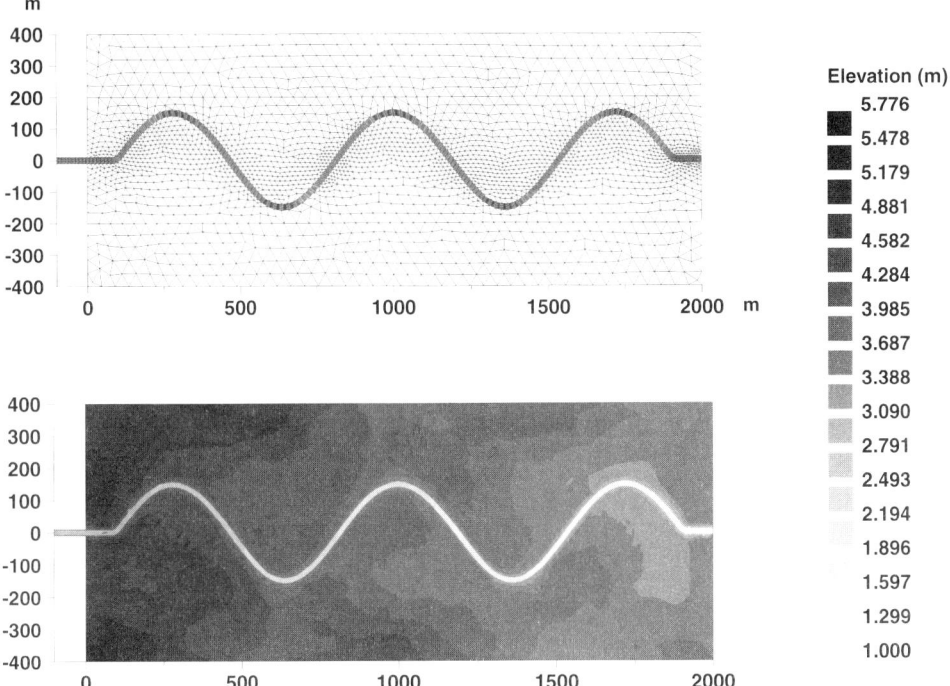

Figure 13.9 Finite element mesh and topography used by Bates et al. (1998b) to study spatial sensitivity to uniform parameter change in a two-dimensional finite element solution of the St Venant equations. The topography for the mesh was generated by fractal analysis of real floodplain topographies

uniform change lies somewhere between the two. This implies that while it would appear straightforward to calibrate such models to predict bulk flow and inundation extent using the relationships contained in Figure 13.10 and Figure 13.11, calibrating and validating a model to replicate local hydraulics is a considerably more difficult proposition. Such sensitivities should perhaps be expected as higher order numerical models are typically based on highly non-linear systems of equations solved in a spatially distributed format. However, they may substantially complicate the interpretation of point scale validation evidence from both laboratory and field experiments as the potential for either non-optimal or equifinal solutions is increased.

In considering all types of model validation against field data something of a glass ceiling would thus appear to exist. While it may only be possible to discriminate and calibrate rather simple models from bulk flow or inundation extent data, fields of point scale data which could overcome this problem may, for the reasons discussed above, be difficult to interpret with higher order models.

13.5 Conclusions

The above discussion has highlighted a number of gaps in the current validation of hydraulic models. While hydraulics is relatively rich in analytical solutions and scale model data, field data

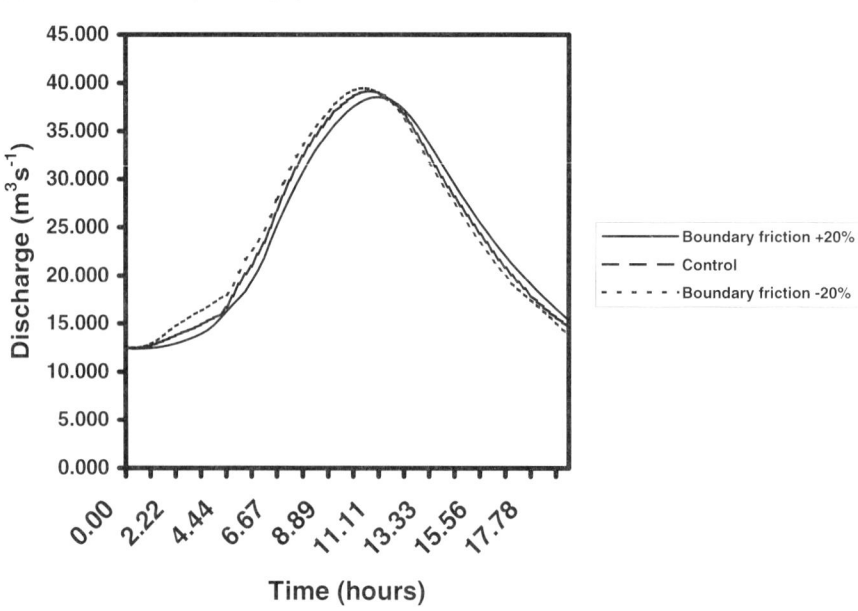

Figure 13.10 Response of bulk outflows predicted by a two-dimensional finite element model of the hypothetical floodplain reach shown in Figure 13.9 to spatially uniform change in boundary friction

sufficient to validate anything other than the most simple models are not routinely available. Where more discriminatory field data exist these can be shown to incorporate uncertainties that significantly limit model identifiability (in the case of inundation extent) or to have been collected at a limited range of scales and for a limited range of environments (in the case of velocity vectors). The discipline would clearly benefit from the collection of a number of benchmark field data sets for a variety of flow situations and scales to supplement analytical and laboratory data sets. However, all validation data sources contain flaws to a greater or lesser extent, and no single source can be relied upon to provide conclusive confirmation/falsification of a model hypothesis. Rather, a variety of validation tests need to be conducted for each type of code, and we are perhaps close to being able to define what these should be. Ayer (1936) takes this one step further and suggests that a statement (in our case a model hypothesis) is only genuine if one knows how to verify or falsify the proposition which it embodies. Thus we cannot consider models to be genuine hypotheses unless we can specify the validation tests they need to be subjected to and what evidence would constitute sufficient grounds for their acceptance or rejection.

We also need to explore whether new technologies can aid the validation process. High frequency VHF radar can be used to estimate surface velocities in coastal waters (A. Lane et al. 1999; Sentchev and Yaremchuk 1999) and can be used in conjunction with numerical models (e.g. Williams et al. 2000). Similarly, the quantitative retrieval of suspended particulate matter concentrations in coastal waters from airborne remote sensing (e.g. Moore et al. 1999) may become a realistic technique for future model validation studies. In general, there are very few studies which attempt field validation of hydraulic model flow velocities and associated trans-

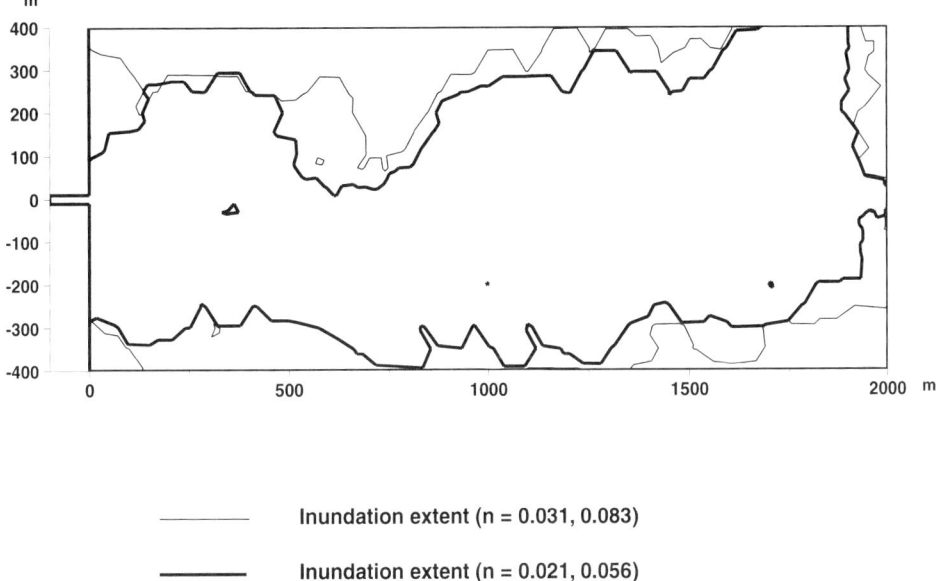

Inundation extent (n = 0.031, 0.083)

Inundation extent (n = 0.021, 0.056)

Figure 13.11   Response of inundation extent predicted by a two-dimensional finite element model of the hypothetical floodplain reach shown in Figure 13.9 to spatially uniform change in boundary friction

port, despite the fact that in management terms it may be the material transported by such flows that is of the greatest significance. This is particularly true of two- and three-dimensional schemes and it is clear that such modelling and appropriate field data collection needs to be more closely integrated. However, a conundrum exists in that codes capable of predicting the velocity fields that drive sediment and pollutant transport in anything but the simplest situations may be structurally very difficult to analyse. As a result there is some danger of reductionism as we may become overly focused on very small scale processes (see the discussion by Lane and Richards 1997).

Despite these gaps in hydraulic model validation/falsification, significant science can still be accomplished. Even a partially validated model may tell us more about processes than continued and unfocused empirical investigation. Models can be used to accomplish a variety of scientific tasks such as testing ideas for reasonableness, organising/checking field data, the design of field investigations and assessing process dominance. Whether validation/falsification is possible thus depends on the science that one is trying to accomplish and, in our view, the process would benefit from a more pragmatic approach. We would consider a model sufficiently validated if the simulations/predictions it makes fulfil a particular objective. Whether a model is acceptable or not is at present a subjective decision (see Konikow and Bredehoeft 1992). We would support recasting this as a practical decision by asking before each study 'what do we want the model to do?'. Thus we need to specify, and more importantly agree, a priori tests a given model needs to pass to be 'valid' for the purpose for which it was conceived. This is not to say that we should be accepting of what Morton (1993) calls *mediating* models that contain assumptions which are false and known to be false and which attempt to bridge the gap between theory (which may be partly qualitative in nature) and quantitative prediction. However, there is

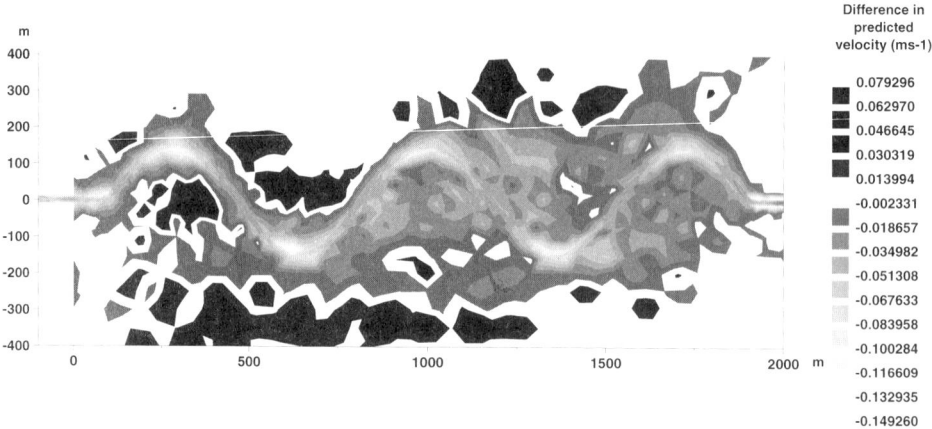

Figure 13.12 Response of local velocity vectors predicted by a two-dimensional finite element model of the hypothetical floodplain reach shown in Figure 13.9 to spatially uniform change in boundary friction

a conundrum here: models must simplify to be useful but in the limit all such simplifications must be a contraction of reality and thus somehow false. Resolution of such issues must surely proceed on the basis of agreed testing and evidence in line with Beven's (1996) call for a 'thoughtful' approach to hydrological modelling.

Lastly, the hydraulic model validation studies that have been attempted thus far are best classed as history matching (see Konikow and Bredehoeft 1992 and Section 13.1), and the predictive ability of such codes has not yet been studied. It is, however, clear that if we use a model in a predictive mode we need to test the model in a like fashion and undertake a post-audit of performance and prediction. This is 'predictive validation' in our original hierarchy (Section 13.1) and as Young (this volume, chapter 7) points out is conditional on possible future falsification by new data. Similarly, Diskin and Simon (1977) note that if we want robust distributed models we need to test them against distributed data. How many such tests are enough must depend on the problem being studied, but could be statistically quantified (see Romanowicz et al. 1996) and also defined on the basis of applied utility. However, neither predictive nor distributed validation has yet been satisfactorily accomplished in hydraulics.

In conclusion, model-building should be considered as no different from any other means of hypothesis testing. Debates over model validation are in reality debates over what constitutes proof and are thus an essential part of scientific activity. It is unlikely that we will attain conclusive answers to such questions, or indeed that we should want to. The definition of proof in particular fields needs to be an ongoing dialogue, continually under revision as a result of new data and ideas. To stop questioning model validation would amount to ending scientific progress. Scientifically, we should be comfortable with such uncertainty and the provisional nature of truth despite legal and political pressures to the contrary.

REFERENCES

Akhavan, R., Ansari, A., Kang, S. and Mangiavacchi, N. 2000. Subgrid-scale interactions in a numerically simulated planar turbulent jet and implications for modelling. *Journal of Fluid Mechanics*, **408**, 83–120.

Ayer, A.J. 1936. *Language, Truth and Logic*. Gollancz, London.

Bastiaans, R.J.M., Rindt, C.C.M., Nieuwstadt, F.T.M. and van Steenhoven, A.A. 2000. Direct and large-eddy simulation of the transition of two- and three-dimensional plane plumes in a confined enclosure. *International Journal of Heat and Mass Transfer*, **43**, 2375–2393.

Bates, P.D. and Anderson, M.G. 1996. A preliminary investigation into the impact of initial conditions on flood inundation predictions using a time/space distributed sensitivity analysis. *Catena*, **26**, 115–134.

Bates, P.D. and De Roo, A.P.J. (2000). A simple raster-based model for floodplain inundation. *Journal of Hydrology*, 236, 54–77.

Bates, P.D. and Hervouet, J-M. 1999. A new method for moving boundary hydrodynamic problems in shallow water. *Proceedings of the Royal Society of London, Series A*, **455**, 3107–3128.

Bates, P.D., Anderson, M.G., Baird, L., Walling, D.E. and Simm, D. 1992. Modelling floodplain flow with a two dimensional finite element scheme. *Earth Surface Processes and Landforms*, **17**, 575–588.

Bates, P.D., Horritt, M., Smith, C. and Mason, D. 1997. Integrating remote sensing observations of flood hydrology and hydraulic modelling. *Hydrological Processes*, **11**, 1777–1795.

Bates, P.D., Stewart, M.D., Siggers, G.B., Smith, C.N., Hervouet, J-M. and Sellin, R.H.J. 1998a. Internal and external validation of a two dimensional finite element model for river flood simulation. *Proceedings of the Institution of Civil Engineers, Water Maritime and Energy*, **130**, 127–141.

Bates, P.D., Horritt, M. and Hervouet, J-M. 1998b. Investigating two dimensional finite element predictions of floodplain inundation using fractal generated topography. *Hydrological Processes*, **12**, 1257–1277

Bates, P.D., Wilson, C.A.M.E., Hervouet, J-M. and Stewart, M.D. 1999. Two dimensional finite element modelling of floodplain flow/Modélisation à deux dimensions par eléments finis d'un ecoulement en plaine. *La Houille Blanche*, **3/4**, 61–67.

Bathurst, J.C. 1986. Physically based distributed modelling of an upland catchment using the Systeme Hydrologique Europeen. *Journal of Hydrology*, **87**, 79–102.

Benkhaldoun, F. and Monthe, L. 1994. An adaptive nine-point finite volume Roe scheme for two dimensional Saint Venant equations. In: Molinaro, P. and Natale, L. (eds), *Modelling Flood Propagation over Initially Dry Areas*. American Society of Civil Engineers, New York, 30–44.

Berbery E.H., Mitchell K.E., Benjamin S., Smirnova, T., Ritchie, H., Hogue, R. and Radeva, E. 1999. Assessment of land-surface energy budgets from regional and global models. *Journal of Geophysical Research*, **104**(D16), 19 329–19 348.

Beven, K.J. 1989. Changing ideas in hydrology: the case of physically based distributed models. *Journal of Hydrology*, **105**, 79–102.

Beven, K.J. 1996. TOPMODEL: a critique. *Hydrological Processes*, **11**, 1069–1085.

Braithwaite, R.B. 1953. *Scientific Explanation*. Cambridge University Press, Cambridge.

Defina, A., D'Alpaos, L. and Matticchio, B. 1994. A new set of equations for very shallow water and partially dry areas suitable to 2D numerical models. In: Molinaro, P. and Natale, L. (eds), *Modelling Flood Propagation over Initially Dry Areas*. American Society of Civil Engineers, New York, 72–81.

DeVriend, H.J. and Geldof, H.J. 1983. Mean flow velocity in short river bends. *American Society of Civil Engineers, Journal of Hydraulic Engineering*, **109**, 991–1011.

Didden, N. and Schott, F. 1992. Seasonal variations in the western tropical Atlantic: surface circulation from Geosat altimetry and WOCE model results. *Journal of Geophysical Research*, **97**(C3), 3529–3541.

Diskin, M.H. and Simon, E. 1977. A procedure for the selection of objective functions for distributed hydrological models. *Journal of Hydrology*, **34**, 129–149.

Ervine, D.A. and MacLeod, A.B. 1999. Modelling a river channel with distant floodbanks. *Proceedings of the Institution of Civil Engineers, Water Maritime and Energy*, **136**, 21–33.

Evangelinos, C., Lucor, D. and Karniadakis, G.E. 2000. DNS-derived force distribution on flexible cylinders subject to vortex-induced vibration. *Journal of Fluids and Structures*, **14**, 429–440.

Ferguson, R.I., Kirkbride, A.D. and Roy, A.G. 1996. Markov analysis of velocity fluctuations in gravel-bed rivers. In: Ashworth, P.J., Bennett, S.J., Best, J.L. and McLelland, S.J. (eds), *Coherent Flow Structures in*

Open Channels. John Wiley, Chichester, 165–183.

Fische-Antze, T., Stoesser, T., Bates, P.D. and Olsen, N.R.B. (in press). 3D numerical modelling of open-channel flow with submerged vegetation. *Journal of Hydraulics Research.*

Fread, D.L. 1984. Flood routing. In: Anderson, M.G. and Burt, T.P. (eds), *Hydrological Forecasting.* John Wiley, Chichester, Chapter 14.

Fread, D.L. 1993. Flood routing. In: Maidment, D.R. (ed.), *Handbook of Applied Hydrology.* Mc-Graw Hill, New York, Chapter 10.

Freer, J., Beven, K. and Ambroise, B. 1996. Bayesian estimation of uncertainty in runoff prediction and the value of data: an application of the GLUE approach. *Water Resources Research,* **32**, 2161–2173.

Ghia, U., Ghia, K.N. and Shin, C.T. 1982. High-Re solutions for incompressible flow using the Navier–Stokes equations and a multigrid method. *Journal of Computational Physics,* **48**, 387–411.

Gregory, K.J. and Walling, D.E. 1973. *Drainage Basin Form and Process.* Edward Arnold, London.

Hanfling, O. 1981. *Logical Positivism.* Blackwell, Oxford.

Hardy, R.J., Bates, P.D. and Anderson, M.G. 2000. Modelling suspended sediment deposition on a fluvial floodplain using a two-dimensional dynamic finite element model. *Journal of Hydrology,* **229**, 202–218.

Harpin, R., Webb, D.R., Whitlaw, C.D., Samuels, P.G. and Wark, J.B. 1995. *Benchmarking of Hydraulic Models.* Stage 1 Final Report to the UK National Rivers Authority, Research and Development Project 508, National Rivers Authority, UK.

Harvey, D. 1969. *Explanation in Geography.* Edward Arnold, London.

Hervouet, J-M. 2000. A high resolution 2-D dam-break model using parallelization. *Hydrological Processes,* 14, 2211–2230.

Hervouet, J-M. and Van Haren, L. 1996. Recent advances in numerical methods for fluid flows. In: Anderson, M.G., Walling, D.E. and Bates, P.D. (eds), *Floodplain Processes.* John Wiley, Chichester, 183–214.

Hicks, F.E., Jin, Y.C. and Steffler, P.M. 1990. Flow near sloped bank in curved channel. *American Society of Civil Engineers, Journal of Hydraulic Engineering,* **116**, 55–70.

Horritt, M.S. 1999. A statistical active contour model for SAR image segmentation. *Image and Vision Computing,* 17, 213–224.

Horritt, M.S. 2000. Development of physically based meshes for two-dimensional models of meandering channel flow. *International Journal of Numerical Methods in Engineering,* 47, 2019–2037.

Ip, J.T.C., Lynch, D.R. and Friedrichs, C.T. 1998. Simulation of estuarine flooding and dewatering with application to Great Bay, New Hampshire. *Estuarine Coastal and Shelf Science,* 47(2), 119–141.

Kawahara, M. and Umetsu, T. 1986. Finite element method for moving boundary problems in river flows. *International Journal of Numerical Methods in Fluids,* 6, 365–386.

Kemp, M.J., Nash, D.F.T. and Anderson, M.G. 1989. On modelling installation designs for soil–water instrumentation, with reference to casagrande type piezometer systems. *Earth Surface Processes and Landforms,* **14**, 375–382.

King, I.P. and Roig, L. 1988. Two dimensional finite element models for floodplains and tidal flats. In: Niki, K. and Kawahara, M. (eds), *Proceedings of an International Conference on Computational Methods in Flow Analysis,* Okayama, Japan, 711–718.

Kodama, T., Wang, S.Y. and Kawahara, M. 1996. Model verification on 3D tidal current analysis in Tokyo Bay. *International Journal for Numerical Methods in Fluids,* **22**, 43–66.

Kohane, R. and Welz, R. 1994. Combined use of FE models for prevention of ecological deterioration of areas next to a river hydropower complex. In: Peter, A., Wittum, G., Meissner, U., Brebbia, C.A., Gray, W.G. and Pinder, G.F. (eds), *Computational Methods in Water Resources X.* Kluwer, The Netherlands, Vol. 1, 59–66.

Knight, D.W. and Sellin, R.H.J. 1987. The SERC Flood Channel Facility. *Journal of the Institute of Water and Environmental Management,* **1**, 198–204.

Knight D.W. and Shiono K. 1996. River channel and floodplain hydraulics. In: Anderson M.G., Walling D.E. and Bates P.D. (eds), *Floodplain Processes.* John Wiley, Chichester, 139–182.

Konikow, L.F. and Bredehoeft, J.D. 1992. Groundwater models cannot be validated. *Advances in Water Resources,* **15**, 75–83.

Lane, A., Knight, P.J. and Player, R.J. 1999. Current measurement technology for near-shore waters. *Coastal Engineering,* 37, 343–368.

Lane, S.N. 1998. Hydraulic modelling in hydrology and geomorphology: a review of high resolution approaches. *Hydrological Processes*, **12**, 1131–1150.

Lane, S.N. and Richards, K.S. 1997. Linking river channel form and process: time, space and causality revisited. *Earth Surface Processes and Landforms*, **22**, 249–260.

Lane, S.N. and Richards, K.S. 1998. High resolution, two-dimensional spatial modelling of flow processes in a multi-thread channel. *Hydrological Processes*, **12**, 1279–1298.

Lane, S.N., Richards, K.S. and Chandler, J.H. 1994. Application of distributed sensitivity analyses to a model of turbulent open channel flow in a turbulent river channel. *Proceedings of the Royal Society of London, Series A*, **477**, 49–63.

Lane, S.N., Richards, K.S. and Chandler, J.H. 1995. Within-reach spatial patters of processes and channel adjustment. In: Hickin, E.J. (ed.), *River Geomorphology*. John Wiley, Chichester, 105–130.

Lane, S.N., Bradbrook, K.F., Richards, K.S., Biron, P.A. and Roy, A.G. 1999. The application of computational fluid dynamics to natural river channels: three-dimensional versus two-dimensional approaches. *Geomorphology*, **29**, 1–20.

Le, H., Moin, P. and Kim, J. 1997. Direct numerical simulation of turbulent flow over a backward-facing step. *Journal of Fluid Mechanics*, **330**, 349–374.

Leclerc, M., Bellemare, J-F., Dumas, G. and Dhatt, G. 1990. A finite element model of estuarine and river flows with moving boundaries. *Advances in Water Resources*, **13**, 158–168.

Li, C.S. 1997. Waveform sampling LiDAR applications in complex terrain. *International Journal of Remote Sensing*, **18**, 2087–2104.

Lopez, F. and Garcia, M. 1997. *Open Channel Flow through Simulated Vegetation: Turbulence Modelling and Sediment Transport*. Hydrosystems Laboratory, Department of Civil and Environmental Engineering, University of Illinois at Urbana-Champaign.

Lynch, D.R. and Gray, W.G. 1980. Finite element simulation of flow deforming regions. *Journal of Computational Physics*, **36**, 135–153.

Lynch, D.R. and Officer, C.B. 1985. Analytic test cases for three-dimensional hydrodynamic models. *International Journal of Numerical Methods in Fluids*, **5**, 529–543.

Marks, K.J. and Bates, P.D. (2000). Integration of high resolution topographic data with floodplain flow models. *Hydrological Processes*, **14**, 2109–2122.

Martinuzzi, R. and Tropea, C. 1993. The flow around surface-mounted, prismatic obstacles in a fully developed channel flow. *Transactions of the American Society of Mechanical Engineers, Journal of Fluids Engineering*, **115**, 85–92.

Moore, G.F., Aiken, J. and Lavender, S.J. 1999. The atmospheric correction of water colour and the quantitative retrieval of suspended particulate matter in Case II waters: application to MERIS. *International Journal of Remote Sensing*, **20**, 1713–1733.

Moramarco, T. and Singh, V.P. 2000. A practical method for analysis of river waves and for kinematic wave routing in natural channel networks. *Hydrological Processes*, **14**, 51–62.

Moramarco, T., Fan, Y. and Bras, R.L. 1999. Analytical solution for channel routing with uniform lateral inflow. *American Society of Civil Engineers, Journal of Hydraulic Engineering*, **125**, 707–713.

Morton, A. 1993. Mathematical models: questions of trustworthiness. *British Journal for the Philosophy of Science*, **44**, 659–674.

Morvan, H., Pender, G., Wright, N.G. and Ervine, D.A. (2000). Three-dimensional modelling of the flow mechanisms in flooded meandering channels. In: Koch, M. and Tönsmann, F. (eds), *Proceedings of the International Symposium on Flood Defence*, Kassel, Germany, 20–23 September 2000. *http://www.uni-kassel.de/fb14/wasserbau/symposium* 2000.

Moulin, C. and Ben Slama, E. 1998. The two-dimensional transport module SUBIEF. Applications to sediment transport and water quality processes. *Hydrological Processes*, **12**, 1183–1195.

Nicholas, A.P. and Smith, G.H.S. 1999. Numerical simulation of three-dimensional flow hydraulics in a braided channel. *Hydrological Processes*, **13**, 913–929.

Nicholas, A.P. and Walling, D.E. 1997. Modelling flood hydraulics and overbank deposition on river floodplains. *Earth Surface Processes and Landforms*, **22**, 59–77.

Olsen, N.R.B. and Kjellesvig, H.M. 1999. Three-dimensional numerical modelling of bed changes in a sand trap. *Journal of Hydraulic Research*, **37**, 189–198.

Olsen, N.R.B. and Stokseth, S. 1995. Three-dimensional numerical modelling of water flow in a river with large bed roughness. *Journal of Hydraulic Research*, **33**, 571–581.

Panchang, V., Chen, W., Xu, B., Schlenker, K., Demirbilek, Z. and Okihiro, M. 2000. Exterior bathymetric effects in elliptic harbor wave models. *American Society of Civil Engineers, Journal of Waterway, Port, Coastal and Ocean Engineering*, **126**, 71–78.

Pasche, E. 1984. *Turbulence Mechanism in Natural Streams and the Possibility of its Mathematical Representation* (in German). Research Report No. 52, Mitteilungen Instituit für Wasserbau und Wasserwirtschaft, RWTH, Aachen, Germany.

Popper, K.R. 1959. *The Logic of Scientific Discovery*. Hutchinson, London.

Rajendran, V.P., Constantinescu, S.G. and Patel, V.C. 1999. Experimental validation of numerical model of flow in pump-intake bays. *American Society of Civil Engineers, Journal of Hydraulic Engineering*, **125**, 1119–1125.

Rastogi, A. and Rodi, W. 1978. Predictions of heat and mass transfer in open channels. *American Society of Civil Engineers, Journal of the Hydraulics Division*, **104**, 397–420.

Reynolds, O. 1895. On the dynamical theory of incompressible viscous fluids and the determination of the criterion. *Philosophical Transactions of the Royal Society*, **186A**, 123–164.

Richards, K. 1990. Real geomorphology. *Earth Surface Processes and Landforms*, **15**, 195–197.

Rodi, W. 1980. *Turbulence Models and their Application in Hydraulics*. IAHR Special Publication, Delft, The Netherlands.

Rodi, W., Ferziger, J.H., Breur, M. and Pourquié, M. 1997. Status of large eddy simulation: results of a workshop. *Transactions of the American Society of Mechanical Engineers, Journal of Fluids Engineering*, **119**, 248–262.

Romanowicz, R., Beven, K.J. and Tawn, J. 1996. Bayesian calibration of flood inundation models. In: Anderson M.G., Walling D.E. and Bates P.D. (eds), *Floodplain Processes*. John Wiley, Chichester, 333–360.

Samuels, P.G. 1990. Cross section location in one-dimensional models. In: White, W.R. (ed.), *International Conference on River Flood Hydraulics*. Wiley, Chichester, 339–350.

Sauvaget, P., David, E., Demmerle, D. and Lefort, P. 2000. Optimum design of large flood relief culverts under the A89 motorway in the Dordogne–Isle confluence plain. *Hydrological Processes*, **14**, 2311–2329.

Sellin R.H.J. and Willets B.B. 1996. Three-dimensional structures, memory and energy dissipation in meandering compound channel flow. In: Anderson M.G., Walling D.E. and Bates P.D. (eds), *Floodplain Processes*. John Wiley, Chichester, 255–298.

Sentchev, A. and Yaremchuk, M. 1999. Tidal motions in the Dover Straits as a variational inverse of the sea level and surface velocity data. *Continental Shelf Research*, **19**, 1905–1932.

Shaw, E.M. 1983. *Hydrology in Practice*, 2nd edition. Van Nostrand Reinhold, London.

Singh, V.P. 1996. Kinematic wave modelling in water resources: surface water hydrology. John Wiley, New York, 1399pp.

Stewart, M.D., Bates, P.D. Anderson, M.G., Price, D.A. and Burt, T.P. 1999. Modelling floods in hydrologically complex lowland river reaches. *Journal of Hydrology*, **223**, 85–106.

Stoesser, T. 1999. Development and validation of a CFD code for simulating turbulent open-channel flows. Internal Report, School of Geographical Sciences, University of Bristol.

Tchamen, G.W. and Kawahita, R.A. 1998. Modelling wetting and drying effects over complex topography. *Hydrological Processes*, **12**, 1151–1183.

Tominaga, A. and Nezu, I. 1991. Turbulent structure in compound open-channel flows. *American Society of Civil Engineers, Journal of Hydraulic Engineering*, **117**, 21–41.

Tsujimoto, T., Shimizu, T. and Okada, T. 1991. Turbulent structure of flow over rigid vegetation-covered bed in open channels. Progress Report 1, Hydraulics Laboratory, Kanazawa University, Japan.

Williams, J.J., MacDonald, N.J., O'Connor, B.A. and Pan, S. 2000. Offshore sand bank dynamics. *Journal of Marine Systems*, **24**, 153–173.

Wood, I.R. and Tong, L. 1989. Dispersion in an open channel with a step in the cross section. *Journal of Hydraulic Research*, **27**, 587–601.

Younis, B.A. 1996. Progress in turbulence modelling for open channel flows. In: Anderson, M.G., Walling, D.E. and Bates, P.D. (eds), *Floodplain Processes*. John Wiley, Chichester, 299–332.

14 Modelling Water Quality Processes in Riverine Systems

R.A. FALCONER, B. LIN AND S.M. KASHEFIPOUR
School of Engineering, University of Wales, Cardiff, UK

14.1 INTRODUCTION

In recent years there has been growing public concern about the quality of water within many riverine systems, particularly in those parts of the world where rivers have become increasingly used as receiving water bodies for discharges of domestic effluents, industrial by-products, agricultural waste and urban drainage. Human and aquatic life is often threatened by the transport of pollutants through riverine systems to coastal waters, and it is therefore not surprising to find that from a water quality point of view, rivers have been studied more extensively and for longer than any other bodies of water (Thomann and Mueller 1987). This is probably due to the fact that many people live close to, or interact with, rivers and streams. Moreover, the large amount of sediment often carried by rivers may result in the deposition of fine sediments along the river basin and in riverine deltas, thereby causing a restriction to navigation and also exacerbating flooding problems. The flux of suspended sediments through riverine systems can also provide an important transport path for highly toxic adsorbents, such as heavy metals and radionuclides etc.

Society's interest in rivers begins with an analysis of the flow characteristics, in the form of the magnitude and duration of discharges, coupled with the chemical and biological characteristics of the river waters. Rivers provide a rich and diverse ecosystem and the riverine basin may therefore be considered from the physical, chemical and biological perspective. The main characteristics of a riverine system include: (i) geometry (including width and depth), (ii) bed slope (including bed roughness and meandering), (iii) velocity of flow, (iv) mixing characteristics (including dispersion and diffusion), (v) water temperature, (vi) suspended solids and morphology. Likewise, from the viewpoint of important chemical and biological characteristics, there are primarily (Thomann and Mueller 1987):

1. *Chemical*: (i) dissolved oxygen, (ii) pH, acidity and alkalinity, (iii) suspended particulate matter, (iv) nitrates and phosphates, and (v) toxic substances (e.g. heavy metals).
2. *Biological*: (i) bacteria and viruses, (ii) fish populations, (iii) rooted aquatic plants, and (iv) algae.

In considering the impact of water pollution and suspended particulate matter on the hydro-environmental and ecological characteristics of a river, it is important to be able to predict mathematically the hydrodynamic, solute and suspended sediment transport processes

Model Validation: Perspectives in Hydrological Science. Edited by M.G. Anderson and P.D. Bates.
© 2001 John Wiley & Sons, Ltd.

in the riverine system. In this context hydroinformatics software tools have become increasingly used by water engineers and environmental managers to assess the impact of changes in the boundary conditions, with such changes including geometric (due to channelisation), morphological (due to sedimentation), hydro-environmental (due to a wastewater discharge), or eco-hydraulic (due to algal blooms).

This chapter is devoted to the establishment of the governing differential equations and parameters describing the hydrodynamic, water quality and sediment transport processes occurring regularly in riverine systems. Details are given of one-, two-, and – to a lesser extent – three-dimensional modelling approaches (1-D, 2-D and 3-D), and with examples being given of the predictive capabilities of such models as developed by the authors for a reach of the River Humber, in the northeast of England.

14.2   HYDRODYNAMIC MODELLING

14.2.1   Governing Equations

Three-Dimensional Flows

The numerical models used by water engineers and environmental managers to predict the flow, water quality and contaminant and sediment transport processes in riverine systems are based on first solving the governing hydrodynamic equations. For a Cartesian coordinate system, with the main body of the flow in the x direction, the corresponding 3-D Reynolds equations for mass and momentum in the flow direction can be written in a general conservative form as (Falconer 1993):

$$\frac{\partial u}{\partial x} + \frac{\partial v}{\partial y} + \frac{\partial w}{\partial z} = 0 \tag{1}$$

$$\underbrace{\frac{\partial u}{\partial t}}_{1} + \underbrace{\frac{\partial u^2}{\partial x} + \frac{\partial uv}{\partial y} + \frac{\partial uw}{\partial z}}_{2} = \underbrace{X}_{3} - \underbrace{\frac{1}{\rho}\frac{\partial P}{\partial x}}_{4} - \underbrace{\frac{\partial \overline{u'u'}}{\partial x} + \frac{\partial \overline{u'v'}}{\partial y} + \frac{\partial \overline{u'w'}}{\partial z}}_{5} \tag{2}$$

where u, v, w = velocity components in x, y, z coordinate directions, respectively, t = time, X = body force in x direction, ρ = fluid density, P = fluid pressure, $\overline{u'u'}, \overline{u'v'}, \overline{u'w'}$ = Reynolds (or apparent) stresses in x direction on x, y, z planes, respectively.

Similar equations to (2) can be written to evaluate the velocity components v and w in the y and z directions, respectively. For the numbered terms in the momentum equation (2), these terms refer to: local acceleration (term 1), advective (or convective) acceleration (2), body force (3), pressure gradient (4) and turbulent shear stresses (5).

In modelling riverine flows in two and three dimensions then the effects of the earth's rotation will need to be included giving, for the body force components:

$$\left.\begin{array}{l} X = 2v\varpi \sin \phi \\ Y = -2u\varpi \sin \\ Z = -g \end{array}\right\} \tag{3}$$

where ϖ = speed of earth's rotation, ϕ = earth's latitude and g = gravitational acceleration. The main effects of the earth's rotation, giving rise to the Coriolis acceleration, are to set up transverse water surface slopes across the river and to enhance the effect of secondary currents and meandering.

For 3-D flow predictions, either the full 3-D governing equations are solved, which leads to a complex numerical formulation to evaluate the pressure P, or, more usually, a hydrostatic pressure distribution is assumed to occur in the vertical (z) direction, and leading to an expression for P of the following form:

$$P(z) = \rho g(H - z) + P_a \tag{4}$$

where H = total depth of flow, z = elevation above the bed and P_a = atmospheric pressure. The corresponding derivative of equation (4), for inclusion in equation (2), gives:

$$\frac{\partial P}{\partial x} = \rho g \frac{\partial H}{\partial x} - \rho g \sin \theta + \frac{\partial P_a}{\partial x} \tag{5}$$

where θ = angle of channel slope. Likewise, a similar representation can be written for the pressure gradient in the y direction. Apart from modelling flows in long rivers, the effects of the atmospheric pressure gradient are generally small and are neglected.

The only unknown terms remaining in equation (2) are the Reynolds stresses, which need to be related to the 3-D velocity field before solving for the water levels and the 3-D velocity components. In solving for these stresses, Boussinesq (Goldstein 1938) proposed that they could be represented in a diffusive manner, giving:

$$\left.\begin{aligned}
-\overline{u'u'} &= v_t \left[\frac{\partial u}{\partial x} + \frac{\partial u}{\partial x}\right] \\
-\overline{u'v'} &= v_t \left[\frac{\partial u}{\partial y} + \frac{\partial v}{\partial x}\right] \\
-\overline{u'w'} &= v_t \left[\frac{\partial u}{\partial z} + \frac{\partial w}{\partial x}\right]
\end{aligned}\right\} \tag{6}$$

where v_t = kinematic eddy viscosity.

To determine the eddy viscosity, this parameter can be obtained in several ways (Falconer and Chen 1996). The simplest approach is to assume a constant value based on field data. However, while this approach may be adequate for predicting velocity distributions in large water bodies, such as coastal basins, lakes or reservoirs, it is not particularly accurate for 3-D model simulations in riverine systems, where such models are generally only used for predicting complex velocity field distributions in the vicinity of structures (such as bridge piers) and short river reaches. Another approach is to apply a zero-equation turbulence model, similar to that prescribed by Prandtl's mixing length hypothesis (Goldstein 1938), wherein:

$$v_t = \ell^2 J \tag{7}$$

where ℓ = a characteristic mixing length and J = magnitude of local velocity gradients in x, y, z directions. Using this approach the mixing length can readily be determined for typical logarithmic type velocity profiles, but for the more complex flow fields where 3-D models are appropriate, then the velocity distribution is unlikely to be primarily logarithmic in form and strong secondary currents often lead to a more complex and less well-defined mixing length.

Hence, for most practical problems where 3-D models are appropriate for riverine simulations, then the turbulent stresses given in equation (2) need to be solved using either a two-equation turbulence model of the k–ε type, or a turbulence model of the algebraic stress type and wherein the Reynolds stress terms are solved directly. For the more usual approach, using either the linear or non-linear k–ε model, the eddy viscosity is defined as:

$$v_t = \frac{C_\mu k^2}{\varepsilon} \tag{8}$$

where C_μ = turbulent model coefficient (= 1.68 from experimental data), k = turbulent kinetic energy and ε = dissipation rate of turbulent kinetic energy. Transport equations are derived for k and ε (Rodi 1984), which, in general, include transport by advection, production and dissipation. At the walls and bed, the wall function is adopted adjacent to the bed, which expresses the velocity in terms of the local friction velocity at the first grid point of the computational domain adjacent to the wall. Likewise at the free surface the velocity components and the turbulent fluctuations normal to the surface, and the normal derivatives of all other variables, are set to be zero except for the rate of turbulent energy dissipation ε. The expression for ε at the first grid point below the free surface is given as (Noat and Rodi 1982):

$$\varepsilon = \frac{C_\mu^{3/4} k^{3/2}}{\kappa} \left(\frac{1}{\gamma} + \frac{1}{0.07H} \right) \tag{9}$$

where γ = distance from surface and κ = von Karman's constant (= 0.41).

Two-Dimensional Flows

For many practical problems where there are significant variations across the streamwise flow direction, for example flow over riverine floodplains and through mangrove forests, it is commonplace for the velocity field to be determined using a 2-D depth integrated numerical model. To determine the hydrodynamic velocity field using a 2-D model, the governing 3-D equations are integrated over the depth, giving for equations (1) and (2) respectively,

$$\frac{\partial H}{\partial t} + \frac{\partial q_x}{\partial x} + \frac{\partial q_y}{\partial y} = 0 \tag{10}$$

$$\frac{\partial q_x}{\partial t} + \beta \left[\frac{\partial u q_x}{\partial x} + \frac{\partial v q_x}{\partial y} \right] = f q_y - gH \frac{\partial H}{\partial x} + gH \sin \theta + \frac{\tau_{xw}}{\rho} - \frac{\tau_{xb}}{\rho}$$
$$+ 2 \frac{\partial}{\partial x} \left[\bar{\varepsilon} \frac{\partial q_x}{\partial x} \right] + \frac{\partial}{\partial y} \left[\bar{\varepsilon} \left(\frac{\partial q_x}{\partial y} + \frac{\partial q_y}{\partial x} \right) \right] \tag{11}$$

where q_x, q_y = discharge per unit width in x, y directions, β = momentum correction factor for non-uniform vertical velocity profile, τ_{xw}, τ_{xb} = surface and bed shear stress components respectively in x-direction, and $\bar{\varepsilon}$ = depth averaged eddy viscosity.

For the momentum correction factor, this parameter can be estimated either from field data, which is preferable, or alternatively by assuming a logarithmic velocity profile to give:

$$\beta = \left[1 + \frac{g}{C^2 \kappa^2} \right] \tag{12}$$

where C = de Chezy bed roughness coefficient.

For the surface wind stress a quadratic friction law is generally assumed, giving:

$$\tau_{xw} = C_s \rho_a W_x W_s \tag{13}$$

where C_s = air–water resistance coefficient, ρ_a = air density, W_x = wind velocity component in x direction and W_s = wind speed. For the resistance coefficient, this is normally specified using a piecewise formulation, such as (Wu 1982), whereby:

$$\left.\begin{array}{ll} C_s = 1.25 \times 10^{-3} W_s^{-0.2} & \text{for } W_s \le 1\,\text{m/s} \\ C_s = 0.5 \times 10^{-3} W_s^{-0.5} & \text{for } 1 < W_s \le 15\,\text{m/s} \\ C_s = 2.6 \times 10^{-3} & \text{for } W_s > 15\,\text{m/s} \end{array}\right\} \tag{14}$$

For the bed friction, this term is also generally represented in the form of a quadratic friction law, as given by:

$$\tau_{xb} = \rho g q_x \frac{V_s}{C^2 H} \tag{15}$$

where V_s = depth averaged fluid speed.

To determine the de Chezy value for the bed roughness, most of the widely used models tend to use the Manning formula, which expresses C in terms of the local depth as follows:

$$C = \frac{H^{1/6}}{n} \tag{16}$$

where n = Manning roughness coefficient, with typical values of n being in the range from 0.012 for smooth lined channelised rivers to 0.04 or more for meandering rivers with vegetation etc.

Although this approach is appropriate for most rivers, the Manning representation assumes that the flow is rough turbulent flow and that the local headloss is dependent only on the size and characteristics of the bed roughness, i.e. form drag dominates totally. However, for low velocity flows on shallow floodplains and wetlands, Reynolds number effects may be significant, reflecting the increased influence of skin friction. This complex hydrodynamic phenomenon can be represented using the more comprehensive friction formulation given by Colebrook–White (Henderson 1966) as:

$$C = -17.715 \log_{10}\left[\frac{k_s}{12H} + \frac{0.282C}{R_e}\right] \tag{17}$$

where k_s = Nikuradse equivalent sand grain roughness and R_e = Reynolds number for riverine systems = $4q_s/v$, where q_s = discharge per unit width along streamline and v = kinematic laminar viscosity. The other advantage in using the Colebrook–White formulation to represent the bed roughness, rather than the Manning formulation, is that the physical roughness parameter k_s can be directly related to the height of bed features, such as ripples or dunes, rather than based on a descriptive representation of the bed characteristics as for the Manning formulation.

In many regions of the world rivers are sited within mangrove forests, or regions of dense vegetation, where the trees may reach heights of up to 40 m and have diameters of up to 1 m, depending upon the age and species of the mangrove. Some species also have dense roots which protrude up to or above the high water line. Past research has shown that natural vegetation can reduce the flow velocity in such rivers considerably due to the additional resistance introduced by the trees and/or vegetation. The effects of this vegetation may be taken into account by enhancing the roughness coefficient in equation (15), without specifying the contribution of vegetation elements any further. However, this approach is unrealistic in that the form drag arising as the fluid flows past the trees or vegetation is represented by an increased bed roughness. In contrast, the authors have been undertaking research on modelling this complex hydrodynamic phenomenon by adding another roughness term to the resistance equation (15) and of a form similar to that for flow around circular cylinders, that is:

$$F_d = \frac{\rho C_d V_s^2 mA}{2g} \tag{18}$$

where C_d = drag coefficient, m = density of trees or vegetation per unit area and A = projected area normal to flow for a single tree or vegetation stem.

Finally, for the depth-averaged eddy viscosity $\bar{\varepsilon}$, this parameter can preferably be estimated from field data of the vertical velocity profile or, assuming bed-generated turbulence dominates over free shear layer turbulence, then a logarithmic velocity profile can be assumed giving (Elder 1959):

$$\bar{\varepsilon} = 0.167\kappa U_* H \tag{19}$$

where U_* = shear velocity ($\sqrt{g V_s}/C$). However, field data by Fischer et al. (1979) showed that the turbulent diffusion coefficient in straight, fairly uniform rivers is generally much higher and is more accurately represented by:

$$\bar{\varepsilon} = 0.15 U_* H \tag{20}$$

For most practical riverine systems, even this value is low compared to measured data recorded in rivers, with values for $\bar{\varepsilon}/(U_* H)$ typically ranging from 0.42 to 1.61.

One-Dimensional Flows

In most model studies of riverine systems the basin may be regarded as a 1-D system, with the longitudinal flow features dominating the system. The governing St Venant equations of motion are obtained by integrating the 3-D equations of motion (i.e. equations (1) and (2)) over the area of flow to give:

$$T\frac{\partial Z}{\partial t} + \frac{\partial Q}{\partial x} = q \tag{21}$$

$$\frac{\partial Q}{\partial t} + \beta \frac{\partial (Q^2/A)}{\partial x} = -gA\frac{\partial Z}{\partial x} - gAS_f \tag{22}$$

where T = surface width of flow, Z = water surface elevation above horizontal datum, Q = discharge, q = lateral inflow or outflow per unit length of river, A = area of flow, and S_f = friction slope, or slope of total energy line, which is given by the following representation:

$$S_f = \frac{Q|Q|}{C^2 A^2 R} \tag{23}$$

where R = hydraulic radius for conveyance segment and C (the de Chezy coefficient) may be estimated from either equations (16) or (17).

In solving equations (21) and (22) numerically to provide time varying values for Z and Q, the authors have developed a new numerical model FASTER (Flow And Solute Transport in Estuaries and Rivers) as outlined in detail in a reference manual (Kashefipour et al. 1999). In the algorithm equation (21) is first substituted into equation (22) to give the momentum equation:

$$\frac{\partial Q}{\partial t} + \frac{2\beta Qq}{A} - \frac{2\beta QT}{A}\frac{\partial Z}{\partial t} - \frac{\beta Q^2}{A^2}\frac{\partial A}{\partial x} = -gA\frac{\partial Z}{\partial x} - g\frac{Q|Q|}{C^2 A^2 R} \tag{24}$$

The terms on the left hand side of equation (24) refer to the local and advective accelerations and the terms on the right hand side refer to the pressure gradient and the bed resistance, respectively.

By re-formulating equation (22) in the form given in equation (24), the governing equations can be double integrated with respect to time (t) and volume (v) over a control volume to provide a highly accurate finite volume solution. The equations are formulated on a staggered grid to provide advantages in treating the typical hydrodynamic boundary conditions that are commonly used in such models.

The implicit finite volume solution of the governing equations is second order accurate in space and time and is unconditionally stable. However, where reasonable precision is required the Courant number, expressed in the form of $C_r = (\Delta t\sqrt{gA/T})/\Delta x$, should be less than five. The scheme remains stable for higher time steps, or Courant numbers, but accuracy reduces particularly at wave peaks.

14.2.2 Initial and Boundary Conditions

Initial Conditions

At the start of simulations, the initial values of the dependent variables must be specified in any numerical model. Moreover, in river models the initial values for lateral inflows or outflows must also be specified at any points and for all time. For a cold start the initial discharges are usually set to zero across the domain for each river reach, and the water surface elevations are set horizontally, and equal to high water for tidal simulations, or the maximum bed elevation plus 0.1 m.

Open Boundary Conditions

Boundary conditions must be specified at the downstream and upstream ends of the river reach. Boundaries may be located at either Z or Q points on the staggered grid, with the elevation Z being specified at the centre of the volume and the discharge Q being specified at the volume face. There are four possible boundary combinations along each reach, as summarised in Table 14.1. Time histories of the relevant boundary variables are specified as the solution progresses.

Internal Boundary Conditions

At a junction in a system of reaches, the end of a reach is defined as an internal boundary condition and is generally a passive boundary condition. Due to the different type of junctions

Table 14.1 Types of channel boundary combinations

Type of reach	Type of boundary	
	Upstream	Downstream
1	Water elevation	Water elevation
2	Discharge	Water elevation
3	Discharge	Discharge
4	Water elevation	Discharge

possible, the assumptions made may be different. Two different types of junctions are shown in Figure 14.1.

For dummy junctions, e.g. Figure 14.1a, due to water elevation compatibility and mass conservation it can be assumed that $Z_1 = Z_2 = Z_J$ and $Q_1 = Q_2$, respectively. For critical flow sections (for Figure 14.1a) the weir equation can be used as the passive boundary, giving $Q_J = F_{weir}(Z_1, Z_2)$. In this type of section mass, but not momentum, is efficiently transferred. In the case of an abrupt expansion or contraction, mass and energy conservation can be applied giving: $Q_1 = Q_2$ and $Z_1 + (V_1^2/2g) = Z_2 + (V_2^2/2g) \pm H_f$. The term H_f, or the friction loss, depends upon the flow direction with respect to abrupt transitions. For multi-reach internal boundary conditions, such as indicated in Figure 14.1b, for mass and energy conservation it can be assumed that $Q_1 = Q_2 + Q_3$ and the energy head in all branches are equal at the junction point, giving $E_1 = E_2 = E_3$, where $E = Z + (\alpha V^2/2g)$, where α = kinetic energy correction factor. If the velocities occurring at a junction are relatively small, then the velocity head term may be neglected and thus the energy conservation equation reduces to $Z_1 = Z_2 = Z_3 = Z_J$.

For determining the values of Z and Q at any point in a reach at the present time step, then the values at the lower and upper ends of the reach must be specified. Thus, at each new time step, the flow conditions at all junctions must first be determined before the solution can be obtained for each individual reach segment. One of the most convenient solution algorithms for determining the reach bounds is the influence line technique described by Daily and Harleman (1972). Details of the use of this approach and the treatment of multiple junctions are given in Kashefipour et al. (1999).

14.3 WATER QUALITY MODELLING

14.3.1 Governing Equations

Three-Dimensional Flow Fields

In modelling numerically the flux of water quality constituents, contaminants or sediments within a riverine system, the conservation of mass equation can first be written in general terms for a 3-D flow field as given by (Harleman 1966):

$$\underbrace{\frac{\partial \phi}{\partial t}}_{1} + \underbrace{\frac{\partial u\phi}{\partial x} + \frac{\partial v\phi}{\partial y} + \frac{\partial w\phi}{\partial z}}_{2} + \underbrace{\frac{\partial}{\partial x}\overline{u'\phi'} + \frac{\partial}{\partial y}\overline{v'\phi'} + \frac{\partial}{\partial z}\overline{w'\phi'}}_{3} = \underbrace{\phi_s + \phi_d + \phi_k}_{4} \qquad (25)$$

where ϕ = time-averaged solute concentration, ϕ_s = source or sink solute input (e.g. an outfall), ϕ_d = solute decay or growth term, and ϕ_k = total kinetic transformation rate for solute. The

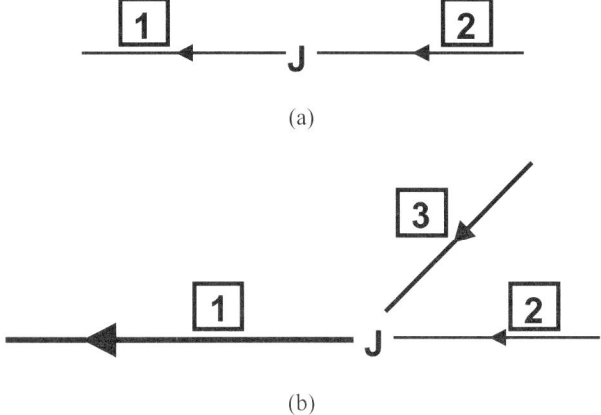

Figure 14.1   Two different types of junctions

individual terms in equation (25) – generally referred to as the advective–diffusion equation – refer to: local effects (term 1), transport by advection (2), turbulence effects (3), and source (or sink), decay (or growth) and kinetic transformation effects (4).

The cross-produced terms $\overline{u'\phi'}$ etc. represent the mass flux of solute due to the turbulent fluctuations and, by analogy with Fick's law of diffusion, it can be assumed that this flux is proportional to the mean concentration gradient and is in the direction of decreasing concentration. Hence, the terms can be written as:

$$\left.\begin{aligned}
\overline{u'\phi'} &= -D_{tx}\frac{\partial\phi}{\partial x} \\[2mm]
\overline{v'\phi'} &= -D_{ty}\frac{\partial\phi}{\partial y} \\[2mm]
\overline{w'\phi'} &= -D_{tz}\frac{\partial\phi}{\partial z}
\end{aligned}\right\} \tag{26}$$

where D_{tx}, D_{ty}, D_{tz} = turbulent diffusion coefficients in x, y, z directions. For riverine flows it is common to assume isotropic turbulence and to approximate the horizontal diffusion terms to the depth mean coefficients as given by Fischer (1973), whereby in the absence of field data, these terms are often equated to:

$$D_{tx} = D_{ty} = 0.15U_*H \tag{27}$$

Likewise, for the vertical diffusion coefficient, in the absence of stratification and field data it is common to assume a linear shear stress distribution and a logarithmic velocity profile giving (Vieira 1993):

$$D_{tz} = U_*\kappa z\left(1 - \frac{z}{H}\right) \tag{28}$$

However, as indicated for hydrodynamic modelling, 3-D models only tend to be used for short riverine reaches and where complex secondary flow features exist. The above transport equation is therefore typically used to predict sediment transport fluxes in the lee of bridge piers etc.

Two-Dimensional Flow Fields

For more practical problems where there are significant variations in the flow field across the river, such as over floodplains or mangrove forests, a 2-D numerical model solution is more commonly used, together with the depth integrated advective–diffusion equation (Falconer 1991):

$$\frac{\partial \phi H}{\partial t} + \frac{\partial \phi q_x}{\partial x} + \frac{\partial \phi q_y}{\partial y} - \frac{\partial}{\partial x}\left[HD_{xx}\frac{\partial \phi}{\partial x} + HD_{xy}\frac{\partial \phi}{\partial y}\right]$$

$$- \frac{\partial}{\partial y}\left[HD_{yx}\frac{\partial \phi}{\partial x} + HD_{yy}\frac{\partial \phi}{\partial y}\right] = H[\phi_s + \phi_d + \phi_k] \qquad (29)$$

where ϕ = depth average solute concentration, ϕ_s, ϕ_d, ϕ_k = depth average concentrates corresponding to ϕ_s, ϕ_d, ϕ_k, in equation (25) and $D_{xx}, D_{xy}, D_{yx}, D_{yy}$ = depth average longitudinal dispersion and turbulent diffusion coefficients in x, y directions.

For the dispersion–diffusion terms, these coefficients can be shown to be of the following form (Preston 1985):

$$\left.\begin{aligned}
D_{xx} &= \frac{(D_\ell U^2 + D_t V^2)H\sqrt{g}}{V_s C} + D_w \\[2mm]
D_{yy} &= \frac{(D_\ell V^2 + D_t U^2)H\sqrt{g}}{V_s C} + D_w \\[2mm]
D_{xy} &= D_{yx} = \frac{(D_\ell - D_t)UVH\sqrt{g}}{V_s C} + D_w
\end{aligned}\right\} \qquad (30)$$

where D_ℓ = depth average longitudinal dispersion constant, D_t = depth average turbulent diffusion constant, D_w = wind-induced dispersion coefficient and U, V = depth average velocities in x, y directions, respectively. For values of D_ℓ and D_t, these coefficients can be preferably obtained from field data, or alternatively minimum values obtained by assuming a logarithmic velocity profile, wherein $D_\ell = 5.93$ (Elder 1959) and $D_t = 0.15$ (Fischer 1973). However, in practical studies these values tend to be rather low (Fischer et al. 1979), with measured values of D_ℓ and D_t ranging from 8.6 to 7500 and 0.42 to 1.61, respectively. In the absence of field data, undertaken in the form of extensive dye dispersion studies, the authors have found that the most accurate results have generally been obtained using their DIVAST (Depth Integrated Velocities And Solute Transport) model with values of $D_\ell = 13.0$ and $D_t = 1.2$.

One-Dimensional Flow Fields

As for hydrodynamic studies, most riverine water quality studies are generally based on using a 1-D numerical model. For this purpose the 3-D advective–diffusion equation (25) is integrated over an arbitrary cross-sectional area A to give:

$$\frac{\partial \phi A}{\partial t} + \frac{\partial \phi Q}{\partial x} - \frac{\partial}{\partial x}\left[AD_x\frac{\partial \phi}{\partial x}\right] = S_s + S_d + S_L \qquad (31)$$

where ϕ = area average solute concentration, S_s = area average source term, S_d = area average decay term and S_L = lateral input (or output).

One of the most significant terms in equation (31) is the longitudinal dispersion coefficient. As indicated from numerous field studies reported in the literature, the dispersion coefficient in natural rivers is dependent upon many hydrodynamic parameters, including: depth, width, velocity and shear velocity (Fischer et al. 1979). However, Kashefipour and Falconer (2000) have applied dimensional analysis procedures to more than 80 data sets in 30 natural rivers in the USA and have shown that the most accurate equation for estimating the longitudinal dispersion coefficient can be expressed as:

$$D_x = \left[7.428 + 1.775 \left(\frac{T}{H}\right)^{0.620} \left(\frac{U_*}{U}\right)^{0.572} \right] HU \left(\frac{U}{U_*}\right) \tag{32}$$

where T = surface width, H = maximum depth and U_* = local shear velocity.

In solving equation (31) numerically a new implicit algorithm has been developed using the authors' model FASTER (Flow And Solute Transport in Estuaries and Rivers). This finite volume-based solution procedure calculates the advection of a concentrate of solute, or suspended sediments at each face of any control volume, by means of a modified form of the highly accurate ULTIMATE QUICKEST scheme (Lin and Falconer 1997). As before, a space staggered grid system is used to solve the finite volume form of equation (31), with the variable ϕ being located at the centre of the control volume.

14.3.2   Initial and Boundary Conditions

Initial Conditions

As for the hydrodynamic model, initial values of the solute concentration ϕ must be specified across the domain and within the model. In general, the concentration is set to zero across the domain at the start of the simulations, or some predetermined concentration when the base concentration is not zero. For example, when the domain is dominated by salt water, then the initial salinity level may typically be set at 35 parts per thousand or similar.

Open Boundary Conditions

For the input of solute concentration levels at the upstream end of a river reach, or through outfalls etc., the concentration has to be specified from known conditions, e.g. where the concentration is zero or at a base value, or preferably from field data. In contrast, at the downstream boundary the concentration is extrapolated from within the domain for outgoing flows (Falconer 1986). For inflow across a downstream boundary, e.g. where the model is applied to an estuary reach and the tide is incoming, then preferably the concentration level is available from field data (e.g. for salinity), or the net outgoing concentration is determined from the ebb tide simulations and a scaling factor is applied to estimate the return current concentration (Falconer 1986).

The concentration at all junctions must be specified at the upstream and downstream ends of each reach. In the model FASTER the advective–diffusion equation is solved explicitly at the junction, with the dispersion and decay terms assumed to be negligible in the immediate vicinity of the junction. Full details of the model and the treatment of junctions and boundary conditions are given in Kashefipour et al. (1999).

14.4 WATER QUALITY PROCESSES

14.4.1 General

In modelling water quality processes in rivers a range of indicator organisms are often modelled, including (Falconer and Chen 1996):

1. *Physical indicators of water quality*:
   (i)   Suspended solids – including inorganics, e.g. sand, and organics, e.g. sewage solids.
   (ii)   Turbidity – important in governing the flow conditions and biological processes for many water quality constituents.
   (iii) Temperature – important in governing the flow conditions and biological processes for many water quality constituents.
   (iv)   Other physical indicators – such as colour, conductivity and radioactivity.
2. *Chemical indicators of water quality*:
   (i)   Dissolved oxygen – which is important for water quality, since it is essential for supporting most forms of aquatic life.
   (ii)   Biochemical oxygen demand – which is a measure of the consumption of bacteria during aerobic degradation of organic matter.
   (iii) Nitrogen – which is an essential nutrient for biological growth and a major constituent of domestic effluents. Nitrogen may be present in a variety of chemical forms, including organic, ammoniacal, nitrite and nitrate nitrogen.
   (iv)   Phosphorus – which is also an essential nutrient for biological growth and which appears exclusively as phosphate in aquatic environments. Phosphate may be present in several forms, including orthophosphate, condensed phosphates (i.e. pyro-, meta- and poly-phosphates) and organically based phosphates.
   (v)   Chlorides – commonly occurring in the form of salinity in estuarine reaches and an indication of sewage pollution in rivers.
   (vi)   Metals – particularly those which are easily dissolved in water and are toxic, e.g. arsenic, cadmium, chromium, lead and mercury. These metals occur mainly as a result of industrial discharges, mine water tailings and agriculture. Although these contaminants are usually only present in low quantities, they can accumulate in the food chain and cause toxicity problems.
3. *Biological indicators of water quality*:
   (i)   Pathogens – including the faecal coliform group, with the principal bacteria being *Escherichia coli.*
   (ii)   Other biological organisms – such as algae.

Some of the main water quality indicator organisms modelled in riverine flows will be outlined below. The formulations outlined herein and included in the FASTER, DIVAST and TRIVAST models are primarily based on the US EPA formulations included in the QUAL2E model (Brown and Barnwell 1987). For convenience in outlining the water quality indicator organism parameters, or the sediment transport characteristics in the transport equation, the advective–diffusion equation is reduced to the area average source term S_d.

Total and Faecal Coliforms (C)

Total and faecal coliform levels are included as indicators of pathogen contamination in surface waters. Both coliform types are generally expressed as first order decay functions giving, for the source term:

$$S_d = -k_5AC \tag{33}$$

where C = concentration of coliforms (in colony counts per 100 ml) and k_5 = coliform die-off rate (per day). The range of values for inclusion in the model are 0.05 to 4.0 per day.

For the coliform die-off rate the parameter depends upon a number of variables including: sunlight, acidity (in the form of pH), sunlight and temperature. For further details see Aver and Neihaus (1993).

Biochemical Oxygen Demand (L)

As for coliforms, BOD is generally expressed in the form of a first order reaction to describe the de-oxygenation of ultimate carbonaceous BOD. The functional relationship expressed in most models takes account of additional BOD removal due to sedimentation, flocculation and scour, giving:

$$S_d = -k_1AL - k_3AL \tag{34}$$

where L = concentration of ultimate carbonaceous BOD (mg/l), k_1 = deoxygenation rate co-efficient, typically in the range 0.02 to 3.4 per day, and k_3 = rate of loss of carbonaceous BOD due to settling etc., with typical values in the range -0.36 to 0.36 per day.

Dissolved Oxygen (O)

The balance of oxygen in a riverine system depends upon the capacity of the river to re-aerate itself. Apart from being dependent upon the advection and diffusion processes, this capacity is a function of major sources and sinks of oxygen, including: atmospheric re-aeration, biochemical oxidation of carbonaceous BOD and nitrogenous organic matter, sediment oxygen demand, algae production and consumption by respiration. Thus, the dissolved oxygen formulation in the 1-D model can be written as:

$$S_d = k_2(O^* - O)A - k_1LA - k_4P - \alpha_5 f_4 \beta_1 N_1 A - \sigma_6 \beta_2 N_2 aA + \alpha_3 \mu aA - \alpha_4 raA \tag{35}$$

where O = concentration of dissolved oxygen (mg/l), k_2 = re-aeration rate in accordance with Fick's law for diffusion (per day), O^* = saturation concentration of dissolved oxygen at local temperature, P = wetted perimeter, k_4 = sediment oxygen uptake in the range 1.5 to 9.8 mg O_2/m^2/day, α_5 = rate of oxygen uptake per unit of ammonia nitrogen oxidation, in the range 3.0 to 4.0 mg O_2/mg N, α_6 = rate of oxygen uptake per unit of nitrite nitrogen oxidation, in the range 1.0 to 1.14 mg O_2/mg N, α_3 = oxygen production per unit of chlorophyll a, and α_4 = oxygen uptake per unit of chlorophyll a in mg O_2/mg chlorophyll a.

For the re-aeration rate k_2, the value is generally assumed to vary spatially in accordance with the O'Connor and Dobbins (1958) relationship given as:

$$k_2 = \frac{\sqrt{D_m V_s}}{H^{3/2}} \tag{36}$$

where k_2 = re-aeration rate in days, V_s = depth average fluid speed, in ft/s, H = local depth in feet, and D_m = molecular diffusion in feet^2/day, given as:

$$D_m = 1.91 \times 10^{-3}(1.037)^{T-20} \tag{37}$$

where T = fluid temperature.

For the dissolved oxygen saturation level within the water column, this decreases with temperature and salinity, with the usual representation in models being that given by Bowie et al. (1985):

$$O^* = 14.6244 - 0.367134T + 0.0043972T^2 - 0.966S + 0.00205ST + 0.002739S^2 \tag{38}$$

Nutrients

Living organisms require about 40 elements occurring naturally in the earth's crust and atmosphere to sustain growth and reproduction. However, when dealing with water quality modelling and eutrophication in rivers, the most important nutrients are generally recognised as being nitrogen and phosphorus as the limiting nutrients (see Figure 14.2) and carbon and silicon to a much lesser extent. The principal external sources of these nutrient inputs are municipal and industrial wastes, agricultural and forest runoff, urban and suburban runoff, and atmospheric fallout. The limiting components will now be discussed individually.

Nitrogen Cycle

In riverine waters there is a stepwise transformation from organic (N_4) to ammonia (N_1) to nitrite (N_2) and nitrate (N_3) nitrogen. This transformation is included in the water quality module outline herein as follows:

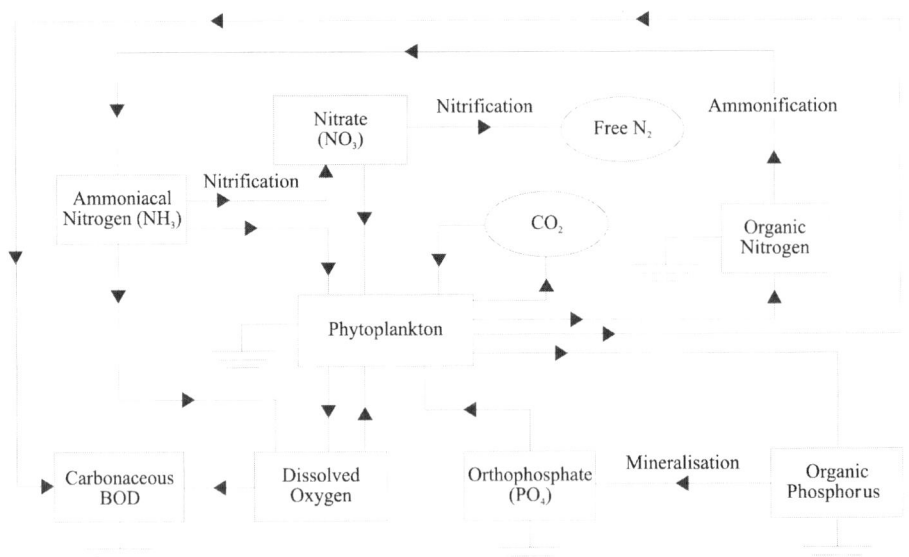

Figure 14.2   Eutrophication cycle including nitrogen, phosphorus and phytoplankton

1. Organic nitrogen (N_4)

$$S_d = -\beta_3 N_4 A - \sigma_4 N_4 A + \alpha_N \rho_N RaA \qquad (39)$$

where N_4 = concentration of organic nitrogen (mg/l), β_3 = rate constant for hydrolysis of organic nitrogen to ammonia nitrogen, in the range 0.02 to 0.4 per day, σ_4 = rate of coefficient for organic nitrogen settling, in the range 0.0001 to 0.1 per day, α_N = ratio of organic nitrogen to chlorophyll a, ρ_N = fraction of dead and respired phytoplankton recycled top organic nitrogen, R = total phytoplankton loss rate, and a = concentration of chlorophyll a (mg/l).

2. Ammonia nitrogen (N_1)

$$S_d = \beta_3 N_4 A - \beta_1 f_4 N_1 A + \alpha_N (1 - \rho_N) RaA - \beta_A \alpha_N \mu aA \qquad (40)$$

where N_1 = concentration of ammonia nitrogen (mg/l), β_1 = rate constant for biological oxidation of ammonia nitrogen in the range 0.1 to 1.0 per day, f_2 = oxygen limiting factor = $O_2/(k_o + O_2)$, where k_o = half-saturation constant, in the range 0.5 to 2.0 mg O_2/l, β_A = ammonia preference factor and μ = gross growth rate (per day) (Bowie et al. 1985).

3. Nitrite nitrogen (N_2)

$$S_d = \beta_1 f_4 N_1 A - \beta_2 N_2 A \qquad (41)$$

where N_2 = concentration of nitrite nitrogen (mg/l) and β_2 = rate constant for oxidation of nitrite nitrogen, in the range 0.2 to 2.0 per day.

4. Nitrate nitrogen (N_3)

$$S_d = \beta_2 N_2 A - \alpha_N \mu (1 - \beta_A) aA \qquad (42)$$

where N_3 = concentration of nitrate nitrogen (mg/l).

In cases where only a one-stage process is modelled for nitrification (i.e. oxidation of ammonia to nitrate directly, rather than to nitrite first and then to nitrate) then this process can be simulated by modelling the conservation of nitrate nitrogen directly, as given by:

$$S_d = f_4 \beta_1 N_1 A - \alpha_N \mu (1 - \beta_A) aA \qquad (43)$$

Phosphorus Cycle (P)

The phosphorus cycle is similar to the nitrogen cycle. Phosphorus in its organic form is produced by algal death, which converts to its dissolved inorganic state for primary production. Phosphorus, when discharged from municipal treatment works, is generally in the dissolved inorganic form and can readily be taken up by algae. Phosphorus is included in the FASTER model in the following form:

1. Organic phosphorus (P_1)

$$S_d = \beta_4 P_1 A - \sigma_5 P_1 A + \alpha_p P_p RaA \qquad (44)$$

where P_1 = concentration of organic phosphorus (mg/l), β_4 = organic phosphorus decay rate, in the range 0.01 to 0.7 per day, σ_5 = organic phosphorus settling rate (per day),

α_p = ratio of phosphorus to chlorophyll a, and P_p = fraction of dead and respired phytoplankton recycled to organic phosphorus.

2. Dissolved phosphorus (P_2)

$$S_d = \beta_4 P_1 A - \sigma_6 P_2 A - \alpha_p \mu a A \qquad (45)$$

where P_2 = concentration of inorganic or dissolved phosphorus (mg/l) and σ_6 = dissolved phosphorus settling rate (per day).

Phytoplankton Algae–Chlorophyll a (a)

Algae are the dominant component of the primary producers in many riverine systems. Algal dynamics is closely linked with nutrient dynamics and dissolved oxygen levels in water quality models, as is the case in the model FASTER. Poor turbidity, taste and odour problems are often caused by algal blooms. The commonly used approach, as included in the present FASTER model, is to aggregate all of the algae into a single constituent, i.e. chlorophyll a. The general governing processes for chlorophyll a can be given as:

$$S_d = (\mu - R - G)aA \qquad (46)$$

where G = settling rate for algae (per day).

Temperature Dependence

All of the rate constants cited herein are temperature dependent, with the range of values given being applicable at the standard temperature (i.e. 20°C) value and then corrected for temperature according to a Streeter–Phelps type formulation, given as:

$$k_T = k_{20} \theta^{(T-20)} \qquad (47)$$

where k_T = value of rate constant for local water temperature (per day), k_{20} = rate constant at standard temperature (i.e. 20°C), and θ = empirical constant for each rate constant, with typical values for θ being given by Brown and Barnwell (1987).

14.5 MODELLING APPLICATIONS

14.5.1 General

The differential flow and water quality equations outlined herein can be solved using a wide range of commercial or research orientated hydroinformatics software tools. The authors have been involved in the development of 3-D (TRIVAST), 2-D (DIVAST) and 1-D (FASTER) flow, water quality and sediment transport models and details are given herein of the application of these models to flow in compound channels and, in particular, to a tidal reach of the Humber Basin and incorporating the use of all three model types.

14.5.2 Flow and Solute Transport in Compound Channels

Natural rivers often consist of a main channel and floodplains. When flow becomes over-bank the fast flow in the main channel will be retarded by the slower moving flows from the floodplain, causing a large lateral exchange of momentum. Recent studies show that a significant influence of secondary flows occurs in the vicinity of the junction between the main channel and floodplain. Although the magnitude of the secondary flow velocity is normally less than 10% of the longitudinal velocity, it modifies the cross-sectional distribution of longitudinal velocity and hence the mixing processes.

The following example shows how 3-D numerical models can be used to predict the flow and solute transport processes in compound channels. The numerical model used in this calculation is a non-linear k–ε model originally developed by Speziale (1987). A staggered grid was used in the model, in which the secondary velocity components were placed at the sides of a computational cell, with all other variables being placed at the centre of the cell. The size of the mesh was set uniformly, except near the boundaries where the mesh sizes were considerably reduced.

Figures 14.3 and 14.4 show a comparison of the model predicted longitudinal and secondary velocity distributions in a two-stage open channel, with the measured velocity being given by Tominaga and Nezu (1991). It can be seen from these results that the pattern of the twin vortices (i.e. one in the main channel and one on the floodplain) at each side of the junction and the free-surface vortex in the left hand side of the channel are well predicted. Nevertheless, the predicted distribution of the longitudinal velocity is smoother than the measured results, although the influence of secondary currents on the main flow distribution can clearly be seen.

This example highlights a detailed modelling study of complex 3-D hydrodynamic processes, which cannot be adequately represented using a 2-D or 1-D approach. This is particularly important for floodplain modelling where lateral fluxes of solutes or sediments are important. Further details of this study are given in Lin and Shiono (1995).

14.5.3 Model Applications to Humber Basin

General

The Humber Basin is a well-mixed river and estuarine system, located along the east coast of northern England, and providing an outlet to the North Sea for the rivers Trent and Ouse and

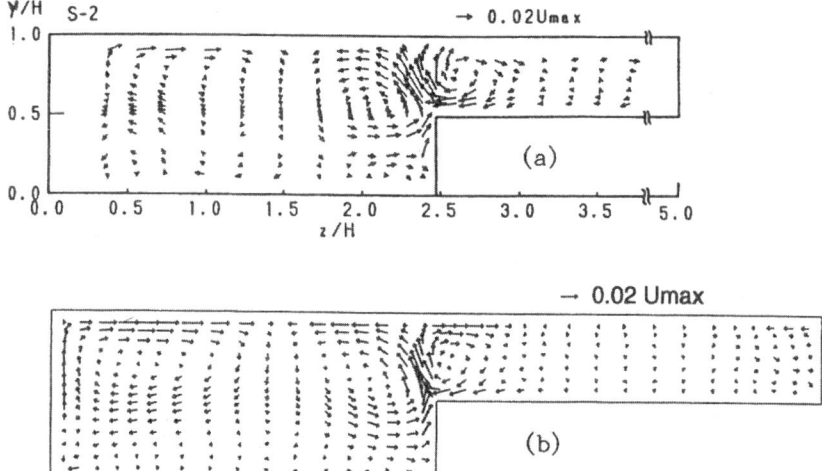

Figure 14.3   Comparison of numerical model predicted and measured secondary velocity distributions. (a) Experiment (Tominaga and Nezu 1991); (b) model prediction

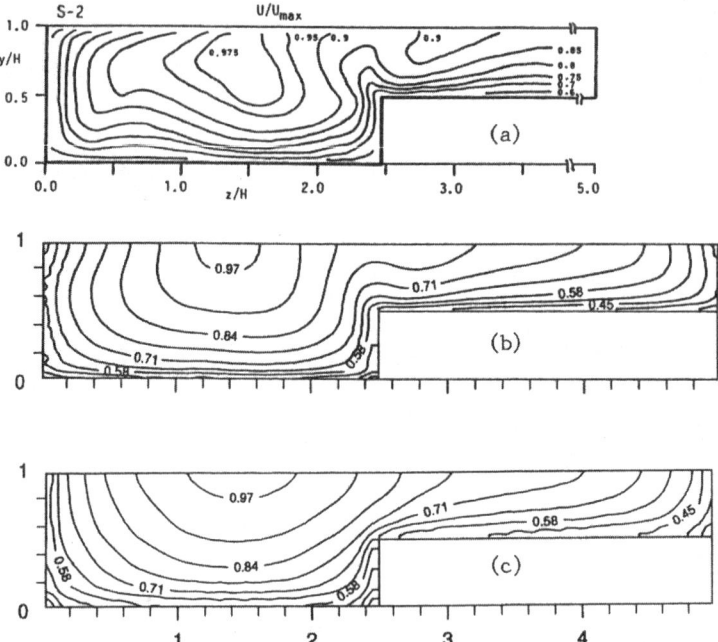

Figure 14.4 Comparison of numerical model predicted and measured main velocity distributions. (a) Experiment (Tominaga and Nezu 1991); (b) non-linear $k-\varepsilon$ model prediction; (c) linear $k-\varepsilon$ model prediction

shipping access to a number of ports, including Hull, Immingham and Grimsby. The management and monitoring of water quality in the Humber Basin is coordinated by the Environment Agency (North East Region).

The tidal currents outside of the estuary are predominantly north–south, and parallel to the Yorkshire and Lincolnshire coastlines. During flood tide the flow is seawards with a band of water, approximately 8 km wide, separating from the main current and flowing into the estuary. This pattern is reversed during the ebb tide, with the seaward boundary (shown in Figure 14.5) being parallel to, and southeast of, Spurn Head and acting effectively as a streamline.

Specification of Models

The finite difference and finite volume models of the Humber Basin were simulated in 2-D and 3-D models using a 500 m grid size horizontally and eight layers vertically in the 3-D model, with the thickness of the top layer being 4 m at mean water level, and with the other layers being 3 m in depth. At the centre of the sides of each grid square a representative depth below Ordnance Datum and the river basin was required. For this purpose bathymetric data given on the Admiralty Chart 109 and additional information provided by Associated British Ports were used, together with a digitising ground modelling software package to generate interpolated depths at the required grid locations. For the 1-D model riverine cross-sections were obtained from the Admiralty Chart and a varying grid size was used, varying from 350 m to 650 m.

In order to drive the hydrodynamic models, water elevation data recorded by Associated

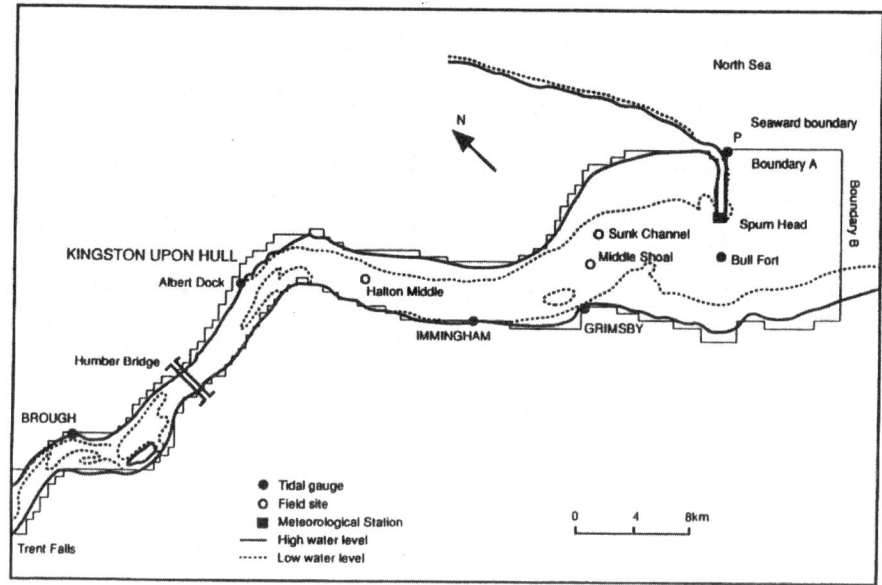

Figure 14.5 Plan of Humber Estuary showing open boundary conditions

British Ports, the Environment Agency (North East Region) and Yorkshire Water plc were used at both the seaward and landward boundaries, i.e. from Spurn Head, or just beyond, to Trent Falls. Although it is generally preferable to avoid using water elevations at both boundaries for such a study, in this case no time varying flow data were available at Trent Falls at the time when these studies were undertaken. Field measurements of water elevations, velocities and various water quality constituents were also available at several sites in between the two open boundaries. Data were available for different tidal ranges at each site, with a complete tidal cycle being monitored at quarter hour intervals.

The measured water elevations were specified in the models at every hour and with the corresponding required values at both open boundaries, for each half time step, being obtained by interpolation using Everett's formula. For the discharges from sewage work etc., mean daily rates were input at the outfall cells, with linear interpolation being used to obtain intermediate values.

For the water quality constituent inputs from the rivers Wharfe, Aire, Don, Trent and Ouse, the boundary values used were obtained by summing up the individual components. For the loadings from the various sewage treatment works at Hull etc., the corresponding values were obtained from Yorkshire Water plc. For the water quality constituent concentrations at the open boundaries full details are given earlier for the treatment of the boundary conditions. However, in summary the ebb tide concentrations along the open boundaries were extrapolated from within the computational domain, whereas for flood tide simulations the boundary values were set to the corresponding open sea conditions.

Apart from the open boundary conditions, initial conditions had to be included in the models for water levels, velocities and water quality constituents at all points across the model domain. The water levels and velocities were generally set to some predetermined constant everywhere, with the water levels being equated to the level at the landward boundary at the start of the first tidal cycle and the velocities initially set to zero. At the open seaward boundary the water levels were varied in a half sinusoidal manner for the first three hours of the first tide, varying from the initial condition as prescribed by the Trent Falls boundary value to the correct boundary value after three hours. This smooth introduction of the water level phase difference between both boundaries reduced the initial disturbances within the domain. For the water quality constituents a mean value was also assumed everywhere at the commencement of all simulations, with the initial value generally being set to the open seaward boundary condition.

Other inputs into the model included the bed friction and surface wind stress. For the bed friction the Darcy–Weisbach resistance formula was used, including Reynolds number effects (see equation (17)), as found to be particularly important for modelling shallow floodplain or wetlands flows with a k_s value of 20 mm being found to generally give fairly good agreement between predicted and measured data across the domain. Although the use of a constant value of k_s across the basin was questionable, no field data were available to justify varying this roughness coefficient. For the surface wind stress a constant mean velocity was assumed to be appropriate, although most of the results reported herein are for a zero wind condition.

Calibration and Verification of Models

In general it is always necessary to be able to check the water levels and velocity field predictions obtained from a numerical model with field data before proceeding with any analysis of the predictive results obtained and the same being true for a laboratory model investigation. For the model studies undertaken for this basin, comparisons were first made, using the 3-D and 2-D models, of the water level and velocity predictions and measurements taken at several sites along the estuary. For the 2-D model predictions a typical set of results are shown in Figure 14.6 for comparisons at Middle Shoal. As can be seen, the corresponding typical comparison shows good agreement between both sets of results. The only site where the agreement was not so good was at 7.5 km from Trent Falls. This was thought to be primarily due to the relatively coarse grid at this site along the estuary failing to represent the geometry and bathymetry of the basin with sufficient accuracy near the landward boundary. For the 3-D model a typical velocity distribution is shown in Figure 14.7 for part of the domain during the flood tide, and for the top and third layers, respectively, with a typical comparison of the velocity profile through the depth being shown in Figure 14.8 at Halton Middle. Again, good agreement was attained between the velocities at the various elevations through the depth and at all of the sites considered.

The 1-D model, FASTER, with the same specifications as the 3-D and 2-D models, was also applied to predict the water surface elevations and velocity distributions for several sites along the Humber Basin. The predicted water surface elevations and velocities for a complete tidal cycle at the site of Sunk Channel are shown as an example in Figures 14.9 and 14.10, respectively, with a comparison of the predictions with corresponding measured data. These results again clearly show good agreement between the predicted and measured data, with further details being given in Kashefipour and Falconer (2000).

For the water quality constituents along the basin, comparisons were made against detailed measurements undertaken by Yorkshire Water plc in 1978, at several sites along the estuary. Predictions were made using the water quality formulations cited previously and comparisons between measured and predicted results along the estuary are shown for salinity, biochemical oxygen demand, organic nitrogen, oxidised nitrogen and dissolved oxygen in Figures 14.11 to 14.15, respectively. As can be seen from Figures 14.11 to 14.15, the agreement between the measured and predicted concentration levels was encouraging for all of the individual constituents considered. Calibration indicated that the following values for the decay rates and temperature correction factors were most appropriate:

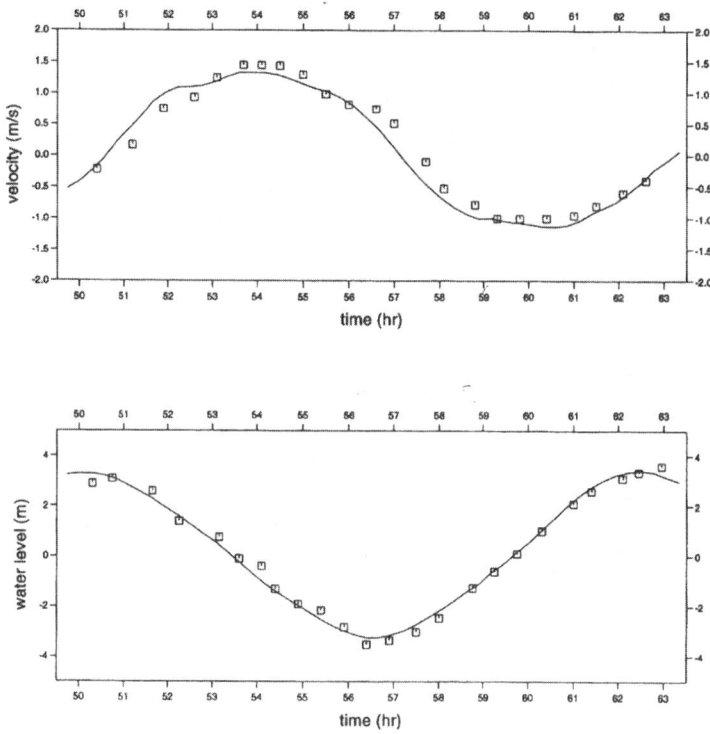

Figure 14.6 Comparison of predicted and measured depth averaged velocities and water elevations for a mid-range tide at Middle Shoal

$k_1 = 0.21\text{d}^{-1}$ – deoxygenation rate coefficient for BOD
$\theta_1 = 1.047$ – temperature correction for BOD deoxygenation
$k_3 = 0.13\text{d}^{-1}$ – rate loss of carbonaceous BOD due to settling
$\theta_3 = 1.08$ – temperature correction for loss of carbonaceous BOD
$\beta_3 = 0.40\text{d}^{-1}$ – hydrolosis of organic nitrogen to ammonia nitrogen
$\theta_3 = 1.047$ – temperature correction for hydrolosis of organic nitrogen
$\sigma_4 = 0.10\text{d}^{-1}$ – rate constant for organic nitrogen settling
$\theta_4 = 1.024$ – temperature correction for organic nitrogen settling
$\beta_1 = 1.00\text{d}^{-1}$ – rate constant for biological oxidation of ammonia nitrogen
$\theta_1 = 1.093$ – temperature correction for biological oxidation of ammonia nitrogen
$k_4 = 4.0\,\text{mg O}_2/\text{m}^2$ – sediment oxygen demand rate
$\sigma_5 = 3.5\,\text{mg O}_2/\text{mg N}_1$ – rate of oxygen uptake per unit of ammonia nitrogen oxidation.

In addition to water quality constituents, sediment transport and heavy metal fluxes have also been modelled extensively in the model applications to the Humber Basin. For the sediment transport predictions, the 3-D advective–diffusion equation (25) was modified to include an extra term on the right hand side of the equation in the form of a net erosion flux, given as:

$$E = \gamma W_f(\phi_e - \phi) \tag{48}$$

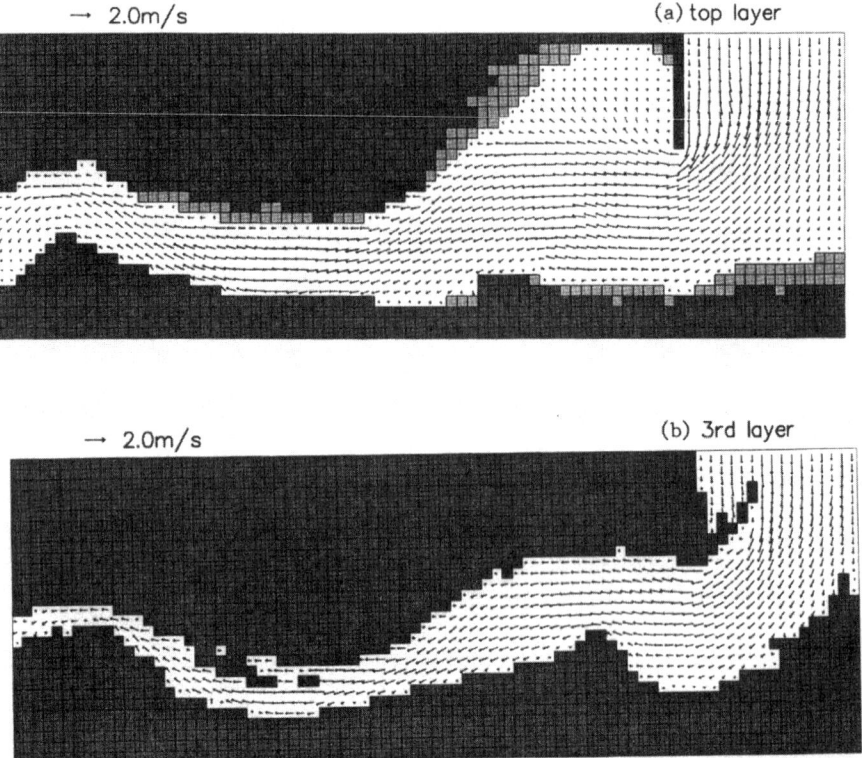

Figure 14.7 Predicted velocity field near mean water level during a flood tide for (a) top layer and (b) third layer

where E = net erosion (or deposition) per unit area of bed, γ = a profile factor given by the ratio of the bed concentration to the mean concentration, W_f = particle settling velocity, ϕ_e = equilibrium concentration, and ϕ = layer, depth or area-averaged suspended sediment concentration for 3-D, 2-D or 1-D model simulations, respectively. Equilibrium conditions are said to occur in the flow when the sediment flux vertically upwards from the bed due to turbulence is in equilibrium with the net sediment flux downwards due to the fall velocity (or gravity). For the equilibrium concentration many sediment transport formulations have been proposed, with the bed load and suspended load formulations by van Rijn (1984a,b) now being among the most widely used in computational models. Full details of van Rijn's formulations for sediment transport in river applications are given in Falconer and Chen (1996).

In modelling sediment transport processes in 1-D river models, Kashefipour and Falconer (2000) have introduced a new empirical equation to predict suspended sediment fluxes, based on using measured data in the River Humber provided by Associated British Ports (formerly British Transport Docks Board). This equation can be applied to estimate S_s and also the suspended concentrations at the boundaries. According to the tidal current direction, when the flow enters the domain the suspended sediment concentration at the boundary can be provided

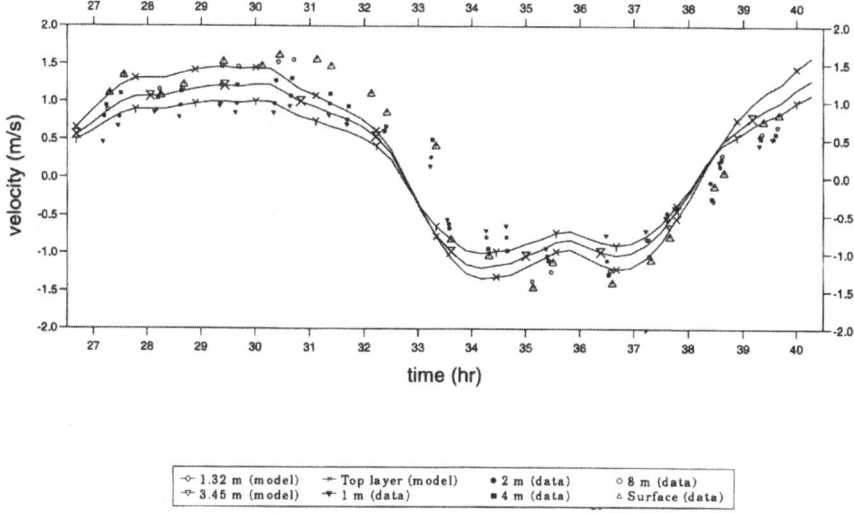

Figure 14.8 Comparison of predicted and measured velocities for a spring tide at Halton Middle

by the following equation, which is another form of the equation presented by Kashefipour and Falconer:

$$\phi_B = 213.3 U^{1.211} \frac{B}{Q} \text{ for } U < 0.45 \text{ m/s}$$

$$\phi_B = 1202.26 U^{2.908} \frac{B}{Q} \text{ for } U \geq 0.45 \text{ m/s}$$

(49)

where: ϕ_B = area average suspended sediment concentration at the boundary (mg/l), U = area average velocity in the direction of the flow (m/s), Q = discharge (m^3/s) and B = bottom width of the channel (m).

The 1-D model with the same specifications as used for the hydrodynamic simulations was applied to predict suspended sediment concentrations for several sites along the Humber Basin. The measured data obtained for two complete tidal cycles, namely mid and spring on 6 July 1978 and 5 September 1979, respectively, were used to calibrate and verify the model predictions. For example, the predicted suspended sediment concentrations and measured data for Sunk Channel are shown in Figures 14.16 and 14.17. These typical comparisons show good agreement between both sets of results. Similar results were obtained for other sites along the Humber River.

Finally, the sediment transport and water quality models have been extended to predict trace metal contaminant levels in riverine basins, including the sorption and desorption between dissolved and particulate forms in the water column and the sediments. The distribution of contaminants between the dissolved and particulate phases has been determined from a series of field and laboratory tests, with the partitioning coefficient K_D – used to define the distribution of the metal between the particulate and dissolved phase – being given as (Turner et al. 1993):

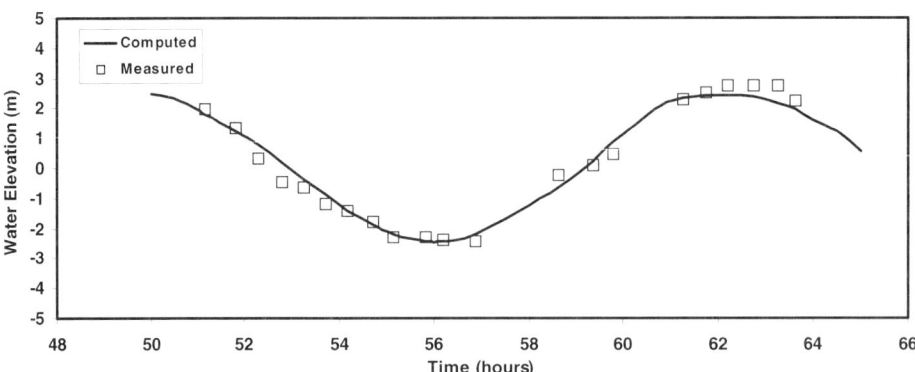

Figure 14.9 Comparison of predicted and measured water surface elevations for a mid-range tide at Sunk Channel

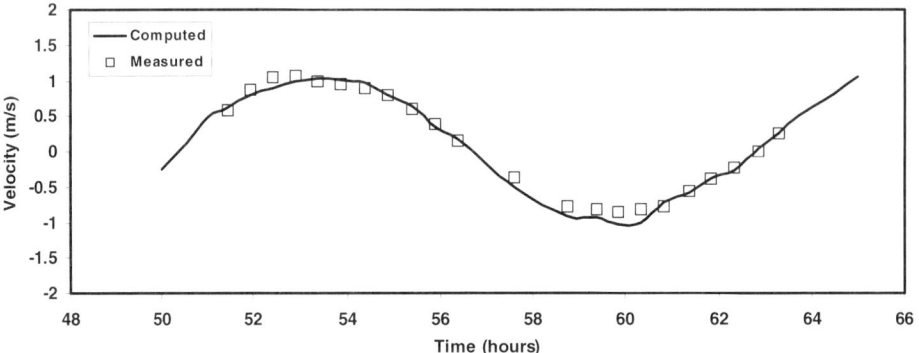

Figure 14.10 Comparison of predicted and measured velocities for a mid-range tide at Sunk Channel

$$K_D = \frac{P}{C} \qquad (50)$$

where P = metal concentration adsorbed on suspended sediments and C = metal concentration in solution, or dissolved, phase. The relationship assumes that the constituents are in equilibrium and that sorption is completely reversible. In estuarine waters salinity has been found to play an important role in the distribution of metals. In seawater the concentration of dissolved trace metals is generally much lower than in fresh water, with this being due to the increased competition from seawater cations for particle sorption sites (Turner et al. 1993). A relationship incorporating salinity has been defined by Bale (1987) giving:

$$\log_e K_D = a \log_e(S + 1) + \log_e K_D^0 \qquad (51)$$

where K_D^0 = partitioning coefficient in fresh water, S = salinity and a = constant. This formula-

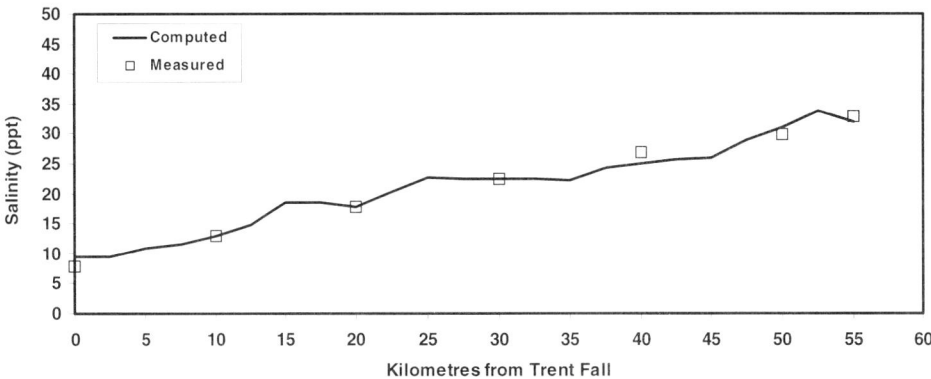

Figure 14.11 Comparison of predicted and measured salinity levels along Basin

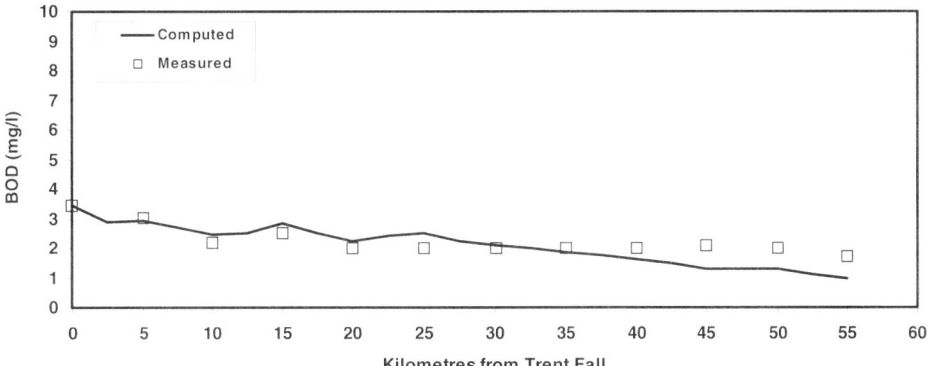

Figure 14.12 Comparison of predicted and measured BOD concentrations along Basin

tion has been added to the models outlined herein for predicting the fate of heavy metals in a number of estuarine and riverine basins, including in particular the Humber and Mersey basins.

In applying this refinement specifically to the Humber Basin, the partition coefficient was included such that the appropriate distribution coefficient was determined for each grid cell and at every time step. The 2-D model DIVAST was applied to simulate the distribution, behaviour and characteristics of the dissolved and particulate metal phases in the Humber for different partition coefficients. These scenarios included: (i) no partitioning of contaminants between the dissolved and adsorbed phases, (ii) partitioning of contaminants with no salinity dependence, and (iii) salinity dependence. Simulations were undertaken for a salinity dependent partitioning coefficient, with a freshwater partitioning coefficient $K_D = 6600\,\text{ml/g}$ and a $= -0.653$, with a constant landward boundary value set at $2.5\,\text{mg/l}$ for dissolved and $16.5\,\text{mg}$ for particulate contaminant concentrations to simulate the input of metal-bearing river waters (Edwards et al. 1987).

Figure 14.18 shows a graphical representation of the dissolved trace metal distribution for a mid-flood tide and after four tidal cycles of simulation. At the landward boundary the high trace metal concentrations occur due to the metal inputs into the upper estuary from the tidal rivers, and the decreasing partition coefficient with increasing salinity results in more trace metals

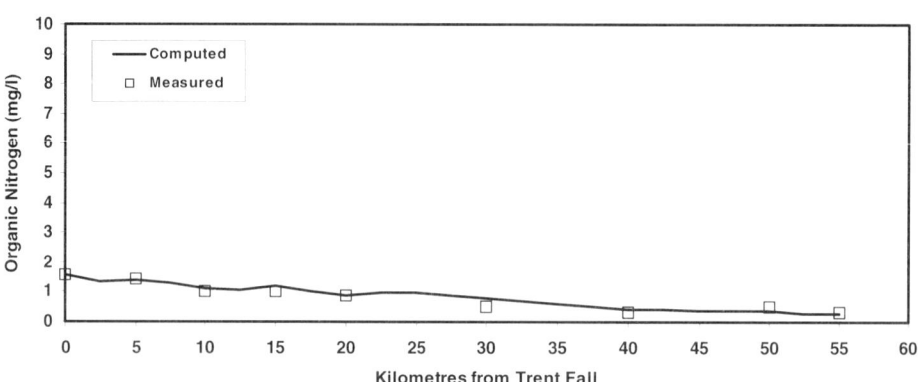

Figure 14.13   Comparison of predicted and measured organic nitrogen levels along Basin

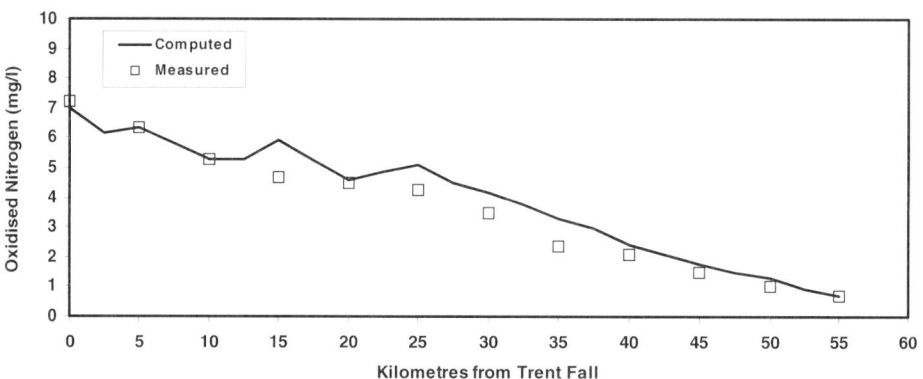

Figure 14.14   Comparison of predicted and measured oxidised nitrogen levels along Basin

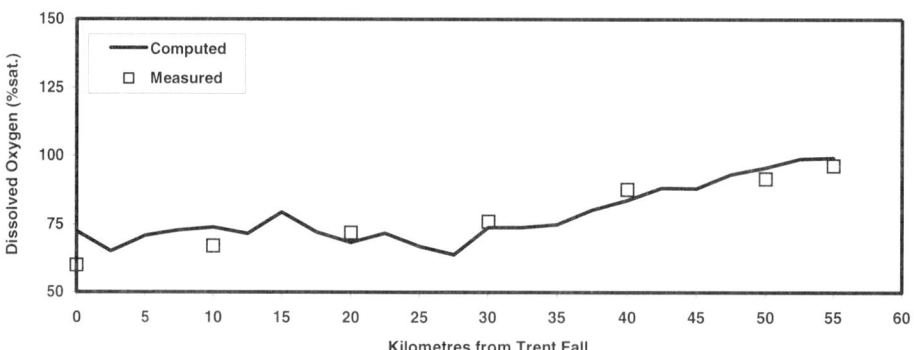

Figure 14.15   Comparison of predicted and measured dissolved oxygen levels along Basin

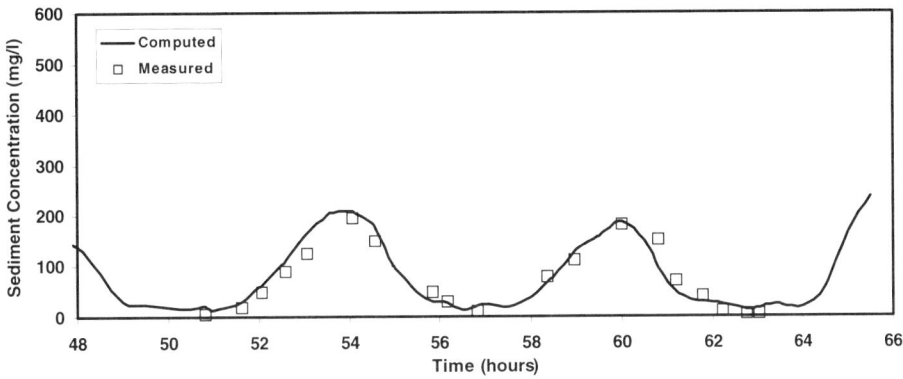

Figure 14.16   Comparison of predicted and measured suspended sediment concentration for a mid-range tide at Sunk Channel

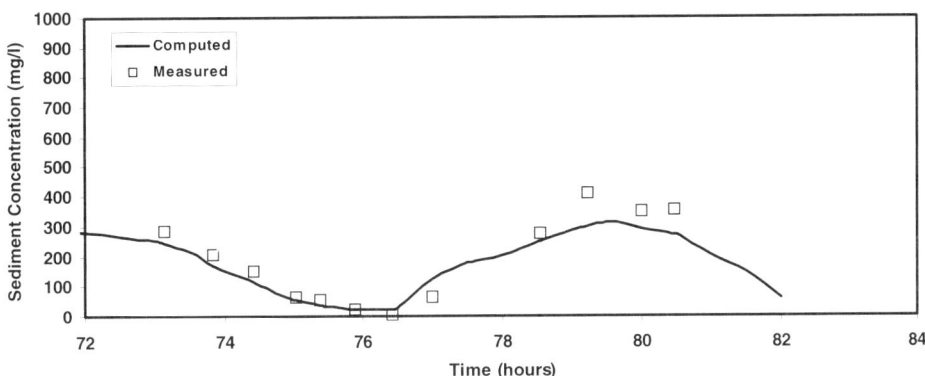

Figure 14.17   Comparison of predicted and measured suspended sediment concentration for a spring-range tide at Sunk Channel

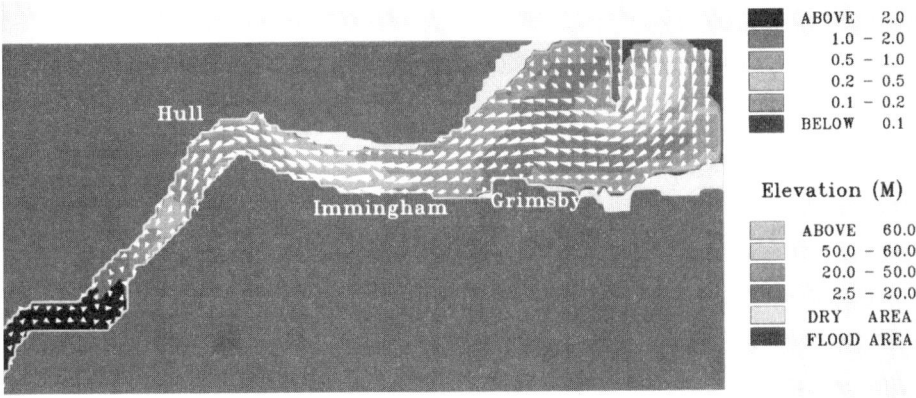

Figure 14.18   Model predicted tracer metal concentration distribution for a mid-flood tide

being predicted in the dissolved form further down the estuary. This modelling approach has been shown to give good predictions of the axial and transverse distributions of trace metals along the Humber Basin (see Ng et al. 1996), and this refinement has now been extended to three-dimensions and is currently being tested for the Mersey.

In comparing the numerically predicted solute distributions with the corresponding field measured data, various statistical methods are commonly used to measure the degree of accuracy. These methods have been used in these model studies of the Humber to varying degrees and are outlined in detail for sediment transport in Kashefipour and Falconer (2000). The various methods used include:

Standard Error:

$$SE = \frac{1}{N} \sum_{i=1}^{N} |S_{mi} - S_{pi}| \tag{52}$$

Root Mean Square:

$$RMS = \frac{1}{N} \sqrt{\sum_{i=1}^{N} (S_{mi} - S_{pi})^2} \tag{53}$$

Average Percentage Error:

$$E = \frac{SE}{\frac{1}{N} \sum_{i=1}^{N} S_{mi}} \times 100\% \tag{54}$$

Slope (b) *and correlation coefficient* (R^2) *of best fit line between model predictions and measured data:*

$$b = S_m/S_p \tag{55}$$

where N = number of each series of data points used, S_m = measured data values and S_p = predicted values. Typical values of SE, RMS, E and b, particularly for suspended sediments, are given in Kashefipour and Falconer (2000).

14.6   CONCLUSIONS

Details are given herein of the structure of hydroinformatics software tools for predicting flow, water quality and sediment and contaminant transport processes in riverine systems, with particular reference being made to the use of such tools by engineers, scientists and environmental managers for environmental impact assessment studies. Details are given of the governing hydrodynamic equations for three-, two- and one-dimensional flow-field predictions and the corresponding advective–diffusion equation for the transport of water quality indicator, sediment and contaminant solutes within the water column. In applying the general advective–transport equation to a range of solutes (including sediments), details are given of the corresponding kinetic reactions and decay/growth rates for the most commonly modelled water quality indicators. Models developed by the authors for predicting 3-D, 2-D and 1-D flow and solute concentration distributions are summarised, together with details being given of the treatment of the boundary and initial conditions within such models.

A refined and highly accurate finite difference scheme has been used to solve the advective–diffusion equation, namely the ULTIMATE QUICKEST scheme, with this scheme producing little numerical diffusion and no undershoot or overshoot in regions where high concentrations exist.

Finally, details are given of the application of the authors' models TRIVAST (3-D), DIVAST (2-D) and FASTER (1-D) to a tidal reach of the Humber Basin, which drains large catchments of England including several major cities, and discharges into the North Sea along the north-eastern coast of England. The water quality indicators modelled include: salinity, temperature, total and faecal coliforms, biochemical oxygen demand, dissolved oxygen, organic, ammoniacal and oxidised nitrogen, and organic and dissolved phosphorus. The models were then extended to predict the fluxes of cohesive and non-cohesive sediments, including bed and suspended load transport, with the 2-D model being further extended to predict heavy metal distributions using dynamic partition coefficients.

All three models have predicted the hydrodynamic features of the basin very accurately, with good agreement being obtained for all of the models between the predicted elevations and velocities and the corresponding field measured data. Likewise, each of the predicted individual water quality constituent distributions was also compared with field data along the basin, with the agreement between the predicted and measured results again being generally encouraging. The three models were also tested for predicting cohesive and non-cohesive sediment fluxes, with good agreement obtained between the models and measured data, and with all three models predicting very similar suspended sediment concentrations. Finally, the 2-D model was extended to predict heavy metal concentrations, with dynamic decay rates, and this novel approach is now being included in the 3-D and 1-D models.

ACKNOWLEDGEMENTS

The numerical model studies reported herein were funded primarily through a number of grants including: the Engineering and Physical Sciences Research Council (grant GR/J/48337), the Natural Environmental Research Council LOIS programme (grant GST/02/0772), and Yorkshire Water plc. The authors are also grateful to the following organisations for providing data: the Environment Agency (North East Region), and ABP Research and Consultancy Ltd.

REFERENCES

Aver, M.T. and Neihaus, S.L. 1993. Modelling faecal coliform bacteria-1. Field and laboratory determination of loss kinetics. *Water Research*, IAWQ, **27**, 693–701.

Bale, A.J. 1987. The characteristics, behaviour and heterogeneous reactivity in estuarine suspended particles in estuaries. PhD Thesis, University of Plymouth.

Bowie, G.L. et al. 1985. *Rates, Constants and Kinetics Formulations in Water Quality Modelling.* Environmental Research Laboratory, US EPA, Athens, GA, Report No. EPA/600/3-85/040.

Brown, L.C. and Barnwell, T.O. Jr 1987. *The Enhanced Stream Water Quality Models QUAL2E and QUAL2E-UNCAS: Documentation and User Manual.* Environmental Research Laboratory, US EPA, Athens, GA, Report No. EPA/600/3-87/007, May.

Daily, J.E. and Harleman, D.R.F. 1972. *Numerical Model for the Prediction of Transient Water Quality in Estuary Networks.* Ralph M. Parsons Laboratory for Hydraulics and Water Resources, MIT Report No. 158.

Edwards, A., Freestone, R. and Urquhart, C. 1987. *The Water Quality of the Humber Estuary*. Report of the Humber Estuary Committee, Yorkshire Water Authority.

Elder, J.W. 1959. The dispersion of marked fluid in a turbulent shear flow. *Journal of Fluid Mechanics*, **5**(4), 544–560.

Falconer, R.A. 1986. Water quality simulation study of a natural harbour. *Journal of Waterway, Port, Coastal and Ocean Engineering*, ASCE, **112**(1), 15–34.

Falconer, R.A. 1991. Review of modelling flow and pollutant transport processes in hydraulic basins. *Proceedings of First International Conference on Water Pollution: Modelling, Measuring and Prediction*, Southampton, UK, Computational Mechanics Publications, 3–23.

Falconer, R.A. 1993. An introduction to nearly horizontal flows. In: Abbott, M.B. and Price, W.A. (eds), *Coastal, Estuarial and Harbour Engineers' Reference Book*. E. & F.N. Spon, London, 27–36.

Falconer, R.A. and Chen, Y. 1996. Modelling sediment transport and water quality processes on tidal floodplains. In: Anderson, M.G., Walling, D.E. and Bates, P.D. (eds), *Floodplain Processes*. John Wiley, Chichester, 361–398.

Fischer, H.B. 1973. Longitudinal dispersion and turbulent mixing in open channel flow. *Annual Review of Fluid Mechanics*, **5**, 59–78.

Fischer, H.B., List, E.J., Koh, R.C.J., Imberger, J. and Brooks, N.H. 1979. *Mixing in Inland and Coastal Waters*. Academic Press, San Diego.

Goldstein, S. 1938. *Modern Development in Fluid Dynamics*, Vol. 1. Oxford University Press, Oxford.

Harleman, D.R.F. 1966. Diffusion processes in stratified flow. In: Ippen, A.T. (ed.), *Estuary and Coastline Hydrodynamics*. McGraw-Hill, New York, 575–597.

Henderson, F.M. 1966. *Open Channel Flow*. Collier-Macmillan, London.

Kashefipour, S.M. and Falconer, R.A. 2000. An improved model for predicting sediment fluxes in estuarine waters. *Fourth International Conference on Hydroinformatics*, 23–27 Iowa, July, 1–8.

Kashefipour, S.M., Falconer, R.A. and Lin, B. 1999. *FASTER Model Reference Manual*. Environmental Water Management Research Centre Report, Cardiff University.

Lin, B. and Falconer, R.A. 1997. Tidal flow and transport modelling using the ULTIMATE QUICKEST scheme. *Journal of Hydraulic Engineering*, ASCE, **123**(4), 303–314.

Lin, B. and Shiono, K. 1995. Numerical modelling of solute transport in compound channel flows. *Journal of Hydraulic Research*, IAHR, **33**(6), 773–788.

Ng, B., Turner, A., Tyler, A.O., Falconer R.A. and Millward, G.E. 1996. Modelling contaminant geochemistry in estuaries. *Water Research*, **30**, 63–74.

Noat, D. and Rodi, W. 1982. Numerical simulations of secondary currents in channel flow. *Journal of Hydraulic Engineering*, ASCE, **108**(HY8), 948–968.

O'Connor, D.J. and Dobbins, W.E. 1958. Mechanism of reaeration in natural streams. *Transactions of the ASCE*, **123**, 641–684.

Preston, R.W. 1985. *The Representation of Dispersion in Two-Dimensional Shallow Water Flow*. Central Electricity Research Laboratories, Report No. TPRD/U278333/N84, May.

Rodi, W. 1984. *Turbulence Models and their Application in Hydraulics*, 2nd edition. International Association for Hydraulics Research, Delft, the Netherlands.

Speziale, C.G. 1987. On non-linear $k - \lambda$ and $k - \varepsilon$ model of turbulence. *Journal of Fluid Mechanics*, **187**, 459–475.

Thomann, R.V. and Mueller, J.A. 1987. *Principles of Surface Water Quality Modelling Control*. Harper Collins, New York.

Tominaga, A. and Nezu, I. 1991. Turbulence structure in open channel with a step in the cross-section. *Journal of Hydraulic Engineering*, ASCE, **117**, 21–41.

Turner, A., Millward, G.E., Bale, A.J. and Morris, A.W. 1993. Application of the K_D concept to the study of trace metal removal and desorption during estuarine mixing. *Estuarine, Coastal and Shelf Science*, **36**, 1–13.

van Rijn, L.C. 1984a. Sediment transport part 1: bed load transport. *Journal of Hydraulic Engineering*, ASCE, **10**, 1431–1456.

van Rijn, L.C. 1984b. Sediment transport part 2: suspended load transport. *Journal of Hydraulic Engineering*, ASCE, **10**, 1613–1641.

Vieira, J.K. 1993. Dispersive processes in two-dimensional models. In: Abbott, M.B. and Price, W.A. (eds),

Coastal, Estuarial and Harbour Engineers' Reference Book. E. & F.N. Spon, London, Chapter 14, 179–190.

Wu, J. 1969. Wind-stress coefficients over sea surface from breeze to hurricane. *Journal of Geophysical Research*, **87**, 9704–9706.

15 Modelling Sediment Entrainment into Suspension, Transport and Deposition in Rivers

MARCELO H. GARCIA

Department of Civil and Environmental Engineering, University of Illinois, USA

15.1 INTRODUCTION

The most common modes of sediment transport in rivers are bedload and suspended load. In the case of bedload, sediment grains roll, slide or saltate over each other, never deviating too far from the bed (Nino and Garcia 1994, 1998). In the case of suspended load, the fluid turbulence comes into play carrying sediment particles well up into the water column (Garcia and Parker 1991, 1993). In both cases, the driving force for sediment transport is the action of gravity on the fluid phase; this force is transmitted to the particles via drag.

The floodplains of most sand-bed rivers often contain copious amounts of silt and clay finer than approximately 50 μm. This material is often called wash load because it moves through the river system without being present in the bed in significant quantities. The wash load has been described 'as those grains that are in perpetual suspension'. However, the wash load in the water column exchanges with the banks and the floodplain rather than with the river bed. Greatly increased wash load, for example, can lead to thickened floodplain deposits, with a consequent increase in bankfull depth.

Alluvial rivers that are free to scour and fill during floods can broadly be divided into two types: sand bed streams and gravel bed streams. Sand bed streams typically have values of median bed sediment size between 0.1 mm and 1 mm. The sediment tends to be relatively well sorted, with values of geometric standard deviation of the bed sediment size varying from 1.1 to 1.5. Gravel bed streams typically have values of median size of the bed sediment exposed on the surface of 15 mm to 200 mm or larger; the substrate is usually finer by a factor of 1.5 to 3. The geometric standard deviation of the substrate sediment size is typically quite large, with values in excess of 3 being quite common. Although gravel and coarser material constitute the dominant sizes, there is usually a substantial amount of sand stored in the interstices of the gravel substrate (Klingeman et al. 1998).

Two dimensionless parameters provide an effective delineator of rivers into the above two types (Garcia 1999, 2000). The first of these is the dimensionless Shields stress defined as:

$$\tau^* = \frac{\tau_b}{\rho g R D} = \frac{HS}{RD}$$

where τ_b is the bed shear stress, g is the gravitational acceleration, ρ is the water density,

Model Validation: Perspectives in Hydrological Science. Edited by M.G. Anderson and P.D. Bates.
© 2001 John Wiley & Sons, Ltd.

$R = (\rho_s/\rho - 1)$ is the submerged specific gravity of the sediment, D is the sediment size (mean diameter), H is the flow depth and S is the stream slope which for steady, uniform flow is the same as the energy gradient. The second of these is a particle Reynolds number R_{ep} defined as:

$$R_{ep} = \frac{\sqrt{gRDD}}{\nu}$$

where ν is the kinematic viscosity of water. This second parameter can be considered as a dimensionless surrogate for grain size. Figure 15.1 shows a plot of the values of the Shields stress evaluated at bankfull flow versus particle Reynolds number for six sets of field data: (a) gravel bed rivers in Wales, UK (Wales); (b) gravel bed rivers in Alberta, Canada (Canada); (c) gravel bed rivers in the Pacific Northwest, USA (Pacific); (d) single-thread sand streams (Sand sing); (e) multiple-thread sand streams (Sand mult); (f) large sand bed rivers (Parana, Missouri etc.); and (g) large-scale laboratory experiments conducted at St Anthony Falls Laboratory (SAFL), University of Minnesota (Parker et al. 1998). There are three curves in the diagram that make it possible to know whether a given bed sediment will go into motion, and if this is the case whether or not the prevailing mode of transport will be in suspension or as bedload. The diagram also can be used to predict what kind of bedforms can be expected. For example, ripples will develop in the presence of a viscous sublayer and fine-grained sediment. If the viscous sublayer is disrupted by coarse sediment particles, then dunes will be the most common type of bedform. The Shields regime diagram also shows a clear distinction between the conditions observed in sand bed rivers and gravel bed rivers at bankfull stage. As one would expect, in gravel bed rivers sediment is predominantly transported as bedload, while in sand bed streams sediment is transported mainly in suspension.

The emphasis here is placed on understanding the mechanics of suspended load transport in rivers, with the goal of providing the knowledge needed to do sound mathematical modelling of sedimentation processes in the hydrologic cycle, particularly in problems aiming at the restoration and naturalisation of streams and rivers. It is hoped also that the fundamental mathematical treatment of suspended load that follows will help in the design of field experiments by pointing out the flow and sediment transport parameters that need to be measured, and will provide also the theoretical foundation needed to validate both analytical and numerical models of sediment entrainment into suspension, transport and deposition in rivers.

While the focus herein is on non-cohesive sediments, most of the material covered also applies to the case of fine-grained, cohesive sediments, the main exception being the sediment fall velocity and bed boundary conditions.

15.2 MASS CONSERVATION OF SUSPENDED SEDIMENT

Suspended sediment differs from bedload sediment in that it may be diffused throughout the vertical column of fluid via turbulence. Here the local instantaneous volume concentration of suspended sediment at point x_i and time t is denoted as $c(x_i, t)$. The Eulerian particle velocity field is denoted as $v_i(x_j, t)$. For fine particles the following approximation holds:

$$v_i = u_i - v_s\delta_{i3} \tag{1a}$$

where u_i is the Eulerian flow velocity field, v_s is the sediment fall velocity in quiescent water, and δ_{i3} is the Kronecker delta. The above velocities are instantaneous values. In the above form of

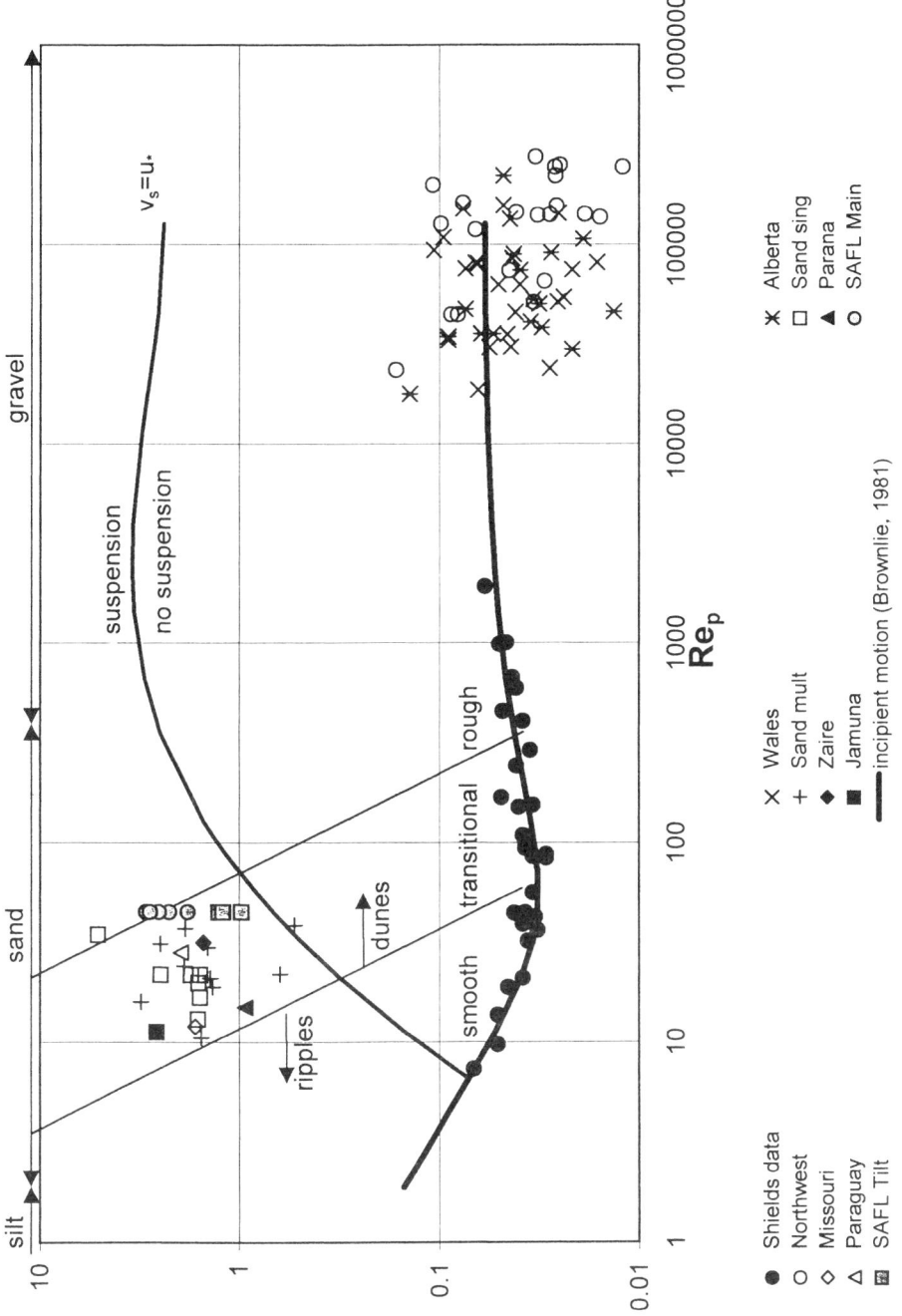

Figure 15.1 Shields regime diagram (Garcia 2000)

the relation, $x_3 = z$ is constrained to be upward vertical. The form for an arbitrary Cartesian system is

$$v_i = u_i - v_s k_i \tag{1b}$$

where k_i is a unit vector in the upward vertical direction. (Note that $k_i = (0, 0, 1) = \delta_{i3}$ in a coordinate system for which x_3 is directed upward vertical.) As long as the suspended sediment under consideration is sufficiently coarse so as not to undergo Brownian motion (i.e. silt or coarser), molecular effects can be neglected. Suspended particles are transported solely by convective fluxes. The volume convective flux F_{si} of suspended sediment is given by the following expression:

$$F_{si} = cv_i \tag{2}$$

Consider an arbitrary volume, fixed in Eulerian space. It has volume V and surface area A. The local unit outward normal vector to the surface is denoted as n_i. The equation of mass balance of suspended sediment can be written in words as

$$\frac{\partial}{\partial t}[\text{mass in volume}] = [\text{net mass inflow rate}] \tag{3}$$

Noting that the mass inflow rate across an area dS is given by $-\rho_s F_{si} n_i dS$, equation (3) translates to

$$\frac{\partial}{\partial t}\left(\iiint_v \rho_s c\, dV\right) = -\rho_s \iint_s F_{si} n_i dS \tag{4}$$

According to the divergence theorem,

$$\iint_s F_{si} n_i dS = \iiint_v \frac{\partial F_{si}}{\partial x_i} dV \tag{5}$$

Between equations (2), (4) and (5), it is found that

$$\rho_s \iiint \left[\frac{\partial c}{\partial t} + \frac{\partial v_i c}{\partial x_i}\right] dV = 0 \tag{6}$$

In so far as the choice of volume V is entirely arbitrary, it can be concluded that

$$\frac{\partial c}{\partial t} + \frac{\partial v_i c}{\partial x_i} = 0 \tag{7}$$

Herein it is assumed that the approximation embodied in equation (1) applies for suspended sediment. This yields the result

$$\frac{\partial c}{\partial t} + \frac{\partial F_{si}}{\partial x_i} = 0 \tag{8a}$$

where

$$F_{si} = (u_i - v_s k_i)c \tag{8b}$$

denotes the volume flux of suspended sediment in the x_i direction. If $x_3 = z$ denotes an upward vertical (or quasi-vertical) coordinate, equations (8a) and (8b) reduce to

$$\frac{\partial c}{\partial t} + \frac{\partial uc}{\partial s} + \frac{\partial vc}{\partial n} + \frac{\partial}{\partial z}[(w - v_s)c] = 0 \tag{8c}$$

Here $u_i = (u, v, w)$ denotes the Eulerian fluid velocity field. Note that here s and n are not necessarily tangential to the bed.

15.3 REYNOLDS AVERAGING

The parameter u_i and c are now decomposed into averages over turbulence and fluctuations about the mean:

$$c = \bar{c} + c'; \quad u_i = \bar{u}_i + u_i' \tag{9a,b}$$

Substituting equations (9a) and (9b) into equation (8), reducing with the equation of fluid continuity and averaging, the following Reynolds conservation equation of suspended sediment is obtained:

$$\frac{\partial c}{\partial t} + \frac{\partial \bar{F}_{si}}{\partial x_i} = 0 \tag{10a}$$

where

$$\bar{F}_{si} = (\bar{u}_i - v_s k_i)\bar{c} + \overline{u_i' c'} \tag{10b}$$

Again, if $x_3 = z$ is upward vertical, equations (10a) and (10b) reduce to

$$\frac{\partial \bar{c}}{\partial t} + \bar{u}\frac{\partial c}{\partial s} + \bar{v}\frac{\partial \bar{c}}{\partial n} + (\bar{w} - v_s)\frac{\partial \bar{c}}{\partial z} = -\frac{\overline{\partial u' c'}}{\partial s} - \frac{\overline{\partial v' c'}}{\partial n} - \frac{\overline{\partial w' c'}}{\partial z} \tag{10c}$$

It is seen from equation (10b) that the mean flux of suspended sediment $\bar{F}_{si}$ is composed of two components, i.e. a mean convective flux and a Reynolds flux. The Reynolds flux $\overline{u_i' c'}$ in the above relation is clearly diffusive in nature. The simplest closure assumption one could make for these terms is

$$\overline{u_i' c'} = -D_d \frac{\partial \bar{c}}{\partial x_i} \tag{11a}$$

In equation (11a), the kinematic eddy diffusivity D_d is assumed to be a scalar quantity. For the case of non-isotropic turbulence, equation (11a) must be generalised to the form

$$\overline{u_i'c'} = -D_{dij}\frac{\partial \bar{c}}{\partial x_j} \tag{11b}$$

Here D_{dij} is a tensor quantity. It is often assumed to represent a diagonal matrix, such that $D_{dij} = 0$ if $i \neq j$, and $D_{d11} \neq D_{d22} \neq D_{d33}$.

15.4 BOUNDARY CONDITIONS

Equation (10), when closed with a Fickian assumption such as equation (11), represents a convective–diffusive equation for suspended sediment. The condition of vanishing flux of suspended sediment across (normal to) the water surface defines the upper boundary condition.

Let n_i^s denote a unit vector normal to the water surface. The boundary condition there takes the form

$$\bar{F}_{si}\big|_{\text{water surface}} \cdot n_i^s = 0 \tag{12}$$

As a special case, uniform steady flow over a flat (when averaged over bedforms) bed is considered. In this case,

$$\bar{u}_i = (\bar{u}(z), 0, 0) \tag{13}$$

If z is furthermore taken to be (nearly) upward vertical from the bed, and the water surface is taken to be (nearly) horizontal, $n_i^s = k_i = (0, 0, 1)$, and equation (12) reduces to

$$\bar{F}_{sz}\big|_{z=H} = 0 \tag{14a}$$

where

$$\bar{F}_{sz} = -v_s\bar{c} + \overline{w'c'} \tag{14b}$$

The boundary conditions at the bed differs from the one at the water surface, in that it must account for entrainment of sediment into the flow from the bed, and deposition from the flow

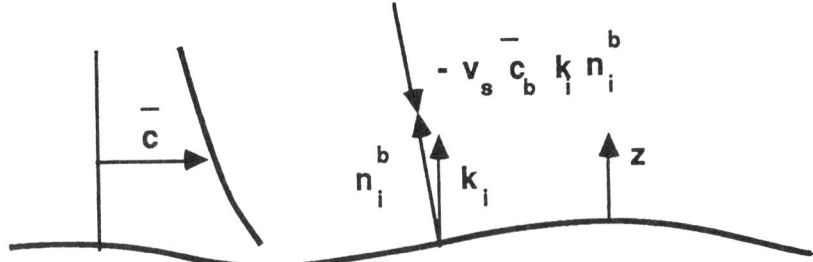

Figure 15.2 Definition diagram for bed sediment boundary conditions

onto the bed. As noted above, the mean flux $\bar{F}_{si}$ can be divided into two components, a mean convective part $(\bar{u}_i - v_s k_i)\bar{c}$ and a Reynolds part $\overline{u_i'c'}$.

Let n_i^b denote a unit upward normal vector at the erodible boundary (bed) as shown in Figure 15.2. As long as porous flow can be neglected, the flow must satisfy at least the following boundary condition:

$$\bar{u}_i|_{bed}n_i^b = 0 \tag{15}$$

It follows from equations (10b) and (15) that the mean flux of suspended sediment onto the bed is given by $-$ D, where

$$D = v_s\bar{c}_b k_i n_i^b \tag{16}$$

denotes the volume rate of *deposition* of suspended sediment per unit time per unit bed area. Here $\bar{c}_b$ denotes a near bed value of $\bar{c}$.

The remaining term in the expression for $\bar{F}_{si}$ is the Reynolds flux due to turbulence. The component of the Reynolds flux of suspended sediment near the bed that is directed upward normal to the bed may be termed the rate of *erosion*, or, more accurately, *entrainment* of bed sediment into suspension per unit bed area per unit time. The entrainment rate E is thus given by

$$E = \overline{u_i'c'}|_{near\ bed} \cdot n_i^b \tag{17}$$

The terminology 'near bed' is employed to avoid possible singular behaviour at the bed (located at $z = 0$).

It is seen from equations (10b) and (15) to (17), that the net upward normal flux of suspended sediment at (or rather just above) the bed is given by

$$\bar{F}_{si}|_{near\ bed} \cdot n_i^b = v_s(E - \bar{c}_b k \cdot n^b) \tag{18}$$

where

$$E \equiv \frac{E}{v_s} \tag{19}$$

denotes a dimensionless rate of entrainment of bed sediment into suspension.

The required bed boundary condition, then, is a specification of E. Typically a relation of the following form is assumed:

$$E = E(\tau_{bs}, \text{other parameters}) \tag{20}$$

where τ_{bs} denotes the boundary shear stress due to skin friction.

In analogy to equations (13) and (14), in the special case of steady, uniform flow over a flat (averaged over bedforms) bed, with z defined to be (nearly) vertical, equation (18) reduces to

$$\bar{F}_{sz}|_{near\ bed} = v_s(E - \bar{c}_b) \tag{21}$$

It is furthermore assumed that an equilibrium steady, uniform suspension has been achieved; it follows that there should be neither net deposition on $(\bar{F}_{sz} < 0)$ or erosion from $(\bar{F}_{sz} > 0)$ the bed.

That is, $\bar{F}_{sz} = 0$, yielding the result

$$E = \bar{c}_b \tag{22}$$

This relation simply states that the entrainment rate equals the deposition rate, so that there is no net normal flux of suspended sediment at the bed.

15.5   RELATION FOR SEDIMENT ENTRAINMENT

A number of relations are available in the literature for estimating the entrainment rate of sediment into suspension E (and thus the reference concentration $\bar{c}_b$ for the equilibrium case). Garcia and Parker (1991) performed a detailed comparison of eight such relations against data. The relations were checked against a carefully selected set of data pertaining to equilibrium suspensions of uniform sand. In this case, it is possible to measure $\bar{c}_b$ directly at some near bed elevation $z = b$, and to equate the result to E according to equation (22).

The data consisted of some 64 sets from ten different sources, all pertaining to laboratory suspensions of uniform sand with a submerged specific gravity R near 1.65. The shear stress due to skin friction τ_{bs}, and the associated shear velocity due to skin friction $u*_s$, given by

$$\tau_{bs} = \rho u*_s^2 \tag{23}$$

were computed using a relation similar to equation (28a) and the following relation for the roughness k_s,

$$k_s = 2 \cdot D \tag{24}$$

The data covered the following ranges:

E: 0.002–0.06
$u*_s/v_s$: 0.70–7.50
H/D: 240–2400
R_{ep}: 3.50–37.00

The range of values of $R_{ep} = (gRD^3)^{1/2}/v$ corresponds to a grain size range from 0.09 mm to 0.44 mm. Except for the somewhat small values of H/D, the values cover a range that includes typical field sand bed streams.

Three of the relations for E performed particularly well and are presented herein. The first of these presented is the relation of Garcia and Parker (1991). The reference level is taken to be 5% of the depth; that is,

$$\frac{b}{H} = 0.05 \tag{25}$$

The relation takes the form,

$$E = \frac{AZ_u^5}{\left(1 + \dfrac{A}{0.3} Z_u^5\right)} \tag{26a}$$

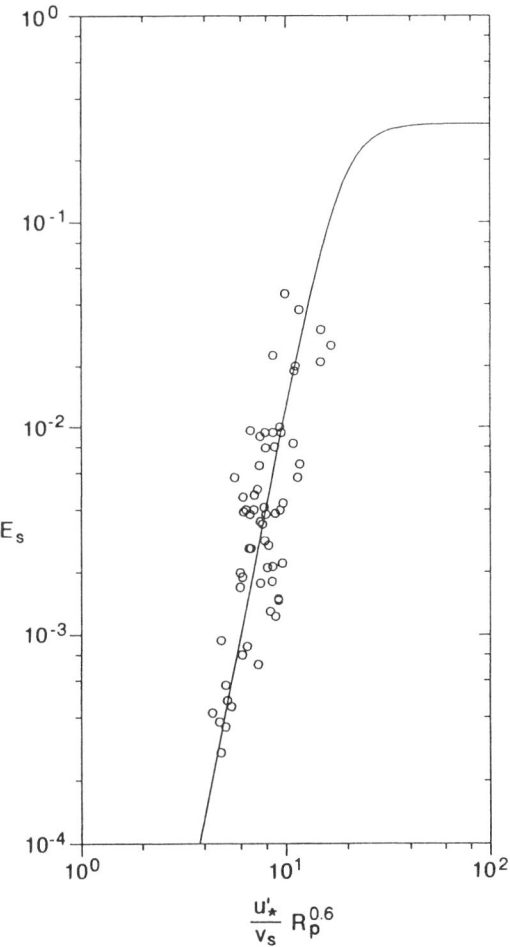

Figure 15.3   Sediment entrainment function of Garcia and Parker (1991)

$$A = 1.3 \cdot 10^{-7}; \quad Z_u = \frac{u*_s}{v_s} R_{ep}^{0.6} \qquad (26b,c)$$

Equation (26a) is compared against the data in Figure 15.3. The relation was modified later to account for the effect of viscous effects on the entrainment of fine-grained, non-cohesive sediment (Garcia and Parker 1993).

A second relation that performed very well is that of Van Rijn (1984). This relation takes the form,

$$E = 0.015 \frac{D}{b} (\tau_s^*/\tau_c^* - 1)^{1.5} R_{ep}^{0.2} \qquad (27)$$

where τ_s^* denotes the Shields stress due to skin friction. Van Rijn computes τ_{bs} from

$$C_{fs} = \frac{1}{\kappa} \ln\left(12\frac{H}{k_s}\right)^{-2} \tag{28a}$$

where for uniform material

$$k_s = 3 \cdot D \tag{28b}$$

In performing the comparison, Garcia and Parker (1991) estimated τ_c^* from a fit to the Shields curve due to Brownlie (1981). This fit can be expressed as

$$\tau_c^* = 0.22 \cdot R_{ep}^{-0.6} + 0.06 \cdot \exp(-17.77 \cdot R_{ep}^{-0.6}) \tag{29}$$

A third relation that performs very well is that of Smith and McLean (1977). It can be expressed as

$$E = 0.65 \frac{\gamma_o(\tau_s^*/\tau_c^* - 1)}{1 + \gamma_o(\tau_s^*/\tau_c^* - 1)} \tag{30}$$

where

$$\gamma_o = 0.0024 \tag{31}$$

The value b at which E is to be evaluated is given by the relation below:

$$b = 26.3 \cdot (\tau_s^*/\tau_c^* - 1) \cdot D + k_s \tag{32}$$

Here k_s denotes the equivalent roughness height for a fixed bed.

More recently, Zyserman and Fredsoe (1994) have proposed a relation very similar to that of Garcia and Parker (1991) which seems to perform quite well in sandy streams.

15.6   EQUILIBRIUM SUSPENSIONS

Consider normal flow in a wide, rectangular open channel. The bed is assumed to be erodible and has no curvature when averaged over bedforms. The z-coordinate is quasi-vertical, implying

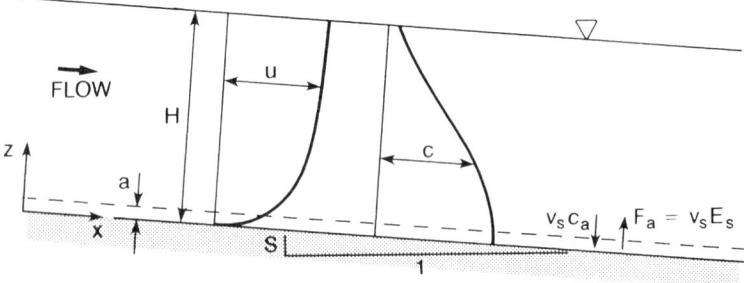

Figure 15.4   Definition diagram for open-channel equilibrium transport conditions

low channel slope S. The suspension is likewise assumed to be in equilibrium. That is, $\bar{c}$ is a function of z along. As illustrated in Figure 15.4, the flow and suspension are uniform in s and n and steady in time, so that equation (10a) reduces to

$$\frac{d\bar{F}_{sz}}{dz} = 0 \tag{33}$$

Equation (33) yields the condition $\bar{F}_{sz}$ = constant. The constant can be evaluated as 0 from equation (14a), i.e. the condition of vanishing normal flux across the water surface. As a result, equation (33) integrates with the aid of equation (14b) to yield

$$\overline{w'c'} - v_s\bar{c} = 0 \tag{34}$$

It is appropriate to close this equation with the assumption of an eddy diffusivity: from equation (11a)

$$\overline{w'c'} = -D_d\frac{d\bar{c}}{dz} \tag{35}$$

so that equation (34) becomes

$$-D_d\frac{d\bar{c}}{dz} - v_s\bar{c} = 0 \tag{36}$$

Equation (36) has a simple physical interpretation. The term $-v_s\bar{c}$ represents the rate of sedimentation of suspended sediment under the influence of gravity; it is always directed downward. If all of the sediment is not to settle out, there must be an upward flux that balances this term. The upward flux is provided by the effect of turbulence, acting to yield a Reynolds flux. According to equation (35), this flux will be directed upward as long as $d\bar{c}/dz < 0$. It follows that the equilibrium suspended sediment concentration decreases for increasing z, so that turbulence diffuses sediment from zones of high concentrations (near the bed) to zones of low concentration (near the water surface).

Further progress requires an assumption for the kinematic eddy diffusivity D_d. The simple approach taken here is that of Rouse (1937). It involves the use of the *Prandtl analogy*. The argument is as follows. Fluid mass, heat, momentum etc. should all diffuse at the same kinematic rate due to turbulence and thus have the same kinematic eddy diffusivity, because each is a property of the fluid particles, and it is the fluid particles that are being transported by Reynolds fluxes. The velocity profile is approximated as logarithmic throughout the depth.

The vertical kinematic eddy viscosity takes the form:

$$D_d = \kappa u*z\left(1 - \frac{z}{H}\right) \tag{37}$$

Equation (37) is now substituted into equation (36), which is then integrated from the nominal bed level to distance z above the bed in z. The resulting form can be cast as

$$\bar{c} = \bar{c}_b\left[\frac{(H - z)/z}{(H - b)/b}\right]^Z \tag{38}$$

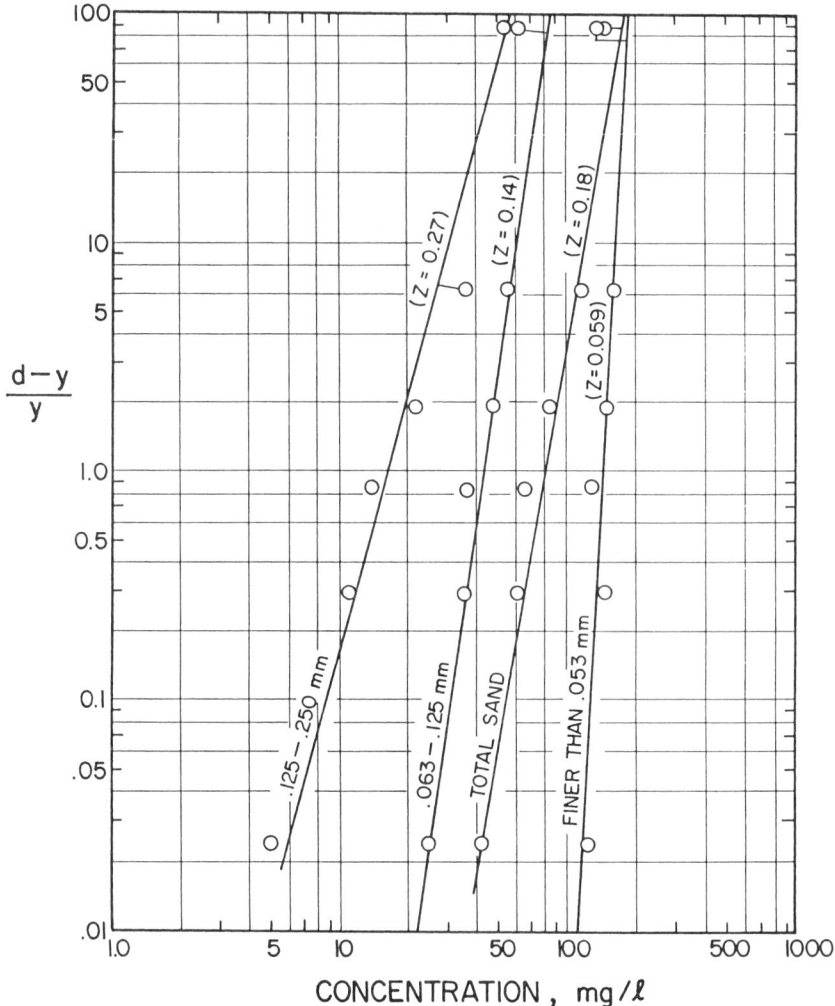

Figure 15.5 Distribution of suspended sediment in the Amazon river at Manacapuru, where the flow depth is 44 m (Vanoni 1980)

where Z denotes the *Rouse number*, a dimensionless number given by

$$Z = \frac{v_s}{\kappa u*} \tag{39}$$

Sediment concentration profiles of suspended sediment observed in the Amazon River, Brazil, are plotted in Rousean form in Figure 15.5 (Vanoni 1980). Despite the wide range of sediment sizes present in the water column, it is cleat that equation (38) provides a good predictor for sediment concentration profiles under equilibrium conditions, provided that the near bed concentration as defined in equation (16) can be estimated.

Note, that from equation (22), $\bar{c}_b$ is equal to the dimensionless sediment entrainment rate E in the case of the present equilibrium suspension. Table 15.1 gives a summary of all the formulations that have been proposed in the literature to predict near bed volumetric sediment concentrations for equilibrium conditions.

15.7   NON-EQUILIBRIUM SUSPENSIONS

The above relations for E can be combined with equations (22) and (38) to provide a complete treatment for the case of equilibrium suspensions. It is of value to investigate the formulation of a typical non-equilibrium suspension.

The classical problem of this type is the *pick-up* problem, illustrated in Figure 15.6. Sediment-free water flows from a non-erodible bed to an erodible bed. Sediment is gradually entrained, such that far downstream an equilibrium suspension is attained. That is, where $\bar{c}_e(z)$ denotes the equilibrium profile,

$$\text{Lim}_{s \to \infty} \bar{c}(s, z) = \bar{c}_e(z) \tag{40}$$

In order to simplify the problem, it is assumed that the composite bed roughness of the non-erodible portion is identical to that of the erodible portion. The suspension is assumed to be developing in the s-direction but to be constant in time and uniform in the transverse direction. The origin of the s-coordinate is taken at the junction between the non-erodible and erodible portions.

In fact, local scour should occur just downstream of the junction. The bed elevation, however, can be taken to be constant over time scales that are long compared to the characteristic time of the flow but short compared to that required for noticeable bed erosion.

The flow is taken to be at equilibrium; thus equation (13) is satisfied. From equations (10a), (10b) and (11a), the equation of sediment mass conservation can be written as

$$\bar{u}\frac{\partial \bar{c}}{\partial s} - v_s\frac{\partial \bar{c}}{\partial z} = \frac{\partial}{\partial z}\left(D_d \frac{\partial \bar{c}}{\partial z}\right) + \frac{\partial}{\partial s}\left(D_d \frac{\partial \bar{c}}{\partial s}\right) \tag{41}$$

The boundary condition at the water surface is given in equations (14) and (11a) as

$$\left[D_d\frac{\partial \bar{c}}{\partial z} + v_s\bar{c}\right]_{z=H} = 0 \tag{42}$$

The condition (22) cannot be used as a bed boundary condition, because it applies solely to equilibrium flow (Parker 1978; Garcia and Parker 1991). The appropriate boundary condition is obtained from equations (11a) and (17):

$$-D_d\frac{\partial \bar{c}}{\partial z}\big|_{z=b} = v_s E \tag{43}$$

The boundary condition at the junction is simply

$$\bar{c}\big|_{z=0} = 0 \tag{44}$$

Table 15.1 Near-bed sediment concentration functions

Formula	Equation	Parameters	Reference height
Einstein (1950)	$c_{ae} = \dfrac{q^*}{23.2(\tau_s^*)^{0.5}}$		$a = 2D_s$
Engelund and Fredsoe (1976, 1982)	$c_{ae} = \dfrac{0.65}{(1 + \lambda_b^{-1})^3}$	$\lambda_b = \left[\dfrac{\tau_s^* - 0.06 - \dfrac{\beta p\pi}{6}}{0.027(R+1)\tau_s^*}\right]^{0.5}$; $p = \left[1 + \left(\dfrac{\dfrac{\beta\pi}{6}}{\tau_s^* - 0.06}\right)^4\right]^{-0.25}$; $\beta = 1.0$	$a = 2D_s$
Smith and McLean (1977)	$c_{ae} = \dfrac{0.65\gamma_o S_o}{1 + \gamma_o S_o}$	$S_o = \dfrac{\tau_s^* - \tau_c^*}{\tau_c^*}$; $\gamma_o = 2.1 \cdot 10^{-3}$	$a = \alpha_o(\tau_s^* - \tau_c^*)D_s + k_s$ $\alpha_o = 26.3$
Itakura and Kishi (1980)	$c_{ae} = k_1\left(k_2 \dfrac{u_*}{v_s}\dfrac{\Omega}{\tau^*} - 1\right)$	$\Omega = \dfrac{\tau^*}{k_3}\left(k_4 + \left[\dfrac{\exp(-A_o^2)}{\int_{A_o}^{\infty} \exp(-\xi^2)d\xi}\right] - 1\right)$; $A_o = \dfrac{k_3}{\tau^*} - k_4$; $a = 0.05H$ $k_1 = 0.008$; $k_2 = 0.14$; $k_3 = 0.143$; $k_4 = 2.0$	
Van Rijn (1984)	$c_{ae} = 0.015 \dfrac{D_s}{a} \dfrac{S_o^{1.5}}{D_*^{0.3}}$	$D_* = D_s\left(\dfrac{gR}{v^2}\right)^{1/3}$	$a = \dfrac{\Delta_b}{2}$ if Δ_b known else $a = k_s \cdot a_{min} = 0.01H$

Celik and Rodi (1984)	$c_{ae} = \dfrac{k_o C_m}{I}$	$C_m = 0.034\left[1 - \left(\dfrac{k_s}{H}\right)^{0.06}\right]\dfrac{u_*^2}{gRH}\,\dfrac{U_m}{v_s}$; $\quad a = 0.05H$ $I = \displaystyle\int_{0.05}^{1}\left(\dfrac{1-\eta}{\eta}\cdot\dfrac{\eta_a}{1-\eta_a}\right)^{v_s/0.4u_*}d\eta$; $\eta = z/H$; $\eta_a = 0.05$; $k_o = 1.13$
Akiyama and Fukushima (1986)	$E_s = 0$; $\quad Z < Z_c$ $E_s = 3\cdot10^{-12}Z^{10}\left(1 - \dfrac{Z_c}{Z}\right)$; $\qquad Z_c < Z < Z_m$ $E_s = 0.3$; $\quad Z > Z_m$	$Z = \dfrac{u_*}{v_s}R_{ep}^{0.5}$; $\quad R_{ep} = \dfrac{\sqrt{RgD_sD_s}}{v}$; $\quad Z_c = 5$; $\quad Z_m = 13.2$ $\quad a = 0.05H$
Garcia and Parker (1991)	$E_s = \dfrac{AZ_u^5}{1 + \dfrac{A}{0.3}Z_u^5}$ or $E_s = AZ_u^5$;	$Z_u = \dfrac{u_*'}{v_s}R_{ep}^n$; $\quad u_*' = \dfrac{g^{0.5}}{C'}U_m$; $\quad a = 0.05H$ $C' = 18\cdot\log\left(\dfrac{12R_b}{3D_s}\right)$; $\quad n = 0.6$; $\quad A = 1.3\cdot10^{-7}$
Zyserman and Fredsoe (1994)	$c_{ae} = \dfrac{0.331(\theta' - 0.045)^{1.75}}{1 + \dfrac{0.331}{0.46}(\theta' - 0.045)^{1.75}}$	$\theta' = \dfrac{(u_*')^2}{(s-1)gD_s}$ $\quad a = 2D_s$

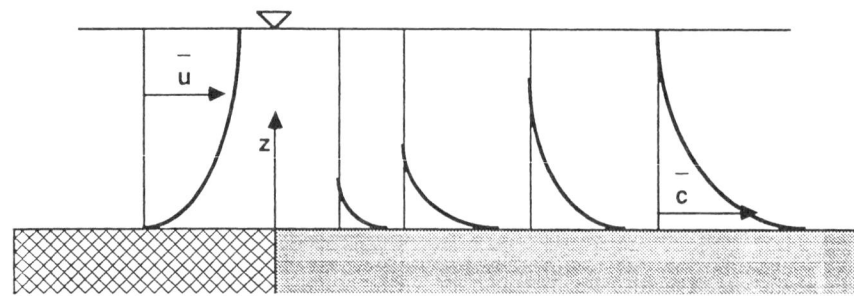

Figure 15.6 Definition diagram of sediment pick-up problem

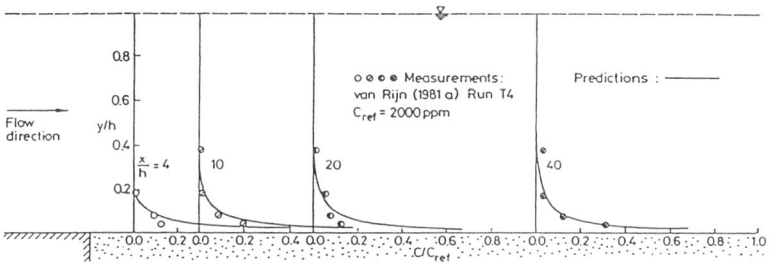

Figure 15.7 Comparison of measured and numerically predicted sediment concentration profiles resulting from net entrainment into suspension (Celik and Rodi 1988)

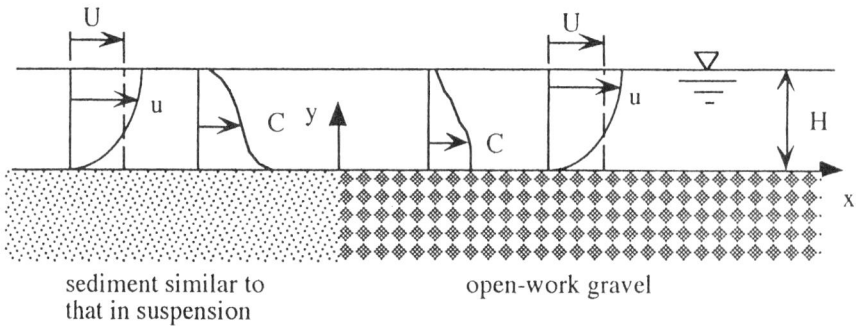

Figure 15.8 Equilibrium sediment-laden flow going from sediment-covered bed to open-work gravel bed (Huang and Garcia 2000)

The pick-up problem has attracted the attention of many authors since the 1930s (e.g. Mei 1968). Two more recent treatments are those due to Van Rijn (1981) and Celik and Rodi (1988). Van Rijn uses the above formulation, with a closure for D_d afforded by a k–ε turbulence model for the flow. A sample prediction from Celik and Rodi is shown in Figure 15.7.

Another classical non-equilibrium case is the *put-down* problem illustrated in Figure 15.8. In this problem, a steady uniform suspension flows from a sediment-covered bed to an open-gravel

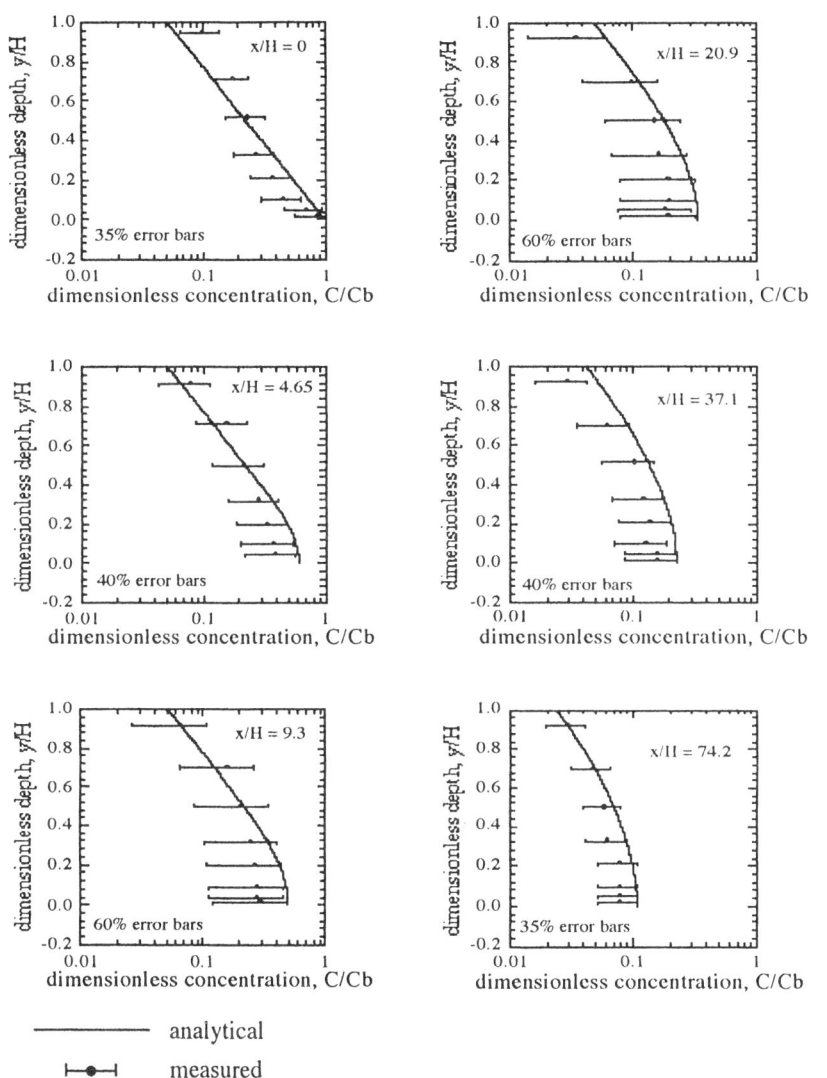

Figure 15.9 Comparison of measured and analytically estimated sediment concentration profiles in a uniform flow over a perforated bottom (Huang and Garcia 2000)

bed. As the sediment deposits on the gravel pores it cannot be resuspended by the flow turbulence and eventually the water column will be free of suspended sediment. Observations have shown that accumulation of fine sediment in the pores of spawning open-work gravel have a detrimental effect on biota. Einstein (1968) studied this problem experimentally in the laboratory and more recently, Huang and Garcia (2000) have found an analytical solution that provides estimates of vertical sediment concentration profiles which compare reasonably well with laboratory observations by Wang and Ribberink (1986), as shown in Figure 15.9.

15.8 GENERALISATION FOR SEDIMENT MIXTURES

Rather few researchers have considered the problem of suspensions of mixed grain sizes. Let the grain size range of interest be divided into N subranges, each with mean size ϕ_j on the phi scale and geometric mean diameter $D_j = 2^{-\phi_j}$, where $j = 1 \ldots N$. Furthermore, let $\bar{c}_j$ denote the volume concentration of sediment in the jth subrange. It follows that

$$\bar{c} = \sum_{j=1}^{N} \bar{c}_j \tag{45}$$

As long as the suspension remains dilute, the equation for mass conservation generalises easily; for example, equations (10a), (10b), and (11) generalise to

$$\frac{\partial \bar{c}_j}{\partial t} + (\bar{u}_i - v_s k_i)\frac{\partial \bar{c}_j}{\partial x_i} = \frac{\partial}{\partial x_i}\left(D_d \frac{\partial \bar{c}_j}{\partial x_i}\right) \tag{46}$$

The boundary condition at the water surface similarly generalises in a straightforward manner: the flux of suspended sediment normal to the water surface should vanish for each grain size range.

The boundary condition at the bed can be formulated by generalising the relations for sediment entrainment rate normal to the bed embodied in equation (17) (reduced with equation (11a)); where E_j denotes the volume entrainment rate for the jth subrange,

$$E_j = \overline{u_i' c_j'}|_{\text{near bed}} n_i^b = -D_d \frac{\partial \bar{c}_j}{\partial z_n}|_{\text{near bed}} \tag{47}$$

In the above relation, z_n is specifically defined to be a coordinate defined upward normal from the bed (averaged over bedforms).

Garcia and Parker (1991) have provided a generalised treatment for E_j in the case of mixtures. Let F_j denote the volume fraction of material in the surface layer of the bed in the jth gain range. In analogy to equation (19), it is assumed that

$$E_j = v_{sj} F_j E(Z_{uj}) \tag{48a}$$

where the functional relation between E and Z_{uj} is given by equation (26a). The parameter Z_{uj} is specified as

$$Z_{uj} = \lambda_m \frac{u*_s}{v_{sj}} R_{epj}^{0.6}\left(\frac{D_j}{D_{50}}\right)^{0.2} \tag{48b}$$

In the above relations, v_{sj} denotes the fall velocity of grain size D_j in quiescent water, D_{50} denotes the median size of the surface material in the bed,

$$R_{epj} = \frac{\sqrt{RgD_j}D_j}{\nu} \tag{48c}$$

and the parameter λ_m is given by

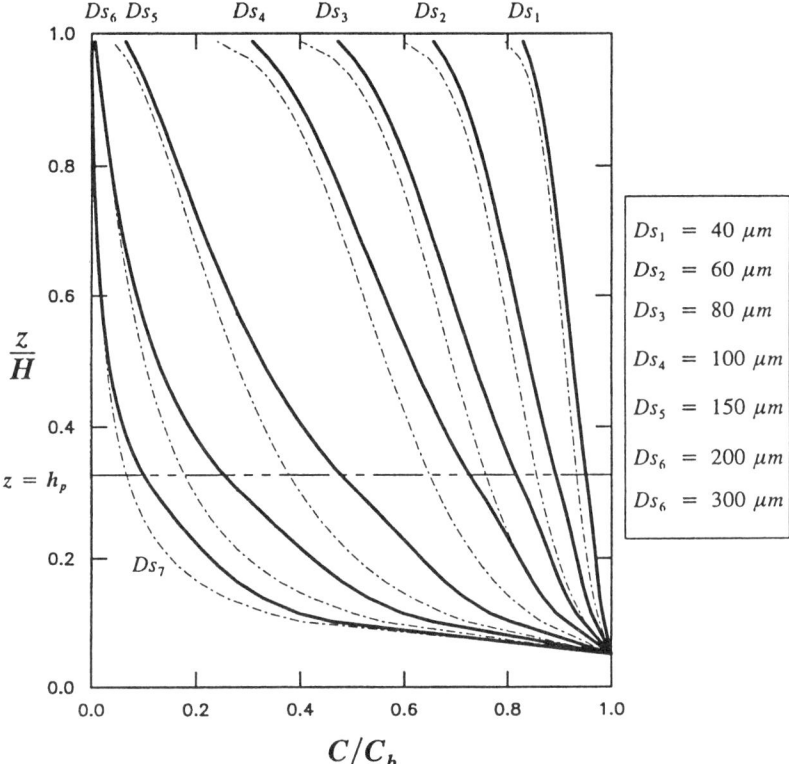

Figure 15.10 Comparison of equilibrium sediment concentration profiles in the presence of vegetation computed with a two-equation turbulence model versus those estimated with the Rousean distribution for different mean particle sizes (Lopez and Garcia 1998)

$$\lambda_m = 1 - 0.288\sigma_\phi \tag{48d}$$

Hence σ_ϕ denotes the arithmetic standard deviation of the bed surface material on the phi scale. No summation in j is implied in the above equations. Comparisons of values estimated with (48a) against observations made under equilibrium conditions in the Niobrara river (USA) indicate good agreement (Garcia and Parker 1991).

15.9 SUSPENDED SEDIMENT TRANSPORT IN VEGETATED CHANNELS

In the last decade, there has been an increasing need to understand sediment processes in wetlands, floodplains and rivers with vegetated banks, by which suspended solids and chemical

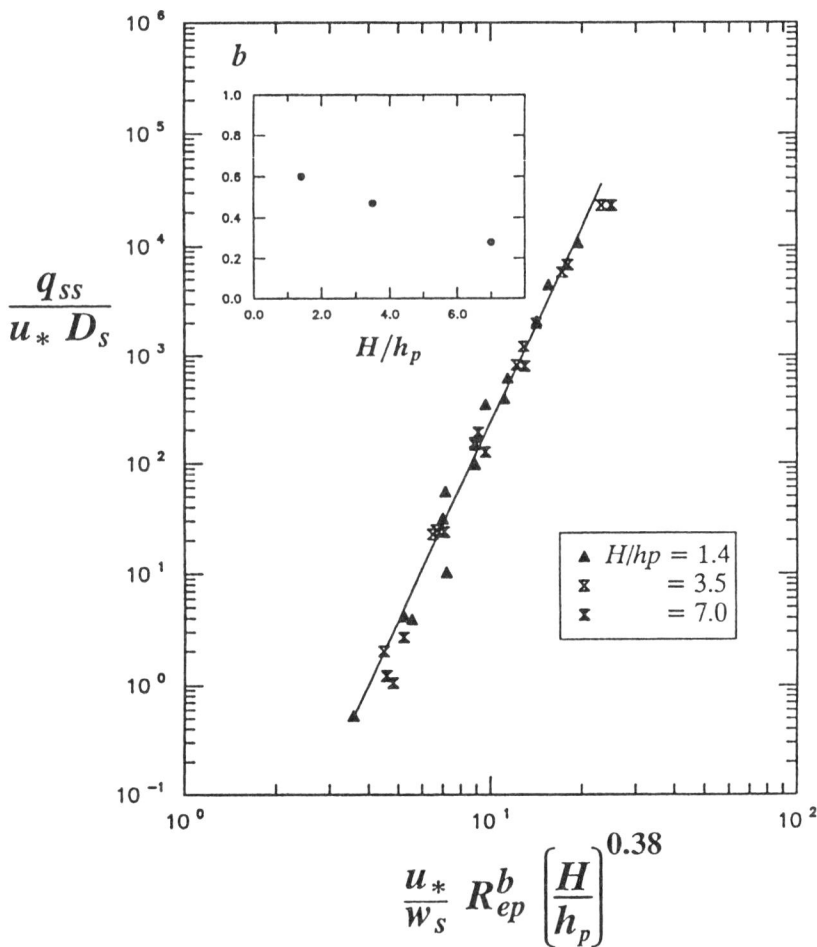

Figure 15.11 Dimensionless volumetric suspended sediment transport rates predicted with a two-equation turbulence model as a function of flow characteristics, particle size and vegetation characteristics (Lopez and Garcia 1998)

contaminants (pesticides, heavy metals etc.) are deposited and retained within a natural or man-made waterway. When properly formulated, mathematical models can be quite useful to assess the role played by vegetation in reducing both the capacity of water flows to transport sediment in suspension and the ability of turbulence to entrain sediment into suspension. Recently, a two-equation turbulence model based on the k–ε closure scheme was developed to simulate the flow and turbulence characteristics of open-channel flows through non-emergent vegetation (López and Garcia 1997). Once the performance of the model was verified, the flow structure of vegetated open channels was studied with the help of numerical experiments. Simulated rigid and flexible plants were used to validate the model. Dimensional analysis allowed identification of the dimensionless parameters that govern suspended sediment transport processes in the presence of vegetation, and thus helped in the design of numerical

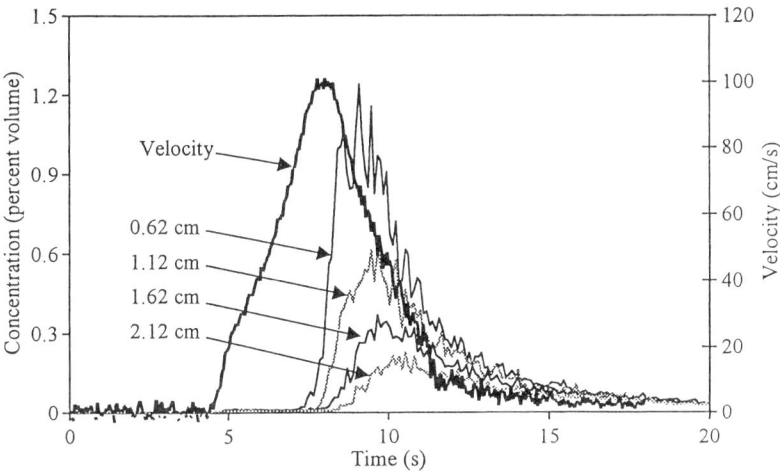

Figure 15.12 Time series of flow velocity and suspended sediment concentrations at different elevations above the sediment bed (Admiraal and Garcia 2000)

experiments to investigate the role of different flow properties, sediment characteristics, and vegetation parameters upon the transport capacity.

The two-equation turbulence model was found to accurately represent the mean flow and turbulence structure of open channels through simulated vegetation, thus providing the necessary information to estimate suspended sediment transport processes (López and Garcia 1998). A reduction of the averaged streamwise momentum transfer toward the bed (i.e. shear stress) induced by the vegetation was identified as the main reason for lower suspended sediment transport capacities in vegetated waterways compared with those observed in non-vegetated channels under similar flow conditions (López and Garcia 1998). Simulated profiles of kinematic eddy viscosity were used to solve the sediment diffusion equation (equation (36)), yielding distributions of relative sediment concentration slightly in excess of the ones predicted by the Rousean formula (Figure 15.10). A power law was found to provide a very good collapse of all the numerically generated data for suspended sediment transport rates in vegetated channels (Figure 15.11).

15.10 SEDIMENT RESUSPENSION BY UNSTEADY FLOWS

Past sedimentation research has focused primarily on steady flows. However, a number of important problems occur in unsteady flows. Examples include sediment transport by floods, boat wakes and flow surges. As shown above, there are a number of entrainment relations available for predicting the upward flux of sediment from a mobile bed, but as flow unsteadiness increases, the reliability of such relations becomes doubtful. Motivated by the need to assess the environmental impact of increasing navigation in the Upper Mississippi River Basin, a set of experiments was conducted in a custom-made rectangular duct located in the Ven Te Chow Hydrosystems Laboratory of the University of Illinois (Garcia et al. 1999). A computer-operated valve governed the velocity of the water in the duct, and the flow velocity, wall shear stress and vertical distribution of suspended sediment were simultaneously measured. An acoustic profiler was used to measure suspended sediment concentrations of two sizes on non-cohesive sediment,

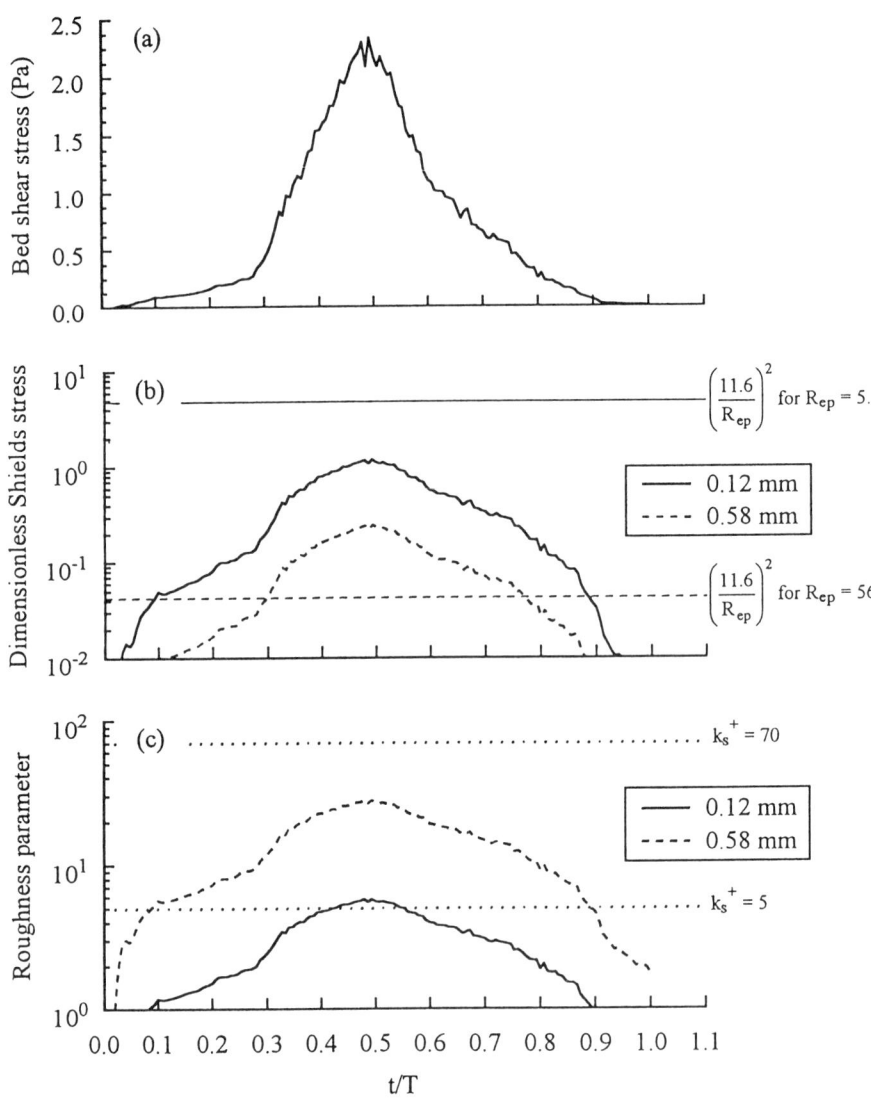

Figure 15.13 (a) Bed shear stress time series imposed on a sand bed by an unsteady flow, (b) corresponding Shields stress time series, and (c) roughness parameter time series for 120 μm and 580 μm sand beds (Admiraal and Garcia 2000)

a coarse sand and a fine sand. Time lags were observed between the bed shear stress and the upward flux (entrainment) of sand from the bed (Admiraal and Garcia 2000). This is shown for one of the tests in Figure 15.12. The phase lags were larger for tests with fine sand than for tests with coarse sand. The finer sediment seems to be more protected from the flow turbulence by the presence of a viscous sublayer, and does not respond to flow changes as readily as the coarser material that is more exposed to the flow turbulence, as depicted in Figure 15.13. The Garcia–Parker relation, equation (26a), was found to predict well sediment entrainment rates for weakly unsteady flows. However, for rapidly accelerating flows, the Garcia–Parker equation had to be corrected to account for time lags.

ACKNOWLEDGEMENTS

This is a by-product of research conducted as part of the project 'Development of an Integrated Scientific and Technological Framework for Stream Naturalization', Grant 98-NCERQA-M1, funded by the Water and Watersheds Program of the United States Environmental Protection Agency. This financial support is gratefully acknowledged. Most of the field data in Figure 15.1 were kindly provided by Gary Parker.

REFERENCES

Admiraal, D. and Garcia, M.H. 2000. Entrainment response of bed sediment to time-varying flows, *Water Resources Research*, **36**(1), 335–348.

Akiyama, J. and Fukushima, Y. 1986. Entrainment of noncohesive sediment into suspension. *3rd International Symposium on River Sedimentation*, Wang S.Y., Shen, H.W. and Ding, L.Z. (eds), University of Mississippi, pp, 804–813.

Brownlie, W.R. 1981. *Prediction of Flow Depth and Sediment Discharge in Open Channels Flows*. Report No. KH-R-34A, Keck Laboratory of Hydraulics and Water Resources, California Institute of Technology.

Celik, I. and Rodi, W. 1988. Modelling suspended sediment transport in nonequilibrium situations. *Journal of Hydraulic Engineering*, **114**(10), 1157–1191.

Einstein, H.A. 1950. The bed load function for sediment transportation in open channels. *Technical Bulletin 1026*, US Department of Agriculture, Soil Conservation Service, Washington, D.C.

Einstein, H.A. 1968. Deposition of suspended particles in a gravel bed. *Journal of the Hydraulics Division, ASCE*, **94**(HY5), 1197–1205.

Engelund, F. and Fredsoe, J. 1976. A sediment transport model for straight alluvial channels. *Nordic Hydrology*, **7**(5), pp 293–306.

Engelund, F. and Fredsoe, J. 1982. Hydraulic theory of alluvial rivers. In: Chow, V.T. (ed.), *Advances in hydroscience*. Academic Press, Inc., San Diego, CA, **13**, pp. 187–215.

Garcia, M.H. 1999. Sedimentation and erosion hydraulics. In: Mays, L. (ed.), *Hydraulic Design Handbook*. McGraw-Hill, New York, 6.1–6.113.

Garcia, M.H. 2000. Discussion of 'The legend of A.F. Shields'. *Journal of Hydraulic Engineering, ASCE*, vol. 126, N9, Sept., 718–720.

Garcia, M.H. and Parker, G. 1991. Entrainment of bed sediment into suspension. *Journal of Hydraulic Engineering*, **117**(4), 414–435.

Garcia, M.H. and Parker, G. 1993. Experiments on the entrainment of sediment into suspension by a dense bottom current. *Journal of Geophysical Research*, **98**(C3), 4793–4807.

Garcia, M.H., Admiraal, D.M. and Rodriguez, J.F. 1999. Laboratory experiments on navigation-induced bed shear stresses and sediment resuspension. *International Journal of Sediment Research*, **14**(2), 303–317.

Huang, X. and Garcia, M.H. 2000. Pollution of gravel spawning grounds by deposition of suspended sediments. *Journal of Environmental Engineering, ASCE*, Vol. 126, N10, October, 963–967.

Itakura, T. and Kishi, T. 1980. Open channel flow with suspended sediments. *Journal of Hydraulics Division, ASCE*, **106**(8), pp. 1325–1343.

Klingeman, P.C., Beschta, R.L., Komar, P.D. and Bradley, J.B. (eds) 1998. *Gravel-Bed Rivers in the Environment*. Water Resources Publications, Colorado.

López, F. and Garcia, M.H. 1997. *Open Channel Flow Through Simulated Vegetation: Turbulence Modelling and Sediment Transport*. Wetlands Research Program Technical Report, WRP-CP-10, US Army Corps of Engineers, Waterways Experiment Station, Vicksburg, MS.

López, F. and Garcia, M.H. 1998. Open-channel flow through simulated vegetation: suspended sediment transport modelling. *Water Resources Research*, **34**(9), 2341–2352.

Mei, C.C. 1998. Nonuniform diffusion of suspended sediment. *Journal of the Hydraulics Division, ASCE*, **95**(HY1), 581–584.

Nino, Y. and Garcia, M.H. 1994. Gravel saltation II: modelling. *Water Resources Research*, AGU, **30**(6), 1915–1924.

Nino, Y. and Garcia, M.H. 1998. Using Lagrangian particle saltation observations for bedload sediment transport modelling. *Hydrological Processes*, **12**, 1197–1218.

Parker, G. 1978. Self-formed straight rivers with equilibrium banks and mobile bed. Part 1. The sand-silt river. *Journal of Fluid Mechanics*, **89**(1), pp. 109–125.

Parker, G., Toro-Escobar, C. and Voigt, R. Jr 1998. *Countermeasures to Protect Bridge Piers from Scour*. NCHRP Report 24-7, vol. 2, December.

Rouse, H. 1937. Modern conceptions of the mechanics of turbulence. *Transactions, ASCE*, **102**, 463–543.

Smith, J.D. and McLean, S.R. 1977. Spatially averaged flow over a wavy surface. *Journal of Geophysical Research*, **83**, 1735–1746.

van Rijn, L.C. 1981. *The Development of Concentration Profiles in a Steady, Uniform Flow without Initial Sediment Load*. Publication No. 255, Delft Hydraulics Laboratory.

van Rijn, L.C. 1984. Sediment transport, Part II. Suspended load transport. *Journal of Hydraulic Engineering*, **110**(11), 1613–1641.

Vanoni, V.A. 1980. *Sediment Studies in the Brazilian Amazon River Basin*. United Nations Development Program, Report No. KH-P-168, Keck Laboratory of Hydraulics and Water Resources, California Institute of Technology.

Wang, Z.B. and Ribberink, J.S. 1986. The validity of a depth-integrated model for suspended sediment transport. *Journal of Hydraulic Research*, IAHR, **24**(1), 53–67.

Zyserman, J.A. and Fredsoe, J. 1994. Data analysis of bed concentration of suspended sediment. *Journal of Hydraulic Engineering*, **120**(9), pp. 1021–1042.

16 The 'Validation' of Hydrodynamic Models: Some Critical Perspectives

STUART N. LANE[1] AND KEITH S. RICHARDS[2]
[1]School of Geography, University of Leeds, UK
[2]Department of Geography, University of Cambridge, UK

16.1 PHILOSOPHICAL ISSUES

The purpose of this chapter is to present a critique of the conventional concept of 'validation' as applied to the assessment and testing of numerical models. A starting point for this is to consider the relationship between the use of the term 'validation', and certain philosophical issues underlying scientific method. In fact, two terms are often used interchangeably in numerical modelling studies: 'validation' and 'verification'. The term 'falsification', by comparison, is rarely encountered. The implications of this use of language are worth exploring. Some authors (e.g. Anderson and Woessner 1992; Flavelle 1992) conclude that 'verification' means ensuring that computer algorithms and code accurately solve the equations upon which the model is based, while 'validation' involves demonstrating that a model is a 'good', 'correct' or 'sufficient' representation of reality. The implied distinction is between internal consistency on the one hand, and external testing on the other, the former being theoretical, logical or abstract, and the latter being empirical and practical. However, this is not a consistent use of the terms. Oreskes et al. (1994), for example, define 'validation' as the process of assessing the fidelity with which a model's closed-system structure represents a natural system without containing detectable flaws, and with internal consistency (a process more theoretical than empirical). In this chapter, the aim is to explore 'conventional(ist)' empirical approaches to model 'validation', and to evaluate these against a range of alternative strategies for what might more appropriately be termed model 'assessment'. The chapter follows Howes and Anderson (1988) in noting that the creation of a numerical model involves several stages. The physics of an environmental process are converted into a mathematical representation or approximation, the mathematical formulation is represented as a set of numerical algorithms, and the algorithms are converted into computer code. At each stage, some form of assessment of the conversion is required, both in general as the model is being developed, and in particular when specific applications are being considered (for which the decisions made in model development may not be uniformly acceptable). A conventional, empirical – indeed, positivist – approach relies on comparison of model predictions with observed data for validation purposes, but this cannot resolve the source of error when the resulting fit is poor, which thus demands the use of other approaches that may be less formalised.

Regardless of how 'validation' and 'verification' are defined, it is clearly impossible to demonstrate the truth of any proposition within a system, except in a closed system where all

Model Validation: Perspectives in Hydrological Science. Edited by M.G. Anderson and P.D. Bates.
© 2001 John Wiley & Sons, Ltd.

possible manifestations of the system are known (Oreskes et al. 1994). The real world is clearly not a closed system, and we do not have perfect knowledge of it. Herein lies a basic problem. 'Validation' or 'verification' of theory (and therefore by extension, of a model based on that theory) is philosophically impossible (Popper 1968; Konikow and Bredehoeft 1992). This is because no amount of empirical testing can ever cover all possible events in time and space, and thus guarantee that the theory (or model) will perform adequately outside the range of observed conditions or events. According to Popper, a theory or model can only ever be falsified. While the theoretical impossibility of demonstrating the truth of any proposition within an open system is clear (e.g. Oreskes et al. 1994), questions arise over the practical possibility of falsification, for similar reasons. It may be the case that confidence in a model's performance and predictive capability is partly a reflection of prior falsification: the model's earlier failures have led to improvements in its representation of reality. However, this will have involved value judgements that are not objectively measurable. As Beck (1987) has cogently argued,

> given . . . field observations of the behaviour of environmental systems, . . . models . . . which have become enormously complex assemblies of very many hypotheses, cannot be effectively falsified. This is partly a function of uncertainty in the field data, certainly a function of current limitations in the method of system identification, and essentially a function, in the event of demonstrating a significant mismatch between the model and observations, of being unable to distinguish which among the multitude of hypotheses have been falsified.
>
> (p. 1396)

The level of complexity in models of environmental systems is such that falsification is: (i) dependent upon what criteria we set as necessary for falsification; (ii) inevitable, given the complexity of the real world that is being modelled; and (iii) of no real use unless it can inform the modeller of exactly why the model is failing. If validation/verification in these conventional terms is philosophically impossible, and falsification practically impossible, where does this leave those activities that are used to increase confidence in model predictions, and which are currently subsumed under the validation, verification or falsification headings? This question has to be asked as, in large measure, the practical purpose behind creating models of environmental processes is to enable extrapolation across space and/or time, and to understand or predict situations where no measurements are available. To be able to do this, it is necessary to assess the extent to which the model is likely to be representative of reality in those new spaces and times. Validation/verification/falsification, regardless of whether or not they are possible, are necessary activities in order to allow models to be used in practical applications.

Given that the question has to be asked, there are various answers. One is that the activities conventionally associated with validation must be undertaken, but that they can only demonstrate a limited degree of validity (Oreskes et al. 1994), which is likely in due course to be shown to be illusory. In this context, model 'validation', in which predictions are checked against field data, is really falsification. The model moves from being shown to be valid for certain sets of conditions, to being invalid and needing amendment as a result of further testing. A different answer, from the sociology of science, might be that mere use of the term 'validation' is a linguistic device to disguise what is for long periods the uncertain status of models. When a model can be said to have been subject to 'validation', there is an implication of truthfulness that is reassuring to those managers and policy-makers who use the predictions of models. This is an example of the hermeneutics of science, as analysed by Markus (1987) in his deconstruction of the legitimation devices behind the structure of scientific papers (e.g. 'materials and methods' sections that commonly fail to provide the basis for the repeatability that justifies their existence).

This critical view of 'validation' leads to a conclusion that 'assessment' is a preferable term for the inclusive range of testing procedures required to develop confidence in the use of a particular model. A positivist mode of testing, based on prediction and comparison with data, is itself subject to considerable uncertainty. It must therefore be supplemented by other procedures. Some of these may seem less rigorous, but only because complex models do not allow the simple verification or falsification process that might be viable in a closed-system form of scientific activity: i.e. more formal approaches are precluded by the nature of what is being modelled. Indeed, the 'softer' forms of theory testing that characterise the social sciences, and that are used in combination, have many parallels in model assessment. These include structural corroboration, referential adequacy, multiplicative replication, and searching for negative evidence (Phillips 1992). This reliance on a combination of different means of model assessment will be considered below, following a review of conventional(ist), data-dependent validation methods. Thus, this chapter seeks to argue and illustrate that evidence of successful prediction in observed spaces and times (conventional validation) cannot provide a sufficient basis for use of a model beyond the set of situations for which the model has been empirically tested. A range of other assessment strategies must be invoked.

16.2 ASSESSMENT BASED ON EMPIRICAL DATA: THE CONVENTIONAL APPROACH

The conventional, positivist approach to model assessment is to run the model to make predictions which are then compared with observational data. It is likely that this use of independent check data will continue to provide a necessary part of a model assessment strategy, as it requires modellers to ask critical questions of their models. If a model has been developed incorrectly with respect to its mathematical representation, numerical solution, boundary condition specification or parameterisation (i.e. it provides an incorrect representation of the system upon which it is based), then this may be reflected in divergence between model predictions and independent observations. Even if independent data are not available, the modeller may implicitly compare model predictions with some expectation. For example, a model may converge on theoretically unrealistic values, and this alone may indicate that the model is invalid. However, comparison with independent check data only provides a general test of validity, but does not help us to understand *why* a model is invalid. On its own, therefore, a predict-and-compare test is an insufficient assessment of validity. When a model *fails* to predict independent data adequately, *something* must be wrong (Luis and McLaughlin 1992). The converse is, unfortunately, not true. When model predictions are correct, the model is not necessarily valid, as it is possible for an invalid model to provide an adequate representation of some aspects of reality. Beven (1989), for example, emphasises that hydrological models may produce the 'right results' for the 'wrong reasons'. A particular problem with the assessment of modern models is that they are capable of generating such richly detailed predictions. Many supply four-dimensional prediction data of very high space–time resolution. The implication is that no measurement strategy can hope to equal the level of system description achieved by the model predictions. As a result, comparison of predictions with observations demands novel field measurement methods and the rigorous control of field sampling. This section assesses some of the main problems in undertaking a comparison of predicted and observed data.

A first issue concerns the relationship between measurement in the field, and the time–space resolution of model predictions. At least part of the deviation between observed and expected values will reflect the fact that measurement procedures will probably not match the scale of discretisation of the model, and the representation effected by the predicted data. There are then several questions relating to the definition of the adequacy of fit, how to measure it, and how to judge the accuracy and precision with which the model represents reality (assuming that reality is somehow embodied in the observed data). The richness of model output data implies that models can generate either or both distributed information about the simulated 'internal' behaviour of the model system or time series of single variables at the outlet of the model's spatial domain. For example, a hydrological model may generate maps of soil moisture as well as overall catchment discharge. Thus, there is an issue concerning the combined use of these two types of output data in the assessment of a model's performance. Finally, a model will often be capable of predicting different kinds of (multivariate) data, and the assessment process can therefore judge the model on its capability of providing adequate simultaneous descriptions of these multivariate outputs: discharge and water quality determinands, for example.

16.2.1   Modelling and Measurement Scales

A particular problem that arises when using empirical data to test a model reflects the difficulty of representing a natural open system using sampled measurements. Such data are in essence being viewed as representations of 'reality'. However, the assumptions made during design of measurement apparatus (e.g. Hacking 1983), during research design (e.g. sampling), and during data reduction and analysis mean that this representation is highly approximate. The 'reality' against which a model prediction is compared can be incomplete at best and seriously blurred at worst. As an example, Lane et al. (1999) demonstrate how some of the poor correspondence between model predictions and field observations of two-dimensional patterns of depth-averaged velocity can be attributed to problems of accurately orienting an electromagnetic current meter into a local coordinate system. Thus, there is no fundamental reason to suppose that the empirical, observed data describe or represent 'reality' with any greater accuracy or precision than the model's predictions. Chorley's observation (1978) that whenever the word 'theory' is mentioned to a geomorphologist, (s)he instinctively reaches for a soil auger, reflects a mind-set that has greater belief and confidence in measurement, but which lacks a firm philosophical foundation for this.

Model predictions are averaged in space–time for discrete areas or volumes (cell-centred or face-centred), over a specified time period. Empirical data may be measured over a different scale from that at which predictions are made by a model (Luis and McLaughlin 1992). Measurements, except those derived from remote sensing, are commonly made at a point. In many hydrological and hydraulic applications, the measurement area or volume is smaller than that of model discretisation scale. Thus, there is an immediate discrepancy in the comparison unless control is exercised in the field measurement process in order to ensure appropriate commonality of data scales. A typical example is that of comparing field-measured flow velocities with those derived using a depth-averaged flow model (such as that discussed below in relation to Table 16.2). Such a model predicts depth-averaged velocities. Thus, depth-averaged velocities will need to be derived from point velocities measured in the field. This will require: (i) an assumption about the vertical velocity profile (it may be assumed to be semi-logarithmic, which is likely also to be the assumption embodied in the flow model); (ii) sampling of point velocities at appropriate height intervals above the bed; (iii) measurements of velocities at those heights for appropriate sampling periods; and (iv) generalisation of these field data using regression analysis. Correspondence between field and model predictions will be limited when departures occur from the initial assumptions during the field data collection, as well as because of errors in the model.

However, the problem is more fundamental than this. Observations always contain information that reflects the behaviour of the natural, open system within which they are made. The model will contain information that reflects decisions by the modeller about model closure during model encoding. Thus, there will inevitably be differences between model predictions and observations as a result of model simplification. In the case of three-dimensional modelling of flow in river confluences, Lane et al. (1999) demonstrated that difficulties of model validation using independent field data arose because the field measurements contain variance associated with the complex, spatial variation in grain position and orientation on the river bed. The model did not seek to predict complex three-dimensional flow structures over the details of a gravel-bed river topography. Indeed, the technology for determination of the necessary topographic information is still not available for rivers in which the suspended sediment load is high. The model's topographic boundary condition was therefore smoothed, and the small-scale roughness was parameterised using a roughness coefficient. However, the empirical data used to assess the model's ability to predict flow structures in a junction zone were 'contaminated' by the *actual* roughness effects, and accordingly there remain fundamental limits to the ability of the data to provide an adequate basis for model validation. At the very least, this means that considerable scatter should be expected in *any* plot of model predictions against observations (e.g. Figure 16.1), and that there may be limits on the extent to which this scatter can be reduced: a certain level of imprecision is inevitable (see below). Some parameters, notably those of higher order (e.g. estimates of turbulent kinetic energy), will show systematic error (i.e. inaccuracies, see below) as the generating processes (e.g. the effects of topographic irregularity upon turbulence) cannot be properly represented in the model. What this implies is that model validation must involve more than simply comparing model predictions with independent empirical data as a model is normally closed to create a system that is fundamentally different to that which has been measured.

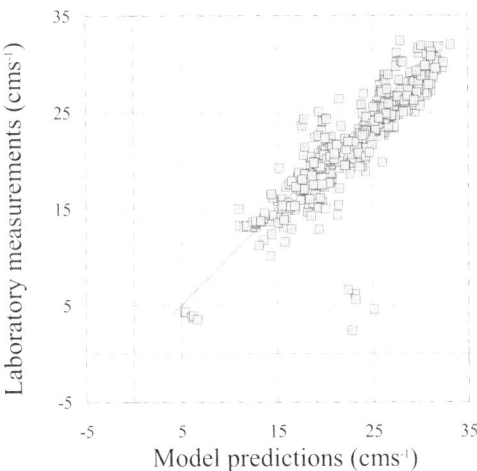

Figure 16.1   An example comparison of observations with model predictions (from Bradbrook et al. 2001) for a confluence of rectangular tributaries, with no bed discordance, and one tributary angled at 30° to the post-confluence flow direction (see Figure 16.3b for geometry). Laboratory measurements are based upon results reported in Biron et al. (1996)

16.2.2 Adequacy of Fit

As noted above, it is necessary to take care in designing a sampling scheme in order to compare a richly detailed set of model predictions with the typically sparse data available from field measurement. Poor choice and sampling of observed data can result in apparently adequate prediction from an invalid model. Thus, in terms of formal comparison with data, a key issue is what constitutes an adequate fit. As Oreskes and Belitz (this volume, chapter 3) note, the model validation literature employs descriptors such as 'adequate' that are subjective and often poorly defined. In addition to poorly defined terms, there are poorly defined methods, particularly in terms of the validation of hydraulic models. For instance, comparison of model predictions with laboratory or field measurements commonly involves plotting the measured and predicted variables on the same axis against a parameter like position or time. Examples of this involve comparison of predicted and observed velocity profiles (e.g. Czernuszenko and Rylov 2000; Meselhe and Sotiropolous 2000; Sanjiv and Marelius 2000; Sofialidis and Prinos 2000) in hydraulic modelling or predicted and observed discharge hydrographs in hydrological modelling. Such comparisons are useful for a qualitative assessment of model performance, and for ready identification of those locations in time or space where the model is not performing well. However, given that the number of observation locations (in space or time) is commonly less than the number of locations for which predictions are available, formal indicators based upon principles of statistical inference must be a crucial part of determining adequacy of fit.

Comparing model predictions with independent observations requires consideration of (following Flavelle 1992): (i) the extent to which variation in the observations is explained by the predictions, which can vary from 0% (imprecise or poor fit) to 100% (precise or good fit); (ii) the extent to which predictions agree with observations, which can vary from perfect equality (accurate or unbiased) to perfect inequality (inaccurate or biased); and (iii) the extent to which the predictions provide sufficiently reliable information for them to be accepted when there are no check data (e.g. at non-measured locations within the model or in situations where boundary conditions or topographic parameters are different to those used when the check data were collected). In addition, it is common for a model to predict the variance of a process well, but to appear to perform badly in a conventional statistical assessment because the timing of events is slightly in error. This is a particular characteristic of hydrograph prediction models. Furthermore, there may be an issue of parsimony, so that an index of model performance is required which makes allowance for the model complexity as represented by the number of adjustable parameters it contains: a model 'efficiency' index.

The consequence of these different aspects of the question of adequacy of fit is that the measurement of adequacy, or more precisely, of goodness of fit, is necessarily a multivariate problem, and no single statistic can be relied upon. In hydrological modelling, the multivariate nature of the problem has been recognised for some time and Table 16.1 summarises a sample of the goodness of fit tests that may be employed for assessment of a hydrological model. The first such objective function (OF_1) is useful if a bias towards the prediction of lower values (lower discharges) is preferred. The second function is the conventional measure of variance accounted for, or explained by the model and is familiar as the coefficient of determination. A normalised sum of squares (Ibbit and O'Donnell 1971) may usefully be employed when there is a need to compare the goodness of fit of simulations of differing length (OF_3). The logarithmic transformation of the sum of squares of errors is a useful statistic to employ when errors are non-Gaussian or there is heteroscedasticy in the errors. Finally, the model efficiency (Nash and Sutcliffe 1970) is a useful means of summarising the success of a model, as it results in negative values in cases where the sum of squared model residuals exceeds variance of the observed time series. In all cases it is important to note that, if the parameters in the model have been determined through

Table 16.1   A variety of objective functions that may be used in the quantitative evaluation of the goodness of fit of a model

Relative mean absolute error:	$$OF_1 = \frac{1}{n} \sum_{i=1}^{n} \left| \frac{y_{obs_i} - y_{pred_i}}{y_{obs_i}} \right|$$
Variance accounted for by the model:	$$OF_2 = \frac{\sum_{i=1}^{n} (y_{obs_i} - y_{pred_i})^2}{\sum_{i=1}^{n} (y_{obs_i} - \bar{y})^2}$$
Normalised sum of squares of errors:	$$OF_3 = \frac{(nF)^{1/2}}{\sum_{i=1}^{n} y_{obs_i}}, \text{ where } F = \sum_{i=1}^{n} (y_{obs_i} - y_{pred_i})$$
Log-transformed sum of squares of errors:	$$OF_4 = \sum_{i=1}^{n} [\log(y_{obs_i}) - \log(y_{pred_i})]^2$$
Model efficiency:	$$OF_5 = 100 \left| \frac{(Fo - F)}{Fo} \right|, \text{ where } Fo = \sum_{i=1}^{n} (y_{obs_i} - \bar{y})^2$$

an optimisation procedure in which one or more of these objective functions are minimised or maximised, they cannot then be used as a test statistic in a validation exercise. What this example shows is that there are a range of possible indicators of model performance (as well as for model optimisation) whose relevance depends not just on the model that is being used, but the use to which the model will be put. It is unlikely that clear and generic rules as to which function is most appropriate will emerge independently from a particular application, and different functions may result in different conclusions about model validity. What is critical, therefore, is careful application of a number of tests, and careful justification of the reasons for choosing those tests. This is rarely seen in the evaluation of either hydrological or hydraulic models.

16.2.3   Precision and Accuracy in Testing Model Predictions

Assessments of both accuracy and precision are important for deciding whether or not a model may be used for predictive purposes. Since perfect representation of all processes is not possible, it is necessary to ascertain whether or not the model provides information of sufficient reliability for its predictions to be accepted in the absence of independent test data. This is a challenging part of 'validation', as it is necessary to determine whether or not both the precision and accuracy of the model are acceptable *and* whether or not the model is reliable in situations where boundary conditions or topographic detail, and even processes, differ. This requires the use of an appropriate selection of the summary statistics outlined in the previous section, coupled with checks that the slope and intercept of a regression line of predicted on observed values are not statistically distinguishable from, respectively, the line of equality (a slope of unity), and a value of zero. Initially, this seems straightforward, involving standard *t*-statistics to evaluate the observed slope and intercept against the null hypothesis values of unity for slope and zero for intercept. Unfortunately, this proves problematic. Figure 16.2 shows two hypothetical assessments of model performance. Both Figures 16.2a and 16.2b have the same slope between model observations and predictions. Figure 16.2a has a more precise level of agreement, and this is reflected in a lower standard error ($\hat{S}b1$) than in Figure 16.2b ($\hat{S}b2$). A comparison of slopes is based upon:

$$t = \frac{b-1}{\hat{S}bi} \tag{1}$$

Equation (1) shows that as a model becomes *more* precise (Figure 16.2a), the value of t rises, and hence it becomes more possible to be more confident that the slope of the relationship is significantly different from the line of equality (that is, the relationship appears to be inaccurate). Thus, at first appearance, it is easier to demonstrate that a more imprecise model (Figure 16.2b) is accurate than to demonstrate that a more precise model (Figure 16.2a) is accurate.

This inverse relationship between the demonstration of precision and accuracy in a regression comparison appears to demonstrate an effect rather similar to Heisenberg's uncertainty principle; the more one criterion appears valid, the more difficult it is to demonstrate validity of the other. However, this situation actually arises from a misunderstanding of the hypotheses that the test upon slopes is addressing. The normal null hypothesis is that the slope is not significantly different from unity. We aim to reject the null hypothesis, and to demonstrate that the model is inaccurate. Hence, tests on the slope (and intercept) of the regression relationship are not to demonstrate that the model is accurate, but to check that the model is not inaccurate. With greater precision (i.e. more explained variance), it is easier to demonstrate that a model is giving us inaccurate predictions, and that there is bias in our model. As Flavelle (1992) notes, tests on the slope of a regression relationship only allow bias to be shown, and not lack of bias. Thus, while tests upon the slope of a relationship can be used to identify an inaccurate model, they cannot be used to demonstrate model accuracy. Returning to the question of sufficiency in a model, this demonstrates that statistics cannot help us to identify what is an acceptable level of accuracy. As some have argued (e.g. Flavelle 1992; Jackson et al. 1992), this must remain a subjective decision.

16.2.4   Model-defined Limits to Precision

It is unlikely that any model will ever contain a full representation, across all spatial and temporal scales, of all processes responsible for observed behaviour at a particular point. Thus,

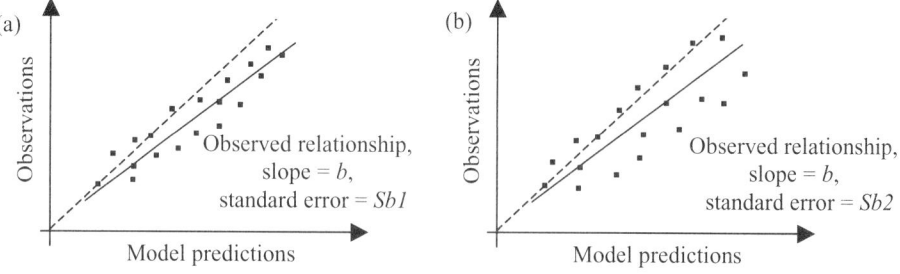

Figure 16.2   Two hypothetical comparisons of model predictions and observations where the comparisons have the same slope (*b*) but different standard errors ($\hat{S}bi$): $\hat{S}b1 < \hat{S}b2$

there will always be some variance unexplained by a model. The larger the amount of unexplained variance, the more imprecise model predictions will be. This is distinct from model accuracy, as an inaccurate model may be quite precise, and an imprecise model may be quite accurate. Lane et al. (1999) illustrate this in a comparison of two-dimensional and three-dimensional models of flow processes in confluences. Table 16.2 shows that in terms of precision, the two-dimensional model performs as well as, if not better than the three-dimensional model: it explains more of the variance of the field observations. However, difference in levels of agreement (the slope parameter in a comparison of predicted and observed values) suggests that the two-dimensional model predictions are less accurate, with the two-dimensional model systematically under-estimating the higher velocities and over-estimating lower velocities. The three-dimensional model (more accurate, less precise) provides a better representation of the general patterns of those velocities than the two-dimensional model (less accurate, more precise). One issue that therefore arises is that of interaction between precision, accuracy and model dimensionality. Lower dimensionality models may be easier to calibrate (and hence to make more precise, or to achieve what Oreskes et al. (1994) call 'forced empirical adequacy'), both because there are fewer parameters, and because sensitivity to those parameters is greater. As model dimensionality is increased, parameter sensitivity may be reduced (e.g. Lane et al. 1999; Bradbrook et al. 2000) making parameterisation less effective, while the difficulty of supplying more detailed boundary conditions and topographic data inevitably reduces model precision. Thus, possible levels of precision are model dependent. The more difficult question is how this relates to model accuracy, and this may be the ultimate determinant of appropriate model scale. If increasing model dimensionality does not increase model accuracy with respect to the predictions that a model makes, then additional dimensionality is unnecessary. This issue may merit further evaluation as part of a notion of multi-criteria model assessment.

16.2.5   Assessment Using both Model System Output and Internal Distributed Predictions

Aside from the above discussion, more general problems emerge in terms of establishing validity using empirical test data, primarily for the reason of equifinality. Oreskes and Belitz (this volume, chapter 3) refer to the common problem that different models, or the same model with different combinations of parameter values, can perform equally well with respect to empirical data (cf. Table 16.2). This has been defined in hydrology as a form of model equifinality (e.g. Beven 1989, 1996), raising the possibility that a model may produce the 'right' results, but for the 'wrong' reasons (Beven 1989): a model must not only predict well, but it should do so for the right reasons. This issue is partly dependent upon the choice of test data. In depth-averaged modelling of floodplain inundation, model evaluation using an outflow catchment hydrograph

Table 16.2 Comparison of model predictions with field observations for a three-dimensional model and a two-dimensional depth-averaged model. Each model was run with identical topographic conditions and boundary conditions where possible. Full details are provided in Lane et al. (1999)

Number of observations	X-direction		Y-direction	
	r^2	Slope	r^2	Slope
3-D model predictions, depth-averaged	0.728	1.041	0.413	0.840
2-D model predictions	0.810	0.735	0.418	0.631

is more likely to produce problems of equifinality than evaluation of model predictions using distributed patterns of floodplain inundation (e.g. Horritt 2000). Analysis in the aggregate, or using averaged quantities, increases the possibility that errors cancel one another. Unfortunately, such spatially distributed data are not always available (e.g. Beven and Binley 1992), although developments in technology (e.g. remote sensing) are allowing spatial mapping of key hydrological (e.g. Horritt and Bates 2000) and hydraulic (e.g. Lane et al. 1998) variables, and hence identification of those locations *within a* model domain where predictions are invalid. For example, Bates et al. (1997) used Landsat TM data to evaluate the spatial patterns of inundation predicted by a two-dimensional finite element model for floodplain flow. This research demonstrated both the real potential for using satellite imagery to assess distributed patterns of floodplain inundation, but also a set of research issues concerned with the integration of model predictions on grids of different geometries and resolutions. At a smaller scale, Lane et al. (1995) demonstrated that a depth-averaged model produced poor predictions in those zones of a braided river where the flow was strongly three-dimensional, even though the model attempted to correct for the effects of depth-averaging upon model predictions. Subsequent research developed a three-dimensional model for those zones, which was shown to result in a more accurate representation of the flow field (Lane et al. 1999). This is the sense in which modelling is an evolutionary activity, in which problems of establishing model validity lead to new ways of viewing and treating the same problem. It also demonstrates how continual iteration between independent data, whether generated in the laboratory or the field, can help the modeller assess the validity of assumptions made during the modelling process. A key criterion is the availability of distributed information.

16.2.6   The Problem of Matching Boundary Conditions Between Model and Prototype

There is an additional set of problems related to differential control of the boundary conditions in the model and the field (or laboratory). The experimental design of laboratory studies of confluences can be difficult, due to the need to provide two combining flows. Best and Roy (1991) and Bradbrook et al. (1998) studied parallel channels with a splitter plate. McLelland et al. (1996) studied a parallel channel, but with a plan angle in each of the steps at the confluence entrance, to simulate the commonly observed morphology associated with avalanche faces when viewed in plan. Biron et al. (1996) were able to create an angled tributary by diverting water upstream and reintroducing it at 30° to the main flow direction further downstream (Figure 16.3b). A number of problems arise. First, the parallel studies are easy to design in the laboratory and easy to represent in numerical meshes. However, most natural channels have angled tributaries, and this has been observed to have important effects upon flow separation (e.g. Best 1987; Biron et al. 1996), and so upon both time-averaged hydrodynamics (e.g. water surface gradients) and turbulence, and hence the flow field within the confluence. Thus, even though Bradbrook et al. (1998) were able to validate a model using information from a parallel channel confluence, significant uncertainties remained over the extent to which the parallel case validates other confluence configurations. For this reason, Bradbrook et al. (2000a) used the data of Biron et al. (1996). This confirmed that the same level of validation was not obtained in the angled confluence case, due to problems with process representation, grid design and boundary condition specification.

However, an additional problem emerged in the Biron et al. (1996) comparison. Specification of the upstream inflow condition was originally based upon the assumption that the tributary entering the main channel was straight (Figure 16.3a). However, the Biron et al. experimental design was based upon siphoning off the tributary flow, running it parallel to the main flow and

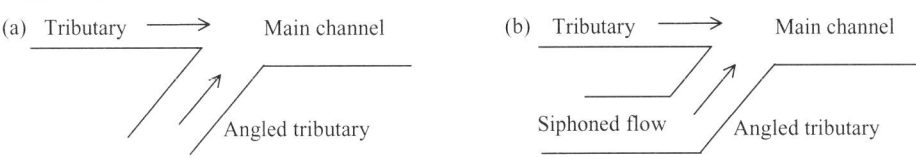

Figure 16.3   The experimental set-up originally modelled numerically (a) and that used in the Biron et al. (1996) laboratory experiments

then curving the tributary flow into the confluence with the main flow (Figure 16.3b). Thus, it contained secondary circulation associated with streamline curvature, which had a major effect upon flow in the confluence zone (Lane et al. 1999). What this shows is that, just as a model involves simplifying assumptions, so experimental observations reflect the ways in which the experiment is designed, constructed and controlled. Thus, even if the experimental data are 'correct', they may still be viewed as 'incorrect' in the terms of the validation exercise for which they are required. A problem clearly arises, then, if these test data (incorrect in the terms of the model validation exercise in progress), are used to reject a model and to accept the alternate hypothesis that the model provides an incorrect representation of reality. What this suggests is the need to explore very carefully the *validity* of the *data* being used for model *assessment*. Ababou et al. (1992) argue that careful thought should be given to the scope of experiments used for model validation, with more trust being placed in those with a broader range of coverage of controlling parameters, such that the generality of the experimental data set is increased.

16.2.7   Multivariate Data Assessment Procedures

Models are often highly complex structures embodying a number of sub-models which cascade their predictions from one to another, and therefore result in progressive multiplication of prediction error. In addition, model parameter values are uncertain, and have wide error bands. This implies that the 'true' parameter value is only known within a certain confidence band, and parameter estimation is stochastic; the value may be considered to be a sample from a probability distribution. Given these sources of error and uncertainty, overall model predictions are generally likely to be highly uncertain. This has been shown through the use of Monte Carlo methods, in which multiple simulations (realisations) are generated for combinations of parameter values drawn from specified parameter probability distributions. A particular example of this approach is that of the 'GLUE' methodology (Generalised Likelihood Uncertainty Estimation), developed by Beven (1989) and Beven and Binley (1992). This analysis shows that uncertainty bounds on model predictions may be extremely wide, implying that little confidence can be placed in model output when the uncertainties embodied within the model are explicitly accounted. The confidence intervals defined by the GLUE procedures are also rarely Gaussian in character, implying that they are difficult to determine without a case-by-case GLUE analysis.

This seems like a counsel of despair, but a key point is that as more information becomes available, the simulated likelihoods can be updated a posteriori. This means that additional data may be used to reduce the uncertainty, by changing the weights given to individual realisations according to their ability to predict the additional phenomena. Thus some realisations of a discharge prediction model may be rejected because they predict unreasonable patterns of shallow groundwater levels, and the removal of these from the complete set of realisations then narrows the confidence band for the predicted time series of discharge. Another example is that hydrochemical or suspended sediment concentrations may be employed to evaluate the likelihood of discharge predictions, to narrow the range of uncertainty. These are illustrations of a more general issue, namely that models abstract a specific part of reality for a particular purpose, although the reality itself involves continuous, interrelated processes. These related processes generate outcomes whose variability provides information on the phenomenon which is the immediate issue for concern. Thus, multivariate information or data can provide considerable assistance in reducing uncertainty in model prediction; this is equivalent to the notion of 'structural corroboration' identified by Phillips (1992) as a tool for evaluating theory. Richards (1996) illustrates this in the context of a case study approach to the modelling of glacier hydrology, where hydrochemical data assist in the interpretation of the validity of discharge predictions from a distributed, physically based meltwater production model.

16.2.8   Conclusion

It seems that a fundamental assumption underlying conventional, positivist, 'validation' of models against empirical data is that, somehow, such data are themselves 'better than the models' as representations or descriptors of some assumed 'reality'. However, if there is one thing that sophisticated four-dimensional model predictions tells us, it is that there is no warrant for this assumption. Empirical data are generally very poor at representing the spatio-temporal dynamics of natural systems, and are therefore a weak basis for effective 'validation' of modern hydrodynamic models. Thus, in place of a conventional positivist validation is a post-positivist model assessment process in which a wide range of different criteria are brought to bear on the question of the capability of models to capture particular process dynamics in a realistic way. This is post-positivist, because it replaces the assumed primacy of quantitative, empirical data with a series of more or less qualitative and theoretical evaluations (Brown 1996). The following section considers the forms taken by some of these evaluations.

16.3   ALTERNATIVE ASPECTS OF MODEL ASSESSMENT

Both Sargent (1982) and Howes and Anderson (1988) have emphasised that model assessment is a more structured and continuous set of procedures than simply an empirical validation against observed data. The complexity of hydrological and hydraulic models means that a wide range of additional approaches to model 'validation' may be required, not just as optional alternatives, but as necessary prerequisites and continual tests (e.g. ASME 1993). Empirical evaluation is an operational procedure, but there are other elements of an overall model assessment which precede any data-driven 'validation'. However, as Figure 16.4 implies, there is a continual cycling through the phases of assessment as models are refined and developed. This section reviews several aspects of these alternative assessment strategies, including: (i) conceptual model assessment; (ii) assessment computational tests and analytical solutions; (iii) assessment using sensitivity analysis; (iv) assessment based on visualisation; and (v) evaluation against professional standards.

16.3.1   From Physical Principles to Mathematical Representation: Conceptual Model Validation

It is well established that it is critical to undertake conceptual model validation (e.g. Usunoff et al. 1992). For a model to have any chance of representing reality, it must include relevant processes within an adequate structure (Usunoff et al. 1992). Concurrently, to make model

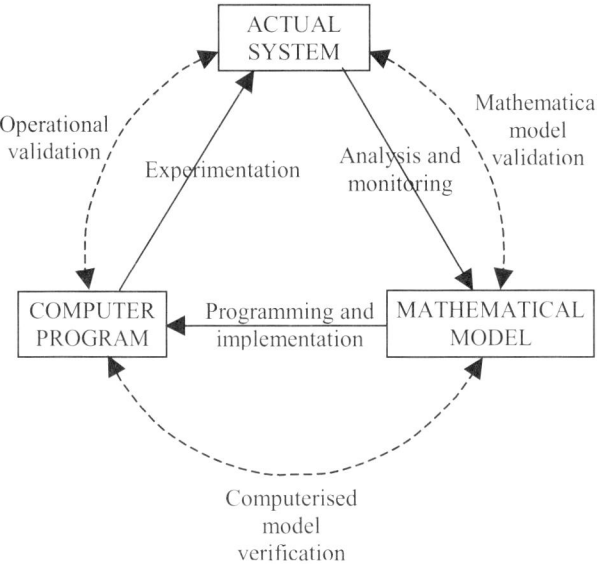

Figure 16.4   A three-stage model evaluation strategy (from Howes and Anderson 1988)

solution possible, both with reference to available boundary conditions and topographic information, and numerical solution issues, it is common to introduce simplifying assumptions, which will also introduce parameters that may or may not have a clear physical meaning. The validity of these assumptions needs to be addressed, both spatially and temporally. As Usunoff et al. note, this makes conceptualisation an especially difficult issue, as much of it depends on the modeller's view of the system under study. Indeed, Luis and McLaughlin (1992) conclude that the assumption that a model structure is correct is convenient but not necessarily justified in most practical applications.

Central to conceptual model evaluation are the assumptions that have been made to transform basic physical principles into a set of mathematical rules for encoding into the model. This process commonly involves: (i) assumptions regarding the spatial dimensionality of the problem; (ii) assumptions regarding the temporal scale of the problem; and (iii) assumptions regarding necessary processes for inclusion in the model. These three categories are commonly interlinked. For instance, in hydraulic modelling, it is common to time-average the governing equations for computational reasons, which means that a model's process inclusion must change, as the effects of turbulence upon mean flow parameters must be modelled. Similarly, the decision to depth-average requires that attention is given to the resultant dispersion terms associated with secondary circulation. Assumptions about (iii) in turn determine the spatial and temporal scales of model enquiry.

An example of conceptual model validation is provided in Lane et al. (1995). This involved an application of a steady-state, depth-averaged rigid-lid model to a divided channel. Five issues of process representation needed consideration (Table 16.3): (i) the acceptability of the assumption of a steady state; (ii) the appropriateness of a rigid-lid treatment; (iii) the suitability of the turbulence closure model; (iv) the acceptability of depth-averaging in general, and also in the context of the corrections made for secondary circulation; and (v) the appropriateness of the bottom shear stress calculation. Table 16.3 illustrates a number of characteristics of conceptual

Table 16.3 An example of conceptual model validation, based upon Lane et al. (1995)

Assumption	Nature of assumption	Evaluation
Steady state (a): Implicit	Implicit to the Reynolds averaging procedure in which the effects of instantaneous velocity fluctuations on momentum transfer and therefore average flow parameters are modelled, although this does not provide information on the velocity fluctuations themselves	These fluctuations represent continual local fluid accelerations and decelerations which may be critical for sediment transport, implying that the assumption of a steady-state in this sense is not appropriate. There was some justification with the intended scale of application of this model to reach-scale dynamics. At this scale, turbulence processes associated with the erosion and deposition of individual particles may cancel each other both across space and through time, particularly when sediment transport rates are high. In practice, there was little alternative, due to computational constraints
Steady state (b): Explicit	A quasi steady-state is assumed in model application; during the time period for which model application is made, the discharge is assumed to be constant. The length of this time period is defined by the rate of change of discharge	This assumption is widely used in modelling studies of natural channels: the approximation of an unsteady flow system by a series of short-term steady flows
Rigid lid	$\nabla \cdot (hu) = 0$ for mass continuity $u \cdot \overline{\nabla u} = -\rho^{-1}\nabla p_s + \underline{f}$ for momentum, i.e. the lid is specified as a horizontal fixed plane, gravity term drops out, and pressure term introduced to momentum equation to account for deviation from a horizontal lid	Limits the extent to which the model can be used for free-surface flows because the computed pressure will be equivalent to the displacement of the free surface, but will then, in part, drive the flow without the surface being displaced. Bernard (1992) suggests that this can be assessed by reference to the Froude number (Fr). This relates the inertia of a unit mass of streamflow to the speed of a shallow gravity wave. For Fr > 1, displacements of the free surface are large and can result in substantial energy loss and resistance to flow. Bernard (1992) notes that a rigid-lid approximation is acceptable wherever the Fr < 0.5, and in the absence of major variations in width and depth, at Fr numbers as large as 0.7. This assumption is not suitable at higher Fr which would imply hydraulic jumps or moving surface waves. Leopold et al. (1960) note that mean Fr rarely exceeds 0.5 in natural, mobile-bed channels. However, the extent to which this is true at all locations *within* a channel needs assessment with reference to specific model applications
Turbulence model	To model the effects of turbulence upon mean flow properties, the model used a two-equation k-ε model based upon Launder and Spalding (1972)	Improvement on other numerical modelling studies which make use of a zero equation turbulence model for the study of natural river channels, but there are still limits on the effectiveness of a two-equation model (e.g. Younis 1992). It is likely that turbulence has

		an important effect on the transport of momentum and therefore mean flow properties. More sophisticated models are more computationally demanding and have only been applied to extremely simple problems in rectangular ducts and channels (e.g. Malin and Younis 1990). For these reasons, this type of model was not used in this study
Depth-averaging	Three-dimensional, Reynolds-averaged flow equations are integrated through the vertical to produce depth-averaged equations	Still a widespread use of one-dimensional models for the study of natural channels. Acceptability of depth-averaging needs to be considered with respect to the relative importance of these vertical motions. Rodi et al. (1981) note that even in the presence of three-dimensional effects, in many rivers the depth-averaging method may be sufficiently accurate for practical purposes, particularly if the width-to-depth ratio is high and changes in bed elevation are not rapid (Bhallamudi and Chaudhury 1991). Depth-averaging in this case attempts to represent the effects of secondary circulation upon the distribution of the mean flow properties
Secondary circulation model	Used to model the dispersion terms introduced by depth-averaging. Approach based upon modelling the z-dependence of the streamwise and out-of-plane velocities, and the effects of centrifugal and frictional forces on a vertical column of water. The dispersion terms introduced by depth-averaging are replaced by a term that represents the shear stress associated with secondary flow. The latter is determined from the streamwise vorticity which includes empirically derived terms for vorticity production, dissipation, diffusion and vortex stretching	A useful representation of secondary circulation effects as it places less emphasis upon the accurate specification of the local radius of curvature of the topography, and more upon iterative estimation of the curvature of the depth-averaged streamlines. However, this is unable to represent the effects of topographic discordance and shear-related turbulence as secondary flow generating mechanisms. In addition to planform streamline curvature, a helical circulation arises from horizontal separation vortices formed in the lee of avalanche faces at the entrance to confluences. This requires a three-dimensional treatment of the problem. Computational difficulties mean that the way to proceed would seem to be based upon using three-dimensional models to improve understanding of secondary circulation effects, and from this to investigate the way in which three-dimensional effects are parameterised in two-dimensional models (e.g. Shimizu et al. 1990)
Bottom stresses	Bottom stresses determined from $C_f \rho \bar{V}_i (\bar{V}_i^2 + \bar{V}_j^2)^{1/2}$ with the bottom friction given by: $C_f = g n^2 h^{-1/3}$. The latter is equivalent to the Darcy–Weisbach friction factor specified in terms of Manning's n	Justification of this treatment of bottom shear stress, based upon a square law relationship, is important if only because depth-averaging implicitly makes it a large contributor to the force balance. The equation for bottom shear stress is used extensively in other depth-averaged models. Use of the square law relationship can only be justified if there is acceptable description of the roughness coefficient, specified in terms of n

model validation. First, central to validation is a grounding in existing knowledge, and this provides support by demonstrating that other modellers have used similar assumptions, or by demonstrating that the assumptions being made in the model are an improvement upon those currently being made in the literature. This is not an argument that a model must be more complex than its predecessors, but more a justification with reference to existing modelling strategies with a presumption that the need for increases in model complexity that follow from changing model assumptions will need to be evaluated at later stages of the modelling process (e.g. through sensitivity analysis). The outcome here is a clear identification of what the model can and cannot achieve, and the spatial and temporal scales for which it may be valid. Second, the conceptual model evaluation can provide a key indication of where the model is likely to fail, and hence where it may be necessary to give particular attention to other aspects of validation. For instance, the secondary circulation correction required by depth-averaging is based upon streamline curvature. Thus, it is particularly important to consider model predictions where vertical flows may be introduced for other reasons, such as zones of flow separation or strong lateral shear.

In summary, conceptual model evaluation is critical for identifying the levels of simplification introduced as a result of assumptions made during model development, whether in terms of process or of scale. This evaluation is thus fundamental to defining the applicable limits of a particular model, the field data needed to initialise and check a model, and hence the way in which the next stages of validation should proceed.

16.3.2   From Mathematics to Computational Algorithms

The process of converting physical–mathematical relationships into computational algorithms and computer code requires consideration of the levels of accuracy required in a numerical solution, the forms of discretisation (finite difference, finite element), and the solution methods to be employed. Once these issues have been resolved, there is a need for a preliminary set of assessments of the validity of this conversion. One way of achieving this is by comparing a numerical solution based on the chosen algorithms with an analytical one, in order to demonstrate the range of conditions for which the model holds. This has been illustrated, for example, by Horritt (2000). Horritt considers the case of a channel that is annular in planform but rectangular in cross-section, noting that this type of channel is a useful basis for study, as it would be expected to contain strongly rotational flows, and hence spatial variation in velocity magnitude and direction and free surface displacement. These are key characteristics for the effective representation of natural open channel flows. Comparison of a range of model solutions (with different meshes and numerical solvers) with the analytical solution allowed basic guidelines for mesh generation and numerical solution for this class of problem. While Horritt (2000) demonstrates that some of these conclusions may be extended for the design of effective meshes for modelling flows in naturally meandering channels, it is also recognised that analytical solutions only provide a test for those cases defined by the assumptions made to allow analytical solutions to be reached (Oreskes et al. 1994; Horritt 2000). In essence, comparison with analytical solutions, as with other forms of model evaluation, should be viewed as one part of a much broader approach.

16.3.3   Error Testing of Computer Code

Once a physical–mathematical model has been converted into a computational tool, there is a requirement to test the programme code in order that confidence can be placed in its solutions of the mathematical representations embodied within it. Both error checking of computer code, and identifying the source of error, can be difficult tasks for a complex model in which several sub-models exist. For example, a distributed hydrological model will embody infiltration, hillslope runoff, and channel routing procedures that must be correctly coded individually, as well as capable of correctly transferring their outputs to one another. Testing computer code is best achieved by modularising the program, then testing each element. One way of undertaking this is to test sub-model programs against relatively simple cases that have already been the subject of analysis in the literature. Replication of published results from other models gives confidence in the validity of the process followed as physical relations are mathematically and algorithmically expressed, then coded. For example, El-Hames and Richards (1998) illustrate a catchment-scale model in which sub-models are tested against a range of pre-existing model applications to simple test cases. A new hillslope runoff model was compared with an existing model (KININF, developed by Lima 1992) on an application to a small experimental plot for which good rainfall–runoff data were available. This in fact illustrates another dimension of model validation, namely the integrated use of laboratory or small-scale experimental facilities to 'prove' a numerical model. If a numerical model is able to replicate the behaviour observed, and measured, within a controlled experimental facility such as a flume or a rainfall–runoff plot, then there can be more confidence in the results obtained from the model when it is applied to uncontrolled field examples (since it is reasonable to discount some potential sources of failure if the model has already performed adequately in a comparison with data generated under controlled conditions).

An additional requirement before a model is used in a wide range of environmental applications is to test the stability of the model, and the numerical limitations encountered. El-Hames and Richards (1995), for example, have examined the numerical difficulty (in terms of the numbers of iterations per time step) associated with the solution of the Richards' equation to determine water redistribution in soils with various hydraulic conductivities and antecedent moisture conditions. This information may then be coupled to that on the stability of numerical solution, which is dependent on the time step and the scale of spatial discretisation; there are well-known criteria for selecting combinations of these constraints that maintain stability, such as the Courant number:

$$\frac{|v|\Delta t}{\Delta x} \leq 1 \tag{2}$$

where v is the rate at which a property moves through a spatial domain which has been discretised with a grid cell dimension of Δx and a time step of Δt. This kind of analysis is important to avoid numerical solutions which become unstable, but there remains a problem of preventing numerical diffusion as a result of rounding, truncation, and error propagation that together combine to cause simulations to drift off target. Of course, this presupposes that such drift is indeed an error, and not a result of inherent non-linearity and sensitivity to initial conditions.

16.3.4   The Role of Sensitivity Analysis

Sensitivity analysis can perform at least five functions as part of a modelling exercise (Lane et al. 1994). It can: (i) determine that, in response to representative variation of model input parameter values and boundary conditions, theoretically realistic model behaviour is experienced (e.g. Howes and Anderson 1988; Lane et al. 1994); (ii) determine that the model is sufficiently sensitive to represent perceived behaviour in the real case (Howes and Anderson 1988); (iii) identify those parameters to which the model is most sensitive (Young et al. 1971; McCuen 1973; Beven 1979) and which must therefore be given most attention in terms of acquisition and quality; (iv) improve model representation through limiting the sensitivity of a particular model component (McCuen 1973); and (v) assess the likely magnitude of error in a model prediction that arises from a particular parameter specification (e.g. McCuen 1973; Lane et al. 1994). Thus, in theory, sensitivity has much to offer the process of assessing that model behaviour is adequate. In practice, undertaking sensitivity analysis as part of a model validation strategy is complicated by a number of issues.

First, most models have many parameters, and hence many more possible parameter combinations. The simplest form of sensitivity analysis is called factor perturbation. This will involve perturbing a model parameter across a realistic range of possible values (assuming that such a realistic range can be identified), while all other parameters are held constant. This mirrors the sort of experimental activity that a flume researcher might adopt, in which experimental response is evaluated in response to certain sets of imposed conditions. However, as the behaviour of the model in response to one parameter depends upon the values taken by other parameters, the parameters are inter-correlated and so this provides a relatively restricted understanding of actual sensitivity. Thus, undertaking simulations for all combinations of parameter values may prove impossibly time demanding. As a result, sensitivity analysis has increasingly made use of stochastic methods, such as the Generalised Likelihood Uncertainty Estimation methodology of Beven and Binley (1992). This is based on the premise that errors in model structure, boundary conditions and calibration data are such that it is only possible to make general statements of likelihood as to the possible values of parameter sets. Beven and Binley note that this approach is especially important for hydrological applications as the number of parameters is high, there is parameter inter-correlation and a wide range of parameter sensitivities.

The second issue that arises is non-linearity in model response to parameter perturbation. Thus, while simple, first order assessments of sensitivity based upon direct differentiation (e.g. Beven 1979) or linear factor perturbation (e.g. Bates and Anderson 1996; Lane et al. 1994) can indicate the general characteristic of model behaviour, the problem of non-linear model response is only addressed if an appropriate analytical framework is provided. The third critical aspect of a sensitivity analysis is that of spatial distribution, already considered in Section 16.2.5. Research has shown that hydrological and hydraulic models can exhibit complex spatial response to uniform factor perturbation due to spatially distributed feedback within the model. For instance, Lane et al. (1994) demonstrated that a depth-averaged model of flow processes in a braided river could exhibit localised zones of velocity *increase* in response to roughness increase. This is intuitively incorrect, as velocity should fall as roughness increases. However, this response was shown to result from the way in which the model had been developed (using a rigid lid) coupled to the non-linear dependence of roughness upon both a friction coefficient and water depth (Lane and Richards 1998). In shallower zones, the relative effect of roughness increases was greater than in the deeper zones. Thus, as roughness was increased flow through the shallower zones decreased. To conserve mass, and because the lid was fixed, this meant that the flow through the deeper zones had to increase. In deeper zones, the contribution of roughness increases to the local force balance was reduced, such that the increased flow resulted in velocity increases. While none of these results were thought to be indicative of real world behaviour, they helped validate model response as, given the assumptions made by the model (i.e. the fixed-lid model closure), intuitively reasonable model behaviour was occurring. Central to reaching this conclusion was a theoretical appraisal of spatially distributed model behaviour. Empirical consideration of any one location in isolation could have caused the model to be deemed invalid, if a positive relationship between roughness and velocity had been observed without comparison with other locations. The main problem with arguing that sensitivity analysis needs to be spatially distributed is that it significantly enhances the complexity of the process. The above example was based upon uniform perturbation of a single parameter. Most parameters in the hydrological and hydraulic literature will display spatial variability and to undertake distributed sensitivity analysis via distributed parameter perturbation will be exceptionally time-consuming. The way forward has to be based upon basic knowledge of parameter groupings in space. For instance, in studies of floodplain flows, it is possible to divide the solution domain into channel and floodplain (e.g. Aronica et al. 1998) as a first step towards spatially distributing parameter values. However, assuming that n parameter values must be considered and there are z spatially delimited zones, the number of possible parameter combinations is n^z, and this for a single parameter. As with multiple parameter combinations in a non-spatially distributed sense, this requires recourse to Monte Carlo simulation if effective progress is to be made.

16.3.5 Model Assessment Using Visualisation

Given that modern models are capable of simulating processes evolving dynamically in four-dimensional space–time, it is clear that conventional empirical testing will often be virtually impossible, and is certainly unlikely to provide reliable approach to either falsification or validation. An increasingly important aspect of model assessment is therefore visualisation of the results of simulations or predictions (Haase et al. 2000). Modelling studies of fundamentally spatio-temporal processes such as climatic phenomena, ocean–atmosphere interactions such as El Niño, or eddy shedding processes at scales from boulders in rivers, through tributary junctions, to ocean currents at continental margins, must be illustrated not as static images but as animations. For example, the dynamics of ocean circulation displayed by the Fine Resolution Antarctic Model (FRAM) can be represented and viewed via an Internet animation at: *http:// www.mth.uea.ac.uk/ocean/fram.html*; but are not readily pictured in a conventional publication (where spatio-temporal structures are represented in a series of static pictures). A similar example is provided at *http://www.geog.leeds.ac.uk/staff/s.lane*. This shows the animation of periodic upwelling in a simple river channel confluence (parallel channels with a step in one of the channels). It also shows a two-dimensional series of unanimated images as published in a conventional journal (Bradbrook et al. 2000). Comparison of the animated and unanimated versions demonstrates just how much additional information can be obtained from an animated simulation, and hence the power of new forms of knowledge dissemination. Validating the data contained in this animation is difficult, as it is associated with a temporally evolving, spatially distributed flow field. With the technology available for that research, time series could only be obtained for singular, sequential measurement locations (i.e. the information was temporally rich, but spatially poor), even though model output was available simultaneously for all grid cells in the model (i.e. it was spatially rich and, in relative terms, temporally poor). Animation clearly allows the modeller to obtain a better feel for the extent to which the model reproduces temporally evolving, spatially distributed processes that can be observed through experimentation or in the field, but not necessarily readily measured. This is a normal part of scientific validation, but one that presents difficulties given that conventional modes of presenting results for peer assessment in the scientific literature may not adequately capture those dynamics that are of interest.

It might be hoped that scientists responsible for simulations such as these will have undertaken an assessment of the validity of the representation of eddy motions illustrated in these visualisations of model output. The model of upwelling in a simple confluence produced 3.1×10^6 data values per second. With current measurement technology, this rate of data collection (in time and space) cannot be achieved. So, how does one warrant a conclusion that a visualisation of model output captures the essentials of the dynamics of the simulated process? There seem few guidelines on what constitutes an acceptable visual representation. However, there are some interesting philosophical implications of the role of visualisation. Conventional empirical validation relies on measurement; visualisation relies on observation and perception. Observation is a more wide-ranging and more strongly theory-laden process than measurement (which should not be interpreted as a statement that measurement is not theory-laden, only that observation is more deeply pervaded by theory). It seems likely that the acceptance of a model on the basis of the evidence of the visualisation of animated model output reflects the observer's conclusion that the model captures the theoretical understanding of the relevant processes that he/she possesses. Thus, theory, understanding and explanatory power are brought to the fore as aspects of model assessment or validation, as observation recovers ground lost to measurement. This trend is associated with an increased sophistication of models, which in turn has led to visualisation as a tool for representing their output. In due course, it may be that quantitative tools will be developed that automatically capture and summarise the structures visible in animated model output. For the moment, however, qualitative and theoretically informed interpretation appears to be the basis on which visualisation may be used in model assessment. And yet, as Rhoads and Thorn (1996, p. 45) note, geomorphologists have tended to adopt sophisticated measurement techniques, because they perceive them to be more reliable than qualitative assessment based upon visual perception. In doing so, they failed to recognise the theoretical content of such measurement techniques, both that arising from the physical theory used to develop the apparatus (Hacking 1983), and that arising from the instrument deployment strategies of the researcher him/herself. As reliance on visualisation increases, the absurdity of this position becomes apparent, and the possibilities of pluralist approaches to model assessment that are qualitative, holistic, theory-laden and based on explanatory power, become that much greater.

16.3.6   Assessment via Basic Principles and Standards

In areas of the natural, environmental and engineering sciences where the use of models is well established, standards for model validation are more explicit. These commonly include reference to factors such as mass balance, mesh performance and numerical accuracy (e.g. Ababou et al. 1992; Jackson et al. 1992). A good example is the 1993 editorial statement by the American Society of Mechanical Engineers' *Journal of Fluid Engineering* (Table 16.4: ASME 1993), with respect to the control of numerical accuracy. This makes interesting reading given the observed tendency of hydrologists and hydraulics researchers to ground their results in model predictions. The ASME benchmark for publication includes only one that involves model assessment using check data, and this is optional and no substitution for a set of other compulsory checks that the modeller must make before a model may be deemed 'fit for purpose'. This certainly runs against the essence of most hydrological and hydraulic modelling approaches where the key target seems to have been goodness of fit, rather than issues surrounding numerical solution. However, with the growing complexity of many numerical approaches to hydrological and hydraulic problems, issues such as mesh resolution and artificial numerical diffusion can only become more important, and it is critical that standards in this respect are introduced in the hydrological application of models.

16.3.7   Conclusion: Integrated Assessment

From the foregoing, it would appear that progress may be made through the adoption of more

Table 16.4   The American Society of Mechanical Engineers *Journal of Fluid Engineering* statement on the Control of Numerical Accuracy

---

1. Authors must be precise in describing the numerical method used; this includes an assessment of the formal order of accuracy of the truncation error introduced by individual terms in the governing equations, such as diffusive terms, source terms, and most importantly, the convective terms. It is not enough to state, for example, that the method is based on a 'conservative finite-volume formulation', giving then a reference to a general CFD textbook.

2. The numerical method used must be at least formally second order accurate in space (based on a Taylor series expansion) for nodes in the interior of the computational grid. The computational expense of second, third, and higher order methods are more expensive (per grid point) than first order schemes, but the computational efficiency of these higher order methods (accuracy per overall cost) is much greater. And, it has been demonstrated many times that, for first order methods, the effect of numerical diffusion on the solution accuracy is devastating.

3. Methods using a blending or switching strategy between first and second order methods (in particular, the well-known 'hybrid', 'power-law', and related exponential schemes) will be viewed as first order methods, unless it can be demonstrated that their inherent numerical diffusion does not swamp or replace important modelled physical diffusion terms. A similar policy applies to methods invoking significant amounts of explicitly added artificial viscosity or diffusivity.

4. Solutions over a range of significantly different grid resolutions should be presented to demonstrate grid-independent or grid-convergent results. This criterion specifically addresses the use of improved grid resolution to systematically evaluate truncation error and accuracy. The use of error estimates based on methods such as Richardson extrapolation or those techniques now used in adaptive grid methods may also be used to demonstrate solution accuracy.

5. Stopping criteria for iterative calculations need to be precisely explained. Estimates must be given for the corresponding convergence error.

6. In time-dependent solutions, temporal accuracy must be demonstrated so that the spurious effects of phase error are shown to be limited. In particular, it should be demonstrated that unphysical oscillations due to numerical dispersion are significantly smaller in amplitude than captured short-wavelength (in time) features of the flow.

7. Clear statements defining the methods used to implement boundary and initial conditions must be presented. Typically, the overall accuracy of a simulation is strongly affected by the implementation and order of the boundary conditions. When appropriate, particular attention should be paid to the treatment of inflow and outflow boundary conditions.

8. In the presentation of an existing algorithm or code, all pertinent references or other publications must be cited in the paper, thus aiding the reader in evaluating the code and its method without the need to redefine details of the methods in the current paper. However, basic features of the code must be outlined according to Item 1, above.

9. Comparison to appropriate analytical or well-established numerical benchmark solutions may be used to demonstrate accuracy for another class of problems. However, in general this does not demonstrate accuracy for another class of problems, especially if any adjustable parameters are involved, as in turbulence modelling.

10. Comparison with reliable experimental results is appropriate, provided experimental uncertainty is established. However, 'reasonable agreement' with experimental data alone will not be enough to justify a given single-grid calculation, especially if adjustable parameters are involved.

---

wide-ranging, complementary, multivariate and holistic treatments of model assessment that: (i) recognise the contributions from conceptual uncertainties, computational uncertainties and data limitations, as well as numerical uncertainties; and (ii) explicitly represent the model as an evolving tool whose current status is transitory, as changes to all aspects of the model occur during an ongoing testing process. Coleman and Stern (1997) provide in part a methodology that may be able do this. They recognise that uncertainties can be divided into those that can be estimated (e.g. numerical uncertainty, parameter uncertainty, boundary condition uncertainty, data uncertainty), and those which cannot (which are labelled 'modelling uncertainty' and refer to those aspects of uncertainty that are fundamental to the decision to model a particular set of processes based upon a particular conceptualisation). Contributions from each of the knowable uncertainties are estimated so that when test data are compared with model predictions, these can be used to set the best possible expected correspondence between model and predictions given known model uncertainties, which sets the scale at which validation is possible (Coleman and Stern 1997). However, this still relies heavily on the assumption that models are best assessed on their capability to generate quantitative output that matches measured data. In fact, similar problems arise here to those discussed in Section 16.2.3 in relation to the question of accuracy, in that the greater the model uncertainty the greater the possibility that empirical test data lie within this uncertainty (Coleman and Stern 1997). However, the strict interpretation here is not the greater the uncertainty the easier it is to validate a model, but rather the greater the uncertainty the more difficult it is to show that the model is invalid. Thus, this sort of analysis is useful for demonstrating those situations where known uncertainties make model validation impossible. Since the richness of model output seems likely to continue to outstrip the volume and quality of empirical data, there are good grounds to suppose that assessment on the basis of visualisation and theoretical, qualitative interpretation and explanatory power will be more widely accepted.

16.4   PHILOSOPHICAL ISSUES RE-VISITED

The issues discussed above suggest that model 'validation', in the context of modern hydrological and hydrodynamic models, is a slippery concept. The word 'validation' itself, with its implications of truth and finality, might usefully be replaced by an 'assessment' which is richer in content, and an ongoing process. One reason for such a conclusion is that there is no single basis for defining model 'success'. As a result of this, Jackson et al. (1992) note that acceptance of a model involves presenting a case for a model that might reasonably lead a technically informed person to consider it to be acceptable. However, as these authors note, different evaluators may have different ideas about what is an acceptable form or level of test criterion. Sceptics may reject a model, while enthusiasts introduce *ad hoc* assumptions or relax the criteria used for predictive success. This latter tendency risks the acceptance of theoretically and empirically inadequate models. A second reason lies in the impossibility of validation and the impracticability of falsification, which together challenge the idea that observational evidence should alone dictate model success. Deviation between observations and predictions will remain one significant part of the scientific research process. There are, however, alternative ways of evaluating model behaviour which allow for both the open-system nature of many modelled environmental systems, and the complex stages through which model development passes as it moves from physical theory through algorithm coding to operational testing. Progress comes not from determining the predictive success of a model, but more from understanding and investigating those situations where the model fails in a variety of different ways (e.g. Kirkby 1996). However, neither can progress rely solely on the argument that explanatory power should determine modelling success; model assessment has to rely on the messy combination of approaches outlined in Sections 16.2 and 16.3 above.

Model assessment is an evolutionary process concerned with establishing the domains of predictive success, and extending those domains through model development. However difficult falsification may be, model prediction failures can increase both methodological and substantive understanding, through the integrated evaluation they allow of model assumptions, computer code or boundary and initial condition data. When combined with sensitivity analysis, this can identify those components of a model that need further attention. While this has a methodological role in improving model performance, it also has a substantive role in identifying those components of the model that matter: that is, those processes that seem to make a difference. If, in a more general sense, model predictions do not conform with existing empirical–theoretical understanding, then real research progress may be made. Such differences may cause existing knowledge to be viewed in a different way and even new modelling, field and experimental activities to be undertaken (numerical simulation exercises, redesign of experimental facilities, measurement of field environments using new techniques). Deviation between predictions and observations may occur even with a valid model, but this too may cause the researcher to view a particular problem in a very different way, and hence design new experimental, field and modelling campaigns that generate new knowledge, and hence enhance understanding (Kirkby 1996). The key criteria in determining whether or not a model contributes to our scientific understanding must be whether or not the model causes us to think differently, to ask new questions of existing data, or to collect new data. As Oreskes et al. (1994) note, the primary value of models is heuristic, providing representations for guiding further study, even though they are not susceptible to proof. Models are most useful when they are used to challenge existing formulations, rather than to validate or verify them.

An obvious criticism here is that many hydrological and hydraulic problems require some degree of prediction. If a model can only be falsified, and as part of this must be subjected to continued testing, should we use a model to predict? A focus upon explanatory power, upon identifying new research questions, and upon new areas for model development and new field or experimental campaigns, seems to miss the real need for practical prediction of real world phenomena. This reflects a key difference between research (into both methodological aspects of model development, and substantive understanding of hydrological and geomorphological processes), with its focus upon understanding and explanation and its essentially scientific objectives, and consultancy (the application of a model to evaluate a particular problem), with its focus upon prediction, which is strictly the application of a particular technology. Given the impossibility of verification, it is for the regulatory community, supported by advice from the scientific community, to determine which models should be used, and the extent to which environmental policy can and should be based upon predictions from models that display forced empirical adequacy (Oreskes et al. 1994). Verification, and only limited validation, may be less than satisfying (Anderson and Woessner 1992), but probably inevitable.

Perhaps the final conclusion must be to return to the peculiar fascination that modellers have with 'validation'. What makes this 'peculiar' is the fact that significantly less attention is given to the problem of validation by those working in laboratory or field contexts, although the issue of validation is just as critical here. The requirement of most numerical models is, through a combination of process simplification and specification of assumptions, that the model provides an adequate representation of reality. But in this respect, a numerical model does not differ from experimental research or a field campaign, both of which employ measurement equipment, sampling and analysis that reflect a priori theoretical knowledge of the subject of our study. For instance, in hydraulics research, it is impossible to achieve simultaneous scaling of the Froude number and the Reynolds number (Peakall et al. 1996), and it is common to relax the Reynolds number on the grounds that the Froude number is thought to exert a critical control upon the dynamics of mobile beds, and hence Froude-scaling is a requirement; this is a form of closure similar to that required in modelling. Similarly, the experimentalist will design the combinations of boundary conditions that will be used as part of a measurement program, in just the same way that the numerical modeller has to decide on the parameter values, and combinations of parameter values, to be used in the model. The findings of a piece of experimental activity will therefore be dependent upon the boundary conditions the experimentalist determines to be important, and how these are combined through experimental design. It is tempting to argue that the direct study of the real world (i.e. fieldwork) overcomes these problems. However, the researcher must again determine here where to study and when, as well as what equipment to use to measure particular phenomena, and how to sample those phenomena. Thus, we can argue that all research activity in hydrology and geomorphology will involve some form of simplification and closure, in which the researcher makes decisions as to what is and is not included in the research activity, and which must necessarily restrict the spatial and temporal relevance of that research. Thus, what makes a model acceptable is a question which must equally be applied to field case studies and laboratory experiments, and which is no more answered in the context of numerical modelling by the practice of making predictions to compare with empirical data than it is in these other forms of scientific enquiry.

REFERENCES

Ababou, R., Sagar, B. and Wittmeyer, G. 1992. Testing procedures for spatially distributed flow models. *Advances in Water Resources*, **15**, 181–198.

Anderson, M.P. and Woessner, W.W. 1992. The role of the postaudit in model validation. *Advances in Water Resources*, **15**, 167–173.

Aronica, G., Hankin, B. and Beven, K. 1998. Uncertainty and equifinality in calibrating distributed roughness coefficients in a flood propagation model with limited data. *Advances in Water Resources*, **22**, 349–365.

ASME, 1993. Statement upon the control of numerical accuracy. *American Society of Mechanical Engineers, Journal of Fluid Engineering*.

Bates, P.D. and Anderson, M.G. 1996. A preliminary investigation into the impact of initial conditions on flood inundation predictions using a time/space distributed sensitivity analysis. *Catena*, **26**, 115–134.

Bates, P.D., Horritt, M.S., Smith, C.N. and Mason, D. 1997. Integrating remote sensing observations of flood hydrology and hydraulic modelling. *Hydrological Processes*, **11**, 1777–1795.

Beck, M.B. 1987. Water quality modelling: a review of the analysis of uncertainty. *Water Resources Research*, **23**, 1393–1442.

Bernard, R.S. 1992. STREMR: numerical model for depth-averaged incompressible flow. Technical report HY-105, US Army Corps Engineers, Waterways Experiment Research Station, Vicksburg, Mississippi.

Best, J.L. 1987. Flow dynamics at river channel confluences: implications for sediment transport and bed morphology. In: Etheridge, F.G., Flores, R.M. and Harvey M.D. (eds), *Recent Developments in Fluvial Sedimentology*, SEPM Special Publication 39, 27–35.

Best, J.L. and Roy, A.G. 1991. Mixing-layer distortion at the confluence of channels of different depth. *Nature*, **350**, 411–413.

Beven, K.J. 1979. A sensitivity analysis of the Penman–Monteith actual evapotranspiration estimates. *Journal of Hydrology*, **44**, 169–190.

Beven, K.J. 1989. Changing ideas in hydrology – the case of physically-based models. *Journal of Hydrology*, **105**, 157–172.

Beven, K.J. 1996. Equifinality and uncertainty in geomorphological modelling. In: Rhoads, B.L. and Thorn, C.E. (eds), *The Scientific Nature of Geomorphology*. John Wiley, Chichester, 289–214.

Beven, K.J. and Binley, A.M. 1992. The future of distributed models: model calibration and uncertainty prediction. *Hydrological Processes*, **6**, 279–298.

Bhallamudi, S.M. and Chudhury, M.H. 1991. Numerical modelling of aggradation and degradation in alluvial channels. *ASCE Journal of Hydraulic Engineering*, **119**, 1145–64.

Biron, P.M., Best, J.L. and Roy, A.G. 1996. Effects of bed discordance on flow dynamics at river channel confluences. *ASCE Journal of Hydraulic Engineering*, **122**(12), 676–682.

Bradbrook, K.F., Biron, P., Lane, S.N., Richards, K.S. and Roy, A.G. 1998. Investigation of controls on secondary circulation and mixing processes in a simple confluence geometry using a three-dimensional numerical model. *Hydrological Processes*, **12**, 1371–1396.

Bradbrook, K.F., Lane, S.N., Richards, K.S., Biron, P.M. and Roy, A.G. 2000. Large eddy simulation of periodic flow characteristics at river channel confluences. *Journal of Hydraulic Research*, **38**, 207–216.

Bradbrook, K.F., Lane, S.N., Richards, K.S., Biron, P.M. and Roy, A.G. 2001. The role of bed discordance at asymmetrical river channel confluences. *ASCE, Journal of Hydraulic Engineering*, in press.

Brown, H.I. 1996. The methodological roles of theory in science. In: Rhoads, B.L. and Thorn, C.E. (eds), *The Scientific Nature of Geomorphology*. John Wiley, Chichester, 3–20.

Chorley, R.J. 1978. Bases for theory in geomorphology. In: Embleton, C., Brunsden, D. and Jones, D.K.C. (eds), *Geomorphology: Present Problems and Future Prospects*. Oxford University Press, Oxford, 1–13.

Coleman, H.W. and Stern, F. 1997. Uncertainties and CFD code validation. *Journal of Fluids Engineering*, **19**, 795–803.

Czernuszenko, W. and Rylov, A.A. 2000. A generalisation of Prandtl's model for 3D open channel flows. *Journal of Hydraulic Research*, **38**, 133–139.

El-Hames, A.S. and Richards, K.S. 1995. Testing the numerical difficulty applying Richards' equation to sandy and clayey soils. *Journal of Hydrology*, **167**, 381–391.

El-Hames A.S. and Richards K.S. 1998. An integrated, physically-based model for arid region flash flood prediction capable of simulating dynamic transmission loss. *Hydrological Processes*, **12**, 1219–1232.

Flavelle, P. 1992. A quantitative measure of model validation and its potential use for regulatory purposes. *Advances in Water Resources*, **15**, 5–13.

Haase, H., Bock, M., Hergenrother, E., Knopfle, C., Koppert, H.J., Schroder, F., Trembilski, A. and Weidenhausen, J. 2000. Meteorology meets computer graphics – a look at a wide range of weather visualisations for diverse audiences. *Computers and Graphics – UK*, **24**, 391–397.

Hacking, I. 1983. *Representing and Intervening: Introductory Topics in the Philosophy of Natural Science*. Cambridge University Press, New York.

Horritt, M.S. 2000. Development of physically based meshes for two-dimensional models of meandering channel flow. *International Journal of Numerical Methods in Engineering*, **47**, 2019–2037.

Horritt, M.S. and Bates, P.D. (in press) Predicting floodplain inundation: raster-based modelling versus the finite element approach. *Hydrological Processes*.

Howes, S. and Anderson, M.G. 1988. Computer simulation in geomorphology. In: Anderson, M.G. (ed.), *Modelling Geomorphological Systems*. John Wiley, Chichester, 421–440.

Ibbit, R.P. and O'Donnell, T. 1971. Fitting methods for conceptual catchment models. *Journal of the Hydraulics Division, American Society of Civil Engineers*, **97**, 1331–1342.

Jackson, C.P., Lever, D.A. and Sumner, P.J. 1992. Validation of transport models for use in repository performance assessments: a view illustrated for INTRAVAL test case 1b. *Advances in Water Resources*, **15**, 33–45.

Kirkby, M.J. 1996. A role for theoretical models in geomorphology? In: Rhoads, B.L. and Thorn, C.E. (eds), *The Scientific Nature of Geomorphology*. John Wiley, Chichester, 257–272.

Konikow, L.F. and Bredehoeft, J.D. 1992. Groundwater models cannot be validated. *Advances in Water Resources*, **15**, 75–83.

Lane, S.N. and Richards, K.S. 1998. Two-dimensional modelling of flow processes in a multi-thread channel. *Hydrological Processes*, **12**, 1279–1298.

Lane, S.N., Richards, K.S. and Chandler, J.H. 1994. Distributed sensitivity analysis in modelling environmental systems. *Proceedings of the Royal Society, Series A*, **447**, 49–63.

Lane, S.N., Richards, K.S. and Chandler, J.H. 1995. Within reach spatial patterns of process and channel adjustment. In: Hickin, E.J. (ed.), *Rivers*, John Wiley, Chichester, 105–130.

Lane, S.N., Biron, P.M., Bradbrook, K.F., Butler, J.B., Chandler, J.H., Crowell, M.D., McLelland, S.J.,

Richards, K.S. and Roy, A.G. 1998. Integrated three-dimensional measurement of river channel topography and flow processes using acoustic doppler velocimetry. *Earth Surface Processes and Landforms*, **23**, 1247–1267.

Lane, S.N., Bradbrook, K.F., Richards, K.S., Biron, P.M. and Roy, A.G. 1999. The application of computational fluid dynamics to natural river channels: three-dimensional versus two-dimensional approaches. *Geomorphology*, **29**, 1–20.

Launder, B.E. and Spalding, D. 1972. Lectures in Mathematical Models of Turbulence, Academic Press, New York.

Leopold, L.B., Bagnold, R.A., Wolman, M.G. and Brush, L.M. 1960. Flow resistance in sinuous or irregular channels. USGS Professional Paper, 282D, 111–35.

Lima, J.T. 1992. Model KININF for overland flow on pervious surfaces. In: Parsons, A.J. and Abrahams, A.D. (eds), *Overland Flow*. University College Press, London, 69–88.

Luis, S.J. and McLaughlin, D. 1992. A stochastic approach to model validation. *Advances in Water Resources*, **15**, 15–32.

Malin, M.R. and Younis, B.A. 1990. Calculation of turbulent buoyant plumes with a Reynolds stress and heat-flux transport closure. *International Journal of Heat and Mass Transfer*, **33**, 2247–64.

Markus, G. 1987. Why is there no hermeneutics of natural sciences? Some preliminary theses. *Science in Context*, **1**, 5–51.

McCuen, R.H. 1973. The role of sensitivity analysis in hydrologic modelling. *Journal of Hydrology*, **18**, 37–53.

McLelland, S.J., Ashworth, P.J. and Best, J.L. 1996. The origin and downstream development of coherent flow structures at channel junctions. In: Ashworth, P.J., Bennett, S., Best, J.L. and McLelland, S.M. (eds), *Coherent Flow Structures in Open Channels*. John Wiley, Chichester, 459–490.

Meselhe, E.A. and Sotiropolous, F. 2000. Three-dimensional numerical model for open-channels with free surface variations. *Journal of Hydraulic Research*, **38**, 115–121.

Nash, J.E. and Sutcliffe, J.V. 1970. River flow forecasting through conceptual models. Part I – a discussion of principles. *Journal of Hydrology*, **83**, 307–335.

Oreskes, N., Shrader-Frechette, K. and Belitz, K. 1994. Verification, validation and confirmation of numerical models in the earth sciences. *Science*, **263**, 641–646.

Peakall, J., Ashworth, P.J. and Best, J.L. 1996. Physical modelling in fluvial geomorphology: principles, applications and unresolved issues. In: Rhoads, B.L. and Thorn, C.E. (eds), *The Scientific Nature of Geomorphology*. Wiley, Chichester, 221–254.

Phillips, D.C. 1992. *The Social Scientist's Bestiary: A Guide to Fabled Threats to, and Defences of, Naturalistic Social Sciences*. Pergamon Press, Oxford.

Popper, K.R. 1968. *The Logic of Scientific Discovery*. Hutchinson, London.

Rhoads, B.L. and Thorn, C.E. 1996. Observation in geomorphology. In: Rhoads, B.L. and Thorn, C.E. (eds), *The Scientific Nature of Geomorphology*. John Wiley, Chichester, 21–56.

Richards, K.S. 1996. Samples and cases: generalisation and explanation in geomorphology. In: Rhoads, B.L. and Thorn, C.E. (eds), *The Scientific Nature of Geomorphology*. John Wiley, Chichester, 171–190.

Rodi, W., Pavlovic, R.N. and Srivatsa, S.K. 1981. Prediction of flow and pollutant spreading in rivers. In: Fischer, H.B. (ed), Transport Models for Inland and Coastal Waters, Academic Press, New York, 542pp.

Sanjiv, S.K. and Marelius, F. 2000. Analysis of flow past submerged vanes. *Journal of Hydraulic Research*, **38**, 65–71.

Sargent, R.G. 1982. Verification and validation of simulation models. In: Cellier, F.E. (ed.), *Progress in Modelling and Simulation*, Academic Press, London, 159–169.

Shimizu, Y., Yamaguchi, H. and Itakura, T. 1990. Three-dimensional computation of flow and bed deformation. *ASCE Journal of Hydraulic Engineering*, **116**, 1090–1106.

Sofialidis, D. and Prinos, P. 2000. Turbulent flow in open channels with smooth and rough floodplains. *Journal of Hydraulics Research*, **37**, 615–640.

Usunoff, E., Carrera, J. and Mousavi, S.F. 1992. An approach to the design of experiments for discriminating among alternative conceptual models. *Advances in Water Resources*, **15**, 199–214.

Young, G.K., Tseng, M.T. and Taylor, R.S. 1971. Estuary water temperature sensitivity to meteorological conditions. *Water Resources Research*, **7**, 1173–1181.

Younis, B.A. 1992. Is Turbulence Modelling of any Use? AIRH Conference, Institution of Civil Engineers, London, April 1992.

17 The Validation of Ice-sheet Models

MARTIN J. SIEGERT[1] AND ANTONY J. PAYNE[2]
[1]*School of Geographical Sciences, University of Bristol, UK*
[2]*Department of Geography, University of Southampton, UK*

17.1 INTRODUCTION

The cryosphere comprises several forms of ice masses, ranging from sea ice a few metres thick over the large areas of high-latitude oceans, to small valley glaciers which are tens to hundreds of metres thick, to huge continental-scale ice sheets several kilometres thick. Around 75% of all fresh water on Earth is currently held as ice. Moreover, around 99% of this fresh water is locked within large ice sheets. During the last Ice Age, the coverage of ice sheets over the mid-latitudes caused so much transfer of water from the oceans that global sea level fell by 120 m. Continental-scale ice sheets affect climate because they alter directly the hypsometry of the Earth's surface and its albedo. They also affect global oceanic conditions by regulating the transfer of fresh water to and from the oceans. In addition, ice sheets are agents of large-scale erosion and deposition. Ice sheets, therefore, act in numerous ways that affect the environment, past, present and future, of our planet. Because of this, the ability to model ice sheets is valuable not only to glaciologists but to Earth science in general. Consequently, the validation of ice-sheet models is necessary if their results are to be of use in scientific research.

Numerical ice-sheet models are usually organised as a two-dimensional grid or mesh (comprising many individual cells) over the area of ice sheet. Calculations of ice flow are made at each cell, based on well-established physical principles. The interaction of ice flux between grid cells describes the time-dependent response of the ice sheet to changes in (1) imposed surface mass balance (the inputs and outputs of the ice sheet) and (2) mechanisms internal to the ice sheet.

Models of large-scale ice flow can be subdivided into two main groups. First, are those used to model ice sheets over continental scales. This category of model has a number of uses. They are perhaps most well known in the context of predicting ice-sheet response to climate change. In particular, much work has been done on the response of the present-day Greenland and Antarctic ice sheets to anthropogenic warming in the coming millennia (Huybrechts and Oerlemans 1990; Huybrechts and de Wolde 1999). However, this type of model can also be used to study the effects of internal interactions within the ice-sheet flow system and their role in large-scale and local flow instability (Huybrechts and Oerlemans 1988; Payne 1995; Payne and Dongelmans 1997; Payne 1999). A third use is in understanding the behaviour of the late Quaternary ice sheets both in terms of their growth and decay (Huybrechts 1990; Siegert and Dowdeswell 1995) and the potential for flow surging (Greve and MacAyeal 1996; Marshall and Clarke 1997).

Ice-sheet models typically comprise three main components. These are the temporal evolution of ice thickness, which also incorporates the areal expansion and contraction of the ice sheet; a

Model Validation: Perspectives in Hydrological Science. Edited by M.G. Anderson and P.D. Bates.
© 2001 John Wiley & Sons, Ltd.

reduced model of how stresses and velocities vary within an ice mass (the key assumption is that longitudinal stresses are minimal and therefore that a local stress balance exists); and the temporal evolution of the internal temperature field of the ice sheet. In addition, there are many subsidiary models for effects such as isostasy (the deflection of underlying lithosphere by the load of the ice sheet); basal slip of the ice over its underlying substrate; the pattern of snow accumulation on and ice/snow melt from the upper ice-sheet surface; and the effect of temperature on the viscosity of ice.

The second category of large-scale ice-flow models is those of ice shelves (floating ice) and fast-flowing ice streams (which are thought to have many of the features of ice shelves). The physics controlling the flow of ice shelves and streams is very much more complex than that controlling ice-sheet flow. In particular, longitudinal and transverse stresses are often the dominant components of the force balance. Stress and velocity fields must therefore be determined simultaneously throughout the entire ice mass (as opposed to the local stress balance discussed above for ice sheets). Longitudinal and transverse stresses are important because ice shelves and streams are characterised by reduced (or, in the case of shelves, zero) basal traction. The assumption that vertical gradients are minimal is often made to simplify the modelling (MacAyeal 1989). The main component of this type of model is the determination of the velocity field for a given ice mass' geometry (MacAyeal 1989). Comparatively little work has been done on the temporal evolution of ice thickness in response to this velocity field (Rommelaere and Ritz 1996). One novel application of this type of model is to determine the basal traction acting on an ice stream given high-resolution data on surface velocity (as has recently become available from satellite-borne radar studies). This requires the inversion of the ice flow model for which control methods are used (MacAyeal 1993; MacAyeal et al. 1996).

A feature of all large-scale ice-flow models is the extreme non-linearity of the underlying equations which they attempt to solve. This is a consequence of the empirical law thought to govern ice flow, in which strain rates vary with the cube of applied stress (Glen 1955). In addition, there are several other features which introduce non-linearity into the system. For instance, the exponential relationship between the viscosity of ice and its temperature (Hooke 1981; Paterson and Budd 1982), and the coupling between ice flow and temperature via frictional heat generation (Clarke et al. 1977).

17.2   MODEL INPUTS – A POTENTIAL SOURCE OF ERROR?

With all numerical models, the quality of model output is dependent on the model input. The most important data used as boundary and forcing conditions in an ice-sheet model are subglacial topography and surface mass balance. For present-day ice sheets the quality of these model inputs is dependent on field data which, for some areas, is noticeably absent. Airborne radar surveying of large ice sheets provides the only viable method of acquiring information on the ice-sheet base at a continental scale. Such surveys are organised as a series of gridded flightlines. Although the technique is sound, there are currently two limitations to data sets collected by radar surveying. The first is that information between flightlines is absent. Interpolation software is used to infer the topography in such regions. The second limitation, which is acute in Antarctica, is that large portions of some ice sheets remain to be surveyed. In Antarctica the subglacial topography in these regions has to estimated from a few seismic measurements and is subject to a high level of potential error. To date, Antarctic ice-sheet models have been run over a topographic DEM based on only a 40% coverage of airborne radar data.

There have been several recent advances to alleviate this problem. Since the 1970s, when radar data from over 400 000 km worth of flightlines were collected over a wide area of Antarctica (albeit at a coarse line spacing), there have been several surveys over much smaller regions with close flightline spacing. It is only recently, however, that these data have been collated and a new subglacial topography produced. This work is being organised in a major international project called BEDMAP (Figure 17.1). However, although the new DEM will be much improved and ice-sheet models will benefit as a result, the DEM will still have problems caused by data absence. More radar work in Antarctica is clearly needed. However, BEDMAP has been designed so that any new data can be automatically submitted to the database, resulting in a continual update of the DEM.

Mass balance of large ice masses is, if anything, more difficult to establish over a wide area than the subglacial conditions. Although it is possible to measure annual mass balance, there is no quick and easy method by which these data can be obtained. This is a real problem for ice-sheet models of Antarctica and Greenland. Because of this, many ice-sheet models are coupled with models of the surface balance. The accuracy of these models is not discussed here, but we note that the accuracy of the ice-sheet model results will be dependent on the validity of the surface balance. We note also that a similar problem exists when the processes of iceberg calving and ice-shelf basal melting are incorporated into ice-sheet models.

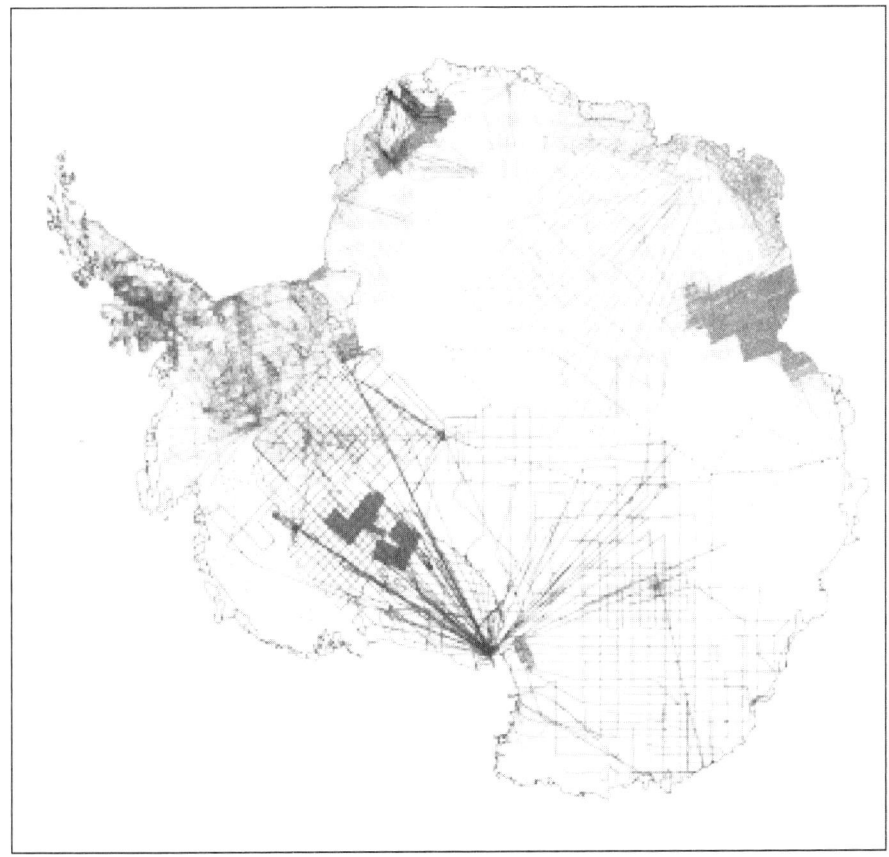

Figure 17.1   BEDMAP airborne radar flightlines in Antarctica (Lythe et al. 2000)

In conclusion, ice-sheet models currently rely on a number of data sets for input and boundary conditions. These data are often far from perfect, and the ice-sheet results that are calculated from them should be viewed with this in mind.

17.3   THE EISMINT EXERCISE

The majority of large-scale ice-flow models have been developed by individual researchers or within small institution-based groups. This do-it-yourself history is in stark contrast to the large community-based models which dominate research in the counterpart fields of meteorology and oceanography. This distinctive approach perhaps arose for two main reasons. First, glaciological studies are undertaken by a wide variety of researchers varying from geographers to geophysicists as well as meteorologists and oceanographers, who have different research aims and are widely scattered (there are very few Departments of Glaciology in the world). The core group necessary to develop a community model has therefore never existed. Second, the level of complexity of large-scale ice-flow models is such that successful model development can be undertaken by a single researcher. This diversity of models is both a strength, in that specific models are honed to specific applications, and a weakness, in that no consensus on the basic behaviour of the models exists. There are approximately 15 distinct models in existence, each developed separately although based largely on the same physical assumptions (see above).

The European Ice Sheet Modelling Initiative (EISMINT) was developed to address this diversity of existing models. It had three main aims. First, to perform a model intercomparison exercise so that a consensus could be reached on the basic predictions of the models. Second, this process would help to identify areas of good practice in modelling. It would also help identify any obvious flaws in individual existing models and provide a useful benchmark to speed the development of future models. Finally, weaknesses in the glaciological theory underlying the present generation of models were identified and addressed by a series of focused workshops. The EISMINT project was funded by the European Science Foundation in two phases which ran from January 1993 to December 1995, and from January 1996 to December 1997. This allowed a total of four meetings specifically addressing intercomparison, as well as workshops on basal processes, climate interactions, oceanic interactions, former ice sheets, lithospheric interactions and ice rheology.

Model intercomparisons were made at three levels. The first level was aimed solely at testing the effects of numerical implementation on model prediction. For this reason, the physics incorporated into the models was tightly constrained, as were the values of the various parameters and boundary conditions employed in the models. The separate origins of many of the existing models has led to a wide variety of numerical techniques being employed. Several finite-element models exist, although the majority of ice-sheet models employ finite differences. Within the latter class, a wide variety of methods are employed to deal with the basic non-linearity of ice flow (see above). So fully implicit, semi-implicit and explicit discretisations all exist, as well as models which incorporate one or both horizontal dimensions, and may or may not include the vertical.

The second level of intercomparison allowed individual modellers to use their preferred values for the various model parameters. The aim of this intercomparison was to determine the effect which the many poorly constrained physical parameters had on the overall prediction of the models. The final level of intercomparison was to model 'real' ice masses through a given climate change scenario. Examples included a long-term steady state experiment; the response to stepped changes in forcing; the response to glacial–interglacial climate change; and the response to future, anthropogenic change. This latter exercise focused on the two present-day ice sheets of

Greenland and Antarctica, as well as the response of valley glaciers. The aim here was to test the predictions of complete models and therefore not to prescribe the details of individual experiments completely. In this way, models which had already been used to generate predictions on ice-sheet response to climate change could be compared fully. The aim of the various levels of intercomparison therefore progressed from details of numerical implementation towards whole-model assessment. It is important to emphasise that the whole EISMINT exercise was driven by model intercomparison and not by testing against real-world data sets. This was principally because of the paucity of real-world data available for such testing (as discussed elsewhere in the chapter).

17.3.1   EISMINT Levels One and Two

The work discussed stems mainly from the International Symposium on Ice Sheet Modelling held in Chamonix during September 1995 and published in *Annals of Glaciology*, 23. This meeting represented the conclusion of the first phase of EISMINT. The volume also contains useful intercomparisons of the way in which isostasy is incorporated into ice-sheet models (Le Meur and Huybrechts 1996); the effects of different basal slip parameterisations (Pattyn 1996); and the effects of numerical discretisation (Lam and Dowdeswell 1996). A Level One intercomparison of existing ice-sheet models is presented by Huybrechts et al. (1996). Their approach attempts to be as flexible as possible and a series of experiments are presented which could be adapted for models with one and two horizontal dimensions, and which may or may not have included internal temperature evolution. The boundary conditions and parameter values used in the models were prescribed as tightly as possible. A major simplification was made in that although internal temperature calculations could be incorporated, their effect on ice viscosity was omitted. The methodology was therefore to reduce the potential differences between models to a minimum so that any observed differences could be interpreted purely in terms of model numerics.

The actual intercomparison exercise progressed through a number of stages in which individual groups resubmitted their results and a consensus set of results gradually emerged. This process was found to be necessary because many of the differences initially identified between models actually arose because of trivial reasons (such as ambiguity in the experiment descriptions). This process was not without its dangers, in that results inevitably tended to converge towards median values as outlying models were modified. The modification process may have been genuine (e.g. in finding coding errors) but may also have been driven by a perceived need to conform with the bulk of models. In this way genuine reasons for difference may have become obscured. This was partially avoided, in some cases, by the existence of certain analytical solutions to the equations describing ice-sheet flow. The design of the experiments recognised the existence of these solutions, which to a large extent can be taken as 'truth'. Unfortunately, the number of analytical solutions available is very limited and apply to highly simplified situations. Nonetheless their use was an important feature of the intercomparison.

In general the models agreed very closely in terms of ice-thickness evolution, internal-temperature evolution and the predicted velocity patterns within the ice mass. Figure 17.2 shows some selected results for one of the experiments. Interpretation of the remaining scatter between models yielded useful information on the effect of spatial discretisation on model prediction. Two groups of models were found to exist within the results based on the horizontal discretisation used to incorporate ice flow. Models which employed a more stable discretisation were found to be less accurate in comparison to a more precise formulation which was, however, less stable numerically. This difference can be interpreted in terms of the amount of spatial averaging implicit in a particular scheme and is discussed further by Hindmarsh and Payne (1996). The

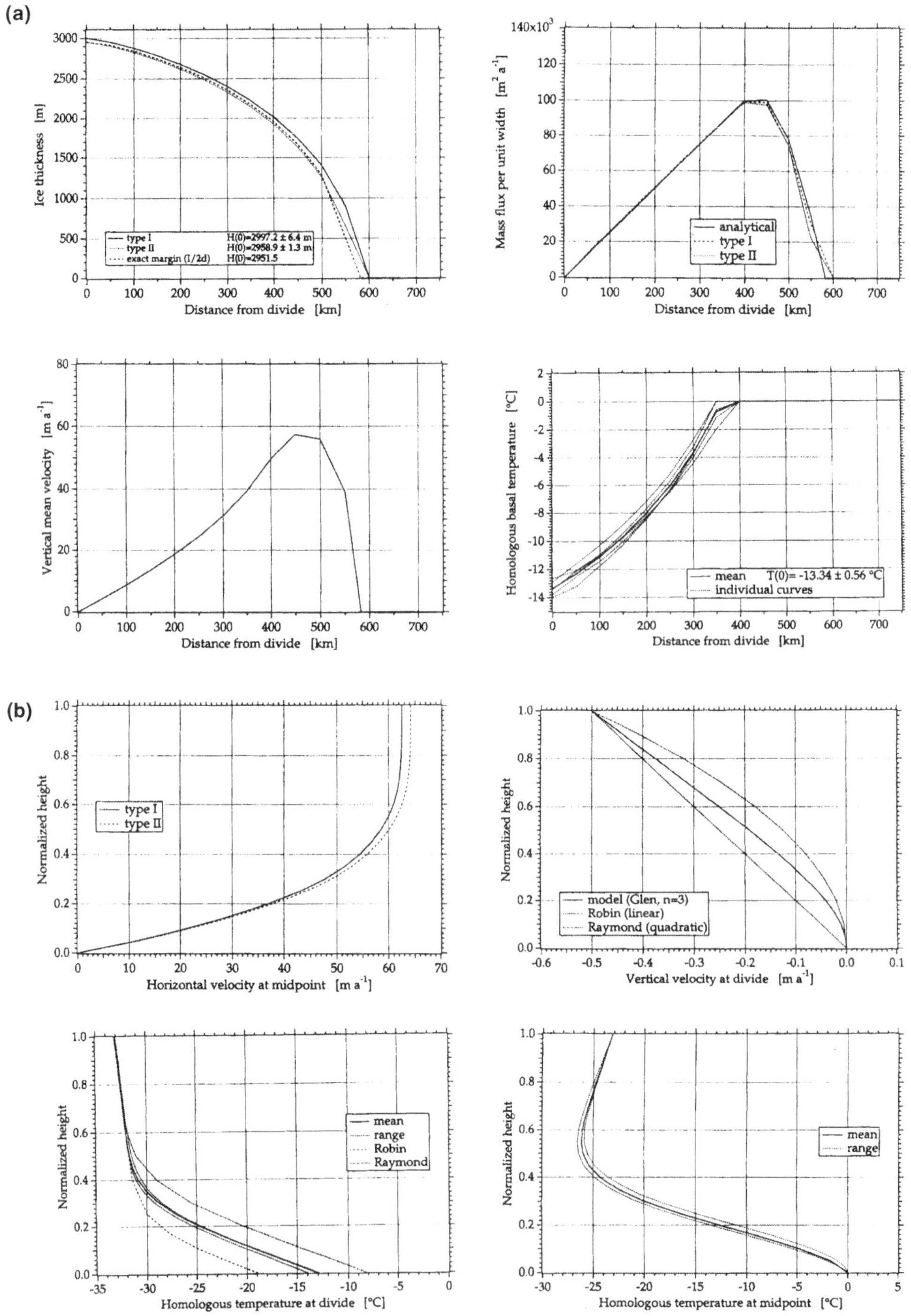
(a)
Ice thickness [m]
Distance from divide [km]
type I H(0)=2997.2 ± 6.4 m
type II H(0)=2958.9 ± 1.3 m
exact margin (I/2d) H(0)=2951.5
Mass flux per unit width [m² a⁻¹]
140x10³
analytical
type I
type II
Vertical mean velocity [m a⁻¹]
Homologous basal temperature [°C]
mean T(0)= -13.34 ± 0.56 °C
individual curves
(b)
Normalized height
Horizontal velocity at midpoint [m a⁻¹]
type I
type II
Vertical velocity at divide [m a⁻¹]
model (Glen, n=3)
Robin (linear)
Raymond (quadratic)
Homologous temperature at divide [°C]
mean
range
Robin
Raymond
Homologous temperature at midpoint [°C]
mean
range

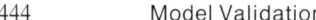

Huybrechts et al. (1996) intercomparison and those discussed below provide useful benchmarks to future glaciological modelling. They provide an effective means of testing a model under development (this is important given the limited number of analytical solutions against which to test) and also highlight potential difficulties in the numerical implementation of the ice-flow equations.

The major shortcoming of the Huybrechts et al. (1996) intercomparison is that the ice-flow and temperature calculations evolved separately. While this eased the complexity of the inter-comparison task, it represents an unrealistic assumption and does not reflect current model usage. A subsequent exercise (Payne et al. 2000) dealt with the effects of flow–temperature coupling. This coupling arises because of the relationship between ice viscosity and temperature (ice near its melting point is up to a thousand times softer than ice at $-50°$ C), and the effect of the frictional heat generated by ice flow (dissipation) on the energy budget of an ice mass.

The experimental design was similar to the Huybrechts et al. (1996) intercomparison so that all boundary conditions were symmetrical. The results showed much more between-model variability than previously. The reason for this is shown in Figure 17.3. Most models shows signs of pattern formation, which, because of the symmetry of the boundary conditions, must be generated internally within the ice-flow equations. One likely mechanism is the interaction between ice temperature and flow, or 'creep instability' (Clarke et al. 1977). This instability arises because of the mutual interaction between ice viscosity and temperature, and ice flow and dissipation. The presence of the patterning is a fairly robust feature of all models but its details vary between models. This implies that the numerical implementation is, to a certain extent, guiding the development of the patterning. The presence of high-frequency noise in some of the model results also implies that some of the models are experiencing numerical difficulties.

This second intercomparison generated more contentious results than the first but is probably more useful. Further work is needed to understand the causes of this patterning and the role which model numerics play in its development. However, two conclusions are possible. First, in more complex simulations of real ice sheets, the existence of this type of instability could well be misinterpreted as being caused by the highly variable bedrock topography and climatological forcing employed in these simulations. The interpretation of the flow fields in simulations of real ice masses should therefore be treated with some caution. Second, if the phenomenon has a physical rather than numerical basis then the patterning could be linked to the development of ice streams. Payne and Baldwin (2000) discuss the patterning further.

17.3.2   EISMINT Level Three

Simulations of the Greenland and Antarctic Ice Sheets, as well as various valley glaciers and the Ross Ice Shelf, Antarctica, comprised the third level of the EISMINT exercise. The aim of this level was to compare model predictions of the past and future evolution of these ice masses with the minimum amount of constraint on the details of the individual models. The details of the Antarctic intercomparisons will not be discussed here because the number of models participating in these exercises was relatively small. The ice-shelf exercise is described in MacAyeal et al. (1996).

Figure 17.2   Intercomparison results for the moving-margin experiments in steady state. (a) ice thickness, mass fluxes, velocities and basal temperatures. (b) Vertical profiles of temperature and velocity for ice-sheet divide and midpoint. From Huybrechts et al. (1996)

Figure 17.3 Predicted steady-state basal temperatures for each model in the intercomparison. Temperatures are in K and uncorrected for pressure melting point variation. The ice-covered area is shaded grey. The units of the x and y axes are kilometres. From Payne et al. (2000)

Nine groups participated in the Greenland intercomparison. Experiments consisted of simulating the past evolution of the ice sheet over the last two glacial–interglacial cycles (the last 250 ka). In addition, response to various climate-change scenarios over the next 500 years was simulated. Each model was prepared so that they used the same parameterisation of ablation and accumulation to air temperature. Realistic boundary conditions were employed for both the underlying bedrock topography and climate change over the period of interest (obtained from the various Greenland Summit ice cores). All models simulated the basic geometry of the ice sheet in a very realistic fashion with the present-day positions of all of the major ice domes and divides being modelled correctly. There was, however, some variability in the predicted basal thermal regime, which implies that thermal regime does not affect ice-sheet form greatly (confirming the results quoted above).

Results highlighted the importance of the parameterisation used to model ice/snow melt (ablation) and precipitation, and their relationships to air temperature. The areal extent of the ice sheet is principally determined by the predicted ablation within the model. There was a certain degree of consensus on the timing and magnitude of areal extent changes over the last 250 ka before present (BP), and the present-day and the last interglacial (the Eemian) were found to represent minima. In contrast, the average thickness of the ice sheet is determined principally by the precipitation rate. This is because the area experiencing ablation only represents a narrow fringe around the ice sheet, while the majority of the ice sheet is affected by accumulation directly. The relationship of air temperature to ablation and precipitation is complex: increased air temperatures imply greater ablation rates but also greater accumulation rates (because of the increased moisture content of warm air). This led to the counter-intuitive result (which was confirmed by all models) that the Last Glacial Maximum (LGM at 21 ka BP) was a period when areal extent was at a maximum (reduced ablation) but the ice sheet was relatively thin (divide thickness 3150 ± 150 m). In contrast, the Eemian ice mass was relatively less extensive but was thicker (divide thickness 3300 ± 100 m). The present-day ice mass appears to be intermediate in terms of both area and thickness (divide thickness 3200 ± 100 m). It is encouraging that the majority of models were able to simulate the various conflicting effects consistently.

The contribution of the world's valley glaciers to future sea-level change is investigated by Oerlemans et al. (1998). Although this work was part of the EISMINT project, it does not address model intercomparison but concentrates on the modelled response of 12 valley glaciers to imposed climate warming. A mass-balance model for each glacier is separately tuned using the past variation of the glacier's length. A variety of linear warming rates are then imposed with and without concurrent changes in precipitation. The differences between the predicted response of individual glaciers are large and created mainly by differences in their hypsometry. Results from the scenarios vary greatly from instances where few glaciers will survive past the year 2100, to a relatively minor (10 to 20%) loss in volume from 1990 levels.

A very useful distinction is made in this study between the static and dynamic responses of a glacier to mass-balance change. The former ignores the effects of a glacier's flow on its response, so that static sensitivity is simply the rate at which mass balance changes with increased air temperature. Dynamic sensitivity incorporates the feedback between a glacier's changing geometry and its mass balance. This distinction has also been used in discussing the response of ice sheets to climate change, where inheritance effects from the LGM must also be incorporated (Huybrechts and de Wolde 1999).

17.3.3 Lessons Learnt from the EISMINT Exercise

Three main scientific conclusions sprang from the exercise. First, the current generation of ice-sheet models are consistent with one another and with the available analytical solutions. The main area of concern lies in their ability to cope with flow/temperature coupling. This means that local details of ice flow may not be modelled correctly; however, the global response of an ice mass to a climate-change signal appears to be modelled robustly. Second, the application of existing models to the present-day ice sheets again reveals a large degree of correspondence, which is largely due to the response being mass-balance driven rather than reflecting changes in internal flow regime. Similarly, models of ice shelves and streams show a large degree of agreement. Third, the main areas of concern reflect the boundaries between different types of ice-flow system, for instance at grounding lines (between grounded ice sheets and floating ice shelves) and at the onset and lateral boundaries of ice streams. These transitional zones fall between the traditional model classification of ice sheet and ice shelf/stream, and are the focus of much recent work. This is because, although small in terms of the areas they occupy, they are thought to be crucial in determining the dynamics of the overall ice-sheet flow system. Very recently, ice-sheet and ice-stream models have been coupled. This is important because ice streams discharge the vast majority of ice lost from the present-day ice sheets. Accurate prediction of an ice sheet's future behaviour must therefore be based on models which incorporate its dynamics correctly (Marshall and Clarke 1997; Hulbe and MacAyeal 1999).

The organisation of the intercomparison itself also yielded some useful lessons. First, it is essential to specify the model experiments as fully as possible. Parameter values and boundary conditions should be stated unambiguously. In addition, the consequences of the way in which different types of model (finite element and finite difference, or two and three dimensional) are to be incorporated should be thought out fully and the potential for comparison to analytical solutions investigated. This saves much time in the preparation and interpretation of the results. Second, the format of the model results (both in terms of units and text formatting) should be specified as completely as possible. Again, thought at the planning stage saves considerable effort during the interpretation stage. The Payne et al. (2000) exercise included ten groups spread worldwide and involved some 3500 sets of data. There is therefore a considerable data management task associated with an intercomparison, especially if contributing groups resubmit revised results. This task is eased considerably by the use of the Internet.

17.4   COMPARING ICE-SHEET MODELS WITH REAL GLACIOLOGICAL DATA

Because most ice-sheet models are based on well-established physical assumptions about the flow of ice, results from different models will be similar, provided they work properly. However, these assumptions, and data used as input to the models, may oversimplify the actual glaciological situation. This results in mismatch between ice-sheet model results and real glaciological measurements.

When comparing model output with glaciological data, the scale of the data sets needs to be similar to the scale of the ice-sheet model output (e.g. averaged over 5 to 20 km). Real data at a finer resolution than this show how ice-sheet models simplify the actual flow of ice sheets. Determining the difference between model output and these real data represents an important way in which to assess the validity of ice-sheet models.

Recent intercomparison of ice-sheet models has utilised the present Greenland Ice Sheet as a test case (see Section 17.3). However, the data used to compare model results is restricted to the generalised surface elevation, a smoothed version of ice thickness and present-day ice extent. At a continental scale, this form of model validation is appropriate. However, as ice-sheet models become more sophisticated, it will be necessary to test their output against a variety of ice-sheet measurements for validation purposes. This section outlines data sets currently available that have yet to be compared fully with ice-sheet model output, but should be used for the future validation of ice-sheet models.

17.4.1 Surface and Sub-ice Morphology, and the Flow of Ice in Central Regions

The advent of satellite altimetry has resulted in the determination of the surface morphology of ice sheets to a high degree of accuracy. At present, most ice-sheet models of Greenland and Antarctica work at a 5 to 20 km scale. These models replicate the broad shape of the ice sheets well. However, real morphology at this or a finer scale cannot be modelled at present. ERS-1 altimetry of the Greenland and Antarctic ice sheets reveals that there is a complex morphology on the ice surface related to the flow of ice.

Airborne and ground-based ice-penetrating radar data allow the identification and measurement of subglacial lakes. Subglacial lakes are identified on radar records by the presence of three distinct characteristics (Figure 17.4). The first is especially strong reflections from the ice-sheet base, which are typically 10–20 dB stronger than adjacent ice–bedrock reflections. The second characteristic is that echoes of constant strength occur along the track, indicative of an interface which is very smooth on the scale of the radar wavelength. The third is a very flat and virtually horizontal character (i.e. mirror-like), with maximum slopes typically less than 1%. By collating

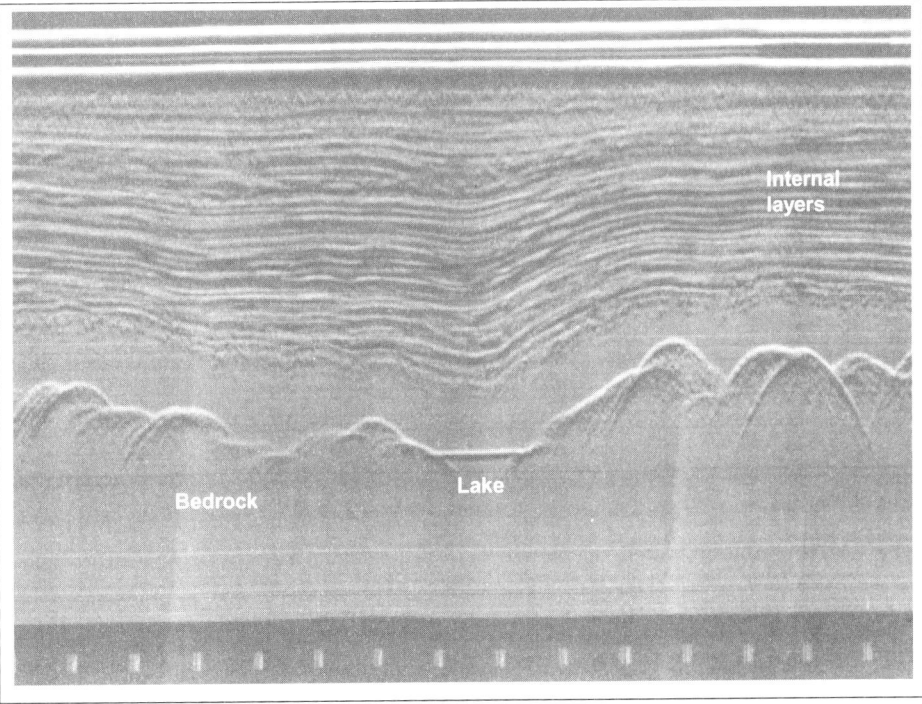

Figure 17.4 Example of a subglacial lake, imaged from SPRI 60 MHz airborne radar data. The lake is about 3.5 km below the ice-sheet surface at Ridge B, and is about 5 km long. The section is aligned orthogonal to ice-sheet flow. Continuous internal layering can be seen across the radar section

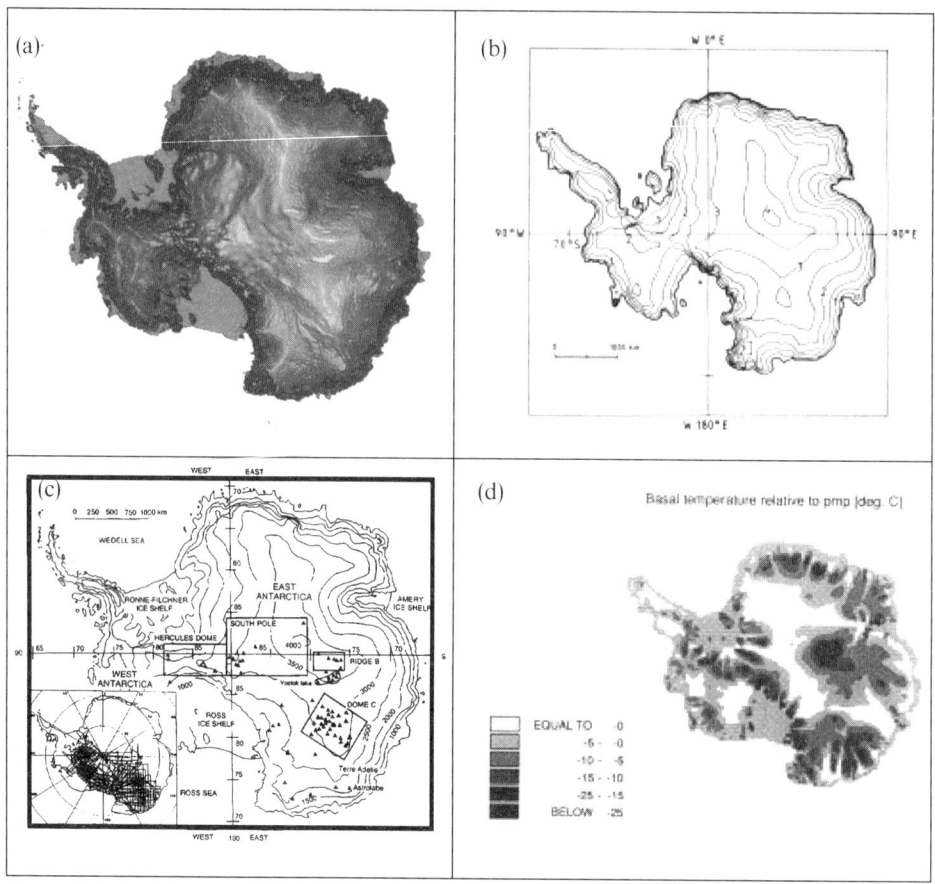

Figure 17.5 (a) ERS-1 ice surface. (b) Huybrechts model of Antarctica (ice-sheet surface) (Huybrechts 1992). (c) Large-scale ice surface (from SPRI folio) and location of subglacial lakes (Dowdeswell and Siegert 1999a). (d) Huybrechts basal thermal regime (Huybrechts 1992)

the surface elevation, bedrock morphology and locations of subglacial lakes, a powerful data set is made to test against ice-sheet model output.

A good example of where ice-sheet model results (Huybrechts 1992) match well with large-scale ice-sheet features, yet poorly with sub-grid cell morphology is at Dome C, East Antarctica (Figure 17.5). The ice surface around Dome C has a generalised surface slope of about 0.08°. However, there are a number of regions where the slope reduces to less than 0.01°. These 'flat' surfaces are caused by ice flow over subglacial trenches and/or subglacial lakes (e.g. Lake Vostok). In either case, the flow of ice is altered from the base-parallel shearing that most ice-sheet models account for, to longitudinal extension. In order to solve this problem, an accurate representation of subglacial topography is required. Although there is a major pro-gramme aimed at establishing an up to date bedrock elevation for Antarctica (BEDMAP; see

Section 17.2), the resolution of these new data is not fine enough to solve the problem of complex flow of ice over subglacial topography highlighted here.

17.4.2   Radar Layering and the Internal Flow of Ice

Ice-sheet models are capable of calculating the flow of ice in three dimensions. Ice-penetrating radar information can be used to verify calculated flowpaths because it reveals internal layering which is assumed to be isochronous (Figure 17.4). All internal layering is caused by changes in electrical properties of ice. However, there are three different ways of changing the electrical properties of ice, yielding three main types of internal layering. The first type of internal layer is when there are changes in the density of ice. This is a dominant process at ice depths less than 700 m. However, below this level the density of ice does not change very much and so internal reflections must be caused by other processes. The second form of layering is caused by the acidity of ice. Layers with an acidity in excess of the normal background level of glacier ice are formed when the aerosol products of ancient volcanic activity are incorporated in the snow chemistry on the former ice surface. This acid snow is subsequently buried by later snow fall to its present-day position, several tens of thousand years later. These acid layers are therefore isochronous surfaces. The third type of layering is where there are changes in the crystallography of ice. Such layering is thought to develop from acidic layers in the presence of enhanced stress across the stoss face of subglacial hills (Fujita et al. 1999).

Internal layers are often continuous across large sections of the Antarctic Ice Sheet (Siegert et al. 1998). Because they are isochronous, their patterns can be used to match the three-dimensional flow of ice calculated in numerical models (Mayer and Siegert 2000). However, as yet very few ice-sheet modelling investigations have utilised this potentially powerful measurement (Whillans 1976).

17.4.3   Location of Subglacial Lakes and Ice-sheet Thermal Conditions

Subglacial lakes are evidence for melting at the ice-sheet base. They can, therefore, be used to validate the thermal conditions calculated by ice-sheet models. Over 70 subglacial lakes have been identified at the base of the 13 million km^2 Antarctic Ice Sheet (Oswald and Robin 1973; Siegert et al. 1996; Dowdeswell and Siegert 1999a). The bulk of the lakes are located in the interior of the Antarctic Ice Sheet, at or close to ice divides in the form of domes and ridges, where ice velocity is low (Dowdeswell and Siegert 1999a). Almost 60% of lakes are found within 200 km of an ice crest, remembering that ice flowlines from divide to coastal margin are often over 1000 km long in Antarctica (Figure 17.5). Only about 15% of subglacial lakes are positioned more than 500 km from an ice divide. At least 16 subglacial lakes occur at locations which are close to the onset of enhanced ice flow, some hundreds of kilometres from the ice-sheet crest. An example is provided by three subglacial lakes near the onset of fast flow into Byrd Glacier (Figure 17.5). Byrd Glacier is fast-flowing and drains a very large interior ice-sheet drainage basin into the Ross Ice Shelf (Drewry 1983). These subglacial lakes are similar in size and depth to the small and probably shallow lakes found in major subglacial basins in the ice-sheet interior.

Huybrechts (1992) used the location of subglacial lakes to verify the broad-scale thermal character of the ice-sheet base across East Antarctica (Figure 17.5). Warm-based conditions were predicted across both the centre of East Antarctica and at the base of ice streams. However, because of the smoothed bedrock topography used as model input, individual lakes, or the outline of specific lake regions were not able to be matched well. There is, therefore, plenty of scope to use subglacial lakes in the future verification of ice-sheet model thermal conditions.

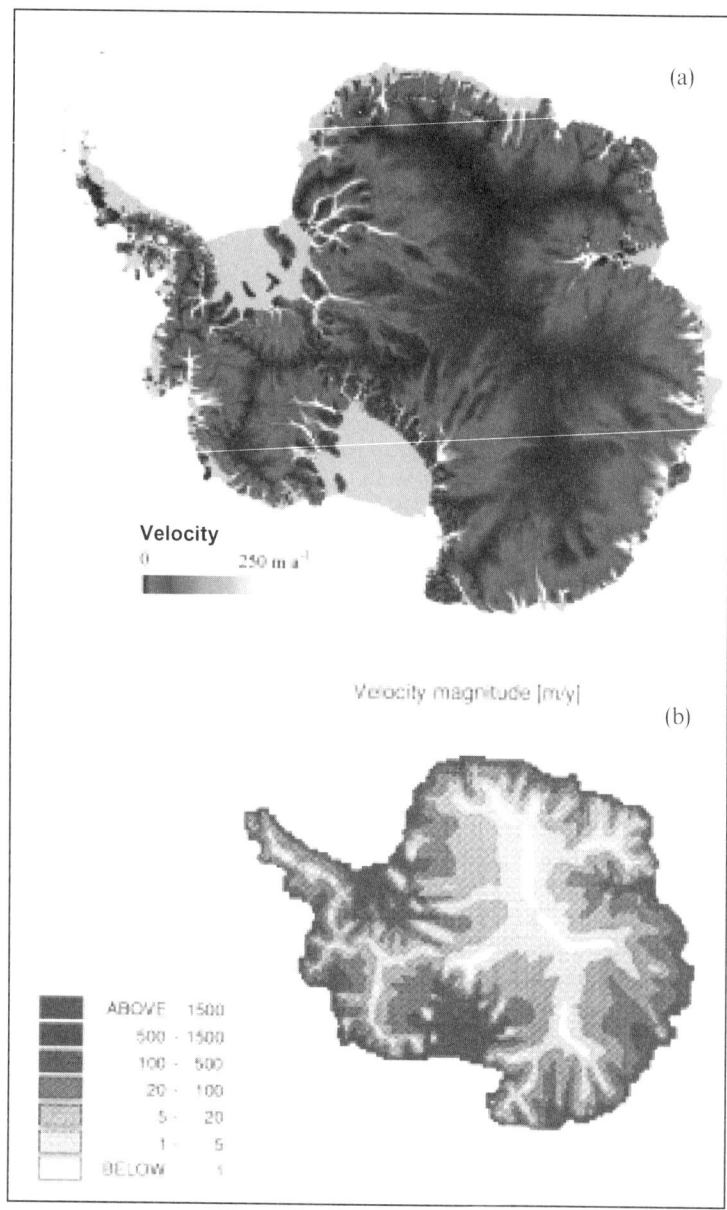

Figure 17.6   (a) Antarctic Ice Sheet flux (Bamber et al. 2000). (b) Huybrechts model of Antarctic ice sheet velocity (Huybrechts 1992)

Further validation of the ice-sheet thermal regime can be made by comparing results with temperature profiles from ice cores. However, this comparison is limited by the small number of deep ice cores and the lack of penetration to the lower layers of the ice sheet in most cases.

17.4.4   Balance Flux Modelling of Ice Sheets

A recent simple model of ice flow in Antarctica shows the flux of ice from the interior of the ice sheet to the onset of ice streams, into the ice streams and onto the ice shelves (Figure 17.6) (Bamber et al. 2000). The model is based on the ERS-1 altimetric ice surface and a new map of ice accumulation in Antarctica. It simply calculates the flux of ice required to maintain the present ice profile assuming the ice sheet in 'steady state'. This model indicates that the transition from slow flowing ice in the interior to fast moving ice in ice streams occurs across the inland region of the ice sheet. In Antarctica, several subglacial lakes have been found at the onset regions of the ice streams, showing clearly that, downstream, the ice-sheet base is warm. Since the model is not based on assumptions of how ice flows, the results could be used to compare numerical ice-sheet model output in future.

17.4.5   Satellite Interferometry

One aspect of ice sheet modelling that is in need of independent data sets for comparison and validation is the calculated velocity of ice. Until recently such data have been scarce over large ice sheets. However, recent Interferometric Synthetic Aperture Radar (InSAR) techniques have been applied to glaciers and ice sheets to reveal the surface velocity continuously across wide areas. The new RADARSAT data have a complete coverage of Antarctica and will be processed in the near future. These data will be a vital addition to the data set used to modulate numerical models of the Antarctic Ice Sheet. Work to date has shown that development of ice streams in West Antarctica seen through InSAR data (Joughin et al. 1999) compare well with Bamber's simple calculation of Antarctic ice flux (Bamber et al. 2000).

InSAR data revealing the surface velocity of ice sheets also has an application in modelling subglacial sliding. In areas where the internal ice-sheet deformation is known to occur in a simple manner, the difference between the InSAR velocity and velocity due to internal deformation of ice will be equivalent to the component due to basal processes. Currently models are incapable of predicting accurately the subglacial sliding and, so, InSAR data may be of use in establishing reliable algorithms for this process.

17.4.6   Summary

Numerical models of modern ice sheets can be compared with a variety of independent data sets at a variety of scales, for validation purposes. At present, most ice-sheet models operate with smoothed representations of topographic inputs and make assumptions about the flow of ice over quite large horizontal distances. Because of this, the spatial resolution over which they can be judged is limited. Modern ice-sheet models do not replicate the real flow of ice over complex subglacial topography.

The temperature calculation in ice-sheet models is vital to the ice-flow parameters. The location of subglacial lakes can be used to validate the thermal component of models of the Antarctic Ice Sheet. Satellite-derived InSAR ice-velocity fields across large areas of the ice surface mark a next phase in the validation of ice flow, and may be of use in constructing algorithms for ice sliding, a reliable solution to which has so far eluded ice-sheet modellers.

17.5   TESTING NUMERICAL RECONSTRUCTIONS OF QUATERNARY ICE SHEETS

The principal type of geological data used to study former ice sheets is the glacial geomorphology of the land's surface and near-surface stratigraphy, which has previously been used to gain information about the extent of the former ice sheets but also potentially holds information on their flow dynamics. Additional information can be obtained from deep sea and ice cores, which have previously been used to obtain estimates of global ice volume variation but are increasingly used to obtain information on the regional dynamics of ice sheets.

There are two principal benefits in studying the behaviour of former ice sheets. The first is that the geological data available contain information relating to periods in excess of the typical time constants associated with ice sheets (10 to 100 ka). Data collected directly from present-day ice sheets relate to only a small fraction of these time scales (at most 100 years) and are inadequate for the tasks of both assessing ice-sheet response to climatic/environmental change and understanding possible natural variation within the ice-sheet system.

The second advantage in studying former ice sheets is that the bed of the ice sheet is directly accessible. It is clear that many of the more important processes controlling ice-sheet behaviour are determined by the interaction of the ice sheet with its underlying substrate. The beds of the present-day ice sheets are only accessible by drilling through the kilometres of overlying ice. Understandably, the number of such drilling projects is extremely limited (the most successful being the Ice Stream B work summarised by Engelhardt et al. (1990)) and the results are hampered by being essentially point measurements. For this reason many basal process theories are based on observations on valley glaciers and may not be appropriate to the expanded scale of ice sheets. In contrast, many areas occupied by former ice sheets offer an almost continuous, coherent spatial record of glacigenic deposits and bedforms which reflect past basal conditions. The problem is the interpretation of this record. On the one hand, time constraint of the processes of formation is poor, while, on the other hand, the spatial dimensions of the ice sheets which formed these bedforms are still under debate.

In the absence of real information on former ice-sheet flow and past climate, numerical models of late Quaternary ice sheets must be validated by proxy records of ice flow, from the geological record. Since field data are *evidence* they can be used to test ice-sheet model results. Disagreement between model results and field data means that the ice-sheet results are most likely to be wrong. Thus, modelling of former ice-sheets is led by field evidence.

Geological data can be divided into groups depending on glacial information they hold. For example, the maximum extent of ice can be established from terminal moraines (on land and in shallow marine areas), ice-proximal deltas within proglacial lakes, and glacigenic sedimentary fans at the mouths of cross-shelf bathymetric troughs. In addition, the nature of ice-sheet decay can be established from similar geological features, especially if there is a dating control, and records of postglacial isostatic uplift (such as raised beaches). The direction of former ice-sheet flow can be established from glacial landforms that are aligned in the flow direction (such as striations, drumlins and flutes). These features also provide information about the subglacial environment since most will only form if the glacier is warm based.

Although subglacial topography can be determined accurately, most other model inputs, like the time-dependent variation in air temperature and ice-surface mass balance, are far less well known and require proxy records to formulate an estimate. However, if the geological record is well established, informed inverse-type procedures in ice-sheet modelling can be used to reconstruct ice sheets and provide information on past climate. One recent example of where this procedure has worked is the reconstruction of the LGM Eurasian ice sheet (Siegert et al. 1999).

(a)   (b)

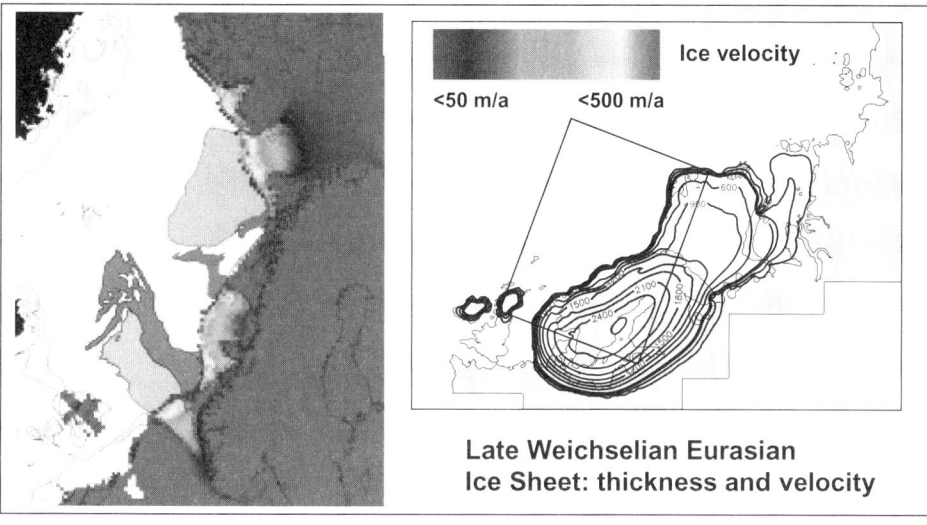

Ice velocity

Figure 17.7 Eurasian Ice Sheet model results (a) Ice sheet velocity and the locations of large sedimentary fans (light grey shade) and major sedimentary slides (dark grey shade). (b) Ice-sheet thickness (contours in 300 m) (Dowdeswell and Siegert 1999b; Siegert et al. 1999; Taylor 1999)

   The Eurasian Arctic includes what are now the Barents and Kara epi-continental seas. In the last ten years a substantial amount of geological information on the late Quaternary ice sheet has been obtained. However, with the absence of ice cores in this area, the paleoclimate required as an ice-sheet input is not known. Siegert et al. (1999) used the inverse-type modelling procedure to reconstruct the ice sheet without the need for an initial detailed climate. The inverse procedure works by forcing the ice-sheet model to match geological data, through intuitive adjustments to the model's climate inputs. Model results are then tested against geological data not used previously in the experiment. The model then predicts the ice-sheet configuration compatible with the geological data, and an associated climate scenario.

   For the Eurasian ice sheet, the maximum extent of the LGM ice sheet has recently been established (Svendsen et al. 1999). Since there are several lines of evidence that indicate ice growth began at around 28 000 years ago, Siegert et al. (1999) were able to force the ice-sheet model to match the geological limits assuming that only 10 000 years was available for ice-sheet build-up. They started with the modern mass-balance conditions for Svalbard and Scandinavia, and reduced the value of the mean annual air temperature to enable the ice sheet to grow. Model results were compared with marine sedimentary evidence for ice-stream activity at the mouths of several bathymetric troughs, and inactivity in some other troughs. The results showed that a large, dynamic ice sheet existed across the Barents Sea, incorporating ice streams within bathymetric troughs which transported sediment to the continental margin fans (Dowdeswell and Siegert 1999b; Siegert et al. 1999; Taylor 2000) (Figure 17.7). However, further east, the ice sheet was much thinner, probably cold based, and slow moving. This reconstruction contrasts markedly from earlier reconstructions of the Eurasian ice sheet, which showed a massive 3-km-thick ice dome centred between the Barents and Kara seas. Geological evidence in conjunction with numerical modelling has therefore falsified this early reconstruction.

Deglaciation of former ice sheets is also possible to model if geological evidence is available. Most ice-sheet models have an isostatic component, and the calculations of postglacial bedrock uplift can be compared with measured rates of uplift.

Ice-sheet modelling has now been performed for most of the former major ice sheets. Numerical results can be interpreted best with the aid of geological data, and the conclusions from this interpretation fed back into the modelling experiments. It should be noted that a circular argument is likely to develop unless some independent geological evidence is withheld for results comparison, rather than model forcing. It should also be noted that the true validation of models of former ice sheets is probably unlikely to ever happen. However, as the geological database increases so the range of plausible ice-sheet scenarios that match with the geological data reduces.

17.6   SUMMARY AND FUTURE DEVELOPMENTS

Ice-sheet modelling allows quantitative predictions of how large ice masses behave and respond to environmental change. They are validated by a combination of intercomparison with existing models and by comparing model results with 'real' information gathered in the field or from remote sensing. Although the large-scale behaviour of ice sheets can be reproduced well by numerical models, there are limitations to the accuracy of numerical ice-sheet modelling at a finer scale. This is because (1) ice-sheet models are derived from assumptions about the flow of ice which may not be accurate at all places in the ice sheet and (2) subglacial topography and other model inputs may be insufficiently well known to allow the model to work at a fine scale.

The latter problem can be solved by obtaining high-resolution geophysical data sets of existing ice sheets to establish the subglacial topography, internal structure and surface elevation of the ice sheet. Such data acquisition has been ongoing for several decades. There are a number of world data-centres where information on glaciers and ice sheets are held, and several ongoing or planned programmes that aim to collate specific types of geophysical information. For example, the BEDMAP programme is establishing a new subglacial topography in Antarctica based on all available airborne radar data. Further, a new map of subglacial topography in Greenland is being compiled (J. Bamber, Bristol Glaciology Centre, personal communication) based on recently acquired NASA airborne radar data. However, even after these subglacial data are combined there will still be areas of both ice sheets that remain to be surveyed by airborne radar. The mapping of internal radar layering allows the establishment of the three-dimensional flow of ice. Conceivably, maps of isochronous surfaces in the ice sheet, based on internal radar layers, could be determined wherever the ice sheet has been surveyed by airborne radar. Information from new satellites such as RADARSAT will shortly result in highly detailed surface topographies of Antarctica and Greenland that supplement existing data, and fill the data holes south and north of 80°. Because of the advances in measuring ice sheets, in the near future numerical models of the Antarctic and Greenland ice sheets will be run with a series of much improved boundary conditions.

A solution to the former problem requires an advance in how the flow of ice is modelled beyond the empirical flow law that most use. The next generation of ice-sheet models may benefit from advances in both these areas. A potential way forward is in advanced numerical techniques that describe the flow of ice without having to make simplifying assumptions. These models are currently under construction by a number of scientists. However, to test them, detailed information about the ice sheet is required from geophysical sources. Such data are currently only available along two-dimensional cross-sections. These cross-sections are often aligned along the line of ice flow. Thus, the next step in ice-sheet modelling may require experiments along two-dimensional ice-sheet flowlines, where a full suite of geophysical data is available, in order to develop numerical techniques that may result in advances to three-dimensional continental-scale modelling.

Ice-sheet models are used to predict how the cryosphere responds to and affects climate change. From climate information gathered from ice cores, there is reason to believe that ice sheets, oceans and the atmosphere interact with each other. It is therefore essential that future ice-sheet models be coupled with models of atmospheric and ocean circulation in order to predict how the combined ice–ocean–atmosphere system behaves.

ACKNOWLEDGEMENTS

We thank Dr C.S.M. Doake for reviewing a draft of this chapter.

REFERENCES

Bamber, J.L., Vaughan, D.G. and Joughin, I. 2000. Widespread complex flow in the interior of the Antarctic Ice Sheet. *Science*, **287**, 1248–1250.

Clarke, G.K.C., Nitsan, U. and Paterson, W.S.B. 1977. Strain heating and creep instability in glaciers and ice sheets. *Reviews of Geophysics and Space Physics*, **15**, 235–247.

Dowdeswell, J.A. and Siegert, M.J. 1999a. The dimensions and topographic setting of Antarctic subglacial lakes and implications for large-scale water storage beneath continental ice sheets. *Geological Society of America, Bulletin*, **111**, 254–263.

Dowdeswell, J.A. and Siegert, M.J. 1999b. Ice-sheet numerical modeling and marine geophysical measurements of glacier-derived sedimentation on the Eurasian Arctic continental margins. *Geological Society of America, Bulletin*, **111**, 1080–1097.

Drewry, D.J. 1983. *Antarctica: Glaciological and Geophysical Folio*. Scott Polar Research Institute, University of Cambridge.

Engelhardt, H., Humphrey, N., Kamb, B. and Fahnstock, M. 1990. Physical conditions at the base of a fast flowing Antarctic ice stream. *Science*, **248**, 57–59.

Fujita, S., Maeno, H., Uratsuka, S., Furukawa, T., Mae, S., Fujii, Y. and Watanabe, O. 1999. Nature of radio-echo layering in the Antarctic ice sheet detected by a two-frequency experiment. *Journal of Geophysical Research*, **104**(B6), 13013–13124.

Glen, J.W. 1955. The creep of polycrystalline ice. *Proceedings of the Royal Society of London, Series A*, **228**, 519–538.

Greve, R. and MacAyeal, D.R. 1996. Dynamic/thermodynamic simulations of Laurentide ice sheet instability. *Annals of Glaciology*, **23**, 328–335.

Hindmarsh, R.C.A. and Payne, A.J. 1996. Time step limits for stable solutions of the ice sheet equation. *Annals of Glaciology*, **23**, 74–85.

Hooke, R.L. 1981. Flow law for polycrystalline ice in glaciers. *Reviews of Geophysics*, **19**, 664–672.

Hulbe, C.L. and MacAyeal, D.R. 1999. A new numerical model of coupled inland ice sheet, ice stream, and ice shelf flow and its application to the West Antarctic Ice Sheet. *Journal of Geophysical Research*, **104**(B11), 25349–25366.

Huybrechts, P. 1990. A 3-D model for the Antarctic Ice Sheet: a sensitivity study on the glacial–interglacial contrast. *Climate Dynamics*, **5**, 79–92.

Huybrechts, P. 1992. The Antarctic ice sheet and environmental change: a three dimensional modelling study. *Reports on Polar Research* (Alfred-Wegener-Institut für Polar und Meeresforschung) 99.

Huybrechts, P. and Wolde, J. de. 1999. The dynamic response of the Greenland and Antarctic ice sheets to multiple-century climatic warming. *Journal of Climate*, **12**, 2169–2188.

Huybrechts, P. and Oerlemans, J. 1988. Evolution of the East Antarctic Ice Sheet: a numerical study of thermomechanical response patterns with changing climate. *Annals of Glaciology*, **11**, 52–59.

Huybrechts, P. and Oerlemans, J. 1990. Response of the Antarctic Ice Sheet to future greenhouse warming. *Climate Dynamics*, **5**, 93–102.

Huybrechts, P., Payne, A.J. and EISMINT Intercomparison Group, 1996. The EISMINT benchmarks for testing ice-sheet models, *Annals of Glaciology*, **23**, 1–12.

Joughin, I., Gray, L., Bindschadler, R., Price, S., Morse, D., Hulbe, C., Mattar, K. and Werner, C. 1999. Tributaries of West Antarctic ice streams revealed by RADARSAT interferometry. *Science*, **286**, 283–286.

Lam, J.K.W. and Dowdeswell, J.A. 1996. Adaptive-grid finite-volume model of glacier-terminus fluctuations. *Annals of Glaciology*, **23**, 86–93.

Le Muer, E. and Huybrechts, P. 1996. A comparison of different ways of dealing with isostasy: examples from modelling the Antarctic Ice Sheet during the last glacial cycle. *Annals of Glaciology*, **23**, 309–317.

Lythe, M.B., Vaughan, D.G. and the BEDMAP consortium. 2000. BEDMAP – bed topography of the Antarctic. 1:10,00,000 map. BAS (Misc.) 9, Cambridge.

MacAyeal, D.R. 1989. Large-scale flow over a viscous basal sediment: theory and application to Ice Stream B, Antarctica. *Journal of Geophysical Research*, **94**(B4), 4071–4087.

MacAyeal, D.R. 1993. A tutorial on the use of control methods in ice-sheet modeling. *Journal of Glaciology*, **39**, 91–98.

MacAyeal, D.R., Bindschadler, R.A. and Scambos, T.A. 1995. Basal friction of Ice Stream E, West Antarctica. *Journal of Glaciology*, **41**, 247–262.

MacAyeal, D.R., Rommelaere, V., Huybrechts, P., Hulbe, C.L., Determan, J. and Ritz, C. 1996. An ice-shelf model test based on the Ross Ice Shelf, Antarctica. *Annals of Glaciology*, **23**, 46–51.

Marshall, S.J. and Clarke, G.K.C. 1997. A continuum mixture model of ice stream thermodynamics in the Laurentide Ice Sheet 1. Theory. *Journal of Geophysical Research*, **102**, 20 599–20 613.

Mayer, C. and Siegert, M.J. 2000. Numerical modelling of ice-sheet dynamics across the Vostok subglacial lake, central East Antarctica. *Journal of Glaciology*, **46**, 197–205.

Oerlemans, J., Anderson, B., Hubbard, A., Huybrechts, P., Jóhannesson, T., Knap, W.H., Schmeits, M., Stroeven, A.P., van de Wal, R.S.W., Wallinga, J. and Zuo, Z. 1998. Modelling the response of glaciers to climate warming, *Climate Dynamics*, **14**, 267–274.

Oswald, G.K.A. and de Q. Robin, G. 1973. Lakes beneath the Antarctic Ice Sheet. *Nature*, **245**, 251–254.

Paterson, W.S.B. and Budd, W.F. 1982. Flow parameters for ice sheet modelling. *Cold Region Science and Technology*, **6**, 175–177.

Pattyn, F. 1996. Numerical modelling of a fast flowing outlet glacier: experiments with different basal conditions. *Annals of Glaciology*, **23**, 237–246.

Payne, A.J. 1995. Limit cycles in the basal thermal regime of ice sheets. *Journal of Geophysical Research*, **100**(B3), 4249–4263.

Payne, A.J. 1999. A thermomechanical model of ice flow in West Antarctica. *Climate Dynamics*, **15**, 115–125.

Payne, A.J. and Baldwin, D.J. 2000. Analysis of the ice flow instabilities identified in the EISMINT intercomparison exercise. *Annals of Glaciology*, **30**, 204–210.

Payne, A.J. and Dongelmans, P.W. 1997. Self organization in the thermomechanical flow of ice sheets. *Journal of Geophysical Research*, **102**, 12 219–12 234.

Payne, A.J. et al. 2000. Results from the EISMINT Phase 2 Simplified Geometry Experiments: the effects of thermomechanical coupling. *Journal of Glaciology*, **46**, 227–238.

Rommelaere, V. and Ritz, C. 1996. A thermomechanical model of ice-shelf flow. *Annals of Glaciology*, **12**, 13–20.

Siegert, M.J. and Dowdeswell, J.A. 1995. Modelling ice-sheet sensitivity to late Weichselian environments in the Svalbard Barents Sea region. *Journal of Quaternary Science*, **10**, 33–43.

Siegert, M.J., Dowdeswell, J.A., Gorman, M.R. and McIntyre, N.F. 1996. An inventory of Antarctic subglacial lakes. *Antarctic Science*, **8**, 281–286.

Siegert, M.J., Hodgkins, R. and Dowdeswell, J.A. 1998. A chronology for the Dome C deep ice-core site through radio-echo layer correlation with the Vostok ice core, Antarctica. *Geophysical Research Letters*, **25**, 1019–1022.

Siegert, M.J., Dowdeswell, J.A. and Melles, M. 1999. Late Weichselian glaciation of the Eurasian High Arctic. *Quaternary Research*, **52**, 273–285.

Svendsen, J.I., Astakov, V.I., Bolshiyanov, D.Yu., Demidov, I., Dowdeswell, J.A., Gataullin, V., Hjort, Ch., Hubberten, H.W., Larsen, E., Mangerud, J., Melles, M., Möller, P., Saarnisto, M. and Siegert, M.J. 1999. Maximum extent of the Eurasian ice sheets in the Barents and Kara Sea region during the Weichselian. *Boreas*, **28**, 234–242.

Taylor, J. 1999. Large-scale sedimentation and ice sheet dynamics in the Polar North Atlantic. Unpublished PhD Thesis, University of Bristol, 2 volumes.

Whillans, I.M. 1976. Radio-echo layers and the recent stability of the West Antarctic Ice Sheet. *Nature*, **264**(5582), 152–155.

18   Discussion of Model Validation in Relation to the Regional and Global Scale

JENS CHRISTIAN REFSGAARD

Geological Survey of Denmark and Greenland, Copenhagen, Denmark

18.1   INTRODUCTION

This chapter describes and discusses the present status of model validation in relation to regional and global scale hydrological modelling. A terminology and methodology for model validation is defined. The different categories of hydrological models are outlined and discussed, both models originating in the traditional hydrological community and land surface parameterisation schemes in the atmospheric models. A review of six selected cases of regional and global scale hydrological models is carried out. The six cases ranging between 110 000 km^2 and 80 million km^2 are characterised by different types of hydrological models, different degrees of coupling to atmospheric models and differences in validation status. On this basis the advantages and limitations of conceptual and physically based model types are discussed. The land surface components in atmospheric models are seen generally not to perform as well for simulation of discharge as the traditional hydrological models. A serious constraint in the validation status of regional and global scale hydrological models is that they generally only are validated against historical discharge measurements at a few stations. To improve the credibility of such models for the prediction of climate change or land use change effects on water resources there is a need to focus more on validation of internal spatial variables. At regional and global scales the only realistic data source for such internal model validation appears to be spatial remote sensing data.

18.1.1   The Need for Large-scale Hydrological Modelling

Hydrological modelling of large areas is relevant for two main purposes, namely water resources management and climate research.

Human activities such as constructions of dams, land reclamation from wetlands, water abstraction from surface and groundwater sources, irrigation, changes in cropping patterns, and de/af-forestation may have significant influences on the water resources in a region. With the increasing pressure on land and water resources and the increased demands for food, electricity and water caused by high population growth, it becomes increasingly important that the available water resources are used in a sustainable manner. Water allocations to individual sectors and to different subregions should be fulfilled in a manner, which is acceptable from an overall quantitative and qualitative perspective.

Model Validation: Perspectives in Hydrological Science. Edited by M.G. Anderson and P.D. Bates.
© 2001 John Wiley & Sons, Ltd.

The need for an integrated water resources management in river basins has frequently been voiced and is commonly recognised as a prerequisite for sustainable development. In this context the need for water resources management on larger river basins is generating a demand for hydrological modelling at large scales. These activities are mostly carried out within the traditional hydrological community, involving both researchers and practitioners.

In addition, General Circulation Models (GCMs) and Limited Area Models (LAMs) that are widely applied at global and regional scale in climate research include hydrological descriptions of large areas. These activities are mostly driven by the meteorological scientific community in cooperation with hydrologists.

18.1.2   The Need for Model Validation

Hydrological models are being developed and applied in increasing number and variety. While the operational hydrological models previously have been primarily of the conceptual type or simpler (WMO 1975, 1986, 1992), the trends in recent years both in the research and in the operational communities have been towards more emphasis on distributed and physically based hydrological models. At the same time contradictions are emerging regarding the various claims of model applicability on the one hand and the lack of validation of these claims on the other hand. Hence, the credibility of the models is often questioned, and sometimes with good reason. This is particularly true with regard to the more complex physically based models such as the SHE (Abbott et al. 1986a,b; Bathurst and O'Connell 1992; Refsgaard and Storm 1995) and the Thales (Grayson et al. 1992a,b).

Whereas much attention during the past three decades has been given to specific procedures for parameter assessment, calibration and, to a lesser extent, validation of lumped models (e.g. Fleming 1975; WMO 1975, 1986, 1992; Klemes 1986; Sorooshian et al. 1993), very limited attention has so far been devoted to the far more complicated tasks in connection with distributed models, where problems related to validation of internal variables and multiple scales also have to be considered (Refsgaard 2000).

Comprehensive progress has been made in atmospheric modelling during the past decades. This includes attempts to validate the models. However, it is only within the last decade or so that significant focus has been put on validation of the land surface part of these models against observed land surface data such as discharges, snow cover and soil moisture contents.

18.1.3   Contribution of this Chapter

The aim of this chapter is to describe and discuss the present status of model validation in relation to regional and global scale hydrological modelling.

18.2   HYDROLOGICAL SPATIAL SCALES AND MODEL CATEGORIES

18.2.1   Definition of Spatial Scales

Hydrological modelling is being carried out at spatial scales ranging from point scale to global scale. A variety of definitions of such scales has been proposed (e.g. Dooge 1986; Becker 1992; Blöschl and Sivapalan 1995). In the present chapter the definitions given in Table 18.1 will be used.

Process descriptions in hydrology are often studied and modelled at the point scale. This usually results in rather complex process descriptions. When moving to larger scales the spatial

Table 18.1   Definition of spatial hydrological modelling scales

Spatial scale	Characteristics	
	Length	Area
Point scale	< 10 cm	
Field or hillslope scale	100 m	
Catchment scale	3–100 km	$10–10^4$ km^2
Regional scale	100–1000 km	$10^4–10^6$ km^2
Continental or global scale	> 1000 km	> 10^6 km^2

variabilities of physical parameters and variables have to be taken into account. This can in principle be done in two ways, either by aggregation or upscaling (Heuvelink and Pebesma 1999). The aggregation strategy implies that the same process equations are used at a larger scale and that the model output at the larger scale is derived after aggregation of either data and/or model runs at the smaller scale. The upscaling strategy implies that process descriptions are somehow integrated to represent the lumped conditions at a larger scale. This generally results in simpler descriptions as expressed by, for example, Dooge (1985):

> the finer scale processes may either be ignored or may be represented by their statistical effects in the large scale description. . . . It has been found in practice that the models based on continuum mechanics are too complex to allow for the spatially variable nature of hydrologic systems to be taken into account in large scale modelling and they have been simplified to such an extent that they became in effect simple conceptual models.

Many hydrological models have their origin and were originally tested at the hillslope or at the field scale, identical to the size of many small experimental catchments. TOPMODEL (Beven et al. 1995) is a representative of a topography driven hillslope model and the first test of the SHE (Bathurst 1986) is an example of a field scale/experimental catchment model.

18.2.2   Model Categories in the Hydrological Community

The two traditional approaches in hydrological modelling are the conceptual and the physically based ones (Refsgaard (1996) and many others). Typical examples of conceptual model codes are the Stanford Watershed Model (Crawford and Linsley 1966), the Sacramento (Burnash 1995), the HBV (Bergström 1995) and the Arno/Xinan-Jiang (Dümenil and Todini 1992). Typical examples of physically based model codes are the MIKE SHE (Abbott et al. 1986a,b; Refsgaard and Storm 1995) and the Thales (Grayson et al. 1992a,b). The conceptual models are used in either a lumped or a semi-distributed mode. A lumped model implies that the catchment is considered as one computational unit, whereas a semi-distributed model uses some kind of distribution, either in subcatchments or in hydrological response units, where areas with the same key characteristics are aggregated to sub-units without considering their actual locations within the catchment. Examples of hydrological response units considered in semi-distributed models are elevation zones, which are relevant for snow modelling, and combinations of soil and vegetation type, which may be relevant for simulation of evapotranspiration (and hence runoff generation). A distributed model, on the other hand, provides a description of catchment processes at georeferenced computational grid points within the catchment. A physically based model may be used as either fully distributed or, in certain cases, semi-distributed.

The fundamental difference between the conceptual and the physically based models lies in their process descriptions and the way spatial variability is treated. The physically based models contain equations which have originally been developed for point scales and which provide detailed descriptions of flows of water, solutes and energy. The variability of catchment characteristics is accounted for explicitly through the variations of hydrological parameter values among the different computational grid points. This approach leaves the variability within a grid as unaccounted for, which in some cases is of minor importance but in other cases may pose a serious constraint. The conceptual models (irrespective of whether they are used in a lumped or a semi-distributed mode) use empirical process descriptions, which have built-in accounting for the spatial variability of catchment characteristics. The conceptual models have been developed for and are very good at describing runoff generation and overall water balances, but generally fail to provide details of water flows and soil moisture storages as well as descriptions of fluxes of solutes and energy.

The conceptual models have typically been applied to catchments ranging from 10 to 10^4 km^2, while the physically based models additionally have been applied at smaller scales for process studies at field and hillslope scale.

18.2.3   Model Categories for Land Surface Parameterisation of Atmospheric Models

During the past decade remarkable progress has been made in modelling the world's climate by use of General Atmospheric Circulation Models (GCMs). The spatial resolution of GCMs are of the order of 10^4 km^2, i.e. larger than the computational units used both for conceptual and, in particular, for physically based hydrological models. In the meteorological and the global change modelling community hydrological models are often denoted 'land surface parameterisation schemes' and form submodels of the larger atmospheric models. An overview of the different methodologies for coupled atmospheric–hydrological modelling adopted in previous studies is given in Table 18.2 and may be summarised as follows:

(a) Land surface parameterisation schemes composed of a simple bucket that neither considers vertical nor horizontal variability of soil moisture. An example of this approach is Manabe (1969).

(b) Land surface parameterisation schemes composed of soil–vegetation–atmosphere-transfer (SVAT) that do not consider sub-grid spatial variability. Examples of this approach are the vertical Biosphere–Atmosphere-Transfer-Scheme (BATS) (Dickinson et al. 1986) and the Simple Biosphere Model (SiB) (Sellers et al. 1986).

(c) Land surface parameterisation schemes composed of SVAT models with statistical representation of sub-grid heterogeneity. An example of such approach is the work by Entekhabi and Eagleson (1989).

(d) Land surface parameterisation schemes based on a traditional lumped conceptual hydrological model. Dümenil and Todini (1992) describe the ARNO rainfall–runoff model for use in an atmospheric model and Rowntree and Lean (1994) report results from using the ARNO scheme in a GCM model. Lumped conceptual models implicitly account for spatial variability in the empirical (conceptual) process equations.

(e) Land surface parameterisation schemes which subdivide the atmospheric model grid into smaller (hydrological) units of variable properties and subsequent aggregation of the hydrological response. This approach is reported by Koster and Suarez (1992) but has not been used much so far. The sub-units may then again be described using the point scale approach (b) or one of the catchment scale approaches (c) or (d).

Table 18.2 Overview of different methodologies for coupled atmospheric–hydrological modelling with special emphasis on description of sub-grid spatial variability and capability of utilising spatial remote sensing data

Type	Land surface parameterisation scheme					Validation possible?		
	Hydrological units per atmospheric grid	Sub-grid variability within atm. grid	Vegetation-unsaturated zone flow description	Subsurface soil water distribution within atmospheric grid	Example	Validation at point scale?	Validation at meso scale	Utilise remote sensing data
Bucket	One	No	Very simple	No	Manabe (1969)	No	No	No
Column SVAT	One	No	Advanced	No	BATS (Dickenson et al. 1986; SiB (Sellers et al. 1986)	Yes	No	No
Stochastic SVAT	One	Statistical representation	Simple	No	Entekhabi and Eagleson (1989)	No	Yes	No
Lumped conceptual hydrological model	One	Implicitly included in hydrological model equations	Simple	No	Dümenil and Todini (1992)	No	Yes	No
Semi-distributed	Some	Atm. model grid subdivided in hydrological sub-units	Medium	No	Koster and Suarez (1992)	(Yes)	Yes	(Yes)
Distributed physically based	Many	Atm. model grid subdivided in hydrological sub-units	Advanced	Yes. Groundwater model	Famiglietti and Wood (1994a,b)	Yes	Yes	Yes

In addition to the above aspect of spatial variability of surface processes the subsurface soil water and groundwater processes often become important in larger catchments. This adds a sixth group to the above model classes:

(f)   Models which in addition to spatial variability of the surface processes incorporate subsurface soil water redistribution within the atmospheric model grid. Very few such models have been reported yet. Famiglietti and Wood (1994a,b) describe a simplified model of this type based on the TOPMODEL concept and an initial application to FIFE data.

Each of the six approaches has their strengths and shortcomings. Approach (a) is very simple and not being used much any longer. Approach (b) is very suitable for a point scale where process models can be validated. However, by ignoring heterogeneity this approach cannot be validated at the catchment scale where its physical realism may be questioned.

Approaches (c) and (d) are generally suitable at the catchment scale provided that the variability assumptions are valid. However, these approaches do not provide information at point scale. Hence, they cannot be validated at point scale and cannot fully utilise spatial information such as remote sensing data which usually are available with a much finer resolution than the model discretisation.

Approach (f), which in its most advanced form comprises a distributed physically based description of land surface processes with a finer resolution in the hydrological model than in the atmospheric model, has the potential for providing the most correct description making full use of all available data. Such an approach is particularly needed, when groundwater flow or water tables play a significant role for land surface processes, e.g. in cases of wetlands and artificially drained areas, as well as in areas where significant human interactions have occurred. However, this approach is not yet fully operational.

Approach (e) may be a compromise enabling both validation at point scale, where good data sets are often available, direct utilisation of remote sensing data, and accounting of spatial variability. The limitation of this approach is that the lateral movements of water within catchments are usually not accounted for. This approach corresponds otherwise to the semi-distributed hydrological models.

The only model type which is fully capable of utilising remote sensing data is the distributed physically based model.

The early versions of land surface components in GCMs are often characterised as simple by hydrologists (Becker 1992; Beven 1995). Beven (1995) points out that although the more detailed and complex models, such as the BATS and the SiB, are sophisticated in their vertical representation of the soil–plant boundary layer system, they are essentially point scale models that are used at the GCM grid scale without accounting for the spatial variability of the process parameters. The key problem today is, according to Laval (1997), how to account for sub-grid scale spatial variability in rainfall, vegetation cover and soil hydraulic parameters.

18.3   VALIDATION IN HYDROLOGICAL MODELLING

18.3.1   Confusion on Terminology

No unique and generally accepted terminology exists at present in the hydrological community. As illustrated below in the few examples from different hydrological subcommunities, many different, and highly contradictory, definitions and methodologies for model validation are presently used.

Konikow (1978) and Anderson and Woessner (1992) use the term verification with respect to the governing equations, the code or the model. According to Konikow (1978) a model is verified 'if its accuracy and predictive capability have been proven to lie within acceptable limits of errors by tests independent of the calibration data'. Tsang (1991) uses the term model verification in the meaning of checking the model's capability to reproduce historical data. Anderson and Woessner (1992) define model validation as tests showing whether the model can predict the future. As opposed to the above others, Flavelle (1992) distinguishes between verification (of computer code) and validation (of site-specific model). Oreskes et al. (1994), using a philosophical framework, state that verification and validation of numerical models of natural systems theoretically is impossible, because natural systems are never closed and because model results are always non-unique. Instead, in their view models can only be 'confirmed'.

18.3.2   Can a Hydrological Model be Validated?

In addition to the terminology confusion the more fundamental question, whether hydrological models at all can be validated, has been discussed at length during the past years. Konikow and Bredehoeft (1992) argued, as exponents of the one perception that the terms validation and verification are misleading and their use should be abandoned: 'the terms validation and verification have little or no place in ground-water science; these terms lead to a false impression of model capability'. The main argument in this respect relate to the anti-positivistic view that a theory (in this case a model) can never be proved to be generally valid, but can in contrary be falsified by just one example. De Marsily et al. (1992) argued in a response to Konikow and Bredehoeft (1992) for a more pragmatic view: 'using the model in a predictive mode and comparing it with new data is not a futile exercise; it makes a lot of sense to us. It does not prove that the model will be correct for all circumstances, it only increases our confidence in its value. We do not want certainty; we will be satisfied with engineering confidence.'

Another example of a scientist critical towards the term model validation is Beven (1996), who, among others, as a reaction against many model developers and users who regularly use the term almost as a rubber stamp, stated that model validation is not possible: 'is there any example of a successfully validated distributed model at the catchment scale?'

A key problem in this discussion is that the term hydrological model often is used in two different senses, namely as a model code (software program) and as a site-specific model including input data and parameter values.

18.3.3   Proposed Framework for Discussion on Model Validation

The terminology and methodology used in the following for discussion of model validation in hydrology is described in details in Refsgaard (2000) based on ideas outlined by Schlesinger et al. (1979), Klemes (1986) and Anderson and Woessner (1992).

The key terms are here defined as follows:

- A *model code* is a generic software program, which can be used for different catchments without modifying the source code.
- A *model* is a site application of a code to a particular catchment, including input data and parameter values.
- A model code can be *verified*. A code verification involves comparison of the numerical solution generated by the code with one or more analytical solutions or with other numerical solutions. Verification ensures that the computer program accurately solves the equations that constitute the mathematical model.

- Model *validation* is here defined as the process of demonstrating that a given site-specific model is capable of making accurate predictions for periods outside a calibration period. A model is said to be validated if its accuracy and predictive capability in the validation period have been proven to lie within acceptable limits or errors.

It is important to notice that the term model validation refers to a site-specific validation of a model. This must not be confused with a more general validation of a model code, which will never be possible.

The validation test scheme proposed below is based on Klemes (1986), who states that a model should be tested to show how well it can perform the kind of task for which it is specifically intended. The four types of test correspond to different situations with regard to whether data are available for calibration and whether the catchment conditions are stationary or the impact of some kind of intervention has to be simulated.

The *split-sample test* is the classical test, being applicable to cases where there is sufficient data for calibration and where the catchment conditions are stationary. The available data record is divided into two parts. A calibration is carried out on one part and then a validation on the other part. Both the calibration and validation exercises should give acceptable results.

The *proxy-basin test* should be applied when there is not sufficient data for a calibration of the catchment in question. If, for example, streamflow has to be predicted in an ungauged catchment Z, two gauged catchments X and Y within the region should be selected. The model should be calibrated on catchment X and validated on catchment Y and vice versa. Only if the two validation results are acceptable and similar can the model command a basic level of credibility with regard to its ability to simulate the streamflow in catchment Z adequately.

The *differential split-sample test* should be applied whenever a model is to be used to simulate flows, groundwater levels and other variables in a given gauged catchment under conditions different from those corresponding to the available data. The test may have several variants depending on the specific nature of the modelling study. If, for example, a simulation of the effects of a change in climate is intended, the test should have the following form. Two periods with different values of the climate variables of interest should be identified in the historical record, such as one with a high average precipitation and the other with a low average precipitation. If the model is intended to simulate streamflow for a wet climate scenario, then it should be calibrated on a dry segment of the historical record and validated on a wet segment. Similar test variants can be defined for the prediction of changes in land use, effects of groundwater abstraction and other such changes. In general, the model should demonstrate an ability to perform through the required transition regime.

The *proxy-basin differential split-sample test* is the most difficult test for a hydrological model, because it deals with cases where there is no data available for calibration and where the model is directed to predicting non-stationary conditions. The test is a combination of the two previous tests.

18.4   REVIEW OF CASES OF REGIONAL AND GLOBAL SCALE HYDROLOGICAL MODELLING

In this section some cases of regional and global hydrological modelling are reviewed. Only cases with model areas larger than 100 000 km^2 and simulation time steps of one day or finer have been considered. Such cases are, according to the definitions shown in Table 18.1, characterised as belonging to the upper end of the regional scale or to the global scale. It should be emphasised that the review is not complete in terms of including all reported cases, but rather selective in presenting typical cases which, altogether, aim at providing a state-of-the-art picture.

The cases are briefly described below in a sequence following descending catchment areas. Key features of the selected cases are given in Table 18.3.

18.4.1   Simulation of Large River Basins with the ECHAM Model

Dümenil and Todini (1992) presented a comparison of the simulated and observed discharge hydrographs for 16 of the largest river basins of the earth. The simulations were made with the ECHAM model, where the Arno scheme (conceptual hydrological approach) is integrated as a runoff component. The spatial resolution is 5.625°, corresponding to a grid size of typically 600 km × 600 km. The model was run for a 20 year period with apparently no calibration carried out for the hydrological component. Thus the entire simulation period serves as a validation test of the category proxy-basin test (see Section 18.3.3 above).

The hydrograph comparisons shown by Dümenil and Todini (1992) are based on data aggregated to monthly values averaged over the 20 year simulation period. The study is very interesting, because it was one of the first attempts of its kind to validate a GCM against discharge data. The test results indicate that there is still significant scope for improvements. For a few of the river basins the results are quite good, while for many others either the annual mean and/or the annual variation deviates substantially from the measured data.

18.4.2   Baltic Sea Basin

The Baltic Sea Basin covers a total land area of 1 600 000 km^2 at the outlet through the Danish Straits. This does not include the surface area of the Baltic Sea itself, 377 000 km^2. The water balance of this basin was simulated by the HBV Baltic Basin Water Balance Model (HBV-Baltic; Graham 1999) based on 25 subbasins ranging from 21 000 km^2 to 144 000 km^2. The model code used was the most recent version of the conceptual HBV (Lindström et al. 1997). The model simulated the precipitation runoff process on a daily basis using daily synoptic precipitation and air temperature data. The 11 years of available monthly runoff record was divided into two periods, with the earlier period being used for model calibration and the remaining period for model validation (split-sample test). As an illustration of a typical spatial discretisation for conceptual hydrological models the subbasin boundaries for the HBV-Baltic model is shown in Figure 18.1. Graham (1999) concludes that the model generally simulates the runoff well for the entire basin while the runoff simulated at individual subbasins, although still quite well, is less accurate. The comparison of simulated and recorded discharge hydrographs shown in Figure 18.2 is at the same good level of agreement as is usually found when conceptual models are applied by experienced hydrologists.

The HBV-Baltic model is used for cooperative research with both meteorological and hydrological modelling within the Baltic Sea Experiment (BALTEX). The objective is to provide off-line analysis for coupled model development and to fill a needed role until truly coupled models become available. Furthermore, the model is suitable for operational applications and will be used to extend runoff records, fill in missing data and perform quality checks on new observations (Graham, 1999).

One example of application is an intercomparison with the land surface component of Max-Planck's global ECHAM4/T106 climate model (Graham et al. 1988). In this case the HBV-Baltic was run with daily precipitation and temperature data from the climate model. The model results in terms of simulated runoff, snow water equivalent and soil moisture deficit were then compared to the equivalent values from running the climate model alone. A key conclusion from this study is that there are fundamental differences in the conceptualisation in the land

Table 18.3 Key characteristics of selected cases of hydrological modelling at regional and global scale

Study area Location	Area (10³ km²)	Purpose	Hydrological model Name	Category	Spatial discretisation	Atmospheric model Fully coupled	De-coupled Provide input	Validation Approach	Status	Reference
16 of the largest river basins globally	80000	Validation of GCM against discharge hydrographs	Arno	Conceptual	360000 km²	ECHAM2	ECHAM2	Proxy-basin Discharge data	Research off-line	Dümenil and Todini (1992)
Baltic Sea	1600	Cooperative research with atmospheric and hydrological models (BALTEX)	HBV	Conceptual	25 subcatchments 21000–144000 km²	—	ECHAM4	Split-sample Discharge data	Research off-line	Graham (1999)
Mackenzie river basin	1600	Research linked to validation of GCM performance	SLURP	Conceptual	25 units average: 64000 km²	—	CCC GCM II	Split-sample Discharge data	Research off-line	Kite et al. (1994)
Senegal river basin	350	Research project. Integration of hydrological models and remote sensing data	MIKE SHE	Physically-based	22000 grids 16 km²	—	—	Split-sample Discharge data Internal validation against satellite data	Research off-line	Andersen (1998)
Yangtze river near Three Gorges	180	Operational flood forecasting	NAM-MIKE 11	Conceptual	15 subcatchments 3000–30000 km²	—	HIRLAM	Split-sample Water level data	Operational Real-time test	DHI (1998)
Odra river basin	110	Cooperative research with atmospheric and hydrological models (BALTEX)	SEWAB	Physically-based	350 grids 324 km²	—	—	Split-sample Discharge data	Research off line	Ruhe et al. (1998)

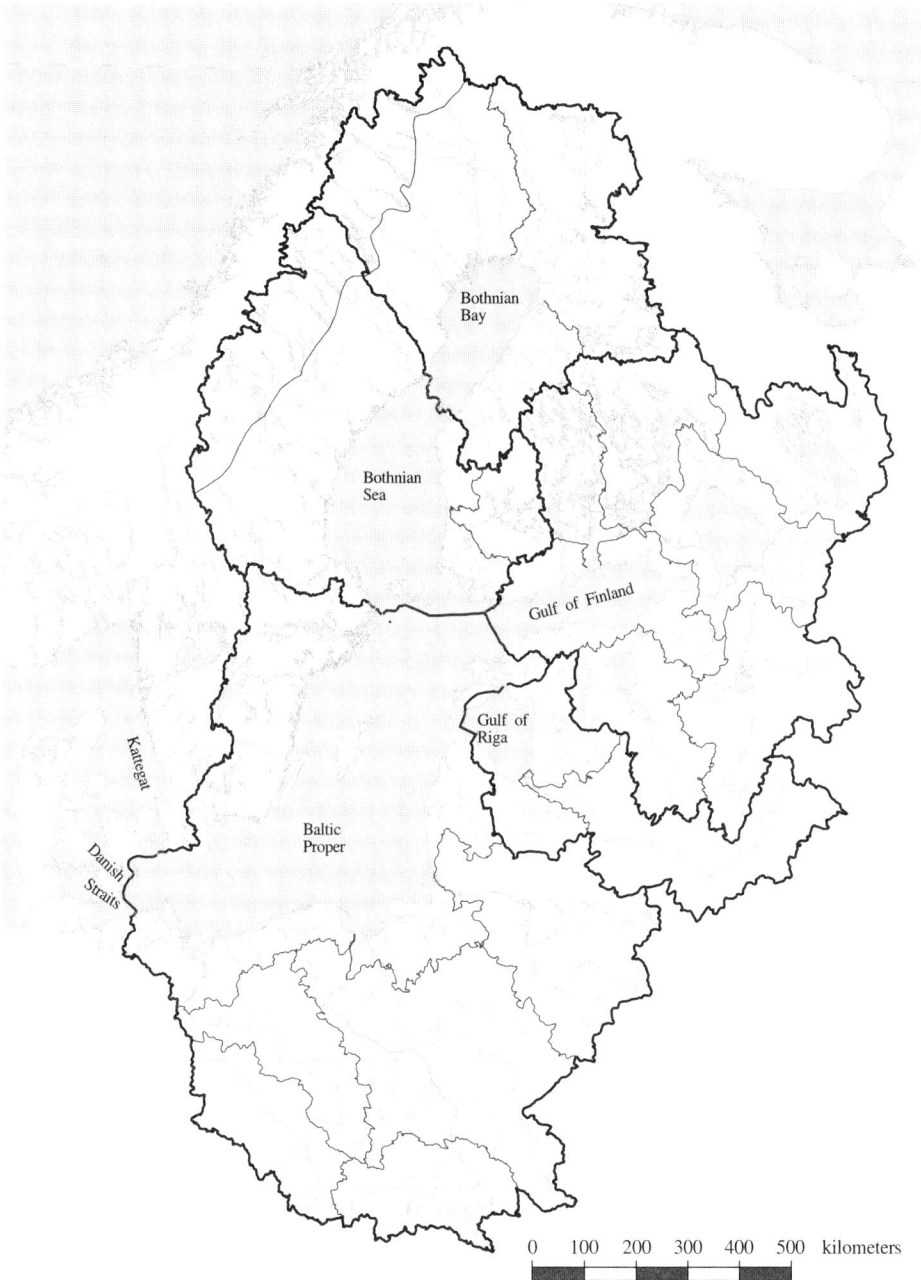

Figure 18.1 Basin boundaries for the HBV-Baltic. The five main Baltic Sea drainage basins are outlined with thick lines. Subbasins are numbered and delineations are shown with thin lines (from Graham 1999)

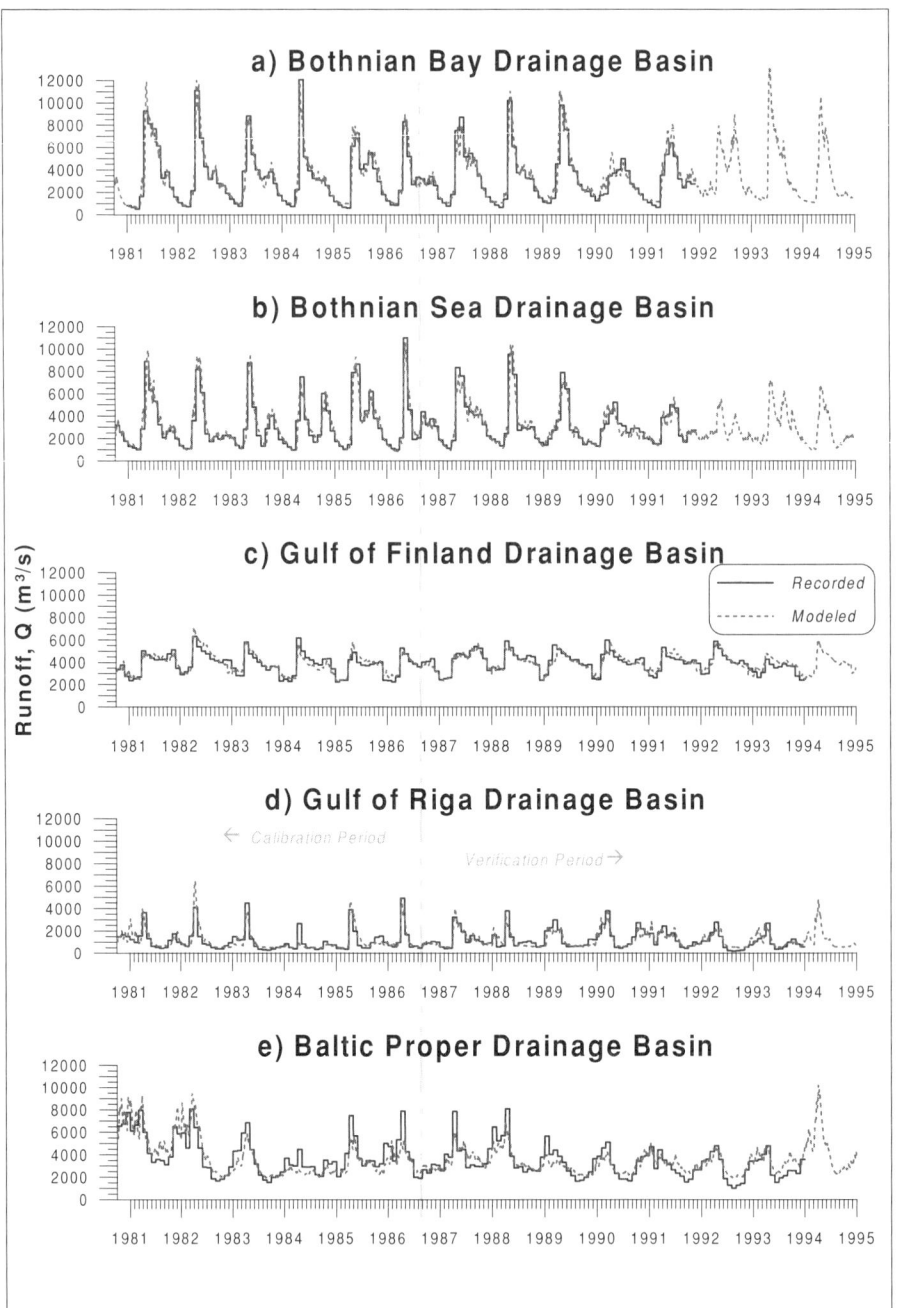

Figure 18.2 Simulated and recorded discharge for the five main Baltic Sea drainage basins for the calibration and validation periods (from Graham 1999)

surface component of the climate model and the hydrological model, resulting in significant differences in model outputs. Another important conclusion is that there appears to be compensating errors in the climate model, e.g. soil moisture is simulated reasonably well although the simulated precipitation shows significant errors. To avoid such compensating errors the authors strongly recommend putting more focus on validation of internal variables.

18.4.3   Mackenzie River Basin

The Mackenzie River Basin in northwestern Canada covers $1\,600\,000\,km^2$. Kite et al. (1994) have combined a hydrological model (SLURP) with a GCM. The SLURP is a semi-distributed hydrological model in which the parameters are related to land cover. At each time step (one day) the model is applied sequentially to a matrix of so-called grouped response units (GRU) and land covers. A GRU is a part of the catchment made up of a number of computational units which may or may not be continuous. In this case the river basin was divided into a 5×5 matrix of land cover classes and GRUs, calculating runoff from each land cover area within each GRU and routing the streamflow downstream from one GRU to the next. The process descriptions in the SLURP may be classified as mainly being of the conceptual type.

The model was calibrated on the basis of daily data from the period 1986–1987, and the period 1988–1990 was used for model validation.

The hydrological model was run on the basis of two different sets of input data, namely observed data from 25 climatological stations and data simulated by the Canadian Climate Centre GCM II (CCC GCM II). The CCC GCM II was operated with a transformed grid of $3.75° \times 3.75°$. It was concluded that using CCC GCM II data as distributed input to SLURP produced a hydrograph closer to the recorded one than using observed climatic data.

Another comparison made was between the discharge simulated by the CCC GCM II alone and the SLURP. Here the results show that using the SLURP with the GCM data produces a much better representation of the recorded flow regime as compared to the CCC GCM II as a stand-alone application.

18.4.4   Senegal River Basin

The Senegal River Basin covers $350\,000\,km^2$ in Guinée, Mali, Senegal and Mauritania in Western Africa. This basin was selected as focus catchment in a research project, which had the overall objective to integrate remote sensing data and hydrological modelling and investigate how and which types of remote sensing data can provide valuable additional information from a hydrological modelling perspective.

The hydrology of the Senegal River Basin has been modelled using a new version of the MIKE SHE code. This new version is the standard version of MIKE SHE (Abbott et al. 1986b; Refsgaard and Storm 1995) with a substitution of the standard complex groundwater module by a simple groundwater module (Christierson 1997). This approach, which for the groundwater component only aims at simulation of interflow and baseflow, is similar to the one successfully tested by Knudsen et al. (1986). The new MIKE SHE version, as illustrated in Figure 18.3 is oriented towards surface water studies such as the coupling of a hydrological model with remote sensing data for the Senegal River Basin, where detailed description of groundwater conditions may not be required.

The model solves the equations of snow melt, interception, evapotranspiration, overland flow, channel flow, unsaturated flow and saturated subsurface flow and in this way describes the major flow processes of the entire land phase of the hydrological cycle. The model is fully

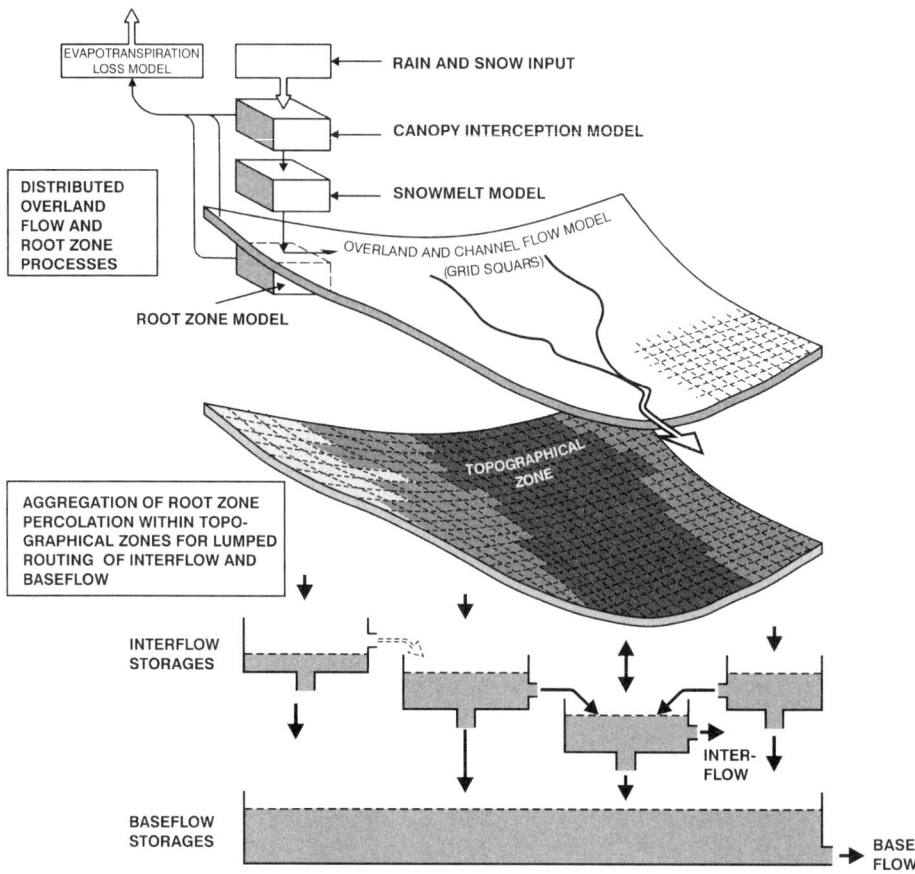

Figure 18.3   Model structure for MIKE SHE version with the simple groundwater module

distributed and physically based except for the new groundwater module which may be characterised as semi-distributed and conceptual.

Previous experience has shown that the linear reservoir approach for the groundwater zone, as applied here, is sufficient for an accurate description of the discharge hydrographs (Refsgaard and Knudsen 1996). The advantage of this approach is that it requires much less geological data, is less demanding with respect to computational requirements and is easier to calibrate.

For the MIKE SHE application to the Senegal River Basin (Andersen 1998) a spatial discretisation of $4 \times 4\,km^2$ was used for the root zone. Daily data on rainfall from 112 stations and discharge data from 11 gauging stations for a ten year period were available for calibration and validation purposes. The first calibration was carried out for a five year period using one of the downstream discharge stations, Bakel, only (Figures 18.4 and 18.5). Subsequently, internal validation was carried out against other discharge stations for the same period. Finally, a split-sample validation test was carried out against data from the other five years that had not been used for calibration. As part of the research activities it is intended to derive information on

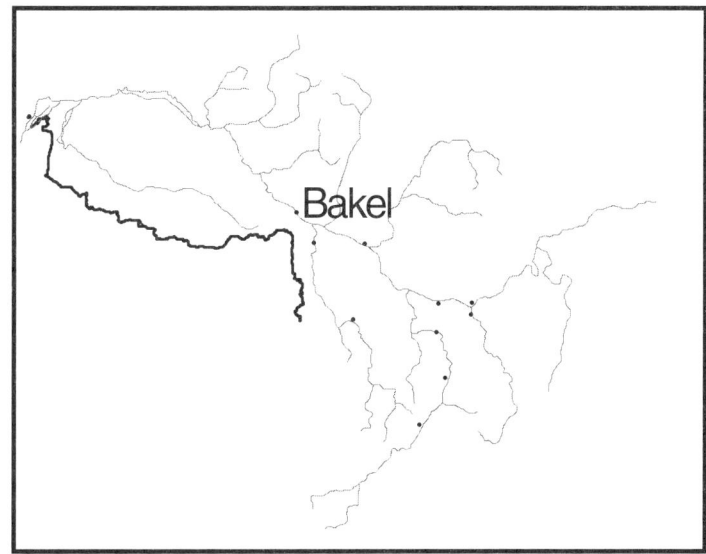

Figure 18.4 The Senegal River Basin and the location of the Bakel discharge station together with other upstream stations (from Andersen 1998)

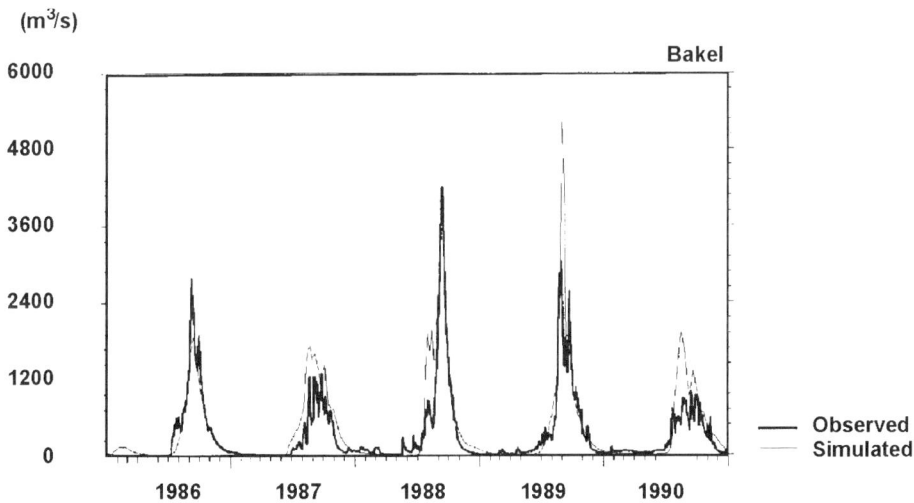

Figure 18.5 The first calibration results from the Bakel discharge station (from Andersen 1998)

soil moisture from satellite data (Sandholt et al. 2000) and compare these data to the soil moisture simulated by the model. Although the remote sensing data do not provide ground truth, this comparison may be considered as a kind of validation test on the model's capability to simulate spatial patterns of soil moisture. Such a test is a soft one, because the remote sensing data contain a considerable amount of uncertainty; however, at such a large scale remote sensing data appear to be the only possibility for conducting just some kind of internal validation tests for a distributed model.

18.4.5 Yangtze River Basin

The Yangtze River Basin covers a total area of $1\,800\,000\,km^2$ in China. Within the framework of a flood forecasting project (DHI 1998) $180\,000\,km^2$ of the Middle Yangtze River Valley, including the city of Wuhan and the Three Gorges Valley, was modelled.

The hydrological model was the conceptual NAM (Nielsen and Hansen 1973), for which the area was divided into 15 subcatchments ranging from 3000 to $30\,000\,km^2$. The NAM subcatchments provided input (lateral flows) to the hydrodynamic routing facilities in MIKE 11 (Havnø et al. 1995).

In a forecasting situation the hydrological model receives rainfall forecasts from a limited area meteorological model, HIRLAM (DMI 1995), which has been established with a spatial discretisation of $0.5°$ (about 50 km). The HIRLAM covers an area of about $26 \times 10^6\,km^2$, i.e. much more than the Yangtze River Basin.

The objective of the modelling was to forecast water levels in the Yangtze river during flood situations. The model was calibrated on historical data and tested operationally in real-life against the catastrophic floods during the 1998 flood season.

18.4.6 Odra River Basin

The Odra River Basin covers $110\,000\,km^2$ mainly in Poland. Within the framework of the BALTEX (Baltic Sea Experiment) Ruhe et al. (1998) have established a grid-related hydrological model capable of coupling to a regional scale atmospheric model for the Odra. The model is based on a land surface scheme SEWAB (Mengelkamp et al. 1997) coupled with a linear horizontal water routing scheme (Lohman et al. 1996). The model is established on a rotated grid with a resolution of $0.167°$ (about 18 km) like the atmospheric model REMO. The hydrological model is run on a 30 minute time step.

The model was calibrated on data for the period 1992–1993 against measured discharge data and subsequently validated against data for the catastrophic 1997 flood event. For the calibration and test the hydrological model was run independent of the atmospheric model. The calibration results were carried out on the basis of 350 precipitation stations, while only 30 stations were available for the 1997 validation test. Ruhe et al. (1998) reports a good agreement between observed and simulated discharges for one upstream gauging station for the calibration period, whereas the 1997 validation test showed a good overall fit with regard to flood volume, but a relatively poor agreement with regard to flood peak size and timing.

18.5 DISCUSSION AND CONCLUSIONS

18.5.1 Which Hydrological Model Type to Use

In the case studies reviewed above successful results have been reported from applications of both conceptual and physically based types of hydrological models. It is noticed also that the test results from the case studies exclusively deal with discharge data, while no validation tests have been carried out for additional objectives such as simulation of spatial patterns of soil moisture, evapotranspiration or runoff generation.

Several intercomparison studies of lumped conceptual, semi-distributed conceptual and distributed physically based models have indicated that their respective capability of predicting catchment runoff are almost identical (e.g. Refsgaard and Knudsen 1996). The question of which model type to use for regional and global scale purposes thus boils down to the question of what the purpose of the hydrological modelling is. The strengths, potentials and limitations of the two main classes of models may be summarised as follows:

- The conceptual models are today ready for operational use for simulation of runoff at the regional and even global scale. Thus for establishing, for example, overall water balances and for operational flood forecasting this model type appear to be adequate and suitable. The conceptual models can easily be semi-coupled with atmospheric models in such a way that the atmospheric model generates input for the hydrological model, but due to incompatible spatial resolution and lack of energy balance description it is not possible to make a suitable feedback mechanism to the atmospheric model.
- The physically based models enable a more detailed description of hydrological processes with a better spatial resolution, which is of great importance for the research on land surface–atmosphere interactions. Thus this model type makes it possible to simulate spatial land surface patterns which are directly compatible with available spatial remote sensing data. Because physically based models contain spatial information on soil moisture and easily enable inclusion of standard SVAT schemes, they can be fully coupled with atmospheric models with mutual feedback mechanisms.

Irrespective of the above general conclusion, it should be noted that the term physically based must be taken with certain reservations when such codes are applied at regional and global scales. MIKE SHE, for instance, is usually categorised as a physically based system. This is strictly speaking only correct if it is applied on an appropriate scale. A number of scale problems arise when the MIKE SHE is used on a regional scale (Refsgaard and Storm 1995). In addition, if there is a considerable uncertainty attached to the basic information, and if the spatially and temporally varying variables (such as groundwater table elevations) cannot be validated against observations, a MIKE SHE model of that particular site, cannot be considered as fully physically based, but will degenerate towards a detailed conceptual model.

In brief, it may be concluded that at present the conceptual models have a major operational potential, whereas the distributed models have the major research potential of interest in regional and global hydrological modelling.

18.5.2   Land Surface Components in Atmospheric Models Versus Hydrological Models

The case studies comprise two examples of comparisons between river discharge simulations by a land surface component in an atmospheric model and by a traditional conceptual hydrological model. Both Kite et al. (1994) and Graham et al. (1998) conclude that there are large differences in the conceptualisation between the atmospheric model and the hydrological model and that the performance of the hydrological model can be assumed to be better than the performance of the atmospheric model. The results from the global scale modelling presented by Dümenil and Todini (1992) support the view that the land surface components in atmospheric models are still lacking predictive capability with regard to discharge simulation as compared to traditional hydrological models.

Considering that atmospheric modelling and hydrological modelling have developed from different positions characterised with different objectives this difference in performance in discharge simulation is not surprising. Thus, atmospheric modelling is still operating at a relatively large spatial resolution, it has traditionally focused on the vertical exchange of vapour and energy to the atmosphere, and it has only recently started to consider lateral movement of water at the land surface and subsurface. In contrast, hydrological models have, by now, focused on discharge simulation for almost four decades.

However, there is a clear distinction in performance between the two classes of atmospheric modelling, namely the GCMs and LAMs. While the performance of the GCMs with the coarse resolution is relatively poor (Dümenil and Todini 1992; Kite et al. 1994) the performance of the LAMs with finer spatial resolutions are relatively much better (Graham et al. 1998; Ruhe et al. 1998). This indicates that a key constraint for the atmospheric models is related to the spatial resolution, which in the case of the GCMs appears to be too coarse for a satisfactory discharge simulation.

18.5.3   Validation Status

As discussed previously in this chapter, model validation tests are often not carried out with a scientific rigour that justifies the often used notion of validated models. This also applies to regional and global scale modelling. Three main problems in this respect will be addressed in the following, namely the question of which validation test scheme to use, the question of internal validation of distributed models and the scaling issue.

The key principle behind the suggested model validation test schemes is that the model should be tested to show how well it can perform the kind of task for which it is specifically intended (Klemes 1986). This results in four different test types corresponding to whether data are available for calibration and whether the catchment conditions are stationary or the impact of some kind of human intervention has to be predicted.

The six cases reported above have all carried out validation tests, mostly the traditional split-sample test. This implies that the models have demonstrated a certain validity for simulating the discharge at given sites, corresponding to the level of accuracy achieved during the validation tests. The models have not been subject to validation tests aiming at demonstrating their possible capabilities for, for example, predicting the effects of climate change or land use change. This is a limitation with regard to the theoretical basis for their practical use to water resources management and climate change studies.

In water resources management one of the important fields of application for a hydrological model is prediction of the effects of land use changes. Many such studies have been reported; however, most of them can be characterised as hypothetical predictions, because the models have not been subject to adequate validation tests (Lørup et al. 1998). The main reason for this lack of adequate validation tests is probably that it is not straightforward due to the non-stationarity of the catchment conditions, in which case the procedure outlined in Section 18.3.3 above prescribes that experience (data) from other catchments that have undergone similar land use changes should have been used in a so-called differential split-sample test.

In climate change studies an important field of application is prediction of the effects of climate change on water resources. For changes which are small as compared to the natural climate variability such tests can be carried out by use of the differential split-sample test outlined in Section 18.3.3 above implying calibration on 'dry' periods and validation on 'wet' periods and vice versa. However, for long-term climate changes, which are so large that they cause changes in land use and vegetation characteristics, validation tests are in practice very difficult to design and conduct (Refsgaard and Knudsen 1996).

It must be realised that the above recommended validation tests are so demanding that many applications today would fail to meet them. This does not imply that these modelling studies are not useful, only that their output should be realised to be somewhat more uncertain than is often stated and that they should not make use of the term 'validated model'.

Another characteristic of the reviewed cases is that the validation tests only focus on discharge data, with the Senegal River Basin case (Sandholt et al. 1999) as the only exception. This finding is in accordance with Refsgaard (2000) who, on the basis of a review of the validation status of distributed models, concluded:

- Distributed models are usually calibrated and validated only against runoff data, while spatial data are seldom available.
- In the few cases where model simulations have been compared with field data on internal variables, these test results are generally of less accuracy than the results of the validation tests against runoff data.
- Authors, who have not been able to test their models' capabilities to predict internal spatial variables, often state that their distributed models provide physically realistic descriptions of spatial patterns of internal variables. While such statements may be correct in many cases, they can, due to lack of data, not be documented and should therefore be expressed more cautiously.

There are probably two main reasons for this lack of internal validation, namely incompatibility between field data and model structure and a general lack of suitable spatial data.

The structure of conceptual models generally poses a serious constraint in comparing outputs from these models with field data, except for discharge data. There are two reasons for this. First, the spatial variability of field variables such as soil moisture is not explicitly accounted for, but implicitly built into the model equations, which therefore do not describe small-scale variations at all. Areas with different soil types, soil depths, vegetation types, climate conditions and depth to groundwater table are in a conceptual model often lumped together to one computational unit, which is usually incompatible with field data. Second, the spatial discretisation in conceptual models usually follows catchment boundaries, while field data often are either point or grid values.

Spatial data of importance for distributed modelling can be categorised in two groups, namely the input data on model parameters such as topography, geology, land use and soil type and data comparable to model variables such as snow cover, soil moisture, leaf area index, root depth, actual evapotranspiration, surface temperature and depth of surface water inundation. Whereas the spatial data describing model parameters often are available in existing databases (although not always in the desired spatial resolution), field data on the model variables are seldom available. The key source of such spatial data, except for experimental catchments, is remote sensing data. Although there is obvious potential for use of remote sensing data, the success stories of relevance to regional and global modelling in operational applications are in practice limited to mapping of land use and snow cover. However, with the recent progress in remote sensing techniques spatial data with a sufficient accuracy may be expected to become operationally available for a number of new fields within the coming years (e.g. Sandholt et al. 1999).

With respect to scaling, it should be emphasised that with the present generation of distributed physically based model codes, which do not contain adequate up- or downscaling methodologies, separate calibration and validation tests have to be carried out every time the grid size is changed (Refsgaard 1997). For conceptual models the heterogeneity of process variables and parameters are implicitly built into the model equations and do not in themselves generate a need for recalibration and new validation tests when the scale is changed. Thus, on the basis of experiences with the conceptual HBV model in Sweden and in the Baltic Sea Basin, Bergström and Graham (1998) conclude that the parameters are relatively stable over a wide range of scales. However, in cases where the input data to the conceptual models change significantly as a consequence of change of scale (e.g. different precipitation or temperature values as an effect of upscaling the topography in mountainous regions) conceptual models also will require new calibration and validation tests.

18.5.4   Key Problems and Challenges

The key problems and research challenges identified in relation to validation of hydrological models at the regional and global scale are:

- To bridge the gap between the hydrological and atmospheric modelling communities.
- To adopt rigorous validation test schemes and disseminate the practical use of such methodology throughout the modelling communities.
- To carry out validation tests on existing models to assess to what extent these models have satisfactorily predictive capabilities to contribute to the solution of practical water management problems.
- To make full use of remote sensing data for calibration and validation of hydrological models in order to enable internal validation of spatially simulated variables.

ACKNOWLEDGEMENTS

This work was carried out within the framework of the project 'Integration of Earth Observation Data in Distributed Agrohydrological Models (INTEO)', which is funded by the Danish research councils under the 'Earth Observation Program' (contract no. 9600668). Phil Graham and Sten Bergström at the Swedish Meteorological and Hydrological Institute are thanked for providing copies of their figures.

REFERENCES

Abbott, M.B., Bathurst, J.C., Cunge, J.A., O'Connel, P.E. and Rasmussen, J. 1986a. An introduction to the European Hydrological System – Systeme Hydrologique Européen 'SHE', 1: History and philosophy of a physically-based distributed modelling system. *Journal of Hydrology*, **87**, 45–59.

Abbott, M.B., Bathurst, J.C., Cunge, J.A., O'Connel, P.E. and Rasmussen, J. 1986b. An introduction to the European Hydrological System – Systeme Hydrologique Européen 'SHE', 2: Structure of a physically-based distributed modelling system. *Journal of Hydrology*, **87**, 61–77.

Andersen, J. 1998. Modelling the Senegal River Basin. In: *Proceedings INTEO-Workshop*, 5–6 October 1998, Dakar, Senegal. Institute of Geography, University of Copenhagen.

Anderson, M.P. and Woessner, W.W. 1992. The role of postaudit in model validation. *Advances in Water Resources*, **15**, 167–173.

Bathurst, J.C. 1986. Physically-based distributed modelling of an upland catchment using the Systeme Hydrologique Européen. *Journal of Hydrology*, **87**, 79–102.

Bathurst, J.C. and O'Connell, P.E. 1992. Future of distributed modelling: the Systeme Hydrologique Européen. *Hydrological Processes*, **6**(3), 265–277.

Becker, A. 1992. Criteria for a hydrologically sound structuring of large scale land surface process models. In: O'Kane, J.P. (ed.), *Advances in Theoretical Hydrology, A Tribute to James Dooge*. European Geophysical Society Series on Hydrological Sciences, 1. Elsevier, Amsterdam, 97–111.

Bergström, S. 1995. The HBV model. In: Singh, V.P. (ed.), *Computer Models of Watershed Hydrology*. Water Resources Publications, Highlands Ranch, CO, 443–476.

Bergström, S. and Graham, L.P. 1998. On the scale problem in hydrological modelling. *Journal of Hydrology*, **211**, 253–265.

Beven, K. 1995. Linking parameters across scales: subgrid parameterization and scale dependent hydrological models. *Hydrological Processes*, **9**, 507–525.

Beven, K. 1996. A discussion on distributed hydrological modelling. In: Abbott, M.B. and Refsgaard, J.C. (eds), *Distributed Hydrological Modelling*. Kluwer Academic Publishers, Dordrecht, 255–278.

Beven, K.J., Lamb, R., Quinn, P., Romanowicz, R. and Freer, R. 1995. TOPMODEL. In: Singh, V.P. (ed.), *Computer Models of Watershed Hydrology*, Water Resources Publications, 627–668.

Blöschl, G. and Sivapalan, M. 1995. Scale issues in hydrological modelling: a review. *Hydrological Processes*, **9**, 251–290.

Burnash, R.J.C. 1995. The NWS river forecast system – catchment modelling. In: Singh, V.P. (ed.), *Computer Models of Watershed Hydrology*, Water Resources Publications, 311–366.

Christierson, B.V. 1997. *WATBAL-MIKE SHE Integration. Linear Reservoir Module*. DHI Report 4710, Danish Hydraulic Institute, Hørsholm, Denmark.

Crawford, N.H. and Linsley, R.K. 1966. *Digital Simulation in Hydrology, Stanford Watershed Model IV*. Department of Civil Engineering, Stanford University, Technical Report 39.

De Marsily, G., Combes, P. and Goblet, P. 1992. Comments on 'Ground-water models cannot be validated', by L.F. Konikow and J.D. Bredehoeft. *Advances in Water Resources*, **15**, 367–369.

DHI, 1998. *Heavy Rains and Flood Forecasting in the Middle Yangtze River Valley*. Final Report, Phase 2 prepared for the World Bank and Changjiang Water Resources Commission by Danish Hydraulic Institute in association with Danish Meteorological Institute and University of Copenhagen.

Dickinson, R.J., Henderson-Sellers, A., Kennedy, P.J. and Wilson, M.F. 1986. *Biosphere Atmosphere Transfer Scheme (BATS) for the NCAR Community Climate Model*. NCAR Technical Note, NCAR, TN275 + STR.

DMI, 1995. *Application of the HIRLAM System in China: Heavy Rain Forecast Experiments in the Yangtze River region*. Danish Meteorological Institute, Copenhagen.

Dooge, J.C. 1985. *Hydrological Modelling and the Parametric Formulation of Hydrological Processes on a Large Scale*. WCP-Publ. Ser. No. 96, WMO/TD-No. 43, Geneva.

Dooge, J.C. 1986. Looking for hydrologic laws. *Water Resources Research*, **22**, 46S–58S.

Dümenil, L. and Todini, E. 1992. A rainfall–runoff scheme for use in the Hamburg climate model. In: O'Kane, J.P. (ed.), *Advances in Theoretical Hydrology, A Tribute to James Dooge*. European Geophysical Society Series on Hydrological Sciences, 1. Elsevier, Amsterdam, 129–157.

Entekhabi, D. and Eagleson, P.S. 1989. Land surface hydrology parameterization for atmospheric general circulation models including subgrid scale spatial variability. *Journal of Climate*, **2**, 816–831.

Famiglietti, J.S. and Wood, E.F. 1994a. Multiscale modelling of spatially variable water and energy balance processes. *Water Resources Research*, **30**(11), 3061–3078.

Famiglietti, J.S. and Wood, E.F. 1994b. Application of multiscale water and energy balance models on a tallgrass prairie. *Water Resources Research*, **30**(11), 3079–3093.

Flavelle, P. 1992. A quantitative measure of model validation and its potential use for regulatory purposes. *Advances in Water Resources*, **15**, 5–13.

Fleming, G. 1975. *Computer Simulation Techniques in Hydrology*. Elsevier, New York.

Graham, L.P. 1999. Modelling runoff to the Baltic Sea. *Ambio*, **28**, 328–334.

Graham, L.P., Bergström, S. and Jacob, D. 1998. A discussion of land parameterization in hydrologic and climate models – example from the Baltic Sea Basin. *Contribution to the Second International Conference on Climate and Water*, Espoo, Finland, 17–20 August.

Grayson, R.B., Moore, I.D. and McHahon, T.A. 1992a. Physically based hydrologic modelling, 1. A terrain-based model for investigative purposes. *Water Resources Research*, **28**(10), 2639–2658.

Grayson, R.B., Moore, I.D. and McHahon, T.A. 1992b. Physically based hydrologic modelling, 2. Is the concept realistic? *Water Resources Research*, **28**(10), 2639–2658.

Havnø, K., Madsen, M.N. and Dørge, J. 1995. MIKE 11 – a generalized river modelling package. In: Singh, V.P. (ed.), *Computer Models of Watershed Hydrology*, Water Resources Publications, 733–782.

Heuvelink, G.B.M. and Pebesma, E.J. 1999. Spatial aggregation and soil process modelling. *Geoderma*, **89**(1–2), 47–65.

Kite, G.W., Dalton, A. and Dion, K. 1994. Simulation of streamflow in a macroscale watershed using general circulation model data. *Water Resources Research*, **30**(5), 1547–1559.

Klemes, V. 1986. Operational testing of hydrological simulation models. *Hydrological Sciences Journal*, **31**, 13–24.

Knudsen, J., Thomsen, A. and Refsgaard, J.C. 1986. WATBAL: a semi-distributed, physically based hydrological modelling system. *Nordic Hydrology*, **17**, 347–362.

Konikow, L.F. 1978. Calibration of groundwater models. In: *Verification of Mathematical and Physical Models in Hydraulic Engineering*. American Society of Civil Engineering, New York, 87–93.

Konikow, L.F. and Bredehoeft, J.D. 1992. Ground-water models cannot be validated. *Advances in Water Resources*, **15**, 75–83.

Koster, R.D. and Suarez, M.J. 1992. Modelling the land surface boundary in climate models as a composite of independent vegetation stands. *Journal of Geophysical Research*, **97**, 2697–2715.

Laval, K. 1997. Hydrological processes in GCMs. In: Sorooshian, S., Gupta, H.V. and Rodda, J.C. (eds), *Land Surface Processes in Hydrology, Trials and Tribulations of Modelling and Measuring*. NATO ASI Series 46, Springer-Verlag, Berlin, 45–61.

Lindström, G., Johansson, B., Persson, M., Gardelin, M. and Bergström, S. 1997. Development and test of the distributed HBV-96 model. *Journal of Hydrology*, **201**, 272–288.

Lohmann, D., Nolte-Holube, R. and Raschke, E. 1996. A large scale horizontal routing model to be coupled to land surface parameterisation schemes. *TELLUS*, **48A**(5), 708–721.

Lørup, J.K., Refsgaard, J.C and Mazvimavi, D. 1998. Assessing the effects of land use change on catchment runoff by combined use of statistical tests and hydrological modelling: case studies from Zimbabwe. *Journal of Hydrology*, **205**, 147–163.

Manabe, S. 1969. Climate and ocean circulation: I The atmospheric circulation and the hydrology of the earth's surface. *Monthly Weather Review*, **97**, 739–774.

Mengelkamp, H-T., Warrach, K. and Raschke, E. 1997. *A Land Surface Scheme for Atmospheric and Hydrologic Models*: SEWAB (Surface Energy and Water Balance). GKSS External Report, available at GKSS Research Center, Geesthacht, Germany.

Nielsen, S.A. and Hansen, E. 1973. Numerical simulation of the rainfall–runoff process on a daily basis. *Nordic Hydrology*, **4**, 171–190.

Oreskes, N., Shrader-Frechette, K. and Belitz, K. 1994. Verification, validation and confirmation of numerical models in the earth sciences. *Science*, **264**, 641–646.

Refsgaard, J.C. 1996. Terminology, modelling protocol and classification of hydrological model codes. In: Abbott, M.B. and Refsgaard, J.C. (eds), *Distributed Hydrological Modelling*. Kluwer Academic Publishers, Dordrecht, 17–39.

Refsgaard, J.C. 1997. Parameterisation, calibration and validation of distributed hydrological models. *Journal of Hydrology*, **198**, 69–97.

Refsgaard, J.C. 2000. Using spatial data in a formal approach to calibration and validation of models. In: Grayson, R. and Blöschl, G. (eds), *Spatial Patterns in Catchment Hydrology: Observations and Modelling*. Cambridge University Press, Cambridge.

Refsgaard, J.C. and Knudsen, J. 1996. Operational validation and intercomparison of different types of hydrological models. *Water Resources Research*, **32**(7), 2189–2202.

Refsgaard, J.C. and Storm, B. 1995. MIKE SHE. In: Singh, V.P. (ed.), *Computer Models of Watershed Hydrology*. Water Resources Publications, Highlands Ranch, CO, 809–846.

Rowntree, P.R. and Lean, J. 1994. Validation of hydrological schemes for climate models against catchment data. *Journal of Hydrology*, **155**, 301–323.

Ruhe, C., Lobmeyr, M., Mengelkamp, H-T. and Warrach, K. 1998. A grid related distributed hydrological model and its application over large river basins. In: Lemmelä, R. and Helenius, N. (eds), *Proceedings of the Second International Conference on Climate and Water*, Espoo, Finland 17–20 August 1998, 863–872.

Sandholt, I., Andersen, J., Dybkjær, G., Lo, M., Refsgaard, J.C., Rasmussen, K. and Jensen, K.H. 1999. Use of remote sensing data in distributed hydrological models: applications in the Senegal River Basin, *Danish Journal of Geography*, **99**, 47–57.

Schlesinger, S., Crosbie, R.E., Gagné, R.E., Innis, G.S., Lalwani, C.S., Loch, J., Sylvester, J., Wright, R.D.,

Kheir, N. and Bartos, D. 1979. Terminology for model credibility. SCS Technical Committee on Model Credibility. *Simulation*, **32**(3), 103–104.

Sellers, P.J., Mintz, Y., Sud, Y.C. and Dalcher, A. 1986. A Simple Biosphere (SiB) model for use within General Circulation Models. *Journal of Atmospheric Sciences*, **43**, 505–531.

Sorooshian, S., Duan, Q. and Gupta, V.K. 1993. Calibration of rainfall–runoff models: application of global optimization to the Sacramento soil moisture accounting model. *Water Resources Research*, **29**, 1185–1194.

Tsang, C-F. 1991. The modelling process and model validation. *Ground Water*, **29**, 825–831.

WMO, 1975. *Intercomparison of Conceptual Models Used in Operational Hydrological Forecasting.* WMO Operational Hydrology Report No. 7, WMO No. 429. World Meteorological Organisation, Geneva.

WMO, 1986. *Intercomparison of Models for Snowmelt Runoff.* WMO Operational Hydrology Report No. 23, WMO No. 646. World Meteorological Organisation, Geneva.

WMO, 1992. *Simulated Real-time Intercomparison of Hydrological Models.* WMO Operational Hydrology Report No. 38, WMO No. 779. World Meteorological Organisation, Geneva.

Author Index

Ababou et al. (1992) 423, 432
Abbott et al. (1986a,b) 117, 462, 463, 473
Admiraal and Garcia (2000) 410
Akaike (1974) 123, 157
Akhavan et al. (2000) 329
Akiyama and Fukushima (1986) 403
Albert and Krajeski (1998) 264, 274
Alley and Emery (1986) 31
Andersen (1980) 92
Andersen (1997) 475
Andersen (1998) 470, 474
Anderson (1968) 29
Anderson and Larson (1996) 267
Anderson and Woessner (1992) 1, 23, 33, 74, 413, 433, 467
Andreas (1987) 265
Arola and Lettenmaier (1996) 278
Aronica et al. (1998) 26, 431
Arons and Colbeck (1995, 1998) 273
Ascher (1981) 27
Ascher (1983) 29
Ascher (1987) 38
Ascher (1989) 35
Ascher (1993) 27, 28
Ascher and Overholt (1983) 27, 29, 38
Aver and Neihaus (1993) 369
Avissar (1992) 278
Avissar and Pielke (1989) 164
Ayer (1936) 325, 350

Bacon (1620) 119
Bader et al. (1954) 276
Badji and Dautrebande (1995) 182
Bair (1994) 46, 74
Bair and Wood (1999) 60
Bale (1987) 380
Bales and Harrington (1995) 261
Bamber et al. (2000) 452, 453
Barker and Young (1985) 85, 100
Barnett (1990) 93, 102
Barton and Bathols (1988) 181
Bassett (1994) 46
Bastet et al. (1999) 316
Bastiaans et al. (2000) 329

Bates and Anderson (1996) 430
Bates and De Roo (2000) 343
Bates and Hervouet (1999) 331, 333, 335
Bates et al. (1992) 342
Bates et al. (1996) 185, 348
Bates et al. (1997) 181, 183, 342, 422
Bates et al. (1998a) 180, 185, 327, 333, 334, 342
Bates et al. (1998b) 187, 348, 349
Bates et al. (1999) 336
Bathurst (1986) 334, 463
Bathurst and O'Connell (1992) 462
Bear (1972) 85
Beck (1987) 47, 414
Beck and Young (1975) 124
Beck et al. (1990) 47
Beck et al. (1997) 1, 23, 27
Becker (1992) 462, 466
Beer and Young (1983) 120, 125
Beljaars and Holtslag (1991) 265
Benkhaldoun and Monthe (1994) 331
Berbery et al. (1999) 334
Bergen (1975) 276
Bergström (1991) 262
Bergström (1995) 463
Bergström and Graham (1999) 480
Bergström et al. (1991) 211
Bernard (1992) 428
Best and Roy (1991) 422
Bethke (1992) 24
Betts et al. (1996) 164
Beven (1979) 430
Beven (1989) 187, 326, 415, 420, 423
Beven (1993) 26, 47, 48
Beven (1995) 49, 466
Beven (1996a) 26, 46, 47, 420
Beven (1996b) 5, 45, 467
Beven (1996c) 352
Beven (1997) 51
Beven (2001a) 8, 49
Beven (2001b) 48, 51
Beven and Binley (1992) 422, 423
Beven and Germann (1981) 239
Beven and Germann (1982) 201, 234
Beven and Kirkby (1979) 117

Beven and Linley (1992) 430
Beven et al. (1995) 51, 463
Bhallamudi and Chaudhury (1991) 429
Billings and Voon (1986) 125
Binley et al. (1996) 45
Biron et al. (1996) 417, 422
Blauvelt (1999) 62, 63
Blöschl (1999) 278
Blöschl and Kirnbauer (1991) 264, 272, 273
Blöschl and Sivapalan (1995) 7, 8, 462
Blöschl et al. (1991a) 278, 280
Blöschl et al. (1991b) 278, 280
Bonansea (1995) 181
Bond et al. (1999) 316
Booltink (1993) 218
Booltink (1994) 197, 295, 218
Booltink and Bouma (1991) 201
Booltink and Verhagen (1996) 209
Booltink et al. (1991) 201, 205, 218
Booltink et al. (1993) 218, 219
Bouma (1984) 201
Bouma and Dekker (1978) 196
Bouma and Hoosbeek (1996) 195
Bouma and Van Lanen (1987) 197, 201
Bouma et al. (1977) 222
Bouma et al. (1996) 202
Bowie et al. (1985) 370, 371
Box and Jenkins (1970) 87–88, 118, 123, 126
Bradbrook et al. (1998) 422
Bradbrook et al. (2000) 431
Bradbrook et al. (2001) 417
Braithwaite (1953) 325
Braun et al. (1994) 262
Bredehoeft (1993) 306
Bredehoeft and Konikow (1992) 32
Bredehoeft and Konikow (1993) 61, 74, 294, 297, 311
Bronstert (1999) 258
Brooks and Corey (1964) 246, 247
Brooks et al. (1994) 46
Brown (1996) 424
Brown and Barnwell (1987) 368, 372
Brown and Laase (1995) 74
Brownlie (1981) 398
Brun and Pahaut (1991) 277
Brun et al. (1989) 263, 272, 275, 276, 278
Brunner (1991) 26
Brunner and Ascher (1992) 25, 26
Burkard et al. (1991) 283
Burke et al. (1998) 169, 176, 178, 179
Burnash (1995) 463

Carrera and Neuman (1986) 85, 89, 102, 106
Carroll (1987) 283

Carroll and Cressie (1996, 1997) 284
Carroll et al. (1995) 273
Cartwright (1989) 14
Cartwright (1999) 17
Celik and Rodi (1988) 403, 404
Chalmers (1990) 46
Chang et al. (1997) 283
Chappell et al. (2000) 130
Chorley (1978) 416
Christierson (1997) 473
Church (1935, 1937) 267
Clarke et al. (1977) 440, 445
Cline and Carroll (1999) 280
Cline et al. (1998a) 278, 284
Cline et al. (1998b) 280
Cohen (1991) 183
Colbeck (1972) 274
Colbeck (1979) 276
Colbeck (1983) 276
Colbeck (1987) 274
Colbeck (1991) 272
Colbeck (1997) 273
Colbeck et al. (1990) 275, 276
Cooley (1977) 311, 312
Cooley (1979) 294, 312
Cooley (1982) 312
Cooley (1983) 312
Cooley (1985) 106
Cooley (1993a) 314, 315
Cooley (1993b) 313, 314
Cooley (1997) 294, 314
Cooley and Hill (1992) 313
Cooley and Naff (1990) 312, 313
Cooley and Vecchia (1987) 313, 314
Cooley et al. (1986) 313
Corr et al. (1995) 182
Crawford and Linsley (1966) 117, 463
Currey (1997) 181
Czernuszenko and Rylov (2000) 418

Daamen (1996) 174
Daamen and Simmonds (1996) 168, 173
Dagan (1985) 105
Daily and Harleman (1972) 364
Daly et al. (1999) 262, 263, 264, 265
Daly et al. (2000) 265
Daniel et al. (1997) 247
Davenport et al. (2000) 187
Davis et al. (1985) 274
Davis et al. (1987) 276
Davis et al. (1992) 45
Davis et al. (1993) 276
Davis et al. (1995) 279, 280, 282
Davis et al. (1997) 279

De Dean (1999) 36
De Lima and Olimpio (1989) 73
De Marsily et al. (1992) 32, 467
De Vriend and Geldof (1983) 338
Dean (1999) 36, 37
Defina et al. (1994) 331, 333
Dekker (1998) 196
Denoth et al. (1979) 274
DHI (1998) 470
Di Pietro and La Folie (1991) 235, 242
Dickinson et al. (1986) 464, 465
Didden and Schott (1992) 334
Dierckx et al. (1986) 211
Dietrich and Chapman (1993) 88, 91, 102
Dietrich and Newsam (1989) 97, 105
Dietrich et al. (1993) 77, 104
Dingman (1984) 234
Diskin and Simon (1997) 352
Dobson et al. (1985) 166
Doering (1965) 205
Domenico and Schwartz (1998) 67
Dooge (1985) 463
Dooge (1986) 462
Douville (1997) 264
Dowdeswell and Siegert (1999a) 450, 451
Dowdeswell and Siegert (1999b) 455
Dozier (1987) 261
Dozier (1992) 261
Drusche et al. (1999) 178
Dullien (1992) 274
Dümenil and Todini (1992) 463, 464, 465, 469, 470,
        477, 478
Dunne (1998) 3, 4
Dupre (1993) 17
Durner et al. (1999) 316

Edwards et al. (1987) 381
Einstein (1950) 402
Einstein (1968) 405
Elder (1959) 362, 366
Elder et al. (1991) 273
Elder et al. (1998) 273
El-Hames and Richards (1995) 427
El-Hames and Richards (1998) 427
Elliott (1980) 92
Elsenheimer (1999) 6
Emmett (1978) 45
Engelhardt et al. (1990) 454
Engelund and Fredsoe (1976, 1982) 402
Entekhabi and Eagleson (1989) 464, 465
Ervine and MacLeod (1999) 330
Essery (1997) 264, 278
Evangelinos et al. (2000) 329
Ewing et al. (1999) 294, 295, 297, 310

Faeh (1997) 257
Falconer (1986) 367
Falconer (1991) 366
Falconer and Chen (1996) 359, 367, 378
Famiglietti and Wood (1994a,b) 465, 466
Fawcett et al. (1995) 5
Feddes et al. (1978) 211
Feddes et al. (1988) 205
Ferguson (1986) 280
Ferguson et al. (1996) 334
Fily et al. (1999) 285
Finke (1993) 197, 205
Finke and Bosma (1993) 2–5, 201, 205, 210
Finke et al. (1996) 202
Finsterle and Faybishenko (1999) 316
Fischer (1973) 365, 366
Fischer et al. (1979) 262, 366, 367
Fischer-Antze et al. (in press) 336, 337, 340
Fischhoff (1982) 27
Flavelle (1992) 413, 418, 420, 467
Flemming (1975) 462
Flood and Gutelius (1997) 187
Flury et al. (1994) 45
Foster et al. (1993) 62, 63
Franklin (1993) 5
Franks and Beven (1997a) 48
Franks and Beven (1997b) 49
Franks et al. (1997) 118, 149
Franks et al. (1998) 48
Fread (1984, 1993) 330
Freer et al. (1996) 48, 327
Freer et al. (1997) 8
Friedland (1990) 71
Fujita et al. (1999) 451
Fuller et al. (1994) 184
Funtowicz and Revetz (1985) 33

Galantowicz et al. (1999) 173, 178, 179
Gamerman (1997) 123
Garcia (1999, 2000) 389
Garcia and Parker (1991) 389, 396, 398, 401, 403,
        406, 407
Garcia and Parker (1993) 389, 397
Garcia et al. (1999) 409
Gerber (1987) 58
Germann (1985) 239, 241, 242
Germann (1987) 196
Germann (1990a,b) 239, 257
Germann and Bürgi (1996) 242
Germann and Di Pietro (1996) 233, 234, 240, 241,
        242, 246, 257, 258
Germann et al. (1997) 235, 240
Ghia et al. (1982) 337, 338
Giere (1988) 13, 16

Giere (1999) 16
Glen (1955) 440
Glendinning (1998) 274
Glendinning and Morris (1999) 276
Goldstein (1938) 359
Goldstein and Smith (1974) 104
Golub and Van Loan (1989) 94
Golub et al. (1979) 102
Goode and Konikow (1990) 311
Goodison (1981) 267
Goodison et al. (1987) 267, 282
Graham (1999) 469, 470, 472
Graham et al. (1988) 469, 477, 478
Graybill (1961) 312
Grayson et al. (1992a) 462, 463
Grayson et al. (1992b) 317, 462, 463
Greenwood (1989) 75
Greenwood et al. (1985) 211
Greenwood et al. (1990) 211
Gregory and Walling (1973) 326
Greve and MacAyeal (1996) 439
Griend et al. (1996) 170
Griffiths (1999) 6
Grossman and Vaughn (1999) 57
Gupta et al. (1985) 46
Gupta et al. (1998) 26, 46
Gutierre and Kouvelis (1991) 29

Haase et al. (2000) 431
Hacking (1983) 416, 432
Hadamard (1932) 78
Hall and Pilkey (1991) 36
Hall et al. (1995) 280
Hallikainen et al. (1985) 165
Hanfling (1981) 325
Hankin and Beven (1998a,b) 26
Hardy et al. (2000) 348
Harleman (1966) 364
Harpin et al. (1995) 341
Harr (1995) 63, 71
Harrington and Bales (1998) 271
Harrington et al. (1995) 280
Hartman et al. (1999) 280
Hartmann (2000) 256
Harvey (1969) 325
Harvey (1989) 156
Harvey et al. (1996) 45
Hassanizadeh and Carrera (1992) 217
Hasselman (1998) 118
Hasselman et al. (1997) 118
Hastie and Tibshirani (1996) 155
Havnø et al. (1995) 476
Hemmerle (1975) 104
Henderson (1966) 361

Henderson (1995) 182
Henderson et al. (1996) 45
Hervouet (2000) 326, 336
Hervouet and Van Haren (1996) 328, 330, 332
Heuvelink and Pebsema (1999) 463
Hewlett and Hibbert (1963) 5
Hicks et al. (1990) 338
Hill (1992) 28
Hill (1998) 294
Hill et al. (1998) 294, 315
Hillel and Gardner (1969) 205
Hillyer and Stakhiv (1997) 37, 38
Hindmarsh and Payne (1996) 443
Hodges (1995) 34, 35
Hogan (1996) 267
Hollenbeck et al. (1996) 177
Hooke (1981) 440
Hoosbeek and Bryant (1992) 195
Hornberger et al. (1985) 119
Horritt (1999) 183, 343
Horritt (2000) 185, 334, 335, 422, 426
Horritt and Bates (2000) 422
Horritt et al. (2000) 184
Houser et al. (1998) 32
Houston (1990) 36
Houston (1991) 36, 38
Howes and Anderson (1988) 413, 424, 425, 527
Howson and Urbach (1993) 46
Huang and Garcia (2000) 404, 405
Huber (1990) 62
Huber (1991) 62
Hughes (1999) 60, 71, 75
Hulbe and MacAyeal (1999) 448
Hutson and Wagenet (1991) 205
Huybrechts (1990) 439
Huybrechts (1992) 450, 451, 452
Huybrechts (1996) 445
Huybrechts and de Wolde (1999) 430, 447
Huybrechts and Oerlemans (1988) 439
Huybrechts and Oerlemans (1990) 439
Huybrechts et al. (1996) 443, 445

Ibbit and O'Donnell (1971) 418
Illangasekare et al. (1990) 271, 274, 278
Iman and Conover (1980) 205, 212
Iman and Conover (1982) 213, 223
Imhoff et al. (1987) 182
Ip et al. (1998) 331
Isham and Kaufmann (1995) 27
Itakura and Kishi (1980) 402
Ivins and Porrill (1994) 183

Jackson and O'Neill (1990) 170, 171
Jackson and Schmugge (1991) 171

Jackson et al. (1992) 420, 432
Jackson et al. (1993) 176
Jackson et al. (1995) 177
Jakeman and Hornberger (1993) 81, 119, 127
Jakeman et al. (1990) 127
Janssen (1993) 216, 217
Janssen (1995a) 197
Janssen (1995b) 216
Janssen et al. (1993) 197, 205, 212, 213, 216
Jarvis et al. (1999) 316
Jin (1989) 172
Jobson and Harbaugh (1999) 295
Johnston and Di Nardo (1997) 134
Jordan (1991) 262, 272, 275, 279
Jordan et al. (1989) 274
Jordan et al. (1999) 262, 272, 275
Jordan et al. (2000) 264
Jordon (1991) 265
Jordon et al. (1999) 265
Jost (1991) 62
Joughin et al. (1999) 453

Kabat et al. (1997) 178
Kannen (1995) 182
Kashefipour and Falconer (2000) 367, 376, 378, 384
Kashefipour et al. (1999) 363, 364, 367
Kattelmann (1986) 272, 274
Kattelmann (1989) 268, 271
Kattelmann and Dozier (1999) 270, 274
Kawahara and Umetsu (1986) 331
Keesman (1989) 216
Keesman (1990) 216
Keesman and van Straaten (1998, 1989) 216
Kemp et al. (1989) 326
Kennedy (1989) 71
Kerr (1998) 175, 179
King and Roig (1988) 333
Kirkby (1976) 119
Kirkby (1996) 432, 433
Kirnbauer et al. (1994) 261
Kite (1995) 278
Kite et al. (1994) 470, 473, 477, 478
Klemes (1982) 32
Klemes (1986) 317, 462, 467, 468, 478
Klingeman et al. (1998) 389
Klute (1986) 199, 201, 210, 218
Knight and Sellin (1987) 334, 338
Knight and Shiono (1996) 331, 342
Knudsen et al. (1986) 473
Koblinsky et al. (1993) 181
Kodama et al. (1996) 348
Kohane and Welz (1994) 332
Koivusalo and Heikinheimo (1999) 273
Kondo and Yamasaki (1990) 263

Konikow (1978) 467
Konikow (1986) 29, 309
Konikow (1990) 311
Konikow (1992) 23, 322
Konikow (1995) 309
Konikow and Bredehoeft (1992) 1, 23, 24, 25, 30,
        32, 33, 123, 351, 352, 467, 414
Konikow and Patten (1985) 29, 31, 32, 38
Konikow and Swain (1990) 30
Konikow et al. (1997) 309, 310
Koshinen et al. (1997) 284
Koster and Suarez (1992) 464, 465
Kouwen and Li (1980) 190
Kuhn (1962) 119, 120
Kyriakidis (1997) 49

Lam and Dowdeswell (1996) 443
Landau and Lifshitz (1989) 84
Lane (1998) 328
Lane and Richards (1997) 351
Lane and Richards (1998) 333, 430
Lane et al. (1994) 348, 427, 430
Lane et al. (1995) 334, 347, 348, 422, 425
Lane et al. (1998) 422
Lane et al. (1999) 330, 347, 348, 350, 416, 417, 420,
        422
Lang and Sidhu (1983) 172
Lapen and Martz (1996) 273
Launder and Spalding (1972) 429
Laval (1997) 466
Lawes et al. (1882) 248
Le et al. (1997) 338
Le Meur and Huybrechts (1996) 443
Le Vine and Karam (1996) 172
Le Vine et al. (1990) 175
Leavesley (1989) 261
Leavesley and Stannard (1990) 279
Leavesley and Striffler (1979) 279
Leclerc et al. (1990) 331
Leonard et al. (1990) 36
Leopold and Maddock (1953) 83
Leopold et al. (1960) 428
Lesaffre et al. (1998) 277
Lewis and Grossman (1999) 66
Li (1997) 333
Li and Islam (1999) 173
Liaw (1986) 151
Lima (1992) 427
Lin (1997) 187
Lin and Falconer (1997) 367
Lin and Shiono (1995) 373
Lindström et al. (1997) 469
Ling et al. (1995) 273
Link and Marks (1999) 279

Lipton and Ward (1997) 284
Liston and Sturm (1998) 278, 284
Liston et al. (1993) 278
Loague and Kyriakidis (1997) 49
Lohmann et al. (1998) 48
Lopez and Garcia (1997) 338, 408
Lopez and Garcia (1998) 407, 408, 409
Lørup et al. (1998) 478
Loth and Graf (1998a,b) 278
Loth et al. (1993) 278
Loth et al. (1998a,b) 272
Luce et al. (1997) 278
Luce et al. (1998) 278
Luce et al. (1999) 280
Luis and McLaughlin (1992) 415, 416, 425
Lynch and Gray (1980) 331, 335
Lynch and Officer (1985) 341
Lythe et al. (2000) 441

MacAyeal (1989) 440
MacAyeal (1993) 440
MacAyeal et al. (1996) 440, 445
Malin and Younis (1990) 429
Manabe (1969) 465
Marks and Bates (2000) 333
Marks et al. (1998) 264
Markus (1987) 414
Marquardt and Snee (1975) 104
Marsh (1999) 262
Marsh and Woo (1984a,b) 270
Marshall (1990) 71
Marshall and Clarke (1997) 439, 448
Marshall and Oglesby (1994) 278
Martinec (1975) 263
Martinec and Rango (1980) 280
Martinec and Rango (1981) 284
Martinuzzi and Tropea (1991) 338
Matson (1994) 62
Mattikalli et al. (1998a) 176, 177
Mattikalli et al. (1998b) 176
Mayer and Siegert (2000) 451
McCombie and McKinley (1993) 74
McCuen (1973) 430
McDonnell (1991) 248
McKay et al. (1979) 205, 212
McLaughlin (1984) 318, 319, 320
McLaughlin and Townley (1996) 45, 89
McLelland et al. (1996) 422
McMadamon et al. (1993) 273
Mdaghri Alaoui (1998) 242
Mdaghri Alaoui et al. (1997) 241, 242
Meadows et al. (1972) 34
Melloh (1999) 263
Melloh et al. (1997) 279

Meselhe and Sotiropolous (2000) 418
Metheny (1998) 72, 73
Moll and Overmars (1990) 183
Mood and Graybill (1963) 296, 315, 317
Moore (1995) 34, 35
Moore et al. (1999) 350
Moramarco and Singh (2000) 335
Moramarco et al. (1999) 335
Morel-Seytoux (1966) 294, 309
Morel-Seytoux (1973) 315
Morel-Seytoux (1983a,b) 296
Morel-Seytoux (1985) 295
Morel-Seytoux (1999) 298, 317
Morel-Seytoux and Khanji (1975) 296
Morel-Seytoux and Nimmo (1999) 296
Morel-Seytoux and Restrepo (1987a,b,c) 295
Morel-Seytoux et al. (1973) 295
Morel-Seytoux et al. (1980) 295
Morris et al. (1997) 272, 273
Morton (1993) 46, 351
Morvan et al. (2000) 348
Moulin and Ben Slama (1998) 336
Mudaliar (1999) 168
Myette et al. (1987) 64, 74

Nachabe and Morel-Seytoux (1995a) 299, 318
Nachabe and Morel-Seytoux (1995b) 318
Narasimhan (1998) 4
Nash and Sutcliffe (1970) 418
National Research Council (1990) 1, 2
Nelder and Mead (1965) 265
Neumann (1992) 32
Newsam (1982) 96, 104, 112
Newsam and Barakat (1985) 104
Ng et al. (1996) 384
Nicholas and Smith (1999) 334, 347, 348
Nicholas and Walling (1997) 348
Nielsen and Hansen (1973) 476
Nigg (2000) 26
Nino and Garcia (1994, 1998) 389
Njoku and Entekhabi (1996) 175
Njoku et al. (1999) 164
Noat and Rodi (1982) 360
Nolin and Dozier (1993) 277
Nolin et al. (1993) 280
Noyelle et al. (1995) 182
NRS (1990) 6

O'Connor and Dobbins (1958) 369
Oerlemanns et al. (1998) 447
Oliver et al. (1994) 182
Olsen and Kjellesvig (1999) 336
Olsen and Stokseth (1995) 336, 337, 340
O'Neill et al. (1996) 171, 172, 176

Oreskes (1998) 1, 23, 35
Oreskes (1999) 33
Oreskes et al. (1994) 1, 23, 25, 74, 123, 124, 295, 413,
    414, 426, 433, 467
Ormsby et al. (1985) 182
Oswald and Robin (1973) 451
Owe and van de Griend (1998) 166

Pacelle (1986) 71, 72
Painter et al. (1998) 277, 280, 285
Panchang et al. (2000) 334, 335
Parker (1978) 401
Parker et al. (1985) 205, 218
Parker et al. (1998) 390
Parkinson and Young (1998) 139, 149
Pasche (1984) 337, 338, 340
Paterson and Budd (1982) 440
Pattyn (1996) 443
Payne (1995) 439
Payne (1999) 439
Payne and Baldwin (2000) 445
Payne and Dongelmans (1997) 439
Payne et al. (2000) 445, 446, 448
Peakall et al. (1996) 433
Pearson et al. (2000) 166
Peccei (1997) 34
Peck et al. (1980) 283
Perla (1982) 276
Perla (1985) 276
Peters (1996) 58
Phillips (1992) 415, 424
Pilkey (1990) 36
Pilkey (1994) 36
Pilkey (1995) 36, 37
Pilkey (1997) 36, 38
Pilkey (2000) 36, 37
Pilkey and Thieler (1992) 37
Pilkey et al. (1993) 36
Pinol et al. (1997) 8
Pomeroy et al. (1997) 284
Popper (1959) 119, 121, 123, 325
Popper (1968) 414
Press et al. (1992) 265
Preston (1985) 366
Prickett and Pettyjohn (1995) 60

Qin and Karnieli (1999) 284

Rajendran et al. (1999) 336
Ramamoorthi (1988) 182
Ramsey (1995) 182, 190
Rango (1993) 261
Rango and Salomonson (1974) 181
Rastogi and Rodi (1978) 329

Rayner (2000) 25, 25, 33
Refsgaard (1996) 4, 463
Refsgaard (1997) 480
Refsgaard (2000) 462, 467, 479
Refsgaard and Knudsen (1996) 474, 477, 479
Refsgaard and Storm (1995) 462, 463, 473, 477
Refsgaard and Storm (1996) 1, 2
Reitz et al. (1979) 165
Remson et al. (1971) 300
Reynolds (1895) 329
Rhoads (1994) 46
Rhoads and Thorn (1996) 432
Richards (1931) 235
Richards (1990) 6, 46, 326
Richards (1994) 6, 46
Richards (1996) 424
Richards et al. (1997) 6, 7
Ritsema (1998) 196
Ritz (1996) 440
Robert et al. (1993, 1995, 1997) 209
Rodi (1980) 330
Rodi (1984) 360
Rodi et al. (1981) 429
Rodi et al. (1997) 330
Romanowicz and Beven (1998) 48
Romanowicz et al. (1996) 48, 352
Rosenthal and Dozier (1996) 280
Roth et al. (1990) 241
Rouse (1937) 399
Rowntree and Lean (1994) 464
Ruhe et al. (1998) 470, 476, 478
Rykiel (1994) 295

Samuels (1990) 330
Sandholt et al. (1999) 479
Sandholt et al. (2000) 479
Sanjiv and Marelius (2000) 418
Sarabandi and Chiu (1997) 168
Sarewitz and Pielke (1999) 25
Sarewitz et al. (2000) 25
Sargent (1982) 424
Sauvaget et al. (2000) 336
Scherrer (1997) 257
Schlesinger et al. (1979) 467
Schmugge (1998) 181
Schmugge and Choudhury (1981) 168
Schroter and Whiteley (1992) 273
Schroter et al. (1991) 273
Schultz (1998) 181
Schultze et al. (1999) 296, 316
Schweppe (1965) 125, 156
Sellers et al. (1986) 464, 465
Sellers et al. (1992) 173
Sellers et al. (1996) 278

Sellin and Willetts (1996) 334, 342, 347
Sen and Stoffa (1995) 46
Senay et al. (2000) 179
Sentchev and Yaremchuk (1999) 250
Shackley et al. (1998) 118
Shannon and Weaver (1962) 79
Shaw (1983) 327
Shaw and Southwell (1941) 57
Shi and Dozier (1995a,b) 284
Shi and Dozier (2000a) 283
Shi and Dozier (2000b) 283
Shih et al. (1997) 276
Shimizu et al. (1990) 429
Shook et al. (1993) 280
Siegert and Dowdeswell (1995) 439
Siegert et al. (1996) 451
Siegert et al. (1998) 451
Siegert et al. (1999) 455
Silberstein et al. (1999) 284
Simon and Kahn (1984) 34
Singh (1996) 330
Singh et al. (1997) 274
Sivapalan (1995) 8
Sivapalan et al. (1997) 8
Slater et al. (2000) 262, 263, 278
Sloan and Moore (1984) 5
Smith (1990) 38
Smith (1997) 181
Smith and McLean (1977) 398, 402
Smith et al. (1995) 183
Sofialidis and Prinos (2000) 418
Solomon (1993) 182
Sommerfeld et al. (1991) 263, 280, 284
Sommerfeld et al. (1994) 271
Sonka et al. (1993) 182
Sorooshian et al. (1993) 462
Speziale (1987) 373
Stafford (1997) 209
Stafford (1998) 75
Stakgold (1967) 90, 91
Stakgold (1968) 90, 91
Steefel and van Cappellen (1998) 1, 23
Steenhuis et al. (1994) 196
Steenhuis et al. (1999) 5
Stehman (1999) 282
Stewart et al. (1999) 332
Stoesser (1999) 336, 338, 340
Sturm and Benson (1997) 276
Sturm et al. (1997) 273, 274
Suarez (1999) 18
Suppes (1969) 13
Suppes (1984) 17
Svendsen et al. (1999) 455

Tarboton et al. (1995) 264, 278
Taylor (1999) 455
Tchamen and Kawahita (1998) 331
Thomann and Mueller (1987) 357
Tierney (1992) 34
Tietje and Tapkenhinrichs (1993) 197, 202
Tikhonov and Arsenin (1977) 104
Tominaga and Nezu (1991) 338, 373, 374
Topp et al. (1980) 199
Tsang (1991) 1, 23, 33, 311, 467
Tsang (1992) 1, 23
Tsujimoto et al. (1991) 338
Turner (1993) 379
Turner et al. (1993) 380
Turpin et al. (1999) 278, 280
Tuteja and Cuname (1997) 274
Tversky and Kahneman (1982) 27, 29, 35

Ulaby et al. (1981) 165
Ulaby et al. (1982) 181
US Bureau of Census (1997) 57
US Environmental Protection Agency (1985) 63
US National Research Council (1990) 6
Usunoff et al. 1992) 424

Van Dam et al. (1990) 199, 210
Van Dyke (1987) 57
Van Genuchten (1980) 205, 210, 218, 219
Van Genuchten et al. (1991) 205
Van Rijn (1981) 404
Van Rijn (1984) 397, 402
Van Rijn (1984a,b) 378
Vanclooster et al. (1994) 211
Vanoni (1980) 400
Vecchia and Cooley (1987) 313
Vereecken (1988) 202, 211
Vereecken et al. (1992) 202
Verhagen and Bouma (1997) 209
Verhagen et al. (1995) 209
Vieria (1993) 365

Wagenet and Hutson (1989) 211, 219
Wagenet et al. (1989) 219
Walker and Goodison (1993) 184
Walland and Simmonds (1996) 263, 278
Wallis et al. (1989) 120, 125
Wang and Choudbury (1981) 168, 170–171
Wang and Ribberink (1986) 405
Wang and Schmugge (1980) 166, 169
Wang et al. (1990) 175, 176
Want et al. (1983) 168
Want et al. (1995) 182
Webster (1992) 240
Wegmuller et al. (1995) 182

Weinstein (1986) 62
Wheater et al. (1993) 117
Whillans (1976) 451
White (1994) 182, 183
Whitehead and Young (1975) 124, 125
Wigmosta et al. (1994) 264, 279, 280
Wigneron et al. (1993) 172
Wigneron et al. (1995) 169, 171, 172
Wilheit (1978) 168, 169, 171
Williams and Shah (1992) 183
Williams et al. (2000) 350
Woesnner and Anderson (1996) 74
Wolfram (1991) 107
Wood (1998) 5
Wood and Tong (1989) 338
Wösten (1987) 201, 207
Wösten and Van der Zee (1993) 201
Wösten and Van Genuchten (1988) 207
Wu (1982) 361

Yang et al. (1997) 264, 278
Yang et al. (1999) 273, 278
Ye et al. (1998) 119, 128, 130
Yeh (1986) 85, 294
Young (1974) 118, 125
Young (1978) 87, 118, 124
Young (1983) 118, 124
Young (1984) 85, 87, 100, 118, 123, 124, 125, 126,
     139, 150, 153, 157
Young (1985) 124
Young (1986) 117, 157

Young (1988) 124, 153
Young (1992) 118, 124
Young (1993) 119, 120, 124, 125, 130, 153, 155
Young (1996) 119
Young (1998a) 118, 199, 124, 125, 153
Young (1998b) 118, 119, 120, 124, 125, 130,
     153
Young (1999a) 118, 139, 149
Young (1999b) 124, 153
Young (2000a) 125, 155, 156
Young (2000b) 125, 128, 130
Young and Beven (1994) 118, 119, 120, 124, 125,
     130, 134, 153, 155
Young and Jakeman (1979, 1980) 124
Young and Lees (1993) 118, 124
Young and Minchin (1991) 118, 124
Young and Pedregal (1998) 120, 124
Young and Pedregal (1999) 104, 120, 124
Young and Tomlin (2000) 120, 139, 149, 150
Young and Wallis (1994) 120, 125
Young et al. (1971) 430
Young et al. (1996) 118, 124, 139, 149
Young et al. (1997) 119, 120, 130
Young et al. (1998) 139, 120, 130
Young et al. (2000) 125
Younis (1992) 429
Younis (1996) 329

Zak et al. (1997) 48
Zill and Cullen (1997) 82
Zyserman and Fredsoe (1994) 398, 403

Subject Index

1-d diffusion problem 106
2-d and 3-d models 376

Aberjona River 64
Ablation patterns 280
Absorption curves 199
Active contour models (snakes) 183
Adequacy of fit 418
ADV (Acoustic Doppler Velocimeters) 334, 347
Advanced information techniques 209
Adversarial legal system 60
ADZ (Aggregated Dead Zone) model 120, 125
Agro chemical breakthrough 196
Agro ecosystem models 228
Airborne radiometric mapping 178
Airborne remote sensing 175
Airy model 33, 34
Amazon river 400
Antecedent soil moisture 246
ARNO rainfall runoff model 464
Atmospheric modelling 278, 464
Average percentage error 384
AVHRR (Advanced Very High Resolution
    Radiometer) 181

Baltic sea basin 469
Basal outflow 263
Basal temperatures 446
BATS (Biosphere Atmosphere Transfer Scheme)
    modelling 464
Bayesian
    Analysis 105
    Framework 48
    Methods 18
Beach processes and erosion 35
Bed friction 186
Bedford-Ouse model (BM) 125, 142, 150, 157, 158
BEDMAP 441, 450, 456
Beers law 170
Best match 96
Biochemical oxygen demand 369
Biomass productions 208
Black box model 148
Boundary condition uncertainty 434
Boundary conditions 333, 394, 422

Boundedness of an operation 88
Brightness temperatures 170
Bulk flow measures 341
Bulk outflows 350
Bulk snow properties 263
Bypass flow 201, 202, 217
Byrd glacier 451

Calibrated parameter space 226
Calibration and ill-conditioning 77
Calibration and verification of models 376
Calibration methodology 317
Calibration 15
Canopy 171
Capillary potential 233
CASI (Compact Airborne Spectrometer
    Instrument) data 187, 190
Catchment storage 145
C-band radiometer 179
CCF (Cross Correlation Function) 133
Chebyshev expansions 110
Chezy bed roughness coefficient 185, 361
Civil lawsuit 59
Civil trial 61
Classical mechanical models 15
Cochella Valley 30, 32
Concentration functions 402
Conceptual error 32, 33, 38
Conceptual models 117, 118
Conceptualisation 33
Conditional proposition 29
Conditional validation 123
Conditioning 49
Conductivity curves 213
Conflicts 75
Confounding of errors 300
Correction factors 15
Counterfactual consequences 20
Courant number 363, 427
Coweeta catchment 5, 121, 129, 130
Critical rationalism 6, 48
Cross examination 60
CRREL energy balance site 264, 274

Darcy's law 45, 60, 222, 274 , 297

Data model calibration 20
Data model vindication 20
Data uncertainty 434
DBM (Data-Based Mechanistic) modelling l, 119, 120, 124–125, 128–130, 131, 134, 141–143, 146, 148, 154, 156
DBMS (DBM Simulation) 128, 142, 146, 150, 157–158
DBMS, IHACRES and BM models 149
DCE (Dichloroethene) 63, 67
Decision making process 23, 37
Degree of ill-conditioning 90
DEM (Digital Elevation Model) 8, 185, 187, 333, 344
Depth averaged velocities 377
Depth averaged velocity profiles 340
Desorption curves 199
Dielectric discontinuity 165
Differential split sample test 468
Discretisation 84, 107
Dislectric constant 172
Dissolved oxygen 369
Distinction making 19
Distributed physically based model 3
Distribution systems 82
DIVAST (2-D), (Depth Integrated Velocities And Solute Transport) model 366, 368, 381, 385
Dominant mode behaviour 150
Dotty plots 51
Draining front see Wetting and draining fronts
Drawdown 299
Dual porosity models 316
Dye flow 270
Dynamic order 119

Earthquake prediction 26
Eastern flevoland 217
ECHAM model 469
ECM (Electromagnetic Current Meter) 347
Economic models 28
Eddy viscosity 359
Effective rainfall 144
Effective rainfall flow model 132
EISMINT (European Ice Sheet Modelling INiTiative) 442–443, 445, 447
Emissivity 166, 168
Environmental systems 79
EPA (Environmental Protection Agency) 295, 368
Equifinality 47, 118
Equilibrium suspensions 398
Error magnification 100
Error testing 426
Errors 297
ERS–1 183, 450, 453

ERS–1 SAR data 343, 345
ESTAR 176
Estimation (optimisation) 19, 123
Eulerian particle velocity field 390
Eurasian ice sheet model 455
Evapo-transpiration 81
Expert witnesses 57, 58, 62, 72

Falsification, 46, 121, 123
FASTER (1-D), (Flow And Solute Transport in Estuaries and Rivers) model 367, 368, 376, 385
Federal rules of evidence 62
Fick's law 297
Field data 64, 72, 73, 75
FIFE experiment 173, 176, 466
Finite volume models 374
FIR (Finite Impulse Response) 126
FIS (Fixed Interval Smoothing) 124, 153
Flood model validation 183
Floodplain deposition 348
Floodplain storage 342
Fluid flows 105
Fluidity 247
Flux predictions 48
Forward and inverse problems 85
Free surface flow in unsaturated soils 248
Fresnel equations 165
Front velocities 249
Froude number 333, 435, 436
Frye rule 62
Functional macropores 256
Future scenarios models 296

GAM (Generalised Additive Modelling) approach 155
Gaussian noise 173
GCM (General Circulation Model) 25, 48, 464
Generic models 296
Geologic systems 309
Geostatistical analysis 32, 203
GEWEX (Global Energy and Water cycle Experiment) 334
GIS (Geographical Information Systems) 181, 197, 209
Global climate change 25
Global optimum 46
Global scale hydrological modelling 468
GLUE (Generalised Likelihood Uncertainty Estimation) procedure 26, 327, 423
GPS (Global Positioning System) 209, 210
Grain size 277
Greenland ice sheet 448
Groundwater levels 226, 227

Groundwater models 310
Groundwater pumping 30

HBV model 463, 469
HEC-RAS model 331
Henry's law 297
Heterogeneity 7, 60
HIRLAM model 476
History matching 352
HMC (Hybrid Metric Conceptual) models 118, 148
HMC model 119, 121, 129, 141, 145, 146
Homogeneous infiltration 196
Horton infiltration excess 44
Humber basin 373, 374, 381
Hydraulic conductivity 69, 101, 200
Hydraulic models 328, 358
Hydraulic properties 175, 205
HYDRCOIN (HYDRologic COde INtercomparison) 309
HYDRO model 336, 337, 338
Hydrologic model 59
Hypothetico- deductive approach 119
Hysterisis 199

Ice flow 439, 445
Ice flow models 440
Identification problems 88
IHACRES model 119, 120, 127–129, 130, 142, 143, 146, 150, 157, 158
Ill-conditioning 7, 77, 78, 83, 100
Ill-posedness 82, 92, 107
Image segmentation 183
Increments error 301
Input and data error 300
Input identification 87, 96
InSAR (Interferometric Synthetic Aperture Radar) 453
Instability and numerical issues 92
Instability of system identification 88
Interpretation errors 301
Intrinsic model errors 302
Inundation extent 181–182, 346, 351
Inverse problems 86, 87, 89, 105, 119
ISIS model 331
Isostasy and continental drift 33
IUH (Infinite dimensional Unit Hydrograph) 126
IV (Instumental Variable) methods 123

Jury 57, 58
Jury verdict 70

k-ε turbulence closure scheme 408
Kalman filter method 173

Kepler's laws 17
Kinematic
    Eddy diffusivity 399
    Eddy viscosity 359
    Wave 239, 242, 256
    Wave approximation 274
KININF model 427
Kriging 283

Landsat MSS 181, 280
Landsat TM (Thermatic Mapper) 183, 285, 422
Large scale hydrological modelling 461
Last glacial maximum 447
Latin hypercube sampling techniques 198, 212, 217
Lay witnesses 62
l-band brightness temperature 178
Leaching 208
LEACHM (Leaching Estimation And CHemistry Model) 205, 207, 211, 219
LEACHW 219, 222
Legal liability 37
LES (Large Eddy Simulation) 329
LiDAR (Light Detection And Ranging) 187, 188, 190, 333
Limits to growth model 34
Little Washita watershed 177
Log transformed sum of squares of error 419
Lower Santa Cruz river 29, 31
Lysimeter drainage 244

Mackenzie River basin 473
Macropore walls 196, 222
Macropores 218
Macroporous soils 196
Mammouth Mountain California 264
Manning roughness coefficient 361
Mass balance 224
Mass conservation of suspended sediment 390
Mathematical model 78
Matrix flow 246
MCMC (Markov Chain Monte Carlo) analysis 134
Measurement 45
Melt water outflow 268
Metric and HMC approaches 119
Metric models 117, 118
Microwave radiation 166
Microwave reflectance techniques 181
MIKE 11 model 331, 476
MIKE SHE model 463, 473, 474, 477
MOCDENSE 310
Model assessment
    Calibration 74, 223
    Code 467
    Development process 304

Efficiency 418, 419
Equations errors 298
Evaluation 25, 425
Intercomparisons 442
Space 52
Validation 217, 225, 294, 428, 432, 462
Modelling
And measurement scales 416
Protocols 306
Scale 3
Models
ARNO 464
BM 142, 150, 157, 158
DBM l, 120, 125, 128–130, 131, 134,
141–143,146, 154–156
DIVAST (2-D) 366, 368, 381, 385
ECHAM 469
FASTER (1-D) 367, 368, 376, 385
GCMs (General Circulation Models) 25, 464
HEC-RAS 331
HIRLAM 476
HMC 121, 129, 141, 145, 146
HYDRO 336, 337, 338
Ice flow models 440
IHACRES 119, 120, 127–129, 130, 142, 143, 146,
150, 157, 158
ISIS 331
KININF 427
LEACHM 205, 207, 211, 219
LEACHW 219, 222
MIKE 11 331, 476
MIKE SHE 463, 473, 474, 477
MODFLOW 295
Pratt 34
QUAL2E 368
RANS 329, 336, 337, 348
SHE 117, 462
SLURP 473
SNTHERM 262, 265, 266, 271, 276, 279,280
SSIIM 337, 340
Stanford watershed 326, 463
TELEMAC- 2D 185
TF 143, 157
THALES 462
TOPMODEL 51, 117, 466
TRIVAST 368, 385
TVP/SDP 152
Models as
Analogies with other real systems 12
Analogy theory 14
Images of data 13
Implicit conditional 29, 31
Instances of theories 13
Tamed data models 14

Tamed theories 12
Models of radiation transport 172
MODFLOW groundwater model 295
Moisture contents 209
Momentum
Correction factor 360
Dissipation 257
Dissipation approach 240, 243, 258
Of infiltration 240
Of flow in soils 235
Monte Carlo
Analysis 220
Realisations 52
Simulation 198, 205, 207, 208, 211, 225, 314, 431
Simulation methods 149
Multi objective optimisation procedure 26
Multiple acceptable models 51

Navier Stokes equations 328, 334
Newton's law of shear 235
Newton's laws of motion 60
Newtonian mechanics 11, 17
Nitrogen cycle 370
NMSS (Non Minimum State Space) 152
NOAA 280
Noise 182
Noise process 132
Non equilibrium suspensions 401
Non identifiability 47
Non parametric estimation 144
Non point source pollutants 203
Non uniqueness 24
Non-linear inverse problems 102
Non-linear model estimation 155
Non-stationary and non-linear model estimation
152
Normalised sum of squares of error 419
Numerical accuracy 433
Numerical and data errors 300
Numerical uncertainty 434
Nutrients 370

Objective functions 419
Odra River basin 476
One dimensional (1-d)
Framework 83
Flow fields 366
Flows 362
Snow process models 262
One step outflow method 218
Open boundary conditions 367
Optimal parameter values 315
Optimised model 49
Optomistic bias 29

Organic phosphorous 371
Overbank flow 338

PACF (Partial Auto Correlation Function) 135,
    136
Parameter
  Estimation error 298
  Estimation from measurements 240
  Uncertainty 434
Parametrical non uniqueness 24
Passive microwave emission 163
Passive microwave remote sencing 164
PCE (Perchloroethene) 63
Pedotransfer functions 202
Peer review 306
Penetration depth 166
Percolation 81
Performance check procedure 301
Phosphorous cycle 371
Physical system 78
Physically based models 175
Physics based models 117
Phytoplamkton algae- chlorophyll 372
Pick up problems 401, 404
Plaintiffs expert 66, 72
Planning studies 305
Point scale data 347
Point scale radiometric measurements 167, 172
Poiseuille's law 244
Porous media 233
Porous media flow 233
Posteriori flow classification 258
Prandtl analogy 399
Pratt model 34
Precision 420
Prediction 25
Predictive error 30
Predictive uncertainty 26
Predictive validation 138
Preferential flow 235, 257
Pressure dominated flow 248
Pressure extractors 218
Pre-trial discovery 58, 64
Principles of passive microwave radiometry 164
Prior information 101, 103
Probablistic flood classification 184
Problems of oversimplification 35
Process based models 49
Process descriptions 48
Profile of equilibrium concept 36
Proxy basin test 468
Public participation 305
Public policy 26, 27
Put down problems 404

QUAL2E model 368
Quaternary ice sheets 454

Radar 283
Radar layering 451
RADARSAT 453, 456
Radiation transport models 168
Radiative transfer model 173
Rainfall, 81, 127, 128, 144
Rainfall flow model 136, 140, 148
RANS (Reynolds Averaged Navier Stokes)
    equations 329
RANS (Reynolds Averaged Navier-Stokes) model
    329, 336, 337, 348
Rate of return analysis 27
Rayleigh jeans approximation 165
Rayleigh quotient 94
Realist approach 6
Reitholzbach lysimeter 245
Relative mean absolute error 419
Remote sensing of flood extent 181
Residual soil water equivalent 265
RETC 207, 208
Retentivity and conductivity characteristics 206
Retentivity curves 212
REV 234
Reynolds averaging 393
Reynolds number 361, 390, 435
Richards equation 222
RIV (Refined Instumental Variable) approach 124
River confluences 417
River flood models 180
River infiltration 68
River Meuse 345
River Thames 180
RMS (Root Mean Squared) error 174, 384
RORASC (ROtated RAndom SCan)
    Cycle 223
    Procedure 216, 225
Roughness 168
Rouse number 400
Rousean distribution 407

Salinity 380
Salt River valley 309
Salt River 29–32
Sand bed rivers 389
Santa Cruz river 309
SAR (Synthetic Aperture Radar)
    Imagery 179, 181, 182, 183, 187, 190
    Imaging systems 182, 343
Satellite interometry 453
Scale hierarchy 196
Scale model comparison 336

Scaling approaches 8
Scientific method 58, 59
Scientific testimony 57
Sediment
    Boundary conditions 394
    Concentration profiles 400, 404, 405
    Mixtures 406
    Transport process 378
Senegal River basin 473,474
Sensitivity analysis 222, 223, 302, 427
Sensitivity and uncertainty analysis 213
Set theory approach 216
Shallow water depth contours 337
Shared vision modelling, 305
SHE (Systeme Hydrolgique Europeen) model 117,
        462
Shields regime 391
Shields stress 397
Simulation and calibration procedures 86
Site specific models 296
Sleepers river 264, 267
SLURP model 473
Snow
    Cover 271
    Cover properties 283
    Density 273
    Depth 266
    Emissivity 265
    Extent 279, 282
    Grain size 275
    Surface temperature 284
    Surface texture 285
    Surface wetness 284
    Temperature profiles 273, 275
    Texture 276
    Water equivalent 263, 264, 269, 280–284
    Wetness 274
SNTHERM
    Model 262, 265, 266, 271, 276, 279, 280
    Predictions 274
    Simulation 267, 268
Soil
    Characteristics 195
    Hydraulic properties 214
    Physical properties 201
    Roughness properties 178
    Survey data 203
    Vegetation-atmosphere models 163
    Vegetation atmospheric transfer 464
    Water content 178
    Water relations 198
Soil moisture
    maps 166, 175, 176, 179
    Retention 198

Variations 251, 253, 258
Solute transport in compound channels 372
Sources of modelling errors 301
Spaceborne remote sensing 177
Spatial
    Awareness 218
    Scales 462
    Variability 218
Spatially distributed snow models 261, 277
Specification of models 374
Spectral reflectance 275
Split sample test 3, 468
SRIV (Simplified RIV) algorithms 124
SSARR 263, 265
SSIIM model 337, 340
St Venant equations 185, 330, 362
Stabilisation 100, 112
Standard error 384
Stanford watershed model 326, 463
Staring class-pedotransfer function 208
Staring series 208, 210
State dependent parameter (sdp) analysis 130
Stochastic
    Calibration 104
    Dynamic modelling 122
    Framework 113
    Residuals 133
Structure and order identification 121
Sub ice morphology 449
Subglacial lakes 449, 451
Subglacial topography 454
Successive forward problems 89
Suction crust infiltrometer 218
Surface infiltration 227
Suspended sediment 357, 395
Suspended sediment transport rates 408
SVAT (Soil Vegetation-Atmosphere Transfer)
        models 465, 477
SWAMP-l radiometer 179
System identification 87, 91, 98
Systematic error and bias 27, 38
Systems transfer function 85

TCE (Trichloroethene) 63, 67
TDR (Time Domain Reflectometry) measurements
        199, 210
Technical advisory committees 305
TELEMAC- 2D model 185
Temperature dependence 372
Temperature index model 268
Temporal and spatial divergence 24
Tensiometer profiles 218
Testimony 70
TF (Transfer Function) models 124, 143, 157

THALES model 462
Theory based calibration 20
Theory based models 19
Theory based vindication 20
Theory vindication 16
Three dimensional flow 66, 358, 364
Tile drains 217
Time series snow maps 281
Time Variable Parameter (TVP) estimates 124
TIN (Triangular Irregular Network) 333
TOPMODEL 51, 117, 466
Total and faecal coliforms 368
Tracer metal concentration 383
Transmissivity 85, 299
Trial process 58
TRIVAST model 368, 385
Truthful estimation 20
Turbulence 185, 399
Turbulence model 408, 409
Turbulent eddy viscosity parameters 186
Turbulent kinetic energy 417
TVP/SDP model 152
τ–ω–h–Q models 176
Two dimensional flow fields 366
Two dimensional numerical hydraulic models 180
Two dimentional flows 360
Types of models 295

Uncertainty
    Analysis 38, 61, 139, 197, 434
    In predictions 51
Unknown or unlikely events 28
UNSODA database 228
Unsteady flows 409
US legal system 58
US Supreme Court 62
USGS (United States Geological Survey) 319

Vadose zone 202, 317
Validating snow models 272
Validating spatially distributed snow models 279
Validation
    And calibration process 184
    Of advanced methods 179
    Of flood routing models 342
    Of hydraulic models 332
    Test 3
Van Genuchten parameters 207
Variances 419
Vector data sources 342
Vegetated channels 407
Vegetation
    Height 169, 189
    Height map 189
Velocities 379
Velocity distributions 374
Velocity vectors 352
Verification 294, 413
Vertical kinematic eddy viscosity 399
Viscosity 235
Visualisation 431
Volume flux density 250

Walnut Gulch Watershed 176
Water
    Balance 227
    Budget 80
    Contents 210
    Elevations 377
    Film 236
    Fluxes 81
    Quality modelling 364
    Stress 211
    Table 64
    Table map 65
Wave model 211
WCRP (World Climate Research Programme) 334
Wetting and draining fronts 237–239
White noise 133
WOCE (World Ocean Circulation Experiment)
        334
Wrong calibration 301

Yangtze River basin 476